NEUROLOGY AND PREGNANCY: PATHOPHYSIOLOGY AND PATIENT CARE

HANDBOOK OF CLINICAL NEUROLOGY

Series Editors

MICHAEL J. AMINOFF, FRANÇOIS BOLLER, AND DICK F. SWAAB

VOLUME 171

ELSEVIER

NEUROLOGY AND PREGNANCY: PATHOPHYSIOLOGY AND PATIENT CARE

Series Editors

MICHAEL J. AMINOFF, FRANÇOIS BOLLER, AND DICK F. SWAAB

Volume Editors

ERIC A.P. STEEGERS, MARILYN J. CIPOLLA, AND ELIZA C. MILLER

VOLUME 171

3rd Series

ELSEVIER

ELSEVIER
Radarweg 29, PO Box 211, 1000 AE Amsterdam, Netherlands
The Boulevard, Langford Lane, Kidlington, Oxford OX5 1GB, United Kingdom
50 Hampshire Street, 5th Floor, Cambridge, MA 02139, United States

Notices
Knowledge and best practice in this field are constantly changing. As new research and experience broaden our understanding, changes in research methods, professional practices, or medical treatment may become necessary.

Practitioners and researchers must always rely on their own experience and knowledge in evaluating and using any information, methods, compounds, or experiments described herein. In using such information or methods they should be mindful of their own safety and the safety of others, including parties for whom they have a professional responsibility.

With respect to any drug or pharmaceutical products identified, readers are advised to check the most current information provided (i) on procedures featured or (ii) by the manufacturer of each product to be administered, to verify the recommended dose or formula, the method and duration of administration, and contraindications. It is the responsibility of practitioners, relying on their own experience and knowledge of their patients, to make diagnoses, to determine dosages and the best treatment for each individual patient, and to take all appropriate safety precautions.

To the fullest extent of the law, neither the Publisher nor the authors, contributors, or editors assume any liability for any injury and/or damage to persons or property as a matter of products liability, negligence or otherwise, or from any use or operation of any methods, products, instructions, or ideas contained in the material herein.

British Library Cataloguing-in-Publication Data
A catalogue record for this book is available from the British Library

Library of Congress Cataloging-in-Publication Data
A catalog record for this book is available from the Library of Congress

ISBN: 978-0-444-64239-4

For information on all Elsevier publications
visit our website at https://www.elsevier.com/books-and-journals

Publisher: Nikki Levy
Editorial Project Manager: Kristi Anderson
Production Project Manager: Punithavathy Govindaradjane
Cover Designer: Alan Studholme

Typeset by SPi Global, India

Handbook of Clinical Neurology 3rd Series

Available titles

Vol. 79, The human hypothalamus: basic and clinical aspects, Part I, D.F. Swaab, ed. ISBN 9780444513571
Vol. 80, The human hypothalamus: basic and clinical aspects, Part II, D.F. Swaab, ed. ISBN 9780444514905
Vol. 81, Pain, F. Cervero and T.S. Jensen, eds. ISBN 9780444519016
Vol. 82, Motor neurone disorders and related diseases, A.A. Eisen and P.J. Shaw, eds. ISBN 9780444518941
Vol. 83, Parkinson's disease and related disorders, Part I, W.C. Koller and E. Melamed, eds. ISBN 9780444519009
Vol. 84, Parkinson's disease and related disorders, Part II, W.C. Koller and E. Melamed, eds. ISBN 9780444528933
Vol. 85, HIV/AIDS and the nervous system, P. Portegies and J. Berger, eds. ISBN 9780444520104
Vol. 86, Myopathies, F.L. Mastaglia and D. Hilton Jones, eds. ISBN 9780444518996
Vol. 87, Malformations of the nervous system, H.B. Sarnat and P. Curatolo, eds. ISBN 9780444518965
Vol. 88, Neuropsychology and behavioural neurology, G. Goldenberg and B.C. Miller, eds. ISBN 9780444518972
Vol. 89, Dementias, C. Duyckaerts and I. Litvan, eds. ISBN 9780444518989
Vol. 90, Disorders of consciousness, G.B. Young and E.F.M. Wijdicks, eds. ISBN 9780444518958
Vol. 91, Neuromuscular junction disorders, A.G. Engel, ed. ISBN 9780444520081
Vol. 92, Stroke – Part I: Basic and epidemiological aspects, M. Fisher, ed. ISBN 9780444520036
Vol. 93, Stroke – Part II: Clinical manifestations and pathogenesis, M. Fisher, ed. ISBN 9780444520043
Vol. 94, Stroke – Part III: Investigations and management, M. Fisher, ed. ISBN 9780444520050
Vol. 95, History of neurology, S. Finger, F. Boller and K.L. Tyler, eds. ISBN 9780444520081
Vol. 96, Bacterial infections of the central nervous system, K.L. Roos and A.R. Tunkel, eds. ISBN 9780444520159
Vol. 97, Headache, G. Nappi and M.A. Moskowitz, eds. ISBN 9780444521392
Vol. 98, Sleep disorders Part I, P. Montagna and S. Chokroverty, eds. ISBN 9780444520067
Vol. 99, Sleep disorders Part II, P. Montagna and S. Chokroverty, eds. ISBN 9780444520074
Vol. 100, Hyperkinetic movement disorders, W.J. Weiner and E. Tolosa, eds. ISBN 9780444520142
Vol. 101, Muscular dystrophies, A. Amato and R.C. Griggs, eds. ISBN 9780080450315
Vol. 102, Neuro-ophthalmology, C. Kennard and R.J. Leigh, eds. ISBN 9780444529039
Vol. 103, Ataxic disorders, S.H. Subramony and A. Durr, eds. ISBN 9780444518927
Vol. 104, Neuro-oncology Part I, W. Grisold and R. Sofietti, eds. ISBN 9780444521385
Vol. 105, Neuro-oncology Part II, W. Grisold and R. Sofietti, eds. ISBN 9780444535023
Vol. 106, Neurobiology of psychiatric disorders, T. Schlaepfer and C.B. Nemeroff, eds. ISBN 9780444520029
Vol. 107, Epilepsy Part I, H. Stefan and W.H. Theodore, eds. ISBN 9780444528988
Vol. 108, Epilepsy Part II, H. Stefan and W.H. Theodore, eds. ISBN 9780444528995
Vol. 109, Spinal cord injury, J. Verhaagen and J.W. McDonald III, eds. ISBN 9780444521378
Vol. 110, Neurological rehabilitation, M. Barnes and D.C. Good, eds. ISBN 9780444529015
Vol. 111, Pediatric neurology Part I, O. Dulac, M. Lassonde and H.B. Sarnat, eds. ISBN 9780444528919
Vol. 112, Pediatric neurology Part II, O. Dulac, M. Lassonde and H.B. Sarnat, eds. ISBN 9780444529107
Vol. 113, Pediatric neurology Part III, O. Dulac, M. Lassonde and H.B. Sarnat, eds. ISBN 9780444595652
Vol. 114, Neuroparasitology and tropical neurology, H.H. Garcia, H.B. Tanowitz and O.H. Del Brutto, eds. ISBN 9780444534903
Vol. 115, Peripheral nerve disorders, G. Said and C. Krarup, eds. ISBN 9780444529022
Vol. 116, Brain stimulation, A.M. Lozano and M. Hallett, eds. ISBN 9780444534972
Vol. 117, Autonomic nervous system, R.M. Buijs and D.F. Swaab, eds. ISBN 9780444534910
Vol. 118, Ethical and legal issues in neurology, J.L. Bernat and H.R. Beresford, eds. ISBN 9780444535016
Vol. 119, Neurologic aspects of systemic disease Part I, J. Biller and J.M. Ferro, eds. ISBN 9780702040863
Vol. 120, Neurologic aspects of systemic disease Part II, J. Biller and J.M. Ferro, eds. ISBN 9780702040870
Vol. 121, Neurologic aspects of systemic disease Part III, J. Biller and J.M. Ferro, eds. ISBN 9780702040887
Vol. 122, Multiple sclerosis and related disorders, D.S. Goodin, ed. ISBN 9780444520012
Vol. 123, Neurovirology, A.C. Tselis and J. Booss, eds. ISBN 9780444534880
Vol. 124, Clinical neuroendocrinology, E. Fliers, M. Korbonits and J.A. Romijn, eds. ISBN 9780444596024
Vol. 125, Alcohol and the nervous system, E.V. Sullivan and A. Pfefferbaum, eds. ISBN 9780444626196

All volumes in the 3rd Series of the **Handbook of Clinical Neurology** are published electronically, on Science Direct: http://www.sciencedirect.com/science/handbooks/00729752.

Foreword

Pregnancy and childbirth directly concern a sizeable portion of the world's population. The nervous system participates in these processes in many ways and makes them possible through hormonal changes. In addition, any treatment of a pregnant patient may have important and often long-term consequences for the developing fetus. Yet, standard neurology textbooks often fail to discuss these topics in any depth. The previous series of the *Handbook of Clinical Neurology* (HCN) had close to 100 entries under the indexed heading of "Pregnancy," but these usually consisted of footnotes added to the discussion of major disorders such as headache or epilepsy. Therefore we are proud to introduce the present two volumes dedicated to various aspects of neurology and pregnancy.

The first volume is dedicated to the neurophysiologic changes that accompany pregnancy, an overview of the neurologic and psychiatric diseases seen during that period, and the approach to the care of these conditions. A chapter entitled "Neuro-obstetrics" emphasizes the need for a multidisciplinary approach to the care of women with neurologic comorbidities, both preconceptionally and throughout pregnancy. Other chapters discuss in detail neuroimaging and radiation exposure during pregnancy as well as the ethical aspects of complicated pregnancies. The volume concludes with a discussion of the diagnosis and management of congenital neurologic diseases during pregnancy.

The second volume deals in detail with neurologic disorders that may accompany pregnancy, such as stroke, trauma, and neurologic complications associated with preeclampsia and eclampsia. There are chapters that discuss compression neuropathies, acute psychiatric illnesses, migraine and other headaches, movement disorders, disorders of sleep, and CNS tumors. Moreover, a chapter is dedicated to the management of chronic neurologic disorders during pregnancy.

We have had the good fortune of working with three experienced and world-renowned volume editors. Prof. Eric A.P. Steegers is a gynecologist and the head of the Department of Obstetrics & Gynecology at the Erasmus University Medical Center in Rotterdam. Dr. Marilyn J. Cipolla is professor in the Department of Neurological Sciences with joint appointments in Obstetrics, Gynecology & Reproductive Sciences and Pharmacology at the University of Vermont Larner College of Medicine, Burlington. Dr. Eliza C. Miller is a vascular neurologist and assistant professor of neurology in the Division of Stroke and Cerebrovascular Disease at Columbia University, New York City. As series editors, we reviewed all the chapters in the volumes and made suggestions for improvement, but we are delighted that the volume editors and chapter authors produced such scholarly and comprehensive accounts of different aspects of the neurology of pregnancy. Hence, we hope that the volumes will appeal to clinicians and neuroscientists alike. Our goal is to provide clinicians with a state-of-the-art reference that summarizes the clinical features and management of the many neurologic manifestations of pregnancy. We also intend to provide basic researchers with the foundations for new approaches to the study of the complex changes that accompany pregnancy.

In addition to the print version, the chapters in the two volumes are also available electronically on Elsevier's ScienceDirect website, which is becoming increasingly popular with readers and will improve the book's accessibility. Indeed, all the volumes in the present series of Handbook are available electronically on this website. This will make them even more accessible to readers and facilitate searches for specific information.

As always, it is a pleasure to thank Elsevier, our publisher, and in particular Michael Parkinson in Scotland, Nikki Levy and Kristi Anderson in San Diego, and Punithavathy Govindaradjane at Elsevier Global Book Production in Chennai, for their assistance in the development and production of the *Handbook of Clinical Neurology*.

Michael J. Aminoff
François Boller
Dick F. Swaab

Preface

Neurology and Pregnancy is a two-volume comprehensive multidisciplinary guide to best practices for neurologists, obstetricians, and other specialists who care for women with acute or chronic neurologic disorders. All aspects of care for the pregnant woman with neurologic illness, such as preconception care, genetic counseling, pregnancy in patients with chronic neurologic disorders, and acute pregnancy-related neurologic complications, are covered. Postpartum care and complications, including lactation concerns, are also addressed, as are the long-term effects of pregnancy and its associated problems on maternal brain health.

In the first volume (vol. 171 of the *Handbook of Clinical Neurology*), the complex neurophysiologic changes in pregnancy are summarized from a basic and translational science perspective. This is followed by a discussion of general issues relevant to the provision of multidisciplinary neurologic care for pregnant patients, including neuroimaging, the principles of neuro-obstetric critical care, and ethical and medicolegal concerns. The final part describes normal fetal cerebral development and summarizes the management of the most common prenatal neurologic disorders.

The second volume (vol. 172 of the *Handbook of Clinical Neurology*) covers acute neurologic disorders presenting in pregnancy, including preeclampsia/eclampsia, ischemic and hemorrhagic stroke, and infections of the central nervous system. This is followed by discussions of chronic neurologic conditions, such as epilepsy, migraine, and multiple sclerosis. As innovative care can be achieved only by the continuous integration of clinical management with advances in research, all chapters provide insights into the newest developments in related translational neuroscience and the implications for the future of the field.

The many contributors to this volume reflect diverse perspectives, from basic and translational researchers to clinicians and clinical researchers in a broad range of specialties. We are grateful to have been given the opportunity to learn and benefit from their extraordinary expertise. Michael Parkinson, the development editor, has thoughtfully provided vital support throughout the process of bringing this work to publication.

We believe that this text will improve the complex multidisciplinary care needed for pregnant women with neurologic disorders, contribute to improved maternal and fetal outcomes, and enhance the long-term health of mothers affected by neurologic disease and also that of their children.

Eric A.P. Steegers
Marilyn J. Cipolla
Eliza C. Miller

Contributors

K. Allegaert
Department of Development and Regeneration; Department of Pharmaceutical and Pharmacological Sciences, KU Leuven, Leuven, Belgium; Department of Clinical Pharmacy, Erasmus University Medical Center, Rotterdam, The Netherlands

A. Anderson
Department of Neurology, Weill Institute for Neurosciences, University of California San Francisco, San Francisco, CA, United States

T. Barghouthi
Department of Neurology, Wake Forest Baptist Medical Center, Winston-Salem, NC, United States

S.L. Berga
Department of Obstetrics and Gynecology, Jacobs School of Medicine and Biomedical Sciences, University at Buffalo, Buffalo, NY, United States

K. Bernstein
Department of Anesthesiology, Columbia University Irving Medical Center, New York, NY, United States

R. Bove
Department of Neurology, Weill Institute for Neurosciences; Department of Neurology, School of Medicine, University of California San Francisco, San Francisco, CA, United States

V.L. Brooks
Department of Physiology and Pharmacology, Oregon Health & Science University, Portland, OR, United States

I.A. Brussé
Department of Obstetrics and Gynecology, Division of Obstetrics and Fetal Medicine, Erasmus University Medical Center, Rotterdam, The Netherlands

M.S. Burn
Department of Obstetrics, Gynecology & Reproductive Sciences, Yale School of Medicine, New Haven, CT, United States

C. Bushnell
Department of Neurology, Wake Forest Baptist Medical Center, Winston-Salem, NC, United States

M.J. Cipolla
Department of Neurological Sciences; Department of Obstetrics, Gynecology & Reproductive Sciences; Department of Pharmacology, University of Vermont Larner College of Medicine, Burlington, VT, United States

J. Dudink
Department of Neonatology, University Medical Center, Utrecht, The Netherlands

M. Figurelle
Department of Neurology, Wake Forest Baptist Medical Center, Winston-Salem, NC, United States

Q. Fu
Institute for Exercise and Environmental Medicine, Texas Health Presbyterian Hospital Dallas; Department of Internal Medicine, University of Texas Southwestern Medical Center, Dallas, TX, United States

D. Garg
Division of Reproductive Endocrinology and Infertility, Department of Obstetrics and Gynecology, University of Utah School of Medicine, Salt Lake City, UT, United States

K. Gordo
Department of Anesthesiology, Columbia University Irving Medical Center, New York, NY, United States

P. Govaert
Department of Neonatology, Erasmus University Medical Center, Rotterdam, The Netherlands; Department of Neonatology, ZNA Middelheim, Antwerp; Department of Rehabilitation and Physical Therapy, Gent University Hospital, Gent, Belgium

D.R. Grattan
Centre for Neuroendocrinology and Department of Anatomy, School of Biomedical Sciences, University of Otago, Dunedin, New Zealand

C.M. Heesch
Department of Biomedical Sciences and Dalton Cardiovascular Research Center, University of Missouri, Columbia, MO, United States

H. Hussey
Department of Anesthesiology, Columbia University Irving Medical Center, New York, NY, United States

P. Hussey
Department of Anesthesiology, Columbia University Irving Medical Center, New York, NY, United States

A.C. Johnson
Department of Neurological Sciences, University of Vermont Larner College of Medicine, Burlington, VT, United States

A.C.M. Kluivers
Department of Obstetrics and Gynecology, Division of Obstetrics and Fetal Medicine, Erasmus University Medical Center, Rotterdam, The Netherlands

S.R. Ladyman
Centre for Neuroendocrinology and Department of Anatomy, School of Biomedical Sciences, University of Otago, Dunedin, New Zealand

R. Landau
Department of Anesthesiology, Columbia University Irving Medical Center, New York, NY, United States

R. Lemley
Department of Neurology, Wake Forest Baptist Medical Center, Winston-Salem, NC, United States

J. Maguire
Department of Neuroscience, Tufts University School of Medicine, Boston, MA, United States

A.M. Mandel
The Neurological Institute, Columbia University Irving Medical Center, New York, NY, United States

C. McCormack
Center for Science and Society, Columbia University, New York, NY, United States

E.C. Miller
Department of Neurology, Columbia University Vagelos College of Physicians and Surgeons; NewYork-Presbyterian Hospital, Columbia University Irving Medical Center, New York, NY, United States

A. Mitchell
Department of Obstetrics and Gynecology, Columbia University Irving Medical Center, New York, NY, United States

C. Monk
Department of Obstetrics and Gynecology; Department of Psychiatry, Columbia University Irving Medical Center; New York State Psychiatric Institute, New York, NY, United States

M.A. O'Neal
Department of Neurology, Brigham and Women's Hospital, Boston, MA, United States

K.M. Prager
Department of Medicine, Columbia University Irving Medical Center, New York, NY, United States

K.N. Sheth
Department of Neurology, Yale School of Medicine, New Haven, CT, United States

S.S. Sheth
Department of Obstetrics, Gynecology & Reproductive Sciences, Yale School of Medicine, New Haven, CT, United States

K. Shetler
Department of Neurology, Columbia University Vagelos College of Physicians and Surgeons, New York, NY, United States

Z. Shi
Department of Physiology and Pharmacology, Oregon Health & Science University, Portland, OR, United States

J. Singh
Department of Neurology, Weill Institute for Neurosciences, University of California San Francisco, San Francisco, CA, United States

D. Smok
Department of Obstetrics Gynecology, Columbia University Vagelos College of Physicians and Surgeons, New York, NY, United States

E.A.P. Steegers
Department of Obstetrics and Gynecology, Erasmus University Medical Center, Rotterdam, The Netherlands

F. Triulzi
Department of Pediatric Neuroradiology, Università Degli Studi di Milano, Milan, Italy

E.E. Whitaker
Department of Anesthesiology; Department of Neurological Sciences, University of Vermont Larner College of Medicine, Burlington, VT, United States

D.M. Zambrano Espinoza
Department of Neurology, Columbia University Vagelos College of Physicians and Surgeons, New York, NY, United States

Contents

Section I

Neurophysiologic changes in pregnancy

Handbook of Clinical Neurology, Vol. 171 (3rd series)
Neurology and Pregnancy: Pathophysiology and Patient Care
E.A.P. Steegers, M.J. Cipolla and E.C. Miller, Editors
https://doi.org/10.1016/B978-0-444-64239-4.00001-1

Chapter 1

Neuroendocrine mechanisms of reproduction

DEEPIKA GARG[1] AND SARAH L. BERGA[2]*

[1]*Division of Reproductive Endocrinology and Infertility, Department of Obstetrics and Gynecology, University of Utah School of Medicine, Salt Lake City, UT, United States*

[2]*Department of Obstetrics and Gynecology, Jacobs School of Medicine and Biomedical Sciences, University at Buffalo, Buffalo, NY, United States*

Abstract

The neuroendocrinology of reproduction focuses on the neuromodulation of gonadotropin-releasing hormone (GnRH), the ontogeny of the hypothalamic–pituitary–gonadal axis, and common reproductive events and conditions, namely, puberty, the menstrual cycle, and disorders of reproductive function. The core concept underpinning the neuroendocrinology of reproduction is neuroregulation of hypothalamic GnRH drive. In both men and women, reproductive function requires that GnRH input elicit appropriate secretion of follicle-stimulating hormone and luteinizing hormone from the anterior pituitary and that the gonads respond to such input appropriately. Moreover, insufficient GnRH drive causes hypothalamic hypogonadism and secondary insufficiency of gonadal sex steroid hormone synthesis and release in both sexes. Alterations in GnRH drive also reflect gonadal conditions such as dysgenesis, hyperandrogenism, gonadotropin mutations, and aging and loss or absence of oocytes or Sertoli cells. The most common cause of insufficient GnRH drive is functional, that is, due to the endocrine effects of psychologic or behavioral variables. Rarely does reduced GnRH drive reflect organic or congenital causes such as developmental defects, brain tumors, or celiac disease. Despite a common neuropathogenesis the heterogeneity of behavioral variables associated with reduced GnRH drive has resulted in a variety of names, including functional hypothalamic amenorrhea, stress-induced anovulation, and psychogenic amenorrhea.

INTRODUCTION

The neuroendocrinology of reproduction focuses primarily on the ontogeny and neuromodulation of the hypothalamic–pituitary–gonadal axis. The core concept underpinning the neuroendocrinology of reproduction is regulation of hypothalamic gonadotropin-releasing hormone (GnRH) drive. In both men and women, reproductive function requires that GnRH input elicit appropriate secretion of follicle-stimulating hormone (FSH) and luteinizing hormone (LH) from the anterior pituitary and that the gonads respond to such input appropriately. We review herein the neuroregulation of GnRH as it relates to common reproductive events and conditions, namely, puberty, the menstrual cycle, and disorders of gonadal function. The neuromodulation of GnRH is complex; a panoply of external and internal factors regulates GnRH drive to align reproductive function with circumstance.

GnRH is the proximate signal to the gonadotrope cells of the anterior lobe of the pituitary. Pulsatile GnRH release into the portal vasculature at the level of the median eminence stimulates the synthesis and secretion of the two key gonadotropins: FSH and LH. These two gonadotropins, in turn, drive gonadal function.

*Correspondence to: Sarah L. Berga, M.D., Jacobs School of Medicine and Biomedical Sciences, University at Buffalo, Department of Obstetrics and Gynecology, Division of Reproductive Endocrinology and Infertility, Buffalo, NY 14203. United States. Tel: +1-336-978-3726, E-mail: slberga@buffalo.edu

In women, gonadotropins regulate oocyte development, folliculogenesis, and the development and maintenance of the corpus luteum. They drive the synthesis of the traditional gonadal sex steroid hormones: estradiol from granulosa cells, androstenedione from theca cells, and progesterone from luteinized granulosa cells. Likewise, in men, LH stimulates the production of testosterone from Leydig cells and FSH stimulates spermatogenesis from Sertoli cells. LH and FSH are the primary, but not only, regulators of gametogenesis. This chapter focuses on understanding the regulation of GnRH drive and how its neuromodulation fosters or impairs reproductive competence.

GnRH is released into the portal circulation as pulses, and both pulse frequency and pulse amplitude gate the pituitary release of gonadotropins. In women, increases and decreases in GnRH drive result in syndromal alterations in reproductive function, the two most common of which are functional hypothalamic hypogonadism/anovulation/amenorrhea (FHH/FHA) and polycystic ovary syndrome (PCOS). There are sex differences in GnRH pulse frequency and amplitude; the GnRH pulse frequency in men is roughly circhoral whereas that in women in the follicular phase is about one pulse every 90 min. The increased GnRH-LH pulse frequency observed in women with PCOS of nearly one pulse every 60 min suggests androgenization of the hypothalamic GnRH "pulse generator" (Berga et al., 1993). In men, reduced GnRH drive results in a spectrum of hypothalamic hypogonadism (HH) that manifests as what is often termed "low testosterone syndrome" and oligoasthenozoospermia. Rarely is the cause of reduced GnRH drive organic or congenital; typically the etiology is functional and related to psychologic and behavioral factors. The most readily recognized condition in women that is attributable to functional slowing of the GnRH "pulse generator" is functional hypothalamic amenorrhea (FHA). Women with FHA have insufficient GnRH drive to the point that they are anovulatory and chronically hypoestrogenic, which leads to amenorrhea and is readily clinically recognized. Other terms for this condition include psychogenic amenorrhea, exercise amenorrhea, functional hypothalamic chronic anovulation, and stress-induced anovulation. This syndrome was originally described as psychogenic or hypothalamic amenorrhea (Reifenstein, 1946). As FHH/FHA occurs due to behaviors and attitudes, it is theoretically reversible. In contrast, organic or congenital causes, such as brain tumors and defective migration of GnRH neurons during fetal development from the olfactory placode into the hypothalamus, are generally not readily reversible. In both men and women, functional and organic causes of HH are diagnoses of exclusion.

THE ROLE OF THE GONADOTROPIN-RELEASING HORMONE (GnRH) "PULSE GENERATOR" IN INITIATION AND MAINTENANCE OF GONADAL FUNCTION

Biosynthesis and action of GnRH

GnRH is synthesized in hypothalamic GnRH neurons and secreted into the hypophyseal portal circulation. The GnRH release pattern regulates the synthesis and release of the pituitary gonadotropins, FSH and LH (Schally et al., 1971). In addition to regulating pituitary gonadotropin secretion, it has been suggested that GnRH works in an autocrine and paracrine manner in different tissues (e.g., ovary, placenta) and plays crucial roles in other important physiologic functions such as implantation (Yu et al., 2011).

When released into the portal circulation, GnRH is a decapeptide; however, it is the cleavage product of the 92-amino-acid precursor peptide called prepro-GnRH. The GnRH gene is located on human chromosome 8 (8p11.2–p21). The prepro-GnRH peptide contains GnRH (10 amino acids), GnRH-associated peptide (56 amino acids), signal sequence (23 amino acids), and a proteolytic processing site (3 amino acids). GnRH-associated peptide can stimulate secretion of gonadotropin and inhibit secretion of prolactin. The pulsatile pattern of hypothalamic GnRH release led to the term "hypothalamic GnRH pulse generator" (Pohl and Knobil, 1982). GnRH neurons form a complex network with their cell bodies located in the infundibular (arcuate) nucleus and the medial preoptic area of hypothalamus (Stopa et al., 1991) and are functionally integrated, with abundant interconnections and connections to other neuronal populations. Of note, GnRH are olfactory neurons that must travel to the hypothalamus from the olfactory placode. One cause of organic forms of HH in both men and women is failure of migration (Layman, 1999; Caronia et al., 2011).

The regulation of GnRH secretion is complex and involves multiple central pathways. Pulsatile GnRH release into the portal circulation is required for gonadotropin secretion from the anterior pituitary, and thus it is critical for fertility and reproduction. Mutations of the GnRH-1 gene in mice render them hypogonadal, and the hypogonadism can be reversed by either GnRH-1 gene therapy or transplantation of GnRH neurons (Mason et al., 1986). In addition, defective or absent GnRH secretion in humans leads to hypogonadotropic hypogonadism (HH), which typically manifests as absence of puberty and thus infertility; puberty and reproductive function are restored by exogenous provision of pulsatile GnRH therapy (Santoro et al., 1986). A genetic cause of premature activation of the GnRH pulse generator, which leads to central precocious puberty in boys and girls, has been described. The mutation in the imprinted gene MKRN3, located at chromosome 15q11–q13, causes a deficiency of the protein product (Abreu et al., 2013). Various factors regulate GnRH secretion, including psychologic stress, sex steroids, and energy balance; behavioral factors are viewed as functional, as contrasted with organic, modulators of the GnRH function.

The actions of GnRH are mediated through its receptors. Two types of GnRH receptors have been identified: GnRH type I and GnRH type II (GnRH-2). Both receptors are structurally and functionally distinct. GnRH mediates its pituitary actions through GnRH type I receptor. Another form of GnRH (GnRH-2) has been discovered in animals, and its gene is located on chromosome 20 (20p13). There is extensive expression of GnRH-2 in the CNS and extra-CNS tissues, and these receptors may contribute to the regulation of reproductive behavior. The physiologic role of GnRH-2 in humans is unclear.

Neuroanatomy of GnRH pulse generator

The term "GnRH pulse generator" refers to the intrinsically pulsatile pattern of GnRH release into the portal circulation and requires that endogenously pulsatile GnRH neurons fire synchronously. Mediators that disrupt, suppress, or disconnect the cross-talk among GnRH neurons are many and include central factors such as glutamate, GABA, kisspeptin, antimüllerian hormone (AMH), and sex steroids, for example, estradiol, progesterone, and androgens. As GnRH neurons lack sex steroid receptors, the actions of estrogens are thought to be mediated by afferent neurons (e.g., those secreting glutamate, GABA, kisspeptin, neuropeptide Y, and others). GnRH is rapidly degraded by proteases after being released in the portal circulation and thus has a very short half-life of 2–4 min. Pulsatile GnRH secretion is a necessary requirement for FSH and LH synthesis and secretion; continuous GnRH release causes profound desensitization of GnRH receptors (Wildt et al., 1981). The demonstration that GnRH receptors are desensitized by continuous, as opposed to pulsatile, GnRH exposure led to the formulation of GnRH agonists and antagonists with longer half-lives. GnRH analogs now comprise an extensive clinical armamentarium used to effect ovarian suppression and prevent endogenous LH surges during ovarian stimulation protocols.

The higher the pulse frequency of GnRH, the more the secretion of LH is favored over the release of FSH; in contrast, slower frequency of GnRH pulses promotes FSH release more than LH release (Gross et al., 1987). The GnRH pulse generator is active during fetal life and then desynchronized until the onset of puberty (Rasmussen et al., 1986; Terasawa, 1995). During infancy and childhood, puberty is inhibited because the GnRH pulse generator is exquisitely sensitive to estrogen negative feedback throughout that stage of human development (Dubois et al., 2016).

Neuromodulation of the GnRH pulse generator

The kisspeptin system is required for normal GnRH secretion and serves as a "gatekeeper" of puberty. The kisspeptin gene product, a 154-amino-acid precursor protein, is a G protein-coupled receptor ligand for its receptor, a seven transmembrane domain, G protein-coupled receptor 54 (GPR54). Within the arcuate nucleus, kisspeptin/neurokinin B/dynorphin (KNDy) neurons secrete kisspeptin, which activates GnRH neurons (Gottsch et al., 2004). In addition to activating GnRH neurons, kisspeptin neurons also form synapses with GnRH neuron terminals in the median eminence, where GnRH release (exocytosis) is stimulated by kisspeptin (Ramaswamy et al., 2008). KNDy neurons and kisspeptin-GPR54 signaling form the final common pathway of the hypothalamic circuitry that governs GnRH pulsatility (McCarthy, 2013). One formulation of the central cascade that regulates GnRH is shown in Fig. 1.1 (Navarro and Kaiser, 2013).

Stressors, defined by behaviors and thoughts that activate hypothalamic–pituitary–adrenal (HPA) axis and the adrenergic and the autonomic nervous systems, lead to a constellation of neuroendocrine alterations, including hypothalamic hypothyroidism, which has the effect of conserving energy (Berga et al., 1989). Since reproduction is highly energy dependent, hypothalamic hypothyroidism is envisioned as an adaptive mechanism in the face of sustained metabolic or psychogenic challenge. The principal "stress system" in the brain also regulates corticotropin-releasing hormone (CRH) via neurons in the hypothalamic paraventricular nuclei, antidiuretic hormone via neurons in supraoptic nuclei, and the locus coeruleus/norepinephrine system in the brainstem via the central sympathetic nervous system (Mastorakos et al., 2006). In addition, metabolic signals mediated through adipocyte-derived leptin play a significant role in regulation of KNDy neurons via an inhibitory GABAergic neuronal pathway (Martin et al., 2014). In a monkey model the administration of the CRH antagonist astressin B reversed the impact of the chronic social stress, subordination, on GABA-A receptor binding in the prefrontal cortex, a site implicated in the regulation of the limbic–HPA, gonadal, and thyroidal axes (Michopoulos et al., 2013a). Alterations in GnRH pulse frequency and

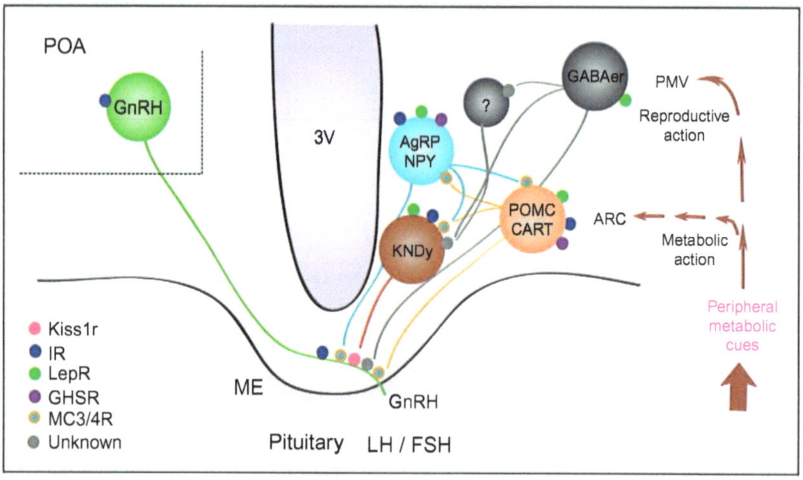

Fig. 1.1. Schematic representation of neural interactions between metabolic and reproductive functions depicting likely sites of action of leptin, insulin, and ghrelin to control GnRH release (Navarro and Kaiser, 2013). Abbreviations: *3V*, third ventricle; *AgRP*, agouti-related peptide; *ARC*, arcuate nucleus; *CART*, cocaine- and amphetamine-regulated transcript; *GABA*, g-aminobutyric acid; *GHSR*, growth hormone secretagogue receptor; *IR*, insulin receptor; *Kiss1r*, kisspeptin receptor; *KNDy*, kisspeptin/neurokinin B/dynorphin; *LepR*, leptin receptor; *MC3r*, melanocortin receptor 3; *MC4r*, melanocortin receptor 4; *ME*, median eminence; *NPY*, neuropeptide Y; *PMV*, ventral premammillary nucleus; *POA*, preoptic area; *POMC*, pro-opiomelanocortin.

amplitude might be in part the result of interposition of glial processes into the synaptic cleft of GnRH neurons, thereby interfering with the density of GnRH-to-GnRH appositions among GnRH neurons and reducing the GnRH cohort size (Witkin et al., 1997).

NEUROENDOCRINOLOGY OF PUBERTY

The pubertal transition is characterized by increasing frequency and amplitude of GnRH pulses and follows a dormant phase during childhood in which GnRH neurons do not synchronously release GnRH. The GnRH pulse generator is active during fetal life and then is quiescent during prepuberty. Changes in the neural cascade allow the GnRH pulse generator to reactivate. Centrally, puberty may be conceptualized as the "mirror image" of functional hypothalamic hypogonadism (FHH).

Gonadotropins and estradiol are low in prepubertal girls. There are important sex differences in the timing of the offset of GnRH drive in infants and the onset of GnRH drive at puberty that are beyond the scope of this chapter. Regardless of sex, however, puberty begins with a nocturnal onset of GnRH pulsatility (Boyar et al., 1974), which initiates gonadal secretion of estradiol in girls and testosterone in boys. During the prepubertal dormant phase the GnRH pulse generator is "hypersensitive" to sex steroid negative feedback. Thus the pubertal transition requires a change in the central sensitivity or the acquisition of central resistance to sex steroid negative feedback (Reiter et al., 1975; Abreu et al., 2013). Genetic, environmental, and nutritional factors interactively modulate feedback sensitivity (Gajdos et al., 2009).

NEUROENDOCRINOLOGY OF THE MENSTRUAL CYCLE

The physiology of the menstrual cycle relies on a temporally precise pattern of GnRH drive and appropriate pituitary and ovarian responses. The phases of the menstrual cycle are termed: (1) follicular, midcycle, and luteal when referring to ovarian events and (2) proliferative and secretory when referring to the endometrial responses to ovarian secretion of estradiol and progesterone.

Follicular phase

The follicular phase commences with the onset of menses and finishes with the initiation of the LH surge. The drop in estradiol and progesterone that accompanies the demise of the corpus luteum from the preceding luteal phase allows FSH to increase relative to LH. The rise in FSH reflects a decrease in its inhibition at the pituitary level by lower estradiol and progesterone levels at the end of the luteal phase. If the increase in FSH is sufficient and there are responsive follicles, folliculogenesis follows. The steady rise in estradiol level across the follicular phase due to progressive follicle growth inhibits FSH release from the pituitary. The decline in FSH in the midfollicular phase prevents initiation of folliculogenesis of smaller follicles that are still FSH dependent. The suppression of FSH allows the lead follicle, which has become FSH independent and LH dependent before the other follicles, to escape demise. The result is generally mono-ovulation (Zeleznik, 2001) in the absence of exogenous use of FSH. Across the follicular phase, both estradiol and GnRH pulse frequency increase progressively (Hall et al., 1992). The duration of the typical follicular phase is about 11–13 days, and it becomes shorter in women as they age. The shorter follicular phase in older women reflects a "premature" rise in FSH during the preceding luteal phase and is likely due to the loss of oocyte number (Santoro et al., 1996) and concomitant decrease in inhibin, which also suppresses FSH.

Midcycle LH surge

As the follicle matures and granulosa cells multiply, estradiol rises exponentially. In the late follicular phase, there is a small rise in progesterone due to granulosa cells acquiring LH receptors and then transforming into luteal cells (Hoff et al., 1983). The accompanying neural events involve a progressive increase in GnRH pulse amplitude and frequency mediated in part by kisspeptin neurons (Fig. 1.1) and eventuate in a massive release of LH termed the LH surge. The LH surge catalyzes proteases in the follicle wall to digest it and in so doing allows for the release of the egg from the follicle through the follicle wall site termed the "stigma." As granulosa cells transform into luteal cells that secrete progesterone, there is a fall in E2 and resumption of negative feedback (Naftolin et al., 2007).

Luteal phase

Gradually rising progesterone secretion from corpus luteum leads to progressive slowing of LH pulses from about one pulse every 60–90 min in the late follicular phase (Filicori et al., 1986) to one pulse every 4 h (Stocco et al., 2007). The corpus luteum also produces inhibin A, the concentrations of which peak in the mid-luteal phase. In contrast, inhibin B secretion is nearly absent during the luteal phase (Welt et al., 1997). Inhibins produced by granulosa and luteal cells inhibit FSH synthesis and release by pituitary gonadotrophs and decrease gonadotropin sensitivity to GnRH (Culler and Negro-Vilar, 1989). In the absence of oocyte fertilization and implantation, there is a gradual decrease in estradiol and progesterone production by the corpus luteum. However, if implantation occurs, the trophoectoderm and then placenta make human chorionic gonadotropin (hCG), which "rescues" the corpus luteum and ensures progesterone production until the luteal–placental shift at around week 9 of gestation (Csapo et al., 1972).

FUNCTIONAL HYPOTHALAMIC AMENORRHEA/ANOVULATION

In women, reduced GnRH drive results in decreased FSH and LH stimulation of ovarian follicles and theca cells that has a spectrum of clinical presentation, including oligomenorrhea, amenorrhea, irregular menstrual intervals, and cycles of short duration that reflect partial or absent folliculogenesis. Full folliculogenesis that results in ovulation requires about 14 days of appropriate GnRH drive. Partial or absent folliculogenesis results from GnRH drive that is intermittently or chronically reduced. In clinical practice, reduced GnRH drive may manifest as failure of or partial response to ovulation induction or controlled ovarian stimulation with clomiphene or letrozole.

The initial demonstration that FHA was due to insufficient GnRH drive was provided by giving exogenous GnRH in a pulsatile manner at a frequency of once every 90 min for 2 weeks and showing that folliculogenesis and pregnancy resulted (Hurley et al., 1983; Miller et al., 1983). Women with FHA have not only reduced GnRH-LH pulse frequency but also decreased FSH concentrations (Berga et al., 1989). Organic forms of congenital HH may predispose to FHA (Caronia et al., 2011), which is typically a form of secondary amenorrhea, but it can manifest as primary amenorrhea. The most dramatic example of FHA is anorexia nervosa. Psychiatric syndromes such as depression, obsessive compulsive disorder, syndromal and nonsyndromal eating disorders, and behavioral factors such as excessive exercise and nutritional deficiencies predispose to FHA, but unlike anorexia nervosa, in which amenorrhea is almost universal, amenorrhea is less common in women with other psychiatric syndromes. Regardless of the combination of predisposing psychiatric and behavioral factors, the pathogenesis of anovulation and menstrual irregularities is similar and due to reduced GnRH drive. In short, FHA/anovulation is a reversible form of reduced GnRH drive that results in anovulation or luteal insufficiency. The behavioral antecedents result in a constellation of neuroendocrine and secondary endocrine disturbances. The complex regulation of hypothalamic function notwithstanding, reduced GnRH drive is best conceptualized as a neuroendocrine adaptation to a medley of metabolic and psychogenic challenges. Indeed, FHA is a constellation of neuroendocrine adaptations that promote survival (allostasis) in the face of metabolic, intrapsychic, and environmental demands that interact synergistically (Berga et al., 1989).

GnRH function is regulated by both inhibitory and stimulatory neuromodulatory signals that align GnRH drive with the internal and external milieu. In women with FHA, there is a significant association between activation of the HPA axis and decline in GnRH drive. Both amenorrheic athletes and nonathletes display elevated levels of serum, cerebrospinal fluid, and urinary cortisol (Biller et al., 1990; Berga et al., 1997; Brundu et al., 2006). Acute nutritional deprivation concomitantly lowers glucose (Fig. 1.2), increases cortisol, and slows LH pulse frequency, the best measure of GnRH pulse frequency (Loucks and Thuma, 2003; Sanders et al., 2018).

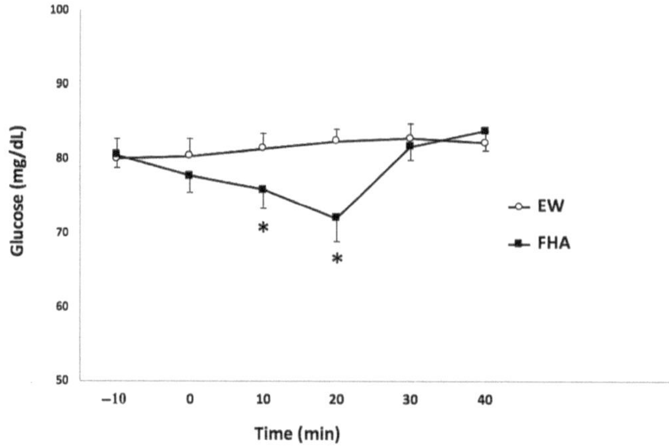

Fig. 1.2. Glucose responses to exercise challenge in women with functional hypothalamic amenorrhea (FHA) and eumenorrheic women (EW) (Sanders et al., 2018).

In short, women with FHA display chronically elevated cortisol levels and reduced LH pulse frequency, and these pathophysiologic changes are accompanied by behavioral patterns such as undernutrition and dysfunctional attitudes that are hypothesized to activate the HPA (Giles and Berga, 1993; Marcus et al., 2001; Berga, 2008). In our monkey model of FHA, seemingly minor stressors synergistically disrupted menstrual cyclicity (Williams et al., 2007). As proof of concept, resumption of ovarian function was achieved by addressing known behavioral concomitants (Berga et al., 2003; Berga and Loucks, 2006; Michopoulos et al., 2013b).

Evidence shows that GnRH pulsatility is modulated by a panoply of functional factors (Fig. 1.1) (Navarro and Kaiser, 2013). Studies performed in monkeys are most relevant to human FHA. Although monkeys are not as sensitive to energetic imbalances and psychologic stressors differ, monkey models have contributed importantly to our understanding of the pathogenesis of stress-induced changes in reproductive function. Challenges that impair reproductive function include nutritional restriction, social instability, chair restraint stress, exercise-related stress, and injection of insulin and endotoxin (Chen et al., 1992; Romero et al., 1993; Shively et al., 1997). Not all female monkeys display equivalent stress sensitivity to the same stressors; stress sensitivity is mediated in part by differences in CRH gene expression with more stress-sensitive female macaques showing elevated CRH in the hypothalamus (Centeno et al., 2007). Ultimately the central neural pathways that mediate stress sensitivity regulate KISS1 expression and sensitivity to estradiol negative feedback (Bethea et al., 2014). Not only are individuals variably sensitive to different stressors, but preexisting stress augments reactivity to subsequent stressors (Bethea et al., 2005; Williams et al., 2007; Berga, 2008).

The impact of stressors on neuroendocrine and reproductive function differs in the presence of gonads. Sex steroids blunt the response of endotoxin, and the response to chair restraint is increased in the presence of ovaries (Chen et al., 1992; Ferin, 1999). In addition the impact of stress is also influenced by follicular versus luteal phases of the menstrual cycle (Ferin, 1999). Responses to stressors also depend on the underlying endocrine or psychologic state of the individual. The neuroendocrine response to exercise was exaggerated in men whose cortisol levels were not suppressed by pretreatment with dexamethasone as compared with those whose cortisol levels were suppressed (Petrides et al., 1994). The preexisting activation of the limbic–HPA axis results in increased reactivity to subsequent stressors. In contrast, during lactational amenorrhea, the response to stress is blunted (Altemus et al., 1995). This differential and variable vulnerability of individuals to physical and psychosocial stressors appears to be mediated in part by altered serotoninergic tone and polymorphisms in the serotonin reuptake transporters (Bethea et al., 2005; Jarrell et al., 2008). In short the response to stressors depends in part on the neuroendocrine status of the individual experiencing the stress.

Pathophysiology

The proximate cause of the reproductive impairment associated with FHA starts with reduced hypothalamic GnRH drive and manifests in the circulation as reduced LH pulse frequency. Folliculogenesis to the point of ovulation requires that GnRH drive be above threshold for approximately 12–14 days. Markedly reduced GnRH drive causes anovulation, while intermittent reductions in GnRH drive cause poor follicular development and luteal phase insufficiency.

FHA is more than an isolated disruption of the HPO axis due to reduced GnRH drive and is associated with a constellation of neuroendocrine secretory aberrations (Berga et al., 1989), including HPA activation and

HP-thyroidal suppression. In FHA, 24-h mean serum cortisol pattern is increased while maintaining a circadian secretion pattern (Berga et al., 1989).

GnRH suppression in response to stress occurs in the absence of ovarian activity; high cortisol levels led to a fall in FSH levels in postmenopausal women who were admitted to the intensive care unit (Spratt et al., 1993). FSH levels continued to decrease after cortisol levels declined as the patients recovered, suggesting a delay between recovery of the HPA axis and restoration of GnRH drive. We found that augmented cortisol secretion is specific to women with FHA and not observed in those with other forms of anovulation, including PCOS (Berga et al., 1997). Eumenorrheic women and those with other causes for anovulation, including PCOS, had similar 24-h secretory patterns for cortisol. Interestingly, women recovering from FHA had normal diurnal cortisol patterns and LH pulse frequencies intermediate between women with eumenorrheic and FHA. Cortisol levels were elevated in women with depression and anorexia nervosa, but not always in bulimia (Mortola et al., 1987; Gwirtsman et al., 1989). Athletic women with amenorrhea had higher cortisol levels and reduced LH pulse frequency as compared with eumenorrheic athletic women (Loucks et al., 1989). Intermittent decreases in LH pulse frequency and slightly increased cortisol levels were noted in eumenorrheic athletes with luteal phase insufficiency. It is unclear how often depression is associated with amenorrhea, and women with FHA did not meet the formal criteria for depression (Giles and Berga, 1993). Higher calorie intake in women with depression and bulimia may reduce HPA activation (Alvero et al., 1998).

It is important to emphasize that GnRH pulse frequency in FHA is not a direct consequence of elevated cortisol per se but rather of the activation of the limbic–HPA axis by behaviors and cognitions. Indeed, continuous hydrocortisone infusions mimicking moderate and severe stress had no impact on secretion of gonadotropin (Samuels et al., 1994). Only when cortisol doses given exogenously exceeded 300% of the physiologic levels was GnRH drive modestly reduced (Saketos et al., 1993). In comparison, in women with FHA, cortisol levels are only 20%–30% higher than those in eumenorrheic women. Women with anorexia nervosa have cortisol elevations about 150% higher than those in eumenorrheic women and cortisol levels decrease to 70% of normal after weight recovery (Gwirtsman et al., 1989). Insulin-induced hypoglycemia activated the limbic–HPA axis and caused a similar decline in LH levels in monkeys pretreated with metyrapone, a cortisol synthesis blocker, and saline (Van Vugt et al., 1997). Similarly, rhesus monkeys with and without adrenalectomy displayed comparable decreases in gonadotropin levels after intravenous infusion of CRH (Xiao et al., 1989). These findings underscore the notion that central activation of the limbic–HPA axis suppresses GnRH drive rather than elevations of cortisol per se.

The central cascade that suppresses GnRH drive and activates the limbic–HPA axis is complex. Women with anorexia nervosa have elevated cerebrospinal fluid levels of immunoreactive CRH (Kaye et al., 1987). In contrast, we found comparable CSF levels of CRH in women with FHA and eumenorrhea indicating disruption of GnRH pulsatility even in the absence of detectable rise in cerebrospinal CRH (Berga et al., 2000). However, women with FHA showed increased CSF cortisol despite "normal" CRH levels, indicating resistance to cortisol feedback inhibition. The main site of inhibition of the HPA axis is hippocampal glucocorticoid receptors. Chronic stress alters these receptors and the feedback set point.

During stress, cortisol levels are elevated and result in metabolic adaptation, presumably to reduce and redirect energy expenditure. This is accomplished in part by lowering thyroxine and triiodothyronine levels and the absence of increased TSH drive; functional and adaptive hypothalamic hypothyroidism is often referred to as "sick euthyroid syndrome" (Samuels et al., 1994).

As noted earlier the preexisting state of the individual gates the response to subsequent challenge (Petrides et al., 1994; Williams et al., 2007). Not surprisingly, recovery from FHA led to marked increases in TSH levels (Berga et al., 1997). This indicates that recovery from FHA comprises a time course expressed initially as a decrease in cortisol levels followed later by recovery of GnRH pulse generator and TRH drive. These observations also suggest that mild hypothalamic hypothyroidism caused by a metabolic stress may alter the endocrine reactivity to subsequent stress. More importantly, these data suggest that the activation of the HPA axis and the HP-thyroidal axis always accompanies FHH/FHA (Berga, 2008). Other metabolic alterations found in FHA include elevated nocturnal secretion of melatonin (Berga et al., 1988), increased growth hormone (GH) (Berga et al., 1989), and low levels of leptin that increase during recovery (Michopoulos et al., 2013b), which collectively reflect metabolic adaptation. Prolactin levels are significantly lower in women with FHA (Berga et al., 1989; Laughlin et al., 1998) and may aid in diagnosis.

Behavioral contributors

The accumulating evidence suggests that all behaviors and attitudes that activate the HPA axis and suppress the thyroidal axis suppress the GnRH pulse generator and cause a spectrum of FHH/FHA. Thus the list of such behaviors and potential variables linked with the development of FHA is likely to be variable and idiosyncratic.

In the clinical setting, it is challenging to identify the specific stressors involved in desynchronizing the GnRH pulse generator and compromising reproductive function. While experimental animal models allow central pathways associated with FHA to be determined, human studies have identified a range of stressors, including exercise, drug use, weight loss, personality disorders, cognitive dissonance, and affective and eating disorders (Shanan et al., 1965; Fries et al., 1974; Vigersky et al., 1977; Berga and Girton, 1989; Giles and Berga, 1993; Marcus et al., 2001). The behavioral heterogeneity of FHA reflects individual valences about what is important, coping skills, metabolic challenge, body habitus, autonomic system, and preexisting levels of stress. Typically, in FHA, there are multiple stressors, apparently minor, such as a combination of poor eating habits, study- or work-related pressures, and relatively increased activity or exercise. However, in the experimental and clinical research settings, subjects have relatively strict inclusion and exclusion criteria to qualify for FHA and thus are not entirely representative of FHA in the clinical setting. In addition, in research settings, investigators attempt to isolate stressors and study "exercise related amenorrhea," "anorexia nervosa," and "idiopathic amenorrhea," thus leading to oversimplification. The neuropathogenesis is similar and behavioral, metabolic, and cognitive variables comingle (Berga, 2008). Oversimplifying reduces the likelihood that patients with FHH/FHA will be counseled about why and how to address the panoply of behavioral interventions.

To identify cognitive, behavioral, and psychologic variables associated with FHA, we compared three groups of women: nine women with FHA not related to excessive exercise, weight loss, drug use, an eating disorder, or an affective disorder; eight with eumenorrhea; and six with an identifiable organic cause of anovulation (Giles and Berga, 1993). All women underwent a structured diagnostic interview using the Schedule for Affective Disorders and Schizophrenia-Lifetime Version (SADS-L). Interestingly, those with amenorrhea due to any cause reported a compromised sense of psychologic equilibrium as conveyed on psychometric inventories but the psychologic domains differed according to cause of amenorrhea (Giles and Berga, 1993). However, women in the FHA group had dysfunctional attitudes and "unrealistic" expectations, such as concomitant perfectionism and high need for social approval (sociotrophy), which we posited might have made it more difficult for them to resolve intrapsychic conflicts between need for approval and desire for attainment and thus more stress sensitive. We used these findings to develop a unique program of cognitive behavior therapy (CBT) designed to ameliorate problematic attitudes and expectations (Berga et al., 2003; Berga and Loucks, 2006; Michopoulos et al., 2013b). These findings demonstrated that women with FHA displayed higher sensitivity to stress at central and neuroendocrine levels and that behavioral interventions reduced stress sensitivity and thereby restored ovulatory ovarian function.

Differential diagnosis

FHA is a diagnosis of exclusion. All other causes of amenorrhea and anovulation must be excluded. Further, anovulation is not always accompanied by amenorrhea. The epidemiology of FHA is not well defined. In a series of 262 patients with adult-onset secondary amenorrhea, hypothalamic suppression was the most common cause (Reindollar et al., 1986). Interestingly the type and severity of stress impact the percentage of women who develop amenorrhea (Drew, 1961). It is difficult to diagnose subtle forms of anovulation and luteal phase deficiency due to mild or intermittent alterations in GnRH drive. However, it is important to understand that these subtle forms augmented by environmental factors are more common than outright amenorrhea and may masquerade as other causes of anovulation, particularly if there is stress-related weight gain (Kaplan and Manuck, 2004). Insufficient GnRH drive can provoke limited follicular development and may lead to menstrual functions in the form of polymenorrhagia or eumenorrhea without the actual ovulation (De Souza et al., 1998). Luteal phase deficiency may indicate an alteration in GnRH drive in women with normal ovarian reserve, but reduced ovarian reserve may also be a cause. It is important for the clinician to remember that responses of gonadotropins to clomiphene and aromatase inhibitors may be reduced as mild stress renders the GnRH pulse generator less responsive to estradiol antagonism.

To determine the functional and hypothalamic cause of decreased ovarian function, all other possible organic causes, such as anatomic anomalies and pregnancy, must be excluded. There is high clinical suspicion for FHA if the LH:FSH ratio is less than 1, estradiol levels are below 60 pg/mL, and progesterone concentrations are below 1 ng/mL. However, a central organic cause should be suspected if there are focal neurologic signs; certain neurologic conditions such as hydrocephalus can present with nonlateralizing and vague symptoms (Lowry et al., 1996). There should be a low threshold to order magnetic resonance imaging (MRI) to rule out any organic causes. Evaluation for celiac disease and autoimmune enteropathy is recommended (Berga, 2010).

Male FHH may present as oligoasthenospermia with reduced or normal gonadotropins. As men do not have obvious clinical symptoms such as amenorrhea, it is more difficult to diagnose FHH in the absence of infertility. Usually, reduced GnRH drive is clinically occult in men and rarely presents with symptoms associated with other causes such as primary HH and anosmia signaling Kallmann syndrome or upper body wasting or mood changes due to isolated HH (Aloi et al., 1997; Nachtigall et al., 1997). The behavioral contributors to FHH have not been as well studied in men as in women. It has often been suggested that men have a "hypothalamus robustus" and are stress resistant, while women possess a "hypothalamus fragilis" and are stress sensitive. While men and women may be variably sensitive to commonplace stressors, it is unlikely that men are impervious to stress. Indeed, most men with low testosterone syndrome may have FHH.

Different behavioral challenges activate somewhat specific neuroendocrine cascades in animal models (Romero et al., 1993). For instance, extreme energy expenditure might activate a combination of metabolic pathways that have the potential to interrupt GnRH drive, while psychogenic challenges might trigger a somewhat different set of neuroendocrine pathways that also has the potential to interrupt GnRH drive. In our monkey model, FHA was highest when they combined metabolic and psychogenic challenges (Williams et al., 2007) and our human data seems to confirm this notion. Women randomized to weight loss alone, running alone, and a combination of the two were studied to determine rates of anovulation and luteal insufficiency; menstrual cycle compromise was highest in the combined group (Bullen et al., 1985). Further, eumenorrheic women with caloric deprivation had reduced LH pulsatility in association with lower serum glucose and increased serum cortisol levels, confirming that the response to stressors results in a constellation of neuroendocrine adaptations (Loucks and Thuma, 2003). While women who were eumenorrheic responded to exercise with an increase in glucose and cortisol, presumably to meet metabolic demand, women with FHA responded to a graded exercise regimen with a small drop in glucose that was associated with much greater increase in serum cortisol as compared with eumenorrheic women (Sanders et al., 2018).

Behavioral management

The behavioral causes of FHH may differ in men and women, but this area has received little investigative attention. It had been assumed that men possess a "hypothalamus robustus" while women have a "hypothalamus fragilis." In men, a prolonged physical and psychologic stress, a war drill, caused a reversible increase in cortisol accompanied by decrease in gonadotropins and testosterone (Opstad, 1992). In women, short-term fasting was associated with suppression of neuroendocrine function and follicular development resulting in anovulation in lean women but not in women of normal weight suggesting increased sensitivity to metabolic challenge in lean women (Alvero et al., 1998). The GnRH pulse generator in men might be more sensitive to energy (food) deficiency than that in women while women may be more sensitive to exercise. Since psychologic challenges may be gender specific, it is difficult to compare psychologic variables in women and men. Common causes of HH include anorexia nervosa or bulimia in 3% of women, are not related to eating disorders in 5% of college-aged women, and show a 55% prevalence of anovulatory cycles or luteal insufficiency in recreational running (Becker et al., 1999).

The key to recovery in women with FHA is to decrease cortisol and thereby increase GnRH drive and restore the thyroidal axis. Recovery may require a prolonged duration of energetic balance and psychologic equilibrium to reverse FHH/FHA. To test the hypothesis that changing attitudes and behavioral modifications will improve and restore ovarian functions, we randomized women with FHA to either CBT or observation for 16 sessions over 20 weeks. A diagnosis of FHA required secondary amenorrhea of more than 6 months duration and the absence of identifiable causes of secondary amenorrhea. Laboratory evaluation included determination of urinary pregnancy test, LH, FSH, estradiol, progesterone, TSH, thyroxine, prolactin, androstenedione, testosterone, and DHEAs. Standard mental health conditions were excluded by structured interview with a trained psychologist. An LH:FSH ratio less than 2 with low estradiol, progesterone, and androgens suggested FHA. CBT sessions focused on understanding nutrition and exercise habits as well as attitudes toward nutrition and exercise, stress management practices, and adopting healthy approaches. Of those randomized to the CBT group, 87% recovered ovarian function in comparison with only 25% of those randomized to observation group (Michopoulos et al., 2013b). CBT not only restored ovarian function but also improved neuroendocrine and metabolic alterations of FHA.

Taken together, existing data indicate that behavioral therapy such as CBT may restore ovarian activity, ameliorate HPA axis activation, and lead to partial recovery of the thyroidal axis in women with FHA. It is important to identify the behavioral antecedents of FHA as recovery depends on diminishing their endocrine effects. The goal of the treatment should be to alter the inciting event or attitudes responsible for initiating or supporting endocrine reactivity. It is crucial to set behavioral goals and have realistic expectations of self and others. Certain behaviors, such as exercise and dieting, may be undertaken to afford a greater sense of control, but they may also unwittingly serve as additional metabolic challenges and enhance the neuroendocrine and autonomic responses to current or subsequent psychogenic and metabolic stressors. Thus focusing on the behaviors and attitudes that fuel FHA are most likely to yield the highest chance for resumption of ovulation and neuroendocrine recovery.

Medical management

FHA is more than an isolated disruption of the GnRH pulse generator and elicits a constellation of neuroendocrine aberrations. The goal of therapy is to reduce stress and its neuroendocrine concomitants (Berga et al., 2003; McEwen, 2006). The usual clinical intervention is oral contraceptives or exogenous hormone replacement therapy in women who wish to avoid pregnancy. The benefit of such management strategies in FHA is not clear. For instance, bone growth demands an anabolic state and the increased metabolic mobilization associated with FHA may provoke catabolism. Furthermore the stimulation of withdrawal bleeding may mask the situation and give a false sense of assurance to both the clinician and the patient.

In 2017 the Endocrine Society published Clinical Practice Guidelines to guide the diagnosis and treatment of FHA (Gordon et al., 2017). The working group highlighted the importance of a multidisciplinary treatment approach, including psychologic, behavioral, dietary, and pharmacologic support, and noted that hormone replacement strategies were highly unlikely to support bone accretion in the presence of metabolic imbalance and hypercortisolism.

Hormone replacement or oral contraceptive therapy has been used by some physicians as a treatment mainstay to prevent osteoporosis. However, bone accretion requires other factors than just the sex steroids (Gordon et al., 2017). Women with anorexia nervosa as well as FHA had lower bone mass than age-matched women with regular cycles; women with anorexia nervosa had lower bone mass than those with FHA and a greater disturbance of neuroendocrine homeostasis (Grinspoon et al., 1999). Women with anorexia nervosa do not respond to exogenous sex steroid administration to compensate for the bone loss (Miller and Klibanski, 1999). In anorexia nervosa as well as FHA, adequate metabolic balance and nutritional factors such as calcium and vitamin D also play an important role in achieving sufficient bone density and may be deficient due to dietary restrictions. Sex steroid therapy alone will not improve ongoing metabolic derangements. Excess cortisol can interfere with osteoblast activity and inhibit bone formation. In women with FHA, both estrogen and androgen are low and this leads to deficient bone formation as well as excessive bone resorption. In addition, cortisol also interferes with estrogen action (Whirledge and Cidlowski, 2013). Since both sex steroids and bisphosphonates primarily work by retarding bone resorption, women with FHA and anorexia nervosa have both reduced bone formation and increased bone resorption. In addition, bisphosphonates are potentially teratogenic. Recovery of regular menses in these women is associated with bone growth, but bone density remains below expected values, possibly reflecting only partial recovery. A small study of oral contraceptive use led to small gains in bone density over placebo; however, cyclic progestin administration led to greater bone loss than placebo over a 12-month intervention period (Hergenroeder et al., 1997). Overall, bone density in FHA depends on genetic factors, degree and duration of hypogonadism, and extent of metabolic compromise.

The long-term health consequences of FHH, hypothyroidism, and hypercortisolism have been difficult to chronicle. In rat models, disruption of hippocampal neurons has been demonstrated (Stein-Behrens et al., 1994). Health consequences such as neuropsychiatric disorders, cognitive decline, and accelerated aging have been associated with elevated cortisol levels (Lupien et al., 1997). Stress has been associated with reduced telomerase activity and shorter telomere length in premenopausal women, indicating accelerated aging (Epel et al., 2004). Additionally, reduced telomerase length was related to higher stress hormones, while lower activity of telomerase was associated with amplified autonomic reactivity to stress (Epel et al., 2006).

Hormone and sex steroid therapy per se is unlikely to be detrimental in FHA, but more than administration of these hormones is required. The effectiveness of psychopharmacologic approaches has not been evaluated in this population but the use of antidepressants can be considered.

The neuromodulatory cascade that responds to stress that eventuates in the constellation of neuroendocrine aberrations described in FHA (Berga et al., 1989) is complex and no one central agent is independently responsible. Our primate model (Michopoulos et al., 2013a) suggested that CRH antagonism may be helpful; however, there is no CRH antagonist available for use in humans. Antidepressants reduce HPA axis activation, but their role in reducing associated metabolic imbalances is unclear. Naltrexone has been used with variable success to restore GnRH drive; the mechanism of action is thought to be antagonism of endogenous opioids (Wildt et al., 1993). Women with FHA given constant infusion of kisspeptin-54 for 10h had a temporary increase in LH pulsatility (Jayasena et al., 2014). However, kisspeptin is not yet available clinically and more research is needed in this area to determine if the response is sustained. Subcutaneous daily recombinant leptin treatment has been investigated for FHA but was only able to restore ovulatory function in less than half of the women treated (Welt et al., 2004).

Thus optimal treatment for FHA is behavioral change that reduces the impact of stressors to thereby ameliorate HPA axis activation and promote metabolic balance. We have shown that CBT has a high likelihood of achieving this goal, including recovery of ovarian function (Berga et al., 2003; Berga and Loucks, 2006; Michopoulos et al., 2013b). Nonpharmacologic therapies, such as stress management, empower individuals by nurturing self-care and competency. These therapies can produce long-term physical and mental health benefits that extend beyond resumption of gonadal function.

Fertility treatment in FHA

Ovulation induction is often recommended for women with FHA who desire fertility. The mechanism of action of ovulation-inducing agents such as clomiphene and letrozole therapy is to elevate FSH levels and initiate folliculogenesis by antagonizing estradiol action. However, women with FHH/FHA are overly sensitive to estradiol and antagonism rarely engenders an increase in FSH sufficient to initiate and sustain folliculogenesis.

Strategies for ovulation induction in FHA include exogenous administration of gonadotropins and exogenous pulsatile GnRH administration (Martin et al., 1993). Women with FHH/FHA may require both LH and FSH to initiate and sustain folliculogenesis with exogenous gonadotropins. Should ovulation induction be done in women with FHH/FHA? Chronic metabolic and psychogenic stress increases the risk of intrauterine growth restriction and preterm delivery (van der Spuy et al., 1988). Even more worrisome is the probability of impaired neuropsychologic development in the offspring. Women with FHA have varying degrees of decrements in thyroxine hormone levels compared with control population (Becker et al., 1999). Children born to women with subclinical hypothyroidism may develop intellectual disability (Thompson et al., 2018). Since the mother is the only source of thyroid hormones during the first trimester and the predominant source in the second trimester, ovulation induction in the presence of metabolic derangements associated with FHA may have a risk for the abnormal neuropsychologic development in the offspring (Morreale de Escobar et al., 2000). In addition, studies suggest that women with FHA exhibit compromised parenting skills. A woman who is chronically stressed is less likely to cope well with the additional stress of caring for a newborn. In summary, ovulation induction in a woman with FHA carries significantly higher risks of preterm delivery, intrauterine growth restriction, and poor neuropsychologic development in the offspring. The preferred plan of action is to address behaviors and cognitions that increase stress sensitivity and promote resumption of ovulatory function before attempting ovulation induction or assisted reproduction.

FHA is theoretically reversible. In one study, 72% of women with FHA recovered in 6 years (Hirvonen, 1977). In another, more than half of the women with FHA who had undergone ovulation induction conceived and became eumenorrheic postpartum (Lewinthal et al., 1987). Hypothalamic recovery likely comprises a temporal pattern of readjustment; namely, HPA axis restoration precedes increased GnRH drive and resolution of hypothalamic hypothyroidism (Michopoulos et al., 2013b). To avoid frustration and unrealistic expectations, patients should be counseled about this timeline and reassured that reproductive recovery takes time following lifestyle alterations (Berga et al., 2003; Berga and Loucks, 2006; Michopoulos et al., 2013b).

POLYCYSTIC OVARY SYNDROME

PCOS is one of the most common causes of anovulation; its presentation includes oligomenorrhea and hyperandrogenism. The definition of PCOS is controversial. In 1990 the National Institutes of Health (NIH) defined PCOS as the presence of chronic anovulation and clinical or biochemical evidence of hyperandrogenism, after exclusion of other etiologies (Zawadzki and Dunaif, 1992). The 2003 Rotterdam European Society for Human Reproduction and Embryology and the American Society for Reproductive Medicine (ESHRE/ASRM) criteria widened the 1990 NIH diagnostic criteria for PCOS and defined it as the presence of at least two of the following criteria: hyperandrogenism, determined by biochemical or clinical signs (excess total testosterone or free testosterone or hirsutism); ovarian dysfunction described by amenorrhea or oligomenorrhea and chronic anovulation; and the presence of polycystic morphology in ovaries (Rotterdam ESHRE/ASRM-Sponsored PCOS Consensus Workshop Group, 2004). The diagnosis of PCOS also requires the exclusion of other related disorders characterized by androgen excess (Azziz et al., 2009).

Pathophysiology

PCOS is conceptualized as anovulation due to hyperandrogenism and is variably accompanied by neuroendocrine aberrations that cause anovulation. One hypothesis is that prenatal androgen excess imprints the fetal GnRH pulse generator. GnRH drive, both pulse frequency and amplitude, is increased in women with PCOS relative to eumenorrheic women (Berga et al., 1993; Kalro et al., 2001). Further, women with PCOS display hypothalamic GnRH resistance to sex steroid feedback (Daniels and Berga, 1997) that is restored by administering an antiandrogen, flutamide (Eagleson et al., 2000), but not by exposure to the insulin-sensitizer metformin (Eagleson et al., 2003). We documented elevations of estradiol, testosterone, and GABA in the cerebrospinal fluid of women with PCOS (Kawwass et al., 2017) and have suggested that increased GABA reflects the altered sex steroid milieu while also increasing GnRH drive (Fig. 1.1). High AMH has been implicated as a cause of increased GnRH pulse frequency (Cimino et al., 2016). Amplified GnRH drive results in increased LH and decreased FSH. Increased LH drive, in turn, pushes and sustains ovarian androgen overproduction while chronically suppressing FSH. In turn, there is insufficient FSH to support folliculogenesis and ovulation. That multiple genetic variants have been linked to PCOS likely explains the heterogeneity of presentation (Day et al., 2018). Expression of this syndrome is influenced by environmental factors such as nutrition and lifestyle (Garg and Merhi, 2015).

Behavioral contributors

PCOS is a complex syndrome and, because of the heterogeneity of presentation, multidisciplinary care may be required to treat the full spectrum of this condition (Bekx et al., 2010). It often presents in adolescence as irregular cycles, hirsutism, insulin resistance, and excessive weight gain. Infertility is another common presentation. PCOS must be distinguished from anovulation due to stress-induced obesity, autoimmune diabetes, attenuated congenital adrenal hyperplasia (CAH), exogenous androgen exposure, and hypercortisolism due to Cushing's syndrome or disease. Treatment priorities depend on determining metabolic status, including insulin resistance, hypertension, dyslipidemia, and treatment goals. Once an accurate diagnosis has been established and concomitant conditions identified, treatment goals include regulation of the menstrual cycle and prevention of endometrial hyperplasia, normalization of nutritional and metabolic status, proiding psychologic support for poor self-image, depression, and anxiety, and treatment of infertility (Bekx et al., 2010).

There are multiple factors that interplay and contribute to the pathogenesis of PCOS, stress being one of them. It has been hypothesized that stress has an important role in regulating the different clinical manifestations in PCOS. Responses to stressors are primarily initiated and maintained by cognitions and behaviors activating the limbic–HPA axis (Smith and Vale, 2006). A study demonstrated exaggerated responses of the sympathetic nervous system, including elevated pulse rate, associated with HPA activation in overweight/obese PCOS patients as compared with lean PCOS patients and age-matched controls (Basu et al., 2018). Metabolic stress aggravates the psychologic, reproductive, and metabolic concomitants of PCOS (Papalou and Diamanti-Kandarakis, 2017).

In addition, obesity is a common feature in women with PCOS; 50%–60% of women with PCOS have an increase in adipose tissue in the abdominal viscera (Legro, 2012). Interestingly, hyperinsulinemia itself causes obesity by having an anabolic effect on fat metabolism (Vilmann et al., 2012). Obesity, glucose, appetite signals, and hyperinsulinemia synergistically influence neuroendocrine function and alter GnRH drive (Fig. 1.1). A vicious cycle ensues in which android fat worsens PCOS by decreasing insulin action and altering neuroendocrine function.

Differential diagnosis

The differential diagnoses of PCOS include other causes of hyperandrogenism such as ovarian hyperthecosis, CAH, Cushing's syndrome, androgen-secreting tumors, and exogenous androgen administration.

Ovarian hyperthecosis comprises most of the cases of hyperandrogenism in postmenopausal women and is due to increased luteinized theca cell mass in the ovarian stroma (Krug and Berga, 2002). When compared with PCOS, hyperthecosis is characteristically associated with more severe biochemical hyperandrogenism and virilization. CAH is an autosomal recessive disorder due to deficiency of enzymes involved in cortisol synthesis in the adrenal cortex. Enzyme deficiency in CAH is mostly due to 21-hydroxylase, which is reported in 90%–95% of cases. Nonclassic 21-hydroxylase deficiency tends to present around puberty with features of hyperandrogenism, menstrual disturbance or hirsutism, acne or infertility. Cushing's syndrome is a rare cause of androgen excess and is due to increased circulating concentrations of cortisol. It can present with variable manifestations, including centripetal adiposity, abdominal striae, facial plethora, abnormal supraclavicular fat pads, and hyperandrogenism (acne, hirsutism, and male pattern baldness) (Howlett et al., 1985). It may also be iatrogenic due to excess steroid exposure. Invariably, Cushing's results in low gonadotropins and hypothalamic amenorrhea.

Rare causes of hyperandrogenism include androgen-secreting tumors of the ovaries or adrenal glands. These tumors typically present with sudden onset and rapidly progressive features of hyperandrogenism. The common virilizing ovarian tumors are Sertoli Leydig cell tumors, granulosa cell tumors, hilar cell tumors, and Brenner tumors. Ovarian androgen-secreting tumors are typically characterized by increased serum testosterone levels. Androgen-secreting adrenal tumors include adrenal adenomas and carcinomas. These tumors present with markedly elevated serum testosterone, as well as urinary 17-ketosteroid levels (Cordera et al., 2003). Imaging techniques such as computerized tomography, MRI, and ultrasound are used to differentiate among these clinical disorders. Administration of anabolic androgen steroids may be used to improve athletic performance or increase libido in women. Their administration can lead to acne, hirsutism, seborrheic changes, and striae and can mimic PCOS (Hartgens and Kuipers, 2004).

Behavioral management

Given the strong association among obesity, insulin resistance, hyperandrogenism, and other manifestations of PCOS, weight loss is recommended for those with an elevated BMI. The Androgen Excess PCOS Society (AE-PCOS) recommends weight loss as one of many lifestyle interventions and as a primary treatment option for obese women with PCOS (Azziz et al., 2009). Several studies have shown that even a small degree of weight loss, as low as 5% of the body weight, helps to reduce the severity of the symptoms (Guzick et al., 1994). Women with this syndrome are at higher risk for metabolic disorders and diabetes, and weight loss may reduce cardiovascular risk and insulin resistance. Women with PCOS should also engage in regular exercise.

Current evidence that affirms these recommendations includes a 2018 study demonstrating that CBT combined with lifestyle modifications, such as exercise and improved nutrition, resulted in higher weight loss and significantly improved quality of life (Cooney et al., 2018). CBT is considerably more effective in assisting with weight loss than lifestyle modifications alone (Brennan et al., 2017). Moran et al. compared the impact of different diets on PCOS-related outcomes. They found slightly more weight loss with a monounsaturated fat-enriched diet, including improved menstrual regularity and superior quality of life with a diet having low glycemic index and reduced depression and improved self-esteem with a diet higher in protein (Moran et al., 2019). It remains to be determined if the central neuroendocrine regulation of appetite and weight homeostasis differs in women with PCOS when compared with these factors in women who are eumenorrheic and, if so, how that understanding would improve treatment approaches.

FUNCTIONAL HYPOTHALAMIC HYPOGONADISM (FHH) IN MEN

The primary signal for gonadal function in men, as in women, is appropriate GnRH drive. HH is characterized by gonadal failure secondary to deficient secretion of GnRH and gonadotropins. The causes are both organic and functional. Stressors are sex specific with men being more sensitive to metabolic and energetic deficiencies than women (Klibanski et al., 1981; MacConnie et al., 1986). Short-term fasting in obese men can decrease serum FSH and testosterone concentrations (Klibanski et al., 1981). As in women, lateralizing neurologic symptoms should prompt an investigation for organic central nervous system causes such as tumors. Kallmann syndrome is uncommon but is generally associated with anosmia as the developmental origin of GnRH neurons is the olfactory placode.

Pathophysiology

In both males and females, gonadal function depends on appropriate GnRH stimulation of pituitary gonadotropes to induce pituitary GnRH receptors and secretion of LH and FSH. Studies in prepubertal children and men with HH have revealed that qualitative and quantitative pituitary responsiveness to GnRH requires prior exposure (Barkan et al., 1985). In addition, GnRH pulse frequency and amplitude regulate pituitary gonadotropin secretion.

Behavioral contributors

Studies have shown reduced serum testosterone levels in response to exercise and undernutrition. Soldiers undergoing 8 weeks of military training, which included intense training exercises and an average daily energy deficit of 1000 kcal, had testosterone levels below the normal range (Nindl et al., 2007). Men undergoing a 5-day war drill showed an increase in cortisol and concomitant decreases in LH, FSH, and testosterone (Opstad, 1992). Gonadotropins and testosterone rebounded only after cortisol returned to baseline. Men who fasted for 48 h showed a dramatic decline in testosterone levels (Cameron et al., 1991). Shemesh et al. reported the case of a 22-year-old man with frank FHH who presented with decreased libido, erectile dysfunction, and oligoasthenozoospermia associated with bodybuilding and severe nutritional fat restriction that responded fully to increased fat intake (Shemesh et al., 2011). However, FHH may be difficult to detect as men may not experience decreased libido or sexual dysfunction even when testosterone levels are low (Spratt et al., 1988); therefore the true incidence of FHH in men is unknown.

Differential diagnosis

Congenital causes of HH in men are characterized by complete or partial lack of pubertal development. Around 50%–60% of individuals affected with the congenital form also exhibit anosmia, defining Kallmann syndrome. Additional phenotypic abnormalities present in patients with Kallmann syndrome may include craniofacial defects, such as cleft lip/palate, neurosensory deafness, digital anomalies such as clinodactyly or syndactyly, renal abnormalities such as unilateral renal agenesis, and neurologic defects such as oculomotor abnormalities and cerebellar ataxia. In Kallmann syndrome, hypogonadism is secondary to GnRH deficiency, which is due to the defective migration of olfactory and GnRH neurons (Mitchell et al., 2011). A growing list of genes encoding neuropeptides and proteins that are associated with the migration of GnRH neurons has been associated with congenital HH, underscoring the complexity and heterogeneity of congenital HH (Silveira et al., 2010).

Acquired causes of HH are mainly structural or functional abnormalities of the hypothalamic–pituitary axis, and many patients have multiple pituitary hormone deficiencies, including gonadotropin deficiency. These disorders include infiltrative conditions of the hypothalamic–pituitary system, such as sarcoidosis, histiocytosis, lymphocytic hypophysitis, or space-occupying lesions, for example, craniopharyngiomas, pituitary adenomas, and other central nervous system tumors (Silveira et al., 2002). Adult-onset isolated HH can be secondary to systemic conditions and exogenous substance use, such as opiates, androgens, psychotropics, and glucocorticoids. It is important to screen for hyperprolactinemia and hypothyroidism. Adult-onset HH can be idiopathic and is a rare disorder characterized by failure of gonadotropin secretion following otherwise normal sexual maturation in men in whom other causes have been excluded (Nachtigall et al., 1997). It presents with severe hypogonadism, along with testosterone levels below 130 ng/dL and no spontaneous reversals (Dwyer et al., 2010). Although more common in women, men may develop anorexia nervosa and other eating disorders; the prevalence was reported in around 5%–15% of the men studied (Lucas et al., 1991).

The differential diagnosis of HH in boys includes constitutional delay, which leads to delay in both growth and puberty. Adolescents with constitutional delay of puberty typically present with delayed growth and bone age. In contrast, patients with congenital HH have normal height during childhood despite the absence of the pubertal growth spurt. Various tests such as daily excretion of FSH and LH in the urine, gonadotropin response to GnRH, and testosterone response to hCG have been proposed but the single measurement of inhibin B was found to have 100% sensibility and specificity to differentiate between constitutional delay of puberty and HH in adolescents (Coutant et al., 2010).

Behavioral management

Like FHH/FHA in women, FHH in men is theoretically reversible with appropriate lifestyle changes. FHH in men and FHA in women represent an adaptive response to combined metabolic and psychogenic stressors. Men may be more sensitive to undernutrition and women to increased expenditure.

A previous study demonstrated that restricted energy intake and physical exertion during training exercises is associated with declines in testosterone and gonadotropins (Opstad, 1992; Friedl et al., 2000). Leptin administration prevented starvation-induced inhibition of the HPG axis (Chan et al., 2003). Importantly, reproductive dysfunction in men with acute energy deficit can be reversed and testosterone levels restored when the triggering event is alleviated (Dwyer et al., 2019).

To date, no specific studies have examined the effect of stress due to prolonged exercise or nutrition deficiency on spermatogenesis or Sertoli cell function (inhibin B levels) in men. Indirect evidence of the altered gonadal status has been suggested in men with FHH (Dwyer et al., 2019).

Medical management

Low testosterone levels in the setting of low or normal gonadotropin levels in men suggest the diagnosis of HH. Rarely, selective LH or FSH deficiency can occur due to inactivating mutations (Layman et al., 2002).

The goals of therapy for HH in adolescents are the initiation and maintenance of normal pubertal development and initiation of fertility when the patient desires. Adult men with symptomatic androgen deficiency are treated with testosterone to improve sexual function and increase muscle mass but this intervention also compromises spermatogenesis and may result in infertility. Testosterone may be administered as intramuscular injections of long-acting testosterone (testosterone cypionate or enanthate) or transdermal formulations, including gel preparations (5–10 g/d), or by the application of testosterone patches on the skin in nongenital areas (Bhasin et al., 2010). The long-term goals of testosterone treatment are to achieve and maintain levels in the normal adult range. However, testosterone therapy is contraindicated in patients desiring fertility, those who have breast or prostate cancer, and those at high risk for prostate cancer, uncontrolled or poorly controlled heart failure, hematocrit more than 50%, severe lower urinary tract symptoms, or untreated severe obstructive sleep apnea (Bhasin et al., 2010).

In men with HH who desire fertility, treatment with gonadotropins may be required for spermatogenesis (Zacharin et al., 2012). The typical gonadotropin regimen includes combination of hCG and FSH, which can be administered via the intrasubcutaneous or intramuscular route. The target is to attain normal testosterone levels, which is usually achieved by 6 months of continuous treatment, and normal spermatogenesis, which is achieved in up to 80% of patients. In patients with incomplete pubertal development and azoospermia, hCG can be started for 6 months followed by adding FSH. The prognosis is better in patients with normally descended testis, large testicular volume, and elevated serum inhibin B levels at the start of the treatment.

THE NEUROENDOCRINOLOGY OF PREGNANCY

The neuroendocrinology of pregnancy entails an interplay among maternal, fetal, and placental endocrine systems. Pregnancy is a profound metabolic challenge that requires not only maternal neuroendocrine adaptations to facilitate the maintenance of pregnancy for the requisite duration but also maternal neuroendocrine signaling to facilitate parturition and lactation. Maternal neuroendocrine responses are required to promote maternal survival during and after labor and delivery. To meet these physiologic challenges, maternal, fetal, and placental neuroendocrine function must be coordinated.

The impact of pregnancy on the maternal endocrine system begins with the production of human hCG from the trophoblast at the time of implantation. Changes in maternal metabolism during pregnancy prioritize fetal growth and include hyperinsulinemia, insulin resistance, increased plasma lipids, and more efficient plasma amino acid transport. There is an immediate increase in maternal metabolism that increases further as pregnancy progresses. Lactation also carries an enormous metabolic demand. Maternal adaptation to the infant's schedule functions as both a psychologic and metabolic stressor. The typical infant does not display circadian rhythms, including a night-asleep, day-awake sleep pattern, until about 12 weeks after birth.

The enlargement of the anterior pituitary gland during pregnancy largely reflects a 10-fold increase in lactotrope size and number. Somatotrope and gonadotrope cell numbers are reduced; corticotrope or thyrotrope numbers are stable (Foyouzi et al., 2004). The marked increase in estrogen levels of placental origin during pregnancy enhances pituitary prolactin synthesis and secretion leading to circulating prolactin levels that are approximately 20 times higher during pregnancy. Amniotic fluid prolactin levels are 10–100 times higher than in the maternal circulation, reflecting decidual production. Serum prolactin levels return to the baseline of nonpregnancy approximately 7 days after delivery in the absence of lactation. With breastfeeding, basal prolactin levels remain elevated but gradually decrease; however, during suckling, there is a brisk rise in prolactin levels within 30 min (Foyouzi et al., 2004). GH levels in the maternal circulation remain unchanged throughout pregnancy. The placenta synthesizes and secretes biologically active GnRH and maternal pituitary gonadotropin production decreases throughout pregnancy. The thyroid gland enlarges to meet metabolic demand, but TSH levels during the first trimester are significantly lower than those in the second and third trimesters or in the nonpregnant state, primarily because of the intrinsic thyrotropic activity of hCG (Glinoer, 1997). Women with hypothyroidism need supplementation because their thyroid gland will be unable to respond to placental hCG with the needed increase in thyroxine (Alexander et al., 2004).

The maternal hypothalamus and pituitary gland and placenta are sources of circulating adrenocorticotropic hormone (ACTH) and CRH during pregnancy (Petraglia et al., 2010). Placental CRH is identical to the hypothalamic CRH in structure, bioactivity, and immunoreactivity (Chan et al., 1988). Exogenous CRH stimulates the release of ACTH from both tissues in a dose-dependent manner. Maternal ACTH levels rise during pregnancy followed by a drop just before parturition and a marked 15-fold increase during the stress of delivery (Petraglia et al., 2010). ACTH concentrations return to the prepregnancy level within 24 h of delivery. CRH concentrations increase due to placental production resulting in high CRH levels in maternal serum during pregnancy, an exponential rise during the "late" third trimester, and a peak during labor (Riley et al., 1991). Both CRH and ACTH levels become undetectable within 24 h after delivery (Goland et al., 1986). Interestingly, while glucocorticoids inhibit the release of maternal CRH and ACTH, they stimulate the expression of placental CRH (Riley et al., 1991). The enhanced cortisol production is due to an increase in the maternal plasma ACTH concentrations and responsiveness of the adrenal cortex to ACTH stimulation during pregnancy (Lindsay and Nieman, 2005). Cortisol secretion follows maternal ACTH secretion and the diurnal rhythm of maternal cortisol is maintained during pregnancy. As a result of the hyperestrogenemia of pregnancy, hepatic production of cortisol-binding globulin is increased, which in turn results in decreased clearance of cortisol and a threefold rise in total plasma cortisol by week 26, when the levels plateau until the onset of labor (Lindsay and Nieman, 2005). Fig. 1.3 shows the integration of maternal, fetal, and placental HPA pathways (Petraglia et al., 2010). Premature activation of maternal and fetal stress pathways by emotional, infectious, and behavioral stressors may result in premature labor (Petraglia et al., 2010). Placental CRH concentrations increase in response to proinflammatory cytokines and glucocorticoids and thus have a pivotal role in the establishment, continuation, and termination of pregnancy (Margioris et al., 1988; Petraglia et al., 2010).

Oxytocin is a peptide neurohormone produced by hypothalamic neurons that project to the posterior lobe

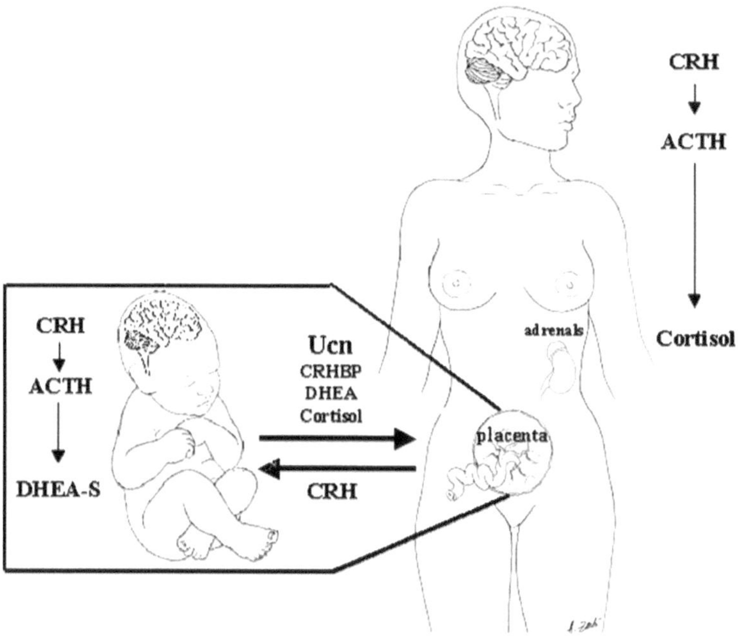

Fig. 1.3. CRH and urocortin (Ucn) secretion in preterm delivery: the placenta represents the major source for CRH, whereas the fetus abundantly secretes Ucn, adrenal DHEA, and CRHBP in maternal circulation (Petraglia et al., 2010).

of the pituitary (Satake et al., 1999). Oxytocin is synthesized by neurons in the supraoptic and periventricular nuclei of the hypothalamus and released as a prepropeptide from the posterior lobe of the pituitary gland (Renaud and Bourque, 1991). Major functions of oxytocin include uterine contractility and milk ejection. Oxytocin levels parallel the increase in maternal serum levels of estradiol and progesterone. The levels increase further with cervical dilation and vaginal distention during labor and delivery, stimulating contraction of the uterine smooth muscles and enhancing fetal expulsion (Zeeman et al., 1997).

SUMMARY

Reproductive function, including pregnancy, parturition, and lactation, requires neuroendocrine adaptation. In particular, gonadal function in both women and men requires appropriate GnRH drive. In women, reduced GnRH drive results in a spectrum of ovarian functions that includes anovulation. Paradoxically, too rapid a GnRH pulse frequency also results in anovulation by raising LH and suppressing FSH. Thus there is a narrow U-shaped curve for GnRH drive in women with both slow and fast GnRH pulse frequencies causing anovulation. In men, there is no known condition associated with a too rapid GnRH pulse frequency. Whether the U-shaped curve in GnRH drive exists for men requires additional investigation. Undoubtedly, mild reductions GnRH drive in men produce an occult phenotype that is often diagnosed at the time of an infertility investigation while more severe suppression of GnRH produces frank hypogonadism and is often termed "low testosterone syndrome." The neuroendocrine concomitants of altered GnRH differ and reflect the underlying pathophysiology. It is important to recognize that reproductive dysfunction is not an isolated neuroendocrine disruption and that many medical conditions have reproductive concomitants. Thus, a thorough appreciation of neuroendocrinology aids both diagnosis and treatment of the full spectrum of reproductive function and dysfunction.

References

Abreu AP, Dauber A, Macedo DB et al. (2013). Central precious puberty caused by mutations in the imprinted gene MKRN3. N Engl J Med 368: 2467–2475.

Alexander EK, Marqusee E, Lawrence J et al. (2004). Timing and magnitude of increases in levothyroxine requirements during pregnancy in women with hypothyroidism. N Engl J Med 351: 241–249.

Aloi JA, Bergendahl M, Iranmanesh A et al. (1997). Pulsatile intravenous gonadotropin-releasing hormone administration averts fasting-induced hypogonadotropism and hypoandrogenemia in healthy, normal weight men. J Clin Endocrinol Metab 82: 1543–1548.

Altemus M, Deuster PA, Galliven E et al. (1995). Suppression of hypothalamic-pituitary-adrenal axis responses to stress in lactating women. J Clin Endocrinol Metab 80: 2954–2959.

Alvero R, Kimzey L, Sebring N et al. (1998). Effects of fasting on neuroendocrine function and follicle development in lean women. J Clin Endocrinol Metab 83: 76–80.

Azziz R, Carmina E, Dewailly D et al. (2009). The androgen excess and PCOS society criteria for the polycystic ovary syndrome: the complete task force report. Fertil Steril 91: 456–488.

Barkan AL, Reame NE, Kelch RP et al. (1985). Idiopathic hypogonadotropic hypogonadism in men: dependence of the hormone responses to gonadotropin-releasing hormone (GnRH) on the magnitude of the endogenous GnRH secretory defect. J Clin Endocrinol Metab 61: 1118–1125.

Basu BR, Chowdhury O, Saha SK (2018). Possible link between stress-related factors and altered body composition in women with polycystic ovarian syndrome. J Hum Reprod Sci 11: 10–18.

Becker AE, Grinspoon SK, Klibanski A et al. (1999). Eating disorders. N Engl J Med 340: 1092–1098.

Bekx MT, Connor EC, Allen DB (2010). Characteristics of adolescents presenting to a multidisciplinary clinic for polycystic ovarian syndrome. J Pediatr Adolesc Gynecol 23: 7–10.

Berga SL (2008). Stress and reproduction: a tale of false dichotomy? Endocrinology 149: 867–868.

Berga SL (2010). Should infertility patients be screened for celiac disease? Gynecol Endocrinol 26: 781–782.

Berga SL, Girton LG (1989). The psychoneuroendocrinology of functional hypothalamic amenorrhea. Psychiatr Clin North Am 12: 105–116.

Berga SL, Loucks TL (2006). Use of cognitive behavior therapy for functional hypothalamic amenorrhea. Ann N Y Acad Sci 1092: 114–129.

Berga SL, Mortola JF, Yen SSC (1988). Amplification of nocturnal melatonin secretion in women with functional hypothalamic amenorrhea. J Clin Endocrinol Metab 66: 242–244.

Berga SL, Mortola JF, Girton L et al. (1989). Neuroendocrine aberrations in women with functional hypothalamic amenorrhea. J Clin Endocrinol Metab 68: 301–308.

Berga SL, Guzick DS, Winters SJ (1993). Increased luteinizing hormone and alpha-subunit secretion in women with hyperandrogenic anovulation. J Clin Endocrinol Metab 77: 895–901.

Berga SL, Daniels TL, Giles DE (1997). Women with functional hypothalamic amenorrhea but not other forms of anovulation display amplified cortisol concentrations. Fertil Steril 67: 1024–1030.

Berga SL, Loucks-Daniels TL, Adler LJ et al. (2000). Cerebrospinal fluid levels of corticotropin-releasing hormone in women with functional hypothalamic amenorrhea. Am J Obstet Gynecol 182: 776–781.

Berga SL, Marcus MD, Loucks TL et al. (2003). Recovery of ovarian activity in women with functional hypothalamic amenorrhea who were treated with cognitive behavior therapy. Fertil Steril 80: 976–981.

Bethea CL, Pau FK, Fox S et al. (2005). Sensitivity to stress-induced reproductive dysfunction linked to activity of the serotonin system. Fertil Steril 83: 148–155.

Bethea CL, Kim A, Reddy AP et al. (2014). Hypothalamic KISS1 expression, gonadotrophin-releasing hormone and neurotransmitter innervation vary with stress and sensitivity in macaques. J Neuroendocrinol 26: 267–281.

Bhasin S, Cunningham GR, Hayes FJ et al. (2010). Testosterone therapy in men with androgen deficiency syndromes: an Endocrine Society Clinical Practice Guideline. J Clin Endocrinol Metab 95: 2536–2559.

Biller BM, Federoff HJ, Koenig JI et al. (1990). Abnormal cortisol secretion and responses to corticotropin-releasing hormone in women with hypothalamic amenorrhea. J Clin Endocrinol Metab 70: 311–317.

Boyar RM, Katz J, Finkelstein JW et al. (1974). Anorexia nervosa. Immaturity of the 24-hour luteinizing hormone secretory pattern. N Engl J Med 291: 861–865.

Brennan L, Teede H, Skouteris H et al. (2017). Lifestyle and behavioral management of polycystic ovary syndrome. J Womens Health (Larchmt) 26: 836–848.

Brundu B, Loucks TL, Adler LJ et al. (2006). Increased cortisol in the cerebrospinal fluid of women with functional hypothalamic amenorrhea. J Clin Endocrinol Metab 91: 1561–1565.

Bullen BA, Skrinar GS, Beitins IZ et al. (1985). Induction of menstrual disorders by strenuous exercise in untrained women. N Engl J Med 312: 1349–1353.

Cameron JL, Weltzin TE, McConaha C et al. (1991). Slowing of pulsatile luteinizing hormone secretion in men after forty-eight hours of fasting. J Clin Endocrinol Metab 73: 35–41.

Caronia LM, Martin C, Welt CK et al. (2011). A genetic basis for functional hypothalamic amenorrhea. N Engl J Med 364: 215–225.

Centeno ML, Sanchez RL, Reddy AP et al. (2007). Corticotropin-releasing hormone and pro-opiomelanocortin gene expression in female monkeys with differences in sensitivity to stress. Neuroendocrinology 86: 277–288.

Chan EC, Thomson M, Madsen G et al. (1988). Differential processing of corticotrophin-releasing hormone by the human placenta and hypothalamus. Biochem Biophys Res Commun 153: 1229–1235.

Chan JL, Heist K, DePaoli AM et al. (2003). The role of falling leptin levels in the neuroendocrine and metabolic adaptation to short-term starvation in healthy men. J Clin Invest 111: 1409–1421.

Chen MD, O'Byrne KT, Chiappini SE et al. (1992). Hypoglycemic 'stress' and gonadotropin-releasing hormone pulse generator activity in the rhesus monkey: role of the ovary. Neuroendocrinology 56: 666–673.

Cimino I, Casoni F, Liu X et al. (2016). Novel role for anti-Müllerian hormone in the regulation of GnRH neuron excitability and hormone secretion. Nat Commun 7: 10055.

Cooney LG, Milman LW, Hantsoo L et al. (2018). Cognitive-behavioral therapy improves weight loss and quality of life in women with polycystic ovary syndrome: a pilot randomized clinical trial. Fertil Steril 110: 161–171.

Cordera F, Grant C, van Heerden J et al. (2003). Androgen-secreting adrenal tumors. Surgery 134: 874–880.

Coutant R, Biette-Demeneix E, Bouvattier C et al. (2010). Baseline inhibin B and anti-Mullerian hormone measurements for diagnosis of hypogonadotropic hypogonadism (HH) in boys with delayed puberty. J Clin Endocrinol Metab 95: 5225–5232.

Csapo AI, Pulkkinen MO, Ruttner B et al. (1972). The significance of the human corpus luteum in pregnancy maintenance. Am J Obstet Gynecol 112: 1061–1067.

Culler MD, Negro-Vilar A (1989). Endogenous inhibin suppresses only basal follicle-stimulating hormone secretion but suppresses all parameters of pulsatile luteinizing hormone secretion in the diestrous female rat. Endocrinology 124: 2944–2953.

Daniels TL, Berga SL (1997). Resistance of gonadotropin releasing hormone drive to sex steroid-induced suppression in hyperandrogenic anovulation. J Clin Endocrinol Metab 82: 4179–4183.

Day F, Karaderi T, Jones MR et al. (2018). Large-scale, genome-wise meta-analysis of polycystic ovary syndrome suggests shared genetic architecture for different diagnosis criteria. PLoS Genet 14: e1007813.

De Souza MJ, Miller BE, Loucks AB et al. (1998). High frequency of luteal phase deficiency and anovulation in recreational women runners: blunted elevation in follicle-stimulating hormone observed during luteal-follicular transition. J Clin Endocrinol Metab 83: 4220–4232.

Drew FL (1961). The epidemiology of secondary amenorrhea. J Chronic Dis 14: 396–407.

Dubois SL, Wolfe A, Radovick S et al. (2016). Estradiol restrains prepubertal gonadotropin secretion in female mice via activation of ER-alpha in kisspeptin neurons. Endocrinology 157: 1546–1554.

Dwyer AA, Hayes FJ, Plummer L et al. (2010). The long-term clinical follow-up and natural history of men with adult-onset Idiopathic hypogonadotropic hypogonadism. J Clin Endocrinol Metab 95: 4235–4243.

Dwyer AA, Chavan NR, Lewkowitz-Shpuntoff H et al. (2019). Functional hypogonadotropic hypogonadism in men: underlying neuroendocrine mechanisms and natural history. J Clin Endocrinol Metab 104: 3403–3414.

Eagleson CA, Gingrich MB, Pastor CL et al. (2000). Polycystic ovarian syndrome: evidence that flutamide restores sensitivity of the gonadotropin-releasing hormone pulse generator to inhibition by estradiol and progesterone. J Clin Endocrinol Metab 85: 4047–4520.

Eagleson CA, Bellows AB, Hu K et al. (2003). Obese patients with polycystic ovary syndrome: evidence that metformin does not restore sensitivity of the gonadotropin-releasing hormone pulse generator to inhibition by ovarian steroids. J Clin Endocrinol Metab 88: 5158–5162.

Epel ES, Blackburn EH, Lin J et al. (2004). Accelerated telomere shortening in response to life stress. Proc Natl Acad Sci U S A 101: 17312–17315.

Epel ES, Lin J, Wilhelm FH et al. (2006). Cell aging in relation to stress arousal and cardiovascular disease risk factors. Psychoneuroendocrinology 31: 277–287.

Ferin M (1999). Stress and the reproductive cycle. J Clin Endocrinol Metab 84: 1768–1774.

Filicori M, Santoro N, Merriam GR et al. (1986). Characterization of the physiological pattern of episodic gonadotropin secretion throughout the human menstrual cycle. J Clin Endocrinol Metab 62: 1136–1144.

Foyouzi N, Frisbaek Y, Norwitz ER (2004). Pituitary gland and pregnancy. Obstet Gynecol Clin North Am 31: 873–892. xi.

Friedl KE, Moore RJ, Hoyt RW et al. (2000). Endocrine markers of semistarvation in healthy lean men in a multistressor environment. J Appl Physiol 88: 1820–1830.

Fries H, Nillius SJ, Pettersson F (1974). Epidemiology of secondary amenorrhea II. A retrospective evaluation of etiology with special regard to psychogenic factors and weight loss. Am J Obstet Gynecol 118: 473–479.

Gajdos ZK, Hirschhorn JN, Palmert MR (2009). What controls the timing of puberty? An update on progress from genetic investigation. Curr Opin Endocrinol Diabetes Obes 16: 16–24.

Garg D, Merhi Z (2015). Advanced glycation end products: link between diet and ovulatory dysfunction in PCOS? Nutrients 7: 10129–10144.

Giles DE, Berga SL (1993). Cognitive and psychiatric correlates of functional hypothalamic amenorrhea: a controlled comparison. Fertil Steril 60: 486–492.

Glinoer D (1997). The regulation of thyroid function in pregnancy: pathways of endocrine adaptation from physiology to pathology. Endocr Rev 18: 404–433.

Goland RS, Wardlaw SL, Stark RI et al. (1986). High levels of corticotropin-releasing hormone immunoactivity in maternal and fetal plasma during pregnancy. J Clin Endocrinol Metab 63: 1199–1203.

Gordon CM, Ackerman KE, Berga SL et al. (2017). Functional hypothalamic amenorrhea: an Endocrine Society clinical practice guideline. J Clin Endocrinol Metab 102: 1413–1439.

Gottsch ML, Cunningham MJ, Smith JT et al. (2004). A role for kisspeptins in the regulation of gonadotropin secretion in the mouse. Endocrinology 145: 4073–4077.

Grinspoon S, Miller K, Coyle C et al. (1999). Severity of osteopenia in estrogen-deficient women with anorexia nervosa and hypothalamic amenorrhea. J Clin Endocrinol Metab 84: 2049–2055.

Gross KM, Matsumoto AM, Bremner WJ (1987). Differential control of luteinizing hormone and follicle-stimulating hormone secretion by luteinizing hormone-releasing hormone pulse frequency in man. J Clin Endocrinol Metab 64: 675–680.

Guzick DS, Wing R, Smith D et al. (1994). Endocrine consequences of weight loss in obese, hyperandrogenic, anovulatory women. Fertil Steril 61: 598–604.

Gwirtsman HE, Kaye WH, George DT et al. (1989). Central and peripheral ACTH and cortisol levels in anorexia nervosa and bulimia. Arch Gen Psychiatry 46: 61–69.

Hall JE, Schoenfeld DA, Martin KA et al. (1992). Hypothalamic gonadotropin releasing hormone secretion and follicle-stimulating hormone dynamics during the luteal-follicular transition. J Clin Endocrinol Metab 74: 600–607.

Hartgens F, Kuipers H (2004). Effects of androgenic-anabolic steroids in athletes. Sports Med 34: 513–554.

Hergenroeder AC, Smith EO, Shypailo R et al. (1997). Bone mineral changes in young women with hypothalamic amenorrhea treated with oral contraceptives, medroxyprogesterone, or placebo over 12 months. Am J Obstet Gynecol 176: 1017–1025.

Hirvonen E (1977). Etiology, clinical features and prognosis in secondary amenorrhea. Int J Fertil 22: 69–76.

Hoff JD, Quigley ME, Yen SSC (1983). Hormonal dynamics at midcycle: a reevaluation. J Clin Endocrinol Metab 57: 792–796.

Howlett TA, Rees LH, Besser GM (1985). Cushing's syndrome. J Clin Endocrinol Metab 14: 911–945.

Hurley DM, Brian RJ, Burger HG (1983). Ovulation induction with subcutaneous pulsatile gonadotropin-releasing hormone: singleton pregnancies in patients with previous multiple pregnancies after gonadotropin therapy. Fertil Steril 40: 575–579.

Jarrell H, Hoffman JB, Kaplan JR et al. (2008). Polymorphisms in the serotonin reuptake transporter gene modify the consequences of social status on metabolic health in female rhesus monkeys. Physiol Behav 93: 807–819.

Jayasena CN, Abbara A, Veldhuis JD et al. (2014). Increasing LH pulsatility in women with hypothalamic amenorrhoea using intravenous infusion of Kisspeptin-54. J Clin Endocrinol Metab 99: E953–E961.

Kalro BN, Loucks TL, Berga SL (2001). Neuromodulation in polycystic ovary syndrome. Obstet Gynecol Clin North Am 28: 35–62.

Kaplan JR, Manuck SB (2004). Ovarian dysfunction, stress, and disease: a primate continuum. ILAR J 45: 89–115.

Kawwass JF, Sanders KM, Loucks TL et al. (2017). Increased cerebrospinal fluid levels of GABA, testosterone and estradiol in women with polycystic ovary syndrome. Hum Reprod 32: 1450–1456.

Kaye WH, Gwirtsman HE, George DT et al. (1987). Elevated cerebrospinal fluid levels of immunoreactive corticotropin-releasing hormone in anorexia nervosa: relation to state of nutrition, adrenal function, and intensity of depression. J Clin Endocrinol Metab 64: 203–208.

Klibanski A, Beitins IZ, Badger T et al. (1981). Reproductive function during fasting in men. J Clin Endocrinol Metab 53: 258–263.

Krug E, Berga SL (2002). Postmenopausal hyperthecosis: functional dysregulation of androgenesis in climacteric ovary. Obstet Gynecol 99: 893–897.

Laughlin GA, Dominguez CE, Yen SSC (1998). Nutritional and endocrine-metabolic aberrations in women with functional hypothalamic amenorrhea. J Clin Endocrinol Metab 83: 25–32.

Layman LC (1999). Genetics of human hypogonadotropic hypogonadism. Am J Med Genet 89: 240–248.

Layman LC, Porto AL, Xie J et al. (2002). FSH beta gene mutations in a female with partial breast development and a male sibling with normal puberty and azoospermia. J Clin Endocrinol Metab 87: 3702–3707.

Legro R (2012). Obesity and PCOS: implications for diagnosis and treatment. Semin Reprod Med 30: 496–506.

Lewinthal D, Corenblum B, Brooks JH et al. (1987). Spontaneous return of menstruation in hypothalamic amenorrhea following gonadotropin-releasing hormone-induced pregnancy. Fertil Steril 47: 870–871.

Lindsay JR, Nieman LK (2005). The hypothalamic-pituitary-adrenal axis in pregnancy: challenges in disease detection and treatment. Endocr Rev 26: 775–799.

Loucks AB, Thuma JR (2003). Luteinizing hormone pulsatility is disrupted at a threshold of energy availability in regularly menstruating women. J Clin Endocrinol Metab 88: 297–311.

Loucks AB, Mortola JF, Girton L et al. (1989). Alterations in the hypothalamic-pituitary-ovarian and the hypothalamic-pituitary-adrenal axes in athletic women. J Clin Endocrinol Metab 68: 402–411.

Lowry DW, Lowry DL, Berga SL et al. (1996). Secondary amenorrhea due to hydrocephalus treated with endoscopic ventriculocisternostomy. Case report. J Neurosurg 85: 1148–1152.

Lucas AR, Beard CM, O'Fallon WM et al. (1991). 50-year trends in the incidence of anorexia nervosa in Rochester, Minn.: a population-based study. Am J Psychiatry 148: 917–922.

Lupien SJ, Gaudreau S, Tchiteya BM et al. (1997). Stress-induced declarative memory impairment in healthy elderly subjects: relationship to cortisol reactivity. J Clin Endocrinol Metab 82: 2070–2075.

MacConnie SE, Barkan A, Lampman RM et al. (1986). Decreased hypothalamic gonadotropin-releasing hormone secretion in male marathon runners. N Engl J Med 315: 411–417.

Marcus MD, Loucks TL, Berga SL (2001). Psychological correlates of functional hypothalamic amenorrhea. Fertil Steril 76: 310–316.

Margioris AN, Grino M, Protos P et al. (1988). Corticotropin-releasing hormone and oxytocin stimulate the release of placental proopiomelanocortin peptides. J Clin Endocrinol Metab 66: 922–926.

Martin KA, Hall JE, Adams JM et al. (1993). Comparison of exogenous gonadotropins and pulsatile gonadotropin-releasing hormone for induction of ovulation in hypogonadotropic amenorrhea. J Clin Endocrinol Metab 77: 125–129.

Martin C, Navarro VM, Simavli S et al. (2014). Leptin-responsive GABAergic neurons regulate fertility through pathways that result in reduced kisspeptinergic tone. J Neurosci 34: 6047–6056.

Mason AJ, Pitts SL, Nikolics K et al. (1986). The hypogonadal mouse: reproductive functions restored by gene therapy. Science 234: 1372–1378.

Mastorakos G, Pavlatou MG, Mizamtsidi M (2006). The hypothalamic-pituitary-adrenal and the hypothalamic-pituitary-gonadal axes interplay. Pediatr Endocrinol Rev 3: 172–181.

McCarthy MM (2013). A piece in the puzzle of puberty. Nat Neurosci 16: 251–253.

McEwen BS (2006). Protective and damaging effects of stress mediators: central role of the brain. Dialogues Clin Neurosci 8: 367–381.

Michopoulos V, Embree M, Reding K et al. (2013a). CRH receptor antagonism reverses the effect of social subordination upon central GABAA receptor binding in estradiol-treated ovariectomized female rhesus monkeys. Neuroscience 250: 300–308.

Michopoulos V, Mancini F, Loucks TL et al. (2013b). Neuroendocrine recovery initiated by cognitive behavioral therapy in women with functional hypothalamic amenorrhea: a randomized, controlled trial. Fertil Steril 99: 2084–2091.

Miller KK, Klibanski A (1999). Clinical review 106: amenorrheic bone loss. J Clin Endocrinol Metab 84: 1775–1783.

Miller DS, Reid RR, Cetel NS et al. (1983). Pulsatile administration of low-dose gonadotropin-releasing hormone: ovulation and pregnancy in women with hypothalamic amenorrhea. JAMA 250: 2937–2941.

Mitchell AL, Dwyer A, Pitteloud N et al. (2011). Genetic basis and variable phenotypic expression of Kallmann syndrome: towards a unifying theory. Trends Endocrinol Metab 22: 249–258.

Moran LJ, Noakes M, Clifton P et al. (2019). Predictors of lifestyle intervention attrition or weight loss success in women with polycystic ovary syndrome who are overweight or obese. Nutrients 11: 492.

Morreale de Escobar G, Obregón MJ, Escobar del Rey F (2000). Is neuropsychological development related to maternal hypothyroidism or to maternal hypothyroxinemia? J Clin Endocrinol Metab 85: 3975–3987.

Mortola JF, Liu J, Gillin CJ et al. (1987). Pulsatile rhythms of adrenocorticotropin (ACTH) and cortisol in women with endogenous depression: evidence for increased ACTH pulse frequency. J Clin Endocrinol Metab 65: 962–968.

Nachtigall LB, Boepple PA, Pralong FP et al. (1997). Adult-onset idiopathic hypogonadotropic hypogonadism—a treatable form of male infertility. N Engl J Med 336: 410–415.

Naftolin F, Garcia-Segura LM, Horvath TL et al. (2007). Estrogen-induced hypothalamic synaptic plasticity and pituitary sensitization in the control of the estrogen-induced gonadotrophin surge. Reprod Sci 14: 101–116.

Navarro VM, Kaiser UB (2013). Metabolic influences on neuroendocrine regulation of reproduction. Curr Opin Endocrinol Diabetes Obes 20: 335–341.

Nindl BC, Barnes BR, Alemnay JA et al. (2007). Physiological consequences of U.S. Army ranger training. Med Sci Sports Exerc 39: 1380–1387.

Opstad PK (1992). Androgenic hormones during prolonged physical stress, sleep, and energy deficiency. J Clin Endocrinol Metab 74: 1176–1183.

Papalou O, Diamanti-Kandarakis E (2017). The role of stress in PCOS. Expert Rev Endocrinol Metab 12: 87–95.

Petraglia F, Imperatore A, Challis JR (2010). Neuroendocrine mechanisms in pregnancy and parturition. Endocr Rev 31: 783–816.

Petrides JS, Mueller GP, Kalogeras KT et al. (1994). Exercise-induced activation of the hypothalamic-pituitary-adrenal axis: marked differences in the sensitivity to glucocorticoid suppression. J Clin Endocrinol Metab 79: 377–383.

Pohl CR, Knobil E (1982). The role of the central nervous system in the control of ovarian function in higher primates. Annu Rev Physiol 44: 583–593.

Ramaswamy S, Guerriero KA, Gibbs RB et al. (2008). Structural interactions between kisspeptin and GnRH neurons in the mediobasal hypothalamus of the male rhesus monkey (Macaca mulatta) as revealed by double immunofluorescence and confocal microscopy. Endocrinology 149: 4387–4395.

Rasmussen DD, Liu JH, Wolf PL et al. (1986). Gonadotropin-releasing hormone neurosecretion in the human hypothalamus: in vitro regulation by dopamine. J Clin Endocrinol Metab 62: 479–483.

Reifenstein EC (1946). Psychogenic or hypothalamic amenorrhea. Med Clin North Am 30: 1103–1114.

Reindollar RH, Novak M, Tho S et al. (1986). Adult-onset amenorrhea: a study of 262 patients. Am J Obstet Gynecol 155: 531–541.

Reiter EO, Kaplan SL, Conte FA et al. (1975). Responsivity of pituitary gonadotropes to luteinizing hormone-releasing factor in idiopathic precocious puberty, precocious thelarche, precocious adrenarche, and in patients treated with medroxyprogesterone acetate. Pediatr Res 9: 111–116.

Renaud LP, Bourque CW (1991). Neurophysiology and neuropharmacology of hypothalamic magnocellular neurons secreting vasopressin and oxytocin. Prog Neurobiol 36: 131–169.

Riley SC, Walton JC, Herlick JM et al. (1991). The localization and distribution of corticotropin-releasing hormone in the human placenta and fetal membranes throughout gestation. J Clin Endocrinol Metab 72: 1001–1007.

Romero LM, Plotsky PM, Sapolsky RM (1993). Patterns of adrenocorticotropin secretagog release with hypoglycemia, novelty, and restraint after colchicine blockade of axonal transport. Endocrinology 132: 199–204.

Rotterdam ESHRE/ASRM-Sponsored PCOS Consensus Workshop Group (2004). Revised 2003 consensus on diagnostic criteria and long-term health risks related to polycystic ovary syndrome. Fertil Steril 81: 19–25.

Saketos M, Sharma N, Santoro NF (1993). Suppression of the hypothalamic-pituitary-ovarian axis in normal women by glucocorticoids. Biol Reprod 49: 1270–1276.

Samuels MH, Luther M, Henry P et al. (1994). Effects of hydrocortisone on pulsatile pituitary glycoprotein secretion. J Clin Endocrinol Metab 78: 211–215.

Sanders KM, Kawwass JF, Loucks T et al. (2018). Heightened cortisol response to exercise challenge in women with functional hypothalamic amenorrhea. Am J Obstet Gynecol 218: 230.e1–230.e6.

Santoro N, Filicori M, Crowley WF (1986). Hypogonadotropic disorders in men and women: diagnosis and therapy with pulsatile gonadotropin-releasing hormone. Endocr Rev 7: 11–23.

Santoro N, Brown JR, Adel T et al. (1996). Characterization of reproductive hormonal dynamics in the perimenopause. J Clin Endocrinol Metab 81: 1495–1501.

Satake H, Takuwa K, Minakata H et al. (1999). Evidence for conservation of the vasopressin/oxytocin superfamily in Annelida. J Biol Chem 274: 5605–5611.

Schally AV, Arimura A, Kastin AJ et al. (1971). Gonadotropin-releasing hormone: one polypeptide regulates secretion of luteinizing and follicle-stimulating hormones. Science 173: 1036–1038.

Shanan J, Brzezinski A, Sulman F et al. (1965). Active coping behavior, anxiety, and cortical steroid excretion in the prediction of transient amenorrhea. Behav Sci 10: 461–465.

Shemesh A, Endevelt R, Levy Y (2011). Reversible nutritional hypogonadism in a 22-year-old man. Am J Med 124: e1–e2.

Shively CA, Laber-Laird K, Anton RF (1997). Behavior and physiology of social stress and depression in female cynomolgus monkeys. Biol Psychiatry 41: 871–882.

Silveira LF, MacColl GS, Bouloux PM (2002). Hypogonadotropic hypogonadism. Semin Reprod Med 20: 327–338.

Silveira LF, Trarbach EB, Latronico AC (2010). Genetics basis for GnRH-dependent pubertal disorders in humans. Mol Cell Endocrinol 324: 30–38.

Smith SM, Vale WW (2006). The role of the hypothalamic-pituitary-adrenal axis in neuroendocrine responses to stress. Dialogues Clin Neurosci 8: 383–395.

Spratt DI, O'Dea LS, Schoenfeld D et al. (1988). Neuroendocrine-gonadal axis in men: frequent sampling of LH, FSH, and testosterone. Am J Physiol 254: E658–E666.

Spratt DI, Longcope C, Cox PM et al. (1993). Differential changes in serum concentrations of androgens and estrogens (in relation with cortisol) in postmenopausal women with acute illness. J Clin Endocrinol Metab 76: 1542–1547.

Stein-Behrens B, Mattson MP, Chang I et al. (1994). Stress exacerbates neuron loss and cytoskeletal pathology in the hippocampus. J Neurosci 14: 5373–5380.

Stocco C, Telleria C, Gibori G (2007). The molecular control of corpus luteum formation, function, and regression. Endocr Rev 28: 117–149.

Stopa EG, Koh ET, Svendsen CN et al. (1991). Computer-assisted mapping of immunoreactive mammalian gonadotropin-releasing hormone in adult human basal forebrain and amygdala. Endocrinology 128: 3199–3207.

Terasawa E (1995). Control of luteinizing hormone-releasing hormone pulse generation in nonhuman primates. Cell Mol Neurobiol 15: 141–164.

Thompson W, Russell G, Baragwanath G et al. (2018). Maternal thyroid hormone insufficiency during pregnancy and risk of neurodevelopmental disorders in offspring: a systematic review and meta-analysis. Clin Endocrinol (Oxf) 88: 575–584.

van der Spuy ZM, Steer PJ, McCusker M et al. (1988). Outcome of pregnancy in underweight women after spontaneous and induced ovulation. Br Med J 296: 962–965.

Van Vugt DA, Piercy J, Farley AE et al. (1997). Luteinizing hormone secretion and corticotropin-releasing factor gene expression in the paraventricular nucleus of rhesus monkeys following cortisol synthesis inhibition. Endocrinology 138: 2249–2258.

Vigersky RA, Andersen AE, Thompson RH et al. (1977). Hypothalamic dysfunction in secondary amenorrhea associated with simple weight loss. N Engl J Med 297: 1141–1145.

Vilmann LS, Thisted E, Baker JL et al. (2012). Development of obesity and polycystic ovary syndrome in adolescents. Horm Res Paediatr 78: 269–278.

Welt CK, Martin KA, Taylor AE et al. (1997). Frequency modulation of follicle-stimulating hormone (FSH) during the luteal-follicular transition: evidence for FSH control of inhibin B in normal women. J Clin Endocrinol Metab 82: 2645–2652.

Welt CK, Chan JL, Bullen J et al. (2004). Recombinant human leptin in women with hypothalamic amenorrhea. N Engl J Med 351: 987–997.

Whirledge S, Cidlowski JA (2013). A role for glucocorticoids in stress-impaired reproduction: beyond the hypothalamus and pituitary. Endocrinology 154: 4450–4468.

Wildt L, Häusler A, Marshall G et al. (1981). Frequency and amplitude of gonadotropin-releasing hormone stimulation and gonadotropin secretion in the rhesus monkey. Endocrinology 109: 376–385.

Wildt L, Leyendecker G, Sir-Petermann T et al. (1993). Treatment with naltrexone in hypothalamic ovarian failure: induction of ovulation and pregnancy. Hum Reprod 8: 350–358.

Williams NI, Berga SL, Cameron JL (2007). Synergism between psychosocial and metabolic stressors: impact on reproductive function in cynomolgus monkeys. Am J Physiol Endocrinol Metab 293: E270–E276.

Witkin JW, O'Sullivan H, Miller R et al. (1997). GnRH perikarya in medial basal hypothalamus of pubertal female rhesus macaque are ensheathed with glia. J Neuroendocrinol 9: 881–885.

Xiao E, Luckhaus J, Niemann W et al. (1989). Acute inhibition of gonadotropin secretion by corticotropin-releasing hormone in the primate: are the adrenal glands involved? Endocrinology 124: 1632–1637.

Yu B, Ruman J, Christman G (2011). The role of peripheral gonadotropin-releasing hormone receptors in female reproduction. Fertil Steril 95: 465–473.

Zacharin M, Sabin MA, Nair VV et al. (2012). Addition of recombinant follicle-stimulating hormone to human chorionic gonadotropin treatment in adolescents and young adults with hypogonadotropic hypogonadism promotes normal testicular growth and may promote early spermatogenesis. Fertil Steril 98: 836–842.

Zawadzki JK, Dunaif A (1992). Diagnostic criteria for polycystic ovary syndrome: toward a rational approach. In: A Dunaif, JR Givens, F Haseltine, GR Merriam (Eds.), Polycystic ovary syndrome. Blackwell Scientific Publications, Cambridge, pp. 377–384.

Zeeman GG, Khan-Dawood FS, Dawood MY (1997). Oxytocin and its receptor in pregnancy and parturition: current concepts and clinical implications. Obstet Gynecol 89: 873–883.

Zeleznik AJ (2001). Follicle selection in primates: "many are called but few are chosen". Biol Reprod 65: 655–659.

Handbook of Clinical Neurology, Vol. 171 (3rd series)
Neurology and Pregnancy: Pathophysiology and Patient Care
E.A.P. Steegers, M.J. Cipolla and E.C. Miller, Editors
https://doi.org/10.1016/B978-0-444-64239-4.00002-3

Chapter 2

Neurophysiological and cognitive changes in pregnancy

DAVID R. GRATTAN* AND SHARON R. LADYMAN

Centre for Neuroendocrinology and Department of Anatomy, School of Biomedical Sciences, University of Otago, Dunedin, New Zealand

Abstract

The hormonal fluctuations in pregnancy drive a wide range of adaptive changes in the maternal brain. These range from specific neurophysiological changes in the patterns of activity of individual neuronal populations, through to complete modification of circuit characteristics leading to fundamental changes in behavior. From a neurologic perspective, the key hormone changes are those of the sex steroids, estradiol and progesterone, secreted first from the ovary and then from the placenta, the adrenal glucocorticoid cortisol, as well as the anterior pituitary peptide hormone prolactin and its pregnancy-specific homolog placental lactogen. All of these hormones are markedly elevated during pregnancy and cross the blood–brain barrier to exert actions on neuronal populations through receptors expressed in specific regions. Many of the hormone-induced changes are in autonomic or homeostatic systems. For example, patterns of oxytocin and prolactin secretion are dramatically altered to support novel physiological functions. Appetite is increased and feedback responses to metabolic hormones such as leptin and insulin are suppressed to promote a positive energy balance. Fundamental physiological systems such as glucose homeostasis and thermoregulation are modified to optimize conditions for fetal development. In addition to these largely autonomic changes, there are also changes in mood, behavior, and higher processes such as cognition. This chapter summarizes the hormonal changes associated with pregnancy and reviews how these changes impact on brain function, drawing on examples from animal research, as well as available information about human pregnancy.

INTRODUCTION

Pregnancy represents one of the most profound physiological challenges faced by an adult. Indeed, a study described the sustained changes in metabolic rate typical of pregnancy as at the limits of human physical capability, similar to that expended in extreme physical activity such as an ultramarathon, but over a longer time frame (Thurber et al., 2019). The hormonal fluctuations associated with pregnancy drive a wide range of adaptive changes to maternal physiology to help the mother cope with these new physiological demands (Napso et al., 2018). Some of the physical changes are obvious. There is enlargement of the uterus to adjust to the growing fetus, and development and differentiation of the mammary glands into highly active secretory tissue for milk production. Underpinning these developments are modifications in the cardiovascular and respiratory systems and changes in glucose homeostasis and fluid balance, in gut morphology to promote energy absorption, and in the way the brain perceives metabolic signals from the body. There are also changes in the skeleton, mobilizing calcium for fetal growth, and in the immune system, changing responses to minimize the possibility of rejection of the fetus. Indeed, there is hardly a system in the body that does not modify its function to adjust to the demands of pregnancy (Napso et al., 2018). The brain is no exception to this, and there are many adaptive changes that occur in the maternal brain that contribute

*Correspondence to: David R. Grattan, Ph.D., Department of Anatomy, University of Otago, P.O. Box 913, Dunedin 9054, New Zealand. Tel: +64-3-4797442, Fax: + 64-3-4797254, E-mail: dave.grattan@otago.ac.nz

to these pregnancy-specific functions, ranging from changes in neurophysiological activity of specific neuronal populations through to plasticity of whole brain circuits to alter maternal mood and behavior.

All of the adaptive changes occurring in the mother during pregnancy are driven by the fluctuating hormonal secretion patterns that are characteristic of this state (Pawluski et al., 2017; Smiley et al., 2019). The mother is exposed to sustained increases in hormones such as progesterone, estradiol, glucocorticoids, and prolactin, maternal hormones that are typically released for discrete periods during her menstrual cycle, but in pregnancy are maintained at high levels for weeks. These hormones are complemented by a catalog of additional pregnancy-specific hormones from the placenta, such as human chorionic gonadotropin (hCG) and chorionic somatomammotropin (or human placental lactogen, hPL). These placental hormones are secreted at specific stages during pregnancy and contribute significantly to driving adaptive change in the mother. As the placenta is essentially fetal tissue, this can be viewed as the fetus overriding the mother's normal endocrine processes and driving changes to promote fetal development, sometimes at a cost to the mother. Many of the hormonal changes that occur during pregnancy impact on the maternal brain. Steroid hormones such as estradiol and progesterone readily access the brain and can influence specific receptors expressed in a wide range of neural centers. Peptide hormones, such as prolactin and hPL, are specifically transported into the brain and, similarly, can act on receptors expressed in numerous brain regions. A particular target of these hormones is the hypothalamus, from where the brain can influence a range of autonomic nervous system targets, thereby influencing many body systems.

In this chapter, we first describe the major changes in pregnancy hormones and the distribution of receptors for these hormones in the brain. We also examine some specific examples of neurophysiological plasticity in the circuits controlling hormone secretion. We then describe some of the hormone-induced adaptive changes that occur in the brain during pregnancy, with a particular focus on metabolic adaptation as this is critical to the body being able to sustain all of these profound changes, and maternal behavior, as the ultimate expression of hormone-induced adaptation of neural circuits. Finally, we examine the effect of pregnancy hormones on "higher" brain function, such as cognition, and evaluate the evidence for the condition known as "mommy brain." While we focus on information known about human pregnancy, where possible, for many of these sections, we necessarily refer to animal studies that provide mechanistic insight. We acknowledge that in some cases it is unknown whether equivalent processes occur in the human maternal brain. As a conclusion, we discuss what happens when the hormone-induced adaptation in the brain is insufficient or goes wrong. To what extent do the neuroendocrine and neurophysiological adaptations of the maternal brain contribute to the risk of neurologic disorders associated with pregnancy and postpartum? It seems likely that the increased risk of mental health problems is a potential cost of the extreme neuronal plasticity required to enable the normal range of adaptive changes that contribute to a healthy pregnancy.

CHANGES IN HORMONES DURING PREGNANCY

Steroid hormones

The key hormonal changes that impact on the maternal brain are those of the sex steroids, estradiol and progesterone, arising first from the maternal ovary and later from the placenta; the glucocorticoids from the maternal adrenal gland; and the protein hormones prolactin from the maternal pituitary gland and its pregnancy-specific homolog hPL from the placenta (Fig. 2.1). In animal studies, mating-induced changes in hormones can begin to impact on neurologic function immediately, such as the mating-induced induction of neurogenesis by the hormone prolactin in mice (Shingo et al., 2003; Larsen and Grattan, 2010). However, there is little clear evidence of such rapid responses in humans. The most important initial change is the maintenance of progesterone secretion from the corpus luteum and the loss of the cyclical changes in sex hormones associated with the menstrual cycle. The postovulatory luteal-phase secretion of progesterone provides a window of time to allow implantation of the fertilized ovum. hCG then serves as a signal enabling maternal recognition of pregnancy. It is secreted very early from the syncytiotrophoblast of the developing embryo and is present in high levels in the blood from around 4 weeks of pregnancy (which is counted from the first day of the last menstrual period or around 2 weeks before ovulation!) and peaking at around 12 weeks. The luteotrophic actions of hCG avert luteal regression, preventing the normal decline in progesterone that would signal the beginning of menstruation, and sustaining progesterone secretion for the first trimester. High levels of progesterone are subsequently maintained throughout the rest of the pregnancy, coming predominantly from the placenta. In almost all species apart from humans, a fall in progesterone occurs immediately prior to parturition. Such a "progesterone withdrawal" has not been observed in humans, but rather a functional progesterone withdrawal has been proposed to occur through changes in progesterone receptor (PR) activity in the myometrium (Zakar and Hertelendy, 2007; Vannuccini et al., 2016).

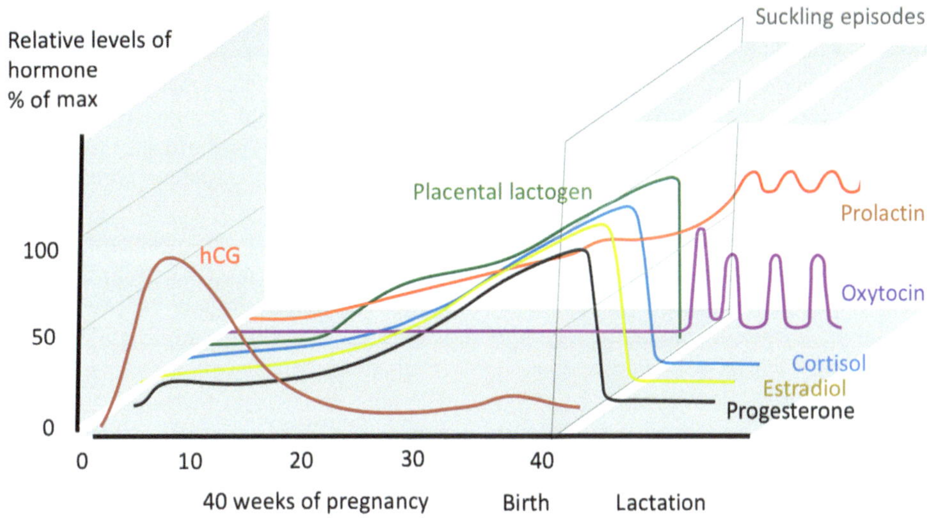

Fig. 2.1. Diagrammatic representation of hormone changes during human pregnancy. Note the abrupt reduction in hormones derived from the placenta (placental lactogen, estradiol, and progesterone) at birth, in contrast to the activation and suckling-induced stimulation of prolactin and oxytocin.

Estradiol-17β (estradiol), the most bioactive form of estrogen, falls after ovulation, and ovarian estradiol secretion remains relatively low throughout the early stages of pregnancy. During the second and third trimesters, however, serum levels of estradiol (and estriol) rise high, with secretion arising from the placenta. By the third trimester, estradiol concentrations in the blood are up to 20-fold higher than ever experienced during the menstrual cycle (Altemus, 2019). Similarly, levels of the glucocorticoid cortisol are elevated during the latter part of pregnancy, likely influenced by corticotrophin-releasing hormone (CRH) that uniquely is expressed by the placenta in primates (Sasaki et al., 1989; Power and Schulkin, 2006; Thomson, 2013). In addition to having a role in stimulating the fetal adrenal, placental CRH acts on the maternal pituitary gland to release adrenocorticotrophic hormone (ACTH), subsequently stimulating cortisol from the adrenal cortex. Unlike hypothalamic CRH, which is inhibited by cortisol, placental CRH is stimulated by cortisol, providing a positive feedback mechanism increasing cortisol secretion as gestation proceeds (Power and Schulkin, 2006). This is a clear example of placental secretions overriding the feedback controls inherent in the maternal endocrine system to drive adaptive changes in the mother. Despite a rise in sex hormone-binding globulin and cortisol-binding globulin during pregnancy, the free levels of these steroid hormones are high, and these steroid hormones readily cross the blood–brain barrier; moreover, their receptors are expressed in a number of brain regions.

Estradiol and PRs are widespread in the brain, being expressed in a number of regions that show pregnancy-specific changes in neurologic function. In particular, they are expressed in the ventromedial and arcuate nucleus of the hypothalamus, throughout the medial preoptic area and bed nucleus of the stria terminalis, the medial amygdala, hippocampus, and cerebral cortex (Simerly et al., 1990; Shughrue et al., 1997a, b; Merchenthaler et al., 2004; Brinton et al., 2008). Glucocorticoid receptors are similarly widespread in the brain, with particular concentrations in the paraventricular nucleus of the hypothalamus and in the hippocampus (Pryce, 2008; Joels, 2018).

Prolactin and placental lactogen

Prolactin levels in the maternal pituitary gland rise progressively throughout pregnancy. Proteomic studies of human pregnancy identified prolactin as one of the top 1% of hormones changing during pregnancy (Romero et al., 2017; Aghaeepour et al., 2018). The increase in prolactin is associated with significant hyperplasia of the pituitary with an increase in lactotroph cell numbers in lactation beginning early in pregnancy (Porter et al., 1990; Scheithauer et al., 1990). Prolactin acts on prolactin receptors expressed widely throughout the hypothalamus, amygdala, and brainstem (Pi and Grattan, 1998a, b; Brown et al., 2010; Kokay et al., 2018). One important consideration is how this relatively large (197–199 amino acids; 21 kDa) polypeptide hormone crosses the blood–brain barrier to gain access to the various regions where the prolactin receptors are expressed. While it is possible that parts of the mediobasal hypothalamus may have an incomplete blood–brain barrier such that hormones can directly access neurons in this area (Schaeffer et al., 2013), it seems unlikely that this is a major mechanism by which prolactin reaches the rest of the brain. Indeed, systemic administration of prolactin has been shown to simultaneously activate neurons throughout the hypothalamus (Brown et al., 2010; Sapsford et al., 2012). There is clear evidence that systemic prolactin crosses the blood–brain barrier through a saturable, carrier-mediated transport system (Walsh et al., 1987). As a result, prolactin levels in the cerebrospinal fluid parallel changes in prolactin in the peripheral circulation (Login and MacLeod, 1977; Nicholson et al., 1980; Grattan and Averill, 1991). Because of the high levels of prolactin receptor expression and prolactin binding seen in the choroid plexus (Walsh et al., 1978, 1990; Pi and Grattan, 1998a, b; Augustine et al., 2003), it had been widely assumed that the prolactin receptor might mediate prolactin entry into the cerebrospinal fluid. However, we have shown that prolactin transport into the brain is independent of the prolactin receptor, occurring just as well in prolactin receptor knockout mice (Brown et al., 2016). Hence, the precise mechanism that translocates prolactin from the blood into the brain remains to be determined.

The placenta serves an important endocrine function in pregnancy and contributes to the altered hormonal profile of a pregnant female. As well as producing the sex steroid hormones described earlier, it also produces a number of important protein hormones, including hPL and a placental variant of growth hormone (hGHv). hPL, the placental analogue of prolactin, is synthesized early in pregnancy and progressively increases throughout pregnancy until parturition (Aghaeepour et al., 2018). By the third trimester, levels of hPL are extremely high, in the order of 1–2 μg/mL (5–10-fold higher than pituitary prolactin). hPL has been shown to act in the brain, presumably crossing the blood–brain barrier by the same process as prolactin. It activates the prolactin receptor with equal affinity to prolactin, and hence the presence of both prolactin and placental lactogen ensures a high level of prolactin receptor activation in the brain throughout pregnancy (both hormones) and into lactation (prolactin). Interestingly, the rise in hPL occurs concurrently with rise in prolactin levels during pregnancy in humans. It would be expected that the high levels of hPL inhibit pituitary prolactin secretion by the activation of the short-loop negative feedback pathway that normally regulates prolactin secretion. The fact that this does not happen suggests that there are adaptive changes in the feedback regulation of prolactin secretion during pregnancy. This is discussed in this chapter.

Changes associated with parturition

Key endocrine events occurring around the time of parturition include progesterone withdrawal, described earlier, and activation of both oxytocin and prolactin secretion (Fig. 2.1). All placentally derived hormones are abruptly lost from the maternal circulation, providing a significant change in signaling to the brain that may have important consequences for postpartum brain function. Oxytocin is released at parturition from the posterior pituitary, when it stimulates uterine contractions for parturition. In the postpartum period, suckling induces prolactin secretion from the anterior pituitary gland, which is required for milk synthesis, and oxytocin is released from the posterior pituitary, which in turn is required for milk ejection from the mammary gland. The high prolactin in the blood at this time continues to cross the blood–brain barrier and has important functions in the brain. Similarly, both the activation of oxytocin neurons at parturition and in response to the suckling stimulus simultaneously induce central release of oxytocin within the hypothalamus (discussed later), where it also has important behavioral effects.

NEUROPHYSIOLOGICAL CHANGES IN THE MATERNAL BRAIN

Oxytocin

Perhaps the most striking and best-characterized neurophysiological change during pregnancy is the change in the pattern of activity of the oxytocin neurons to enable parturition and milk ejection during lactation (Brown, 2016). Oxytocin secretion changes from a relatively low basal release in the nonpregnant state to produce large pulses during parturition (Otsuki et al., 1983; de Geest et al., 1985). The changes in oxytocin secretion occur concurrently with an increase in uterine sensitivity to oxytocin, driven by a marked upregulation of oxytocin receptors in the myometrium (Fuchs et al., 1984). The combination of these effects triggers the rhythmic waves of uterine contractions required for parturition. Oxytocin secretion is then further enhanced by a positive feedback action known as the Ferguson reflex (Ferguson, 1941; Douglas et al., 2001), whereby stretch of the cervix caused by the initial contractions and by the entry of the baby's head into the birth canal induces a spinal somatosensory reflex to promote further oxytocin secretion.

Oxytocin neurons are found in two hypothalamic nuclei, the supraoptic (SON) and paraventricular (PVN) nuclei. While some oxytocin neurons within the PVN project locally within the hypothalamus, many in this nucleus and all in the SON project to the posterior pituitary gland and release oxytocin into the blood. These large neurosecretory neurons are known as "magnocellular" neurons and produce all of the oxytocin involved in stimulating uterine contractions and milk ejection.

The smaller, locally projecting neurons are known as "parvicellular" or small neurons and are involved in releasing oxytocin as a neuropeptide within the brain, regulating a number of behaviors (including maternal behavior; see later) (Neumann, 2008). Interestingly, neuroanatomical studies have found that some of the magnocellular oxytocin neurons have collateral branches of their axon that project into central regions such as the central amygdala and nucleus accumbens and can contribute to the modulation of certain oxytocin-sensitive behaviors (Knobloch et al., 2012). Moreover, while initially a controversial concept, it is now well accepted that the magnocellular oxytocin neurons secrete oxytocin and other molecules from their dendrites, providing a significant source of oxytocin within the central nervous system (Ludwig, 1998).

Oxytocin is synthesized in the neuronal cell bodies in the hypothalamus and transported down the axons to be stored in secretory granules within the nerve terminals in the posterior lobe of the pituitary gland. Secretion of oxytocin from these terminals is essentially identical to neurotransmitter release in other neurons. Activity of the neurons in response to excitatory afferent inputs (for example, from cervical stretch) produces action potentials that pass down the axons, triggering an influx of calcium into the nerve terminal to induce exocytosis of the secretory granules. Thus the pattern of oxytocin secretion is essentially driven by the pattern of activity of the population of oxytocin neurons. Adaptive changes occurring during pregnancy and lactation result in a dramatic change in the neurophysiology of this population of neurons, moving from a relatively slow rate of action potential firing (1–5 spikes per second) to producing high-frequency bursts of activity (up to 100 spikes per second) that last only for a few seconds, but reoccur with increasing frequency as parturition proceeds (Summerlee, 1981). Each burst triggers a wave of oxytocin secretion to drive a uterine contraction (Bicknell, 1988). This bursting activity of the oxytocin neurons continues throughout the period of lactation, when the suckling stimulus provided by the offspring stimulates bursting activity of the oxytocin neurons to induce milk letdown and ejection. Remarkably, while initially triggered specifically by the suckling stimulus, once lactation is established, these milk ejection bursts can become a conditioned stimulus, whereby they will occur in response to stimuli that the mother associates with feeding, such as the baby crying. This is an adaptive change, enabling milk letdown to be initiated prior to the actual suckling, such that milk is immediately available when the baby starts to suckle.

The mechanisms driving the plasticity in oxytocin neurons during pregnancy have been extensively studied (see reviews Brunton and Russell, 2008; Hatton and Wang, 2008; Theodosis et al., 2008; Brown, 2016) (see Fig. 2.2). Changes in functional characteristics of ion channels in the neurons alter the pattern of electrical activity in the neurons (Brown, 2016). There are also marked increases in the afferent inputs into these cells, with glutamatergic, GABAergic, and noradrenergic innervation of oxytocin neurons all increased during lactation (Gies and Theodosis, 1994; El Majdoubi et al., 1996; Michaloudi et al., 1997). The cells also show remarkable structural changes, with retraction of astrocyte coverings allowing increased cell–cell contact between dendrites of adjacent neurons (Perlmutter et al., 1984; Hatton, 1999) and the establishment of intercellular junctions, helping coordinate the activity of the neurons across the population (Theodosis et al., 1981). This structural plasticity is mediated by oxytocin itself, with local release from the somato-dendrite compartment influencing surrounding dendrites (Theodosis et al., 1986). While this level of cellular interaction can account for coordination of oxytocin neurons within a nucleus, generation of a burst of secretion from the posterior pituitary actually involves synchronization across all of the magnocellular neurons across four distinct nuclei (SON and PVN, bilateral), likely requiring internuclei neuronal connections (Brown, 2016).

The hormonal signals that induce the plasticity in oxytocin neuronal activity remain to be determined. Circulating estrogen and progesterone concentrations increase during pregnancy (Brunton and Russell, 2008), but oxytocin neurons do not express estrogen receptor α (ERα) (Peter et al., 1990) and express little ERβ (Shughrue et al., 2002). Nor do they express the classic PRs, even at the end of pregnancy (Francis et al., 2002). While these data suggest that steroid modulation of the oxytocin system must be mediated indirectly, through steroid-sensitive afferent neurons, it is also possible that both hormones influence oxytocin neurons through nonclassic mechanisms. Oxytocin neurons express the membrane estrogen receptor GPR30 (Brailoiu et al., 2007), which might underpin the rapid effects of estrogen to increase oxytocin neuron excitability (Brown et al., 2008). Similarly, progesterone can exert rapid inhibitory actions on oxytocin neurons (Negoro et al., 1973a, b), which is mediated by allosteric modulation of GABA$_A$ receptors after conversion to allopregnanolone (Brussaard et al., 1997). Indeed, progesterone administration can delay birth when given at the end of pregnancy (Antonijevic et al., 2000).

As well as the steroid hormones, however, there is building evidence for regulation of the oxytocin neurons by peptide hormones. In particular, prolactin and hPL have been shown to influence oxytocin neurons.

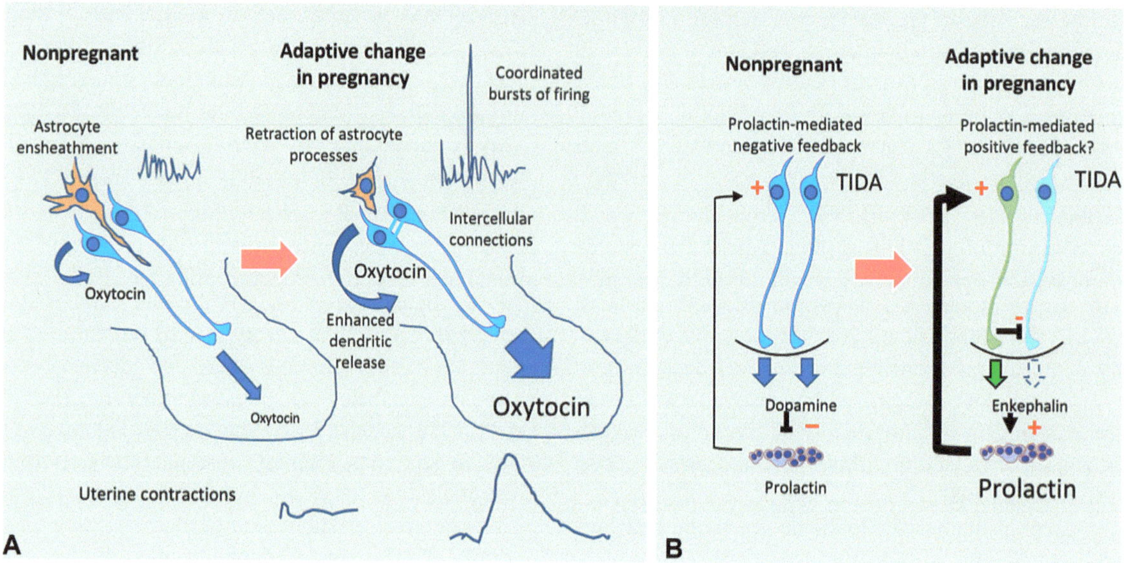

Fig. 2.2. Plasticity in neurophysiological activity of (A) oxytocin and (B) tuberoinfundibular dopamine (TIDA) neurons during pregnancy, resulting in pregnancy- and/or lactation-specific patterns of oxytocin and prolactin secretion, respectively. For oxytocin, there is a shift from low-frequency activity to synchronized bursts of high-frequency resulting in pulses of oxytocin in the blood to drive uterine contractions during birth and milk ejection during lactation. This is controlled by a number of adaptive changes, including withdrawal of astrocytic process surrounding the cells, establishment of cell-to-cell connections, enhanced dendritic release of oxytocin, as well as increased afferent stimulation (not drawn). For prolactin, the normal short-loop feedback mechanism by which it stimulates dopamine release from TIDA neurons is altered. Chronically elevated placental lactogen in pregnancy drives a phenotypic switch in these cells, such that they stop making dopamine and instead make met-enkephalin, which has a stimulatory effect on prolactin secretion. As prolactin continues to stimulate the neurons in pregnancy and lactation, the short-loop feedback essentially operates as positive feedback, promoting further prolactin secretion.

Oxytocin neurons express prolactin receptor mRNA (Kokay et al., 2006) and prolactin administration increases STAT5 phosphorylation in oxytocin neurons (Sapsford et al., 2012; Augustine et al., 2017). Chronic prolactin receptor activation increases oxytocin mRNA expression (Donner and Neumann, 2009), and prolactin increases the number of oxytocin-positive neurons detectable by immunohistochemistry (Augustine et al., 2017). Hence, it appears that the high placental lactogen in mid- and late pregnancy might promote oxytocin synthesis required to cope with the secretory demands of birth and/or lactation. Prolactin treatment prevents the dramatic decrease in oxytocin mRNA expression after removing the suckling stimulus (Ghosh and Sladek, 1995). In contrast to this stimulatory action of prolactin on oxytocin synthesis, prolactin rapidly inhibits the activity of oxytocin neurons in virgin rats (Kokay et al., 2006). This effect is altered in late pregnancy and lactation, however, with the inhibitory effect of prolactin lost in most neurons, and prolactin administration to lactating rats actually increases the firing rate of ~20% of oxytocin neurons (Augustine et al., 2017). This suggests that when these hormones must be secreted together for milk synthesis (prolactin) and milk ejection (oxytocin), prolactin actually supports oxytocin secretion (Grattan et al., 2008; Brown, 2016).

Prolactin inhibition of oxytocin neurons appears to be mediated by activation of potassium channels (Sirzen-Zelenskaya et al., 2011). The excitation that emerges during late pregnancy, however, might be underpinned by the development of a depolarizing action of prolactin in lactation. Prolactin activates transient receptor potential vanilloid 1 (TRPV1) channels in sensory neurons (Patil et al., 2013). SON neurons express TRPV1 channels (Sharif Naeini et al., 2006) and TRPV1 mRNA is upregulated in the SON in lactation (Augustine et al., 2016). Hence, increased TRPV1 channel expression could switch prolactin-induced oxytocin neuron inhibition to excitation in lactation. Another possibility is that prolactin excitation of oxytocin neurons is mediated by afferent inputs. Hypothalamic arcuate nucleus dopamine neurons express prolactin receptors (Kokay and Grattan, 2005) and are excited by prolactin (Romano et al., 2013); these neurons inhibit PVN neurons via coreleased GABA (Zhang and van den Pol, 2016). Given that GABA inhibition of oxytocin neurons switches to excitation in lactation (Lee et al., 2015), prolactin might excite oxytocin neurons in lactating rats via stimulation of GABA transmission from these arcuate nucleus dopamine neurons.

Tuberoinfundibular dopamine neurons and prolactin

The establishment of hyperprolactinemia during pregnancy and lactation provides another well-characterized example of neurophysiological plasticity within a specific neuronal population, this time the tuberoinfundibular dopamine (TIDA) neurons that control prolactin secretion. Prolactin secretion is normally under inhibitory control of the hypothalamus through actions of dopamine released from the TIDA neurons, which are located in the arcuate nucleus of the hypothalamus and project to the median eminence. Dopamine released in the median eminence is transported in the pituitary portal blood to act on dopamine D2 receptors on lactotrophs in the anterior pituitary (Mansour et al., 1990). The observation that mice lacking the dopamine D2 receptor are hyperprolactinemic (Kelly et al., 1997; Saiardi et al., 1997) clearly demonstrates the critical role of dopamine in suppressing endogenous prolactin secretion. Prolactin does not have a specific endocrine target, and as such, it lacks the classic hormone-mediated negative feedback pathway described for all other anterior pituitary hormones. Nevertheless, it is still regulated in a negative feedback manner, with prolactin itself providing the afferent signal in a process known as short-loop feedback through prolactin receptors on TIDA neurons (Lerant and Freeman, 1998; Grattan, 2001; Kokay and Grattan, 2005) (see Fig. 2.2). Electrophysiologic data has demonstrated rapid actions of prolactin on the electrical activity of TIDA neurons in mice (Brown et al., 2012; Romano et al., 2013) or rats (Lyons et al., 2012). In addition to these rapid actions, prolactin also exerts longer-term actions to stimulate dopamine synthesis (Grattan & LeTissier, 2015).

To permit the sustained elevations in prolactin secretion characteristic of pregnancy and lactation, there is a remarkable adaptive change in this short-loop feedback system (Grattan et al., 2008). The mechanisms mediating this adaptive response are only starting to be elucidated. Despite the presence of placental lactogen and rising levels of prolactin, the TIDA neurons apparently no longer release dopamine, rendering the short-loop negative feedback system functionally inactive (Ben-Jonathan et al., 1980; Demarest et al., 1983; Grattan and Averill, 1995; Fliestra and Voogt, 1997). The original perception that hypothalamic dopamine neurons show a “loss of response” to prolactin at this time has proved to be incorrect. In fact, prolactin receptor expression in the dopamine neurons is maintained (Kokay and Grattan, 2005), and acute electrophysiologic responses to prolactin persist in lactation (Romano et al., 2013). Downstream of the prolactin receptor, however, there is a change in the cellular response. Serine 40 phosphorylation of tyrosine hydroxylase in the median eminence is decreased (Feher et al., 2010; Romano et al., 2013), resulting in a reduction in the activity of this enzyme (and a reduction in dopamine synthesis). Strikingly, there is a disconnection between neuronal firing and dopamine release at the median eminence. Even though the electrophysiologic response to prolactin is unchanged, there is no longer detectable release of dopamine in the median eminence (Romano et al., 2013). At the same time as the loss of dopamine secretion, there is an increase in met-enkephalin expression in the dopamine neurons (Merchenthaler, 1993; Szabo et al., 2011; Yip et al., 2019). Hence, the neurons essentially change their phenotype, moving from being dopaminergic to enkephalinergic (Fig. 2.2). As they still respond electrophysiologically to prolactin, they may be mediating a completely different function of prolactin in the brain during pregnancy and lactation.

The hormone changes driving this adaptive plasticity in prolactin feedback are becoming better understood. There is good evidence that the rising levels of estradiol present during pregnancy can promote prolactin secretion, both through inhibition of TIDA neurons and a direct stimulatory action in the pituitary gland. Progesterone withdrawal at the end of pregnancy also promotes prolactin secretion. However, the key signal in triggering the adaptive modifications in the TIDA neurons appears to be placental lactogen, acting through the prolactin receptor itself. As the placental lactogen is not under inhibitory regulation, it essentially bypasses the normal short-loop feedback and provides chronic activation of the prolactin receptors on TIDA neurons. This appears to be critical for the induction of enkephalin in those neurons, as the knockout of prolactin receptors in these neurons prevents this adaptive change (Yip et al., 2019). After birth, the suckling stimulus provides an additional mechanism to stimulate prolactin secretion. The somatosensory stimulation induces a profound suppression of dopamine turnover in TIDA neurons (Selmanoff and Wise, 1981; Demarest et al., 1983; Selmanoff and Gregerson, 1985) and a reduction in tyrosine hydroxylase mRNA levels (Wang et al., 1993). The suckling-induced release of prolactin is also likely to involve one or more prolactin-releasing factors released from the hypothalamus and acting in the pituitary gland. While a number of candidates have been investigated, the nature of this “PRF” remains to be determined.

These two examples show how the hormonal changes of pregnancy can profoundly affect the neurophysiological functioning of individual populations of neurons, in this case resulting in changed patterns of hormone secretion, that subsequently have further effects on other neurons systems. Next we evaluate how some complex neuronal circuits are altered by these hormones, inducing important adaptive changes in maternal physiology.

ADAPTIVE CHANGES IN CENTRAL CONTROL OF MATERNAL PHYSIOLOGY

Central adaptations of maternal energy homeostasis

As mentioned earlier, pregnancy is a metabolically demanding state in which the mother must supply all the energy requirements of the growing fetus, meet her own energy demands, and prepare for the subsequent metabolic demands of lactation. Pregnant females have increased appetite and gain weight over and above the weight gain attributed to fetal growth, leading to the common perception of "eating for two" during pregnancy (Cripps and Williams, 1975; Shirley, 1984; Ladyman and Grattan, 2004; Ladyman et al., 2018). Adaptations include growth and development of mammary glands, changes in glucose homeostasis, and the deposition of fat. Importantly, many of these changes in metabolic process are predictive, establishing a positive energy balance in anticipation of future metabolic demand. This requires a hormone-induced modification of central nervous system circuits that normally tightly control food intake and energy homeostasis.

The hypothalamic circuitry regulating energy homeostasis is complex and has been reviewed in depth elsewhere (e.g. Flak and Myers Jr., 2016). For the purposes of this chapter, we focus on key pathways that might be involved in the hormone-induced adaptive responses during pregnancy. The arcuate nucleus within the hypothalamus is central to appetite regulation, while the ventromedial nucleus (VMN) has classically been described as a "satiety" area as lesions to this area lead to hyperphagia and obesity (reviewed in Elmquist et al., 1999). The PVN is thought to be a site where energy homeostasis information is integrated, and it has bidirectional connections projecting with key neuronal populations in the arcuate nucleus. Within the arcuate nucleus, two of the most well-characterized neuronal populations involved in the regulation of energy balance are the anorectic proopiomelanocortin (POMC) neurons and the orexigenic neuropeptide Y/agouti-related peptide (NPY/AgRP) neurons. POMC is the precursor peptide for a number of different protein products, but it is the downstream actions of its protein product alpha-melanocyte-stimulating hormone (αMSH), which acts through melanocortin receptors (MCRs), that mediate the satiety actions of this neuronal population. Both NPY and AgRP can stimulate food intake, and AgRP is thought to act as an endogenous antagonist of MCRs. Fasting increases NPY and AgRP mRNA while decreasing POMC levels. The adipose tissue-derived hormone leptin provides a critical feedback signal, integrating the activity of these neuronal populations to match the peripheral energy stores. Leptin is normally secretion in proportion to body adipose levels and acts in the hypothalamus, particularly the arcuate nucleus, to suppress food intake and therefore maintain body fat at a reasonably constant level. It usually inhibits the orexigenic NPY/AgRP neurons and stimulates anorexigenic POMC neurons, providing an overall signal to reduce food intake. Leptin is clearly an important regulator of energy homeostasis in both animal models and humans, and a deficiency of leptin is associated with extreme obesity and hyperphagia (Campfield et al., 1995; Halaas et al., 1995; Pelleymounter et al., 1995; Farooqi et al., 2002).

Pregnancy is associated with an increase in both fat mass and leptin levels, as well as an increase in appetite (Hardie et al., 1997; Masuzaki et al., 1997; Amico et al., 1998; Garcia et al., 2000; Ehrhardt et al., 2001; Ladyman and Grattan, 2004; Castracane et al., 2005; Ladyman et al., 2012; Menchetti et al., 2015; Saleri et al., 2015; Gustafson et al., 2019; Szczesna et al., 2019). This is somewhat paradoxical, as the elevated leptin should result in reduced food intake, leading to the hypothesis that pregnancy is a state of relative leptin insensitivity. Studies examining the satiety response to leptin in rodents have demonstrated a marked reduction in leptin-induced suppression of food intake during pregnancy compared with nonpregnant controls (Ladyman and Grattan, 2004; Trujillo et al., 2011; Ladyman et al., 2012). In particular, central administration of leptin in pregnant rats has no effect on food intake (Ladyman and Grattan, 2004). Consistent with this, despite rising leptin levels during pregnancy, POMC mRNA is relatively constant (Mann et al., 1997; Douglas et al., 2002; Rocha et al., 2003; Ladyman et al., 2009) or reduced (Pazos et al., 2013), in alignment with the increase in food intake. Similarly, NPY and AgRP are either constant or increased during pregnancy (Garcia et al., 2003; Rocha et al., 2003; Ladyman et al., 2009; Pazos et al., 2013) as would be expected during a time of increased appetite. Neither of these neuronal populations is apparently responding normally to the rising levels of leptin. Similar changes have been detected in the CSF from pregnant women, with elevated AgRP levels, while POMC levels are unchanged from nonpregnant values (Page-Wilson et al., 2013), despite markedly increased circulating leptin during pregnancy.

As well as the loss of appetite-suppressing effects of leptin during pregnancy, a number of other functions of leptin are also impaired. Leptin action in the brain contributes to the regulation of glucose metabolism (Tups et al., 2017), independent of its effects on body weight and food intake (Morton, 2007; Denroche et al., 2012; Fujikawa and Coppari, 2015). In female rodents, acute leptin treatment leads to impaired glucose tolerance and reduced insulin secretion (Harris, 2014; Ladyman and Grattan, 2016). In contrast, leptin administration has no effect on glucose tolerance or insulin secretion during pregnancy (Ladyman and Grattan, 2016). Leptin can also increase sympathetic nerve activity (SNA) through its effects in the hypothalamus (Shi et al., 2015), and it had been hypothesized that the increased leptin in pregnancy may contribute to the increased SNA that is part of the adaptations in fluid balance and cardiovascular regulation during pregnancy (Reyes et al., 2018). However, it has been shown that, during pregnancy, leptin no longer increases SNA (Shi et al., 2019), representing yet another central function of leptin that is attenuated during pregnancy.

A number of mechanisms underlying pregnancy-induced leptin insensitivity have been characterized (see Fig. 2.3). First, transport of leptin across the blood–brain barrier seems to be impaired, meaning that despite high leptin concentrations in blood, access to the brain is reduced (Gustafson et al., 2019). An increase in levels of leptin-binding proteins has been detected in plasma from pregnant mice (Cumin et al., 1996; Gavrilova et al., 1997) and rats (Seeber et al., 2002). In human pregnancy, the CSF:plasma leptin ratio is lower than in the nonpregnant state, suggesting impaired access of leptin into the brain (Page-Wilson et al., 2013; Andersson-Hall et al., 2018). There is some debate as to whether plasma levels of the short form of the receptor (a predicted binding factor) increase during human pregnancy, as some studies show an increase (Page-Wilson et al., 2013; Andersson-Hall et al., 2018) while others show no change (Lewandowski et al., 1999; Nuamah et al., 2003). Alongside impaired access to the brain, leptin activation of first-order target neurons is attenuated during pregnancy. In rats, central administration of leptin mid-pregnancy, bypassing any impaired transport into the brain, elicits markedly reduced responses in pregnancy, particularly in the VMN (Ladyman and Grattan, 2004, 2005, 2016). This is associated with a reduction in leptin receptor mRNA in the VMN during pregnancy (Ladyman and Grattan, 2005; Szczesna et al., 2019). Several studies have shown an upregulation of SOCS3 in the hypothalamus during pregnancy (Tang et al., 2008; Trujillo et al., 2011; Zampieri et al., 2015; Szczesna et al., 2019). SOCS3 is a key negative regulator in the leptin-responsive JAK/STAT3 signaling pathway and is predicted to reduce leptin action. Transgenic mice with a specific deletion of SOCS3 in leptin receptor-expressing cells demonstrate lower pregnancy-induced weight gain and food intake and have increased sensitivity in the VMH to leptin during pregnancy (Zampieri et al., 2015). Interestingly, mice with a specific deletion of SOCS3 in VMN cells also show increased leptin sensitivity in the VMH during pregnancy, but they do not show lower pregnancy-induced weight gain or food intake (Ramos-Lobo et al., 2017), suggesting that changes in leptin sensitivity in this area during pregnancy may not be responsible for major changes in food intake. VMN-specific SOCS3-KO were less glucose tolerant during pregnancy though, and hence the decrease in leptin sensitivity in this area during pregnancy may contribute to the normal pregnancy-induced adaptations in glucose homeostasis (Ramos-Lobo et al., 2017).

While responses to leptin are clearly reduced in pregnancy, even bypassing the leptin-responsive circuitry in the hypothalamus, by administering the downstream mediator of leptin action, αMSH, does not lead to a suppression of food intake in pregnancy as it does in nonpregnant females (Ladyman, 2008; Ladyman et al., 2009, 2016). Similarly, responses to other satiety factors are also reduced. Cholecystokinin (CCK) is a gut-derived satiety factor that is secreted in response to stretch in the stomach and acts acutely to terminate feeding and limit meal size (Moran, 2000). Interestingly, in rodent studies, increased food intake during pregnancy is associated with increases in meal size rather than increased meal frequency (Strubbe and Gorissen, 1980; Ladyman et al., 2018) suggesting a delayed termination of feeding during a meal. Consistent with this concept, an injection of CCK does not acutely suppress food intake during pregnancy as it does in nonpregnant females (Ladyman et al., 2011). Overall, evidence indicates that there are many sites in the leptin signaling pathway that are attenuated during pregnancy to prevent the effects of leptin (Fig. 2.3).

Leptin is not the only metabolic factor that is functionally suppressed during pregnancy. As part of the metabolic adaptations of pregnancy, there are substantial changes in the regulation of maternal glucose homeostasis, including increased insulin secretion (Baeyens et al., 2016). Pregnancy-induced increases in insulin secretion are necessary for glucose regulation at this time, ensuring there is a constant supply of glucose to the fetus. Insulin has other actions in the body, however, particularly in the brain where it regulates body weight and energy homeostasis as well as affecting mood and memory (Bruning et al., 2000; Lee et al., 2016). In particular, insulin action in the brain has been shown to suppress food intake and hepatic glucose output (Strubbe and Steffens, 1975; Woods et al., 1979; McGowan et al., 1992; Bruning et al., 2000; Obici et al., 2002a, b; Clegg et al., 2003; Brown et al., 2006; Rojas and Schwartz, 2014; Inoue, 2016; Tups et al., 2017). Pregnancy is associated with hyperphagia and increased or normal hepatic output

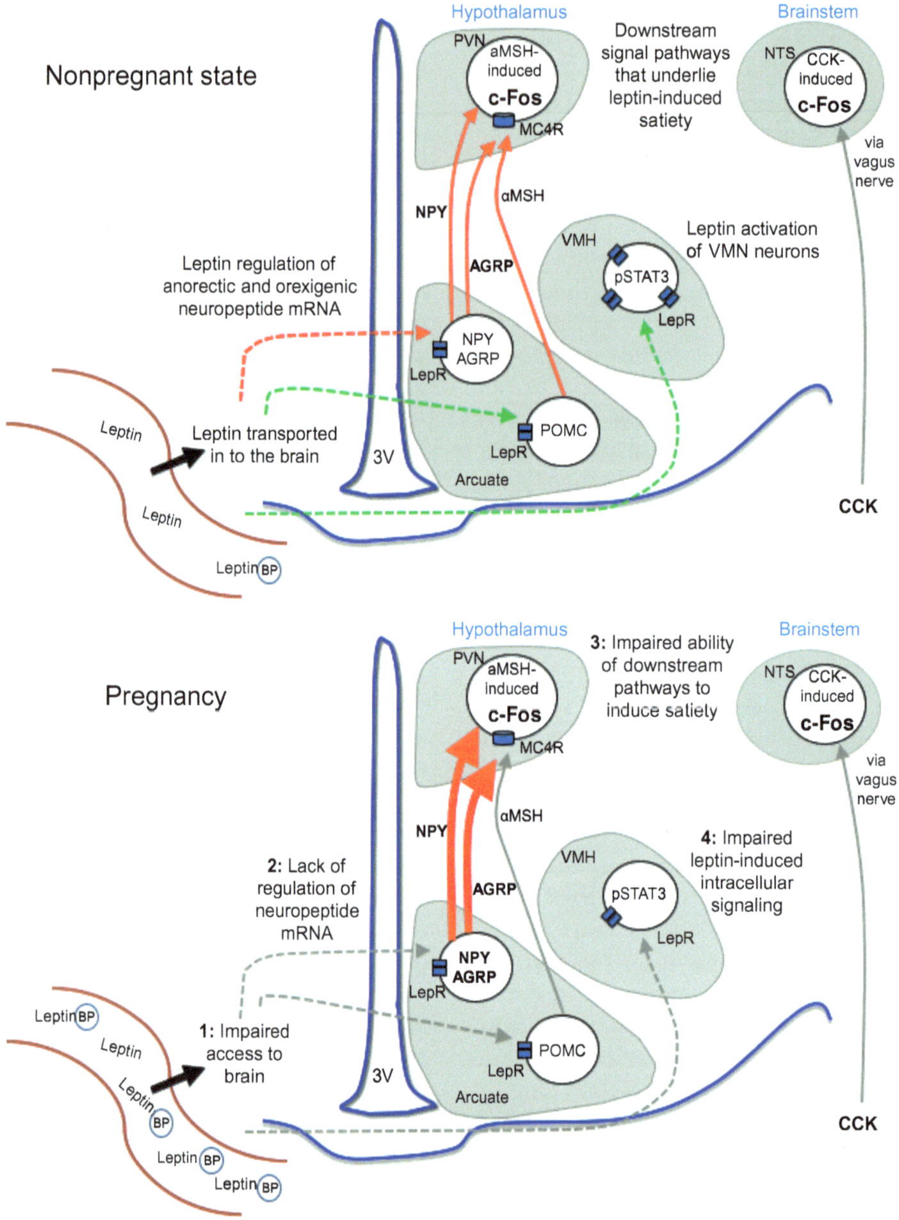

Fig. 2.3. Mechanisms underlying central pregnancy-induced leptin insensitivity. In nonpregnant adults the hormone leptin acts as part of a feedback loop to regulate body fat and appetite. The brain, in particular the hypothalamus, is a key site in leptin-induced satiety. Leptin is secreted by adipocytes into the blood, and then it gains access to the brain by a saturable transport mechanism. Within the arcuate nucleus, leptin suppresses mRNA for the orexigenic neuropeptides NPY and AGRP and stimulates anorectic POMC mRNA. Further downstream, the protein product of POMC mRNA, αMSH, acts in the PVN and other regions to induce the satiety effects of leptin. Leptin also activates neurons in the VMN leading to increased pSTAT3 levels. CCK is an acute-acting hormone that is secreted in response to ingestion of food and acts to terminate a meal. Pregnancy is a state of central leptin insensitivity, which facilitates increased food intake and fat mass. During pregnancy a number of mechanisms have been identified that lead to changes in leptin sensitivity; these include: (1) impaired access of leptin into the brain, (2) a lack of regulation of neuropeptide mRNA levels by leptin, (3) impaired ability of leptin-sensitive downstream pathways to induced satiety, and (4) reduced leptin receptor and attenuated ability of leptin to induce activation of intracellular signaling pathways.

(Carrara et al., 2012; Catalano, 2014) despite raised insulin concentrations, suggesting that like leptin, the brain becomes insensitive to the actions of insulin during pregnancy. Insulin-induced activation of neurons in the hypothalamus, specifically in the areas that regulate energy homeostasis such as the arcuate nucleus and VMN, is attenuated during pregnancy (Ladyman and Grattan, 2017). Insulin-induced increases in SNA are also suppressed in pregnant rats, demonstrating a functional insensitivity to central insulin action during pregnancy (Shi et al., 2019).

Growth hormone (GH) secretion also increases during pregnancy (Kelly et al., 1976; Eriksson et al., 1989; Lacroix et al., 2002; El-Kasti et al., 2008; Gatford et al., 2017) and is important in driving metabolic adaptations in the brain. Transgenic mice with a specific deletion of GH receptor (GH-R) from leptin receptor-expressing cells have increased sensitivity of the VMN to leptin during pregnancy, suggesting that GH plays a role in the generation of VMN-specific leptin insensitivity during pregnancy (Teixeira et al., 2019). Mice with a complete neuronal deletion of GH-R had a subtly different phenotype, with attenuation of pregnancy-induced changes in food intake and fat mass, but no change in leptin sensitivity in the VMN (Teixeira et al., 2019). There is more to learn about the role of GH in mediating pregnancy-specific adaptations.

Hypothalamic fatty acid metabolism is another key pathway in the regulation of energy homeostasis (Lopez et al., 2007) that is attenuated during pregnancy (Martinez de Morentin et al., 2015). Lipids and central de novo production of fatty acids are thought to be involved in overall nutrient sensing in the hypothalamus (reviewed in Lopez et al., 2007). One of the central catalytic enzyme involved in de novo fatty acid synthesis is fatty acid synthase (FAS) and central inhibition of FAS reduces food intake, a process that appears to require the accumulation of malonyl-CoA (Lopez et al., 2007). Pregnancy is associated with increased levels of hypothalamic malonyl-CoA, which should reduce food intake, but does not do so in pregnancy suggesting that the hypothalamus has a reduced capacity to respond (Martinez de Morentin et al., 2015). Interestingly, FAS mRNA is specifically reduced during pregnancy in the VMN, but not in other regions of the hypothalamus (Martinez de Morentin et al., 2015), again likely to drive region-specific changes in response to anorectic molecules.

All of these metabolic adaptations are likely to be driven by the hormonal changes during pregnancy, although there may be species differences in the specific mechanism involved. Most work analyzing these central actions of pregnancy hormones has been in rodents and particularly focused on the development of leptin insensitivity. In pregnant rats, the chronic high levels of placental lactogens, which act on the prolactin receptor, seem to be well positioned to cause this leptin insensitivity. First, chronic prolactin receptor activation in the brain has been shown to lead to a state of leptin insensitivity in female rats (Naef and Woodside, 2007). Using pseudopregnant rats, chronic central prolactin infusion to mimic placental lactogen secretion also leads to the development of a leptin-resistant like state (Augustine and Grattan, 2008) associated with reduced leptin receptor mRNA in the VMN and impaired leptin-induced activation of STAT3 (Augustine et al., 2019). The mechanisms underlying how high prolactin receptor activation leads to leptin insensitivity is currently unknown, and as there is little evidence for colocalization of prolactin receptors and leptin receptors in the brain, it is unlikely this is due to a direct interaction of prolactin on leptin-responsive cells (Ladyman et al., 2010; Nagaishi et al., 2014). It seems unlikely that increase in estradiol during pregnancy is involved in suppressing leptin sensitivity during pregnancy, as increasing levels of estrogen drive increases in leptin sensitivity (Clegg et al., 2006) and estrogen deficiency leads to leptin insensitivity (Ainslie et al., 2001). However, estrogen has been shown to decrease central sensitivity to insulin (Clegg et al., 2006) and may contribute to the suppression of central insulin sensitivity during pregnancy. Administration of progesterone in either OVX or intact rats can increase food intake and body weight without affecting plasma leptin levels suggesting progesterone could potentially drive changes in leptin sensitivity (Hervey and Hervey, 1967; Wade, 1975; Chen et al., 1996; Grueso et al., 2001). However, chronic progesterone administration is not enough by itself to drive a suppression in leptin sensitivity in female rats (Augustine and Grattan, 2008).

Overall, it is clear that there are multiple adaptations in maternal brain that establish a positive energy balance to help the mother prepare for the metabolic demands of the growing fetus and for changes in her own body to adapt to the challenges of both pregnancy and subsequent lactation. A key feature is the attenuation in the sensitivity to many factors that would normally act to suppress appetite, overriding the normal feedback systems that regulate appetite in the nonpregnant adult mammal, and promoting fat deposition and weight gain. While the exact cellular and molecular mechanisms underlying these changes are yet to be fully elucidated, and may be different in different species, the multiple potentially redundant mechanisms involved highlight the importance of establishing a positive energy balance during pregnancy, presumably selected through the course of evolution. While this can be considered an adaptive response, for humans in the present obesogenic environment, this is quite maladaptive. Many women are coming into pregnancy overweight or obese and run the risk of gaining excessive weight during pregnancy. This is associated with increased risk of many pregnancy-related complications (Marchi et al., 2015; Goldstein et al., 2017; LifeCycle Project-Maternal Obesity et al., 2019). Indeed, obesity during pregnancy is now described as the "most common clinical risk factor encountered in obstetric practice" (Krishnamoorthy et al., 2006). It is essential that clinicians assessing weight gain in pregnant women are aware that almost all homeostatic regulators of food intake are suppressed, and likely there are also changes in energy expenditure to promote weight gain. Under these circumstances, voluntary limitation of food intake will be extremely difficult.

Central modulation of body temperature and neuroimmune responses

In humans and commonly studied rodent models such as rats and mice, body temperature is initially increased in pregnancy, but as gestation advances, it declines such that in late pregnancy, core body temperature is lower than that in the prepregnancy state (Fewell, 1995; Hartgill et al., 2011; Gamo et al., 2013). Pregnancy is also associated with reduced circadian variation in body temperature, likely related to changes in activity and the constant heat produced by the metabolic processes of the growing fetus (Wharfe et al., 2016). Changes in body temperature across pregnancy, particularly in late pregnancy, are hypothesized to be due to a regulated shift in the set point for body temperature induced by the hormonal environment of pregnancy (Eliason and Fewell, 1997). Alongside this are a number of changes in the response of body temperature to different stressors. For example, cold-induced thermogenesis is impaired in late pregnancy, and late pregnant rats are less able to maintain body temperature in a cold environment compared with virgins (Imai-Matsumara et al., 1990). Psychologic stressors can increase body temperature (Moltz, 1993); however, this becomes attenuated as pregnancy advances (Fewell and Tang, 1997). The febrile response to an immune challenge is also attenuated during late pregnancy (Kasting et al., 1978; Martin et al., 1995; Tsang et al., 2006). Whether this is the same in late pregnant women is unknown, but the third trimester is associated with increased severity of infections suggestive of a decrease in immune defense at this time (Robinson and Klein, 2012). The suppression of the febrile response to an innate immune challenge has been the most well studied of these thermoregulatory responses during pregnancy and describes yet another system within the maternal brain that undergoes major adaptations during pregnancy. This is an adaptive response, because during pregnancy, elevated body temperature is teratogenic (Edwards, 2006; Ravanelli et al., 2019).

The innate immune response is one component of the immune system that targets infectious pathogens and mounts a defense against these pathogens. This response includes driving a fever in an attempt to make the environment less hospitable to the invading pathogens. Administration of lipopolysaccharide (LPS), a cell-wall component of gram-negative bacteria, is a commonly used experimental protocol to induce an innate immune response to study the physiological processes involved in this process, especially the induction of fever (Roth and Blatteis, 2014). As part of this immune response, there is a release of proinflammatory cytokines from innate immune cells after identifying pathogen recognition receptors (reviewed in Evans et al., 2015). These proinflammatory cytokines, of which IL-6 is particularly important in driving the fever response, act in many places, including the brain, to promote production of prostaglandins (Roth and Blatteis, 2014; Garami et al., 2018). Prostaglandin E2 (PGE2) is a key prostaglandin in fever response, which acts on thermoregulatory neurons within the median preoptic nucleus of the hypothalamus to drive the thermogenic changes involved in generating a fever. Overall, the fever is promoted by heat conservation, such as peripheral vasoconstriction, and heat generation, such as increase in brown adipose tissue and heat-seeking behavior (reviewed in Garami et al., 2018). The role of an attenuated fever response during pregnancy is not well understood, but it is hypothesized that since increases in temperature may be detrimental to the fetus, the suppression of fever may protect the fetus. Indeed, a meta-analysis reports an increase in negative consequences for offspring after fever during pregnancy (Dreier et al., 2014).

It is unlikely that major changes in plasma levels of proinflammatory or antiinflammatory cytokines in response to LPS underlie the attenuated response in pregnancy, as LPS induces a similar plasma profile of cytokines in virgin and late pregnant rats (Mouihate et al., 2005), and the downstream signal transduction pathways also seem to be conserved in virgin or late pregnant rats (Mouihate et al., 2005; Harre et al., 2006). More work has shown that there may be regional specificity to the attenuated LPS-induced IL-1 beta and IL-6 mRNA in various brain regions during late pregnancy compared with virgin rats. In the hypothalamus, including the thermoregulating preoptic area, and the hippocampus, LPS-induced increases in IL-6 and IL-1 beta mRNA are detected in both pregnant and virgin rats, but this increase was lower in pregnancy (Sherer et al., 2017). There is also a decrease in the number of microglia, the resident macrophage cells of the brain that produce IL-6 (Lee et al., 2002), in various brain regions during pregnancy, including the medial prefrontal cortex, the dorsal hippocampus, and the amygdala (Haim et al., 2017). This decrease in microglia may contribute to the reduction in proinflammatory cytokine mRNA in the brain in response to LPS. One key mechanistic finding is that in late pregnancy, there is an attenuation of COX2 (a key enzyme involved in the formation of prostaglandins) and PGE2 production in response to LPS, suggesting a suppression of this fever-inducing pathway (Imai-Matsumura et al., 2002; Mouihate et al., 2002). Furthermore, central administration of PGE2 induces an increase in temperature alongside increased BAT thermogenesis and increased sympathetic drive to BAT in virgin rats but not in late pregnant rats, suggesting that along with decreased production of PGE, the brain also becomes insensitive to the actions of PGE2 (Chen et al., 1999).

Similar to other adaptations of the maternal brain, it is likely that the hormonal milieu of pregnancy drives these changes in the febrile response. Treating OVX rats with estrogen and progesterone to mimic levels seen during the preovulatory LH surge leads to an attenuated febrile response after LPS injection and also a reduction in hypothalamic COX2 levels (Mouihate and Pittman, 2003), indicating the ability of these sex steroids to regulate the febrile response. Moreover, progesterone can directly influence the firing patterns of preoptic thermosensitive neurons (Nakayama et al., 1975). However, treating OVX virgin rats with estrogen and progesterone to specifically mimic the hormonal profile of late pregnancy (days 17–20 of gestation) did not affect the febrile response to LPS (Finley et al., 2015). Both oxytocin and glucocorticoids can suppress proinflammatory factors that mediate the fever response (Drew and Chavis, 2000; Yuan et al., 2016), but given that at this time oxytocin is not evaluated in the brain (Landgraf et al., 1992) and that there is a damping of the HPA response in an immune activation (Brunton and Russell, 2008), there is little evidence to support a major role for these hormones in this process during late pregnancy (reviewed in Barrientos et al., 2019). Prolactin/placental lactogen, another hormone that increases during pregnancy, has been shown to lead to a reduction of body temperature when male rats are made hyperprolactinemic by grafting of donor pituitaries to the renal capsule, suggesting the prolactin could also contribute to the effects of lowering of body temperature during pregnancy (Drago and Amir, 1984).

Suppression of the stress response

Despite the overall increase in basal glucocorticoid levels, described earlier, the acute stress response is markedly attenuated during late pregnancy (Neumann et al., 1998) and lactation (Lightman, 1992). This primarily results from changes to the activity of the hypothalamic–pituitary–adrenal (HPA) axis (Brunton and Russell, 2008). Like the loss of the febrile response, it is thought that the major function of the hyporesponsive stress axis is adaptive, preventing exposure of the fetus to high levels of glucocorticoid hormones that might have an adverse effect on fetal development. Under nonpregnant conditions, activation of the CRH and arginine vasopressin (AVP) neurons in the hypothalamic PVN stimulates the secretion of ACTH from the anterior pituitary corticotrophs. ACTH stimulates the synthesis and secretion of glucocorticoids from the adrenal gland cortex, which in turn act as the effectors of the stress response at target organs. During late pregnancy and lactation, the ACTH and corticosterone responses to stress are significantly diminished as a result of reduced central drive for CRH neuron activation (Brunton and Russell, 2008). Animal studies have shown a marked reduction in *Crh* mRNA expression under basal (unstressed) conditions in both late pregnant rats (Shanks et al., 1999) and mice (Gustafson et al., 2017), as well as reduced c-Fos expression in the PVN, which together are indicative of reduced neuronal activation in this nucleus (da Costa et al., 1996; Woodside and Amir, 1997). This phenomenon also occurs in humans, however, with a number of clinical studies showing that the HPA axis is also less sensitive to stress in pregnant women (Schulte et al., 1990; Hartikaninen-Sorri, 1991; Kammerer et al., 2002).

The most likely mediator of the reduced stress response is the progesterone metabolite allopregnanolone. High levels of progesterone reaching the brain can be metabolized by 5α-reductase to form this neuroactive steroid. Administration of allopregnanolone suppresses HPA axis responses to stress in nonpregnant rats (Patchev et al., 1996; Brunton et al., 2009), while suppressing allopregnanolone production in pregnant rats with a 5α-reductase inhibitor substantially restores HPA axis responses to physical and emotional stressors (Brunton et al., 2009). The 5α-reductase enzyme is markedly upregulated in the brain during late pregnancy (Brunton et al., 2009), promoting the conversion of progesterone to allopregnanolone. Allopregnanolone does not bind to classical intracellular steroid receptors such as the PR, but rather acts by allosteric modulation of the $GABA_A$ receptors, potentiating GABA signaling (Brunton et al., 2014). CRH neurons in the PVN are under direct inhibitory GABAergic control (Miklos and Kovacs, 2002), and thus it seems likely that increased allopregnanolone in the maternal brain in pregnancy may enhance inhibitory inputs to the CRH neurons, thus suppressing HPA axis stress responses.

The anterior pituitary hormone prolactin has also been proposed as a potential mediator of HPA axis suppression during pregnancy and lactation. Hyperprolactinemia causes a similar decrease in the acute stress response (Carter and Lightman, 1987; Torner et al., 2001), an effect that could be prevented by blockade of prolactin signaling (Torner et al., 2001). As described earlier, prolactin (and its pregnancy-specific homolog, hPL) is chronically elevated during pregnancy and lactation. Chronic administration of prolactin to nonpregnant rats suppresses the HPA axis response to stress (Donner et al., 2007) and blocking prolactin receptor activation during lactation increases the ACTH response to stress (Torner et al., 2002). While acute (24-h) inhibition of prolactin secretion in the lactating mouse is not sufficient to reverse the suppressive effect of this physiological state upon *Crh* mRNA expression under nonstressed conditions (Gustafson et al., 2017), the potential for prolactin to act as a restraint on the stress-induced activation of the HPA axis in mice has not yet been determined. Centrally acting prolactin has also been shown to inhibit the formation of stress-induced gastric ulcers and prevent stress-induced hypocalcemia, although these actions have been associated with acute activation, rather than inhibition, of CRH neurons (Drago et al., 1985; Fujikawa et al., 2004, 2005).

Lactation is normally associated with marked reductions in anxiety (Picazo and Fernandez-Guasti, 1993; Toufexis et al., 1999). Anxiety is a complex neurologic trait, but it has been linked with hyperactivity of the HPA stress axis. In both males and nonpregnant female rats, prolactin has dose-dependent anxiolytic actions (Torner et al., 2001). As observed for prolactin-induced suppression of the stress response, the anxiolytic action of prolactin was prevented by blockade of the prolactin receptor, resulting in increased anxiety-related behavior. Hence, prolactin acts as an endogenous anxiolytic, with the ability to reduce the magnitude of the acute stress response. Elevated levels of prolactin during lactation, therefore, coupled with increased expression of prolactin receptors in the PVN, may result in prolactin contributing to the suppression of the stress response during lactation, or perhaps altering the stress response to respond only to factors relevant to the lactating state.

ADAPTIVE CHANGES IN BEHAVIOR

Maternal behavior

Arguably, the ultimate expression of the hormone-induced changes in neurophysiological functioning of a number of neuronal populations is the expression of maternal behavior. This is a stereotypical suite of behaviors expressed in animals, and it is likely to be the neurobiologic basis for the emotional and mood changes experienced in human mothers. Mammalian offspring require parental support for nutrition, protection, and survival. Specialized neural circuits receive sensory, cortical, and hormonal input to generate a coordinated change in maternal behavior (Smiley et al., 2019), which can be defined as the collection of behaviors by a postpartum woman that maximizes the likelihood that her offspring will survive (Bridges, 2015), including adaptations such as the metabolic changes described earlier, ensuring the mother is prepared to provide nutrition to the offspring. In rodents, more overt, young-directed responses include nest-building, retrieval of young, licking/grooming, crouching over the pups, and nursing, as well as increased maternal aggression to protect offspring. Arguably, one of the most important systems supporting parental behavior are the reward circuits involving midbrain dopaminergic neurons that have evolved to promote activities related to parental care that might otherwise be considered costly to the mother.

There have been a number of reviews describing the known neuronal circuitry underlying maternal behavior. The medial preoptic area (MPOA) of the hypothalamus has long been identified as the core integrating structure, and lesions of the MPOA block the expression of maternal behavior (Numan, 1974; Numan et al., 1977, 1988; Jacobson et al., 1980). Modern molecular genetic strategies have refined our understanding by allowing manipulation of individual populations of neurons. For example, targeted ablation of galanin neuron within the MPOA resulted in impaired maternal responses (Wu et al., 2014), highlighting a critical role for this particular subset of neurons. Similarly, ablation of dopamine neurons in a rostral part of the MPOA also impaired maternal behavior, whereas optogenetic stimulation of these cells enhanced maternal care, potentially through enhanced oxytocin secretion (Scott et al., 2015). A critical feature of the maternal behavior network is its ability to engage with reward mechanisms in the brain to motivate the mother to engage in these potentially costly behaviors. Interactions with offspring are highly rewarding for a postpartum female, with a place-preference paradigm showing that mother rats find rats pups more rewarding than cocaine (Mattson and Morrell, 2005). Importantly, the MPOA was highlighted as the only area where greater activation of neurons was seen with a preference for pup-associated versus cocaine-associated cues (Lee et al., 2000). The MPOA provides a key output to the midbrain reward systems through activation of midbrain dopamine projections to the nucleus accumbens (Numan et al., 2005; Numan and Stolzenberg, 2009; Stolzenberg and Numan, 2011). Disruption of this pathway causes significant impairments in maternal behavior (Terkel et al., 1979; Numan and Smith, 1984; Numan et al., 1990). In addition, it is now apparent that much more widespread connections from the MPOA to other brain regions are also involved (Numan and Sheehan, 1997), forming a distributed network regulating maternal behavior.

Multiple human fMRI neuroimaging studies have shown activation of the midbrain reward areas during maternal interactions with offspring. For example, there is increased activity in the dopaminergic reward circuit when subjects view baby images or hear sounds of crying babies (Lonstein et al., 2015). Infant features such as wide eyes, large, round heads, and small noses elicit positive emotions in adults (Lonstein et al., 2015). When participants could control the length of time they viewed images of infants, women demonstrated increased motivation to extend the time they were viewing infant-related stimuli (Charles et al., 2013). In addition to brain areas such as the amygdala and nucleus accumbens, which show activation patterns that match their proposed functions in rodents, many higher-order regions, such as prefrontal cortex, anterior cingulate cortex, and the orbital frontal cortex, are also found to be involved with maternal care (Rutherford et al., 2011). For example, the orbital frontal cortex, an area known to be involved in hedonic stimuli processing, shows increased activity when viewing image stimuli of one's own baby and is positively correlated with improved mood (Lonstein et al., 2015). Together, these studies back the experimental animal data and suggest a highly conserved neuronal network supporting reward and motivation during maternal behavior.

Hormonal regulation of maternal behavior

Convincing evidence for the hormonal basis of maternal behavior was first provided by parabiosis experiments where blood from late pregnant animals was transfused into nonpregnant females, resulting in the rapid induction of maternal care (Terkel et al., 1972). Subsequent studies have defined the hormonal changes involved, including increased levels in circulating estradiol, progesterone, and lactogenic hormones (prolactin and placental lactogens), clearly demonstrating the neuroendocrine interactions underlying this behavior. As described earlier, these same hormones play an integral role in establishing the physiological adaptions in the maternal body to support the growing fetus during pregnancy. These hormonal changes are facilitated by the placenta itself, in addition to maternal changes in the hypothalamic control and release of hormones such as prolactin and oxytocin from the pituitary gland.

The role of the sex steroids in modulating maternal behavior has been the subject of extensive research. Estradiol was first demonstrated to stimulate the onset of maternal behavior, with estradiol administration reducing the latency to demonstrate young-directed maternal behavior in both virgin and pregnancy-terminated primigravid rats (Siegel and Rosenblatt, 1975a, b). These actions of estradiol are mediated through the estrogen receptor alpha (ERα), with ERα knockout virgin female mice displaying deficits in pup retrieval behavior (Ogawa et al., 1998). The MPOA is a key site of estradiol action, with estradiol benzoate implants directly into the MPOA inducing maternal behavior (Numan et al., 1977), and knockdown of ERα expression specifically in the MPOA suppressing postpartum retrieval and nursing behavior (Ribeiro et al., 2012). In contrast to the actions of estradiol, administration of progesterone on its own is ineffective in reducing the latency of virgin female rats to respond maternally to pups (Bridges, 1984). Instead, progesterone acts to modulate the actions of other hormones on maternal behavior. When it is chronically coadministered with estradiol, followed by acute progesterone removal, progesterone further increases the capacity of estradiol to stimulate maternal behavior (Bridges, 1984). Because progesterone's actions on maternal behavior are indirect, the neural sites of progesterone action have been more difficult to define. However, the MPOA also appears to be an important target, as sustained progesterone blocks the estrogen-stimulated display of maternal behavior (Sheehan and Numan, 2002). Progesterone administration in virgin female rats also suppresses prolactin receptor mRNA expression in the MPOA (Bridges and Hays, 2005), and prolactin is a key regulator of maternal behavior, as discussed later. From these studies in rodents, progesterone has emerged as a negative regulator of maternal behavior, highlighting the necessity of progesterone withdrawal prior to parturition for appropriately timed expression of maternal behavior after birth.

Evidence of a role for the lactogenic hormones in maternal behavior was first suggested in 1935, with repeated injections of the "newly discovered" prolactin inducing maternal behavior in virgin female rats (Riddle et al., 1935). In hypophysectomized rats, estradiol and progesterone administration fails to induce maternal behavior (Bridges et al., 1985), indicating that an estrogen-stimulated pituitary-derived factor was required for the expression of maternal behavior. The addition of either prolactin or an ectopic prolactin-secreting pituitary graft was sufficient to significantly reduce the latency of inexperienced virgin female rats to respond maternally, thus confirming the importance of prolactin in maternal behavior (Bridges et al., 1985). Subsequently, it was shown that central infusions of prolactin in virgin mice were able to replicate these findings at concentrations that were ineffective when administered peripherally (Bridges et al., 1990). As placental lactogen shows a similar potency for inducing maternal behavior in virgin rats (Bridges et al., 1997), it is likely to be the primary ligand in modulating prolactin receptor-expressing neural circuits until maternal prolactin levels rise in late pregnancy and postpartum. Like prolactin, placental lactogens readily cross the blood–brain barrier and activate prolactin receptors throughout the brain. Placental lactogen was observed to bind to the choroid plexus and hypothalamus in the maternal brain (Pihoker et al., 1993) and is likely responsible for high levels of phosphorylated signal transducer and activator of transcription 5 (pSTAT5) expression triggered downstream of the prolactin receptor (Brown et al., 2010) detected throughout the brain during pregnancy (Salais-Lopez et al., 2017) compared with nonpregnant females. Generation of prolactin receptor knockout mice confirmed that prolactin was involved in maternal behavior in nonpregnant mice (Lucas et al., 1998). However, because of the essential role of prolactin and placental lactogen in sustaining ovarian progesterone production during pregnancy in rodents (Stocco et al., 2007), prolactin receptor knockout mice are infertile, and the role of prolactin in normal, postpartum maternal behavior could not be assessed. A study of the conditional deletion of prolactin receptors specifically from the MPOA of adult female mice confirmed that prolactin action was required for postpartum maternal behavior, with the mothers abandoning their litters in the absence of prolactin action in the MPOA (Brown et al., 2017).

Oxytocin is also an important peptide in the control of maternal behavior. Circulating oxytocin does not cross the blood–brain barrier in significant concentrations (Leng and Russell, 2016) and, thus, is not involved in behavioral responses (Rosenblatt, 1969; Herrenkohl and Rosenberg, 1974). However, as described earlier, magnocellular oxytocin neurons in the SON and PVN have collateral branches that project centrally to release oxytocin within the brain (Son et al., 2013). Moreover, the parvicellular oxytocin neurons in the PVN that project centrally are also activated during parturition and in response to suckling (Ludwig and Stern, 2015). Finally, oxytocin is also secreted within the SON, PVN, and in surrounding regions from the dendrites of the oxytocin neurons (Lambert and Kinsley, 1993; Rossoni et al., 2008; Ludwig and Stern, 2015). Dendritically released oxytocin can diffuse through the extracellular space within the brain and activate oxytocin receptors in regions that do not receive axonal projections from oxytocin neurons. Collectively, there are multiple pathways by which oxytocin is released within the brain around the time of parturition and throughout lactation. Administration of central oxytocin to estrogen-primed virgin rats rapidly induced the onset of maternal behavior (Pedersen and Prange Jr., 1979; Pedersen et al., 1982). It appears that oxytocin is primarily involved in facilitating the onset of maternal behavior at parturition, rather than being important for the maintenance of these behaviors through the postpartum period (Yoshihara et al., 2017). Lesions to the PVN during late pregnancy severely impact maternal behavior, whereas lesions in the early postpartum period have limited effect (Numan et al., 1985; Insel and Harbaugh, 1989). During parturition, there are also high levels of oxytocin binding in the MPOA and midbrain (Pedersen et al., 1994), and infusion of an oxytocin antagonist into either of these sites from late pregnancy blocks subsequent maternal behavior in postpartum rats (Pedersen et al., 1994). During the postpartum period, oxytocin has more subtle modulatory effects, with central administration of an oxytocin antagonist reducing the time dams spend licking pups and in the kyphotic nursing posture (Pedersen and Boccia, 2003).

In humans, the surge of oxytocin released during parturition and breastfeeding is thought to be associated with central release of oxytocin, which is important for the mother bonding with her baby (Rilling, 2013). Certain polymorphisms in the oxytocin gene are associated with increased infant-directed vocalizations and care (Galea et al., 1997), while other polymorphisms in the oxytocin receptor gene have been linked to empathy and stress reactivity, greater sensitivity during infant interactions, and infant-directed vocalizations (Rodrigues et al., 2009; Feldman, 2016). Although there are many positive associations with oxytocin and parental care, consistent with many animal findings, oxytocin is also associated with prosocial behavior in humans and other animals generally, and thus it is difficult to tease apart prosocial effects versus specific parental care effects.

There is evidence that oxytocin and prolactin may interact in the regulation of maternal behavior. Oxytocin can act as a prolactin-releasing factor in some circumstances (Egli et al., 2006; Kennett and McKee, 2012), potentially providing a mechanism whereby peripheral oxytocin could exert indirect actions in the central nervous system. Conversely, oxytocin neurons express the prolactin receptor (Townsend et al., 2005; Kokay et al., 2006; Sirzen-Zelenskaya et al., 2011). While prolactin acutely suppresses the firing of these neurons, this inhibitory effect is lost in late pregnancy, allowing activation of these neurons in the presence of elevated prolactin (Augustine et al., 2017). Prolactin also stimulates oxytocin synthesis. The precise role of this interdependent, mutual regulation of prolactin and oxytocin in maternal behavior is unclear, but it is an important consideration when evaluating experiments that have manipulated one or other of these hormones.

Maternal aggression

In addition to the nurturing behavior toward their offspring during the postpartum period, females show reduced anxiety and increased aggression toward intruders to protect their offspring from perceived danger. Two specific hypothalamic regions seem to play critical but opposing roles in modulating postpartum aggressive behavior: the PVN that suppresses aggression, and

VMN that promotes aggression. Lesions of the VMN attenuate maternal aggression (Hansen, 1989). Estrogen receptor-expressing neurons in the ventrolateral VMN show increased activity during postpartum aggression (Hashikawa et al., 2017), and chemogenetic inhibition reduces both the duration and frequency of attacks on intruders by a postpartum dam (Hashikawa et al., 2017). Although these studies show that estrogen receptor-expressing cells in the VMN are important for maternal aggression, a role for the female sex steroids themselves in modulating aggression has yet to be confirmed. The PVN also plays an important role in modulating maternal aggression. The impact of PVN lesions on this behavior was dependent on the extent and localization of damage, with larger lesions reducing aggression toward intruders (Consiglio and Lucion, 1996), whereas more targeted lesion that restricted damage to the parvocellular region of PVN caused an increase in aggressive behavior (Giovenardi et al., 1998). Activation of parvocellular neurons in the PVN, therefore, is important for limiting the display of aggressive behavior. Within this region, oxytocin has emerged as an important neuroendocrine modulator, and blocking oxytocin action increases aggression (Giovenardi et al., 1998; Lubin et al., 2003). In addition to oxytocin, there is evidence suggesting vasopressin is also involved in modulating levels of maternal aggression, although these effects have been different between species. In mice, vasopressin 1b receptor, but not 1a, knockout animals displayed reduced levels of maternal aggression (Wersinger et al., 2007a, b), whereas in rats, central administration of an AVP receptor 1a antagonist led to increased maternal aggression (Nephew and Bridges, 2008).

COGNITIVE CHANGES

With such widespread actions of pregnancy hormones on brain function, it is not surprising that higher brain functions might also be affected. Certainly, the popular science perception of pregnancy is one of marked changes in cognitive function. "Pregnancy brain," "mommy brain," "baby brain," and "mumnesia" are just a few of the commonly used terms to describe the forgetfulness or fogginess that is self-reported by many pregnant women, and researchers have also coined their own term: "benign encephalopathy of pregnancy" (Poser et al., 1986). Whether there is any real evidence for cognitive changes, however, is an open question, and this concept is not strongly supported by animal studies, which, if anything, show the opposite. Here we summarize the evidence and discuss the possible neurologic basis for cognitive changes in pregnancy.

In self-reported studies, the range of cognitive issues women experience during pregnancy includes increased forgetfulness, trouble reading and concentrating, confusion, inability to maintain attention, and lack of coordination (Parsons and Redman, 1991; Casey et al., 1999; Brett and Baxendale, 2001). Many women also report that impaired cognitive function experienced during pregnancy led to an inability to perform daily tasks adequately (Brett and Baxendale, 2001). A meta-analysis of studies focusing on more objective measures of cognition did find a significant decline of cognitive functioning in pregnant women compared with nonpregnant women (Davies et al., 2018). Memory was the key function in which impairments were detected, and the third trimester showed the most decline (Davies et al., 2018). Verbal memory recall appears to be particularly negatively impacted by pregnancy (Henry and Rendell, 2007; Glynn, 2010). Measures of executive functioning and attention did not reveal any significant differences between nonpregnant and pregnant states in this meta-analysis (Davies et al., 2018). Interestingly, one study has even shown a correlation between fetal sex and memory decline, with pregnant women carrying a female fetus showing greater impaired memory in working and spatial memory tasks than those with a male fetus (Vanston and Watson, 2005). The effects of pregnancy on memory decline may accumulate in successive pregnancies, as multiparity has been correlated with diminishing performance in verbal memory recall during pregnancy (Glynn, 2012). Perhaps the most surprising finding of this meta-analysis was just how few studies have been done that objectively examine cognitive function during pregnancy, with only 20 studies that had a total of 709 pregnant women meeting the criteria for inclusion in the meta-analysis. Interestingly, while changes in cognition were detected during pregnancy, it should be noted that the performances in the tests to measure aspects of memory and executive function were still within the normal ranges of general cognitive functioning and memory (Davies et al., 2018). It is possible that the tasks used to measure cognitive function and memory have a bias toward skills that are not particularly ethologically relevant for mothers and hence are unable to clarify whether changes in aspects of cognition are an actual decline or a shift in how functioning has been optimized for the upcoming task of nurturing offspring (Macbeth and Luine, 2010).

While the objective and subjective data indicate impaired cognitive changes during pregnancy, particularly when it comes to memory, the cause is yet to be defined. One of the key questions is whether a decline in cognitive function is a secondary result of a change in some other physiological process, such as poor sleep or hypoglycemia. Sleep quantity and quality can be greatly affected by pregnancy (Sedov et al., 2018), and poor sleep can negatively influence cognition (Goel et al., 2009; Killgore, 2010). Initial self-reporting studies suggested that poor sleep was not associated with decline in cognition (Poser et al., 1986). Data using objective testing measures have also shown no correlation between sleep, anxiety, or depression levels on memory test performance in pregnant women (Keenan et al., 1998). Indeed, while sleep loss is a predictor of reported memory loss in pregnant and postpartum women, it does not impact on actual memory test performance suggesting that sleep deficiency impacts on perception of forgetfulness but not on actual memory functioning (Casey et al., 1999). One hypothesis is that the cognitive decline in pregnancy is part of the trade-off to meet the high metabolic demands of this time. In particular, the decline in memory and attention during late pregnancy may be due to a shifting of energy partitioning toward the growing fetus and away from the requirements of the maternal brain (Ziomkiewicz et al., 2019). Brain function is energetically costly to humans, utilizing a large component of the body's glucose resources. Glucose is also the key energy source needed for the developing offspring during pregnancy. Periods of hypoglycemia are not uncommon in late pregnancy (Naik et al., 2017), and hypoglycemia itself is associated with impaired cognition, especially memory, attention, and verbal fluency in nonpregnant diabetic patients (Holmes et al., 1984; McAulay et al., 2001; Sommerfield et al., 2003). Whether diminished cognition function is associated with incidence of hypoglycemia in late pregnancy has yet to be tested (Ziomkiewicz et al., 2019). An alternative hypothesis is that cognition apparently declines because the specific tasks being considered are not a key necessity for the woman to adapt to her current state. This theory suggests that cognition changes are part of an adaptation of the brain to motherhood that reorganizes cognitive functioning to focus on skills needed for the nurturing of offspring (Anderson and Rutherford, 2012).

While evidence does support a subtle but detectable change in cognition during human pregnancy, whether a change in cognition is discernible in animal models is not fully clear. During early and mid-pregnancy, rats perform better than controls in the Morris water maze, a task that assesses spatial learning and memory (Vorhees and Williams, 2006), but performance declines in the last week of pregnancy (Galea et al., 2000). Other studies have seen no decline in performance in the Morris water maze during pregnancy and an increase in working memory and cognitive flexibility, as shown by pregnant rats being quicker to locate the platform when it had changed position in the task (Bodensteiner et al., 2006). In an object placement task, which specifically assesses spatial memory and discrimination, pregnant rats performed better than virgin controls suggesting better spatial memory (Macbeth et al., 2008). Overall, there is little work investigating the changes in cognition during pregnancy in rodent models, and more work has focused on the postpartum and beyond (Pawluski et al., 2016). What evidence there is suggests that cognition is not greatly impaired in rats during pregnancy. Given that in humans it is verbal memory recall that seems to be the aspect of cognition in which decline has been most readily detected, it begs the question of what is the equivalent function in rodents and how it can that be assessed.

During pregnancy the maternal brain is in a highly plastic state, as the female transitions into motherhood and shifts her behavioral focus toward that of caring for and nurturing her offspring. Understanding how the changes in the brain may influence cognition at this time is in its infancy. Highlighting the idea of the pregnant brain as being in a transitional phase, perhaps undergoing preparation for motherhood, MRI studies have shown that pregnancy is associated with a reduction in brain volume in humans, which is reversed by 6 months postpartum (Oatridge et al., 2002). Another study has demonstrated that pregnant women experience a decrease in gray matter, particularly in regions involved in social processes, and this has been shown to persist for at least 2 years following parturition (Hoekzema et al., 2017). The reduction in gray matter is thought to represent a process of the brain becoming more specialized for the task of nurturing offspring by way of synaptic pruning (Hoekzema et al., 2017). This hypothesis would suggest greater social cognition in the females following pregnancy, and there is a small amount of evidence to support this, with pregnant women showing enhanced emotion and face recognition (Pearson et al., 2009; Anderson and Rutherford, 2011, 2012). This may, however, occur at the expense of other cognitive function, such as the deficits in memory observed during pregnancy (Davies et al., 2018), as suggested in the "cognitive reorganization" theory of pregnancy described earlier (Anderson and Rutherford, 2012). It is likely that increased social cognitive abilities allow for better mother–infant bonding and promote the mother's ability to recognize the needs of her offspring thus conferring an advantage to the survival of offspring.

The hippocampus is a brain region that plays a key role in memory, emotions, and learning. In particular, it is integral to the consolidation of information from short-term memory to long-term memory and in spatial memory. Therefore the hippocampus has been a focus for investigating changes in plasticity of the maternal brain during pregnancy, although similar to behavioral changes, there has been more of an emphasis on changes in lactation and in the long term after going through pregnancy and lactation (termed "reproductive experience") rather than specifically during pregnancy (Pawluski et al., 2016). Hippocampal volume is reduced in late pregnant rats at a similar time to when spatial memory is also impaired (Galea et al., 2000) suggesting changes in structure at the same time as changes in function. Interestingly, work examining the long-term effects of reproductive experience has indicated that long after they have weaned off their offspring, reproductively experienced female rats show better spatial learning and memory than age-matched virgin rats (Gatewood et al., 2005) suggesting there is a long-term benefit in memory function of having been pregnant and nurturing offspring.

Morphologic changes in neurons in the hippocampus have been detected during pregnancy, but these are complex and region specific. Late pregnancy is associated with a decrease in complexity of CA3 pyramidal neurons, with less dendritic intersections (Pawluski et al., 2012), while after weaning off pups, the postpartum hippocampal CA3 and CA1 pyramidal neurons both show less complexity, as measured by decreased dendritic lengths and intersections (Pawluski and Galea, 2006). In rats, late pregnancy is associated with increased spine density in the apical CA1 region of the hippocampus, and this increase is maintained during lactation (Kinsley et al., 2006). Increases in spine density allow for additional synaptic inputs and are thought to be a key site for increases of synaptic efficacy in the networks involved in learning, memory, and motivation (Gipson and Olive, 2017; Sheppard et al., 2019). Estradiol and progesterone can stimulate proliferation of dendritic spines in the hippocampus (Woolley and McEwen, 1993; Sheppard et al., 2019) and mimic the high estrogen and progesterone in the later stages of pregnancy in OVX rats leading to a similar increase in CA1 neuronal dendritic spines, strongly supporting the idea that these sex hormones mediate the pregnancy-induced increase in spine density (Kinsley et al., 2006). Directly after birth, however, hippocampal dendritic spine density has been shown to be similar to that in virgins (Frankfurt et al., 2011) and may represent an acute, transitional decline in spine density between the increases observed during pregnancy and later in lactation. Parturition is a time of major hormonal changes, and since dendritic spine density can change within hours estrogen and progesterone fluctuating (Woolley and McEwen, 1993), it is possible that this decline in hippocampal dendritic spine density is due to the major change in sex steroids along with many other hormones in the period of time directly surrounding giving birth. Interestingly, the degree of increase in hippocampal spine density in multiparous rats has been detected to be positively correlated to the number of male pups they have given birth to (Pawluski and Galea, 2006), perhaps lending support to the idea of less cognitive decline in women pregnant with a male fetus (Vanston and Watson, 2005).

It is hypothesized that the increased dendritic spine density in the hippocampus and other regions during pregnancy and lactation contributes to the increase in skills such as navigation and spatial cognitive abilities that are required for the mother to be efficient at finding food to avoid leaving pups unattended for long periods (Kinsley et al., 1999; Lambert et al., 2005). Indeed, reproductive experience in rats leads to better performances at maze tasks (Kinsley et al., 1999) and increases efficiency as predators in terms of speed of catching crickets (Kinsley and Lambert, 2008). Postpartum rats also show better abilities in tasks measuring behavioral flexibility compared with virgin controls along with increased dendritic spines in the medial prefrontal cortex (Leuner and Gould, 2010), a region of the brain associated with attention-shifting behavior (Dalley et al., 2004). This increase in cognitive flexibility in the postpartum is dependent on the hormone oxytocin, as pup removal or oxytocin antagonism negates the improvements in performance (Albin-Brooks et al., 2017). Oxytocin has been shown to improve long-term spatial memory in virgin mice and activate a signaling cascade involved in memory and learning, implicating oxytocin as a modulator of long-term cognitive changes in the maternal brain (Tomizawa et al., 2003). This is not a mechanism of cognitive change during pregnancy, a time of mostly low oxytocin levels, but it is likely to play a role in the effects of lactation on cognition.

The hippocampus is one of the few brain regions where neurogenesis is detected in adulthood. Neurogenesis in the dentate gyrus of the hippocampus has been implicated in cognitive processes and the reorganization of hippocampal circuits that regulate learning, memory, and cognitive function (Toda et al., 2019). Whether there is any change in cell proliferation in the hippocampus during pregnancy is unclear as results from various studies using rodent models are not consistent. Many studies in rats or mice suggest no change (Banasr et al., 2001; Shingo et al., 2003; Pawluski et al., 2010, 2011) while others detect decreased levels of cell proliferation, particularly in late gestation (Rolls et al., 2008; Pawluski et al., 2015; Eid et al., 2019). Work in pregnant guinea pigs has shown increased cell proliferation in the hippocampus during late pregnancy (Wan et al., 2019). The inconsistences in the results from the various studies may indicate that if there is pregnancy-induced change in neurogenesis, then it may be very sensitive to the environmental differences between studies, which can impact on hippocampal neurogenesis (Toda et al., 2019), or that the timing during pregnancy is very specific.

Overall, a decline in memory is the most robust finding regarding changes in cognition during human pregnancies, but this is yet to be confirmed in animal models. Whether this is dependent on the tests available to examine memory in rats and mice or due to species differences remains unknown. Evidence from rodents, however, suggests plasticity in the brain, particularly in the hippocampus with its roles in memory and learning, during pregnancy, and more so in the postpartum and long term after reproductive experience that may be involved in the transition to motherhood.

SUMMARY AND CONCLUDING REMARKS

In this chapter, we have summarized the major changes in hormones that are associated with pregnancy and highlighted how many of these hormones can act in the maternal brain. The hormonal changers of pregnancy drive a wide range of adaptive changes in the mother, helping her cope with the demands of pregnancy and prepare for the subsequent requirements of lactation. Almost every physiological system in the mother is modified in some way. In the brain, there is significant plasticity, whether examined at the level of neurophysiological activity of individual populations of neurons, such as the oxytocin neurons or the TIDA neurons that control prolactin secretion, or at the level of complex neuronal circuits, such as those controlling metabolism, memory, and parental behavior. With such hormone-induced plasticity occurring over a prolonged period, it is perhaps not surprising that sometimes things do not always work optimally. Pregnancy and the postpartum period are associated with increased risk of mental health disorders, such as anxiety and depression. Perhaps this is a cost of the high levels of adaptive plasticity occurring in the maternal brain.

The changing steroid hormones have been most heavily studied for a potential role in mood disorders. Potential imbalances in these hormones could lead to maladaptive maternal mood and behavior. Since levels of both estrogen and progesterone drop precipitously after delivery of the placenta, a "steroid withdrawal" hypothesis has been proposed, which has received support for contributing to postpartum mood disorders in animal work. However, it is important to recognize that many of the mental health problems associated with pregnancy emerge during the pregnancy, and an episode of depression or anxiety before birth is the largest predictor of a postpartum mood disorder (Pawluski et al., 2017). In humans, fluctuating estrogen levels are associated with depressed mood and increased serotonin transporter in the neocortex of women. This likely results in decreased central serotonin levels, which for some time have been implicated in depression (Brummelte and Galea, 2016). Hence, changes in levels of sex steroids may be critical in modulating mood during pregnancy. Increased glucocorticoids are also strongly associated with increased stress, and long-term overactive HPA responses and stress have been implicated in depression. While the HPA axis is physiologically attenuated in non-depressed mothers during pregnancy, it seems clear that alterations in the HPA axis accompany peripartum mental health disorders (Field et al., 2006; Seth et al., 2016; Garcia-Leal et al., 2017).

Animal studies have shown that both oxytocin and prolactin are anxiolytic, and hence these peptide hormones may also be implicated in mood disorders in pregnancy and postpartum. In women, oxytocin released during breastfeeding can inhibit ACTH secretions, which may lead to decreased CORT levels in response to stressful stimuli, such as her baby crying (Swain et al., 2011). Women with low oxytocin during the last trimester are more at risk for postpartum depression than women with higher oxytocin (Brummelte and Galea, 2016). Use of oxytocin during labor (possibly indicative of a lack of endogenous oxytocin release) is positively correlated with higher depressive, anxious, and somatization symptoms 2 months postpartum (Kroll-Desrosiers et al., 2017). Like oxytocin, prolactin is also thought to be a protective factor against stress effects during pregnancy and lactation (Drago et al., 1989). Interestingly, evidence suggests that women who suffer from one complication are more likely to also experience additional complications. For example, women with gestational diabetes have significantly higher rates of anxiety and depression (Silverman et al., 2017) and higher rates of preeclampsia (Weissgerber and Mudd, 2015), and women with depression during pregnancy are more likely to experience preterm birth or have SGA babies (Venkatesh et al., 2016). Such comorbidity is consistent with the hypothesis that, despite the different systems involved, there is a common etiology for different pregnancy complications (such as changes in hormones or hormone signaling).

A greater understanding of the normal hormone-induced adaptations in the maternal brain should be the primary goal of researchers working in the field of pregnancy. At present, too many of the studies evaluating stress and environmental challenges during pregnancy are focusing on the outcomes in terms of fetal development and paying little attention to what is happening to the mother. The more we can learn about the remarkable adaptations of the maternal brain, the more likely we will be to successfully identify and treat neurologic problems of pregnancy and ensure a healthy outcome for both mother and baby.

References

Aghaeepour N, Lehallier B, Baca Q et al. (2018). A proteomic clock of human pregnancy. Am J Obstet Gynecol 218: 347.e1–347.e14.

Ainslie DA, Morris MJ, Wittert G et al. (2001). Estrogen deficiency causes central leptin insensitivity and increased hypothalamic neuropeptide Y. Int J Obes Relat Metab Disord 25: 1680–1688.

Albin-Brooks C, Nealer C, Sabihi S et al. (2017). The influence of offspring, parity, and oxytocin on cognitive flexibility during the postpartum period. Horm Behav 89: 130–136.

Altemus M (2019). Neuroendocrine networks and functionality. Med Clin North Am 103: 601–612.

Amico JA, Thomas A, Crowley RS et al. (1998). Concentrations of leptin in the serum of pregnant, lactating, and cycling rats and of leptin messenger ribonucleic acid in rat placental tissue. Life Sci 63: 1387–1395.

Anderson MV, Rutherford MD (2011). Recognition of novel faces after single exposure is enhanced during pregnancy. Evol Psychol 9: 47–60.

Anderson MV, Rutherford MD (2012). Cognitive reorganization during pregnancy and the postpartum period: an evolutionary perspective. Evol Psychol 10: 659–687.

Andersson-Hall U, Svedin P, Andreasson U et al. (2018). Central and peripheral leptin and agouti-related protein during and after pregnancy in relation to weight change. Clin Endocrinol (Oxf) 88: 263–271.

Antonijevic IA, Russell JA, Bicknell RJ et al. (2000). Effect of progesterone on the activation of neurones of the supraoptic nucleus during parturition. J Reprod Fertil 120: 367–376.

Augustine RA, Grattan DR (2008). Induction of central leptin resistance in hyperphagic pseudopregnant rats by chronic prolactin infusion. Endocrinology 149: 1049–1055.

Augustine RA, Kokay IC, Andrews ZB et al. (2003). Quantitation of prolactin receptor mRNA in the maternal rat brain during pregnancy and lactation. J Mol Endocrinol 31: 221–232.

Augustine RA, Bouwer GT, Seymour AJ et al. (2016). Reproductive regulation of gene expression in the hypothalamic supraoptic and paraventricular nuclei. J Neuroendocrinol 28: e12350.

Augustine RA, Ladyman SR, Bouwer GT et al. (2017). Prolactin regulation of oxytocin neurone activity in pregnancy and lactation. J Physiol 595: 3591–3605.

Augustine RA, Knowles PJ, Khant Aung Z et al. (2019). Impaired hypothalamic leptin sensitivity in pseudopregnant rats treated with chronic prolactin to mimic pregnancy. J Neuroendocrinol 31: e12702.

Baeyens L, Hindi S, Sorenson RL et al. (2016). Beta-cell adaptation in pregnancy. Diabetes Obes Metab 18: 63–70.

Banasr M, Hery M, Brezun JM et al. (2001). Serotonin mediates oestrogen stimulation of cell proliferation in the adult dentate gyrus. Eur J Neurosci 14: 1417–1424.

Barrientos RM, Brunton PJ, Lenz KM et al. (2019). Neuroimmunology of the female brain across the lifespan: plasticity to psychopathology. Brain Behav Immun 79: 39–55.

Ben-Jonathan N, Neill MA, Arbogast LA et al. (1980). Dopamine in hypophysial portal blood: relationship to circulating prolactin in pregnant and lactating rats. Endocrinology 106: 690–696.

Bicknell RJ (1988). Optimizing release from peptide hormone secretory nerve terminals. J Exp Biol 139: 51–65.

Bodensteiner KJ, Cain P, Ray AS et al. (2006). Effects of pregnancy on spatial cognition in female Hooded Long-Evans rats. Horm Behav 49: 303–314.

Brailoiu GC, Dun SL, Brailoiu E et al. (2007). Nesfatin-1: distribution and interaction with a G protein-coupled receptor in the rat brain. Endocrinology 148: 5088–5094.

Brett M, Baxendale S (2001). Motherhood and memory: a review. Psychoneuroendocrinology 26: 339–362.

Bridges RS (1984). A quantitative analysis of the roles of dosage, sequence, and duration of estradiol and progesterone exposure in the regulation of maternal behavior in the rat. Endocrinology 114: 930–940.

Bridges RS (2015). Neuroendocrine regulation of maternal behavior. Front Neuroendocrinol 36: 178–196.

Bridges RS, Hays LE (2005). Steroid-induced alterations in mRNA expression of the long form of the prolactin receptor in the medial preoptic area of female rats: effects of exposure to a pregnancy-like regimen of progesterone and estradiol. Brain Res Mol Brain Res 140: 10–16.

Bridges RS, DiBiase R, Loundes DD et al. (1985). Prolactin stimulation of maternal behavior in female rats. Science 227: 782–784.

Bridges RS, Numan M, Ronsheim PM et al. (1990). Central prolactin infusions stimulate maternal behavior in steroid-treated, nulliparous female rats. Proc Natl Acad Sci U S A 87: 8003–8007.

Bridges RS, Robertson MC, Shiu RP et al. (1997). Central lactogenic regulation of maternal behavior in rats: steroid dependence, hormone specificity, and behavioral potencies of rat prolactin and rat placental lactogen I. Endocrinology 138: 756–763.

Brinton RD, Thompson RF, Foy MR et al. (2008). Progesterone receptors: form and function in brain. Front Neuroendocrinol 29: 313–339.

Brown CH (2016). Magnocellular neurons and posterior pituitary function. Compr Physiol 6: 1701–1741.

Brown LM, Clegg DJ, Benoit SC et al. (2006). Intraventricular insulin and leptin reduce food intake and body weight in C57BL/6J mice. Physiol Behav 89: 687–691.

Brown CH, Brunton PJ, Russell JA (2008). Rapid estradiol-17beta modulation of opioid actions on the electrical and secretory activity of rat oxytocin neurons in vivo. Neurochem Res 33: 614–623.

Brown RS, Kokay IC, Herbison AE et al. (2010). Distribution of prolactin-responsive neurons in the mouse forebrain. J Comp Neurol 518: 92–102.

Brown RS, Piet R, Herbison AE et al. (2012). Differential actions of prolactin on electrical activity and intracellular signal transduction in hypothalamic neurons. Endocrinology 153: 2375–2384.

Brown RS, Wyatt AK, Herbison RE et al. (2016). Prolactin transport into mouse brain is independent of prolactin receptor. FASEB J 30: 1002–1010.

Brown RSE, Aoki M, Ladyman SR et al. (2017). Prolactin action in the medial preoptic area is necessary for postpartum maternal nursing behavior. Proc Natl Acad Sci U S A 114: 10779–10784.

Brummelte S, Galea LA (2016). Postpartum depression: etiology, treatment and consequences for maternal care. Horm Behav 77: 153–166.

Bruning JC, Gautam D, Burks DJ et al. (2000). Role of brain insulin receptor in control of body weight and reproduction. Science 289: 2122–2125.

Brunton PJ, Russell JA (2008). The expectant brain: adapting for motherhood. Nat Rev Neurosci 9: 11–25.

Brunton PJ, McKay AJ, Ochedalski T et al. (2009). Central opioid inhibition of neuroendocrine stress responses in pregnancy in the rat is induced by the neurosteroid allopregnanolone. J Neurosci 29: 6449–6460.

Brunton PJ, Russell JA, Hirst JJ (2014). Allopregnanolone in the brain: protecting pregnancy and birth outcomes. Prog Neurobiol 113: 106–136.

Brussaard AB, Kits KS, Baker RE et al. (1997). Plasticity in fast synaptic inhibition of adult oxytocin neurons caused by switch in GABA(A) receptor subunit expression. Neuron 19: 1103–1114.

Campfield LA, Smith FJ, Guisez Y et al. (1995). Recombinant mouse OB protein: evidence for a peripheral signal linking adiposity and central neural networks. Science 269: 546–549.

Carrara MA, Batista MR, Saruhashi TR et al. (2012). Coexistence of insulin resistance and increased glucose tolerance in pregnant rats: a physiological mechanism for glucose maintenance. Life Sci 90: 831–837.

Carter DA, Lightman SL (1987). Oxytocin responses to stress in lactating and hyperprolactinaemic rats. Neuroendocrinology 46: 532–537.

Casey P, Huntsdale C, Angus G et al. (1999). Memory in pregnancy. II: implicit, incidental, explicit, semantic, short-term, working and prospective memory in primigravid, multigravid and postpartum women. J Psychosom Obstet Gynaecol 20: 158–164.

Castracane VD, Hendrickx AG, Henson MC (2005). Serum leptin in nonpregnant and pregnant women and in old and new world nonhuman primates. Exp Biol Med (Maywood) 230: 251–254.

Catalano PM (2014). Trying to understand gestational diabetes. Diabet Med 31: 273–281.

Charles NE, Alexander GM, Saenz J (2013). Motivational value and slaience of images of infants. Evol Hum Behav 34: 373–381.

Chen SW, Rodriguez L, Davies MF et al. (1996). The hyperphagic effect of 3 alpha-hydroxylated pregnane steroids in male rats. Pharmacol Biochem Behav 53: 777–782.

Chen X, Hirasawa M, Takahashi Y et al. (1999). Suppression of PGE(2) fever at near term: reduced thermogenesis but not enhanced vasopressin antipyresis. Am J Physiol 277: R354–R361.

Clegg DJ, Riedy CA, Smith KA et al. (2003). Differential sensitivity to central leptin and insulin in male and female rats. Diabetes 52: 682–687.

Clegg DJ, Brown LM, Woods SC et al. (2006). Gonadal hormones determine sensitivity to central leptin and insulin. Diabetes 55: 978–987.

Consiglio AR, Lucion AB (1996). Lesion of hypothalamic paraventricular nucleus and maternal aggressive behavior in female rats. Physiol Behav 59: 591–596.

Cripps AW, Williams VJ (1975). The effect of pregnancy and lactation on food intake, gastrointestinal anatomy and the absorptive capacity of the small intestine in the albino rat. Br J Nutr 33: 17–32.

Cumin F, Baum HP, Levens N (1996). Leptin is cleared from the circulation primarily by the kidney. Int J Obes Relat Metab Disord 20: 1120–1126.

da Costa AP, Wood S, Ingram CD et al. (1996). Region-specific reduction in stress-induced c-fos mRNA expression during pregnancy and lactation. Brain Res 742: 177–184.

Dalley JW, Cardinal RN, Robbins TW (2004). Prefrontal executive and cognitive functions in rodents: neural and neurochemical substrates. Neurosci Biobehav Rev 28: 771–784.

Davies SJ, Lum JA, Skouteris H et al. (2018). Cognitive impairment during pregnancy: a meta-analysis. Med J Aust 208: 35–40.

de Geest K, Thiery M, Piron-Possuyt G et al. (1985). Plasma oxytocin in human pregnancy and parturition. J Perinat Med 13: 3–13.

Demarest KT, McKay DW, Riegle GD et al. (1983). Biochemical indices of tuberoinfundibular dopaminergic neuronal activity during lactation: a lack of response to prolactin. Neuroendocrinology 36: 130–137.

Denroche HC, Huynh FK, Kieffer TJ (2012). The role of leptin in glucose homeostasis. J Diabetes Investig 3: 115–129.

Donner N, Neumann ID (2009). Effects of chronic intracerebral prolactin on the oxytocinergic and vasopressinergic system of virgin ovariectomized rats. Neuroendocrinology 90: 315–322.

Donner N, Bredewold R, Maloumby R et al. (2007). Chronic intracerebral prolactin attenuates neuronal stress circuitries in virgin rats. Eur J Neurosci 25: 1804–1814.

Douglas A, Scullion S, Antonijevic I et al. (2001). Uterine contractile activity stimulates supraoptic neurons in term pregnant rats via a noradrenergic pathway. Endocrinology 142: 633–644.

Douglas AJ, Bicknell RJ, Leng G et al. (2002). Beta-endorphin cells in the arcuate nucleus: projections to the supraoptic nucleus and changes in expression during pregnancy and parturition. J Neuroendocrinol 14: 768–777.

Drago F, Amir S (1984). Effects of hyperprolactinaemia on core temperature of the rat. Brain Res Bull 12: 355–358.

Drago F, Continella G, Conforto G et al. (1985). Prolactin inhibits the development of stress-induced ulcers in the rat. Life Sci 36: 191–197.

Drago F, D'Agata V, Iacona T et al. (1989). Prolactin as a protective factor in stress-induced biological changes. J Clin Lab Anal 3: 340–344.

Dreier JW, Andersen AM, Berg-Beckhoff G (2014). Systematic review and meta-analyses: fever in pregnancy and health impacts in the offspring. Pediatrics 133: e674–e688.

Drew PD, Chavis JA (2000). Inhibition of microglial cell activation by cortisol. Brain Res Bull 52: 391–396.

Edwards MJ (2006). Review: hyperthermia and fever during pregnancy. Birth Defects Res A Clin Mol Teratol 76: 507–516.

Egli M, Bertram R, Toporikova N et al. (2006). Prolactin secretory rhythm of mated rats induced by a single injection of oxytocin. Am J Physiol Endocrinol Metab 290: E566–E572.

Ehrhardt RA, Slepetis RM, Bell AW et al. (2001). Maternal leptin is elevated during pregnancy in sheep. Domest Anim Endocrinol 21: 85–96.

Eid RS, Chaiton JA, Lieblich SE et al. (2019). Early and late effects of maternal experience on hippocampal neurogenesis, microglia, and the circulating cytokine milieu. Neurobiol Aging 78: 1–17.

El Majdoubi M, Poulain DA, Theodosis DT (1996). The glutamatergic innervation of oxytocin- and vasopressin-secreting neurons in the rat supraoptic nucleus and its contribution to lactation-induced synaptic plasticity. Eur J Neurosci 8: 1377–1389.

Eliason HL, Fewell JE (1997). Thermoregulatory control during pregnancy and lactation in rats. J Appl Physiol (1985) 83: 837–844.

El-Kasti MM, Christian HC, Huerta-Ocampo I et al. (2008). The pregnancy-induced increase in baseline circulating growth hormone in rats is not induced by ghrelin. J Neuroendocrinol 20: 309–322.

Elmquist JK, Elias CF, Saper CB (1999). From lesions to leptin: hypothalamic control of food intake and body weight. Neuron 22: 221–232.

Eriksson L, Frankenne F, Eden S et al. (1989). Growth hormone 24-h serum profiles during pregnancy—lack of pulsatility for the secretion of the placental variant. Br J Obstet Gynaecol 96: 949–953.

Evans SS, Repasky EA, Fisher DT (2015). Fever and the thermal regulation of immunity: the immune system feels the heat. Nat Rev Immunol 15: 335–349.

Farooqi IS, Matarese G, Lord GM et al. (2002). Beneficial effects of leptin on obesity, T cell hyporesponsiveness, and neuroendocrine/metabolic dysfunction of human congenital leptin deficiency. J Clin Invest 110: 1093–1103.

Feher P, Olah M, Bodnar I et al. (2010). Dephosphorylation/inactivation of tyrosine hydroxylase at the median eminence of the hypothalamus is required for suckling-induced prolactin and adrenocorticotrop hormone responses. Brain Res Bull 82: 141–145.

Feldman R (2016). The neurobiology of mammalian parenting and the biosocial context of human caregiving. Horm Behav 77: 3–17.

Ferguson JKW (1941). A study of the motility of the intact uterus at term. Surg Gynecol Obstet 73: 359–366.

Fewell JE (1995). Body temperature regulation in rats near term of pregnancy. Can J Physiol Pharmacol 73: 364–368.

Fewell JE, Tang PA (1997). Pregnancy alters body-core temperature response to a simulated open field in rats. J Appl Physiol (1985) 82: 1406–1410.

Field T, Diego M, Hernandez-Reif M (2006). Prenatal depression effects on the fetus and newborn: a review. Infant Behav Dev 29: 445–455.

Finley C, Zhang C, Fewell JE (2015). Sex steroid levels near the term of pregnancy do not alter lipopolysaccharide-induced fever in oophorectomized rats. Exp Physiol 100: 323–330.

Flak JN, Myers Jr MG (2016). Minireview: CNS mechanisms of leptin action. Mol Endocrinol 30: 3–12.

Fliestra RJ, Voogt JL (1997). Lactogenic hormones of the placenta and pituitary inhibit suckling-induced prolactin (PRL) release but not the ante-partum PRL surge. Proc Soc Exp Biol Med 214: 258–264.

Francis K, Meddle SL, Bishop VR et al. (2002). Progesterone receptor expression in the pregnant and parturient rat hypothalamus and brainstem. Brain Res 927: 18–26.

Frankfurt M, Salas-Ramirez K, Friedman E et al. (2011). Cocaine alters dendritic spine density in cortical and subcortical brain regions of the postpartum and virgin female rat. Synapse 65: 955–961.

Fuchs AR, Fuchs F, Husslein P et al. (1984). Oxytocin receptors in the human uterus during pregnancy and parturition. Am J Obstet Gynecol 150: 734–741.

Fujikawa T, Coppari R (2015). Living without insulin: the role of leptin signaling in the hypothalamus. Front Neurosci 9: 108.

Fujikawa T, Soya H, Tamashiro KL et al. (2004). Prolactin prevents acute stress-induced hypocalcemia and ulcerogenesis by acting in the brain of rat. Endocrinology 145: 2006–2013.

Fujikawa T, Tamura K, Kawase T et al. (2005). Prolactin receptor knockdown in the rat paraventricular nucleus by a morpholino-antisense oligonucleotide causes hypocalcemia and stress gastric erosion. Endocrinology 146: 3471–3480.

Galea LA, McEwen BS, Tanapat P et al. (1997). Sex differences in dendritic atrophy of CA3 pyramidal neurons in response to chronic restraint stress. Neuroscience 81: 689–697.

Galea LA, Ormerod BK, Sampath S et al. (2000). Spatial working memory and hippocampal size across pregnancy in rats. Horm Behav 37: 86–95.

Gamo Y, Bernard A, Mitchell SE et al. (2013). Limits to sustained energy intake. XVI body temperature and physical activity of female mice during pregnancy. J Exp Biol 216: 2328–2338.

Garami A, Steiner AA, Romanovsky AA (2018). Fever and hypothermia in systemic inflammation. Handb Clin Neurol 157: 565–597.

Garcia MD, Casanueva FF, Dieguez C et al. (2000). Gestational profile of leptin messenger ribonucleic acid (mRNA) content in the placenta and adipose tissue in the rat, and regulation of the mRNA levels of the leptin receptor subtypes in the hypothalamus during pregnancy and lactation. Biol Reprod 62: 698–703.

Garcia MC, Lopez M, Gualillo O et al. (2003). Hypothalamic levels of NPY, MCH, and prepro-orexin mRNA during pregnancy and lactation in the rat: role of prolactin. FASEB J 17: 1392–1400.

Garcia-Leal C, De Rezende MG, Corsi-Zuelli F et al. (2017). The functioning of the hypothalamic-pituitary-adrenal (HPA) axis in postpartum depressive states: a systematic review. Expert Rev Endocrinol Metab 12: 341–353.

Gatewood JD, Morgan MD, Eaton M et al. (2005). Motherhood mitigates aging-related decrements in learning and memory and positively affects brain aging in the rat. Brain Res Bull 66: 91–98.

Gatford KL, Muhlhausler BS, Huang L et al. (2017). Rising maternal circulating GH during murine pregnancy suggests placental regulation. Endocr Connect 6: 260–266.

Gavrilova O, Barr V, Marcus-Samuels B et al. (1997). Hyperleptinemia of pregnancy associated with the appearance of a circulating form of the leptin receptor. J Biol Chem 272: 30546–30551.

Ghosh R, Sladek CD (1995). Prolactin modulates oxytocin mRNA during lactation by its action on the hypothalamo-neurohypophyseal axis. Brain Res 672: 24–28.

Gies U, Theodosis DT (1994). Synaptic plasticity in the rat supraoptic nucleus during lactation involves GABA innervation and oxytocin neurons: a quantitative immunocytochemical analysis. J Neurosci 14: 2861–2869.

Giovenardi M, Padoin MJ, Cadore LP et al. (1998). Hypothalamic paraventricular nucleus modulates maternal aggression in rats: effects of ibotenic acid lesion and oxytocin antisense. Physiol Behav 63: 351–359.

Gipson CD, Olive MF (2017). Structural and functional plasticity of dendritic spines—root or result of behavior? Genes Brain Behav 16: 101–117.

Glynn LM (2010). Giving birth to a new brain: hormone exposures of pregnancy influence human memory. Psychoneuroendocrinology 35: 1148–1155.

Glynn LM (2012). Increasing parity is associated with cumulative effects on memory. J Womens Health (Larchmt) 21: 1038–1045.

Goel N, Rao H, Durmer JS et al. (2009). Neurocognitive consequences of sleep deprivation. Semin Neurol 29: 320–339.

Goldstein RF, Abell SK, Ranasinha S et al. (2017). Association of gestational weight gain with maternal and infant outcomes: a systematic review and meta-analysis. JAMA 317: 2207–2225.

Grattan DR (2001). The actions of prolactin in the brain during pregnancy and lactation. Prog Brain Res 133: 153–171.

Grattan DR, Averill RL (1991). Intrahypothalamic pituitary grafts elevate prolactin in the cerebrospinal fluid and attenuate prolactin release following ether stress. Proc Soc Exp Biol Med 196: 42–46.

Grattan DR, Averill RL (1995). Absence of short-loop autoregulation of prolactin during late pregnancy in the rat. Brain Res Bull 36: 413–416.

Grattan DR, LeTissier P (2015). Hypothalamic control of prolactin secretion, and the multiple reproductive functions of prolactin. In: TM Plant, AJ Zelesnik (Eds.), Knobil and Neill's Physiology of Reproduction, fourth ed.. vol. 1. Elsevier, pp. 469–526.

Grattan DR, Steyn FJ, Kokay IC et al. (2008). Pregnancy-induced adaptation in the neuroendocrine control of prolactin secretion. J Neuroendocrinol 20: 497–507.

Grueso E, Rocha M, Puerta M (2001). Plasma and cerebrospinal fluid leptin levels are maintained despite enhanced food intake in progesterone-treated rats. Eur J Endocrinol 144: 659–665.

Gustafson P, Bunn SJ, Grattan DR (2017). The role of prolactin in the suppression of Crh mRNA expression during pregnancy and lactation in the mouse. J Neuroendocrinol 29: e12511.

Gustafson P, Ladyman SR, Brown RSE (2019). Suppression of leptin transport into the brain contributes to leptin resistance during pregnancy in the mouse. Endocrinology 160: 880–890.

Haim A, Julian D, Albin-Brooks C et al. (2017). A survey of neuroimmune changes in pregnant and postpartum female rats. Brain Behav Immun 59: 67–78.

Halaas JL, Gajiwala KS, Maffei M et al. (1995). Weight-reducing effects of the plasma protein encoded by the obese gene. Science 269: 543–546.

Hansen S (1989). Medial hypothalamic involvement in maternal aggression of rats. Behav Neurosci 103: 1035–1046.

Hardie L, Trayhurn P, Abramovich D et al. (1997). Circulating leptin in women: a longitudinal study in the menstrual cycle and during pregnancy. Clin Endocrinol (Oxf) 47: 101–106.

Harre EM, Mouihate A, Pittman QJ (2006). Attenuation of fever at near term: is interleukin-6-STAT3 signalling altered? J Neuroendocrinol 18: 57–63.

Harris RB (2014). Direct and indirect effects of leptin on adipocyte metabolism. Biochim Biophys Acta 1842: 414–423.

Hartgill TW, Bergersen TK, Pirhonen J (2011). Core body temperature and the thermoneutral zone: a longitudinal study of normal human pregnancy. Acta Physiol (Oxf) 201: 467–474.

Hartikaninen-Sorri AL (1991). Smoking and childbirth. Acta Obstet Gynecol Scand 70: 103–104.

Hashikawa K, Hashikawa Y, Tremblay R et al. (2017). Esr1(+) cells in the ventromedial hypothalamus control female aggression. Nat Neurosci 20: 1580–1590.

Hatton GI (1999). Astroglial modulation of neurotransmitter/peptide release from the neurohypophysis: present status. J Chem Neuroanat 16: 203–221.

Hatton GI, Wang YF (2008). Neural mechanisms underlying the milk ejection burst and reflex. Prog Brain Res 170: 155–166.

Henry JD, Rendell PG (2007). A review of the impact of pregnancy on memory function. J Clin Exp Neuropsychol 29: 793–803.

Herrenkohl LR, Rosenberg PA (1974). Effects of hypothalamic deafferentation late in gestation on lactation and nursing behavior in the rat. Horm Behav 5: 33–41.

Hervey E, Hervey GR (1967). The effects of progesterone on body weight and composition in the rat. J Endocrinol 37: 361–381.

Hoekzema E, Barba-Muller E, Pozzobon C et al. (2017). Pregnancy leads to long-lasting changes in human brain structure. Nat Neurosci 20: 287–296.

Holmes CS, Koepke KM, Thompson RG et al. (1984). Verbal fluency and naming performance in type I diabetes at different blood glucose concentrations. Diabetes Care 7: 454–459.

Imai-Matsumara K, Matsumura K, Morimoto A et al. (1990). Suppression of cold-induced thermogenesis in full-term pregnant rats. J Physiol 425: 271–281.

Imai-Matsumura K, Matsumura K, Terao A et al. (2002). Attenuated fever in pregnant rats is associated with blunted syntheses of brain cyclooxygenase-2 and PGE2. Am J Physiol Regul Integr Comp Physiol 283: R1346–R1353.

Inoue H (2016). Central insulin-mediated regulation of hepatic glucose production [review]. Endocr J 63: 1–7.

Insel TR, Harbaugh CR (1989). Lesions of the hypothalamic paraventricular nucleus disrupt the initiation of maternal behavior. Physiol Behav 45: 1033–1041.

Jacobson CD, Terkel J, Gorski RA et al. (1980). Effects of small medial preoptic area lesions on maternal behavior: retrieving and nest building in the rat. Brain Res 194: 471–478.

Joels M (2018). Corticosteroids and the brain. J Endocrinol 238: R121–R130.

Kammerer M, Adams D, Castelberg Bv BV et al. (2002). Pregnant women become insensitive to cold stress. BMC Pregnancy Childbirth 2: 8.

Kasting NW, Veale WL, Cooper KE (1978). Suppression of fever at term of pregnancy. Nature 271: 245–246.

Keenan PA, Yaldoo DT, Stress ME et al. (1998). Explicit memory in pregnant women. Am J Obstet Gynecol 179: 731–737.

Kelly PA, Tsushima T, Shiu RP et al. (1976). Lactogenic and growth hormone-like activities in pregnancy determined by radioreceptor assays. Endocrinology 99: 765–774.

Kelly MA, Rubinstein M, Asa SL et al. (1997). Pituitary lactotroph hyperplasia and chronic hyperprolactinemia in dopamine D2 receptor-deficient mice. Neuron 19: 103–113.

Kennett JE, McKee DT (2012). Oxytocin: an emerging regulator of prolactin secretion in the female rat. J Neuroendocrinol 24: 403–412.

Killgore WD (2010). Effects of sleep deprivation on cognition. Prog Brain Res 185: 105–129.

Kinsley CH, Lambert KG (2008). Reproduction-induced neuroplasticity: natural behavioural and neuronal alterations associated with the production and care of offspring. J Neuroendocrinol 20: 515–525.

Kinsley CH, Madonia L, Gifford GW et al. (1999). Motherhood improves learning and memory. Nature 402: 137–138.

Kinsley CH, Trainer R, Stafisso-Sandoz G et al. (2006). Motherhood and the hormones of pregnancy modify concentrations of hippocampal neuronal dendritic spines. Horm Behav 49: 131–142.

Knobloch HS, Charlet A, Hoffmann LC et al. (2012). Evoked axonal oxytocin release in the central amygdala attenuates fear response. Neuron 73: 553–566.

Kokay IC, Grattan DR (2005). Expression of mRNA for prolactin receptor (long form) in dopamine and pro-opiomelanocortin neurones in the arcuate nucleus of non-pregnant and lactating rats. J Neuroendocrinol 17: 827–835.

Kokay IC, Bull PM, Davis RL et al. (2006). Expression of the long form of the prolactin receptor in magnocellular oxytocin neurons is associated with specific prolactin regulation of oxytocin neurons. Am J Physiol Regul Integr Comp Physiol 290: R1216–R1225.

Kokay IC, Wyatt A, Phillipps HR et al. (2018). Analysis of prolactin receptor expression in the murine brain using a novel prolactin receptor reporter mouse. J Neuroendocrinol 30: e12634.

Krishnamoorthy U, Schram CM, Hill SR (2006). Maternal obesity in pregnancy: is it time for meaningful research to inform preventive and management strategies? BJOG 113: 1134–1140.

Kroll-Desrosiers AR, Nephew BC, Babb JA et al. (2017). Association of peripartum synthetic oxytocin administration and depressive and anxiety disorders within the first postpartum year. Depress Anxiety 34: 137–146.

Lacroix MC, Guibourdenche J, Frendo JL et al. (2002). Human placental growth hormone—a review. Placenta 23: S87–S94.

Ladyman SR (2008). Leptin resistance during pregnancy in the rat. J Neuroendocrinol 20: 269–277.

Ladyman SR, Grattan DR (2004). Region-specific reduction in leptin-induced phosphorylation of signal transducer and activator of transcription-3 (STAT3) in the rat hypothalamus is associated with leptin resistance during pregnancy. Endocrinology 145: 3704–3711.

Ladyman SR, Grattan DR (2005). Suppression of leptin receptor messenger ribonucleic acid and leptin responsiveness in the ventromedial nucleus of the hypothalamus during pregnancy in the rat. Endocrinology 146: 3868–3874.

Ladyman SR, Grattan DR (2016). Central effects of leptin on glucose homeostasis are modified during pregnancy in the rat. J Neuroendocrinol 28: e12431.

Ladyman SR, Grattan DR (2017). Region-specific suppression of hypothalamic responses to insulin to adapt to elevated maternal insulin secretion during pregnancy. Endocrinology 158: 4257–4269.

Ladyman SR, Tups A, Augustine RA et al. (2009). Loss of hypothalamic response to leptin during pregnancy associated with development of melanocortin resistance. J Neuroendocrinol 21: 449–456.

Ladyman SR, Augustine RA, Grattan DR (2010). Hormone interactions regulating energy balance during pregnancy. J Neuroendocrinol 22: 805–817.

Ladyman SR, Sapsford TJ, Grattan DR (2011). Loss of acute satiety response to cholecystokinin in pregnant rats. J Neuroendocrinol 23: 1091–1098.

Ladyman SR, Fieldwick DM, Grattan DR (2012). Suppression of leptin-induced hypothalamic JAK/STAT signalling and feeding response during pregnancy in the mouse. Reproduction 144: 83–90.

Ladyman SR, Augustine RA, Scherf E et al. (2016). Attenuated hypothalamic responses to alpha-melanocyte stimulating hormone during pregnancy in the rat. J Physiol 594: 1087–1101.

Ladyman SR, Carter KM, Grattan DR (2018). Energy homeostasis and running wheel activity during pregnancy in the mouse. Physiol Behav 194: 83–94.

Lambert KG, Kinsley CH (1993). Sex differences and gonadal hormones influence susceptibility to the activity-stress paradigm. Physiol Behav 53: 1085–1090.

Lambert KG, Berry AE, Griffins G et al. (2005). Pup exposure differentially enhances foraging ability in primiparous and nulliparous rats. Physiol Behav 84: 799–806.

Landgraf R, Neumann I, Russell JA et al. (1992). Push-pull perfusion and microdialysis studies of central oxytocin and vasopressin release in freely moving rats during pregnancy, parturition, and lactation. Ann N Y Acad Sci 652: 326–339.

Larsen CM, Grattan DR (2010). Prolactin-induced mitogenesis in the subventricular zone of the maternal brain during early pregnancy is essential for normal postpartum behavioral responses in the mother. Endocrinology 151: 3805–3814.

Lee A, Clancy S, Fleming AS (2000). Mother rats bar-press for pups: effects of lesions of the mpoa and limbic sites on maternal behavior and operant responding for pup-reinforcement. Behav Brain Res 108: 215–231.

Lee YB, Nagai A, Kim SU (2002). Cytokines, chemokines, and cytokine receptors in human microglia. J Neurosci Res 69: 94–103.

Lee SW, Kim YB, Kim JS et al. (2015). GABAergic inhibition is weakened or converted into excitation in the oxytocin and vasopressin neurons of the lactating rat. Mol Brain 8: 34.

Lee SH, Zabolotny JM, Huang H et al. (2016). Insulin in the nervous system and the mind: functions in metabolism, memory, and mood. Mol Metab 5: 589–601.

Leng G, Russell JA (2016). The peptide oxytocin antagonist F-792, when given systemically, does not act centrally in lactating rats. J Neuroendocrinol 28: e12331.

Lerant A, Freeman ME (1998). Ovarian steroids differentially regulate the expression of prolactin receptors in neuroendocrine dopaminergic neuron populations—a double-label confocal microscopic study. Brain Res 802: 141–154.

Leuner B, Gould E (2010). Dendritic growth in medial prefrontal cortex and cognitive flexibility are enhanced during the postpartum period. J Neurosci 30: 13499–13503.

Lewandowski K, Horn R, O'Callaghan CJ et al. (1999). Free leptin, bound leptin, and soluble leptin receptor in normal and diabetic pregnancies. J Clin Endocrinol Metab 84: 300–306.

LifeCycle Project-Maternal Obesity, Childhood Outcomes Study Group, Voerman E et al. (2019). Association of gestational weight gain with adverse maternal and infant outcomes. JAMA 321: 1702–1715.

Lightman SL (1992). Alterations in hypothalamic-pituitary responsiveness during lactation. Ann N Y Acad Sci 652: 340–346.

Login IS, MacLeod RM (1977). Prolactin in human and rat serum and cerebrospinal fluid. Brain Res 132: 477–483.

Lonstein JS, Levy F, Fleming AS (2015). Common and divergent psychobiological mechanisms underlying maternal behaviors in non-human and human mammals. Horm Behav 73: 156–185.

Lopez M, Lelliott CJ, Vidal-Puig A (2007). Hypothalamic fatty acid metabolism: a housekeeping pathway that regulates food intake. Bioessays 29: 248–261.

Lubin DA, Elliott JC, Black MC et al. (2003). An oxytocin antagonist infused into the central nucleus of the amygdala increases maternal aggressive behavior. Behav Neurosci 117: 195–201.

Lucas BK, Ormandy CJ, Binart N et al. (1998). Null mutation of the prolactin receptor gene produces a defect in maternal behavior. Endocrinology 139: 4102–4107.

Ludwig M (1998). Dendritic release of vasopressin and oxytocin. J Neuroendocrinol 10: 881–895.

Ludwig M, Stern J (2015). Multiple signalling modalities mediated by dendritic exocytosis of oxytocin and vasopressin. Philos Trans R Soc Lond B Biol Sci 370: \rstb.2014.0182.

Lyons DJ, Hellysaz A, Broberger C (2012). Prolactin regulates tuberoinfundibular dopamine neuron discharge pattern: novel feedback control mechanisms in the lactotrophic axis. J Neurosci 32: 8074–8083.

Macbeth AH, Luine VN (2010). Changes in anxiety and cognition due to reproductive experience: a review of data from rodent and human mothers. Neurosci Biobehav Rev 34: 452–467.

Macbeth AH, Gautreaux C, Luine VN (2008). Pregnant rats show enhanced spatial memory, decreased anxiety, and altered levels of monoaminergic neurotransmitters. Brain Res 1241: 136–147.

Mann PE, Rubin BS, Bridges RS (1997). Differential proopiomelanocortin gene expression in the medial basal hypothalamus of rats during pregnancy and lactation. Brain Res Mol Brain Res 46: 9–16.

Mansour A, Meador-Woodruff JH, Bunzow JR et al. (1990). Localization of dopamine D2 receptor mRNA and D1 and D2 receptor binding in the rat brain and pituitary: an in situ hybridization-receptor autoradiographic analysis. J Neurosci 10: 2587–2600.

Marchi J, Berg M, Dencker A et al. (2015). Risks associated with obesity in pregnancy, for the mother and baby: a systematic review of reviews. Obes Rev 16: 621–638.

Martin SM, Malkinson TJ, Veale WL et al. (1995). Fever in pregnant, parturient, and lactating rats. Am J Physiol 268: R919–R923.

Martinez de Morentin PB, Lage R, Gonzalez-Garcia I et al. (2015). Pregnancy induces resistance to the anorectic effect of hypothalamic malonyl-CoA and the thermogenic effect of hypothalamic AMPK inhibition in female rats. Endocrinology 156: 947–960.

Masuzaki H, Ogawa Y, Sagawa N et al. (1997). Nonadipose tissue production of leptin: leptin as a novel placenta-derived hormone in humans. Nat Med 3: 1029–1033.

Mattson BJ, Morrell JI (2005). Preference for cocaine- versus pup-associated cues differentially activates neurons expressing either Fos or cocaine- and amphetamine-regulated transcript in lactating, maternal rodents. Neuroscience 135: 315–328.

McAulay V, Deary IJ, Ferguson SC et al. (2001). Acute hypoglycemia in humans causes attentional dysfunction while nonverbal intelligence is preserved. Diabetes Care 24: 1745–1750.

McGowan MK, Andrews KM, Grossman SP (1992). Chronic intrahypothalamic infusions of insulin or insulin antibodies alter body weight and food intake in the rat. Physiol Behav 51: 753–766.

Menchetti L, Brecchia G, Canali C et al. (2015). Food restriction during pregnancy in rabbits: effects on hormones and metabolites involved in energy homeostasis and metabolic programming. Res Vet Sci 98: 7–12.

Merchenthaler I (1993). Induction of enkephalin in tuberoinfundibular dopaminergic neurons during lactation. Endocrinology 133: 2645–2651.

Merchenthaler I, Lane MV, Numan S et al. (2004). Distribution of estrogen receptor alpha and beta in the mouse central nervous system: in vivo autoradiographic and immunocytochemical analyses. J Comp Neurol 473: 270–291.

Michaloudi HC, el Majdoubi M, Poulain DA et al. (1997). The noradrenergic innervation of identified hypothalamic magnocellular somata and its contribution to lactation-induced synaptic plasticity. J Neuroendocrinol 9: 17–23.

Miklos IH, Kovacs KJ (2002). GABAergic innervation of corticotropin-releasing hormone (CRH)-secreting parvocellular neurons and its plasticity as demonstrated by quantitative immunoelectron microscopy. Neuroscience 113: 581–592.

Moltz H (1993). Fever: causes and consequences. Neurosci Biobehav Rev 17: 237–269.

Moran TH (2000). Cholecystokinin and satiety: current perspectives. Nutrition 16: 858–865.

Morton GJ (2007). Hypothalamic leptin regulation of energy homeostasis and glucose metabolism. J Physiol 583: 437–443.

Mouihate A, Pittman QJ (2003). Neuroimmune response to endogenous and exogenous pyrogens is differently modulated by sex steroids. Endocrinology 144: 2454–2460.

Mouihate A, Clerget-Froidevaux MS, Nakamura K et al. (2002). Suppression of fever at near term is associated with reduced COX-2 protein expression in rat hypothalamus. Am J Physiol Regul Integr Comp Physiol 283: R800–R805.

Mouihate A, Ellis S, Harre EM et al. (2005). Fever suppression in near-term pregnant rats is dissociated from LPS-activated signaling pathways. Am J Physiol Regul Integr Comp Physiol 289: R1265–R1272.

Naef L, Woodside B (2007). Prolactin/leptin interactions in the control of food intake in rats. Endocrinology 148: 5977–5983.

Nagaishi VS, Cardinali LI, Zampieri TT et al. (2014). Possible crosstalk between leptin and prolactin during pregnancy. Neuroscience 259: 71–83.

Naik D, Hesarghatta Shyamasunder A, Doddabelavangala Mruthyunjaya M et al. (2017). Masked hypoglycemia in pregnancy. J Diabetes 9: 778–786.

Nakayama T, Suzuki M, Ishizuka N (1975). Action of progesterone on preoptic thermosensitive neurones. Nature 258: 80.

Napso T, Yong HEJ, Lopez-Tello J et al. (2018). The role of placental hormones in mediating maternal adaptations to support pregnancy and lactation. Front Physiol 9: 1091.

Negoro H, Visessuwan S, Holland RC (1973a). Reflex activation of paraventricular nucleus units during the reproductive cycle and in ovariectomized rats treated with oestrogen or progesterone. J Endocrinol 59: 559–567.

Negoro H, Visessuwan S, Holland RC (1973b). Unit activity in the paraventricular nucleus of female rats at different stages of the reproductive cycle and after ovariectomy, with or without oestrogen or progesterone treatment. J Endocrinol 59: 545–558.

Nephew BC, Bridges RS (2008). Central actions of arginine vasopressin and a V1a receptor antagonist on maternal aggression, maternal behavior, and grooming in lactating rats. Pharmacol Biochem Behav 91: 77–83.

Neumann ID (2008). Brain oxytocin: a key regulator of emotional and social behaviours in both females and males. J Neuroendocrinol 20: 858–865.

Neumann ID, Johnstone HA, Hatzinger M et al. (1998). Attenuated neuroendocrine responses to emotional and physical stressors in pregnant rats involve adenohypophysial changes. J Physiol 508 (Pt 1): 289–300.

Nicholson G, Greeley GH, Jr. Humm J et al. (1980). Prolactin in cerebrospinal fluid: a probable site of prolactin autoregulation. Brain Res 190: 447–457.

Nuamah MA, Sagawa N, Yura S et al. (2003). Free-to-total leptin ratio in maternal plasma is constant throughout human pregnancy. Endocr J 50: 421–428.

Numan M (1974). Medial preoptic area and maternal behavior in the female rat. J Comp Physiol Psychol 87: 746–759.

Numan M, Sheehan TP (1997). Neuroanatomical circuitry for mammalian maternal behavior. Ann N Y Acad Sci 807: 101–125.

Numan M, Smith HG (1984). Maternal behavior in rats: evidence for the involvement of preoptic projections to the ventral tegmental area. Behav Neurosci 98: 712–727.

Numan M, Stolzenberg DS (2009). Medial preoptic area interactions with dopamine neural systems in the control of the onset and maintenance of maternal behavior in rats. Front Neuroendocrinol 30: 46–64.

Numan M, Rosenblatt JS, Komisaruk BR (1977). Medial preoptic area and onset of maternal behavior in the rat. J Comp Physiol Psychol 91: 146–164.

Numan M, Morrell JI, Pfaff DW (1985). Anatomical identification of neurons in selected brain regions associated with maternal behavior deficits induced by knife cuts of the lateral hypothalamus in rats. J Comp Neurol 237: 552–564.

Numan M, Corodimas KP, Numan MJ et al. (1988). Axon-sparing lesions of the preoptic region and substantia innominata disrupt maternal behavior in rats. Behav Neurosci 102: 381–396.

Numan M, McSparren J, Numan MJ (1990). Dorsolateral connections of the medial preoptic area and maternal behavior in rats. Behav Neurosci 104: 964–979.

Numan M, Numan MJ, Pliakou N et al. (2005). The effects of D1 or D2 dopamine receptor antagonism in the medial preoptic area, ventral pallidum, or nucleus accumbens on the maternal retrieval response and other aspects of maternal behavior in rats. Behav Neurosci 119: 1588–1604.

Oatridge A, Holdcroft A, Saeed N et al. (2002). Change in brain size during and after pregnancy: study in healthy women and women with preeclampsia. Am J Neuroradiol 23: 19–26.

Obici S, Feng Z, Karkanias G et al. (2002a). Decreasing hypothalamic insulin receptors causes hyperphagia and insulin resistance in rats. Nat Neurosci 5: 566–572.

Obici S, Zhang BB, Karkanias G et al. (2002b). Hypothalamic insulin signaling is required for inhibition of glucose production. Nat Med 8: 1376–1382.

Ogawa S, Eng V, Taylor J et al. (1998). Roles of estrogen receptor-alpha gene expression in reproduction-related behaviors in female mice. Endocrinology 139: 5070–5081.

Otsuki Y, Yamaji K, Fujita M et al. (1983). Serial plasma oxytocin levels during pregnancy and labor. Acta Obstet Gynecol Scand 62: 15–18.

Page-Wilson G, Reitman-Ivashkov E, Meece K et al. (2013). Cerebrospinal fluid levels of leptin, proopiomelanocortin, and agouti-related protein in human pregnancy: evidence for leptin resistance. J Clin Endocrinol Metab 98: 264–271.

Parsons C, Redman S (1991). Self-reported cognitive change during pregnancy. Aust J Adv Nurs 9: 20–29.

Patchev VK, Hassan AH, Holsboer DF et al. (1996). The neurosteroid tetrahydroprogesterone attenuates the endocrine response to stress and exerts glucocorticoid-like effects on vasopressin gene transcription in the rat hypothalamus. Neuropsychopharmacology 15: 533–540.

Patil MJ, Ruparel SB, Henry MA et al. (2013). Prolactin regulates TRPV1, TRPA1, and TRPM8 in sensory neurons in a sex-dependent manner: contribution of prolactin receptor to inflammatory pain. Am J Physiol Endocrinol Metab 305: E1154–E1164.

Pawluski JL, Galea LA (2006). Hippocampal morphology is differentially affected by reproductive experience in the mother. J Neurobiol 66: 71–81.

Pawluski JL, Barakauskas VE, Galea LA (2010). Pregnancy decreases oestrogen receptor alpha expression and pyknosis, but not cell proliferation or survival, in the hippocampus. J Neuroendocrinol 22: 248–257.

Pawluski JL, van den Hove DL, Rayen I et al. (2011). Stress and the pregnant female: impact on hippocampal cell proliferation, but not affective-like behaviors. Horm Behav 59: 572–580.

Pawluski JL, Valenca A, Santos AI et al. (2012). Pregnancy or stress decrease complexity of CA3 pyramidal neurons in the hippocampus of adult female rats. Neuroscience 227: 201–210.

Pawluski JL, Csaszar E, Savage E et al. (2015). Effects of stress early in gestation on hippocampal neurogenesis and glucocorticoid receptor density in pregnant rats. Neuroscience 290: 379–388.

Pawluski JL, Lambert KG, Kinsley CH (2016). Neuroplasticity in the maternal hippocampus: relation to cognition and effects of repeated stress. Horm Behav 77: 86–97.

Pawluski JL, Lonstein JS, Fleming AS (2017). The neurobiology of postpartum anxiety and depression. Trends Neurosci 40: 106–120.

Pazos P, Lima L, Casanueva FF et al. (2013). Interleukin 6 deficiency modulates the hypothalamic expression of energy balance regulating peptides during pregnancy in mice. PLoS One 8: e72339.

Pearson RM, Lightman SL, Evans J (2009). Emotional sensitivity for motherhood: late pregnancy is associated with enhanced accuracy to encode emotional faces. Horm Behav 56: 557–563.

Pedersen CA, Boccia ML (2003). Oxytocin antagonism alters rat dams' oral grooming and upright posturing over pups. Physiol Behav 80: 233–241.

Pedersen CA, Prange AJ Jr. (1979). Induction of maternal behavior in virgin rats after intracerebroventricular administration of oxytocin. Proc Natl Acad Sci U S A 76: 6661–6665.

Pedersen CA, Ascher JA, Monroe YL et al. (1982). Oxytocin induces maternal behavior in virgin female rats. Science 216: 648–650.

Pedersen CA, Caldwell JD, Walker C et al. (1994). Oxytocin activates the postpartum onset of rat maternal behavior in the ventral tegmental and medial preoptic areas. Behav Neurosci 108: 1163–1171.

Pelleymounter MA, Cullen MJ, Baker MB et al. (1995). Effects of the obese gene product on body weight regulation in ob/ob mice. Science 269: 540–543.

Perlmutter LS, Tweedle CD, Hatton GI (1984). Neuronal/glial plasticity in the supraoptic dendritic zone: dendritic bundling and double synapse formation at parturition. Neuroscience 13: 769–779.

Peter J, Burbach H, Adan RA et al. (1990). Regulation of the rat oxytocin gene by estradiol. J Neuroendocrinol 2: 633–639.

Pi XJ, Grattan DR (1998a). Differential expression of the two forms of prolactin receptor mRNA within microdissected hypothalamic nuclei of the rat. Brain Res Mol Brain Res 59: 1–12.

Pi XJ, Grattan DR (1998b). Distribution of prolactin receptor immunoreactivity in the brain of estrogen-treated, ovariectomized rats. J Comp Neurol 394: 462–474.

Picazo O, Fernandez-Guasti A (1993). Changes in experimental anxiety during pregnancy and lactation. Physiol Behav 54: 295–299.

Pihoker C, Robertson MC, Freemark M (1993). Rat placental lactogen-I binds to the choroid plexus and hypothalamus of the pregnant rat. J Endocrinol 139: 235–242.

Porter TE, Hill JB, Wiles CD et al. (1990). Is the mammosomatotrope a transitional cell for the functional interconversion of growth hormone- and prolactin-secreting cells? Suggestive evidence from virgin, gestating, and lactating rats. Endocrinology 127: 2789–2794.

Poser CM, Kassirer MR, Peyser JM (1986). Benign encephalopathy of pregnancy. Preliminary clinical observations. Acta Neurol Scand 73: 39–43.

Power ML, Schulkin J (2006). Functions of corticotropin-releasing hormone in anthropoid primates: from brain to placenta. Am J Hum Biol 18: 431–447.

Pryce CR (2008). Postnatal ontogeny of expression of the corticosteroid receptor genes in mammalian brains: inter-species and intra-species differences. Brain Res Rev 57: 596–605.

Ramos-Lobo AM, Teixeira PDS, Furigo IC et al. (2017). SOCS3 ablation in SF1 cells causes modest metabolic effects during pregnancy and lactation. Neuroscience 365: 114–124.

Ravanelli N, Casasola W, English T et al. (2019). Heat stress and fetal risk. Environmental limits for exercise and passive heat stress during pregnancy: a systematic review with best evidence synthesis. Br J Sports Med 53: 799–805.

Reyes LM, Usselman CW, Davenport MH et al. (2018). Sympathetic nervous system regulation in human normotensive and hypertensive pregnancies. Hypertension 71: 793–803.

Ribeiro AC, Musatov S, Shteyler A et al. (2012). siRNA silencing of estrogen receptor-alpha expression specifically in medial preoptic area neurons abolishes maternal care in female mice. Proc Natl Acad Sci U S A 109: 16324–16329.

Riddle O, Lahr EL, Bates RW (1935). Maternal behavior induced in virgin rats by prolactin. Exp Biol Med 32: 730–734.

Rilling JK (2013). The neural and hormonal bases of human parental care. Neuropsychologia 51: 731–747.

Robinson DP, Klein SL (2012). Pregnancy and pregnancy-associated hormones alter immune responses and disease pathogenesis. Horm Behav 62: 263–271.

Rocha M, Bing C, Williams G et al. (2003). Pregnancy-induced hyperphagia is associated with increased gene expression of hypothalamic agouti-related peptide in rats. Regul Pept 114: 159–165.

Rodrigues SM, Saslow LR, Garcia N et al. (2009). Oxytocin receptor genetic variation relates to empathy and stress reactivity in humans. Proc Natl Acad Sci U S A 106: 21437–21441.

Rojas JM, Schwartz MW (2014). Control of hepatic glucose metabolism by islet and brain. Diabetes Obes Metab 16: 33–40.

Rolls A, Schori H, London A et al. (2008). Decrease in hippocampal neurogenesis during pregnancy: a link to immunity. Mol Psychiatry 13: 468–469.

Romano N, Yip SH, Hodson DJ et al. (2013). Plasticity of hypothalamic dopamine neurons during lactation results in dissociation of electrical activity and release. J Neurosci 33: 4424–4433.

Romero R, Erez O, Maymon E et al. (2017). The maternal plasma proteome changes as a function of gestational age in normal pregnancy: a longitudinal study. Am J Obstet Gynecol 217: 67.e1–67.e21.

Rosenblatt JS (1969). The development of maternal responsiveness in the rat. Am J Orthopsychiatry 39: 36–56.

Rossoni E, Feng J, Tirozzi B et al. (2008). Emergent synchronous bursting of oxytocin neuronal network. PLoS Comput Biol 4: e1000123.

Roth J, Blatteis CM (2014). Mechanisms of fever production and lysis: lessons from experimental LPS fever. Compr Physiol 4: 1563–1604.

Rutherford HJ, Williams SK, Moy S et al. (2011). Disruption of maternal parenting circuitry by addictive process: rewiring of reward and stress systems. Front Psych 2: 37.

Saiardi A, Bozzi Y, Baik JH et al. (1997). Antiproliferative role of dopamine: loss of D2 receptors causes hormonal dysfunction and pituitary hyperplasia. Neuron 19: 115–126.

Salais-Lopez H, Lanuza E, Agustin-Pavon C et al. (2017). Tuning the brain for motherhood: prolactin-like central signalling in virgin, pregnant, and lactating female mice. Brain Struct Funct 222: 895–921.

Saleri R, Sabbioni A, Cavalli V et al. (2015). Monitoring blood plasma leptin and lactogenic hormones in pregnant sows. Animal 9: 629–634.

Sapsford TJ, Kokay IC, Ostberg L et al. (2012). Differential sensitivity of specific neuronal populations of the rat hypothalamus to prolactin action. J Comp Neurol 520: 1062–1077.

Sasaki A, Shinkawa O, Yoshinaga K (1989). Placental corticotropin-releasing hormone may be a stimulator of maternal pituitary adrenocorticotropic hormone secretion in humans. J Clin Invest 84: 1997–2001.

Schaeffer M, Langlet F, Lafont C et al. (2013). Rapid sensing of circulating ghrelin by hypothalamic appetite-modifying neurons. Proc Natl Acad Sci U S A 110: 1512–1517.

Scheithauer BW, Sano T, Kovacs KT et al. (1990). The pituitary gland in pregnancy: a clinicopathologic and immunohistochemical study of 69 cases. Mayo Clin Proc 65: 461–474.

Schulte HM, Weisner D, Allolio B (1990). The corticotrophin releasing hormone test in late pregnancy: lack of adrenocorticotrophin and cortisol response. Clin Endocrinol (Oxf) 33: 99–106.

Scott N, Prigge M, Yizhar O et al. (2015). A sexually dimorphic hypothalamic circuit controls maternal care and oxytocin secretion. Nature 525: 519–522.

Sedov ID, Cameron EE, Madigan S et al. (2018). Sleep quality during pregnancy: a meta-analysis. Sleep Med Rev 38: 168–176.

Seeber RM, Smith JT, Waddell BJ (2002). Plasma leptin-binding activity and hypothalamic leptin receptor expression during pregnancy and lactation in the rat. Biol Reprod 66: 1762–1767.

Selmanoff M, Gregerson KA (1985). Suckling decreases dopamine turnover in both medial and lateral aspects of the median eminence in the rat. Neurosci Lett 57: 25–30.

Selmanoff M, Wise PM (1981). Decreased dopamine turnover in the median eminence in response to suckling in the lactating rat. Brain Res 212: 101–115.

Seth S, Lewis AJ, Galbally M (2016). Perinatal maternal depression and cortisol function in pregnancy and the postpartum period: a systematic literature review. BMC Pregnancy Childbirth 16: 124.

Shanks N, Windle RJ, Perks P et al. (1999). The hypothalamic-pituitary-adrenal axis response to endotoxin is attenuated during lactation. J Neuroendocrinol 11: 857–865.

Sharif Naeini R, Witty MF, Seguela P et al. (2006). An N-terminal variant of Trpv1 channel is required for osmosensory transduction. Nat Neurosci 9: 93–98.

Sheehan T, Numan M (2002). Estrogen, progesterone, and pregnancy termination alter neural activity in brain regions that control maternal behavior in rats. Neuroendocrinology 75: 12–23.

Sheppard PAS, Choleris E, Galea LAM (2019). Structural plasticity of the hippocampus in response to estrogens in female rodents. Mol Brain 12: 22.

Sherer ML, Posillico CK, Schwarz JM (2017). An examination of changes in maternal neuroimmune function during pregnancy and the postpartum period. Brain Behav Immun 66: 201–209.

Shi Z, Li B, Brooks VL (2015). Role of the paraventricular nucleus of the hypothalamus in the sympathoexcitatory effects of leptin. Hypertension 66: 1034–1041.

Shi Z, Hansen KM, Bullock KM et al. (2019). Resistance to the sympathoexcitatory effects of insulin and leptin in late pregnant rats. J Physiol 597: 4087–4100.

Shingo T, Gregg C, Enwere E et al. (2003). Pregnancy-stimulated neurogenesis in the adult female forebrain mediated by prolactin. Science 299: 117–120.

Shirley B (1984). The food intake of rats during pregnancy and lactation. Lab Anim Sci 34: 169–172.

Shughrue PJ, Lane MV, Merchenthaler I (1997a). Comparative distribution of estrogen receptor-alpha and -beta mRNA in the rat central nervous system. J Comp Neurol 388: 507–525.

Shughrue PJ, Lane MV, Merchenthaler I (1997b). Regulation of progesterone receptor messenger ribonucleic acid in the rat medial preoptic nucleus by estrogenic and antiestrogenic compounds: an in situ hybridization study. Endocrinology 138: 5476–5484.

Shughrue PJ, Dellovade TL, Merchenthaler I (2002). Estrogen modulates oxytocin gene expression in regions of the rat supraoptic and paraventricular nuclei that contain estrogen receptor-beta. Prog Brain Res 139: 15–29.

Siegel HI, Rosenblatt JS (1975a). Estrogen-induced maternal behavior in hysterectomized-overiectomized virgin rats. Physiol Behav 14: 465–471.

Siegel HI, Rosenblatt JS (1975b). Hormonal basis of hysterectomy-induced maternal behavior during pregnancy in the rat. Horm Behav 6: 211–222.

Silverman ME, Reichenberg A, Savitz DA et al. (2017). The risk factors for postpartum depression: a population-based study. Depress Anxiety 34: 178–187.

Simerly RB, Chang C, Muramatsu M et al. (1990). Distribution of androgen and estrogen receptor mRNA-containing cells in the rat brain: an in situ hybridization study. J Comp Neurol 294: 76–95.

Sirzen-Zelenskaya A, Gonzalez-Iglesias AE, Boutet de Monvel J et al. (2011). Prolactin induces a hyperpolarising current in rat paraventricular oxytocinergic neurones. J Neuroendocrinol 23: 883–893.

Smiley KO, Ladyman SR, Gustafson P et al. (2019). Neuroendocrinology and adaptive physiology of maternal care. Curr Top Behav Neurosci 43: 161–210.

Sommerfield AJ, Deary IJ, McAulay V et al. (2003). Moderate hypoglycemia impairs multiple memory functions in healthy adults. Neuropsychology 17: 125–132.

Son SJ, Filosa JA, Potapenko ES et al. (2013). Dendritic peptide release mediates interpopulation crosstalk between neurosecretory and preautonomic networks. Neuron 78: 1036–1049.

Stocco C, Telleria C, Gibori G (2007). The molecular control of corpus luteum formation, function, and regression. Endocr Rev 28: 117–149.

Stolzenberg DS, Numan M (2011). Hypothalamic interaction with the mesolimbic DA system in the control of the maternal and sexual behaviors in rats. Neurosci Biobehav Rev 35: 826–847.

Strubbe JH, Gorissen J (1980). Meal patterning in the lactating rat. Physiol Behav 25: 775–777.

Strubbe JH, Steffens AB (1975). Rapid insulin release after ingestion of a meal in the unanesthetized rat. Am J Physiol 229: 1019–1022.

Summerlee AJ (1981). Extracellular recordings from oxytocin neurones during the expulsive phase of birth in unanaesthetized rats. J Physiol 321: 1–9.

Swain JE, Kim P, Ho SS (2011). Neuroendocrinology of parental response to baby-cry. J Neuroendocrinol 23: 1036–1041.

Szabo FK, Le WW, Snyder NS et al. (2011). Comparison of the temporal programs regulating tyrosine hydroxylase and enkephalin expressions in TIDA neurons of lactating rats following pup removal and then pup return. J Mol Neurosci 45: 110–118.

Szczesna M, Kirsz K, Misztal T et al. (2019). Downregulation of LRb in VMH/DMH during the second half of gestation and upregulation of SOCS-3 in ARC in late-pregnant ewes—implications for leptin resistance. Gen Comp Endocrinol 274: 73–79.

Tang GB, Cui JG, Wang DH (2008). Hypothalamic suppressor-of-cytokine-signalling 3 mRNA is elevated and pro-opiomelanocortin mRNA is reduced during pregnancy in Brandt's voles (Lasiopodomys brandtii). J Neuroendocrinol 20: 1038–1044.

Teixeira PDS, Couto GC, Furigo IC et al. (2019). Central growth hormone action regulates metabolism during pregnancy. Am J Physiol Endocrinol Metab 317: E925–E940.

Terkel J, Blake CA, Sawyer CH (1972). Serum prolactin levels in lactating rats after suckling or exposure to ether. Endocrinology 91: 49–53.

Terkel J, Bridges RS, Sawyer CH (1979). Effects of transecting lateral neural connections of the medial preoptic area on maternal behavior in the rat: nest building, pup retrieval and prolactin secretion. Brain Res 169: 369–380.

Theodosis DT, Poulain DA, Vincent JD (1981). Possible morphological bases for synchronisation of neuronal firing in the rat supraoptic nucleus during lactation. Neuroscience 6: 919–929.

Theodosis DT, Montagnese C, Rodriguez F et al. (1986). Oxytocin induces morphological plasticity in the adult hypothalamo-neurohypophysial system. Nature 322: 738–740.

Theodosis DT, Poulain DA, Oliet SH (2008). Activity-dependent structural and functional plasticity of astrocyte-neuron interactions. Physiol Rev 88: 983–1008.

Thomson M (2013). The physiological roles of placental corticotropin releasing hormone in pregnancy and childbirth. J Physiol Biochem 69: 559–573.

Thurber C, Dugas LR, Ocobock C et al. (2019). Extreme events reveal an alimentary limit on sustained maximal human energy expenditure. Sci Adv 5: eaaw0341.

Toda T, Parylak SL, Linker SB et al. (2019). The role of adult hippocampal neurogenesis in brain health and disease. Mol Psychiatry 24: 67–87.

Tomizawa K, Iga N, Lu YF et al. (2003). Oxytocin improves long-lasting spatial memory during motherhood through MAP kinase cascade. Nat Neurosci 6: 384–390.

Torner L, Toschi N, Pohlinger A et al. (2001). Anxiolytic and anti-stress effects of brain prolactin: improved efficacy of antisense targeting of the prolactin receptor by molecular modeling. J Neurosci 21: 3207–3214.

Torner L, Toschi N, Nava G et al. (2002). Increased hypothalamic expression of prolactin in lactation: involvement in behavioural and neuroendocrine stress responses. Eur J Neurosci 15: 1381–1389.

Toufexis DJ, Rochford J, Walker CD (1999). Lactation-induced reduction in rats' acoustic startle is associated with changes in noradrenergic neurotransmission. Behav Neurosci 113: 176–184.

Townsend J, Cave BJ, Norman MR et al. (2005). Effects of prolactin on hypothalamic supraoptic neurones: evidence for modulation of STAT5 expression and electrical activity. Neuro Endocrinol Lett 26: 125–130.

Trujillo ML, Spuch C, Carro E et al. (2011). Hyperphagia and central mechanisms for leptin resistance during pregnancy. Endocrinology 152: 1355–1365.

Tsang M, Fewell JE, Moore SL (2006). LPS induced hypothermia in pregnant rats: a regulated thermoregulatory response. Physiol Behav 89: 235–240.

Tups A, Benzler J, Sergi D et al. (2017). Central regulation of glucose homeostasis. Compr Physiol 7: 741–764.

Vannuccini S, Bocchi C, Severi FM et al. (2016). Endocrinology of human parturition. Ann Endocrinol (Paris) 77: 105–113.

Vanston CM, Watson NV (2005). Selective and persistent effect of foetal sex on cognition in pregnant women. Neuroreport 16: 779–782.

Venkatesh KK, Nadel H, Blewett D et al. (2016). Implementation of universal screening for depression during pregnancy: feasibility and impact on obstetric care. Am J Obstet Gynecol 215 (517): e511–e518.

Vorhees CV, Williams MT (2006). Morris water maze: procedures for assessing spatial and related forms of learning and memory. Nat Protoc 1: 848–858.

Wade GN (1975). Some effects of ovarian hormones on food intake and body weight in female rats. J Comp Physiol Psychol 88: 183–193.

Walsh RJ, Posner BI, Kopriwa BM et al. (1978). Prolactin binding sites in the rat brain. Science 201: 1041–1043.

Walsh RJ, Slaby FJ, Posner BI (1987). A receptor-mediated mechanism for the transport of prolactin from blood to cerebrospinal fluid. Endocrinology 120: 1846–1850.

Walsh RJ, Mangurian LP, Posner BI (1990). Prolactin receptors in the primate choroid plexus. J Anat 168: 137–141.

Wan L, Tu T, Zhang QL et al. (2019). Pregnancy promotes maternal hippocampal neurogenesis in Guinea pigs. Neural Plast 2019: 5765284.

Wang HJ, Hoffman GE, Smith MS (1993). Suppressed tyrosine hydroxylase gene expression in the tuberoinfundibular dopaminergic system during lactation. Endocrinology 133: 1657–1663.

Weissgerber TL, Mudd LM (2015). Preeclampsia and diabetes. Curr Diab Rep 15: 9.

Wersinger SR, Caldwell HK, Christiansen M et al. (2007a). Disruption of the vasopressin 1b receptor gene impairs the attack component of aggressive behavior in mice. Genes Brain Behav 6: 653–660.

Wersinger SR, Caldwell HK, Martinez L et al. (2007b). Vasopressin 1a receptor knockout mice have a subtle olfactory deficit but normal aggression. Genes Brain Behav 6: 540–551.

Wharfe MD, Wyrwoll CS, Waddell BJ et al. (2016). Pregnancy suppresses the daily rhythmicity of core body temperature and adipose metabolic gene expression in the mouse. Endocrinology 157: 3320–3331.

Woods SC, Lotter EC, McKay LD et al. (1979). Chronic intracerebroventricular infusion of insulin reduces food intake and body weight of baboons. Nature 282: 503–505.

Woodside B, Amir S (1997). Lactation reduces Fos induction in the paraventricular and supraoptic nuclei of the hypothalamus after urethane administration in rats. Brain Res 752: 319–323.

Woolley CS, McEwen BS (1993). Roles of estradiol and progesterone in regulation of hippocampal dendritic spine density during the estrous cycle in the rat. J Comp Neurol 336: 293–306.

Wu Z, Autry AE, Bergan JF et al. (2014). Galanin neurons in the medial preoptic area govern parental behaviour. Nature 509: 325–330.

Yip SH, Romano N, Gustafson P et al. (2019). Elevated prolactin during pregnancy drives a phenotypic switch in mouse hypothalamic dopaminergic neurons. Cell Rep 26: 1787–1799.e1785.

Yoshihara C, Numan M, Kuroda KO (2017). Oxytocin and parental behaviors. Curr Top Behav Neurosci 35: 119–153.

Yuan L, Liu S, Bai X et al. (2016). Oxytocin inhibits lipopolysaccharide-induced inflammation in microglial cells and attenuates microglial activation in lipopolysaccharide-treated mice. J Neuroinflammation 13: 77.

Zakar T, Hertelendy F (2007). Progesterone withdrawal: key to parturition. Am J Obstet Gynecol 196: 289–296.

Zampieri TT, Ramos-Lobo AM, Furigo IC et al. (2015). SOCS3 deficiency in leptin receptor-expressing cells mitigates the development of pregnancy-induced metabolic changes. Mol Metab 4: 237–245.

Zhang X, van den Pol AN (2016). Hypothalamic arcuate nucleus tyrosine hydroxylase neurons play orexigenic role in energy homeostasis. Nat Neurosci 19: 1341–1347.

Ziomkiewicz A, Wichary S, Jasienska G (2019). Cognitive costs of reproduction: life-history trade-offs explain cognitive decline during pregnancy in women. Biol Rev Camb Philos Soc 94: 1105–1115.

Handbook of Clinical Neurology, Vol. 171 (3rd series)
Neurology and Pregnancy: Pathophysiology and Patient Care
E.A.P. Steegers, M.J. Cipolla and E.C. Miller, Editors
https://doi.org/10.1016/B978-0-444-64239-4.00003-5

Chapter 3

Adaptations in autonomic nervous system regulation in normal and hypertensive pregnancy

VIRGINIA L. BROOKS[1]*, QI FU[2,3], ZHIGANG SHI[1], AND CHERYL M. HEESCH[4]

[1]*Department of Physiology and Pharmacology, Oregon Health & Science University, Portland, OR, United States*

[2]*Institute for Exercise and Environmental Medicine, Texas Health Presbyterian Hospital Dallas, Dallas, TX, United States*

[3]*Department of Internal Medicine, University of Texas Southwestern Medical Center, Dallas, TX, United States*

[4]*Department of Biomedical Sciences and Dalton Cardiovascular Research Center, University of Missouri, Columbia, MO, United States*

Abstract

There is an increase in basal sympathetic nerve activity (SNA) during normal pregnancy; this counteracts profound primary vasodilation. However, pregnancy also impairs baroreflex control of heart rate and SNA, contributing to increased mortality secondary to peripartum hemorrhage. Pregnancy-induced hypertensive disorders evoke even greater elevations in SNA, which likely contribute to the hypertension. Information concerning mechanisms is limited. In normal pregnancy, increased angiotensin II acts centrally to support elevated SNA. Hypothalamic sites, including the subfornical organ, paraventricular nucleus, and arcuate nucleus, are likely (but unproven) targets. Moreover, no definitive mechanisms for exaggerated sympathoexcitation in hypertensive pregnancy have been identified. In addition, normal pregnancy increases gamma aminobutyric acid inhibition of the rostral ventrolateral medulla (RVLM), a key brainstem site that transmits excitatory inputs to spinal sympathetic preganglionic neurons. Accumulated evidence supports a major role for locally increased production and actions of the neurosteroid allopregnanolone as one mechanism. A consequence is suppression of baroreflex function, but increased basal SNA indicates that excitatory influences predominate in the RVLM. However, many questions remain regarding other sites and factors that support increased SNA during normal pregnancy and, more importantly, the mechanisms underlying excessive sympathoexcitation in life-threatening hypertensive pregnancy disorders such as preeclampsia.

INTRODUCTION

Pregnancy is a fascinating state, especially because of the multitude of homeostatic changes that transpire to support the developing fetus. One such change is a surge in sympathetic nerve activity (SNA), which increases even further in pregnancy-induced hypertensive disorders such as preeclampsia. However, unfortunately, only limited information has been obtained about the mechanisms in both normal and hypertensive pregnancy. The purpose of this chapter is to summarize the current knowledge, with an eye toward stimulating further research.

LONGITUDINAL CHANGES IN HEMODYNAMICS AND HORMONES

Pregnancy induces rapidly developing and profound changes in cardiovascular function and hormonal levels to support the developing fetus (Fig. 3.1). The primary

*Correspondence to: Virginia L. Brooks, Ph.D., Department of Physiology and Pharmacology, Oregon Health & Science University, 3181 SW Sam Jackson Park Rd., Portland, OR 97239, United States. Tel: +1-503-494-5843, Fax: +1-503-494-4352, E-mail: brooksv@ohsu.edu

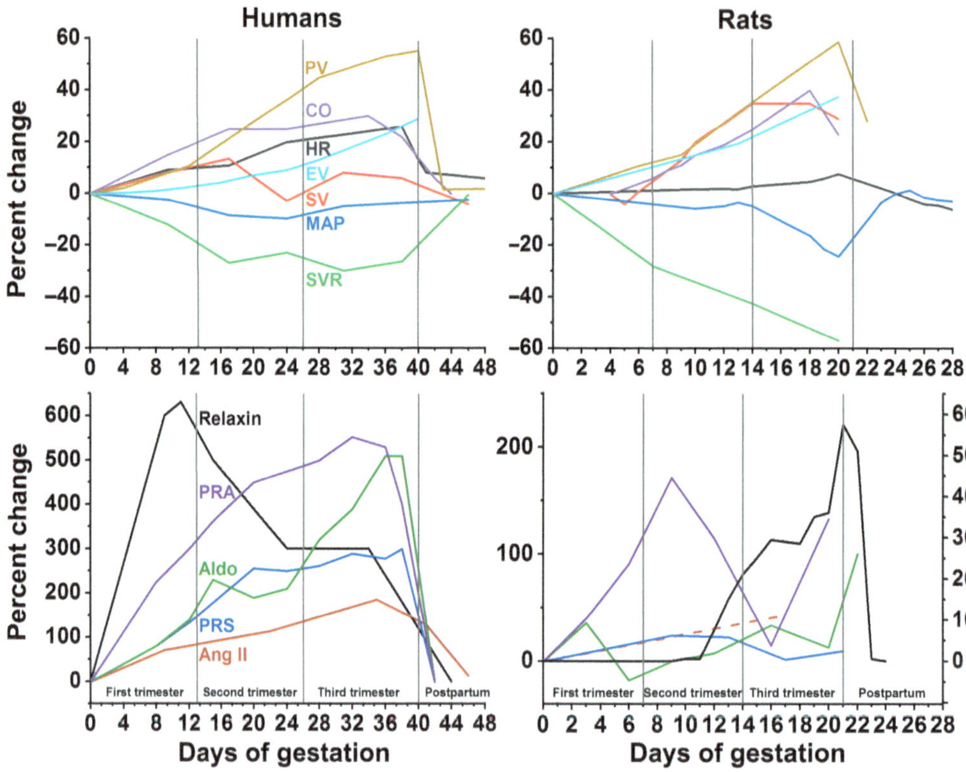

Fig. 3.1. Longitudinal changes in hemodynamic variables and hormones during pregnancy in humans and rats. Hemodynamic variables are shown on the top row, and hormonal values on the bottom row. Human data (*left column*) were adapted from Meah, V.L., Cockcroft, J.R., Backx, K., et al., 2016. Cardiac output and related haemodynamics during pregnancy: a series of meta-analyses. Heart 102, 518–526; de Haas, S., Ghossein-Doha, C., van Kuijk, S.M., et al., 2017. Physiological adaptation of maternal plasma volume during pregnancy: a systematic review and meta-analysis. Ultrasound Obstet Gynecol 49, 177–187; Vricella, L.K., 2017. Emerging understanding and measurement of plasma volume expansion in pregnancy. Am J Clin Nutr 106, 1620s–1625s; Bett, G.C., 2016. Hormones and sex differences: changes in cardiac electrophysiology with pregnancy. Clin Sci (Lond) 130, 747–759; Wilson, M., Morganti, A.A., Zervoudakis, I., et al., 1980. Blood pressure, the renin–aldosterone system and sex steroids throughout normal pregnancy. Am J Med 68, 97–104; Baker, P.N., Broughton Pipkin, F., Symonds, E.M., 1990. Platelet angiotensin II binding and plasma renin concentration, plasma renin substrate and plasma angiotensin II in human pregnancy. Clin Sci (Lond) 79, 403–408; rat data (*right column*) were extracted from Brooks, V.L., Mulvaney, J.M., Azar, A.S., et al., 2010b. Pregnancy impairs baroreflex control of heart rate in rats: role of insulin sensitivity. Am J Physiol Regul Integr Comp Physiol 298, R419–R426; Slangen, B.F., Out, I.C., Verkeste, C.M., et al., 1996. Hemodynamic changes in early pregnancy in chronically instrumented, conscious rats. Am J Physiol Heart Circ Physiol 270, H1779–H1784; Cadnapaphornchai, M.A., Ohara, M., Morris, K.G., Jr., et al., 2001. Chronic NOS inhibition reverses systemic vasodilation and glomerular hyperfiltration in pregnancy. Am J Physiol Renal Physiol 280, F592–598; Nadel, A.S., Ballermann, B.J., Andersen, S., et al., 1988. Interrelationships among arterial peptides, renin, and blood volume in pregnant rats. Am J Physiol Regul Integr Comp Physiol 254, R793–R800; Del Valle, G.O., Mosher, M.D., Conrad, K.P., 1993. Serum immunoreactive erythropoietin and red blood cell mass during pregnancy in conscious rats. Am J Physiol Regul Integr Comp Physiol 265, R399–R403; Sherwood, O.D., 2004. Relaxin's physiological roles and other diverse actions. Endocr Rev 25, 205–234; Fowler, Jr., W.L., Johnson, J.A., Kurz, K.D., et al., 1981. Role of the renin–angiotensin system in maintaining arterial pressure in conscious pregnant rats. Endocrinology 109, 290–295; Garland, H.O., Atherton, J.C., Baylis, C., et al., 1987. Hormone profiles for progesterone, oestradiol, prolactin, plasma renin activity, aldosterone and corticosterone during pregnancy and pseudopregnancy in two strains of the rat: correlation with renal studies. J Endocrinol 113, 435–444; Barron, W.M., 1987. Volume homeostasis during pregnancy in the rat. Am J Kidney Dis 9, 296–302; Gao, H., Tanchico, D.T., Yallampalli, U., et al., 2016. A low-protein diet enhances angiotensin II production in the lung of pregnant rats but not nonpregnant rats. J Pregnancy 2016, 4293431.

initial event is systemic vasodilation, which begins during the luteal phase of the menstrual cycle in women and thus is anticipatory of fetal and maternal needs. Considerable work from the Conrad lab and others indicates that relaxin, produced by the corpus luteum, is a major contributor to the early decrease in systemic vascular resistance, via increases in endothelial-derived nitric oxide and vasodilatory prostaglandins (Conrad, 2011; Leo et al., 2017). Simultaneously, the vasculature resists vasoconstriction induced by angiotensin II (AngII) and sympathetically released norepinephrine (Jarvis et al., 2012; Intapad et al., 2014; Fu, 2018; Reyes et al., 2018a; Spradley, 2019) to aid and abet primary vasodilation.

The dramatic early vasodilation (Chapman et al., 1997) decreases the effective arterial blood volume, which then rapidly stimulates fluid-retaining factors, such as the renin–angiotensin–aldosterone system (RAAS) and the sympathetic nervous system (Jarvis et al., 2012; Tkachenko et al., 2014) (see below). Sodium retention ensues, despite concurrent increases in the glomerular filtration rate and renal blood flow (Tkachenko et al., 2014). This is accompanied by even greater thirst/drinking and vasopressin-mediated water retention, also due in part to arterial underfilling (nonosmotic stimulation of vasopressin release) and relaxin acting centrally, such that plasma osmolality is reset to a lower level (Conrad, 2011; Tkachenko et al., 2014). As a result, plasma volume increases rapidly (Tkachenko et al., 2014; de Haas et al., 2017). Cardiac output rises in parallel due in part to increased stroke volume (increased preload and decreased afterload), increased heart rate (HR), and eccentric cardiac hypertrophy (Meah et al., 2016; Fu, 2018), and the arterial pressure falls because the decrease in systemic vascular resistance exceeds the increase in cardiac output (Tkachenko et al., 2014; Meah et al., 2016; Fu, 2018). Thus, by the end of the first trimester in women, a hyperdynamic circulation, including increases in cardiac output, HR, and plasma volume, is clearly evident and driven largely by systemic vasodilation and an enlarged vascular container that triggers salt and water retention.

Many of these initial changes are maintained or progress throughout human pregnancy (Fig. 3.1). The decrease in systemic vascular resistance, while initially induced largely by corpus luteal products such as relaxin, may be further developed and sustained by additional factors, including placental hormones and the vasodilatory sex steroid estrogen (Conrad and Baker, 2013; Fu, 2018). Plasma volume expansion, driven by additional increments in AngII, aldosterone, and SNA, thus continues concurrently with a delayed but significant increase in red cell volume and therefore blood volume (Tkachenko et al., 2014; Vricella, 2017); cardiac output increases further until just before delivery, when slight decreases have been observed (Meah et al., 2016). As shown in Fig. 3.1, almost all hemodynamic and hormonal factors normalize rapidly after delivery (Hunter and Robson, 1992; Mabie et al., 1994; Gilson et al., 1997; Meah et al., 2016).

Rats have been studied frequently to address mechanistic hypotheses related to maternal adaptations to pregnancy, so we also include an overview of what is known about the longitudinal changes in hemodynamics and hormones in this species (Fig. 3.1). Implantation does not occur until 6 days following fertilization, nearly the end of the first trimester of the 21-day rat gestational period. Soon thereafter (~8 days of gestation), relaxin becomes detectable and, unlike in human pregnancy, rapidly rises to reach a peak just before delivery (Sherwood et al., 1980; Sherwood, 2004; Conrad, 2011). As in humans, this decreases systemic vascular resistance (Gilson et al., 1992; Conrad, 2011), which leads to increases in PRA (Fowler Jr. et al., 1981; Nadel et al., 1988), AngII, sodium and water retention (Atherton et al., 1982), plasma and blood volumes (Atherton et al., 1982; Nadel et al., 1988), cardiac output (Gilson et al., 1992; Slangen et al., 1996; Cadnapaphornchai et al., 2001), and HR (Brooks et al., 2010b). Decreases in MAP are evident midgestation, and MAP falls even more just before delivery (Fowler Jr. et al., 1981; Brooks et al., 2010b).

PREGNANCY INCREASES SNA

Numerous studies of the changes in indirect indices of SNA during human pregnancy have been published. While measurements of plasma norepinephrine levels have been inconsistent, this can be explained by the hyperdynamic circulation that can alter its volume of distribution and clearance (Fu, 2018). However, measurements of HR variability in pregnant women generally suggest a gradual increase in cardiac sympathetic activity, which begins early in gestation, and also a decrease in parasympathetic nerve activity (Lucini et al., 1999; Kuo et al., 2000; Avery et al., 2001; Rang et al., 2002; Kolovetsiou-Kreiner et al., 2018). Similar observations have been made in pregnant rats (Cohen et al., 1988). The greater decrease in HR following pharmacologic blockade of beta adrenergic receptors in pregnant animals further suggests that pregnancy increases cardiac sympathetic tone (Brooks et al., 1997; Lumbers and Yu, 1999).

Direct measurements of SNA have been more consistent. In women, increases in muscle SNA (MSNA) bursts (per minute and per 100 heart beats) have been detected early in pregnancy (5–6 weeks of gestation) (Jarvis et al., 2012; Hissen et al., 2017; Reyes et al., 2018a) (Fig. 3.2). This initial rise develops further during human gestation to reach a peak before delivery (Greenwood et al., 2001; Fischer et al., 2004; Usselman et al., 2015a; Reyes et al., 2018a; Schmidt et al., 2018) (Fig. 3.2). Subtle racial differences have been observed; the MSNA increase in pregnant Asian women is smaller than that seen in pregnant Caucasian women (Okada et al., 2015). Greenwood and colleagues detected an increase in single-unit MSNA per 100 heart beats in late pregnancy and suggested that this reflects enhanced central drive. Using action

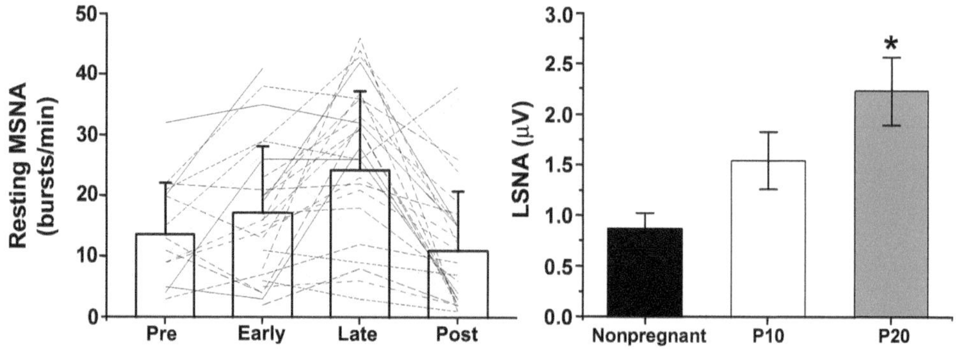

Fig. 3.2. Pregnancy induces a slowly developing sympathoexcitation in both women (left) and rats (right). Left: Data from Okada, Y., Best, S.A., Jarvis, S.S., et al., 2015. Asian women have attenuated sympathetic activation but enhanced renal–adrenal responses during pregnancy compared to Caucasian women. J Physiol 593, 1159–1168; Right: Shi and Brooks, unpublished data. Nonpregnant: n = 20; Gestation day 10 (P10):n = 6; Gestational day 20 (P20): n = 12.

potential waveform detection analysis, Schmidt and colleagues also found that MSNA bursts and integrated activity were increased as was the total number of action potentials per minute or per 100 heart beats, in association with a reduced interspike interval. However, the number of single action potentials per burst was similar in pregnant and nonpregnant women (Schmidt et al., 2018). These authors speculated that the overall enhanced activity was due to decreased central baroreflex gating and decreased baroreflex gain.

In anesthetized late pregnant rats, the activity of sympathetic nerves innervating several organs increases, including splanchnic SNA (SSNA), lumbar SNA (LSNA) (Shi et al., 2015a), and cardiac SNA (Cohen et al., 1988). Elevations in renal SNA (RSNA) have also been reported (Masilamani and Heesch, 1997; Hines et al., 2007), albeit inconsistently (O'Hagan and Casey, 1998; Shi et al., 2015a). Nevertheless, despite insignificant changes in baseline RSNA, blockade of the arcuate nucleus (ArcN) decreased RSNA in late pregnant (but not virgin) rats (Shi et al., 2015a), suggesting that the increment in RSNA is relatively small and not always statistically detectable. Unlike in women, no longitudinal studies have been published, but preliminary observations suggest that pregnancy produces a progressive increase in LSNA in rats, as in women (Fig. 3.2).

Mechanisms of increased basal SNA during pregnancy

Brain sites and neurocircuitry underlying increased basal sympathoexcitation during pregnancy

Only limited information is available concerning the brain nuclei that drive the elevated SNA induced by pregnancy. Nevertheless, the hypothalamic paraventricular nucleus (PVN) is clearly one such site. In nonpregnant female and male rats, the PVN is not significantly involved in supporting ongoing sympathetic activity (Dampney et al., 2018), due largely to the substantial tonic inhibition conferred by multiple inputs, including gamma aminobutyric acid (GABA), nitric oxide, and neuropeptide Y (NPY). In contrast, nonspecific blockade of the PVN (with the $GABA_A$ agonist muscimol) decreases LSNA in anesthetized pregnant (but not nonpregnant) rats, indicating that the PVN supports increased basal sympathetic activity during pregnancy (Shi et al., 2015a). One mechanism that may contribute to the elevated basal activity of PVN presympathetic neurons is a reduction in tonic inhibition. Indeed, while inhibition of PVN ionotropic $GABA_A$ receptors (Dampney et al., 2018) or NPY Y1 and/or Y5 receptors (Cassaglia et al., 2014; Shi et al., 2017) increases SNA in males and nonpregnant females, reflecting this tonic inhibition, these blockers have no effect in pregnant animals; the tonic inhibition by GABA (Kvochina et al., 2009a; Page et al., 2011) and NPY (Shi et al., 2015a) inputs is decreased or lost. Moreover, pregnancy decreases the expression and activity of nitric oxide synthase in the PVN (Heesch et al., 2009). As PVN nitric oxide inhibits SNA by acting presynaptically to inhibit GABA release (Zhang and Patel, 1998), the parallel decreases in GABAergic tone and nitric oxide activity in pregnant rats may be linked.

Moreover, as PVN disinhibition via blockade of PVN $GABA_A$ receptors or inhibitory neuromodulators, such as NPY and nitric oxide, unveils increased SNA drive in nonpregnant animals, PVN presympathetic neurons must also receive tonic excitatory inputs. Thus pregnancy may also elevate basal SNA via increased excitation of the PVN. One potential source of excitatory inputs is the subfornical organ (SFO), which may be activated by circulating hormones, such as AngII or relaxin (see later). Another site is the ArcN, since it projects heavily to the PVN, and ArcN blockade lowers LSNA in pregnant rats (Shi et al., 2015a). ArcN inhibitory NPY neurons and excitatory pro-opiomelanocortin (POMC) neurons, which release α-melanocyte stimulating hormone (α-MSH), converge onto PVN presympathetic neurons (Cassaglia et al., 2014). As suppression of tonic NPY inhibition is required before α-MSH can excite PVN presympathetic neurons (Shi et al., 2015b), and pregnancy decreases PVN NPY inhibition, we tested if α-MSH contributes to basal sympathetic tone. Indeed, blockade of PVN α-MSH MC3/4 receptors lowers SNA in pregnant but not nonpregnant rats (Shi et al., 2015a), indicating that ArcN POMC neurons are one source of excitation of PVN presympathetic neurons during pregnancy.

Thus current evidence supports a role for the PVN, SFO, and ArcN in the pregnancy-induced increases in basal SNA. We will now address possible factors that activate these sites.

Potential factors that activate central sympathetic regulatory regions during pregnancy

Gonadal steroids

Multiple studies of the variations in MSNA during the menstrual cycle have been conducted; however, the results are conflicting. To address this, four labs coalesced their data (Carter et al., 2013) to test the hypothesis that the inconsistent results can be explained by variable and nonparallel surges in estrogen (E2) and progesterone (P), the opposing effects of E2 (sympathoinhibitory) and P (sympathoexcitatory) on SNA (Charkoudian, 2001), and the different phases of the cycle that had been probed. Indeed, the changes in MSNA were best (negatively) correlated with the changes in E2 and the changes in the E2-to-P ratio. Thus it appears that the anticipatory increase in MSNA that has been detected during the mid-luteal phase is due to an increase in progesterone relative to estrogen.

Given that pregnancy elicits rather sizable increases in both E2 and P, the potential contribution of these hormones to the increases in MSNA warrants consideration. Reyes et al. reported a significant direct correlation between E2 and MSNA (Reyes et al., 2018b); however, since E2 is sympathoinhibitory, this may reflect estrogen-induced vasodilation, which indirectly evokes increases in MSNA. Others report no relationship between E2 and MSNA or a weak relationship between P and MSNA ($P=0.08$) (Jarvis et al., 2012). However, no published studies have examined if the increment in MSNA is directly related to a change in the P-to-E2 ratio during pregnancy. In contrast to this hypothesis, the P-to-E2 ratio decreases during pregnancy (Chapman et al., 1998; Okada et al., 2015). Moreover, a reanalysis of our data (Okada et al., 2015) revealed that the increment in MSNA is not related to the changes in E2 or the E2-to-P ratio. Nevertheless, again, a weak relationship between △MSNA and △P was observed ($P=0.08$). This relationship may suggest a contribution of P; however, the counteracting effects of high E2 levels and the parallel increases in the progesterone metabolite and neurosteroid allopregnanolone, which also inhibits SNA (see later), counters this suggestion. Therefore other factors must be involved.

Metabolic hormones: Insulin and leptin

Pregnancy increases the levels of a myriad other hormones, including those that contribute to metabolism and the production of nutrients necessary for fetal development and growth. Two such hormones are insulin and leptin, which also act centrally to increase SNA. Plasma leptin levels increase in pregnant women (Hardie et al., 1997; Sattar et al., 1998; Franco-Sena et al., 2015) and rats (Herrera et al., 2000; Seeber et al., 2002), secondary to both excess adiposity and uterine production and release of leptin. Pregnancy induces insulin resistance in women (Stanley et al., 1998; Lain and Catalano, 2007), rats (Leturque et al., 1984; Rossi et al., 1993; Munoz et al., 1995), and rabbits (Daubert et al., 2007). In women, insulin resistance eventually leads to increments in fasting plasma insulin levels by the end of gestation (Lain and Catalano, 2007; Perichart-Perera et al., 2017; Benaim et al., 2019). In experimental animals, insulin levels are elevated in nonfasted and glucose-challenged experimental animals (Munoz et al., 1995). However, pentobarbital anesthetized or fasted, conscious pregnant animals generally do not exhibit significant increases in plasma or brain insulin (Leturque et al., 1984; Daubert et al., 2007; Shi et al., 2019). Insulin and leptin act in the ArcN to increase SNA by suppressing tonic NPY inhibition and increasing α-MSH excitation in the PVN (Cassaglia et al., 2011, 2016; Ward et al., 2011; Shi and Brooks, 2015; Shi et al., 2015b), similarly to pregnancy. However, the sympathoexcitatory responses to intracerebroventricular (i.c.v.) insulin or leptin are completely eliminated in late pregnant rats, and blockade of ArcN insulin receptors (the site of action of insulin to increase SNA) fails to decrease LSNA (Shi et al., 2019). Collectively, these results indicate that neither hormone likely contributes to pregnancy-induced sympathoexcitation. Because sympathetic activation stimulates glucose uptake by muscle (Nonogaki, 2000), this central resistance to insulin and leptin, in parallel with systemic resistance, aids in the maintenance of sufficiently high glucose concentrations in plasma to ensure adequate delivery to the fetus.

Volume-retaining hormones (RAAS and vasopressin)

As shown in Fig. 3.1, the longitudinal profile of the increase in plasma volume and associated fluid-retaining hormones, such as the RAAS, are closely matched with the increases in SNA in both pregnant women and rats. Therefore a feasible hypothesis is that sympathoexcitatory hormones, such as AngII or aldosterone, contribute to pregnancy-induced sympathoexcitation. In support, the increase in MSNA in pregnant women has been found to be well correlated to the increases in direct renin and aldosterone (Jarvis et al., 2012). In women studied in the third trimester, MSNA was directly related to the increases in vasopressin (Charkoudian et al., 2017; Reyes et al., 2018b). As vasopressin is sympathoinhibitory, this relationship may indirectly reflect the actions of other volume-retaining hormones such as AngII or relaxin, both of which stimulate vasopressin secretion.

More direct evidence has been obtained in animal studies. Compared with nonpregnant females, pregnant animals are more dependent on AngII for maintenance of arterial pressure. Since the vasculature becomes resistant to AngII-induced vasoconstriction, this dependency likely includes a central mechanism. Indeed, Hines and Porter reported that while intravenous (i.v.) AngII increases blood pressure less in pregnant rats, the pressor response to i.c.v. AngII is greater, due in part to greater activation of the sympathetic nervous system (Hines and Porter, 1989). More importantly, O'Hagan et al. showed that either i.c.v. or combined i.c.v. + i.v. administration of losartan, an AT_1 receptor (AT1R) antagonist, shifted baroreflex relationships between arterial pressure and RSNA to the left more in pregnant compared with nonpregnant conscious rabbits (O'Hagan et al., 2001). As a result, RSNA was reduced at any given pressure after AT1R blockade in pregnant animals. However, the central site(s) of action of AngII to increase SNA are unclear. The SFO is a circumventricular organ devoid of a blood–brain barrier and a major site of action of circulating AngII. In pregnant rats, the pressor and sympathoexcitatory responses to nanoinjections of AngII into the SFO are markedly enhanced, and AT1R in the PVN are obligatory in this sympathoexcitatory pathway (Kvochina et al., 2009b). Pregnant rats also exhibit potentiated responses to nanoinjections of AngII into the PVN (Heesch et al., 2010) and ArcN (Shi and Brooks, unpublished observations). In parallel, AT1aR expression is increased in the SFO (Coldren et al., 2015), PVN (Heesch et al., 2010), and ArcN (Shi and Brooks, unpublished observation). Thus circulating AngII, through actions at the SFO, and also the brain RAS acting in the PVN and ArcN, may support elevated basal SNA during pregnancy, at least in experimental animals.

The sympathoexcitatory actions of AngII during pregnancy may be amplified by relaxin. Studies in both male (McKinley et al., 2004) and female rats (Coldren et al., 2015) have revealed that circulating relaxin increases SNA, through an AngII-dependent action in the SFO and subsequent activation of spinally projecting PVN-projecting neurons. Immunohistochemistry experiments revealed that relaxin receptor protein is present in the SFO of both nonpregnant and pregnant rats, and mRNA for relaxin in the SFO is not altered in pregnancy, raising the possibility that high circulating relaxin levels may contribute to pregnancy-induced sympathoexcitation. However, intracarotid artery relaxin infusion did not acutely activate the SFO or spinally projecting neurons in the PVN of term pregnant rats (Coldren et al., 2015). One possible explanation for lack of effect in pregnant rats could be downregulation of relaxin receptors due to chronic elevation of endogenous levels. However, different from most G-protein-coupled receptors, relaxin receptors are resistant to desensitization and responsiveness is maintained during long-term exposure to the agonist (Callander et al., 2009). For example, consistent with long-term actions of relaxin, overexpression of preprorelaxin in the region of the SFO in nonpregnant rats is associated with increased plasma vasopressin for at least 21 days (Silvertown et al., 2005), and endogenous relaxin contributes to central resetting of osmolality regulation throughout pregnancy (McKinley et al., 2004). Thus it is unlikely that CNS relaxin receptors become desensitized in pregnant rats. Alternatively, relaxin receptors in the SFO may be maximally occupied by high levels of endogenous relaxin, or downstream mechanisms, for example, upregulation of brain AT1R, may be already optimally activated by endogenous relaxin in term pregnant rats. Thus relaxin may be another hormone that acts centrally (in the SFO) to support increased SNA during pregnancy; however, additional experiments are required to substantiate this hypothesis.

PREGNANCY IMPAIRS BAROREFLEX CONTROL OF HR AND SNA

For the most part, hemodynamic and blood volume changes that occur during uncomplicated pregnancy (Fig. 3.1) are adaptive, contribute to optimal fetal development, and are well tolerated by pregnant women. However, one potentially deleterious occurrence is a profound decrease in the ability of the arterial baroreflex to compensate for hypotensive challenges. Compared with the nonpregnant state, pregnant women are more prone to orthostatic hypotension. Of greater concern is the fact that pregnancy is associated with a decreased ability to maintain cardiac output and peripheral resistance during blood loss (Clow et al., 2003). Substantial blood loss during delivery and the peripartum period, accompanied by severe hypotension, is a major contributor to maternal morbidity and mortality (Brooks et al., 2010a; Reyes et al., 2018a). Since the arterial baroreflex is an important feedback system for buffering changes in arterial pressure under a variety of physiologic challenges, numerous studies have focused on evaluating arterial baroreflex function in both normotensive and hypertensive pregnancy.

Early studies in anesthetized rabbits first noted that baroreflex-mediated peripheral vasoconstriction in response to arterial baroreceptor unloading (low carotid sinus pressure) was attenuated in pregnant compared with nonpregnant animals (Humphreys and Joels, 1974; Humphreys and Joels, 1982). Both animal and human studies have reported suppressed vascular responsiveness to alpha adrenergic agonists and decreased neurovascular transduction in pregnancy (Conrad and Russ, 1992; Jarvis et al., 2012; Fu, 2018; Reyes et al., 2018a; Spradley, 2019), which could contribute to attenuated baroreflex-mediated vasoconstriction. However, subsequent to the initial observations by Humphreys and Joels (1974, 1982) many studies have provided convincing evidence for an important neural component to pregnancy-associated suppression of baroreflex function.

The well-described brainstem pathway subserving arterial baroreflex function is shown in Fig. 3.3. In response to an increase in arterial pressure, the discharge of afferent fibers from peripheral baroreceptors is increased and second-order neurons in the nucleus tractus solitarii (NTS) are activated. Baroreceptor-sensitive NTS neurons project to and excite GABAergic interneurons in the caudal ventral lateral medulla (CVLM), which in turn project to and inhibit sympathetic premotor neurons in the rostral ventrolateral medulla (RVLM) (Dampney, 1994; Guyenet, 2006). Both presympathetic neurons in the RVLM and the arterial baroreceptors are active at baseline arterial blood pressure, which allows for reciprocal compensatory changes in efferent SNA to the heart and blood vessels in response to both increases (described earlier) and decreases in arterial blood pressure (baroreceptor unloading). Second-order neurons within the NTS also project to and excite cardiac vagal preganglionic neurons in the nucleus ambiguous,

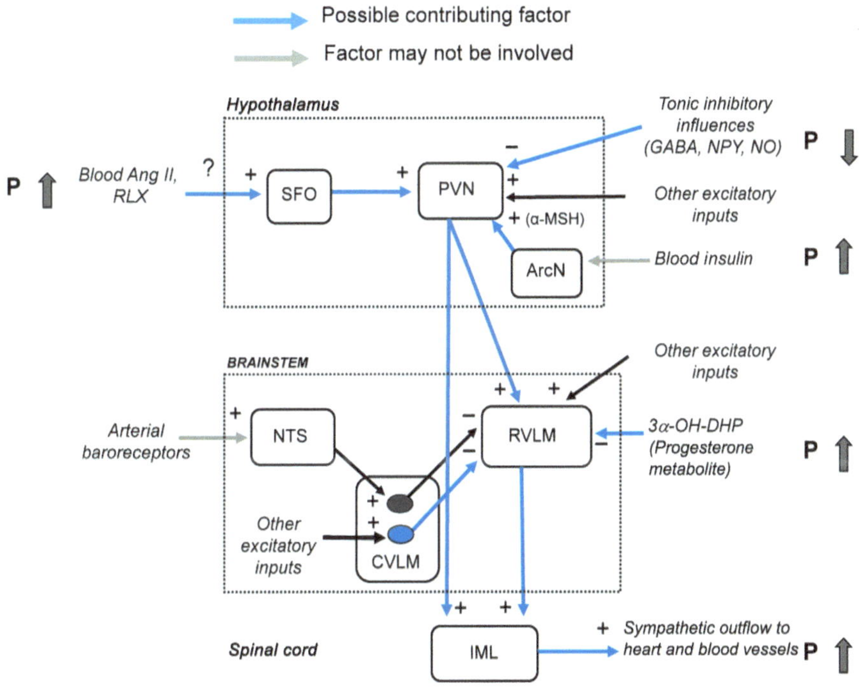

Fig. 3.3. Pathways for control of efferent sympathetic outflow. Major hypothalamic and brainstem sites involved in regulation of preganglionic sympathetic nerve activity in the spinal cord are shown. The direction of changes in hormones or inputs in pregnancy (P) are indicated by vertical *gray up and down arrows*. Circulating angiotensin II (AngII) and relaxin are known to activate paraventricular nucleus (PVN) projecting neurons in the subfornical organ (SFO), and insulin activates α-MSH neurons in the arcuate nucleus, which then project to and excite neurons in the PVN. PVN neurons excite preganglionic sympathetic neurons in the intermediolateral cell column of the spinal cord (IML) through either direct or indirect (synapse in rostral ventrolateral medulla, RVLM) projections. The arterial baroreflex pathway is shown in the bottom half of the diagram. Afferent baroreceptor input activates nucleus tractus solitarius (NTS) neurons, which in turn activate barosensitive GABAergic neurons in the caudal ventrolateral medulla (CVLM) projecting to and inhibiting ongoing activity in presympathetic RVLM neurons. In addition, there is a baroreceptor-independent inhibitory input to RVLM neurons arising from the CVLM that receives excitatory inputs from unknown brain regions. 3α-Hydroxy-dihydroprogesterone (3α-OH-DHP, a metabolite of progesterone) potentiates GABAergic inhibition of the RVLM. "+" indicates an excitatory effect and "−" indicates an inhibitory effect. *Horizontal arrows* indicate pathways where experimental results support (*blue*) or refute (*gray*) a possible role in adaptations during pregnancy.

comprising the parasympathetic nervous system component for baroreflex control of HR (not shown).

Glutamate is the primary excitatory transmitter, and GABA is the primary inhibitory transmitter within the brain regions shown in Fig. 3.3. In addition to the baroreflex pathway through the CVLM, there are baroreflex-independent GABAergic projections primarily from the CVLM to the RVLM that inhibit sympathetic outflow and can influence overall reflex responses (Schreihofer and Guyenet, 2002; Guyenet, 2006).

In general, arterial baroreflex function in animals and humans adapts to changes in behavioral, psychologic, or physiologic conditions, such that the reflex continuously adjusts (resets) to accommodate the particular state and maintain appropriate buffering capacity of the reflex. For example, when baseline arterial pressure increases or decreases for periods as short as 5 min, both the baroreceptor and baroreflex function curves for control of SNA and HR shift toward the prevailing pressure, with no change in sensitivity to increments in arterial pressure. This parallel shift of arterial baroreflex function curves is termed peripheral pressure-dependent resetting and maintains buffering capacity over a broader range of baseline arterial pressures (Chapleau et al., 1989). Other examples of baroreflex resetting clearly involve a central nervous system (CNS) component. For example, during psychologic stress and exercise, "central command" from the cortex and forebrain contributes to resetting of the baroreflex (Dampney, 2016). In the case of exercise, afferent input from sensory receptors in the exercising muscle also plays a role. Arterial pressure, HR, and sympathetic outflow to nonexercising muscles increase progressively with increased exercise intensity, yet baroreflex sensitivity to increments in pressure is maintained due to a rightward shift of the curve toward elevated arterial pressure and an upward shift toward higher SNA (Ogoh et al., 2007; Mueller et al., 2017). Thus under a variety of states, the arterial baroreflex continues to regulate SNA but resets to accommodate the particular behavioral condition.

Studies in pregnant humans and experimental animals, using a variety of approaches, have documented depressed baroreflex control of multiple efferents, including HR, renal, splanchnic, and muscle SNA, and hormones such as vasopressin and ACTH/glucocorticoids (Fu and Levine, 2009; Brooks et al., 2010a; Reyes et al., 2018a; Shi et al., 2019). To evaluate the full range of arterial baroreflex responses, experiments in animal models infused vasoactive drugs to generate complete sigmoidal baroreflex relationships. Fig. 3.4 shows data from groups of conscious nonpregnant and pregnant rats demonstrating baroreflex control of HR (Brooks et al., 2010b) and RSNA (Masilamani and Heesch, 1997). To reduce variability in absolute nerve activity among animals, nerve activity for each animal is most often expressed as a percentage of the resting level (defined as 100%), and data is interpreted

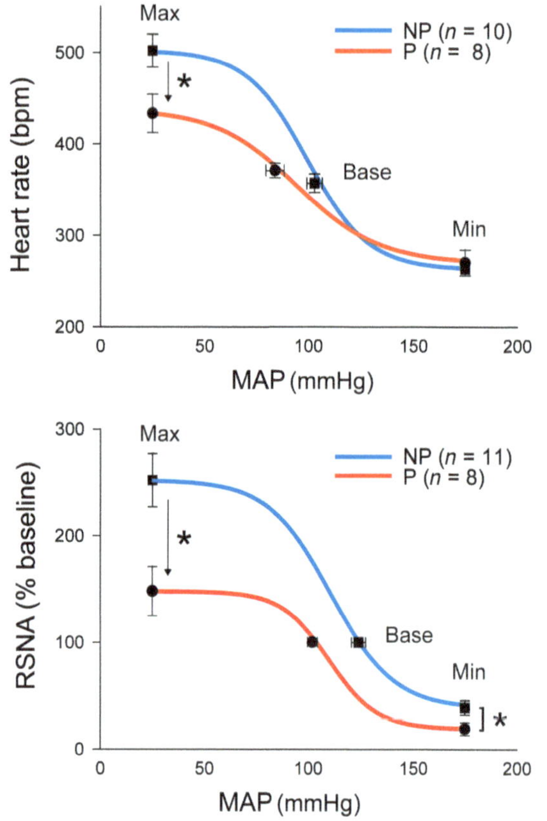

Fig. 3.4. Pregnancy impairs baroreflex control of heart rate (bpm = beats per min) and renal sympathetic nerve activity (RSNA) in conscious rats. Mean arterial pressure (MAP), heart rate, and RSNA were recorded over a range of arterial pressures obtained by infusions of phenylephrine or nitroprusside. Sigmoidal relationships between MAP and heart rate (top) or RSNA (bottom) were determined by fitting a four-parameter equation to data points, and curve parameters were statistically compared between late pregnant (P) and nonpregnant (NP) rats. The uppermost curve in each graph above represents data from NP rats. *Symbols on curves* labeled Base indicate baseline values of MAP, HR, and RSNA (100%). Major effects of pregnancy included increased baseline heart rate, decreased baseline arterial pressure, and decreased maximum heart rate and RSNA attained during hypotension (Max). In addition, pregnancy decreased the minimum RSNA achieved during hypertension (Min). Cardiac baroreflex gain was decreased and the slope (gain) of RSNA responses to decreases in arterial pressure was also decreased. * $P < 0.05$. Redrawn with permission from Brooks, V.L., Mulvaney, J.M., Azar, A.S., et al., 2010b. Pregnancy impairs baroreflex control of heart rate in rats: role of insulin sensitivity. Am J Physiol Regul Integr Comp Physiol 298, R419–R426; Masilamani, S., Heesch, C.M., 1997. Effects of pregnancy and progesterone metabolites on arterial baroreflex in conscious rats. Am J Physiol Regul Integr Comp Physiol 272, R924–R934.

regarding the functional capacity of the baroreflex to compensate for perturbations around resting nerve activity. Three aspects of these curves are particularly affected in pregnancy: the baroreflex function curve is shifted to

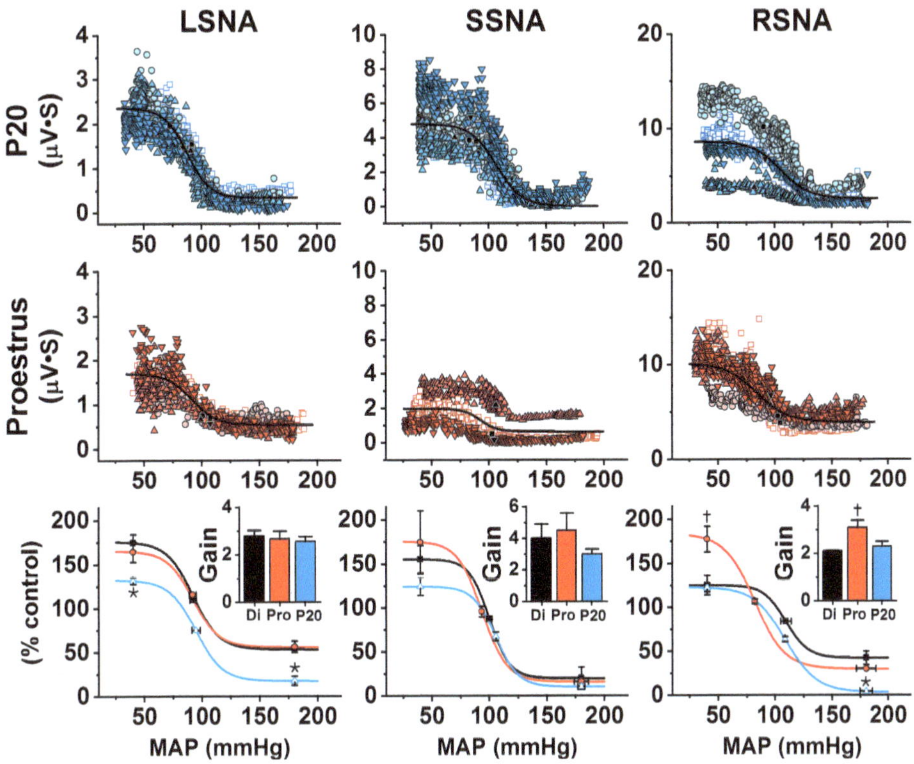

Fig. 3.5. Pregnancy increases basal SNA (in rats) closer to the baroreflex maximum. Comparison of baroreflex control of LSNA, SSNA, and RSNA expressed as raw SNA (results from individual experiments, each shown as different symbols; $n=3$–4 per group) and as % control [grouped data; $n=8$–10 (LSNA), $n=4$–5 (SSNA), $n=5$ (RSNA) in each group]. In the top two rows, baseline MAP and SNA are illustrated as *closed black symbols*. In the bottom row, points ± SEM are the mean sigmoidal baroreflex parameters calculated as previously described (Shi and Brooks, 2015; Shi et al., 2019). Statistical comparisons (*t*-test) of the individual sigmoidal fit parameters between late pregnant (*top row*) and proestrus (*middle row*) revealed that in pregnant rats the baroreflex maximum for LSNA and SSNA was elevated ($P<0.05$), the baroreflex minimum for LSNA and RSNA was suppressed ($P<0.05$), and the baroreflex gain for LSNA and SSNA was increased ($P<0.05$). In the bottom row, $^{*}P<0.05$, late pregnant (P20) lower than diestrus (Di) and proestrus (Pro); $^{\dagger}P<0.05$, proestrus (Pro) is higher than late pregnant (P20) and diestrus (Di). Data adapted from Shi, Z., Brooks, V.L., 2015. Leptin differentially increases sympathetic nerve activity and its baroreflex regulation in female rats: role of oestrogen. J Physiol 593, 1633–1647; Shi, Z., Hansen, K.M., Bullock, K.M., et al., 2019. Resistance to the sympathoexcitatory effects of insulin and leptin in late pregnant rats. J Physiol 597, 4087–4100.

operate around the lower arterial pressure of pregnant rats; during hypotension the maximal increase above baseline for HR or SNA is blunted; and the sensitivity or gain (slope) of the curve around baseline arterial pressure is decreased. In general, responses to a hypertensive challenge appear to be well maintained or even augmented in pregnant animals (Fig. 3.4). Similar observations have been made for effects of pregnancy on baroreflex function curves in rabbits (O'Hagan et al., 2001; Daubert et al., 2007) and sheep (Lumbers and Yu, 1999). These data are consistent with previous reports in conscious rats (Conrad and Russ, 1992) and women (Leduc et al., 1991), indicating that bradycardia in response to i.v. vasoconstrictors is well maintained in term pregnancy. Pharmacologic blockade of the sympathetic or parasympathetic nervous systems has revealed that the sympathetic contribution to baroreflex-mediated tachycardia is particularly blunted in pregnant animals (Brooks et al., 1997; Lumbers and Yu, 1999).

As mentioned earlier, in reports from animal studies, SNA data is standardized (as a percentage of baseline or maximum nerve activity), and it is evident that pregnant animals exhibit less baroreflex excitatory reserve in response to hypotensive challenges (Crandall and Heesch, 1990; Brooks et al., 1995; Masilamani and Heesch, 1997; O'Hagan et al., 2001; Shi et al., 2019). However, standardization before comparisons does not take into account differences in baseline SNA between nonpregnant and pregnant animals. Therefore we replotted normalized data, from publications (Shi and Brooks, 2015; Shi et al., 2019) assessing baroreflex control of LSNA, SSNA, and RSNA in anesthetized pregnant and nonpregnant rats, as absolute SNA. As shown in Fig. 3.5, the basal levels of LSNA, RSNA, and SSNA are much closer to the maximal level elicited by baroreceptor unloading in pregnant rats. Moreover, the reduced baroreflex maximum seen with normalized data (Fig. 3.5) is likely due to the higher position of the baseline SNA on the curve since, if anything, the absolute level during hypotension is higher during pregnancy, at least for LSNA and SSNA. A study in women provides a similar interpretation. Passive postural changes from supine to head-up tilt were used to unload both low- (cardiopulmonary) and high-pressure (arterial) baroreceptors (Okada et al., 2015). Head-up tilt progressively increased absolute MSNA (bursts per minute), which was elevated at all postures in both early and late pregnant women, compared with nonpregnant controls (Fig. 3.6A). Expressed as change from supine values, the increase in MSNA due to tilt was attenuated in late pregnancy, likely due to high baseline (supine) values, closer to the baroreflex maximum (Fig. 3.6B).

Most studies evaluating pregnancy-induced changes in baroreflex function in women have used noninvasive methods, including assessment of HR and blood pressure variability (spectral analysis) or spontaneous baroreflex sensitivity, and are restricted to the segment of baroreflex relationships close to basal values

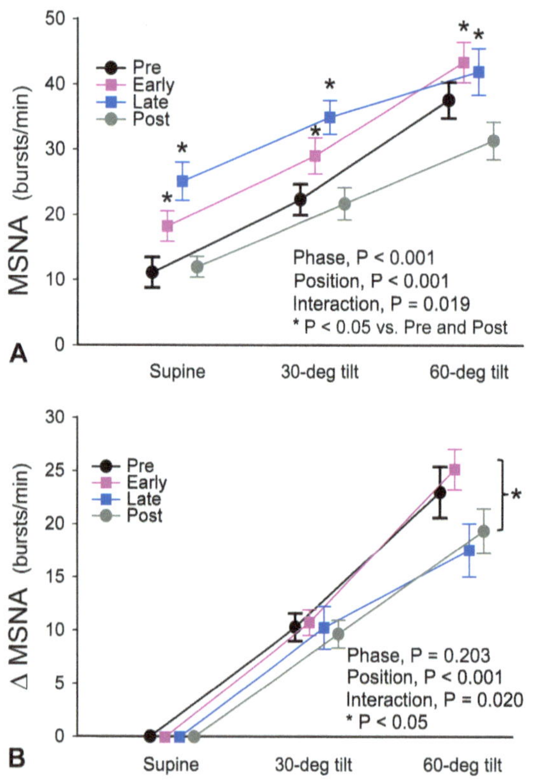

Fig. 3.6. Pregnancy increases basal MSNA (in women) closer to the maximum elicited by head-up tilt. Data adapted from Okada, Y., Best, S.A., Jarvis, S.S., et al., 2015. Asian women have attenuated sympathetic activation but enhanced renal–adrenal responses during pregnancy compared to Caucasian women. J Physiol 593, 1159–1168.

(Rang et al., 2002; Reyes et al., 2018a). Spectral analyses in both the time and frequency domains provide indirect indices related to the relative contributions of the sympathetic and parasympathetic nervous system, and most studies suggest that pregnancy is associated with a relative increase in sympathetic influences (Rang et al., 2002). Spontaneous cardiac baroreflex sensitivity is determined by identifying brief baroreflex-mediated ramps in which spontaneous fluctuations in systolic pressure are associated with simultaneous increases or decreases in pulse interval, reflecting primarily rapid reflex changes in parasympathetic nerve activity to the heart. Using these indirect methods, most (Voss et al., 2000; Silver et al., 2001; Kolovetsiou-Kreiner et al., 2018), but not all (Usselman et al., 2015a), studies in women, and also in experimental animals (Avery et al., 2001; Brooks et al., 2010b), suggest that pregnancy also depresses the parasympathetic component of cardiac baroreflex function. In addition, Usselman et al. (2015a) reported that late pregnancy was associated with decreased spontaneous baroreflex sensitivity for control of efferent MSNA in women. However, a limitation of evaluating spontaneous baroreflex gain based on natural fluctuations in arterial pressure and SNA around baseline is that the elevated baseline MSNA observed in pregnant women is also closer to the baroreflex maximum (Fig. 3.6). Thus, again, the reduced gain may reflect the falling slope of the relationship as it approaches the baroreflex maximum.

As discussed earlier, most experimental evidence suggests that baroreflex function is altered in late pregnancy and attenuated baroreflex compensation is especially evident during baroreceptor unloading (Figs. 3.4 and 3.6), where responses are dependent on removal of barosensitive GABAergic input to the RVLM (Fig. 3.3). Alterations of baroreflex function in pregnancy could be due to changes in afferent baroreceptor input or any number of CNS adaptions that might affect final sympathetic outflow. To test the integrity of efferent sympathetic pathways, responses to excitatory stimuli that do not rely on disinhibition of the RVLM to increase SNA, such as the exercise pressor reflex (isometric hand grip test) or the cold pressor test, have been evaluated in women (Rang et al., 2002). Responses to these tests appear to be preserved in pregnancy. The exercise pressor reflex is initiated by increased discharge in afferent fibers from exercising muscle that project to and activate presympathetic pathways in the brain, ultimately leading to increased glutamatergic excitation of the RVLM (Mueller et al., 2017). Increases in HR and BP following hand grip exercise are not different between pregnant and nonpregnant women (Rang et al., 2002). Extreme cold applied to a limb (cold pressor test) activates excitatory pain pathways, including synapses in the spinal cord and hypothalamic sites, and glutamatergic excitation of the RVLM contributes to the resultant sympathoexcitation (Nakamura et al., 2008). Elevations in MSNA during a cold pressor test are either unchanged (Schobel et al., 1996) or augmented (Usselman et al., 2015b) in term pregnant compared with nonpregnant women. Thus it appears that pregnancy differentially suppresses responses to baroreflex-mediated disinhibition while direct excitation of the RVLM is preserved.

Potential mechanisms for attenuated baroreflex function in pregnancy

Afferent limb

Studies in anesthetized rats in which discharge of the aortic depressor nerve was evaluated in response to brief incremental increases and decreases in arterial pressure revealed that the baroreceptor discharge curve was reset to operate around the lower baseline arterial pressure of pregnancy (peripheral pressure-dependent resetting). However, the change in afferent baroreceptor discharge in response to increments in pressure was preserved, so that baroreceptor sensitivity or gain was not different between nonpregnant and near term pregnant rats (Hines, 2000; Laiprasert et al., 2001). Thus preservation of arterial baroreceptor afferent gain around the prevailing arterial pressure suggests that the decrease in baroreflex sensitivity, and specifically attenuated increases in SNA during hypotension, that occurs during pregnancy must be due, to a large extent, to changes within the brain.

Descending pathways

Descending projections from higher brain centers have the potential to impact arterial baroreflex function by modulating neuronal activity at the brainstem sites. In this regard, hypothalamic nuclei, such as the PVN, have reciprocal connections with the major brainstem nuclei that comprise the medullary baroreflex pathway (Guyenet, 2006; Card and Sved, 2011; Dampney et al., 2018). As discussed earlier, adaptations at the level of the SFO, PVN, and ArcN in the hypothalamus may contribute to increased baseline SNA in pregnant rats in part via suppression of tonic GABA and NPY inhibition. However, while in nonpregnant female rats blockade of $GABA_A$ receptors in the PVN augments arterial baroreflex-mediated increases in HR in response to hypotension and maximum cardiac baroreflex gain is increased, in late pregnant rats disinhibition of the PVN only slightly improved cardiac baroreflex function (Page et al., 2011). These data suggest that there is less (not more) GABAergic restraint of the baroreflex in pregnancy and that GABAergic mechanisms in the PVN do not play a major role in the attenuation of baroreflex-mediated sympathoexcitation in pregnancy (Page et al., 2011). The circulating hormones insulin and leptin have CNS sympathoexcitatory effects (Shi et al., 2019), and, as discussed earlier, both are elevated in pregnancy (Kirwan et al., 2002; Trujillo et al., 2011). In nonpregnant rats (Pricher et al., 2008; Cassaglia et al., 2011; Li et al., 2013) and humans (Young et al., 2010), i.c.v. or i.v. insulin or leptin augment baroreflex-mediated tachycardia and sympathoexcitation during hypotension. However, late pregnant rats become resistant to the effects of i.c.v. insulin and leptin, such that baseline SNA is unchanged and baroreflex control of LSNA remains suppressed compared with nonpregnant controls (Shi et al., 2019). Whether a decrease in brain insulin or leptin concentrations or actions elicits parallel decreases in SNA baroreflex function is unknown. Nevertheless, i.c.v. infusion of insulin improved HR baroreflex gain (Azar and Brooks, 2011; Shi et al., 2019), although other baroreflex parameters were not altered, suggesting only a small contribution of reduced insulin action in the brain. Thus although hypothalamic mechanisms clearly alter control of SNA in nonpregnant rats, studies to date suggest that descending inputs from the hypothalamus are not major contributors to attenuated arterial baroreflex-mediated sympathoexcitation in pregnancy. Studies evaluating potential mechanisms related to the medullary baroreflex pathway are described next.

Medullary baroreflex pathway

To determine potential sites within the medullary baroreflex pathway that may be altered by pregnancy, Curtis et al. (1999) evaluated expression of Fos protein as a marker for neuronal activation in the brainstem of term pregnant compared with nonpregnant rats. In conscious rats, a sustained decrease in arterial blood pressure, to a level that would unload arterial baroreceptors and result in maximum baroreflex-mediated sympathoexcitation, resulted in increased Fos expression in the NTS, CVLM, and RVLM in both groups. However, only in the RVLM was the increase in Fos expression less in pregnant compared with nonpregnant rats. Interestingly, while both noncatecholaminergic and catecholaminergic (C1) cells in the RVLM contribute to sympathetic outflow (Schreihofer and Guyenet, 2000; Guyenet, 2006), the effects of pregnancy to suppress activation of RVLM neurons during hypotension was specific to the noncatecholaminergic neuronal phenotype in this region (Curtis et al., 1999). Since the RVLM provides excitatory input to sympathetic preganglionic neurons in the spinal cord, these results suggest that alterations at the level of the RVLM likely contribute to the attenuated baroreflex-mediated increases in SNA in pregnant rats (Fig. 3.3).

Increased inhibition in the RVLM

Determining the endogenous factors that result in decreased activation of presympathetic neurons in the RVLM of pregnant animals during hypotension presents a challenge, as pregnancy is associated with changes in a multitude of humoral and inflammatory factors, which could potentially influence excitability of presympathetic neurons in this region (Brooks et al., 2010a). As discussed earlier, one explanation for impaired responses in the RVLM is that the ongoing level of activity is closer to the maximum baroreflex-induced level. In addition, inhibitory influences may be increased. Before discussing potential factors, it is important to consider that sympathoexcitation due to baroreceptor unloading (decreased arterial pressure) is mediated by withdrawal of baseline GABAergic tone within the medullary baroreflex pathway (disinhibition) (Guyenet, 2006). The RVLM receives a substantial baroreflex-independent GABAergic input, a large portion of which originates in the CVLM (Schreihofer and Guyenet, 2002) (Fig. 3.3), although other GABAergic influences, including GABA interneurons within the RVLM, may contribute (Schreihofer and Guyenet, 2002; Guyenet, 2006). Thus although the effect of pregnancy is manifested as blunted baroreflex-mediated increases in HR and SNA, it is possible that mechanisms unrelated to the traditional arterial baroreflex pathway could contribute to attenuated sympathoexcitation during hypotension.

Since the final sympathetic outflow emanating from the RVLM at any given time is determined by the balance of excitatory and inhibitory influences, either decreased baseline excitatory drive to barosensitive neurons or increased baseline baroreflex-independent inhibition could contribute to the observation of decreased activation of the RVLM at low arterial pressure in pregnant rats (Curtis et al., 1999). To test the hypothesis that the RVLM of pregnant rats is under greater tonic baroreflex-independent GABAergic inhibition, responses to bilateral blockade of $GABA_A$ receptors in the RVLM were tested in anesthetized pregnant and nonpregnant rats in which the arterial baroreceptors had been surgically denervated. Blockade of $GABA_A$ receptors in the RVLM with nanoinjection of bicuculline produced greater increases in mean arterial pressure and RSNA in pregnant compared with nonpregnant rats (Fig. 3.7) (Kvochina et al., 2007). Thus it appears that pregnancy is associated with augmented tonic GABAergic inhibition of the RVLM, and in the absence of this GABAergic tone, underlying excitatory influences are greater in pregnant rats. Endogenous AngII within the RVLM is excitatory (Allen et al., 2009), and prior blockade of AngII AT1Rs in the RVLM decreased responses to RVLM bicuculline in both groups. However, augmented responses to

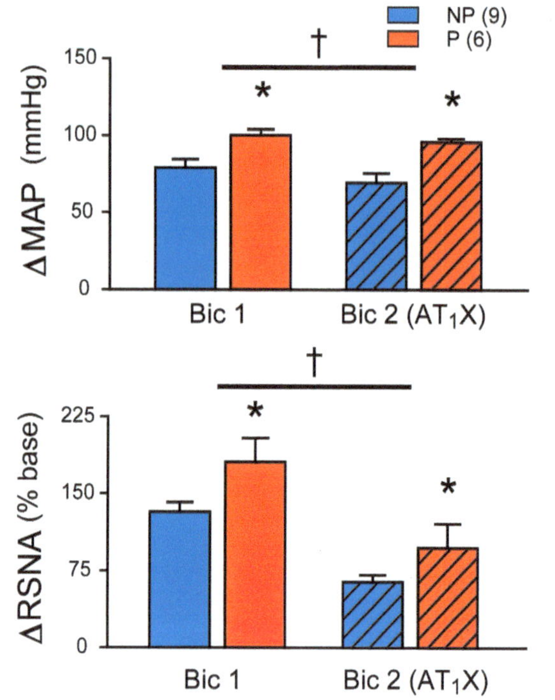

Fig. 3.7. Baroreflex independent $GABA_A$ inhibition of the rostral ventrolateral medulla (RVLM) is increased in late pregnant rats. In sinoaortic denervated rats, increases in mean arterial pressure (Δ MAP) and renal sympathetic nerve activity (Δ RSNA) in response to bilateral blockade of $GABA_A$ receptors (bicuculline, Bic) in the RVLM was greater in P compared with NP rats. Prior blockade of AngII AT_1 receptors (L158,809; AT1X) in the RVLM attenuated responses to Bic, but differences between NP and P rats persisted. $^*P \leq 0.05$, greater than in NP rats. $^†P \leq 0.05$, main effect of AT1X. Modified with permission from Kvochina, L., Hasser, E.M., Heesch, C.M., 2007. Pregnancy increases baroreflex independent GABAergic inhibition of the RVLM in rats. Am J Physiol Regul Integr Comp Physiol 293, R2295–R2305.

bicuculline persisted in pregnant rats (Fig. 3.7). Thus underlying excitatory influences in addition to AngII must contribute to observed differences in responses to RVLM $GABA_A$ blockade between pregnant and nonpregnant rats.

Potential role of neurosteroid metabolite of progesterone

With regard to the mechanisms for increased $GABA_A$ influence in the RVLM, it is possible that, independent of arterial baroreflex inputs, GABA release in the RVLM is greater in pregnant animals. However, neither the source of drive for nonbaroreflex GABAergic inputs to the RVLM nor the potential signal that might increase that input during pregnancy has been elucidated. Another possibility, which would not require increased GABAergic input to the RVLM, would be an increase in positive modulation and/or upregulation of $GABA_A$ receptors within the RVLM so that the effectiveness of existing GABAergic inputs would be amplified in pregnancy. In vitro studies have found that the major metabolite of progesterone, 3α-hydroxy-dihydroprogesterone (3α-OH-DHP, also called allopregnanolone), binds stereospecifically to a unique site on the $GABA_A$ receptor complex and serves as a potent endogenous positive modulator of CNS $GABA_A$ receptors (Mody, 2005). In addition, longer-term exposure to 3α-OH-DHP preferentially increases expression of the $GABA_A$ receptor δ subunit (Sanna et al., 2009). Tonic GABAergic inhibition is mediated mainly by extrasynaptic $GABA_A$ receptors containing the δ subunit, which is the most important receptor subunit in conferring high sensitivity to 3α-OH-DHP in native $GABA_A$ receptors (Mody, 2005; Belelli et al., 2006; Biggio et al., 2009; Maguire and Mody, 2009). Thus the major metabolite of progesterone has both rapid nongenomic effects and longer-term genomic effects, which would potentiate GABAergic inhibition.

Along with progesterone, plasma levels of the neurosteroid metabolite 3α-OH-DHP increase during pregnancy in women (Pennell et al., 2015) and rats (Biggio et al., 2009). 3α-OH-DHP is highly lipid soluble and thus circulating levels have access to the CNS. In addition, synthetic enzymes for 3α-OH-DHP are present within the brain, and it can be synthesized from circulating progesterone or de novo from cholesterol (Stoffel-Wagner, 2001). While circulating levels of 3α-OH-DHP increase in parallel with progesterone, 3α-OH-DHP levels in the brain continue to increase throughout pregnancy after progesterone levels have peaked (Concas et al., 1998). In fact, in certain brain regions, levels of 3α-OH-DHP are several fold higher than circulating levels, and local regulation of the synthetic enzymes for converting progesterone to 3α-OH-DHP within the brain appears to be a major factor in determining brain region, and even cell type specific, actions of this neurosteroid (Belelli and Lambert, 2005). Enzymes responsible for synthesis of 3α-OH-DHP from progesterone, 5α-reductase and 3α-OH steroid oxidoreductase, are present in the medulla (Khanna et al., 1995; Li et al., 1997), and ovarian hormones that are elevated in pregnancy can regulate 3α-OH-DHP synthesis in brain tissue (Mitev et al., 2003). Importantly, messenger RNA for 5α-reductase, the rate-limiting enzyme for synthesis of 3α-OH-DHP, is increased in the RVLM of term pregnant rats (Heesch and Burcks, 2009).

It has been proposed that elevated levels of 3α-OH-DHP in the brain in pregnancy ameliorate responses to stress (Brunton et al., 2014). In women, inadequate elevation of this neurosteroid (and/or the rapid decline postpartum) may contribute to perinatal depressive disorders (Brunton et al., 2014; MacKenzie and Maguire, 2014). In support of a role for the neurosteroid metabolite of progesterone in regulation of SNA, initial experiments demonstrated that i.v. administration of 3α-OH-DHP to either anesthetized (Heesch and Rogers, 1995) or conscious nonpregnant female rats (Masilamani and Heesch, 1997) suppressed arterial baroreflex-mediated sympathoexcitation, while having no effect on afferent baroreceptor discharge (Laiprasert et al., 2001). In these experiments, i.v. administered 3α-OH-DHP would have access to multiple CNS sites. Additional experiments assessed baroreflex function before and after nanoinjections of 3α-OH-DHP into major nuclei in the medullary baroreflex pathway. Nanoinjection of 3α-OH-DHP, but not the inactive isomer 3β-OH-DHP, into the RVLM decreased maximum sympathoexcitation due to baroreceptor unloading (Heesch, 2011) (Fig. 3.8), whereas nanoinjections into the NTS and CVLM were without effect (Brooks et al., 2010a). Thus, similar to the effects of pregnancy to specifically suppress activation of neurons in the RVLM during hypotension (Curtis et al., 1999), acute effects of the neurosteroid metabolite of progesterone to suppress baroreflex-mediated increases in SNA also appear to be specific to the RVLM.

However, well maintained or slightly potentiated baroreflex-mediated sympathoinhibition in response to a hypertensive challenge has been observed in pregnant rats (Crandall and Heesch, 1990; Masilamani and Heesch, 1997; Shi et al., 2019) (Figs. 3.4 and 3.5). Since baroreflex-mediated sympathoinhibition is due to incremental increases in GABAergic input to the RVLM, positive modulation of $GABA_A$ receptors by 3α-OH-DHP at the level of the RVLM would be consistent with enhanced baroreflex-mediated sympathoinhibition. Indeed, 3α-OH-DHP (but not 3β-OH-DHP), administered either intravenously or directly into the RVLM of nonpregnant female rats, decreased the pressure threshold for baroreflex-mediated inhibition of single-unit neuronal discharge in spinally projecting RVLM neurons (Laiprasert et al., 1998). These results suggest that sensitivity of RVLM neurons to endogenously released GABA is increased acutely by 3α-OH-DHP.

Taken together, experiments in nonpregnant female rats provide convincing evidence that exogenous administration of 3α-OH-DHP, the neurosteroid metabolite of progesterone, mimics the effects of pregnancy to suppress baroreflex-mediated sympathoexcitation during hypotension, while sympathoinhibition in response to increased blood pressure is preserved. In addition, the effects of circulating 3α-OH-DHP are likely through an action in the RVLM, since local administration of the neurosteroid into the RVLM, but not other brain regions, produced results similar to i.v. administration. The remaining question was whether elevated endogenous levels of this neurosteroid metabolite of

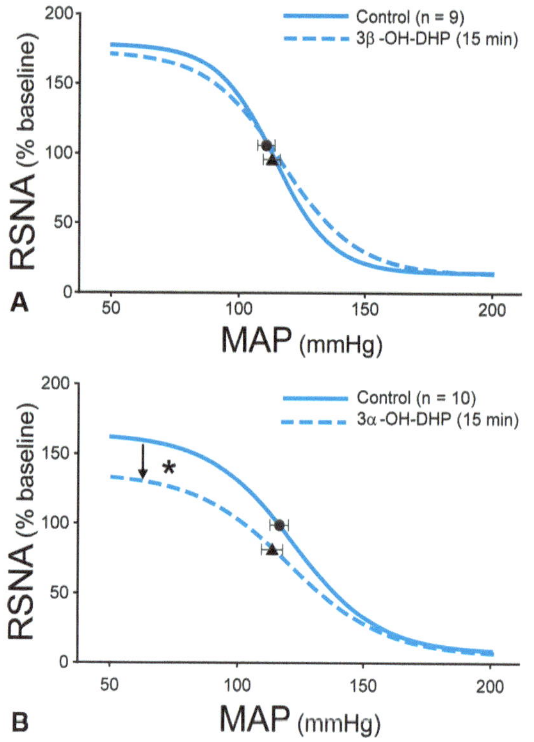

Fig. 3.8. In the RVLM the progesterone metabolite 3α-OH-DHP, but not the inactive isomer 3β-OH-DHP, attenuates arterial baroreflex sympathoexcitation. Baroreflex curves before (control) and 15 min after bilateral nanoinjection of either the α- or β-isomer of 3-OH-DHP into the RVLM of anesthetized rats were constructed using a four-parameter sigmoidal curve formula and the curve coefficients were compared. (A) Nanoinjection of the inactive isomer 3β-OH-DHP into the RVLM had no effect on arterial baroreflex function. (B) Within 15 min following bilateral nanoinjection of 3α-OH-DHP into the RVLM, baroreflex sympathoexcitation was attenuated (decreased maximum RSNA during hypotension). RSNA, renal sympathetic nerve activity; MAP, mean arterial pressure; filled symbols represent resting MAP for each curve; *compared with control curve. Adapted from Heesch, C.M., 2011. Neurosteroid modulation of arterial baroreflex function in the rostral ventrolateral medulla. Auton Neurosci 161, 28–33 with permission.

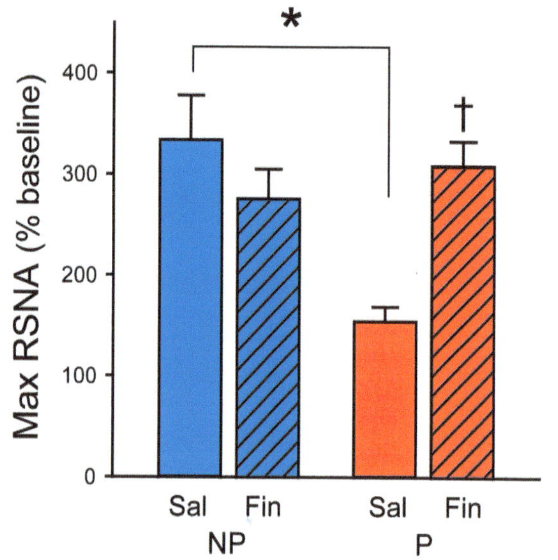

Fig. 3.9. Chronic blockade of 3α-OH-DHP synthesis with the 5α-reductase inhibitor finasteride restores baroreflex-mediated sympathoexcitation in conscious pregnant rats. Baroreflex function curves were constructed (similar to Fig. 3.6) in four groups: nonpregnant (NP) and late pregnant (P) rats treated during the last trimester of pregnancy with either saline (Sal) or finasteride (Fin) ($n=5$ each group). Maximum RSNA during hypotension is shown. In Sal-treated rats, pregnancy reduced baroreflex-mediated sympathoexcitation (*). Finasteride did not change baroreflex function in NP rats, but in P rats maximum RSNA during hypotension was restored to levels not different from those achieved in NP rats. †Pregnant rats: Fin > Sal, $P < 0.05$. Data from Phaup, J.G., Hasser, E.M., Heesch, C.M., 2013. The neurosteroid metabolite of progesterone, 3α-OH-dihydroprogesterone (3α-OH-DHP), is required for attenuated baroreflex mediated sympathoexcitation in pregnancy. FASEB J 27, 1118.38.

progesterone in pregnancy contribute to attenuated baroreflex-mediated sympathoexcitation. If so, chronic inhibition of 3α-OH-DHP synthesis should restore sympathoexcitation during hypotension in pregnant rats. During the last trimester of pregnancy the rate-limiting enzyme for synthesis of 3α-OH-DHP, 5-alpha reductase, was inhibited by daily subcutaneous injections of finasteride, while control rats received subcutaneous saline. Similar to previous studies (Fig. 3.4), the major effect of pregnancy was suppression of maximum baroreflex-mediated sympathoexcitation, and this effect was reversed by chronic finasteride treatment (Phaup et al., 2013) (Fig. 3.9). Taken together, these studies suggest that, likely due to an action in the RVLM, increased endogenous 3α-OH-DHP contributes to the observed effects of pregnancy on baroreflex function.

Potential mechanisms by which pregnancy increases basal SNA and induces sympathoexcitation, yet impairs responses to baroreceptor unloading

The review of literature earlier in the chapter indicates that multiple neural and humoral adaptations during pregnancy likely contribute to observed changes in control of sympathetic outflow. Although many questions remain, experimental evidence to date suggests that mechanisms responsible for maintaining elevated baseline SNA and direct activation of RVLM neurons and those responsible for attenuated baroreflex-mediated sympathoexcitation in pregnancy involve different CNS modalities. One possibility is that, separate from increases in SNA due to disinhibition of the RVLM through the arterial baroreflex, sympathoexcitatory pathways that do not include the RVLM could contribute to directly induced increases in SNA. In addition to excitatory projections to the RVLM, several brain regions, including the PVN, dorsal medial hypothalamus (DMH), and A5 catecholaminergic neurons in the pons, have monosynaptic glutamatergic projections directly to preganglionic neurons in the spinal cord and have been implicated in increased SNA under a variety of conditions (Guyenet, 2006; Card and Sved, 2011). Although there is a paucity of information on a role of these pathways in pregnancy, the PVN contributes to elevated baseline SNA in pregnant rats (Shi et al., 2015a), and it is plausible that a direct PVN to spinal cord pathway could be involved. Importantly, whether the RVLM contributes to or supports the increased basal SNA during pregnancy has not been tested directly. Second, the explanation may be related to different roles of presympathetic cell phenotypes within the RVLM. A series of experiments in male rats evaluated effects of selective lesion of spinally projecting catecholaminergic neurons and found that in C1-lesioned rats, the remaining noncatecholaminergic spinally projecting neurons in the RVLM maintained basal sympathetic tone and, although slightly attenuated, arterial baroreflex function was preserved. In contrast, active sympathoexcitation due to stimulation of the arterial chemoreflex or electrical stimulation of the RVLM was greatly reduced in the absence of the RVLM C1 neurons (Guyenet et al., 2001). Again, little information is available regarding pregnancy. However, during a hypotensive challenge, decreased activation of RVLM neurons in pregnant rats was specific to the noncatecholaminergic neuronal phenotype (Curtis et al., 1999). Noncatecholaminergic cells in the RVLM predominate in arterial baroreflex responses, and their activation during hypotension is limited in pregnancy, presumably related to tonic GABAergic inhibition due to modulation by the progesterone metabolite 3α-OH-DHP. In women, direct sympathoexcitation initiated by the exercise pressor reflex and the cold pressor test are not attenuated by pregnancy, and it is tempting to speculate that, similar to the prominent role of RVLM C1 neurons in direct sympathoexcitation in animals, responses requiring direct activation of the RVLM may primarily involve C1 cells in humans as well. A third possibility to consider is that absolute baseline SNA is elevated in pregnancy. While noncatecholaminergic neurons in the RVLM are sufficient to maintain baseline SNA in normal male rats (Guyenet et al., 2001), it is possible that in pregnancy both RVLM C1 and non-C1 neurons assume a greater role. If so, greater activation of C1 neurons in the RVLM could contribute to both increased baseline SNA and augmented responses to stimuli such as the cold pressor test (Usselman et al., 2015b), but a parallel elevation in non-C1 activity would bring SNA closer to the baroreflex maximum (which is suppressed by 3α-OH-DHP).

Other reflexes altered in pregnancy

Atrial volume reflex

In addition to the arterial baroreflex, there are other important reflexes that contribute to homeostatic control and are altered in pregnancy. In particular, stretch receptors in the right atrium have a critical role in blood volume regulation. Activation of these receptors results in a rather selective reflex inhibition of RSNA (Lovick et al., 1993; Yang and Coote, 2003). The central pathways mediating this reflex have not been fully delineated but include afferent projections from stretch receptors in the atria to the NTS and a critical synapse in the PVN (Lovick et al., 1993; Yang and Coote, 2003). At the level of the PVN, GABAergic interneurons are activated and result in inhibition of other PVN neurons that directly or indirectly regulate renal sympathetic outflow (Deng and Kaufman, 1995; Yang and Coote, 2003). Similar to the arterial baroreflex, reflex responses to atrial stretch are blunted by pregnancy (Kaufman and Deng, 1993; Deng and Kaufman, 1995; Hines and Mifflin, 1995; Hines and Hodgson, 2000). However, different from arterial baroreceptor afferent input that is well maintained in pregnancy, in response to atrial stretch, discharge of the high-frequency subgroup of cardiac vagal afferent fibers is greatly reduced (Deng and Kaufman, 1995; Hines and Hodgson, 2000; Storey and Kaufman, 2004) and activation of neurons in the PVN is blunted (Deng and Kaufman, 1995). Thus depressed function of the atrial volume reflex may contribute to maintenance of the expanded blood volume in pregnancy. Interestingly, acute i.v. administration of the progesterone metabolite 3α-OH-DHP to nonpregnant female rats decreased activation of the PVN during atrial distension, mimicking the effects of pregnancy (Storey and Kaufman, 2004).

Arterial chemoreflex

Resting ventilation and responsiveness to hypoxia are increased in pregnant women (Moore et al., 1986; Usselman et al., 2015a). Experiments in cats verified increased ventilatory responses to hypoxia and further demonstrated that sensitivity of the carotid body chemoreceptors to incremental hypoxia was augmented in pregnancy (Hannhart et al., 1989). In male rats, hypoxic stimulation of carotid body chemoreceptors increases afferent discharge in the carotid sinus nerve and not only mediates increased ventilation but also increases efferent

SNA (Guyenet, 2006). Although efferent SNA in response to hypoxia has not been evaluated in pregnant women or animals, it is possible that augmented chemoreflex function in pregnancy could contribute to the effects of pregnancy on control of SNA.

PREECLAMPSIA ELICITS FURTHER INCREASES IN SNA

Preeclampsia is clinically defined as new onset of hypertension with proteinuria and/or end-organ dysfunction after the 20th week of gestation in a previously normotensive woman (Davey and MacGillivray, 1988; Roberts et al., 2003). According to the latest definition, new-onset thrombocytopenia, renal insufficiency, neurologic complications, liver involvement, and fetal growth restriction may substitute for new-onset proteinuria (Tranquilli et al., 2014). This disorder affects approximately 5% of all pregnancies worldwide (Sibai et al., 2005; Hernandez-Diaz et al., 2009), and it is one of the leading causes of maternal and fetal morbidity and mortality (Lyall and Greer, 1996; Granger et al., 2001a, b; Panchal et al., 2001; Roberts et al., 2003). Despite the critical importance of the problem, therapy for preeclampsia has not changed substantially in over 50 years (Noris et al., 2005; Frishman et al., 2006). There is no definite cure other than delivery (Steegers et al., 2010). The primary reason for the lack of progress in therapy is the failure to identify key underlying mechanisms.

Preeclampsia and sympathetic overactivity

One global theory that has been proposed to explain the morbid nature of preeclampsia is that a hyperadrenergic state may be involved in the pathophysiology of the disorder (Schobel et al., 1996; Greenwood et al., 1998, 2003). In support of this notion, previous microneurographic studies have demonstrated greater resting MSNA in preeclamptic women compared with normotensive pregnant women toward the end of their gestation when clinical signs and symptoms of the complication most commonly become manifest (Schobel et al., 1996; Greenwood et al., 1998, 2001, 2003; Fischer et al., 2004). It is suggested that the increases in peripheral vascular resistance and blood pressure that characterize preeclampsia are mediated, at least in part, by a substantial increase in MSNA (Schobel et al., 1996). Indeed, animal work has indicated that reduced uterine perfusion pressure (RUPP), causing placental ischemia-induced hypertension (a rat model of human preeclampsia), involves adrenergic receptor signaling to promote increases in blood pressure (Spradley et al., 2018). Moreover, the sympatholytic therapy with methyldopa in women with preeclampsia has been found to decrease blood pressure through a reduction in central sympathetic outflow (Roberts and Redman, 1993). Sympathetic activity and blood pressure decrease in parallel after delivery in preeclamptic women, which is further suggestive of a neurogenic component in preeclampsia (Schobel et al., 1996).

The key question that must be addressed is whether sympathetic overactivity can be detected *early* in pregnancy, or whether it occurs only at term when maternal symptoms of preeclampsia most commonly become manifest. A prospective longitudinal study showed that throughout early (within 8 weeks of gestation) and late (between 32 and 36 weeks) pregnancy, as well as postpartum (6–10 weeks after delivery), resting MSNA was markedly augmented in women who developed hypertension at term (Badrov et al., 2019). These results indicate sympathetic overactivity may be a biomarker for predicting the development of preeclampsia and other gestational hypertensive disorders. However, some researchers have proposed that preeclampsia manifests only when both sympathetic overactivity and impaired vasodilatory function occur (Fischer et al., 2004). Whether this is true or not remains to be determined. Interestingly, the magnitude of sympathetic overactivity recorded as multiunit MSNA during late pregnancy was found to be similar in women with preeclampsia and in those with gestational hypertension, but single-unit MSNA was lower in preeclamptic women, suggesting the central mechanisms for sympathetic overactivity underlying these two conditions may be different (Greenwood et al., 2003).

Potential mechanisms for sympathetic overactivity in preeclampsia

The pathophysiologic mechanisms for sympathetic overactivity in preeclampsia are unknown. While there is nothing definitive or proven, the following possibilities may be considered: these include, but are not limited to, impaired sympathetic baroreflex function, overactivation of the RAAS, increased maternal AngII type I receptor agonistic autoantibody (AT_1R-AA), inflammation, activation of the endothelin system, and dysregulation of the natriuretic peptide system (Fig. 3.10).

THE SYMPATHETIC BAROREFLEX

Sympathetic overactivity in preeclampsia could be attributed to an impairment of baroreflex function (e.g., reduced sympathetic baroreflex sensitivity), leading to a decrease in baroreflex-mediated inhibitory restraint on central sympathetic outflow (Ekholm et al., 1994; Schobel et al., 1996). Surprisingly, there is little information available about the sympathetic baroreflex even during normal (uncomplicated) pregnancy in humans.

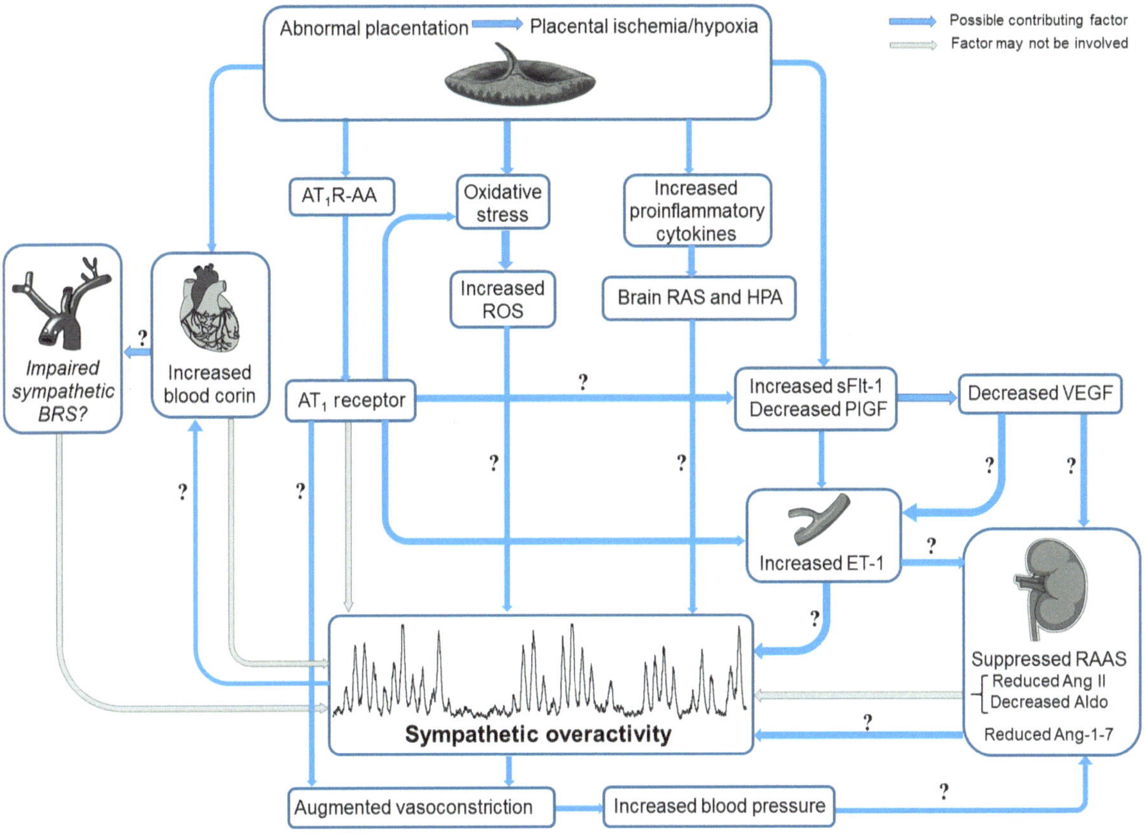

Fig. 3.10. Proposed pathophysiologic mechanisms for sympathetic overactivity in preeclampsia. Aldo, aldosterone; Ang-(1–7), angiotensin-(1–7); AngII, angiotensin II; AT_1R-AA, angiotensin II type I receptor agonistic autoantibody; BRS, baroreflex sensitivity; ET-1, endothelin-1; HPA, the hypothalamic–pituitary–adrenal axis; PIGF, placental growth factor; RAAS, renin–angiotensin–aldosterone system; ROS, reactive oxygen species; RSA, renin–angiotensin system; sFlt-1, soluble fms-like tyrosine kinase 1; VEGF, vascular endothelial growth factor.

One cross-sectional study found that sympathetic baroreflex sensitivity was reduced in normotensive pregnant women during the third trimester of gestation relative to healthy nonpregnant women despite the fact that blood pressure was similar between the groups (Usselman et al., 2015a). In contrast, a longitudinal case report involving a series of repeated measures over time showed that although sympathetic baroreflex sensitivity was variable throughout gestation, the majority of the values indicated a greater sensitivity in normal pregnancy compared with prepregnancy level (Hissen et al., 2017). These preliminary observations need to be confirmed in more pregnant women.

Animal research demonstrated that baroreflex control of RSNA was reset to a higher blood pressure level while the sensitivity of the sympathetic baroreflex (the slope of the baroreflex curve) was not different between pregnant rats with RUPP compared with normal pregnant rats (Hines et al., 2007) (Fig. 3.11). Whether this is true even in humans is unknown. However, MSNA responses to physiologic stimulations such as the Valsalva maneuver, a cold pressor test, and handgrip exercise were found to be similar in women with preeclampsia vs women with normotensive pregnancies (Schobel et al., 1996; Greenwood et al., 1998). These results indicate that baroreflex control of SNA may be intact in preeclampsia. Future studies should consider direct measurement of sympathetic baroreflex function in preeclamptic women.

RAAS

Normal pregnancy is associated with activation of the RAAS, which allows an aldosterone-dependent blood volume expansion (Verdonk et al., 2014). Upregulations of AngII and aldosterone concentrations elicited by the RAAS activation were found to increase MSNA in nonpregnant and pregnant individuals (Matsukawa et al., 1991; Kontak et al., 2010; Jarvis et al., 2012). However, women with preeclampsia display suppression of the RAAS and a reduction in aldosterone concentration (Brown et al., 1994, 1997; Verdonk et al., 2014), indicating the RAAS is not a contributing factor for sympathetic overactivity. These findings also suggest that the RAAS may not be the cause of hypertension in preeclampsia,

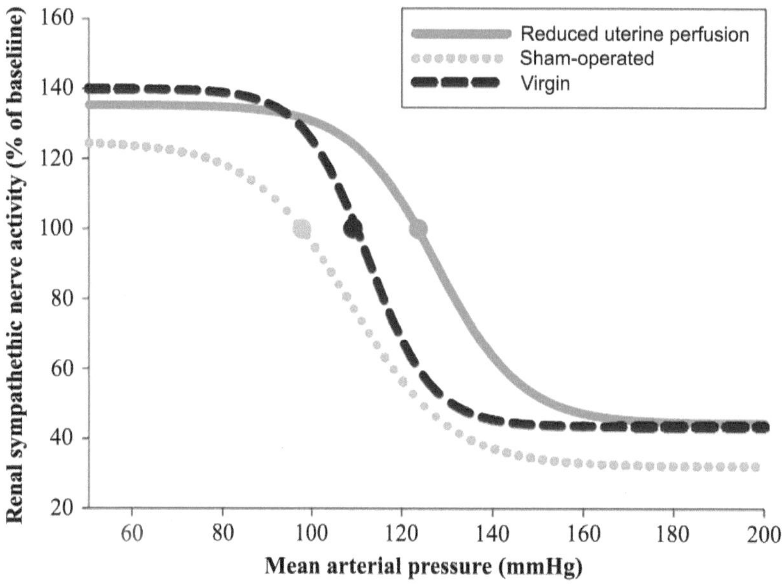

Fig. 3.11. The sympathetic baroreflex curves illustrating the relationship between mean arterial pressure and renal sympathetic nerve activity in nonpregnant virgin rats, normal pregnant sham-operated rats, and pregnant rats with reduced uterine perfusion. *Filled circles* represent the resting mean arterial pressure for each group. Adapted with permission from Hines, T., Beauchamp, D., Rice, C., 2007. Baroreflex control of sympathetic nerve activity in hypertensive pregnant rats with reduced uterine perfusion. Hypertens Pregnancy 26, 303–314.

but rather that its suppression may be the consequence of increased blood pressure (Powe et al., 2011; Saleh et al., 2016). Other possible mechanisms for the suppressed RAAS in preeclampsia may include, but are not limited to, increased levels of atrial natriuretic peptide (ANP), a decrease in vascular endothelial growth factor (VEGF), and an increase in endothelin-1 (ET-1) (Gennari-Moser et al., 2013; Verdonk et al., 2014, 2015; Volpe et al., 2014). It is likely that the suppressed RAAS is counterintuitive in preeclampsia, given the reduced circulating volume in this disorder (Saleh et al., 2016).

Ang-(1–7) is a bioactive component of the RAAS, which has depressor, vasodilatory, and antihypertensive actions (Brosnihan et al., 2003; Valdes et al., 2006; Yamaleyeva et al., 2014). Animal studies have shown that augmentation of central Ang-(1–7) inhibits sympathetic outflow (Gironacci et al., 2004; Kar et al., 2011; Yu et al., 2019). The plasma level of Ang-(1–7) was found to increase by 51% in normotensive pregnant women, but it was significantly decreased in preeclamptic women (Merrill et al., 2002; Chen et al., 2019). Thus the decrease in Ang-(1–7) might contribute, in part, to sympathetic overactivity in preeclampsia.

AT_1R-AA

Preeclampsia could be accompanied by the presence of a maternal circulating AT_1R-AA (Wallukat et al., 1999; Siddiqui et al., 2010). A meta-analysis has concluded that AT_1R-AA positivity may be a valuable indicator for poorer prognosis in preeclampsia (Lei et al., 2016). In contrast, some studies found no difference in AT_1R-AA levels throughout gestation in both women with preeclampsia and women with normotensive pregnancies (Aggarwal et al., 2017). The reasons behind these contradictory results are unclear. More research is needed to confirm the findings of Aggarwal et al. (2017).

AT_1R-AA is functionally similar to AngII, and the latter could increase SNA through the AT_1Rs in the CNS (Stein et al., 1984; Dechend et al., 2000, 2004, 2005, 2006; Ruzicka et al., 2013). Yet, there has been no evidence showing a direct central effect of AT_1R-AA on sympathetic activity in animals or humans. Nonetheless, AT_1R-AA might cause sympathetic overactivity through other possible mechanisms or indirect pathways. For instance, it was found that AT_1R-AA increased the production of reactive oxygen species and decreased the bioavailability of nitric oxide through activation of the AT1Rs in the peripheral vasculature and the CNS (Dechend et al., 2003; Ariza et al., 2007; Parrish et al., 2011; Brewer et al., 2013), which might lead to overactivation of the sympathetic nervous system (Puzserova and Bernatova, 2016; Yan et al., 2017; Haspula and Clark, 2018). In addition, AT_1R-AA was reported to upregulate soluble fms-like tyrosine kinase 1 (sFlt-1) and inactivate VEGF (Zhou et al., 2008), both of which might result in an increase in ET-1 (LaMarca et al., 2009). Previous studies have indicated a sympathoexcitatory effect of endogenous ET-1 in humans (Bruno et al., 2011). AT_1R-AA is associated with increased AngII sensitivity (Wallukat et al., 1999, 2003; Herse et al., 2009; LaMarca et al., 2009; Parrish et al., 2011), leading to augmentation of vasoconstrictor responsiveness.

Inflammation

Preeclampsia is associated with increased proinflammatory, relative to antiinflammatory, cytokines (Catarino et al., 2012; Kucuk et al., 2012; Southcombe et al., 2015; Kalagiri et al., 2016; Cornelius et al., 2019; Zak and Soucek, 2019). Studies in RUPP rats strongly implicate inflammatory processes in the pathogenesis of preeclampsia. For example, adoptive transfer of proinflammatory immune cells into normal pregnant rats replicates many of the features of preeclampsia, including an elevation in proinflammatory cytokines (Harmon et al., 2016; Cornelius et al., 2019). In rats, proinflammatory cytokines act within the SFO and hypothalamus to cause centrally mediated sympathetic overactivity by activating the brain renin–angiotensin system and the hypothalamic–pituitary–adrenal axis in hypertension and other cardiovascular abnormalities (Kang et al., 2008; Wei et al., 2015, 2018; Yu et al., 2017, 2018; Haspula and Clark, 2018). Whether sympathetic overactivity in preeclamptic women is attributed to the increases in proinflammatory cytokines needs to be investigated.

The endothelin system

The endothelin system plays an important role in cardiovascular homeostasis through its direct vascular effects, as well as neural regulation of vasomotor sympathetic tone (Mosqueda-Garcia et al., 1993). ET-1 is a vasoconstrictor and mitogenic peptide produced by endothelial cells (Bruno et al., 2011). Animal research suggests that ET-1 can stimulate the sympathetic nervous system by activation of the subtype A receptor (Gulati et al., 1997; Rossi et al., 1997, 2008; Mortensen, 1999; Nakamura et al., 1999; Rossi and Maliszewska-Scislo, 2008). It also acts in carotid bodies and cervical superior and nodose ganglia, causing changes in the baroreflex control of SNA (Mortensen, 1999). Microneurographic studies in (nonpregnant) humans found that endogenous ET-1 had a sympathoexcitatory effect, especially in individuals with hypertension (Bruno et al., 2011).

Compared with women with normotensive pregnancies, those with preeclampsia were found to have a two- to threefold rise of circulating ET-1; in addition, there was a positive correlation between the severity of preeclampsia and the level of ET-1 (Bernardi et al., 2008; Aggarwal et al., 2012; Verdonk et al., 2015; Saleh et al., 2016). However, so far, there has been no direct evidence showing the impact of ET-1 on SNA in preeclamptic women. Verdonk et al. suggested that ET-1 is not only an independent determinant of both the blood pressure increase and proteinuria in preeclampsia but also a renin suppressor (Verdonk et al., 2015).

The natriuretic peptide system

The natriuretic peptide system, along with the sympathetic nervous system, has been recognized as a key neural–humoral system to maintain overall cardiovascular homeostasis (Volpe, 2014). In general, the beneficial physiologic actions of the natriuretic peptide system are counter-regulatory to those of the sympathetic nervous system (Volpe et al., 1987, 1988; Luchner and Schunkert, 2004). Natriuretic peptides interfere with autonomic and baroreflex control of the circulation, leading to sympathetic inhibition and/or parasympathetic activation (Volpe et al., 1987, 1988; Floras, 1990; Luchner and Schunkert, 2004).

As preeclampsia is associated with sympathetic overactivity, a decrease in maternal natriuretic peptide level would be expected in preeclamptic women. Surprisingly, many studies have shown increased plasma levels of ANP and brain natriuretic peptide (BNP) in preeclampsia (Itoh et al., 1993; Furuhashi et al., 1994; Stepan et al., 1999; Borghi et al., 2000; Tihtonen et al., 2007; Ringholm et al., 2011; Giannubilo et al., 2017), which may be attributed to the strain on the heart caused by high afterload rather than the function of the heart expressed as stroke index or cardiac index (Tihtonen et al., 2007). It is suggested that ANP and BNP may be rather a sequel to preeclamptic pathophysiologic changes and not play an important role as the etiological factor of preeclampsia (Furuhashi et al., 1994). It is possible that increased maternal ANP and BNP levels are a compensatory response to sympathetic overactivity and augmented vasoconstriction and may contribute to the suppressed RAAS in preeclampsia.

Research has demonstrated that the placenta released neprilysin, a widely expressed membrane-bound metalloprotease that binds and cleaves a variety of peptides, including vasodilators, natriuretics, and diuretics, is increased in preeclampsia (Gill et al., 2019), presumably due to the increases in ANP and BNP levels.

Corin is a transmembrane protease discovered in the heart where it converts pro-ANP to active ANP (Yan et al., 1999; Li et al., 2017). Its expression has also been detected in the pregnant uterus, which appears to play an important role in promoting trophoblast invasion and spiral artery remodeling (Cui et al., 2012). New research implies that corin may be involved in the pathogenesis of preeclampsia and other hypertensive disorders in pregnancy (Cui et al., 2012; Kaitu'u-Lino et al., 2013; Khalil et al., 2015; Liu et al., 2015; Miyazaki et al., 2016; Gu et al., 2018; Badrov et al., 2019). It was found that uterine *Corin* messenger RNA and protein levels were significantly lower, while plasma corin content was considerably higher in preeclamptic women compared with nonpregnant or normotensive pregnant women (Cui et al., 2012). It is possible that increased corin content in the blood, which is probably derived from the heart, is a compensatory response to the decreased corin expression in the uterus and/or to the increase in blood pressure in women with preeclampsia. A prospective longitudinal study showed that the change in circulating blood corin from early to late pregnancy was significantly greater in women who developed hypertension at term compared with those who had normotensive pregnancies (Badrov et al., 2019). More importantly, changes in corin from early to late pregnancy were related to blood pressure in late pregnancy, while MSNA in early pregnancy was related to changes in corin from early to late pregnancy (Badrov et al., 2019) (Fig. 3.12). It may be that the greater increase in maternal corin content in women who had hypertension at term is an adaptive response to the further increase in sympathetic activity and oncoming high blood pressure. Conversely, the greater increase in corin might contribute, at least in part, to sympathetic overactivity and, thereby, the development of preeclampsia.

SUMMARY AND FUTURE DIRECTIONS

Normal pregnancy increases basal SNA, which is homeostatically appropriate, but impairs baroreflex control of HR and SNA, which can be harmful. Pregnancy-induced hypertensive disorders evoke even greater elevations in SNA, which likely contribute to the hypertensive state. Only limited information is available concerning the mechanisms. At term, the PVN and ArcN are two hypothalamic nuclei that support the basal sympathoexcitation. In normal pregnancy, increased AngII, by binding to AT1R, acts centrally to support elevated basal SNA. The SFO, ArcN, and PVN are likely (but unproven) targets of the activated RAAS. However, no definitive mechanisms for exaggerated sympathoexcitation that occurs with pregnancy-induced hypertensive disorders have been identified. In addition, normal pregnancy clearly induces increased GABAergic inhibition of the RVLM, and accumulated evidence supports a major role for the locally increased production and actions of 3α-OH-DHP as one mechanism. A consequence is suppression of baroreflex function. The ultimate level of basal SNA is determined by the integration of all excitatory and inhibitory inputs; the clearly apparent increases in SNA suggest that excitatory influences predominate in the RVLM, but this possibility requires direct experimentation.

Thus while some progress has been made in our understanding of the mechanisms by which normal and hypertensive pregnancy alters SNA, HR, and control of SNA and HR, even more remains to be established. This is a challenging quest because the homeostatic changes in normal pregnancy, and the pathophysiologic impact of preeclampsia, are complex and also because the study of pregnant women often precludes direct experimentation. Nevertheless, the following are future areas of research:

1. The degree of basal sympathoexcitation is greater in normal and hypertensive pregnancy than many other physiologic and pathophysiologic states, suggesting

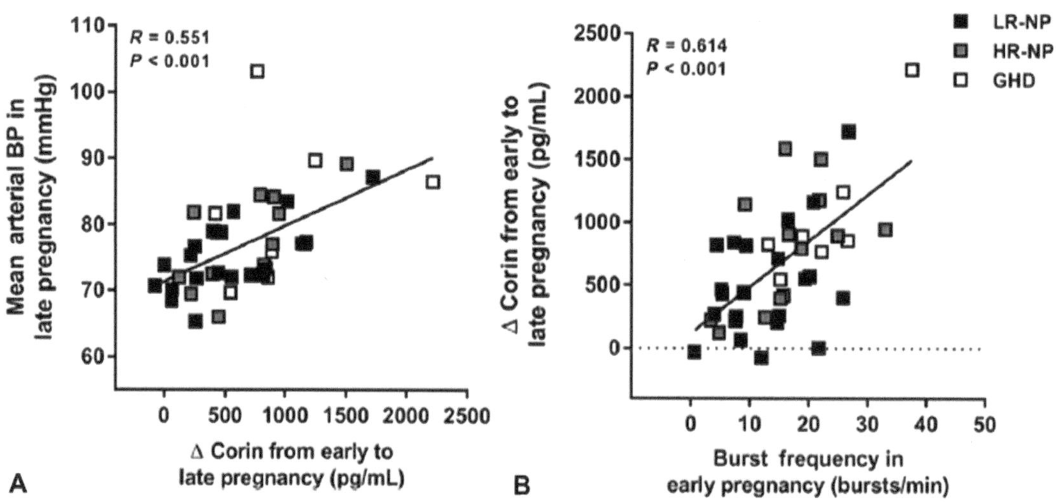

Fig. 3.12. Relationship between mean arterial pressure in late pregnancy and the change in corin from early to late pregnancy (A), and between the change in corin and muscle sympathetic nerve activity in early pregnancy (B), in low-risk normal pregnancy (LR-NP), high-risk normal pregnancy (HR-NP), and women who developed gestational hypertensive disorders (GHD). Adapted with permission from Badrov, M.B., Park, S.Y., Yoo, J.K., et al., 2019. Role of corin in blood pressure regulation in normotensive and hypertensive pregnancy. Hypertension 73, 432–439.

that factors besides the RAAS contribute. Other potential hormonal or nonhormonal factors that have not been directly investigated include aldosterone, progesterone, inflammation (cytokines), and endothelin.

2. Similarly, the high SNA suggests that additional brain sites may contribute, such as the DMH and other CVOs.
3. The study of rodent preeclampsia models, such as the RUPP rat or mouse, may suggest testable hypotheses for hypertensive pregnancy-induced sympathoexcitation in women, especially given the advent of sophisticated approaches for select activation or inhibition of specific brain cell types, such as optogenetics, chemogenetics, and GCaMP6.
4. Experiments in rats have demonstrated profound effects of the neurosteroid metabolite of progesterone, 3α-OH-DHP, on control of RSNA. However, its effects on SNA in other circulatory beds remain to be evaluated. Even less is known about a potential role for 3α-OH-DHP in modulation of SNA in human pregnancy. Given its effects to dampen responses to stress, evaluation of 3α-OH-DHP levels and MSNA in studies in pregnant women could provide important insights.

ACKNOWLEDGMENTS

We gratefully acknowledge the technical assistance of Jennifer Wang and J. Glenn Phaup. This work was supported in part by NIH grants HL088552 (vlb), HL128181 (vlb), HL088184 (qf), HL142605 (qf), HL36245 (cmh), HL098602 (cmh), and AHA grants 13GRNT16990064 (qf) and 15POST23040042 (zs).

References

Aggarwal PK, Chandel N, Jain V et al. (2012). The relationship between circulating endothelin-1, soluble fms-like tyrosine kinase-1 and soluble endoglin in preeclampsia. J Hum Hypertens 26: 236–241.

Aggarwal S, Sunderland N, Thornton C et al. (2017). A longitudinal analysis of angiotensin II type 1 receptor antibody and angiogenic markers in pregnancy. Am J Obstet Gynecol 216: 170 e171–170 e178.

Allen AM, O'Callaghan EL, Chen D et al. (2009). Central neural regulation of cardiovascular function by angiotensin: a focus on the rostral ventrolateral medulla. Neuroendocrinology 89: 361–369.

Ariza AC, Bobadilla NA, Halhali A (2007). Endothelin 1 and angiotensin II in preeclampsia. Rev Invest Clin 59: 48–56.

Atherton JC, Dark JM, Garland HO et al. (1982). Changes in water and electrolyte balance, plasma volume and composition during pregnancy in the rat. J Physiol 330: 81–93.

Avery ND, Wolfe LA, Amara CE et al. (2001). Effects of human pregnancy on cardiac autonomic function above and below the ventilatory threshold. J Appl Physiol 90: 321–328.

Azar AS, Brooks VL (2011). Impaired baroreflex gain during pregnancy in conscious rats: role of brain insulin. Hypertension 57: 283–288.

Badrov MB, Park SY, Yoo JK et al. (2019). Role of corin in blood pressure regulation in normotensive and hypertensive pregnancy. Hypertension 73: 432–439.

Belelli D, Lambert JJ (2005). Neurosteroids: endogenous regulators of the $GABA_A$ receptor. Nat Rev Neurosci 6: 565–575.

Belelli D, Herd MB, Mitchell EA et al. (2006). Neuroactive steroids and inhibitory neurotransmission: mechanisms of action and physiological relevance. Neuroscience 138: 821–829.

Benaim C, Cocate PG, de Barros EG et al. (2019). Longitudinal association of 25-hydroxyvitamin D with adipokines and markers of glucose metabolism among Brazilian pregnant women. Br J Nutr 121: 42–54.

Bernardi F, Constantino L, Machado R et al. (2008). Plasma nitric oxide, endothelin-1, arginase and superoxide dismutase in pre-eclamptic women. J Obstet Gynaecol Res 34: 957–963.

Biggio G, Cristina Mostallino M, Follesa P et al. (2009). GABA(A) receptor function and gene expression during pregnancy and postpartum. Int Rev Neurobiol 85: 73–94.

Borghi C, Esposti DD, Immordino V et al. (2000). Relationship of systemic hemodynamics, left ventricular structure and function, and plasma natriuretic peptide concentrations during pregnancy complicated by preeclampsia. Am J Obstet Gynecol 183: 140–147.

Brewer J, Liu R, Lu Y et al. (2013). Endothelin-1, oxidative stress, and endogenous angiotensin II: mechanisms of angiotensin II type I receptor autoantibody-enhanced renal and blood pressure response during pregnancy. Hypertension 62: 886–892.

Brooks VL, Quesnell RR, Cumbee SR et al. (1995). Pregnancy attenuates activity of the baroreceptor reflex. Clin Exp Pharmacol Physiol 22: 152–156.

Brooks VL, Kane CM, Van Winkle DM (1997). Altered heart rate baroreflex during pregnancy: role of sympathetic and parasympathetic nervous systems. Am J Physiol Regul Integr Comp Physiol 273: R960–R966.

Brooks VL, Dampney RAL, Heesch CM (2010a). Pregnancy and the endocrine regulation of the baroreceptor reflex. Am J Physiol Regul Integr Comp Physiol 299: R439–R451.

Brooks VL, Mulvaney JM, Azar AS et al. (2010b). Pregnancy impairs baroreflex control of heart rate in rats: role of insulin sensitivity. Am J Physiol Regul Integr Comp Physiol 298: R419–R426.

Brosnihan KB, Neves LA, Joyner J et al. (2003). Enhanced renal immunocytochemical expression of ANG-(1-7) and ACE2 during pregnancy. Hypertension 42: 749–753.

Brown MA, Reiter L, Rodger A et al. (1994). Impaired renin stimulation in pre-eclampsia. Clin Sci (Lond) 86: 575–581.

Brown MA, Wang J, Whitworth JA (1997). The renin–angiotensin–aldosterone system in pre-eclampsia. Clin Exp Hypertens 19: 713–726.

Bruno RM, Sudano I, Ghiadoni L et al. (2011). Interactions between sympathetic nervous system and endogenous endothelin in patients with essential hypertension. Hypertension 57: 79–84.

Brunton PJ, Russell JA, Hirst JJ (2014). Allopregnanolone in the brain: protecting pregnancy and birth outcomes. Prog Neurobiol 113: 106–136.

Cadnapaphornchai MA, Ohara M, Morris Jr KG et al. (2001). Chronic NOS inhibition reverses systemic vasodilation and glomerular hyperfiltration in pregnancy. Am J Physiol Renal Physiol 280: F592–F598.

Callander GE, Thomas WG, Bathgate RAD (2009). Prolonged RXFP1 and RXFP2 signalling can be explained by poor internalization and a lack of beta-arrestin recruitment. Am J Physiol Cell Physiol 296: C1058–C1066.

Card JP, Sved AF (2011). Central autonomic pathways. In: IJ Llewellyn-Smith, AJM Verberne (Eds.), Central regulation of autonomic functions, second edn. Oxford University Press, Inc., New York, NY, pp. 3–22.

Carter JR, Fu Q, Minson CT et al. (2013). Ovarian cycle and sympathoexcitation in premenopausal women. Hypertension 61: 395–399.

Cassaglia PA, Hermes SM, Aicher SA et al. (2011). Insulin acts in the arcuate nucleus to increase lumbar sympathetic nerve activity and baroreflex function in rats. J Physiol 589: 1643–1662.

Cassaglia PA, Shi Z, Li B et al. (2014). Neuropeptide Y acts in the paraventricular nucleus to suppress sympathetic nerve activity and its baroreflex regulation. J Physiol 592: 1655–1675.

Cassaglia PA, Shi Z, Brooks VL (2016). Insulin increases sympathetic nerve activity in part by suppression of tonic inhibitory neuropeptide Y inputs into the paraventricular nucleus in female rats. Am J Physiol Regul Integr Comp Physiol 311: R97–r103.

Catarino C, Santos-Silva A, Belo L et al. (2012). Inflammatory disturbances in preeclampsia: relationship between maternal and umbilical cord blood. J Pregnancy 2012: 684384.

Chapleau MW, Hajduczok G, Abboud FM (1989). Peripheral and central mechanisms of baroreflex resetting. Clin Exp Pharmacol Physiol Suppl 15: 31–43.

Chapman AB, Zamudio S, Woodmansee W et al. (1997). Systemic and renal hemodynamic changes in the luteal phase of the menstrual cycle mimic early pregnancy. Am J Physiol Renal Physiol 273: F777–F782.

Chapman AB, Abraham WT, Zamudio S et al. (1998). Temporal relationships between hormonal and hemodynamic changes in early human pregnancy. Kidney Int 54: 2056–2063.

Charkoudian N (2001). Influences of female reproductive hormones on sympathetic control of the circulation in humans. Clin Auton Res 11: 295–301.

Charkoudian N, Usselman CW, Skow RJ et al. (2017). Muscle sympathetic nerve activity and volume-regulating factors in healthy pregnant and nonpregnant women. Am J Physiol Heart Circ Physiol 313: H782–H787.

Chen G, Jin X, Zhang L et al. (2019). Decreased Ang-(1–7) and downregulated intrarenal RAS may contribute to the direct podocyte injury with proteinuria in preeclampsia. Reprod Sci 26: 1146–1157.

Clow KA, Giraud GD, Ogden BE et al. (2003). Pregnancy alters hemodynamic responses to hemorrhage in conscious rabbits. Am J Physiol Heart Circ Physiol 284: H1110–H1118.

Cohen WR, Galen LH, Vega-Rich M et al. (1988). Cardiac sympathetic activity during rat pregnancy. Metabolism 37: 771–777.

Coldren KM, Brown R, Hasser EM et al. (2015). Relaxin increases sympathetic nerve activity and activates spinally projecting neurons in the paraventricular nucleus of nonpregnant, but not pregnant, rats. Am J Physiol Regul Integr Comp Physiol 309: R1553–R1568.

Concas A, Mostallino MC, Porcu P et al. (1998). Role of brain allopregnanolone in the plasticity of g-aminobutyric acid type A receptor in rat brain during pregnancy and after delivery. Proc Natl Acad Sci U S A 95: 13284–13289.

Conrad KP (2011). Maternal vasodilation in pregnancy: the emerging role of relaxin. Am J Physiol Regul Integr Comp Physiol 301: R267–R275.

Conrad KP, Baker VL (2013). Corpus luteal contribution to maternal pregnancy physiology and outcomes in assisted reproductive technologies. Am J Physiol Regul Integr Comp Physiol 304: R69–R72.

Conrad KP, Russ RD (1992). Augmentation of baroreflex-mediated bradycardia in conscious pregnant rats. Am J Physiol Regul Integr Comp Physiol 262: R472–R477.

Cornelius DC, Cottrell J, Amaral LM et al. (2019). Inflammatory mediators: a causal link to hypertension during preeclampsia. Br J Pharmacol 176: 1914–1921.

Crandall ME, Heesch CM (1990). Baroreflex control of sympathetic outflow in pregnant rats: effects of captopril. Am J Physiol Regul Integr Comp Physiol 258: R1417–R1423.

Cui Y, Wang W, Dong N et al. (2012). Role of corin in trophoblast invasion and uterine spiral artery remodelling in pregnancy. Nature 484: 246–250.

Curtis KS, Cunningham JT, Heesch CM (1999). Fos expression in brain stem nuclei of pregnant rats after hydralazine-induced hypotension. Am J Physiol Regul Integr Comp Physiol 277: R532–R540.

Dampney RAL (1994). Functional organization of central pathways regulating the cardiovascular system. Physiol Rev 74: 323–364.

Dampney RA (2016). Central neural control of the cardiovascular system: current perspectives. Adv Physiol Educ 40: 283–296.

Dampney RA, Michelini LC, Li DP et al. (2018). Regulation of sympathetic vasomotor activity by the hypothalamic paraventricular nucleus in normotensive and hypertensive states. Am J Physiol Heart Circ Physiol 315: H1200–h1214.

Daubert DL, Chung MY, Brooks VL (2007). Insulin resistance and impaired baroreflex gain during pregnancy. Am J Physiol Regul Integr Comp Physiol 292: R2188–R2195.

Davey DA, MacGillivray I (1988). The classification and definition of the hypertensive disorders of pregnancy. Am J Obstet Gynecol 158: 892–898.

de Haas S, Ghossein-Doha C, van Kuijk SM et al. (2017). Physiological adaptation of maternal plasma volume during pregnancy: a systematic review and meta-analysis. Ultrasound Obstet Gynecol 49: 177–187.

Dechend R, Homuth V, Wallukat G et al. (2000). AT(1) receptor agonistic antibodies from preeclamptic patients cause vascular cells to express tissue factor. Circulation 101: 2382–2387.

Dechend R, Viedt C, Muller DN et al. (2003). AT1 receptor agonistic antibodies from preeclamptic patients stimulate NADPH oxidase. Circulation 107: 1632–1639.

Dechend R, Muller DN, Wallukat G et al. (2004). AT1 receptor agonistic antibodies, hypertension, and preeclampsia. Semin Nephrol 24: 571–579.

Dechend R, Muller DN, Wallukat G et al. (2005). Activating auto-antibodies against the AT1 receptor in preeclampsia. Autoimmun Rev 4: 61–65.

Dechend R, Homuth V, Wallukat G et al. (2006). Agonistic antibodies directed at the angiotensin II, AT1 receptor in preeclampsia. J Soc Gynecol Investig 13: 79–86.

Deng Y, Kaufman S (1995). Effect of pregnancy on activation of central pathways following atrial distension. Am J Physiol Regul Integr Comp Physiol 269: R552–R556.

Ekholm EM, Piha SJ, Erkkola RU et al. (1994). Autonomic cardiovascular reflexes in pregnancy. A longitudinal study. Clin Auton Res 4: 161–165.

Fischer T, Schobel HP, Frank H et al. (2004). Pregnancy-induced sympathetic overactivity: a precursor of preeclampsia. Eur J Clin Invest 34: 443–448.

Floras JS (1990). Sympathoinhibitory effects of atrial natriuretic factor in normal humans. Circulation 81: 1860–1873.

Fowler Jr WL, Johnson JA, Kurz KD et al. (1981). Role of the renin–angiotensin system in maintaining arterial pressure in conscious pregnant rats. Endocrinology 109: 290–295.

Franco-Sena AB, de Oliveira LC, de Jesus Pereira Pinto T et al. (2015). Factors associated with prospective leptin concentrations throughout pregnancy in pregestational normal weight, overweight and obese women. Clin Endocrinol (Oxf) 82: 127–135.

Frishman WH, Veresh M, Schlocker SJ et al. (2006). Pathophysiology and medical management of systemic hypertension in preeclampsia. Curr Hypertens Rep 8: 502–511.

Fu Q (2018). Hemodynamic and electrocardiographic aspects of uncomplicated singleton pregnancy. Adv Exp Med Biol 1065: 413–431.

Fu Q, Levine BD (2009). Autonomic circulatory control during pregnancy in humans. Semin Reprod Med 27: 330–337.

Furuhashi N, Kimura H, Nagae H et al. (1994). Brain natriuretic peptide and atrial natriuretic peptide levels in normal pregnancy and preeclampsia. Gynecol Obstet Invest 38: 73–77.

Gennari-Moser C, Khankin EV, Escher G et al. (2013). Vascular endothelial growth factor-A and aldosterone: relevance to normal pregnancy and preeclampsia. Hypertension 61: 1111–1117.

Giannubilo SR, Pasculli A, Tidu E et al. (2017). Relationship between maternal hemodynamics and plasma natriuretic peptide concentrations during pregnancy complicated by preeclampsia and fetal growth restriction. J Perinatol 37: 484–487.

Gill M, Motta-Mejia C, Kandzija N et al. (2019). Placental syncytiotrophoblast-derived extracellular vesicles carry active NEP (Neprilysin) and are increased in preeclampsia. Hypertension 73: 1112–1119.

Gilson GJ, Mosher MD, Conrad KP (1992). Systemic hemodynamics and oxygen transport during pregnancy in chronically instrumented, conscious rats. Am J Physiol Heart Circ Physiol 263: H1911–H1918.

Gilson GJ, Samaan S, Crawford MH et al. (1997). Changes in hemodynamics, ventricular remodeling, and ventricular contractility during normal pregnancy: a longitudinal study. Obstet Gynecol 89: 957–962.

Gironacci MM, Valera MS, Yujnovsky I et al. (2004). Angiotensin-(1–7) inhibitory mechanism of norepinephrine release in hypertensive rats. Hypertension 44: 783–787.

Granger JP, Alexander BT, Bennett WA et al. (2001a). Pathophysiology of pregnancy-induced hypertension. Am J Hypertens 14: 178S–185S.

Granger JP, Alexander BT, Llinas MT et al. (2001b). Pathophysiology of hypertension during preeclampsia linking placental ischemia with endothelial dysfunction. Hypertension 38: 718–722.

Greenwood JP, Stoker JB, Walker JJ et al. (1998). Sympathetic nerve discharge in normal pregnancy and pregnancy-induced hypertension. J Hypertens 16: 617–624.

Greenwood JP, Scott EM, Stoker JB et al. (2001). Sympathetic neural mechanisms in normal and hypertensive pregnancy in humans. Circulation 104: 2200–2204.

Greenwood JP, Scott EM, Walker JJ et al. (2003). The magnitude of sympathetic hyperactivity in pregnancy-induced hypertension and preeclampsia. Am J Hypertens 16: 194–199.

Gu Y, Thompson D, Xu J et al. (2018). Aberrant pro-atrial natriuretic peptide/corin/natriuretic peptide receptor signaling is present in maternal vascular endothelium in preeclampsia. Pregnancy Hypertens 11: 1–6.

Gulati A, Rebello S, Kumar A (1997). Role of sympathetic nervous system in cardiovascular effects of centrally administered endothelin-1 in rats. Am J Physiol Heart Circ Physiol 273: H1177–H1186.

Guyenet PG (2006). The sympathetic control of blood pressure. Nat Rev Neurosci 7: 335–346.

Guyenet PG, Schreihoffer AM, Stornetta RL (2001). Regulation of sympathetic tone and arterial pressure by the rostral ventrolateral medulla after depletion of C1 cells in rats. Ann N Y Acad Sci 940: 259–269.

Hannhart B, Pickett CK, Weil JV et al. (1989). Influence of pregnancy on ventilatory and carotid body neural output responsiveness to hypoxia in cats. J Appl Physiol 67: 797–803.

Hardie L, Trayhurn P, Abramovich D et al. (1997). Circulating leptin in women: a longitudinal study in the menstrual cycle and during pregnancy. Clin Endocrinol (Oxf) 47: 101–106.

Harmon AC, Cornelius DC, Amaral LM et al. (2016). The role of inflammation in the pathology of preeclampsia. Clin Sci (Lond) 130: 409–419.

Haspula D, Clark MA (2018). Neuroinflammation and sympathetic overactivity: mechanisms and implications in hypertension. Auton Neurosci 210: 10–17.

Heesch CM (2011). Neurosteroid modulation of arterial baroreflex function in the rostral ventrolateral medulla. Auton Neurosci 161: 28–33.

Heesch CM, Burcks S (2009). 5a-Reductase mRNA is increased in the rostral ventrolateral medulla (RVLM) of term pregnant rats. FASEB J 23: 958.19.

Heesch CM, Rogers RC (1995). Effects of pregnancy and progesterone metabolites on regulation of sympathetic outflow. Clin Exp Pharmacol Physiol 22: 136–142.

Heesch CM, Zheng H, Foley CM et al. (2009). Nitric oxide synthase activity and expression are decreased in the paraventricular nucleus of pregnant rats. Brain Res 1251: 140–150.

Heesch CM, Kvochina L, Burcks S (2010). Increased sympathoexcitation to angiotensin II (A II) and AT1 receptor mRNA in the paraventricular nucleus (PVN) of pregnant rats. FASEB J 24: 1019.10.

Hernandez-Diaz S, Toh S, Cnattingius S (2009). Risk of preeclampsia in first and subsequent pregnancies: prospective cohort study. BMJ 338: b2255.

Herrera E, Lasuncion MA, Huerta L et al. (2000). Plasma leptin levels in rat mother and offspring during pregnancy and lactation. Biol Neonate 78: 315–320.

Herse F, Verlohren S, Wenzel K et al. (2009). Prevalence of agonistic autoantibodies against the angiotensin II type 1 receptor and soluble fms-like tyrosine kinase 1 in a gestational age-matched case study. Hypertension 53: 393–398.

Hines T (2000). Baroreceptor afferent discharge in the pregnant rat. Am J Physiol Heart Circ Physiol 278: R1433–R1440.

Hines T, Hodgson TM (2000). Pregnancy alters cardiac receptor afferent discharge in rats. Am J Physiol Regul Integr Comp Physiol 278: R149–R156.

Hines T, Mifflin SW (1995). Gestational effects on volume-sensitive cardiopulmonary receptor reflexes in the rat. Am J Physiol Regul Integr Comp Physiol 268: R736–R743.

Hines T, Porter JP (1989). Role for central angiotensin II in control of blood pressure during pregnancy. Am J Physiol Regul Integr Comp Physiol 26: R1457–R1461.

Hines T, Beauchamp D, Rice C (2007). Baroreflex control of sympathetic nerve activity in hypertensive pregnant rats with reduced uterine perfusion. Hypertens Pregnancy 26: 303–314.

Hissen SL, El Sayed K, Macefield VG et al. (2017). Muscle sympathetic nerve activity peaks in the first trimester in healthy pregnancy: a longitudinal case study. Clin Auton Res 27: 401–406.

Humphreys PW, Joels N (1974). The carotid sinus baroreceptor reflex in the pregnant rabbit. J Physiol 239: 89–102.

Humphreys PW, Joels N (1982). The response of the hind-limb vascular bed of the rabbit to sympathetic stimulation and its modification by pregnancy. J Physiol 330: 475–488.

Hunter S, Robson SC (1992). Adaptation of the maternal heart in pregnancy. Br Heart J 68: 540–543.

Intapad S, Warrington JP, Spradley FT et al. (2014). Reduced uterine perfusion pressure induces hypertension in the pregnant mouse. Am J Physiol Regul Integr Comp Physiol 307: R1353–R1357.

Itoh H, Sagawa N, Mori T et al. (1993). Plasma brain natriuretic peptide level in pregnant women with pregnancy-induced hypertension. Obstet Gynecol 82: 71–77.

Jarvis SS, Shibata S, Bivens TB et al. (2012). Sympathetic activation during early pregnancy in humans. J Physiol 590: 3535–3543.

Kaitu'u-Lino TJ, Ye L, Tuohey L et al. (2013). Corin, an enzyme with a putative role in spiral artery remodeling, is up-regulated in late secretory endometrium and first trimester decidua. Hum Reprod 28: 1172–1180.

Kalagiri RR, Carder T, Choudhury S et al. (2016). Inflammation in complicated pregnancy and its outcome. Am J Perinatol 33: 1337–1356.

Kang YM, Zhang ZH, Xue B et al. (2008). Inhibition of brain proinflammatory cytokine synthesis reduces hypothalamic excitation in rats with ischemia-induced heart failure. Am J Physiol Heart Circ Physiol 295: H227–H236.

Kar S, Gao L, Belatti DA et al. (2011). Central angiotensin (1–7) enhances baroreflex gain in conscious rabbits with heart failure. Hypertension 58: 627–634.

Kaufman S, Deng Y (1993). Renal response to atrial stretch during pregnancy in conscious rats. Am J Physiol Regul Integr Comp Physiol 265: R902–R906.

Khalil A, Maiz N, Garcia-Mandujano R et al. (2015). Longitudinal changes in maternal corin and mid-regional proatrial natriuretic peptide in women at risk of preeclampsia. Ultrasound Obstet Gynecol 45: 190–198.

Khanna M, Qin K-N, Cheng K-C (1995). Distribution of 3 alpha-hydroxysteroid dehydrogenase in rat brain and molecular cloning of multiple cDNAs encoding structurally related proteins in humans. J Steroid Biochem Mol Biol 53: 41–46.

Kirwan JP, Hauguel-De Mouzon S, Lepercq J et al. (2002). TNF-alpha is a predictor of insulin resistance in human pregnancy. Diabetes 51: 2207–2213.

Kolovetsiou-Kreiner V, Moertl MG, Papousek I et al. (2018). Maternal cardiovascular and endothelial function from first trimester to postpartum. PLoS One 13: e0197748.

Kontak AC, Wang Z, Arbique D et al. (2010). Reversible sympathetic overactivity in hypertensive patients with primary aldosteronism. J Clin Endocrinol Metab 95: 4756–4761.

Kucuk M, Sezer SD, Yenisey C et al. (2012). Comparison of interleukin-6 levels in maternal and umbilical cord blood in early- and late-onset preeclampsia. Gynecol Endocrinol 28: 640–643.

Kuo CD, Chen GY, Yang MJ et al. (2000). Biphasic changes in autonomic nervous activity during pregnancy. Br J Anaesth 84: 323–329.

Kvochina L, Hasser EM, Heesch CM (2007). Pregnancy increases baroreflex independent GABAergic inhibition of the RVLM in rats. Am J Physiol Regul Integr Comp Physiol 293: R2295–R2305.

Kvochina L, Hasser EM, Heesch CM (2009a). Pregnancy decreases GABAergic inhibition of the hypothalamic paraventricular nucleus. Physiol Behav 97: 171–179.

Kvochina L, Phaup JG, Hasser EM et al. (2009b). AT1 receptors in the paraventricular nucleus (PVN) are required for sympathoexcitation due to angiotensin II in the subfornical organ of nonpregnant and pregnant rats. FASEB J 23: 792.2.

Lain KY, Catalano PM (2007). Metabolic changes in pregnancy. Clin Obstet Gynecol 50: 938–948.

Laiprasert JD, Rogers RC, Heesch CM (1998). Neurosteroid modulation of arterial baroreflex-sensitive neurons in rat rostral ventrolateral medulla. Am J Physiol Regul Integr Comp Physiol 274: R903–R911.

Laiprasert JD, Hamlin RL, Heesch CM (2001). Afferent baroreceptor discharge in pregnant rats. Am J Physiol Heart Circ Physiol 281: H2456–H2462.

LaMarca B, Parrish M, Ray LF et al. (2009). Hypertension in response to autoantibodies to the angiotensin II type I receptor (AT1-AA) in pregnant rats: role of endothelin-1. Hypertension 54: 905–909.

Leduc L, Wasserstrum N, Spillman T et al. (1991). Baroreflex function in normal pregnancy. Am J Obstet Gynecol 165: 886–890.

Lei J, Li Y, Zhang S et al. (2016). The prognostic role of angiotensin II type 1 receptor autoantibody in non-gravid hypertension and pre-eclampsia: a meta-analysis and our studies. Medicine (Baltimore) 95: e3494.

Leo CH, Jelinic M, Ng HH et al. (2017). Vascular actions of relaxin: nitric oxide and beyond. Br J Pharmacol 174: 1002–1014.

Leturque A, Burnol AF, Ferre P et al. (1984). Pregnancy-induced insulin resistance in the rat: assessment by glucose clamp technique. Am J Physiol Regul Integr Comp Physiol 246: E25–E31.

Li X, Bertics PJ, Karavalas HJ (1997). Regional distribution of cytosolic and particulate 5alpha-dihydroprogesterone 3alpha-hydroxysteroid oxidoreductases in female rat brain. J Steroid Biochem Mol Biol 60: 311–318.

Li B, Shi Z, Cassaglia PA et al. (2013). Leptin acts in the forebrain to differentially influence baroreflex control of lumbar, renal, and splanchnic sympathetic nerve activity and heart rate. Hypertension 61: 812–819.

Li H, Zhang Y, Wu Q (2017). Role of corin in the regulation of blood pressure. Curr Opin Nephrol Hypertens 26: 67–73.

Liu Y, Hu J, Yu Q et al. (2015). Increased serum soluble corin in mid pregnancy is associated with hypertensive disorders of pregnancy. J Womens Health (Larchmt) 24: 572–577.

Lovick TA, Malpas S, Mahony MT (1993). Renal vasodilatation in response to acute volume load is attenuated following lesions of parvocellular neurones in the paraventricular nucleus in rats. J Auton Nerv Syst 43: 247–256.

Luchner A, Schunkert H (2004). Interactions between the sympathetic nervous system and the cardiac natriuretic peptide system. Cardiovasc Res 63: 443–449.

Lucini D, Strappazzon P, Vecchia LD et al. (1999). Cardiac autonomic adjustments to normal human pregnancy: insight from spectral analysis of R-R interval and systolic arterial pressure variability. J Hypertens 17: 1899–1904.

Lumbers ER, Yu ZY (1999). A method for determining baroreflex-mediated sympathetic and parasympathetic control of the heart in pregnant and non-pregnant sheep. J Physiol 515: 555–566.

Lyall F, Greer IA (1996). The vascular endothelium in normal pregnancy and pre-eclampsia. Rev Reprod 1: 107–116.

Mabie WC, DiSessa TG, Crocker LG et al. (1994). A longitudinal study of cardiac output in normal human pregnancy. Am J Obstet Gynecol 170: 849–856.

MacKenzie G, Maguire J (2014). The role of ovarian hormone-derived neurosteroids on the regulation of GABAA receptors in affective disorders. Psychopharmacology (Berl) 231: 3333–3342.

Maguire J, Mody I (2009). Steroid hormone fluctuations and GABA(A)R plasticity. Psychoneuroendocrinology 34: S84–S90.

Masilamani S, Heesch CM (1997). Effects of pregnancy and progesterone metabolites on arterial baroreflex in conscious rats. Am J Physiol Regul Integr Comp Physiol 272: R924–R934.

Matsukawa T, Gotoh E, Minamisawa K et al. (1991). Effects of intravenous infusions of angiotensin II on muscle sympathetic nerve activity in humans. Am J Physiol 261: R690–R696.

McKinley MJ, Mathai ML, McAllen RM et al. (2004). Vasopressin secretion: osmotic and hormonal regulation by the lamina terminalis. J Neuroendocrinol 16: 340–347.

Meah VL, Cockcroft JR, Backx K et al. (2016). Cardiac output and related haemodynamics during pregnancy: a series of meta-analyses. Heart 102: 518–526.

Merrill DC, Karoly M, Chen K et al. (2002). Angiotensin-(1–7) in normal and preeclamptic pregnancy. Endocrine 18: 239–245.

Mitev YA, Darwish M, Wolf SS et al. (2003). Gender differences in the regulation of 3 alpha-hydroxysteroid dehydrogenase in rat brain and sensitivity to neurosteroid-mediated stress protection. Neuroscience 120: 541–549.

Miyazaki J, Nishizawa H, Kambayashi A et al. (2016). Increased levels of soluble corin in pre-eclampsia and fetal growth restriction. Placenta 48: 20–25.

Mody I (2005). Aspects of the homeostatic plasticity of $GABA_A$ receptor-mediated inhibition. J Physiol 562: 37–46.

Moore LG, Brodeur P, Chumbe O et al. (1986). Maternal hypoxic ventilatory response, ventilation, and infant birth weight at 4,300m. J Appl Physiol 60: 1401–1406.

Mortensen LH (1999). Endothelin and the central and peripheral nervous systems: a decade of endothelin research. Clin Exp Pharmacol Physiol 26: 980–984.

Mosqueda-Garcia R, Inagami T, Appalsamy M et al. (1993). Endothelin as a neuropeptide. Cardiovascular effects in the brainstem of normotensive rats. Circ Res 72: 20–35.

Mueller PJ, Clifford PS, Crandall CG et al. (2017). Integration of central and peripheral regulation of the circulation during exercise: acute and chronic adaptations. Compr Physiol 8: 103–151.

Munoz C, Lopez-Luna P, Herrera E (1995). Glucose and insulin tolerance tests in the rat on different days of gestation. Biol Neonate 68: 282–291.

Nadel AS, Ballermann BJ, Andersen S et al. (1988). Interrelationships among arterial peptides, renin, and blood volume in pregnant rats. Am J Physiol Regul Integr Comp Physiol 254: R793–R800.

Nakamura K, Sasaki S, Moriguchi J et al. (1999). Central effects of endothelin and its antagonists on sympathetic and cardiovascular regulation in SHR-SP. J Cardiovasc Pharmacol 33: 876–882.

Nakamura T, Kawabe K, Sapru HN (2008). Cold pressor test in the rat: medullary and spinal pathways and neurotransmitters. Am J Physiol Heart Circ Physiol 295: H1780–H1787.

Nonogaki K (2000). New insights into sympathetic regulation of glucose and fat metabolism. Diabetologia 43: 533–549.

Noris M, Perico N, Remuzzi G (2005). Mechanisms of disease: pre-eclampsia. Nat Clin Pract Nephrol 1: 98–114. quiz 120.

Ogoh S, Fisher JP, Raven PB et al. (2007). Arterial baroreflex control of muscle sympathetic nerve activity in the transition from rest to steady-state dynamic exercise in humans. Am J Physiol Heart Circ Physiol 293: H2202–H2209.

O'Hagan KP, Casey SM (1998). Arterial baroreflex during pregnancy and renal sympathetic activity during parturition in rabbits. Am J Physiol Heart Circ Physiol 274: H1635–H1642.

O'Hagan K, Skogg K, Stevenson J (2001). AT_1 receptor block does not affect arterial baroreflex during pregnancy in rabbits. Am J Physiol Heart Circ Physiol 280: H1996–H2005.

Okada Y, Best SA, Jarvis SS et al. (2015). Asian women have attenuated sympathetic activation but enhanced renal–adrenal responses during pregnancy compared to Caucasian women. J Physiol 593: 1159–1168.

Page MC, Cassaglia PA, Brooks VL (2011). GABA in the paraventricular nucleus tonically suppresses baroreflex function: alterations during pregnancy. Am J Physiol Regul Integr Comp Physiol 300: R1452–R1458.

Panchal S, Arria AM, Labhsetwar SA (2001). Maternal mortality during hospital admission for delivery: a retrospective analysis using a state-maintained database. Anesth Analg 93: 134–141.

Parrish MR, Wallace K, Tam Tam KB et al. (2011). Hypertension in response to AT1-AA: role of reactive oxygen species in pregnancy-induced hypertension. Am J Hypertens 24: 835–840.

Pennell KD, Woodin MA, Pennell PB (2015). Quantification of neurosteroids during pregnancy using selective ion monitoring mass spectrometry. Steroids 95: 24–31.

Perichart-Perera O, Munoz-Manrique C, Reyes-Lopez A et al. (2017). Metabolic markers during pregnancy and their association with maternal and newborn weight status. PLoS One 12: e0180874.

Phaup JG, Hasser EM, Heesch CM (2013). The neurosteroid metabolite of progesterone, 3α-OH-dihydroprogesterone (3α-OH-DHP), is required for attenuated baroreflex mediated sympathoexcitation in pregnancy. FASEB J 27: 1118.38.

Powe CE, Levine RJ, Karumanchi SA (2011). Preeclampsia, a disease of the maternal endothelium: the role of antiangiogenic factors and implications for later cardiovascular disease. Circulation 123: 2856–2869.

Pricher MP, Freeman KL, Brooks VL (2008). Insulin in the brain increases gain of baroreflex control of heart rate and lumbar sympathetic nerve activity. Hypertension 51: 514–520.

Puzserova A, Bernatova I (2016). Blood pressure regulation in stress: focus on nitric oxide-dependent mechanisms. Physiol Res 65: S309–S342.

Rang S, Wolf H, Montfrans GA et al. (2002). Non-invasive assessment of autonomic cardiovascular control in normal human pregnancy and pregnancy- associated hypertensive disorders: a review. J Hypertens 20: 2111–2119.

Reyes LM, Usselman CW, Davenport MH et al. (2018a). Sympathetic nervous system regulation in human normotensive and hypertensive pregnancies. Hypertension 71: 793–803.

Reyes LM, Usselman CW, Skow RJ et al. (2018b). Sympathetic neurovascular regulation during pregnancy: a longitudinal case series study. Exp Physiol 103: 318–323.

Ringholm L, Pedersen-Bjergaard U, Thorsteinsson B et al. (2011). Atrial natriuretic peptide (ANP) in early pregnancy is associated with development of preeclampsia in type 1 diabetes. Diabetes Res Clin Pract 93: e106–e109.

Roberts JM, Redman CW (1993). Pre-eclampsia: more than pregnancy-induced hypertension. Lancet 341: 1447–1451.

Roberts JM, Pearson G, Cutler J et al. (2003). Summary of the NHLBI working group on research on hypertension during pregnancy. Hypertension 41: 437–445.

Rossi NF, Maliszewska-Scislo M (2008). Role of paraventricular nucleus vasopressin V1A receptors in the response to endothelin 1 activation of the subfornical organ in the rat. J Physiol Pharmacol 59: 47–59.

Rossi G, Sherwin RS, Penzias AS et al. (1993). Temporal changes in insulin resistance and secretion in 24-h-fasted conscious pregnant rats. Am J Physiol Endocrinol Metab 265: E845–E851.

Rossi NF, O'Leary DS, Chen H (1997). Mechanisms of centrally administered ET-1-induced increases in systemic arterial pressure and AVP secretion. Am J Physiol Endocrinol Metab 272: E126–E132.

Rossi NF, Maliszewska-Scislo M, Chen H (2008). Central endothelin: effects on vasopressin and the arterial baroreflex in doxorubicin heart failure rats. Can J Physiol Pharmacol 86: 343–352.

Ruzicka M, Floras JS, McReynolds AJ et al. (2013). Do high doses of AT(1)-receptor blockers attenuate central sympathetic outflow in humans with chronic heart failure? Clin Sci (Lond) 124: 589–595.

Saleh L, Verdonk K, Visser W et al. (2016). The emerging role of endothelin-1 in the pathogenesis of pre-eclampsia. Ther Adv Cardiovasc Dis 10: 282–293.

Sanna E, Mostallino MC, Murru L et al. (2009). Changes in expression and function of extrasynaptic GABAA receptors in the rat hippocampus during pregnancy and after delivery. J Neurosci 29: 1755–1765.

Sattar N, Greer IA, Pirwani I et al. (1998). Leptin levels in pregnancy: marker for fat accumulation and mobilization? Acta Obstet Gynecol Scand 77: 278–283.
Schmidt SML, Usselman CW, Martinek E et al. (2018). Activity of muscle sympathetic neurons during normotensive pregnancy. Am J Physiol Regul Integr Comp Physiol 314: R153–R160.
Schobel HP, Fischer T, Heuszer K et al. (1996). Preeclampsia—a state of sympathetic overactivity. N Engl J Med 335: 1480–1485.
Schreihofer AM, Guyenet PC (2000). Sympathetic reflexes after depletion of bulbospinal catecholaminergic neurons with anti-DbH-saporin. Am J Physiol Regul Integr Comp Physiol 279: R729–R742.
Schreihofer AM, Guyenet PG (2002). The baroreflex and beyond: control of sympathetic vasomotor tone by GABAergic neurons in the ventrolateral medulla. Clin Exp Pharmacol Physiol 29: 514–521.
Seeber RM, Smith JT, Waddell BJ (2002). Plasma leptin-binding activity and hypothalamic leptin receptor expression during pregnancy and lactation in the rat. Biol Reprod 66: 1762–1767.
Sherwood OD (2004). Relaxin's physiological roles and other diverse actions. Endocr Rev 25: 205–234.
Sherwood OD, Crnekovic VE, Gordon WL et al. (1980). Radioimmunoassay of relaxin throughout pregnancy and during parturition in the rat. Endocrinology 107: 691–698.
Shi Z, Brooks VL (2015). Leptin differentially increases sympathetic nerve activity and its baroreflex regulation in female rats: role of oestrogen. J Physiol 593: 1633–1647.
Shi Z, Cassaglia PA, Gotthardt LC et al. (2015a). Hypothalamic paraventricular and arcuate nuclei contribute to elevated sympathetic nerve activity in pregnant rats: roles of neuropeptide Y and alpha-melanocyte-stimulating hormone. Hypertension 66: 1191–1198.
Shi Z, Li B, Brooks VL (2015b). Role of the paraventricular nucleus of the hypothalamus in the sympathoexcitatory effects of leptin. Hypertension 66: 1034–1041.
Shi Z, Madden CJ, Brooks VL (2017). Arcuate neuropeptide Y inhibits sympathetic nerve activity via multiple neuropathways. J Clin Invest 127: 2868–2880.
Shi Z, Hansen KM, Bullock KM et al. (2019). Resistance to the sympathoexcitatory effects of insulin and leptin in late pregnant rats. J Physiol 597: 4087–4100.
Sibai B, Dekker G, Kupferminc M (2005). Pre-eclampsia. Lancet 365: 785–799.
Siddiqui AH, Irani RA, Blackwell SC et al. (2010). Angiotensin receptor agonistic autoantibody is highly prevalent in preeclampsia: correlation with disease severity. Hypertension 55: 386–393.
Silver HM, Tahvanainen KU, Kuusela TA et al. (2001). Comparison of vagal baroreflex function in nonpregnant women and in women with normal pregnancy, preeclampsia, or gestational hypertension. Am J Obstet Gynecol 184: 1189–1195.
Silvertown J, Fraser R, Poterski R et al. (2005). Central effects of long-term relaxin expression in the rat. Ann N Y Acad Sci 1041: 216–222.
Slangen BF, Out IC, Verkeste CM et al. (1996). Hemodynamic changes in early pregnancy in chronically instrumented, conscious rats. Am J Physiol Heart Circ Physiol 270: H1779–H1784.
Southcombe JH, Redman CW, Sargent IL et al. (2015). Interleukin-1 family cytokines and their regulatory proteins in normal pregnancy and pre-eclampsia. Clin Exp Immunol 181: 480–490.
Spradley FT (2019). Sympathetic nervous system control of vascular function and blood pressure during pregnancy and preeclampsia. J Hypertens 37: 476–487.
Spradley FT, Ge Y, Haynes BP et al. (2018). Adrenergic receptor blockade attenuates placental ischemia-induced hypertension. Physiol Rep 6: e13814.
Stanley K, Fraser R, Bruce C (1998). Physiological changes in insulin resistance in human pregnancy: longitudinal study with the hyperinsulinaemic euglycaemic clamp technique. Br J Obstet Gynaecol 105: 756–759.
Steegers EA, von Dadelszen P, Duvekot JJ et al. (2010). Pre-eclampsia. Lancet 376: 631–644.
Stein RD, Stephenson RB, Weaver LC (1984). Central actions of angiotensin II oppose baroreceptor-induced sympathoinhibition. Am J Physiol Regul Integr Comp Physiol 246: R13–R19.
Stepan H, Faber R, Walther D et al. (1999). C-type natriuretic peptide levels in women with gestational hypertension and preeclampsia. Obstet Gynecol 93: 199–202.
Stoffel-Wagner B (2001). Neurosteroid metabolism in the human brain. Eur J Endocrinol 145: 669–679.
Storey E, Kaufman S (2004). Effect of pregnancy and 5alpha-pregnan-3alpha-ol-20-one on atrial receptor afferent discharge in rats. Am J Physiol Regul Integr Comp Physiol 287: R1427–R1433.
Tihtonen KM, Koobi T, Vuolteenaho O et al. (2007). Natriuretic peptides and hemodynamics in preeclampsia. Am J Obstet Gynecol 196: e321–e327.
Tkachenko O, Shchekochikhin D, Schrier RW (2014). Hormones and hemodynamics in pregnancy. Int J Endocrinol Metab 12: e14098.
Tranquilli AL, Dekker G, Magee L et al. (2014). The classification, diagnosis and management of the hypertensive disorders of pregnancy: a revised statement from the ISSHP. Pregnancy Hypertens 4: 97–104.
Trujillo ML, Spuch C, Carro E et al. (2011). Hyperphagia and central mechanisms for leptin resistance during pregnancy. Endocrinology 152: 1355–1365.
Usselman CW, Skow RJ, Matenchuk BA et al. (2015a). Sympathetic baroreflex gain in normotensive pregnant women. J Appl Physiol (1985) 119: 468–474.
Usselman CW, Wakefield PK, Skow RJ et al. (2015b). Regulation of sympathetic nerve activity during the cold pressor test in normotensive pregnant and nonpregnant women. Hypertension 66: 858–864.
Valdes G, Neves LA, Anton L et al. (2006). Distribution of angiotensin-(1–7) and ACE2 in human placentas of normal and pathological pregnancies. Placenta 27: 200–207.
Verdonk K, Visser W, Van Den Meiracker AH et al. (2014). The renin–angiotensin–aldosterone system in pre-eclampsia:

the delicate balance between good and bad. Clin Sci (Lond) 126: 537–544.

Verdonk K, Saleh L, Lankhorst S et al. (2015). Association studies suggest a key role for endothelin-1 in the pathogenesis of preeclampsia and the accompanying renin–angiotensin–aldosterone system suppression. Hypertension 65: 1316–1323.

Volpe M (2014). Natriuretic peptides and cardio-renal disease. Int J Cardiol 176: 630–639.

Volpe M, Cuocolo A, Vecchione F et al. (1987). Vagal mediation of the effects of atrial natriuretic factor on blood pressure and arterial baroreflexes in the rabbit. Circ Res 60: 747–755.

Volpe M, De Luca N, Bigazzi MC et al. (1988). Atrial natriuretic factor potentiates forearm reflex vasoconstriction induced by cardiopulmonary receptor deactivation in man. Circulation 77: 849–855.

Volpe M, Rubattu S, Burnett Jr J (2014). Natriuretic peptides in cardiovascular diseases: current use and perspectives. Eur Heart J 35: 419–425.

Voss A, Malberg H, Schumann A et al. (2000). Baroreflex sensitivity, heart rate, and blood pressure variability in normal pregnancy. Am J Hypertens 13: 1218–1225.

Vricella LK (2017). Emerging understanding and measurement of plasma volume expansion in pregnancy. Am J Clin Nutr 106: 1620s–1625s.

Wallukat G, Homuth V, Fischer T et al. (1999). Patients with preeclampsia develop agonistic autoantibodies against the angiotensin AT1 receptor. J Clin Invest 103: 945–952.

Wallukat G, Neichel D, Nissen E et al. (2003). Agonistic autoantibodies directed against the angiotensin II AT1 receptor in patients with preeclampsia. Can J Physiol Pharmacol 81: 79–83.

Ward KR, Bardgett JF, Wolfgang L et al. (2011). Sympathetic response to insulin is mediated by melanocortin 3/4 receptors in the hypothalamic paraventricular nucleus. Hypertension 57: 435–441.

Wei SG, Yu Y, Zhang ZH et al. (2015). Proinflammatory cytokines upregulate sympathoexcitatory mechanisms in the subfornical organ of the rat. Hypertension 65: 1126–1133.

Wei SG, Yu Y, Felder RB (2018). Blood-borne interleukin-1beta acts on the subfornical organ to upregulate the sympathoexcitatory milieu of the hypothalamic paraventricular nucleus. Am J Physiol Regul Integr Comp Physiol 314: R447–R458.

Yamaleyeva LM, Merrill DC, Ebert TJ et al. (2014). Hemodynamic responses to angiotensin-(1–7) in women in their third trimester of pregnancy. Hypertens Pregnancy 33: 375–388.

Yan W, Sheng N, Seto M et al. (1999). Corin, a mosaic transmembrane serine protease encoded by a novel cDNA from human heart. J Biol Chem 274: 14926–14935.

Yan JQ, Huang F, Hao F et al. (2017). Oxidative stress in the rostral ventrolateral medulla contributes to cardiovascular regulation in preeclampsia. Front Physiol 8: 772.

Yang Z, Coote JH (2003). Role of GABA and NO in the paraventricular nucleus-mediated reflex inhibition of renal sympathetic nerve activity following stimulation of right arterial receptors in the rat. Exp Physiol 88: 335–342.

Young CN, Deo SH, Chaudhary K et al. (2010). Insulin enhances the gain of arterial baroreflex control of muscle sympathetic nerve activity in humans. J Physiol 588: 3593–3603.

Yu Y, Wei SG, Weiss RM et al. (2017). TNF-alpha receptor 1 knockdown in the subfornical organ ameliorates sympathetic excitation and cardiac hemodynamics in heart failure rats. Am J Physiol Heart Circ Physiol 313: H744–H756.

Yu Y, Wei SG, Weiss RM et al. (2018). Angiotensin II type 1a receptors in the subfornical organ modulate neuroinflammation in the hypothalamic paraventricular nucleus in heart failure rats. Neuroscience 381: 46–58.

Yu XJ, Miao YW, Li HB et al. (2019). Blockade of endogenous angiotensin-(1–7) in hypothalamic paraventricular nucleus attenuates high salt-induced sympathoexcitation and hypertension. Neurosci Bull 35: 47–56.

Zak P, Soucek M (2019). Correlation of tumor necrosis factor alpha, interleukin 6 and interleukin 10 with blood pressure, risk of preeclampsia and low birth weight in gestational diabetes. Physiol Res 68: 395–408.

Zhang K, Patel KP (1998). Effect of nitric oxide within the paraventricular nucleus on renal sympathetic nerve discharge: role of GABA. Am J Physiol Regul Integr Comp Physiol 275: R728–R734.

Zhou CC, Ahmad S, Mi T et al. (2008). Autoantibody from women with preeclampsia induces soluble Fms-like tyrosine kinase-1 production via angiotensin type 1 receptor and calcineurin/nuclear factor of activated T-cells signaling. Hypertension 51: 1010–1019.

Handbook of Clinical Neurology, Vol. 171 (3rd series)
Neurology and Pregnancy: Pathophysiology and Patient Care
E.A.P. Steegers, M.J. Cipolla and E.C. Miller, Editors
https://doi.org/10.1016/B978-0-444-64239-4.00004-7

Chapter 4

Physiology of the cerebrovascular adaptation to pregnancy

ABBIE C. JOHNSON*

Department of Neurological Sciences, University of Vermont Larner College of Medicine, Burlington, VT, United States

Abstract

The adaptation of the cerebral circulation to pregnancy is unique compared with other organs and circulatory systems, because the brain requires relatively constant blood flow and water and solute composition to maintain homeostasis. Thus, a major adaptation of the maternal cerebrovasculature to pregnancy is to maintain normalcy in the face of expanded plasma volume, increased cardiac output, and high levels of permeability factors. In this chapter, the effect of pregnancy on critical functions of the cerebral circulation is discussed, including changes occurring at the endothelium and blood–brain barrier (BBB), which protect the maternal brain from changes in BBB permeability. Further, pregnancy-induced changes in the structure and function of cerebral arteries, arterioles, and veins will be discussed as they relate to cerebral vascular resistance, hemodynamics, and cerebral blood flow autoregulation.

INTRODUCTION

The brain is an organ of high metabolic demand that consumes ~20% of the body's oxygen at rest, despite comprising only 2% of body weight (Siegel, 1999). Importantly, the brain has a relatively narrow capacity to tolerate changes in ion and water homeostasis, and blood flow (Siegel, 1999). It is also unique in that it is enclosed within a rigid skull (e.g., Monro-Kellie doctrine), and therefore increased vascular permeability or volume could result in detrimentally increased intracranial pressure that can cause serious neurologic symptoms, brain herniation, and even death. Thus, there is a need to maintain tight control of cerebral blood flow (CBF) and water flux in the face of the 40%–50% increase in plasma volume and cardiac output occurring during pregnancy, as well as a decline in systemic vascular resistance that occurs to maintain a healthy blood pressure (Clapp 3rd and Capeless, 1997). In contrast to other organs outside the central nervous system that undergo substantial increases in both perfusion and transvascular filtration during pregnancy, including the uterus, kidney, and heart, the cerebral circulation must resist these adaptations to counterbalance global hemodynamic changes and maintain the delicate microenvironment of the brain. Thus, the adaptation of the brain and cerebral circulation to pregnancy appears to be to maintain normalcy despite substantial hormonal and cardiovascular changes in almost every other organ (Johnson and Cipolla, 2015). Throughout this chapter, the effect of pregnancy on several aspects of cerebral circulation will be discussed, including the cerebral endothelium and blood–brain barrier (BBB), the structure and function of the cerebrovasculature, hemodynamics, and CBF autoregulation. Further, in the context of these cerebrovascular adaptations, the risk for neurologic complications that can arise upon failure of these adaptations to occur, such as during preeclampsia, will be considered.

VASOMOTOR RESPONSES TO CIRCULATING FACTORS

One of the most important adaptations of the cerebral circulation during pregnancy is counteracting the effects

*Correspondence to: Abbie C. Johnson, Department of Neurological Sciences, University of Vermont Larner College of Medicine, 149 Beaumont Ave., HSRF 416 Burlington, VT, United States. Tel: +1-802-656-4231, Fax: + 1-802-656-5678, E-mail: abbie.johnson@med.uvm.edu

of circulating vasoactive factors. During this period, large amounts of hormones are secreted from the placenta, ovaries, and brain into the maternal circulation, including pro- and anti-inflammatory cytokines, chemokines, and steroids and growth factors (Aagaard-Tillery et al., 2006; Szarka et al., 2010). These factors are critical for the development and survival of the fetus and adaptation of other organ systems needed for a successful pregnancy. Cerebral arteries uniquely adapt during pregnancy to oppose an increase in circulating vasoconstrictors present late in gestation. Exposure to plasma from pregnant women causes vasoconstriction of posterior cerebral arteries in nonpregnant rats; however, this vasoconstrictive effect is absent in arteries of pregnant rats (Amburgey et al., 2010b). This suggests that the cerebral circulation adapts to combat vasoconstrictors present late in gestation. This may be due to either the development of resistance to vasoconstrictors circulating during pregnancy or increased sensitivity to vasodilators also present during pregnancy. Interestingly, this finding was specific to the cerebral vasculature, as the effect of plasma from pregnant women was not seen in mesenteric arteries (Amburgey et al., 2010b). Thus, the adaptation of the cerebral circulation during pregnancy is unique compared with that of other organ systems. The exact mechanism by which the cerebral vasculature resists the vasoconstrictive effect of circulating factors in pregnancy remains unclear, but it may involve receptor downregulation or changes in the influence of the endothelium on vascular tone in response to plasma (Duckles and Krause, 2007; Chan et al., 2010). Regardless, this adaptation of the cerebral circulation likely occurs to prevent the cerebrovasculature from constricting in response to circulating factors and may help maintain physiologic levels of cerebrovascular resistance and blood flow to the brain during pregnancy.

THE CEREBRAL ENDOTHELIUM AND BBB

Vasomotor responses are not the only feature of the cerebral circulation that adapts to pregnancy. The cerebral endothelium that forms the BBB is a complex interface between systemically circulating factors and the delicate microenvironment of the brain. These endothelial cells contain specialized high electrical resistance tight junctions and lack fenestrations (Ueno, 2007; Zlokovic, 2008). BBB tight junctions limit the passage of blood constituents into the brain parenchyma by preventing paracellular transport and are highly protective of the brain milieu (Wahl et al., 1988; Rubin and Staddon, 1999). Cerebral endothelial cells also have a low rate of pinocytosis, which limits the amount of transcellular transport, reinforcing the overall function of the BBB (Reese and Karnovsky, 1967; Brightman and Reese, 1969; Fenstermacher et al., 1988). Pregnancy does not affect mRNA expression of the primary tight junction proteins of the BBB, including claudin-1, claudin-5, occludin, and zona occludens-1, as mRNA expression is similar to the nonpregnant state (Cipolla et al., 2011). In addition, paracellular and transcellular transport at the BBB remain unchanged during normal pregnancy, as BBB permeability to solutes does not increase (Cipolla et al., 2011). Further, hydraulic conductivity, an important parameter that determines water movement through the vessel wall in response to hydrostatic pressure, is normally very low in cerebral endothelial cells due to the high electrical resistance tight junctions and low pinocytotic activity (Roberts et al., 2009). Similar to paracellular and transcellular permeability, hydraulic conductivity is unchanged during pregnancy (Cipolla et al., 2012b), but it increases in response to plasma from preeclamptic women due to elevated circulating oxidized low-density lipoprotein (Amburgey et al., 2010a; Schreurs et al., 2013). Increased hydraulic conductivity during preeclampsia could promote vasogenic edema formation and neurologic symptoms such as seizures, as increased BBB permeability has been linked to several pathologic states, including eclampsia and epilepsy (Oby and Janigro, 2006; Marchi et al., 2007, 2011).

Pregnancy is a state marked by increased circulating permeability factors, including several that are known to promote BBB permeability (Brown et al., 1997; Evans et al., 1997), yet it is remarkable that no such changes in permeability have been measured. For example, vascular endothelial growth factor (VEGF), a cytokine originally named vascular permeability factor, is secreted by the placenta and elevated during pregnancy in the uteroplacental unit and in the maternal circulation (Brown et al., 1997; Evans et al., 1997; Charnock-Jones et al., 2004; Amburgey et al., 2010a). VEGF interacts with its receptors VEGFR1 (or FMS-like tyrosine kinase receptor 1, Flt-1) and VEGFR2 (or fetal liver kinase 1, Flk-1) located on the vascular endothelium to initiate several critical physiologic processes involved in angiogenesis, vascular growth, and endothelial cell survival (Dvorak, 2002; Shibuya, 2013). In addition to VEGF, placental growth factor (PlGF), a member of the VEGF family, is also elevated during pregnancy and contributes to angiogenesis in the uterus and placenta (Krauss et al., 2004). Most notably, VEGF and PlGF are potent vasodilators that increase peripheral microvascular permeability to serum proteins and macromolecules, considered a primary step in preparation for angiogenesis (Feng et al., 1996; Dobrogowska et al., 1998; Dvorak, 2002; Oura et al., 2003). Further, both VEGF and PlGF have been shown to increase BBB permeability (Schreurs et al., 2012). Interestingly, despite elevated circulating VEGF and PlGF during pregnancy, exposure of cerebral vessels to pregnant plasma or serum did not increase BBB permeability (Cipolla et al., 2012b; Schreurs et al., 2012). Moreover, VEGF receptor expression in cerebral arteries does not appear to change during pregnancy, suggesting the lack of effect of VEGF and PlGF is not due to downregulation of VEGFR1/2 or neuropilin (Schreurs et al., 2012). In fact, plasma from late-pregnant rats prevented VEGF-induced increases in BBB permeability (Schreurs et al., 2012), likely due to increased levels of soluble (s)Flt-1. The selective binding of VEGF and PlGF to sFlt-1 is important for regulating their bioavailability, thus limiting the permeability-promoting effects at the BBB during pregnancy (Schreurs et al., 2012). The prevention of circulating permeability factors from increasing BBB permeability is an important adaptation during pregnancy to help maintain brain homeostasis.

Although paracellular and transcellular permeability of the BBB appears to remain intact during pregnancy, efflux transporters present at the BBB are an important regulatory mechanism controlling passage of serum factors into the brain, and this appears to be gestationally regulated. Specifically, p-glycoprotein (Pgp) is an important efflux transporter at the BBB that extricates steroids, cytokines, and chemokines as well as many pharmacologic agents that can pass through the BBB, essentially acting as a gatekeeper to the central nervous system (Begley, 2004; Ueno, 2007). It is an obstacle in administration of therapeutics to the brain, making it difficult for pharmacologic interventions to be delivered in patients with epilepsy, brain tumors, and HIV (Begley, 2004). The role of Pgp in restricting drug delivery to the brain during pregnancy has been investigated in pregnant rodents and nonhuman primates. Pgp protein expression was elevated at the BBB mid-gestation, but returned to prepregnancy levels by late gestation in mice (Coles et al., 2009). Further, Pgp expression and basal activity in cerebral capillaries late in gestation was similar to that seen in the nonpregnant state in rats (Johnson et al., 2018). Protein expression of multidrug resistance-associated protein 1 (Mrp1), another efflux transporter, was also elevated at the BBB mid-pregnancy, and it remained high late in gestation in mice (Coles et al., 2009). A study using positron emission tomography scanning to investigate Pgp activity at the BBB across gestation in nonhuman primates reported that Pgp activity increased with gestational age (Chung et al., 2010). This is in contrast to what has been reported to occur in rats (Johnson et al., 2018); however, methodological or species differences could account for the conflicting findings. Regardless, efflux transporters appear to be gestationally regulated, potentially in response to the increase in circulating factors during pregnancy. Overall, increases in efflux transporter expression and/or activity across gestation are likely a critical adaptation of the BBB to prevent passage of circulating factors into the brain during pregnancy.

Preservation of BBB properties and adaptation of efflux transporters in the face of elevated circulating permeability factors during pregnancy appear to be highly protective and may be central to the prevention of eclamptic seizures. Seizure-provoking serum constituents are found to be circulating late in gestation, highlighting the importance of healthy BBB function. Serum from late-pregnant, but not nonpregnant rats caused hyperexcitability of hippocampal neuronal networks in cultured slices, measured by evoked field potentials (Cipolla et al., 2012b). The increase in excitability was due to serum factors causing neuroinflammation via activation of microglia and secretion of tumor necrosis factor α (Riazi et al., 2008; Cipolla et al., 2012b). However, under normal conditions, microglia are not active in the brain during pregnancy (Johnson et al., 2015). In addition, the brain is not likely to be exposed to circulating serum factors due to the protective nature of the BBB (Ueno, 2007; Cipolla, 2013; Johnson et al., 2015). Importantly, the maternal brain appears to be in a state of increased excitability and seizure susceptibility (Binder et al., 2004; Quick and Cipolla, 2005; Verkman et al., 2006; Maguire et al., 2009; Johnson et al., 2015); however, the vast majority of women do not have seizures during pregnancy, which appears to be due to the BBB. A study investigating the role of the BBB in seizure during pregnancy using chronically implanted EEG electrodes in pregnant and nonpregnant rats provided proof-of-principle evidence that BBB efflux transporters are critical in seizure prevention during pregnancy (Johnson et al., 2018). Spontaneous seizures occurred in response to acute inhibition of efflux transporters that were more pronounced in the hippocampus, a brain region central to learning and memory that is particularly susceptible to seizures, than in the cerebral cortex during pregnancy (Johnson et al., 2018). These seizures were not due to neuroinflammation or regional differences in Pgp expression or basal activity. Rather, latency to seizure onset correlated with circulating levels of estradiol, a neurosteroid that is elevated during pregnancy, is a substrate of efflux transporters, and has acute proconvulsive actions in the hippocampus (O'Leary et al., 1991; Wong and Moss, 1991; Woolley, 1998; Reddy, 2003; Johnson et al., 2018). In the same study, circulating factors during pregnancy were found to exert an inhibitory effect on Pgp activity in isolated capillaries in the hippocampus, but not in the cerebral cortex (Johnson et al., 2018). Thus it appears that BBB efflux transporters are critical in limiting the passage of harmful seizure-provoking serum constituents into the hyperexcitable maternal brain. Further, serum constituents present late in pregnancy affect Pgp activity and have the ability to promptly initiate seizures when their passage into the brain is unregulated (Johnson et al., 2018). Failure of the efflux transporters to adapt and expel high levels of circulating factors

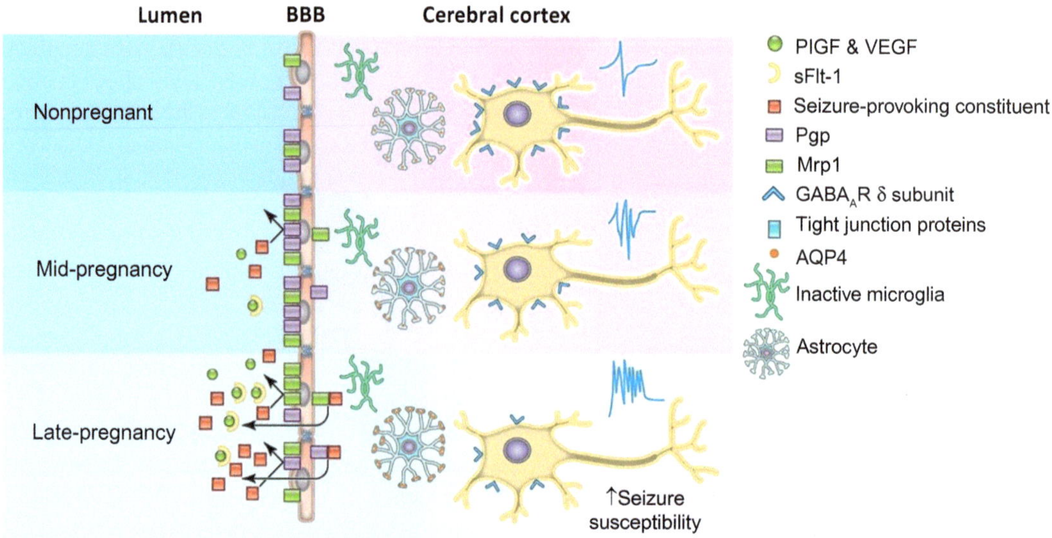

Fig. 4.1. Adaptive response to pregnancy at the blood–brain barrier (BBB). Throughout gestation, the maternal circulation is flooded with many serum factors, including permeability factors and seizure-provoking constituents (Cipolla et al., 2012b). Simultaneously, expression and/or activity of main efflux transporters Pgp and Mrp1 increase at the BBB, potentially in response to increased circulating factors; this serves to protect the brain from these factors (Coles et al., 2009; Chung et al., 2010). *Black arrows* indicate expulsion of circulating factors by Pgp and Mrp1 during pregnancy and protection of the brain from such seizure-provoking factors. Circulating permeability factors VEGF and PlGF increase during pregnancy but are unable to exert an effect on BBB permeability due to concurrent increases in sFlt-1 that limits their bioavailability (Brown et al., 1997; Shibuya, 2013). Pregnancy does not affect tight junction proteins or paracellular permeability of the BBB (Cipolla et al., 2011). Centrally, microglia remain in an inactive state; however, aquaporin-4 (AQP4) expression on astrocytes is increased over the course of gestation (Quick and Cipolla, 2005; Johnson et al., 2014). Further, inhibitory neurotransmitter GABA$_A$ receptor subunits are downregulated in the cerebral cortex and hippocampus, resulting in increased neuronal excitability (Maguire et al., 2009; Johnson et al., 2014). Reproduced from Johnson, A.C., Cipolla, M.J., 2015. The cerebral circulation during pregnancy: adapting to preserve normalcy. Physiology (Bethesda), 30, 139–147.

during pregnancy may be one mechanism by which de novo seizures occur during pregnancy and preeclampsia. Fig. 4.1 summarizes our current understanding of the adaptation of the BBB during normal pregnancy.

CBF AUTOREGULATION AND HEMODYNAMICS

Unlike other organs, the brain lacks energy storage and therefore requires constant blood supply. The cerebral circulation ultimately functions to deliver oxygen, glucose, and nutrient-rich blood to, and remove metabolic waste from, the central nervous system, which is crucial to ensure proper brain function. It is therefore not surprising that blood flow autoregulation is well developed in the brain. CBF autoregulation is the intrinsic property of the brain to maintain relatively constant blood flow in the face of changes in blood pressure (Harper, 1966; Hayman et al., 1981). In normal healthy adults, CBF autoregulation operates between ~60 and 160 mmHg (Lassen, 1959; McHenry Jr. et al., 1974). Pregnancy appears to extend both the upper and lower limits of the CBF autoregulatory curve. In normal pregnant rats, the limits of the CBF autoregulatory curve were investigated by either inducing controlled hemorrhage to lower blood pressure or using phenylephrine infusion to raise blood pressure acutely, together with continuous CBF measurements using laser Doppler flowmetry (Cipolla et al., 2012a; Chapman et al., 2013). These studies found that compared with the nonpregnant state, both the lower and upper limits of CBF autoregulation were shifted (Cipolla et al., 2012a; Chapman et al., 2013). Fig. 4.2 shows the extended autoregulatory curve in the pregnant state compared with the range in the nonpregnant state (Lassen, 1959; Paulson et al., 1990). Taken together, the effect of pregnancy on CBF autoregulation appears to be protective, making the maternal brain better prepared to maintain blood flow in the face of both acute hypotension and hypertension. However, these studies used animal models of pregnancy as such measurements in pregnant women are challenging and potentially dangerous. Studies in humans using noninvasive techniques to measure dynamic autoregulation during normal pregnancy have found improved or no change in CBF

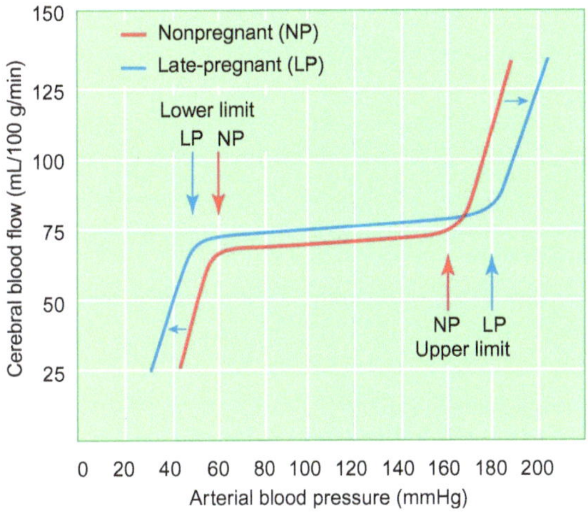

Fig. 4.2. Cerebral blood flow (CBF) autoregulatory curves in the pregnant and nonpregnant states. Within the arterial pressure range of ~60–160 mmHg, CBF is relatively constant due to cerebral arteries and arterioles that constrict in response to increased pressure and dilate in response to decreased pressure (Lassen, 1959; McHenry Jr. et al., 1974). At pressures in excess of the upper limit of autoregulation, forced dilatation of cerebral arteries and arterioles occurs, causing hyperperfusion and a linear relationship between flow and pressure (Lassen, 1959). At pressures below the lower limit of autoregulation, flow drops in response to decreased pressure when vasodilation of cerebral vessels cannot compensate to maintain flow (Lassen, 1959). The drop of CBF during hypotension can cause ischemic brain injury. Impaired autoregulation of CBF occurs at both pressure extremes and flow becomes linearly related to pressure. Pregnancy extends the CBF autoregulatory curve at both ends, shifting both the upper and lower limits modestly (Cipolla et al., 2012a; Chapman et al., 2013). The leftward shift of the lower limit may be due to the enhanced vasodilation of cerebral arteries that occurs during pregnancy in response to decreased intravascular pressure, whereas the outward remodeling of penetrating arterioles may contribute to the rightward shift of the upper limit. Overall, the extension of autoregulation to higher and lower pressures may be protective of the brain especially during parturition. There is no significant change in basal CBF in the pregnant state when compared with the nonpregnant state, that is, an increase of ~10% globally (Buelke-Sam et al., 1982; Cipolla et al., 2011). Reproduced from Johnson, A.C., Cipolla, M.J., 2015. The cerebral circulation during pregnancy: adapting to preserve normalcy. Physiology (Bethesda), 30, 139–147.

autoregulation when compared with nonpregnant women (Janzarik et al., 2014; Van Veen et al., 2016). To our knowledge, no study has determined the limits of CBF autoregulation during human pregnancy, but it is important to understand because of the risk of acute hypotensive and hypertensive episodes, especially during parturition. For example, hemorrhage can occur during parturition causing hypotension, or parturition can lead to increased sympathetic discharge resulting in an acute elevation in blood pressure (Pickering, 2003). Thus, extension of the CBF autoregulatory curve may be a protective adaptation to better maintain cerebral perfusion to the maternal brain during rapid and robust changes in blood pressure.

Numerous studies have measured changes in basal CBF during pregnancy, employing several techniques, such as ultrasonography and magnetic resonance (MR). Early human studies using inhalation of a gaseous mixture of nitrous oxide, oxygen, and nitrogen and the Fick principle to assess CBF, oxygen delivery, and metabolism in the brain reported no differences in CBF between the pregnant and nonpregnant states (McCall, 1948, 1949, 1953; McCall and Taylor, 1952). Transcranial Doppler (TCD) studies that measured CBF velocity in cerebral arteries in pregnant women across gestation revealed CBF velocity decreased during gestation (Williams and Wilson, 1994; Serra-Serra et al., 1997; Belfort et al., 2001; Batur Caglayan et al., 2019). However, changes in vascular resistance and blood flow calculated from these measurements may not accurately reflect CBF due to the lack of information about vessel diameter (Kontos, 1989). A case report measuring longitudinal cerebrovascular reactivity via TCD across gestation in a healthy pregnant patient reported resting CBF velocity did not change over the course of gestation (Steinback et al., 2015). However, cerebrovascular reactivity to elevated CO_2 increased progressively from prepregnancy to the end of the third trimester, suggesting an improved cerebrovascular reserve during pregnancy (Steinback et al., 2015). Using dual-beam angle-independent digital Doppler ultrasonography, diameter and blood flow volume of the internal carotid artery were measured during pregnancy. This study found that CBF increased ~20% across gestation based on these measurements (Nevo et al., 2010). In contrast, a study using MR reported an ~20% decrease in CBF during pregnancy; however, this was in comparison with postpartum, and not prepregnancy values (Zeeman et al., 2003). Studies have used microspheres to measure absolute CBF in late-pregnant rats and found a nonsignificant 5%–10% increase in cortical CBF compared with the nonpregnant state (Buelke-Sam et al., 1982; Cipolla et al., 2011). Further, absolute CBF in the hippocampus was found to be similar between pregnant and nonpregnant rats as measured via hydrogen clearance (Johnson and Cipolla, 2017). Thus, there appears to be conflicting evidence regarding the effect of normal pregnancy on CBF, potentially due to variability in methodology and study populations. Overall, while studies reporting the effect of pregnancy on CBF in women are contradictory, the use of animals, which allowed invasive measurements, suggests CBF remains similar in the pregnant and nonpregnant states.

FUNCTION AND STRUCTURE OF THE CEREBROVASCULATURE

Understanding changes occurring in the structure and function of cerebral arteries and arterioles during pregnancy may shed some light on potential changes in vascular resistance that may in turn drive changes in CBF and autoregulation. Cerebral arteries and arterioles exist in a state of partial constriction and thus have basal myogenic tone. A major contributor to basal tone in the cerebral circulation is the myogenic response of vascular smooth muscle cells (Bayliss, 1902; Mackenzie et al., 1979). Pregnancy does not appear to affect myogenic tone in cerebral pial arteries or penetrating brain arterioles (Chan et al., 2010; Cipolla et al., 2011; Johnson and Cipolla, 2018). While myogenic tone refers to the degree of constriction of a vessel relative to its passive diameter at constant pressure, myogenic reactivity refers to the dynamic response of cerebral arteries and arterioles to changes in intravascular pressure (Kontos et al., 1978). This response is a main contributor to CBF autoregulation by increasing and decreasing cerebrovascular resistance in response to changes in intravascular pressure and appears to be different in the pregnant vs nonpregnant state (Kontos et al., 1978; Faraci et al., 1987a; Faraci and Heistad, 1990; Cipolla et al., 2004; Chapman et al., 2013). In isolated cerebral arteries from pregnant rats, increased intravascular pressure caused forced dilatation at lower pressures when compared with arteries from nonpregnant rats, suggesting a lower capacity to maintain cerebrovascular resistance in the face of increased pressure during pregnancy (Cipolla et al., 2004). However, the CBF autoregulatory curve is shifted to higher and not lower pressures during pregnancy, suggesting other contributors to CBF autoregulation may act in a compensatory way during pregnancy. In contrast, cerebral arteries from pregnant rats had an exaggerated myogenic dilatory response to decreased intravascular pressure, with enhanced vasodilation compared with arteries from nonpregnant rats (Chapman et al., 2013). This may be a protective mechanism by which the maternal cerebrovasculature responds to reductions in intravascular pressure to better maintain blood flow and may contribute to the leftward shift of the lower limit of the autoregulatory curve (Fig. 4.2).

The vasculature of many organ systems, particularly within the uteroplacental circulation, change structurally to accommodate the physiologic adaptation of normal pregnancy (Osol and Mandala, 2009). There is evidence that the cerebral vasculature also structurally remodels during pregnancy in a selective manner that could affect CBF. Structural remodeling describes changes in lumen diameter and vascular wall thickness in response to physiologic or pathologic stimuli (Martinez-Lemus et al., 2009). It can be directed outward or inward, depending on whether the lumen diameter increases or decreases (Mulvany, 1999). Further, remodeling can be hypo-, hyper-, or eutrophic depending on whether the vessel wall thickness decreases, increases, or remains the same, respectively (Mulvany, 1999). During rat pregnancy, no changes in lumen diameter or wall thickness have been measured in cerebral pial arteries, suggesting pregnancy-induced remodeling does not occur in the pial vasculature (Chan et al., 2010). However, brain parenchymal arterioles, precapillary resistance vessels that branch off pial vessels, perfuse the cerebral cortex, and control local blood flow, undergo outward hypotrophic remodeling during pregnancy, resulting in larger vascular lumens and thinner vessel walls than in the nonpregnant state (Cipolla et al., 2011; Johnson and Cipolla, 2018). Despite the structural remodeling of parenchymal arterioles during pregnancy, the ability of these arterioles to conduct local vasodilation upstream to decrease segmental vascular resistance and increase local perfusion in response to neuronal activity remained intact during normal pregnancy (Johnson and Cipolla, 2018). Thus, vascular contributions to neurovascular coupling that matches CBF to metabolic demand to maintain neuronal and glial health appear to be unaffected during pregnancy.

The selective remodeling of parenchymal arterioles during pregnancy is through arteriogenesis and driven by peroxisome proliferator-activated receptor gamma (PPARγ) activation by the hormone (ser)relaxin (Chan and Cipolla, 2011). Interestingly, the primary relaxin receptor is not expressed in parenchymal arterioles (Chan and Cipolla, 2011). However, circulating relaxin appears to cross the BBB and is thought to activate PPARγ on astrocytes and neurons, which in turn exerts a paracrine effect on parenchymal arterioles to drive outward remodeling (Chan and Cipolla, 2011), as illustrated in Fig. 4.3. In contrast to parenchymal arterioles, pial arteries are not intimately associated with brain parenchymal cell types such as astrocytes and neurons, and this may explain the selective effect of pregnancy-induced remodeling on parenchymal arterioles. This process of arteriogenesis is unique compared with angiogenesis; however, angiogenesis also occurs in the brain during pregnancy. In fact, in this state, capillary density increases in the posterior cerebral cortex, which may increase blood supply to neuroendocrine glands that are enlarged and/or involved in production and secretion of pregnancy hormones (Gonzalez et al., 1988). This increase in capillary density has also been linked to increased activation of PPARγ by relaxin (Chan and Cipolla, 2011; Cipolla et al., 2011). Thus, while pregnancy does not affect the structure of the pial vasculature, it seems to have an outward hypotrophic remodeling

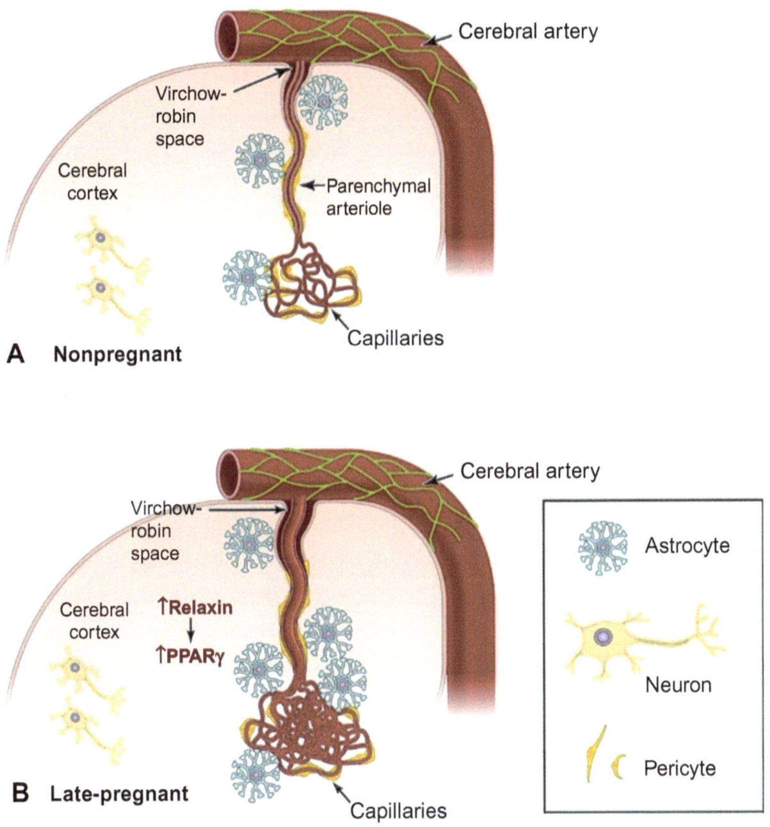

Fig. 4.3. Structural changes of cerebral arteries and arterioles during pregnancy. The cerebral arteries on the brain surface undergo little to no change in lumen diameter or vessel wall thickness during pregnancy. However, intraparenchymal arterioles that branch off the pial surface arteries and perfuse the brain parenchyma undergo outward hypotrophic remodeling resulting in larger vascular lumens and thinner vessel walls compared with the nonpregnant state (Cipolla et al., 2011). Structural remodeling of cerebral arterioles is through a mechanism involving increased circulating relaxin that crosses the BBB and activates PPARγ on astrocytes and neurons, production of VEGF, and activation of matrix metalloproteinases (Chan and Cipolla, 2011; Cipolla et al., 2011). In addition, capillary density increases in the posterior brain region during pregnancy through a similar mechanism (Cipolla et al., 2011). These changes may be due to the increased demand of blood supply to neuroendocrine glands that enlarge during pregnancy (e.g., the pituitary gland) (Gonzalez et al., 1988) or areas involved in production and/or secretion of pregnancy hormones. The increase in vascular volume of the cerebral circulation during pregnancy combined with expanded plasma volume may underlie the susceptibility to edema formation during acute hypertension and other forms of brain injury such as seizure. Reproduced from Johnson, A.C., Cipolla, M.J., 2015. The cerebral circulation during pregnancy: adapting to preserve normalcy. Physiology (Bethesda), 30, 139–147.

effect on parenchymal arterioles, which may contribute to the extension of the CBF autoregulatory curve. Further, this outward remodeling may increase cerebrovascular reserve, which could contribute to the increased reactivity to CO_2 that occurs over the course of gestation, as reported in a case of a healthy pregnant patient (Steinback et al., 2015).

During pregnancy, outward remodeling of parenchymal arterioles and increased capillary density coupled with the approximate 10% hemodilution that occurs could decrease cerebrovascular resistance and increase CBF (Gordon, 2007). Despite these vascular and hemodynamic changes, pregnancy has little effect on cerebrovascular resistance when measured under normotensive conditions in the rat (Cipolla et al., 2011). Pregnancy-induced thinner walls and increased wall stress could decrease small vessel resistance and predispose parenchymal arterioles to increased permeability or rupture under pathologic conditions such as acute hypertension (Cipolla et al., 2011). However, the myogenic tone of parenchymal arterioles was sustained at elevated intravascular pressure, suggesting these arterioles are capable of maintaining cerebral vascular resistance in response to elevated pressure despite the structural remodeling occurring in pregnancy (Johnson and Cipolla, 2018). Further, the substantial contribution of large cerebral arteries to vascular resistance is unique to cerebral circulation (Faraci and Heistad, 1990). In fact, large extracranial and intracranial cerebral arteries contribute ~50% of cerebrovascular resistance (Faraci et al., 1987b; Faraci and Heistad, 1990). Importantly, although downstream arterioles undergo structural changes that could affect small vessel resistance, the pial vasculature does not change structurally during pregnancy and may compensate to maintain normal vascular resistance (Fig. 4.3). Moreover, this pregnancy-induced remodeling appears to be unique to the cerebral cortex. Arterioles perfusing the hippocampus undergo hypotrophic remodeling during pregnancy, but this remodeling does not result in larger lumen diameters (Johnson and Cipolla, 2017). Further, capillary density in the hippocampus is similar in the pregnant and nonpregnant states (Johnson et al., 2018). Thus pregnancy-induced vascular remodeling seems to be selective not only to the level of the vascular tree (pial vs. parenchymal), but also to regional vascular territories (cerebral cortex vs. hippocampus). Implications for these regional differences in regards to hemodynamic changes and mechanisms by which selective remodeling may be occurring remain unclear and require further investigation.

The structural remodeling of cerebral arterioles may be important when considering hypertensive pathologies of pregnancy, such as preeclampsia. Under experimental conditions of acute hypertension, cerebrovascular resistance of cerebral arteries was decreased in pregnant rats, leading to autoregulatory breakthrough and ~40% increase in CBF (Cipolla et al., 2011). The decrease in cerebrovascular resistance with autoregulatory breakthrough was further associated with increased BBB permeability due to greater hydrostatic pressure on the microcirculation (Cipolla et al., 2011). Thus, while the BBB seems to remain intact during normal pregnancy under physiologic conditions, it also appears to be at greater risk of injury during pathologic states such as acute hypertension. Under conditions of elevated intravascular pressure when large arteries become ineffective in regulating CBF due to forced dilatation of myogenic tone, outward hypotrophic remodeling of parenchymal arterioles may predispose the microcirculation to injury by transmitting high hydrostatic pressure downstream. Outward hypotrophic remodeling of brain arterioles could also contribute to the maternal brain being more sensitive to vasogenic edema formation following acute hypertension (Euser and Cipolla, 2007; Cipolla et al., 2012a). The increase in capillary density may further contribute to the susceptibility of the brain to hypertension-induced vasogenic edema during pregnancy by increasing the potential sites of BBB disruption. In addition, pregnancy both prevents and reverses remodeling of cerebral arteries, which occurs in response to chronic hypertension (Cipolla et al., 2006, 2008; Aukes et al., 2007). In the nonpregnant state, chronic hypertension leads to inward hypertrophic remodeling of cerebral arteries, resulting in smaller lumen diameters and thicker vascular walls (Baumbach and Heistad, 1989; Cipolla et al., 2006). This is considered a protective adaptation by which cerebrovascular resistance is increased, thus shielding the microcirculation from elevated arterial blood pressure (Cipolla et al., 2006; Chan et al., 2010). The prevention and/or reversal of this remodeling during pregnancy may be related to the downregulation of angiotensin II type 1 receptor (AT1R) expression in the cerebral circulation, which also occurs during pregnancy (Chan et al., 2010). While the reduction in AT1R expression may be a physiologic adaptation to normal pregnancy, reversal and/or prevention of hypertensive remodeling may make the cerebral microcirculation even more susceptible to injury during states associated with hypertension. This may be important during preeclampsia, where acute elevations in blood pressure are thought to lead to cerebral vasogenic edema formation and subsequent neurologic complications, including eclamptic seizures (Cipolla, 2007).

The adaptation of parenchymal arterioles during pregnancy may be further implicated in long-lasting neurologic complications associated with preeclampsia and eclampsia. Lack of spiral artery remodeling during pregnancy is a feature seen in some women with preeclampsia; this may indicate that a type of small vessel disease occurs during this condition (Egbor et al., 2006). Parenchymal arterioles are the primary vessels involved in small vessel disease in the brain (Rincon and Wright, 2014). Women with preeclampsia seem to be prone to white matter lesions later in life, which may indicate the presence of cerebral small vessel disease (Aukes et al., 2009, 2012; Wiegman et al., 2014). It is possible that the failure of parenchymal arterioles to outward remodel during preeclampsia underlies the potential for white matter lesions later in life. In fact, a study investigating the effect of experimental preeclampsia on the structure and function of parenchymal arterioles that supply white matter and deep gray structures reported that pregnancy-induced remodeling of parenchymal arterioles was absent in experimental preeclampsia (Johnson and Cipolla, 2018). Further, conducted vasodilation that contributes to neurovascular coupling and functional hyperemia was enhanced in normal pregnancy and impaired in experimental preeclampsia (Johnson and Cipolla, 2018). Thus, the lack of remodeling and/or prolonged and persistent impaired function of these arterioles that are central to the pathophysiology of white matter lesions could contribute to white matter disease and cognitive impairment after pregnancies complicated by preeclampsia. We speculate that the lack of adaptation of small vessels in the uterine circulation may be similar to the lack of adaptation of small vessels in the brain, both of which may be occurring in women with preeclampsia. However, the association between small vessel disease of the brain that occurs later in life in women who had preeclampsia with impaired spiral artery remodeling during pregnancy is speculative, and further studies are needed to understand these processes. Moreover, whether the maladaptation of these two vascular beds share similar pathogenesis during preeclampsia remains unclear and requires further investigation.

Pregnancy is a state of increased risk for cerebral venous thrombosis, stroke, and cerebral hemorrhage, all of which may be due to pregnancy-induced changes in the cerebral venous system (Cantu and Barinagarrementeria, 1993; Saposnik et al., 2011; Tate and Bushnell, 2011). Many pregnancy-related changes in the cerebrovasculature, which primarily focused on the arterial compartment of cerebral circulation, have been discussed. However, 70%–80% of cerebral blood volume is contained within the venous system in the brain (An and Lin, 2002). Although much less is known about how pregnancy affects venous outflow and venous pressure that can affect cerebral perfusion and hemodynamics, cerebral venous thrombosis is thought to constitute a substantial proportion of pregnancy-related strokes (McLean and Cushman, 2016). A prospective study classified the cerebrovascular complications in pregnant women from 2006 to 2008 in Chennai, India. Of the ~39,000 deliveries studied, 26 women, none of whom had a history of hypertension, seizures, diabetes, renal disease, smoking, or use of oral contraceptives, were diagnosed with various cerebrovascular complications (Prabhu, 2013). Furthermore, of these 26 women, 16 (62%) were diagnosed with cerebral venous thrombosis, 4 (15%) had intracerebral hemorrhage, and 6 (23%) had arterial infarcts (Prabhu, 2013). Although this study suggests stroke during pregnancy seems to be primarily due to venous rather than arterial complications, the reported incidence of stroke and stroke subtype during pregnancy seems to vary in different parts of the world (Tate and Bushnell, 2011). Regardless, pregnancy may affect the venous compartment of the cerebral vasculature such that the maternal brain is at greater risk for cerebral venous thrombosis and hemorrhage.

The hypercoagulable state of pregnancy is believed to be a primary contributor to cerebrovascular complications during pregnancy (Cantu and Barinagarrementeria, 1993; Saposnik et al., 2011; Bushnell et al., 2014). However, structural changes in the venous system could further contribute to pathologies, including cerebral venous thrombosis and intracranial hemorrhage. In fact, pregnancy does not only affect the structure of cerebral parenchymal arterioles, but also appears to induce remodeling of the venous system. A study reported that the cerebral vein of Galen underwent outward hypotrophic remodeling and had decreased myogenic tone during normal pregnancy, which would likely decrease venous vascular resistance (Van Der Wijk et al., 2013). The larger lumen diameter and decreased vascular resistance could promote venous pooling and stenosis that, in the setting of increased procoagulant factors during pregnancy, could, in turn, lead to venous thrombosis. Further, the hypotrophic remodeling results in a thinner vessel wall, thereby increasing wall tension and wall stress of cerebral veins during pregnancy, which may increase the susceptibility to rupture and cerebral hemorrhage. Together, the venous remodeling reported to occur during pregnancy may increase the risk for both cerebral venous thrombosis and intracerebral hemorrhage that make up a substantial proportion of maternal mortality and cerebrovascular complications during this period. Moreover, cerebral venous remodeling may represent a mechanism by which preeclampsia, a state of exaggerated hypercoagulation, may further promote cerebral venous thrombosis and contribute to the 6–12-fold increased risk for stroke compared with normal pregnancy (Ros et al., 2002; Scott et al., 2012). Whether the outward hypotrophic venous remodeling has mechanisms similar to that seen in the outward hypotrophic arteriolar remodeling occurring in the maternal brain, and how preeclampsia may affect this remodeling, remains unclear and requires further investigation.

CONCLUSIONS

Pregnancy is associated with many adaptations of the cerebral circulation, including changes in receptor and transporter activity and keeping increased permeability factors in check to maintain brain homeostasis and protect against increases in BBB permeability. Structural and functional changes occur in certain segments of the cerebral vasculature; however, CBF and cerebrovascular resistance appear unchanged under normotensive conditions. The CBF autoregulatory curve appears to extend in the pregnant state, protecting the maternal brain from acute and drastic fluctuations in blood pressure. It is remarkable that BBB permeability and CBF are affected so minimally during pregnancy, especially in the face of substantially increased factors that have direct effects on vascular filtration and flow in many other organ systems. However, this supports the principle that the adaptation of the cerebral circulation to normal pregnancy functions to maintain essential oxygen and nutrient delivery and waste removal similar to that seen in the nonpregnant state, especially in the face of tremendous systemic hemodynamic changes associated with pregnancy.

References

Aagaard-Tillery KM, Silver R, Dalton J (2006). Immunology of normal pregnancy. Semin Fetal Neonatal Med 11: 279–295.

Amburgey OA, Chapman AC, May V et al. (2010a). Plasma from preeclamptic women increases blood–brain barrier permeability: role of vascular endothelial growth factor signaling. Hypertension 56: 1003–1008.

Amburgey OA, Reeves SA, Bernstein IM et al. (2010b). Resistance artery adaptation to pregnancy counteracts the vasoconstricting influence of plasma from normal pregnant women. Reprod Sci 17: 29–39.

An H, Lin W (2002). Cerebral venous and arterial blood volumes can be estimated separately in humans using magnetic resonance imaging. Magn Reson Med 48: 583–588.

Aukes AM, Vitullo L, Zeeman GG et al. (2007). Pregnancy prevents hypertensive remodeling and decreases myogenic reactivity in posterior cerebral arteries from Dahl salt-sensitive rats: a role in eclampsia? Am J Physiol Heart Circ Physiol 292: H1071–H1076.

Aukes AM, De Groot JC, Aarnoudse JG et al. (2009). Brain lesions several years after eclampsia. Am J Obstet Gynecol 200: e1–e5.

Aukes AM, De Groot JC, Wiegman MJ et al. (2012). Long-term cerebral imaging after pre-eclampsia. BJOG 119: 1117–1122.

Batur Caglayan HZ, Nazliel B, Cinar M et al. (2019). Assessment of maternal cerebral blood flow velocity by transcranial Doppler ultrasound before delivery and in the early postpartum period. J Matern Fetal Neonatal Med 32: 584–589.

Baumbach GL, Heistad DD (1989). Remodeling of cerebral arterioles in chronic hypertension. Hypertension 13: 968–972.

Bayliss WM (1902). On the local reactions of the arterial wall to changes of internal pressure. J Physiol 28: 220–231.

Begley DJ (2004). ABC transporters and the blood–brain barrier. Curr Pharm Des 10: 1295–1312.

Belfort MA, Tooke-Miller C, Allen Jr JC et al. (2001). Changes in flow velocity, resistance indices, and cerebral perfusion pressure in the maternal middle cerebral artery distribution during normal pregnancy. Acta Obstet Gynecol Scand 80: 104–112.

Binder DK, Oshio K, Ma T et al. (2004). Increased seizure threshold in mice lacking aquaporin-4 water channels. Neuroreport 15: 259–262.

Brightman MW, Reese TS (1969). Junctions between intimately apposed cell membranes in the vertebrate brain. J Cell Biol 40: 648–677.

Brown LF, Detmar M, Claffey K et al. (1997). Vascular permeability factor/vascular endothelial growth factor: a multifunctional angiogenic cytokine. EXS 79: 233–269.

Buelke-Sam J, Nelson CJ, Byrd RA et al. (1982). Blood flow during pregnancy in the rat: I. Flow patterns to maternal organs. Teratology 26: 269–277.

Bushnell C, McCullough LD, Awad IA et al. (2014). Guidelines for the prevention of stroke in women: a statement for healthcare professionals from the American Heart Association/American Stroke Association. Stroke 45: 1545–1588.

Cantu C, Barinagarrementeria F (1993). Cerebral venous thrombosis associated with pregnancy and puerperium. Review of 67 cases. Stroke 24: 1880–1884.

Chan SL, Cipolla MJ (2011). Relaxin causes selective outward remodeling of brain parenchymal arterioles via activation of peroxisome proliferator-activated receptor-gamma. FASEB J 25: 3229–3239.

Chan SL, Chapman AC, Sweet JG et al. (2010). Effect of PPARgamma inhibition during pregnancy on posterior cerebral artery function and structure. Front Physiol 1: 130.

Chapman AC, Cipolla MJ, Chan SL (2013). Effect of pregnancy and nitric oxide on the myogenic vasodilation of posterior cerebral arteries and the lower limit of cerebral blood flow autoregulation. Reprod Sci 20: 1046–1054.

Charnock-Jones DS, Kaufmann P, Mayhew TM (2004). Aspects of human fetoplacental vasculogenesis and angiogenesis. I. Molecular regulation. Placenta 25: 103–113.

Chung FS, Eyal S, Muzi M et al. (2010). Positron emission tomography imaging of tissue P-glycoprotein activity during pregnancy in the non-human primate. Br J Pharmacol 159: 394–404.

Cipolla MJ (2007). Cerebrovascular function in pregnancy and eclampsia. Hypertension 50: 14–24.

Cipolla MJ (2013). The adaptation of the cerebral circulation to pregnancy: mechanisms and consequences. J Cereb Blood Flow Metab 33: 465–478.

Cipolla MJ, Vitullo L, McKinnon J (2004). Cerebral artery reactivity changes during pregnancy and the postpartum period: a role in eclampsia? Am J Physiol Heart Circ Physiol 286: H2127–H2132.

Cipolla MJ, Delance N, Vitullo L (2006). Pregnancy prevents hypertensive remodeling of cerebral arteries: a potential role in the development of eclampsia. Hypertension 47: 619–626.

Cipolla MJ, Smith J, Bishop N et al. (2008). Pregnancy reverses hypertensive remodeling of cerebral arteries. Hypertension 51: 1052–1057.

Cipolla MJ, Sweet JG, Chan SL (2011). Cerebral vascular adaptation to pregnancy and its role in the neurological complications of eclampsia. J Appl Physiol 110: 329–339.

Cipolla MJ, Bishop N, Chan SL (2012a). Effect of pregnancy on autoregulation of cerebral blood flow in anterior versus posterior cerebrum. Hypertension 60: 705–711.

Cipolla MJ, Pusic AD, Grinberg YY et al. (2012b). Pregnant serum induces neuroinflammation and seizure activity via TNFalpha. Exp Neurol 234: 398–404.

Clapp 3rd JF, Capeless E (1997). Cardiovascular function before, during, and after the first and subsequent pregnancies. Am J Cardiol 80: 1469–1473.

Coles LD, Lee IJ, Hassan HE et al. (2009). Distribution of saquinavir, methadone, and buprenorphine in maternal brain, placenta, and fetus during two different gestational stages of pregnancy in mice. J Pharm Sci 98: 2832–2846.

Dobrogowska DH, Lossinsky AS, Tarnawski M et al. (1998). Increased blood–brain barrier permeability and endothelial abnormalities induced by vascular endothelial growth factor. J Neurocytol 27: 163–173.

Duckles SP, Krause DN (2007). Cerebrovascular effects of oestrogen: multiplicity of action. Clin Exp Pharmacol Physiol 34: 801–808.

Dvorak HF (2002). Vascular permeability factor/vascular endothelial growth factor: a critical cytokine in tumor angiogenesis and a potential target for diagnosis and therapy. J Clin Oncol 20: 4368–4380.

Egbor M, Ansari T, Morris N et al. (2006). Morphometric placental villous and vascular abnormalities in early- and late-onset pre-eclampsia with and without fetal growth restriction. BJOG 113: 580–589.

Euser AG, Cipolla MJ (2007). Cerebral blood flow autoregulation and edema formation during pregnancy in anesthetized rats. Hypertension 49: 334–340.

Evans P, Wheeler T, Anthony F et al. (1997). Maternal serum vascular endothelial growth factor during early pregnancy. Clin Sci (Lond) 92: 567–571.

Faraci FM, Heistad DD (1990). Regulation of large cerebral arteries and cerebral microvascular pressure. Circ Res 66: 8–17.

Faraci FM, Heistad DD, Mayhan WG (1987a). Role of large arteries in regulation of blood flow to brain stem in cats. J Physiol 387: 115–123.

Faraci FM, Mayhan WG, Heistad DD (1987b). Segmental vascular responses to acute hypertension in cerebrum and brain stem. Am J Physiol 252: H738–H742.

Feng D, Nagy JA, Hipp J et al. (1996). Vesiculo-vacuolar organelles and the regulation of venule permeability to macromolecules by vascular permeability factor, histamine, and serotonin. J Exp Med 183: 1981–1986.

Fenstermacher J, Gross P, Sposito N et al. (1988). Structural and functional variations in capillary systems within the brain. Ann N Y Acad Sci 529: 21–30.

Gonzalez JG, Elizondo G, Saldivar D et al. (1988). Pituitary gland growth during normal pregnancy: an in vivo study using magnetic resonance imaging. Am J Med 85: 217–220.

Gordon MC (2007). Maternal physiology. In: SG Gabbe, JR Niebyl, JL Simpson (Eds.), Obstetrics: normal and problem pregnancies, fifth edn. Churchill Liverstone 42–65.

Harper AM (1966). Autoregulation of cerebral blood flow: influence of the arterial blood pressure on the blood flow through the cerebral cortex. J Neurol Neurosurg Psychiatry 29: 398–403.

Hayman LA, Berman SA, Hinck VC (1981). Correlation of CT cerebral vascular territories with function: II. Posterior cerebral artery. Am J Roentgenol 137: 13–19.

Janzarik WG, Ehlers E, Ehmann R et al. (2014). Dynamic cerebral autoregulation in pregnancy and the risk of preeclampsia. Hypertension 63: 161–166.

Johnson AC, Cipolla MJ (2015). The cerebral circulation during pregnancy: adapting to preserve normalcy. Physiology (Bethesda) 30: 139–147.

Johnson AC, Cipolla MJ (2017). Altered hippocampal arteriole structure and function in a rat model of preeclampsia: potential role in impaired seizure-induced hyperemia. J Cereb Blood Flow Metab 37 (8): 2857–2869. https://doi.org/10.1177/0271678X16676287.

Johnson AC, Cipolla MJ (2018). Impaired function of cerebral parenchymal arterioles in experimental preeclampsia. Reprod Sci 25: 82A.

Johnson A, Tremble S, Cipolla M (2014). Decreased seizure threshold during pregnancy and experimental preeclampsia: roles for $GABA_A$ receptors and microglial activation. Reprod Sci 21: 100A.

Johnson AC, Nagle KJ, Tremble SM et al. (2015). The contribution of normal pregnancy to eclampsia. PLoS One 10: e0133953.

Johnson AC, Hammer ES, Sakkaki S et al. (2018). Inhibition of blood–brain barrier efflux transporters promotes seizure in pregnant rats: role of circulating factors. Brain Behav Immun 67: 13–23.

Kontos HA (1989). Validity of cerebral arterial blood flow calculations from velocity measurements. Stroke 20: 1–3.

Kontos HA, Wei EP, Navari RM et al. (1978). Responses of cerebral arteries and arterioles to acute hypotension and hypertension. Am J Physiol 234: H371–H383.

Krauss T, Pauer HU, Augustin HG (2004). Prospective analysis of placenta growth factor (PlGF) concentrations in the plasma of women with normal pregnancy and pregnancies complicated by preeclampsia. Hypertens Pregnancy 23: 101–111.

Lassen NA (1959). Cerebral blood flow and oxygen consumption in man. Physiol Rev 39: 183–238.

Mackenzie ET, Farrar JK, Fitch W et al. (1979). Effects of hemorrhagic hypotension on the cerebral circulation. I. Cerebral blood flow and pial arteriolar caliber. Stroke 10: 711–718.

Maguire J, Ferando I, Simonsen C et al. (2009). Excitability changes related to GABAA receptor plasticity during pregnancy. J Neurosci 29: 9592–9601.

Marchi N, Angelov L, Masaryk T et al. (2007). Seizure-promoting effect of blood–brain barrier disruption. Epilepsia 48: 732–742.

Marchi N, Tierney W, Alexopoulos AV et al. (2011). The etiological role of blood–brain barrier dysfunction in seizure disorders. Cardiovasc Psychiatry Neurol 2011: 482415.

Martinez-Lemus LA, Hill MA, Meininger GA (2009). The plastic nature of the vascular wall: a continuum of remodeling events contributing to control of arteriolar diameter and structure. Physiology (Bethesda) 24: 45–57.

McCall M (1948). Cerebral blood flow and metabolism in normal and toxemic pregnancy. Am J Med Sci 216: 596.

McCall M (1949). Cerebral blood flow and metabolism in toxemias of pregnancy. Surg Gynecol Obstet 89: 715–721, illust.

McCall M (1953). Cerebral circulation and metabolism in toxemia of pregnancy; observations on the effects of veratrum viride and apresoline (1-hydrazinophthalazine). Am J Obstet Gynecol 66: 1015–1030.

McCall ML, Taylor HW (1952). The effects of morphine sulfate on cerebral circulation and metabolism in normal and toxemic pregnant women. Am J Obstet Gynecol 64: 1131–1136.

McHenry Jr LC, West JW, Cooper ES et al. (1974). Cerebral autoregulation in man. Stroke 5: 695–706.

McLean K, Cushman M (2016). Venous thromboembolism and stroke in pregnancy. Hematology Am Soc Hematol Educ Program 2016: 243–250.

Mulvany MJ (1999). Vascular remodelling of resistance vessels: can we define this? Cardiovasc Res 41: 9–13.

Nevo O, Soustiel JF, Thaler I (2010). Maternal cerebral blood flow during normal pregnancy: a cross-sectional study. Am J Obstet Gynecol 203: e1–e6.

Oby E, Janigro D (2006). The blood–brain barrier and epilepsy. Epilepsia 47: 1761–1774.

O'Leary P, Boyne P, Flett P et al. (1991). Longitudinal assessment of changes in reproductive hormones during normal pregnancy. Clin Chem 37: 667–672.

Osol G, Mandala M (2009). Maternal uterine vascular remodeling during pregnancy. Physiology (Bethesda) 24: 58–71.

Oura H, Bertoncini J, Velasco P et al. (2003). A critical role of placental growth factor in the induction of inflammation and edema formation. Blood 101: 560–567.

Paulson OB, Strandgaard S, Edvinsson L (1990). Cerebral autoregulation. Cerebrovasc Brain Metab Rev 2: 161–192.

Pickering TG (2003). Effects of stress and behavioral interventions in hypertension. Pain and blood pressure. J Clin Hypertens (Greenwich) 5: 359–361.

Prabhu TR (2013). Cerebrovascular complications in pregnancy and puerperium. J Obstet Gynaecol India 63: 108–111.

Quick AM, Cipolla MJ (2005). Pregnancy-induced up-regulation of aquaporin-4 protein in brain and its role in eclampsia. FASEB J 19: 170–175.

Reddy DS (2003). Pharmacology of endogenous neuroactive steroids. Crit Rev Neurobiol 15: 197–234.

Reese TS, Karnovsky MJ (1967). Fine structural localization of a blood–brain barrier to exogenous peroxidase. J Cell Biol 34: 207–217.

Riazi K, Galic MA, Kuzmiski JB et al. (2008). Microglial activation and TNFalpha production mediate altered CNS excitability following peripheral inflammation. Proc Natl Acad Sci U S A 105: 17151–17156.

Rincon F, Wright CB (2014). Current pathophysiological concepts in cerebral small vessel disease. Front Aging Neurosci 6: 24.

Roberts TJ, Chapman AC, Cipolla MJ (2009). PPAR-gamma agonist rosiglitazone reverses increased cerebral venous hydraulic conductivity during hypertension. Am J Physiol Heart Circ Physiol 297: H1347–H1353.

Ros HS, Lichtenstein P, Bellocco R et al. (2002). Pulmonary embolism and stroke in relation to pregnancy: how can high-risk women be identified? Am J Obstet Gynecol 186: 198–203.

Rubin LL, Staddon JM (1999). The cell biology of the blood–brain barrier. Annu Rev Neurosci 22: 11–28.

Saposnik G, Barinagarrementeria F, Brown Jr RD et al. (2011). Diagnosis and management of cerebral venous thrombosis: a statement for healthcare professionals from the American Heart Association/American Stroke Association. Stroke 42: 1158–1192.

Schreurs MP, Houston EM, May V et al. (2012). The adaptation of the blood–brain barrier to vascular endothelial growth factor and placental growth factor during pregnancy. FASEB J 26: 355–362.

Schreurs MP, Hubel CA, Bernstein IM et al. (2013). Increased oxidized low-density lipoprotein causes blood–brain barrier disruption in early-onset preeclampsia through LOX-1. FASEB J 27: 1254–1263.

Scott CA, Bewley S, Rudd A et al. (2012). Incidence, risk factors, management, and outcomes of stroke in pregnancy. Obstet Gynecol 120: 318–324.

Serra-Serra V, Kyle PM, Chandran R et al. (1997). Maternal middle cerebral artery velocimetry in normal pregnancy and postpartum. Br J Obstet Gynaecol 104: 904–909.

Shibuya M (2013). Vascular endothelial growth factor and its receptor system: physiological functions in angiogenesis and pathological roles in various diseases. J Biochem 153: 13–19.

Siegel GJ (1999). Basic neurochemistry: molecular, cellular, and medical aspects, Lippincott Williams & Wilkins, Philadelphia.

Steinback CD, King EC, Davenport MH (2015). Longitudinal cerebrovascular reactivity during pregnancy: a case study. Appl Physiol Nutr Metab 40: 636–639.

Szarka A, Rigo Jr J, Lazar L et al. (2010). Circulating cytokines, chemokines and adhesion molecules in normal pregnancy and preeclampsia determined by multiplex suspension array. BMC Immunol 11: 59.

Tate J, Bushnell C (2011). Pregnancy and stroke risk in women. Womens Health (Lond) 7: 363–374.

Ueno M (2007). Molecular anatomy of the brain endothelial barrier: an overview of the distributional features. Curr Med Chem 14: 1199–1206.

Van Der Wijk AE, Schreurs MP, Cipolla MJ (2013). Pregnancy causes diminished myogenic tone and outward hypotrophic remodeling of the cerebral vein of Galen. J Cereb Blood Flow Metab 33: 542–549.

Van Veen TR, Panerai RB, Haeri S et al. (2016). Changes in cerebral autoregulation in the second half of pregnancy and compared to non-pregnant controls. Pregnancy Hypertens 6: 380–383.

Verkman AS, Binder DK, Bloch O et al. (2006). Three distinct roles of aquaporin-4 in brain function revealed by knockout mice. Biochim Biophys Acta 1758: 1085–1093.

Wahl M, Unterberg A, Baethmann A et al. (1988). Mediators of blood–brain barrier dysfunction and formation of vasogenic brain edema. J Cereb Blood Flow Metab 8: 621–634.

Wiegman MJ, Zeeman GG, Aukes AM et al. (2014). Regional distribution of cerebral white matter lesions years after preeclampsia and eclampsia. Obstet Gynecol 123: 790–795.

Williams K, Wilson S (1994). Maternal middle cerebral artery blood flow velocity variation with gestational age. Obstet Gynecol 84: 445–448.

Wong M, Moss RL (1991). Electrophysiological evidence for a rapid membrane action of the gonadal steroid, 17 beta-estradiol, on CA1 pyramidal neurons of the rat hippocampus. Brain Res 543: 148–152.

Woolley CS (1998). Estrogen-mediated structural and functional synaptic plasticity in the female rat hippocampus. Horm Behav 34: 140–148.

Zeeman GG, Hatab M, Twickler DM (2003). Maternal cerebral blood flow changes in pregnancy. Am J Obstet Gynecol 189: 968–972.

Zlokovic BV (2008). The blood–brain barrier in health and chronic neurodegenerative disorders. Neuron 57: 178–201.

Handbook of Clinical Neurology, Vol. 171 (3rd series)
Neurology and Pregnancy: Pathophysiology and Patient Care
E.A.P. Steegers, M.J. Cipolla and E.C. Miller, Editors
https://doi.org/10.1016/B978-0-444-64239-4.00005-9

Chapter 5

Neurobiology of maternal mental illness

JAMIE MAGUIRE[1], CLARE McCORMACK[2], ANIKA MITCHELL[3], AND CATHERINE MONK[3,4,5]*

[1]*Department of Neuroscience, Tufts University School of Medicine, Boston, MA, United States*

[2]*Center for Science and Society, Columbia University, New York, NY, United States*

[3]*Department of Obstetrics and Gynecology, Columbia University Irving Medical Center, New York, NY, United States*

[4]*Department of Psychiatry, Columbia University Irving Medical Center, New York, NY, United States*

[5]*New York State Psychiatric Institute, New York, NY, United States*

Abstract

This chapter provides an overview of current research discoveries beginning to uncover the neurobiology of maternal mental illness. Results are described according to standard diagnostic categories (specifically, perinatal depression, perinatal anxiety and OCD, postpartum psychosis and bipolar disorder, and trauma and posttraumatic stress disorder), yet we aim to put this approach in context with the introduction of a classification model for psychiatric research, the research domain criteria, gaining traction in basic and clinical translational fields. We first review a new area of study, the neuroplasticity of the pregnant and postpartum brain, as work here has relevance for understanding the pathophysiology of mental disorders and may provide clues to changes in brain functioning that are related to compromised parenting in the context of postpartum depression. We next provide background information on neuroendocrine and immune changes during pregnancy and, to a lesser extent, the postpartum period, as alterations in these systems are significantly implicated in underlying neurobiology of mental illness for peripartum women. Our discussion of the major mental illnesses for pregnant and postpartum women includes neuroendocrine changes, neuroinflammation, and neurotransmitter alterations, as well as circuit dysfunction. Overall, remarkable progress has been made in identifying variations in neurobiology (and related systems) involved in maternal mental illness; yet, it is clear that, as classified with standard diagnostic systems, these are heterogeneous disorders and there is individual variability in the alterations in neurobiology for the same illness.

INTRODUCTION

The incidence of maternal mental illness is staggering—indeed, depression and anxiety combined are considered the most common complication of pregnancy and childbirth nationwide (Mathematica, 2019), comparable with or dwarfing rates of gestational diabetes at 6%–9% and preterm birth at 8% (Centers for Disease Control, 2019). Specifically, prevalence rates for depression (major and minor combined) are estimated at 8.5%–11% during pregnancy and 6.5%–12.9% in the first year postpartum in one study (Gaynes et al., 2005) and at 18.4% during pregnancy and 19.2% in the first 3 months postpartum in another (Yonkers et al., 2009, 2011). Across their life course, 20% of adult women in the United States will experience a major depressive episode at some point, which is twice the rate for men (Yonkers et al., 2011); accordingly, there is debate as to whether pregnancy, the postpartum, and transition to parenting periods are

*Correspondence to: Catherine Monk, Ph.D., Professor of Medical Psychology, Departments of Obstetrics and Gynecology, and Psychiatry, Columbia University Irving Medical Center, Research Scientist VI, New York State Psychiatric Institute, 630 West 168th Street, PH1540H, New York, NY 10032, United States. Tel: +1-917-543-6031, Fax: +1-212-305-2171, E-mail: cem31@cumc.columbia.edu

particular times of vulnerability, with some epidemiologic studies supporting this assertion (Munk-Olsen et al., 2006) and others not (Vesga-Lopez et al., 2008), except for an approximately 1.2% uptick in depression rates postpartum compared with nonchildbearing periods (Vesga-Lopez et al., 2008). This debate, and the somewhat varying prevalence statistics for depression during pregnancy and the postpartum period, reflects and highlights factors in the standard, categorical approach to diagnosing discrete mental conditions that are relevant to deciphering the neurobiology of maternal mental illness during the peripartum period: (1) There is heterogeneity in symptom profiles within the same diagnostic classification (Putnam et al., 2015, 2017) undermining reliability in diagnosing and potentially contributing to differences in who is counted as depressed (and as having other conditions); indeed psychiatric disorders are frequently comorbid, e.g., depression is often comorbid with anxiety (Simon et al., 2004) and mood disorders are also highly comorbid with alcohol use disorders (Shivani et al., 2002) and substance use disorders (Quello et al., 2005). (2) Reproductive steroids regulate mood although changes in them appear to trigger affective dysregulation in some susceptible women and not in others (Schiller et al., 2016) and at some time periods of psychobiologic transition and not others; for example, some women are depressed in pregnancy but not postpartum, others become depressed postpartum without obvious precursors in the prenatal period (Putnam et al., 2015, 2017).

With respect to anxiety disorders, such as specific phobias, social anxiety disorder, generalized anxiety disorder (GAD), posttraumatic stress disorder (PTSD), panic disorder, and obsessive compulsive disorder, a 12-month prevalence was reported to be 14.9% in nonpregnant women, 13% in past year pregnant women, and 12.3% in postpartum women (Vesga-Lopez et al., 2008). Other data estimates GAD occurring in 8.5%–10.5% of women during pregnancy and 4.4%–10.8% postpartum (Misri et al., 2015). Perinatal obsessive compulsive disorder has been less well studied, but approximately 2.6% of postpartum women have been shown to meet the diagnostic criteria for OCD and 5.4% exhibited subsyndromal levels (Wenzel et al., 2003). Preexisting OCD symptoms may become exacerbated during this time (Williams and Koran, 1997; Maina et al., 1999; Labad et al., 2005; Uguz et al., 2007; Abramowitz et al., 2003; Fairbrother and Abramowitz, 2007; Forray et al., 2010; for a review, see Zambaldi et al., 2009). There is suggestive evidence that depression in the perinatal period is characterized by significant anxiety (Putnam et al., 2017), which traditional psychiatric diagnosis does not capture except as comorbid depression and anxiety. Specifically, obsessive compulsive symptoms are commonly reported in patients with postpartum depression (Wisner et al., 1999a), and high rates of postpartum depression have been observed in women with OCD (Williams and Koran, 1997; Labad et al., 2005). For example, it has been reported that 57% of women with postpartum depression exhibit symptoms of OCD (Wisner et al., 1999b). Rates of bipolar disorder and psychosis are lower: 12-month prevalence of bipolar disorder was reported in 2.3% of nonpregnant women, 2.8% of past year pregnant women, and 2.9% of postpartum women (Vesga-Lopez et al., 2008). The point prevalence rate of schizophrenia in women is 1.2% (Mcgrath et al., 2008) while puerperal psychosis incidence estimates range from 0.89 to 2.6 in 1000 women, and one study reported prevalence of postpartum psychosis in 5 in 1000 women (Vanderkruik et al., 2017).

For all of these mental conditions, social determinants of health significantly contribute to illness risk: in poverty samples, postpartum depression is nearly twice the rate as average prevalence (Chung et al., 2004; Chaudron and Nirodi, 2010) and adverse childhood experiences increase the risk of psychiatric hospitalization postpartum (Meltzer-Brody et al., 2018), underscoring the role of social factors in psychiatric conditions and the complexity of mapping their etiology on biologic as well as psychologic levels (Kendler, 2005, 2019).

In contrast to the typical practice of neurology that draws on a physical exam, standardized neurologic tests, as well as biologic measures for diagnostic specificity, MRI to identify brain lesions in MS or EEG to define seizure disorder, psychiatry primarily relies on self-report of symptoms using standardized questionnaires and patient observation in the absence of any biologically based test (except to rule out other conditions, such as blood test for hypo- or hyperthyroid functioning and, infrequently, lumber puncture for anti-NMDA receptor encephalitis). Leading psychiatrists are highly critical of the latest version of the American Psychiatric Association's Diagnostic and Statistical Manual (DSM)-5, worrying that there is a "lack of validity" to the classification system, calling it "at best a dictionary" or even "a scientific nightmare" (Insel, 2013; Scull, 2015; Cahalan, 2019). When DSM-3, the third version, was published in 1980, it dramatically improved the reliability and hence seeming validity of the diagnostic system by more systematically cataloging symptoms and moving away from conjectures on etiology; yet, the diagnostic categories in that version, and in today's fifth, were based on consensus regarding clinical symptom constellations, given the lack of knowledge about the brain's structure and functioning (Cuthbert and Insel, 2010).

With the emergence of spectacular new neuroscience methodologies and discoveries beginning in the 2000s, psychiatric researchers viewed the DSM as hampering progress because mental illness syndromes based on self-reports of symptoms were artificially siloed and diagnoses unmoored to discrete components of brain functioning—for example, the common dimension of cognitive control is dysregulated in eating disorders and OCD, yet studying their neurobiology happens in isolation (Marsh et al., 2009). These limitations in classification and ontology were viewed as slowing the discovery of pathophysiologic mechanisms, biobehavioral markers, risk prediction tools, and preventive/treatment interventions. In its 2008 Strategic Plan, the National Institute of Mental Health (NIMH) announced an initiative to "develop, for research purposes, new ways of classifying mental disorders based on dimensions of observable behavior and neurobiological measures" (National Institute of Mental Health, 2008). Dubbed the research domain criteria (RDoC), it is a research framework for new approaches to understanding and treating mental disorders. This still-evolving conceptual framework—a new kind of taxonomy for mental disorders for research—is designed to integrate various types of information across multiple levels of analysis, including genetics, molecules, cells, circuitry, behavior, physiology, and self-report, to characterize functioning across six dimensions: negative valance systems, positive valence systems, cognitive systems, systems for social processes, arousal/regulatory systems, and sensorimotor systems (National Institute of Mental Health, 2008; Insel et al., 2010; Casey et al., 2013). For example, by using an RDoC framework, instead of grouping research participants based on symptom-defined diagnostic criteria, they might be selected through an operationalized component of a dimension, such as a particular pattern of activation in reward circuitry under positive valance systems in the context of hormonal changes, which more likely will reveal mechanisms informing how cognition and behavior emerge and are constrained (Cuthbert and Insel, 2010). Importantly, the RDoC model does not privilege biology over other levels of description (self-report); it aims for integration across levels for a data-driven approach to describing psychopathology that includes knowledge of the brain, genes, behavior, and subjective experience. RDoC also does not address etiology; while variation in a brain circuit may help explain disturbed functioning, and a specific allele part of the variation, what *caused* a psychopathologic outcome, with a capital "*C*," is not considered. For etiology, development and life-course perspectives are needed. Specifically, in the words of former NIMH director Thomas Insel, "it is time to rethink mental disorders, recognizing that these are disorders of brain circuits likely caused by developmental processes shaped by a complex interplay of genetics and experience" (Insel and Wang, 2010).

In this chapter, we provide an up-to-date understanding of the neurobiology of maternal mental disorders, acknowledging that the majority of findings are the outcomes of studies using the standard diagnostic categories from the DSM or the International Statistical Classification of Diseases and Related Health Problems (ICD), a medical classification list developed by the World Health Organization, which is similar to DSM. As a consequence, the chapter is divided into sections based on standard diagnostic categories. When possible, we highlight research consonant with an RDoC approach, as well as at points consider etiology with a particular focus on those risk factors occurring early, across the lifespan, and related to brain–behavior outcomes. We begin with a brief overview of changes in the brain, neuroendocrinology, and immune activity during pregnancy and the postpartum periods as they inform an understanding of differences potentially related to pathology. We proceed to review research findings that are beginning to explain the neurobiology of maternal mental illness.

NEUROPLASTICITY IN PREGNANCY

Pregnancy and the transition to parenting can be conceptualized as a developmental period in adulthood involving dramatic shifts in cognition and behavior in addition to physical changes. Evidence from human and animal studies demonstrates that the brain undergoes significant changes in structure and function during the peripartum period, which are generally thought to be adaptive with respect to facilitating maternal orientation toward the mother–infant dyad (Barba-Müller et al., 2019). Methodologically, there are challenges in studying these phenomena, most notably that pregnancy is often considered an exclusion factor for research-based magnetic resonance imaging (MRI), the dominant research tool for examining brain structure and function. Nevertheless, translational studies have begun to provide a picture of the healthy human "parental brain," a necessary step to determining whether different trajectories of neural changes in the peripartum may contribute to risk or resilience for psychopathology. This section provides a brief overview of evidence related to neuroplasticity in the perinatal period.

The mammalian brain demonstrates a striking degree of plasticity in both structure and function in the peripartum, which are thought to promote caregiving-related cognitions and behaviors and enable a mother to effectively adapt to the challenges of this time (Kinsley and Amory-Meyer, 2011; Kim et al., 2016; Barba-Müller et al., 2019). Foundational studies in rodents demonstrate that neural circuits underlying maternal behavior are triggered by endocrine signals during pregnancy and subsequently maintained by experiences of caregiving for pups, with the medial preoptic area consistently identified as a key driver of caregiving behaviors (Numan, 2007; Kohl et al., 2017). Significant plasticity in the hippocampus also occurs during pregnancy in mammals and contributes to cognitive changes across the reproductive life course (Pawluski et al., 2016). While few studies have investigated structural brain changes in humans over this period, there is emerging consensus for global and specific reductions in gray matter volume (GMV) in the postnatal period compared with prepregnancy. A comparison of structural brain changes from pre- to postpregnancy with changes in the adolescent brain revealed very similar patterns of cerebral morphometric changes, including reductions in volume and cortical thickness and surface area, as well as reductions in sulcal depth and increases in sulcal width, suggesting pregnancy and adolescence are both periods characterized by hormonally driven cortical adaptations (Carmona et al., 2019). Extending earlier findings that total brain size reduced from pre- to postpregnancy stages (Oatridge et al., 2002), a longitudinal study additionally demonstrated that pregnancy was associated with GMV reductions in particular areas (Hoekzema et al., 2017). Specifically, compared with prepregnancy, voxel-wise GMV reductions were observed in the superior temporal sulcus, medial and inferior frontal cortex, fusiform areas, and the hippocampus (Hoekzema et al., 2017). These GMV reductions were interpreted as being adaptive in terms of parenting as greater reductions were predictive of more optimal self-reported mother–infant attachment. Data from the same sample also suggests that volumetric changes in reward-sensitive brain regions occur from pre- to postpregnancy, including volume reduction in the right ventral striatum. The degree of volume reduction was associated with increased functional activation of the same region in response to infant cues, leading authors to infer that such anatomical changes over the peripartum facilitate increased responsiveness of a mother's neural reward circuit to the infant (Hoekzema et al., 2020).

Though these studies suggest a reduction in brain volume from pre- to postpregnancy, there is less certainty about whether these changes persist beyond the peripartum. Despite reductions in brain and GMV often being considered characteristics of aging, a study, interestingly, showed that maternal brains appeared significantly younger at 4–6 weeks postpartum (based on a well-validated anatomical "brain age index," calculated via machine learning) compared with 1–2 days after childbirth, suggesting a "rejuvenation" across early postpartum weeks (Luders et al., 2018). In another study, however, most observed reductions in GMV persisted until at least 2 years postpartum; only the hippocampus had returned to baseline volume by 2 years postpartum (Hoekzema et al., 2017). In contrast, Kim et al. (2010) reported increases in GMV in superior, middle, and inferior prefrontal cortex, parietal lobe, and lymbic and cerebellar structures from 2–4 weeks to 3–4 months postpartum, and also that higher degree of change was associated with more positive maternal perceptions of infants (Kim et al., 2010). These findings were replicated in another study using a whole-brain voxel-based approach across similar postnatal time points, with the additional finding of gray matter increase in the precuneus, middle occipital gyrus, and caudate (Luders et al., 2020). Together, these findings point toward a reorganization of brain tissue across the postpartum, perhaps associated with the evolving adjustment to motherhood and the development of multifarious caregiving behaviors.

In addition to structural transformations, the maternal brain has been shown to exhibit functional neuroplasticity over this time period, believed to support mothers as they transition to motherhood and cope with parenting stressors (Lonstein et al., 2015). Rodent models demonstrate that pregnancy is associated with cognitive and behavioral changes facilitating caregiving, including suppression of aversive responses to pups and improved foraging and spatial reasoning ability (Kinsley et al., 1999), promoted by enhanced plasticity in hippocampal neurons (Lambert et al., 2005), and exposure to pups additionally alters neurogenesis in the hippocampus (Pawluski and Galea, 2007). In humans, high-order social cognitive ability becomes increasingly essential during motherhood as mothers must adapt to become attuned and responsive to the needs of their new baby and for successfully raising offspring in a complex social environment (Anderson and Rutherford, 2012). Converging evidence from neuroimaging studies shows heightened emotional response and activation of theory of mind regions in healthy postpartum women (Leibenluft et al., 2004; Swain, 2011; Gingnell et al., 2015) and that inputs from these regions activate regions in the midbrain and striatum, which are critically involved in caregiving motivation. Subsequently, inputs from these regions are processed by middle and lateral prefrontal regions involved in emotion regulation and attention, suggesting their role in organizing maternal response to infant distress sounds, which includes regulating her own emotions and enabling sensitive responses to these cues (Rutherford et al., 2015; Kim et al., 2016). These brain changes confer benefits to the offspring as they facilitate sensitive caregiving by adaptive orientation in motivation and reward to be directed toward the infant; however, it is possible that the same brain changes could contribute to deleterious effects on maternal psychologic well-being for some women. Cárdenas, Kujawa and Humphreys propose a model in which hyporesponsive brain changes may be associated with decreased risk for psychopathology, and hyperresponsive changes with increased risk (Cárdenas et al., 2019).

Relatively little is known about how individual differences such as psychopathology may shape functional brain changes related to the transition to parenting, with the postpartum period of particular interest given potential downstream behavioral effects on interactions with the child. A review of the literature on neural responses to infants among depressed mothers identified only five fMRI studies, which, taken together, suggested heightened amygdala reactivity and aberrant connectivity in response to emotional infant stimuli, as well as lower connectivity between the amygdala, insula, and orbitofrontal cortex; these findings indicate that compared with healthy mothers, depressed mothers have a dampened reward response and reduced empathy to infant stimuli (Bjertrup et al., 2019). Adaptive changes to the parental brain can also be affected by maternal distress: subjective stress is associated with reduced activity in the right medial prefrontal cortex (mPFC) and bilateral midfrontal cortex in response to infant cry, and also with less positive perceptions of parenting (Kim et al., 2011).

Considering the growing evidence pointing to the emergence of a "parental brain" that underlies caregiving behavior and adaptation to motherhood, further investigation into the influence of maternal psychopathology on these neural adaptations likely will underscore the dyadic nature of maternal mental illness as affecting both mother and child (Barrett and Fleming, 2011; Werner et al., 2015). In attempting to understand how psychopathology may emerge or be exacerbated in the peripartum, consideration of the potential negative consequences of these otherwise adaptive changes to the maternal brain, as well as individual differences in the degree of brain change across this period, could reveal neural bases for perinatal mental illness.

NEUROENDOCRINOLOGY AND IMMUNE CHANGES IN PREGNANCY

Pregnancy is a time of profound physiologic changes that serve to meet the demands of this period and ensure protection of the fetus. Moreover, pregnancy, parturition, and lactation are sequential periods characterized by neuroendocrine and inflammatory changes required for the development and maintenance of the fetus and developing infant.

A coordinated sequence of neuroendocrine changes must occur to establish and successfully maintain a healthy pregnancy. These changes in hormone levels coordinate adaptive changes throughout the peripartum period, including parturition and lactation (Russell et al., 2001). In early pregnancy, there is a rapid rise in chorionic gonadotropin (Venning, 1955; Betz and Fane, 2019), which is produced by the trophoblast, specialized cells of the placenta, and functions to maintain the corpus luteum and promote the secretion of progesterone and estrogen (Venning, 1955). Progesterone levels increase during the first 3 months of pregnancy and remain elevated throughout late gestation. After delivery, the levels decline rapidly. Estrogen follows a similar pattern with levels increasing slowly during the first 3 months of pregnancy and rising more dramatically during late pregnancy (Venning, 1955).

There is also a close relationship between the hypothalamic–pituitary–adrenal (HPA) axis and the reproductive system. Initially, there is a rise in the secretion of glucocorticoids, which returns to normal levels after the first trimester and is followed by another increase during the last trimester (Venning, 1955). Typically, corticotropin-releasing hormone (CRH) is produced in neurons in the paraventricular nucleus of the hypothalamus and released into the hypophyseal portal system to trigger the release of adrenocorticotropic hormone (ACTH) from the pituitary gland and then cortisol from the adrenal cortex. During pregnancy, CRH is produced in the placenta and serves to coordinate the physiology of pregnancy and parturition. Immediately after delivery, glucocorticoid levels rapidly decline, becoming undetectable 24 h following delivery (Petraglia et al., 2010). In contrast to the negative feedback regulation of the HPA axis, glucocorticoids positively regulate CRH produced by the placenta, at the same time suppressing stress reactivity at the level of the hypothalamus during the peripartum period (Petraglia et al., 2010).

Neuroendocrine changes associated with parturition are distinct from those that occur throughout pregnancy. Oxytocin levels increase at the onset of labor, peaking at the time of delivery. However, the dynamics of oxytocin levels are a point of contention due to challenges in measurements as different approaches are used and there are other issues regarding the accuracy of the measurements (Prevost et al., 2014). Oxytocin is thought to play a role in both the initiation of labor and the expulsive phase of delivery (Blanks and Thornton, 2003; Prevost et al., 2014), and its importance in parturition is supported by the evidence that oxytocin administration is sufficient to induce labor.

Neuroendocrine changes associated with lactation are also distinct from those associated with either pregnancy or parturition. A marked neuroendocrine adaptation occurring during pregnancy involves alleviation of the mechanisms limiting prolactin secretion, resulting in hyperprolactinemia during pregnancy and lactation. Elevated prolactin levels play an important role in the development and function of the mammary gland. Prolactin secretion is stimulated by copulation and serves to protect the corpus luteum, a mass of cells that forms in the ovary and produces progesterone during early pregnancy. Prolactin secretion also increases during mid-pregnancy, which inhibits hypothalamic prolactin release. During late pregnancy, prolactin levels rise before parturition. Following parturition, prolactin secretion is tightly linked to suckling. In addition to neuroendocrine changes responsible for the physiologic transformations associated with pregnancy, changes in hormone levels, such as those of prolactin and oxytocin, have also been proposed to play a role in maternal behaviors (Grattan and Kokay, 2008).

Pregnancy is a time marked by changes in immune function. It was initially thought to represent a time of immune suppression necessary to allow implantation and growth of the fetus. This concept was supported by the evidence that progesterone exerts immunosuppressive actions. However, we now understand that this is not the case, but rather the immune system is active and tightly controlled during normal pregnancy (Mor et al., 2011). This period is essential for the conservation of the species, and, therefore it is very reasonable to posit that the immune system would function to protect the mother and the offspring. What has become abundantly clear is that the function of the immune system is unique during the peripartum period.

Pregnancy is characterized as a period of heightened recognition, communication, trafficking, and repair (Mor et al., 2011). During normal pregnancy, a number of immune cells, including macrophages, natural killer (NK) cells, and regulatory T cells (Treg), invade the decidua and accumulate around the trophoblast. These immune cells appear to play a supportive role, since their depletion results in the termination of pregnancy due to the deleterious effects on placental development, implantation, or decidual formation in humans and animal models (Croy et al., 2000; Abrahams et al., 2004; Hanna et al., 2006; El Costa et al., 2008).

Pregnancy has been proposed to involve three distinct immunologic phases: (1) implantation, placentation, and as such the first and early second trimester of pregnancy requiring a proinflammatory response; (2) fetal growth and development requiring a supportive immune environment and an antiinflammatory state; and (3) parturition requiring strong immune support. During the first trimester when the blastocyst is implanted in the epithelial lining of the uterus, immune cells function to repair tissue and remove cellular debris (Abrahams et al., 2004; Koga and Mor, 2010); thus the first trimester of pregnancy represents a proinflammatory phase (Mor et al., 2011). The second immunologic phase of pregnancy is a period of fetal growth and development that involves cooperation between the physiology of the mother and the fetus, requiring an antiinflammatory state. Finally, the third immunologic phase of pregnancy occurs in preparation for delivery. At the time of parturition, there is a recruitment of immune cells to promote contraction of the uterus and expulsion of the fetus and placenta. Thus it is apparent that pregnancy is a dynamic period of immune function, which is tightly controlled to support the development of the fetus and result in a successful pregnancy.

The unique function of the immune system during pregnancy is evident from the fact that pregnant women have been demonstrated to respond differently to insults. For obvious ethical reasons, these studies largely rely on ex vivo stimulation measures. For example, stimulation of peripheral blood mononuclear cells with lipopolysaccharide (LPS) increases interleukin-6 (IL-6), tumor necrosis factor-α (TNF-α), and interleukin-1 (IL-1) during late pregnancy compared with early pregnancy, which return to prepregnancy levels postpartum (Gillespie et al., 2016). Further, many autoimmune diseases remit during pregnancy and recur postpartum, suggesting altered immune function during pregnancy. Animal models have been utilized to explore the relationship between pregnancy and inflammatory responses and have demonstrated a diminished response during pregnancy compared with nonpregnant animals (Vizi et al., 2001; Faas et al., 2003; Fofie et al., 2005; Mcclure et al., 2005; Aguilar-Valles et al., 2007; Cui et al., 2009). Thus throughout the peripartum period, the response of the immune system depends on the type of insult and stage of pregnancy.

The inflammatory changes associated with the peripartum period are partially influenced by the neuroendocrine changes, suggesting cooperation between these systems (Bränn et al., 2019). For example, progesterone is known to exert antiinflammatory effects and glucocorticoids also exert regulatory effects on the immune system. Conversely, inflammatory changes can induce neuroendocrine changes, particularly through effects on the HPA axis function. These maternal adaptations involve changes in brain function, driven by neuroendocrine and immune changes, altering stress reactivity, ingestive behaviors, emotional processing, and homeostatic mechanisms, fostering the emergence of complex maternal behaviors and, possibly, the potential for dysfunction in brain and behavior levels.

PERINATAL DEPRESSION

In DSM-5, peripartum depression is classified under the category of major depressive disorder with a specification that symptoms must occur during pregnancy or within the first 4 weeks following delivery. The symptom presentation of patients with postpartum depression resembles those of major depressive disorder and is often comorbid with anxiety disorders (Wisner et al., 2013). Although it is unclear if the symptom presentation is distinct from that of major depressive disorder, it is clear that the symptom onset is unique, which may indicate differences in the underlying neurobiology of disease. Further, postpartum depression also exhibits unique psychosocial risk factors. The heterogeneity in perinatal depression has fueled a discussion about whether there are actually different types of perinatal depression included under this umbrella, which, as indicated previously, has complicated diagnosis, clinical trials, and potentially advancement in understanding the neurobiology of the disorder(s).

Information gained from both the clinical population and preclinical studies has provided insight into the potential mechanisms contributing to the underlying neurobiology of peripartum depression, which includes neuroendocrine changes, neuroinflammation, neurotransmitter alterations, circuit dysfunction, and the involvement of genetics and epigenetics (Pawluski, 2019; Payne and Maguire, 2019). Neuroendocrine changes, ranging from ovarian hormones to glucocorticoids, have been implicated in perinatal depression. The potential involvement of reproductive hormones in perinatal depression has received a great deal of attention, given the symptom onset at the time of hormone withdrawal and the higher incidence of depression in females relative to males (Weissman and Klerman, 1977). However, consistent changes in the levels of reproductive hormones have not been observed in women with perinatal depression (for review, see Schiller et al., 2015). This could be an indication that reproductive hormones do not play a major role in the underlying neurobiology of perinatal depression or could reflect the heterogeneity in the patient population making these associations extremely difficult to detect reliably. Interestingly, there is evidence that women with postpartum depression may exhibit altered sensitivity to gonadal steroids. A seminal study demonstrated that withdrawal from estradiol and progesterone induced depressive symptoms only in patients with a history of postpartum depression (Bloch et al., 2000). These findings suggest that, although it may be difficult to resolve absolute differences in reproductive hormone levels associated with perinatal depression, differences in sensitivity to neuroendocrine changes may be associated with perinatal depression.

Preclinical models also have been employed to examine the impact of changes in reproductive hormones on postpartum depression-like behaviors. Withdrawal from ovarian hormones (progesterone and escalating doses of estradiol) using a hormone-simulated pregnancy model has been shown to induce depression-like behaviors in rats (Stoffel and Craft, 2004). Similarly, estradiol withdrawal in an estrogen withdrawal model of postpartum depression was shown to increase anhedonia (Schiller et al., 2013). Progesterone withdrawal also has been demonstrated to induce depression-like behaviors in rodents (Beckley and Finn, 2007). These preclinical studies are useful for directly demonstrating a relationship between ovarian hormone withdrawal and depression symptoms, with potential relevance to peripartum depression.

Changes in neurosteroid levels also have been implicated in the underlying neurobiology of perinatal depression. Allopregnanolone, a metabolite of progesterone with neuroactive capabilities, has been demonstrated to exert anxiolytic and antidepressant effects (for review, see Schüle et al., 2014), and withdrawal from neurosteroids is capable of inducing depression-like symptoms in rodents (Beckley and Finn, 2007). Reduced allopregnanolone levels have been associated with the risk of developing postpartum depression (Osborne et al., 2017) and correlated with symptom severity in postpartum depression (Hellgren et al., 2014), although other studies failed to demonstrate an association between allopregnanolone levels and postpartum depression (Amin et al., 2006; Deligiannidis et al., 2013). Perhaps the best evidence of a role for neurosteroids in postpartum depression comes from the FDA approval of an allopregnanolone-based treatment for postpartum depression (Scott, 2019), which followed extensive preclinical studies.

Allopregnanolone exerts its neuroactive effects largely through positive allosteric modulation of δ subunit-containing $GABA_A$ receptors ($GABA_ARs$). The δ subunit of the $GABA_AR$ confers neurosteroid sensitivity (Belelli and Lambert, 2005) and, therefore, allopregnanolone is capable of enhancing the tonic GABAergic inhibition mediated by these receptors (Farrant and Nusser, 2005; Belelli et al., 2009). Mice that lack these neurosteroid-sensitive receptors ($Gabrd^{-/-}$ mice) exhibit abnormal postpartum behaviors, including depression-like behaviors and deficits in maternal care (Maguire and Mody, 2008). This preclinical model of postpartum depression-like behaviors was the first to test the therapeutic potential of positive allosteric modulators (PAMs) of $GABA_AR$ δ subunit-containing receptors (Maguire and Mody, 2008). These data provided the foundation for the use of GABA PAMs for the treatment of postpartum depression (Mody, 2019).

Additional neuroendocrine changes implicated in postpartum depression involve the stress hormones, CRH, and cortisol. Stress is a major risk factor for perinatal depression (Righetti-Veltema et al., 1998; Robertson et al., 2004; O'hara and Wisner, 2014) and positively correlates with depression severity scores (Paykel et al., 1980; O'hara et al., 1984; O'hara, 1986). Previous adverse life events also are a risk factor for perinatal depression and similarly correlate with the severity of depression symptoms (O'hara et al., 1984, 1991; Barnet et al., 1996; Guintivano et al., 2018; Meltzer-Brody et al., 2018).

Consistent with the proposed role for stress and stress hormones in the underlying neurobiology of postpartum depression, stress and stress hormones have been shown to induce postpartum depression-like behaviors in preclinical models (for review, see Perani and Slattery, 2014; Pawluski et al., 2017). Chronic stress administered throughout pregnancy induces deficits in maternal care and depression- and anxiety-like behaviors in postpartum dams (Maestripieri et al., 1991; Pardon et al., 2000; Brummelte and Galea, 2010; Nephew and Bridges, 2011; Carini et al., 2013; Murgatroyd et al., 2015; Maguire and Mody, 2016). Exogenous corticosterone administration during pregnancy or lactation is also sufficient to induce maternal care deficits and depression- and anxiety-like behaviors in preclinical models (Brummelte and Galea, 2010; Workman et al., 2013). Elevated levels of CRH during the perinatal period were proposed to form a useful diagnostic criterion for postpartum depression (Yim et al., 2009); however, other studies did not find this association using a covariate-adjusted comparison (Meltzer-Brody et al., 2011). Consistent with previous adverse life events as a risk factor for postpartum depression, early life stress in rodents induces deficits in maternal care and increases depression-like behaviors during the postpartum period (Murgatroyd et al., 2015). These data demonstrate that some of the risk factors for postpartum depression observed in the clinic can be employed to establish useful preclinical models and underscore the need for a life-course perspective when considering etiology, and potentially treatment. In a seminal paper, Nemeroff et al. showed that women with chronic depression responded differently to intervention based on childhood adversity: those with trauma backgrounds responded better to psychotherapy alone than medication alone and did not show a noticeable improvement when provided medication and psychotherapy; those without a trauma background showed the highest levels of remission with medication or medication and psychotherapy (Nemeroff et al., 2003). Similar research on treatment outcomes for women with peripartum depression has not yet been conducted; such work could lead to improved understanding of the neurobiology of peripartum depression and would be consistent with an RDoC approach by, for example, grouping pregnant or postpartum women based on trauma-related difficulties on the dimension of social processes and proceeding with biologically oriented studies related to variation in HPA axis regulation.

Inflammatory changes also have been implicated in perinatal depression. As previously described, from conception through lactation, the peripartum period involves marked changes in immune function. These changes are necessary to facilitate fetal growth and development as well as protect the mother and fetus during this critical time, and they have been implicated in peripartum depression. Altered immune function associated with postpartum depression has largely relied on changes in the expression of cytokines and the number of immune cells in women with postpartum depression. The most robust and reproducible inflammatory changes observed are increased levels of IL-6, IL-1β, and TNF-α associated with postpartum depression, which correlate with depression scores (Corwin and Pajer, 2008; Cassidy-Bushrow et al., 2012; Fransson et al., 2012; Osborne and Monk, 2013) (see Osborne et al., 2019, for some contrasting findings). Additional studies have shown a negative correlation between T cell number and postpartum depression symptoms (Hucklebridge et al., 1994) and decreased levels of IFN-γ levels and a lower IFN-γ:IL-10 ratio (Groer and Morgan, 2007). The kynurenine pathway, which, when dysregulated or overactive, can lead to immune system activation, also has been implicated in postpartum depression with increased kynurenine positively correlated with postpartum depression and depression severity scores (Maes et al., 2002). Inflammatory changes are highly variable, particularly during the peripartum period, and therefore further studies are required to definitively demonstrate inflammatory changes associated with the underlying neurobiology of postpartum depression. The lack of study in preclinical models limits the mechanistic insights that can be drawn between immune function and postpartum depression. Given the dramatic immune changes that occur throughout the peripartum period and the appreciation for inflammatory changes impacting neuronal processes, this topic warrants further investigation. Moreover, the neuroendocrine system impacts inflammatory processes and, therefore, there may be bidirectional influences between these proposed mechanisms that require additional study.

Network-level changes also have been implicated in the underlying neurobiology of postpartum depression, largely as a result of imaging studies in patients with peripartum depression. These changes represent variations in the activity in interconnected groups of neurons that coordinate activity and have been implicated in mediating transitions between behavioral states, including mood (Ochsner et al., 2009). Network activity implicated in mediating mood states have been shown to be disrupted in patients with depression, including patients with major depressive disorder as well as women with postpartum depression (Greicius et al., 2007; Kaiser et al., 2016; Brakowski et al., 2017). For example, changes in the default mode network (i.e. in the absence of external stimuli) have been observed in patients with postpartum depression, specifically, evidence of reduced connectivity of the anterior cingulate cortex, amygdala, hippocampus, and dorsolateral prefrontal cortex (Deligiannidis et al., 2013). The amygdala has particular relevance to depression given that this region has been previously implicated in mood and emotion (Dick et al., 1996; Fossati, 2012). Further evidence suggests a role for the amygdala in depression, including the observed reduction in amygdala volume in patients with major depressive disorder (Hamilton et al., 2008) and neuroanatomical changes in the basolateral amygdala associated with depression (Rubinow et al., 2016). A reduction in connectivity between the amygdala and multiple brain regions, including the ventrolateral prefrontal cortex, insula, caudate, middle and superior temporal regions, occipital cortex, and cerebellum (Ramasubbu et al., 2014; Tang et al., 2019), and increased resting-state activity in limbic areas, including the amygdala (Drevets, 2000; Faymonville et al., 2003), also have been observed in major depressive disorder. Our current understanding surrounding the function of the amygdala in reward, anxiety, and valence processing (Janak and Tye, 2015) suggests that these pathological features involving the amygdala (summarized earlier) are associated with the primary pathology of major depressive disorder. Further, similar gene expression changes have been identified in patients and experimental models of depression (Labonté et al., 2017), including molecular changes in the amygdala, which have been implicated in vulnerability to major depressive disorder (Sibille et al., 2009). The majority of these studies have focused on network-level changes associated with major depressive disorder given that there are fewer studies specifically investigating network-level changes associated with postpartum depression. However, abnormalities in similar networks have been observed in patients with major depressive disorder and postpartum depression. For example, network dysregulation in patients with postpartum depression also involves the amygdala. Postpartum depression has been associated with reduced connectivity between the amygdala and other brain regions, including the insular cortex (Wonch et al., 2016) and the posterior cingulate cortex (Chase et al., 2014). Functional magnetic resonance imaging (fMRI) studies have also been performed to evaluate changes in neural circuits in response to infant-relevant stimuli. For example, hypoactivation of the amygdala is reproducibly observed in response to infant-relevant stimuli in patients with postpartum depression compared with healthy mothers (Moses-Kolko et al., 2010; Laurent and Ablow, 2011, 2013; Wonch et al., 2016), a finding that builds on the emerging data on the parental brain as it is moderated by individual differences in mood states. These data suggest changes in network engagement, potentially specifically related to adaptation for child rearing, associated with postpartum depression. Elucidation of the networks involved in perinatal depression may provide further insight into the underlying neurobiological mechanisms and potential therapeutic strategies to restore healthy network function. Studies are just beginning to explore network-level changes associated with perinatal depression in preclinical models informed by the RDoC approach, which is essential for establishing the mechanisms of network dysfunction associated with perinatal depression and treatment strategies targeted at restoration of healthy network function.

Finally, it is worth noting that in addition to the neuroendocrine, inflammatory, and network-level changes highlighted in this chapter, aberrant plasticity in neurotransmitter systems and genetic and epigenetic changes also have been proposed to contribute to peripartum depression (Pawluski et al., 2017; Payne and Maguire, 2019).

PERINATAL ANXIETY AND OCD

Peripartum anxiety is similar to GAD in that symptoms include excessive, uncontrollable worry and it is associated with three or more of the following six symptoms: restlessness or feeling keyed up or on edge, being easily fatigued, difficulty concentrating or mind going blank, irritability, muscle tension, or sleep disturbance; these symptoms cause clinically significant distress or impairment in social, occupational, or other important areas of functioning. However, peripartum anxiety also has unique features, including commonly held worries that revolve around fears of fetal well-being, maternal wellness, illness in the partner, and/or parental mortality (Misri et al., 2015). One of the limitations in the diagnosis of peripartum GAD is that the physical symptoms of GAD, including fatigue, irritability, difficulty concentrating, tension, and insomnia, may be mistaken for normal consequences of pregnancy and/or new parenthood. Despite the high incidence of anxiety during the peripartum period, the DSM specifier "with peripartum onset" is currently only designated for mood disorders. The lack of a perinatal-specific screening tool for GAD is a significant barrier to diagnosis and treatment. For example, the DSM criteria for GAD specify that symptoms occur for 6 months, which may not be an appropriate time frame for perinatal anxiety, and it has been suggested that 1 month may be more appropriate (Misri et al., 2015).

OCD is an anxiety disorder involving obsessions or compulsions in which recurrent and persistent thoughts or impulses are intrusive and/or inappropriate and cause anxiety or distress. OCD symptoms during the postpartum period commonly involve obsessive and intrusive thoughts about harming the newborn (Sichel et al., 1995; Jennings et al., 1999) and associated checking compulsions (Sichel et al., 1995; Arnold, 1999).

The etiology of peripartum anxiety and OCD are not well understood but are thought to involve genetic, environmental, immunological, and hormonal factors (Brandes et al., 2004). Similar to peripartum depression, the timing of symptom presentation has implicated neuroendocrine factors in the underlying neurobiology of peripartum anxiety and OCD, including rapid changes in oxytocin, estrogen, and progesterone (Stein et al., 1993; Leckman et al., 1994; Young et al., 2001; Bartz and Hollander, 2008); in particular, oxytocin has been implicated in postpartum anxiety given the evidence that it reduces this anxiety (Lonstein et al., 2014).

Due to the timing of symptom onset, changes in neurosteroids have been implicated in postpartum anxiety based on the dramatic alterations in neurosteroid levels during the peripartum period and their well-established anxiolytic effects (Smith et al., 2007). Further, steroid hormone withdrawal increases anxiety-like behaviors in preclinical models (Smith et al., 1998, 2004, 2006; Bitran and Smith, 2005) (for review, see Pawluski, 2019). Imaging studies in postpartum women with OCD symptoms show differential activation of the superior and inferior temporal gyri, orbital frontal cortex, insula, and the mPFC during a psychosocial stressor task (Lord et al., 2012). Preclinical studies have provided information about potential brain regions involved in postpartum anxiety, including the mPFC, bed nucleus of the stria terminalis, and periaqueductal gray (Figueira et al., 2008; Smith et al., 2013; Sabihi et al., 2014; Klampfl et al., 2016). Preclinical and clinical studies have demonstrated changes in GABAergic, norepinephrine, serotonin, CRH, and oxytocin signaling associated with postpartum anxiety (Toufexis et al., 1999; Figueira et al., 2008; Sekiyama et al., 2013; Smith et al., 2013; Stuebe et al., 2013; Klampfl et al., 2014, 2016; Lonstein et al., 2014; Sabihi et al., 2014; Deligiannidis et al., 2016).

Similar to peripartum depression, stress is a risk factor for peripartum anxiety and OCD (Mckeon et al., 1984; Britton, 2008; Dennis et al., 2016), and women with postpartum OCD exhibit increased stress reactivity (Lord et al., 2012). Stress-based preclinical models have been shown to induce anxiety-like behaviors during the postpartum period (Maestripieri et al., 1991) (for review, see Hillerer et al., 2011). For example, chronic restraint stress throughout pregnancy has been shown to increase anxiety-like behaviors in postpartum dams (Maestripieri et al., 1991).

Neuroimaging studies implicate the cortico-striato-thalamo-cortical circuits, which control movement selection and initiation, reinforcement, and reward, in the pathophysiology of OCD. Thus these imaging studies in patients demonstrate unique features compared with peripartum anxiety. However, to our knowledge, imaging studies have not been performed to specifically investigate brain regions involved in perinatal OCD. Abnormalities in the dopaminergic system also have been suggested to play a role in OCD based on the proposed role for the basal ganglia (Welter et al., 2011). Further, serotonin has been implicated in the underlying neurobiology of OCD due to the successful treatment of some individuals with SSRIs (Koran and Saxena, 2000; Kellner, 2010; Pittenger and Bloch, 2014). Glutamatergic signaling also has been specified in OCD processes, largely supported by mutations in the glutamate transporter gene SLCL1A1 identified in association with OCD (Stewart et al., 2013) and treatment with glutamatergic agents (Coric et al., 2005; Grant et al., 2007; Kushner et al., 2007; Wilhelm et al., 2008; Haghighi et al., 2013). As indicated, similar to other peripartum mood disorders, neuroendocrine changes have also been implicated in the underlying neurobiology of peripartum OCD. In particular, ovarian hormones have been implicated in peripartum OCD due to the evidence of symptom manifestation related to alterations in the hormonal profile in women (Maina et al., 1999; Abramowitz et al., 2003; Labad et al., 2005). Although OCD is more challenging to study in preclinical models, compulsivity, stereotypy, and perseverance have been applied as a useful outcome measure with potential relevance to OCD (Alonso et al., 2015), an approach very consistent with RDoC research. However, to our knowledge, no preclinical studies on peripartum OCD have been undertaken to date, which is necessary to understand the underlying neurobiology.

The limited number of studies in animal models is beginning to provide mechanistic insights into postpartum anxiety and OCD. However, the lack of clinical and preclinical studies hinders our understanding of the underlying neurobiology of these disorders. The differences in symptomology suggest that there may be unique features of peripartum anxiety and OCD. To fully appreciate the mechanisms contributing to peripartum anxiety and OCD, additional clinical and preclinical studies are required.

POSTPARTUM PSYCHOSIS AND BIPOLAR DISORDER

In the perinatal period, severe mental illness may occur either as a continuation of preexisting illness or as a new disorder with onset following childbirth. Although not recognized by the DSM or ICD classification systems, "postpartum psychosis" or "puerperal psychosis" are terms commonly used in clinical practice to refer to severe episodes of mania, psychotic depression, or psychosis in the postpartum. Primiparity is a consistently reported obstetric risk factor for postpartum psychosis (Blackmore et al., 2006; Bergink et al., 2011a, 2018), suggesting sensitivity to immune and endocrine changes, as well as psychosocial stress associated with the transition to parenting.

Dysregulated immune activation is associated with acute first-onset psychosis. These patients show elevated monocyte levels but not the normal elevation of T cells seen in healthy postpartum women (Bergink et al., 2013). There is an association between first-onset postpartum psychosis and preeclampsia, a common condition in pregnancy with immune-related pathology (Bergink et al., 2015), as well as autoimmune thyroiditis (Bergink et al., 2011b, 2018).

There also is a clear association between childbirth and onset of bipolar disorder (Munk-Olsen et al., 2012). For women with established bipolar disorder prior to pregnancy, there is a very high risk of relapse and severe mood episodes in the immediate postpartum (Di Florio et al., 2013), with an estimated 35% relapse rate in the postpartum overall and higher rates among those who were medication-naïve during pregnancy (Wesseloo et al., 2016). As with postpartum psychosis, risk for bipolar affective episodes is higher among primiparous women (Blackmore et al., 2006; Di Florio et al., 2014; Munk-Olsen et al., 2014), a finding that remains significant after accounting for the possibility that women experiencing a postpartum episode could be less likely to choose to have additional children (Di Florio et al., 2014). In addition to assisting in the identification of higher-risk women, this finding has implications for candidate neurobiological pathways that may explain associations between parity and severe mood episodes, with endocrine and immunological changes related to first childbirth being potential triggers.

Some women with bipolar disorder may be especially sensitive to the rapid hormonal shift that takes place in the immediate peripartum (described earlier in this chapter), with the role of estrogen being a particular area of interest. In a small study, the occurrence of affective psychosis following childbirth was associated with increased sensitivity of dopamine receptors in the hypothalamus, mediated by the decline in estrogen following parturition (Wieck et al., 1991). The role of estrogen in bipolar affective episodes is further evidenced by converging results of several small studies in nonpregnant women demonstrating that tamoxifen, a selective estrogen receptor modulator, significantly reduces manic symptoms (Meinhard et al., 2014).

Though not extensively studied, oxytocin also may play a role in bipolar affective episodes in the postpartum. Noting commonalities between bipolar disorder and endometriosis, including comorbidity, personality traits, and dysregulated oxytocinergic activity, Dinsdale and Crespi (2017) hypothesize that oxytocin jointly contributes to both conditions. It is highly elevated during parturition, and elevated oxytocinergic activity has been reported in bipolar disorder, particularly during manic episodes (Turan et al., 2013; Lien et al., 2016), and the possibility that oxytocin contributes to bipolar affective episodes in the postpartum is an avenue for future research.

As part of a broader effort to characterize genetic risk for psychiatric disorders (Cross-Disorder Group of the Psychiatric Genomics Consortium, 2013), evidence has emerged for heritability of postpartum psychiatric illness. It has been hypothesized that the occurrence of a psychiatric episode in the peripartum could be a marker of a more heritable form of bipolar disorder (Jones and Craddock, 2007), supported by studies showing familial aggregation of postpartum psychiatric episodes in bipolar disorder patients (Payne et al., 2008) and those with psychiatric disorders in general (Bauer et al., 2019). Moreover, there is no difference between bipolar patients with and without histories of postpartum psychosis on psychological factors generally associated with bipolar disorder, including cognitive style, temperament, or personality characteristics (Perry et al., 2019). Although there is currently insufficient evidence for the use of genetic risk scores as a clinical tool, there is evidence for shared genetic mechanisms among postpartum psychiatric disorders. A multinational study applying polygenic risk scores to postpartum depression showed that genetic risk factors for postpartum depression overlapped with those for bipolar disorder, a stronger association than that seen between bipolar disorder and major depression (Byrne et al., 2014). However, a study using data from Danish population-based registers did not show an association between bipolar disorder polygenic risk scores and postpartum psychiatric episode (defined by hospital admission or prescribed psychotropic medication; predominantly depressive episodes) (Bauer et al., 2019). Further research into genetic risk scores specifically for postpartum psychiatric illness, including bipolar disorder, may help efforts to identify women most at risk.

There are few neuroimaging studies of postpartum affective psychosis. In a study of women at risk for postpartum psychosis (defined as having a diagnosis of bipolar disorder, schizoaffective disorder, prior history of postpartum psychotic episode, or postpartum psychosis in a first-degree relative), structural brain differences were reported between those who did experience an episode of postpartum psychosis. Specifically, there was reduced volume of the parahippocampal gyrus, anterior cingulate cortex, postcentral gyrus, and superior temporal gyrus, patterns of morphology that are largely consistent with general vulnerability to psychosis (Fusté et al., 2017). Comparing these findings with GMV in the peripartum in healthy women, the authors hypothesized that when the typical brain changes occur in the postpartum (Kim et al., 2010) (increase in GMV of frontal lobe and midbrain areas), this could have a protective effect against postpartum psychosis episodes among those at risk. Overall, this is an area in need of further investigation.

TRAUMA AND POSTTRAUMATIC STRESS DISORDER

Altered neurobiology in those with PTSD could be an underlying risk factor for general psychopathology in the peripartum. History of exposure to traumatic events is one of the most common predictors of perinatal psychiatric disorders (Vesga-Lopez et al., 2008). Pregnancy may be a particularly stressful and even traumatic experience for some women, especially those who have significant histories of past trauma (Seng et al., 2010; Muzik et al., 2016). In the United States, risk factors for PTSD during pregnancy include African American racial identity, teen pregnancy, and socioeconomic factors such as education up to the high-school level or less and living in poverty (Seng et al., 2009). During pregnancy, the odds of meeting the criteria for a PTSD diagnosis were highest when the most severe trauma was abuse, followed by prior reproductive trauma (Seng et al., 2009).

Conceptually, childbirth is distinct from other kinds of traumatic events because it is viewed by many women as a positive experience. However, the stress of childbirth itself can be a trigger for PTSD symptoms, possibly more so when the past trauma was interpersonal and thus likely reactivated in the anticipation or engagement of parenting an infant (Moleman et al., 1992; Muzik et al., 2016). A meta-analysis of global studies showed the main vulnerability factors for pregnancy-related PTSD were subjective negative birth experiences, depression during pregnancy, operative birth, fear of childbirth, health complications in pregnancy, prepregnancy history of PTSD, and lack of social support (Ayers et al., 2016). Another systemic review confirmed these vulnerability factors and additionally identified postpartum depression followed by perceived quality of interpersonal interactions with treating medical staff as having the largest association with risk for postnatal PTSD (Grekin and O'hara, 2014). Although preexisting psychopathology is generally a risk factor for PTSD in the postpartum, Muzik et al. (2016) reported only a subset (25%) of women with a lifetime PTSD diagnosis experienced worsening of symptoms in the postpartum; the most salient predictors of this trajectory were negative birth experiences and labor anxiety. The remainder of the sample either experienced no change in PTSD symptoms or, in some cases, it even improved.

Research regarding the neurobiology of PTSD specifically in the peripartum is scarce. The mechanisms of associations between traumatic birth experiences and PTSD symptomatology may differ from those occurring outside of pregnancy due to the physiological changes following childbirth, such as those in the neuroendocrine and immune systems, as well as alterations in oxytocin regulation. The possible therapeutic use of oxytocin for nonpregnant PTSD patients has been examined (Kirsch et al., 2005; Olff et al., 2010), and intranasal administration has been shown to boost treatment response (Koch et al., 2016). Oxytocin may contribute to the extinction of traumatic memories (Sack et al., 2017), which could partially explain the observed temporary memory changes and adaptive extinction of aversive memories seen following childbirth (Brett and Baxendale, 2001). However, a systematic review suggests that this effect of oxytocin, and its anxiolytic properties, may exist only in cases of less severe emotional trauma; for victims of recent traumatic events and/or more severe trauma history, oxytocin may even increase anxiety (Donadon et al., 2018); this is consistent with the heterogeneity of responses to childbirth among PTSD patients (Muzik et al., 2016).

Estrogen also may influence neural reactivity to threat, with elevated estrogen outside pregnancy being associated with greater fear processing in an fMRI study (Hwang et al., 2015). However, lower estrogen levels have also been associated with altered fear inhibition and extinction (Glover et al., 2012, 2013); together, these findings raise the possibility of an inverted U-shaped relationship between estrogen and risk for PTSD (Stevens et al., 2016), although further studies are needed to examine this and determine the role of estrogen in the etiology of PTSD in the peripartum.

DISCUSSION

Science is moving rapidly in its understanding of the neurobiological substrate of maternal mental illness. These new discoveries will lead to new interventions, and, importantly, some of them may not be direct pharmacological agents. For example, Bhattacharyya et al. identified disturbed bioenergetics as a component of major depressive disorder and differences in metabolomic change patterns in those who remitted versus those who did not in response to cognitive behavioral therapy (Bhattacharyya et al., 2019). With greater knowledge of the neurobiology of maternal mental illness, a precision medicine approach, what works for whom and when, will be possible. An initiative to collect a global sample of women with postpartum depression via an app-based platform and thereby ensure sufficient power for genome-wide association studies will greatly contribute to a precision medicine knowledge base (Guintivano et al., 2019). There also is a relevant policy discussion: a call to action and NIH task force focused on how to ethically include pregnant and lactating women in more research (PRGLAC, 2019) with a specific initiative at the National Institute of Child Health and Development (https://www.nichd.nih.gov/About/Advisory/PRGLAC), sometimes known as "Protect Pregnant Women through Research, Not from It." This plan for culture and compliance changes should lead to more research and scientific breakthroughs related to the neurobiology and treatment of maternal mental illness.

References

Abrahams VM, Kim YM, Straszewski SL et al. (2004). Macrophages and apoptotic cell clearance during pregnancy. Am J Reprod Immunol 51: 275–282.

Abramowitz JS, Franklin ME, Schwartz SA et al. (2003). Symptom presentation and outcome of cognitive-behavioral therapy for obsessive-compulsive disorder. J Consult Clin Psychol 71: 1049–1057.

Aguilar-Valles A, Poole S, Mistry Y et al. (2007). Attenuated fever in rats during late pregnancy is linked to suppressed interleukin-6 production after localized inflammation with turpentine. J Physiol 583: 391–403.

Alonso P, Cuadras D, Gabriëls L et al. (2015). Deep brain stimulation for obsessive-compulsive disorder: a meta-analysis of treatment outcome and predictors of response. PLoS One 10: e0133591.

Amin Z, Mason GF, Cavus I et al. (2006). The interaction of neuroactive steroids and GABA in the development of neuropsychiatric disorders in women. Pharmacol Biochem Behav 84: 635–643.

Anderson MV, Rutherford MD (2012). Cognitive reorganization during pregnancy and the postpartum period: an evolutionary perspective. Evol Psychol 10: 147470491 201000402.

Arnold LM (1999). A case series of women with postpartum-onset obsessive-compulsive disorder. Prim Care Companion J Clin Psychiatry 1: 103–108.

Ayers S, Bond R, Bertullies S et al. (2016). The aetiology of post-traumatic stress following childbirth: a meta-analysis and theoretical framework. Psychol Med 46: 1121–1134.

Barba-Müller E, Craddock S, Carmona S et al. (2019). Brain plasticity in pregnancy and the postpartum period: links to maternal caregiving and mental health. Arch Womens Ment Health 22: 289–299.

Barnet B, Joffe A, Duggan AK et al. (1996). Depressive symptoms, stress, and social support in pregnant and postpartum adolescents. Arch Pediatr Adolesc Med 150: 64–69.

Barrett J, Fleming AS (2011). Annual research review: all mothers are not created equal: neural and psychobiological perspectives on mothering and the importance of individual differences. J Child Psychol Psychiatry 52: 368–397.

Bartz JA, Hollander E (2008). Oxytocin and experimental therapeutics in autism spectrum disorders. Prog Brain Res 170: 451–462.

Bauer AE, Liu X, Byrne EM et al. (2019). Genetic risk scores for major psychiatric disorders and the risk of postpartum psychiatric disorders. Transl Psychiatry 9: 288.

Beckley EH, Finn DA (2007). Inhibition of progesterone metabolism mimics the effect of progesterone withdrawal on forced swim test immobility. Pharmacol Biochem Behav 87: 412–419.

Belelli D, Lambert JJ (2005). Neurosteroids: endogenous regulators of the GABA(A) receptor. Nat Rev Neurosci 6: 565–575.

Belelli D, Harrison NL, Maguire J et al. (2009). Extrasynaptic GABAA receptors: form, pharmacology, and function. J Neurosci 29: 12757–12763.

Bergink V, Den Berg Lambregtse-Van M, Koorengevel KM et al. (2011a). First-onset psychosis occurring in the postpartum period: a prospective cohort study. J Clin Psychiatry 72: 1531–1537.

Bergink V, Kushner SA, Pop V et al. (2011b). Prevalence of autoimmune thyroid dysfunction in postpartum psychosis. Br J Psychiatry 198: 264–268.

Bergink V, Burgerhout KM, Weigelt K et al. (2013). Immune system dysregulation in first-onset postpartum psychosis. Biol Psychiatry 73: 1000–1007.

Bergink V, Laursen TM, Johannsen BMW et al. (2015). Pre-eclampsia and first-onset postpartum psychiatric episodes: a Danish population-based cohort study. Psychol Med 45: 3481–3489.

Bergink V, Pop VJM, Nielsen PR et al. (2018). Comorbidity of autoimmune thyroid disorders and psychiatric disorders during the postpartum period: a Danish nationwide register-based cohort study. Psychol Med 48: 1291–1298.

Betz D, Fane K (2019). Human chorionic gonadotropin (HCG). In: StatPearls [Internet], StatPearls Publishing.

Bhattacharyya S, Dunlop BW, Mahmoudiandehkordi S et al. (2019). Pilot study of metabolomic clusters as state markers of major depression and outcomes to CBT treatment. Front Neurosci 13: 926.

Bitran D, Smith SS (2005). Termination of pseudopregnancy in the rat produces an anxiogenic-like response that is associated with an increase in benzodiazepine receptor binding density and a decrease in GABA-stimulated chloride influx in the hippocampus. Brain Res Bull 64: 511–518.

Bjertrup AJ, Friis NK, Miskowiak KW (2019). The maternal brain: neural responses to infants in mothers with and without mood disorder. Neurosci Biobehav Rev 107: 196–207.

Blackmore ER, Jones I, Doshi M et al. (2006). Obstetric variables associated with bipolar affective puerperal psychosis. Br J Psychiatry 188: 32–36.

Blanks AM, Thornton S (2003). The role of oxytocin in parturition. BJOG 110: 46–51.

Bloch M, Schmidt PJ, Danaceau M et al. (2000). Effects of gonadal steroids in women with a history of postpartum depression. Am J Psychiatry 157: 924–930.

Brakowski J, Spinelli S, Dörig N et al. (2017). Resting state brain network function in major depression—depression symptomatology, antidepressant treatment effects, future research. J Psychiatr Res 92: 147–159.

Brandes M, Soares C, Cohen L (2004). Postpartum onset obsessive-compulsive disorder: diagnosis and management. Arch Womens Ment Health 7: 99–110.

Bränn E, Edvinsson Å, Punga AR et al. (2019). Inflammatory and anti-inflammatory markers in plasma: from late pregnancy to early postpartum. Sci Rep 9 (1): 1–10.

Brett M, Baxendale S (2001). Motherhood and memory: a review. Psychoneuroendocrinology 26: 339–362.

Britton JR (2008). Maternal anxiety: course and antecedents during the early postpartum period. Depress Anxiety 25: 793–800.

Brummelte S, Galea LAM (2010). Chronic corticosterone during pregnancy and postpartum affects maternal care, cell proliferation and depressive-like behavior in the dam. Horm Behav 58: 769–779.

Byrne EM, Carrillo-Roa T, Penninx BWJH et al. (2014). Applying polygenic risk scores to postpartum depression. Arch Womens Ment Health 17: 519–528.

Cahalan S (2019). The great pretender, Hachette Book Group, Grand Central Publishing.

Cárdenas EF, Kujawa A, Humphreys KL (2019). Neurobiological changes during the peripartum period: implications for health and behavior. Soc Cogn Affect Neurosci nsz091.

Carini LM, Murgatroyd CA, Nephew BC (2013). Using chronic social stress to model postpartum depression in lactating rodents. J Vis Exp 76: e50324.

Carmona S, Martínez-García M, Paternina-Die M et al. (2019). Pregnancy and adolescence entail similar neuroanatomical adaptations: a comparative analysis of cerebral morphometric changes. Hum Brain Mapp 40: 2143–2152.

Casey BJ, Craddock N, Cuthbert BN et al. (2013). DSM-5 and RDoC: progress in psychiatry research? Nat Rev Neurosci 14: 810–814.

Cassidy-Bushrow AE, Peters RM, Johnson DA et al. (2012). Association of depressive symptoms with inflammatory biomarkers among pregnant African-American women. J Reprod Immunol 94: 202–209.

Centers for Disease Control (2019). Vital statistics rapid release, Report No. 007, May 2019 births: provisional data for 2018, National Vital Statistics System.

Chase HW, Moses-Kolko EL, Zevallos C et al. (2014). Disrupted posterior cingulate-amygdala connectivity in postpartum depressed women as measured with resting BOLD fMRI. Soc Cogn Affect Neurosci 9: 1069–1075.

Chaudron LH, Nirodi N (2010). The obsessive-compulsive spectrum in the perinatal period: a prospective pilot study. Arch Womens Ment Health 13: 403–410.

Chung EK, Mccollum KF, Elo IT et al. (2004). Maternal depressive symptoms and infant health practices among low-income women. Pediatrics 113: e523–e529.

Coric V, Taskiran S, Pittenger C et al. (2005). Riluzole augmentation in treatment-resistant obsessive-compulsive disorder: an open-label trial. Biol Psychiatry 58: 424–428.

Corwin EJ, Pajer K (2008). The psychoneuroimmunology of postpartum depression. J Womens Health 17: 1529–1534.

Cross-Disorder Group of the Psychiatric Genomics Consortium (2013). Identification of risk loci with shared effects on five major psychiatric disorders: a genome-wide analysis. Lancet 381: 1371–1379.

Croy BA, Ashkar AA, Minhas K et al. (2000). Can murine uterine natural killer cells give insights into the pathogenesis of preeclampsia? J Soc Gynecol Investig 7: 12–20.

Cui K, Ashdown H, Luheshi GN et al. (2009). Effects of prenatal immune activation on hippocampal neurogenesis in the rat. Schizophr Res 113: 288–297.

Cuthbert B, Insel T (2010). Classification issues in women's mental health: clinical utility and etiological mechanisms. Arch Womens Ment Health 13: 57–59.

Deligiannidis KM, Sikoglu EM, Shaffer SA et al. (2013). GABAergic neuroactive steroids and resting-state functional connectivity in postpartum depression: a preliminary study. J Psychiatr Res 47: 816–828.

Deligiannidis KM, Kroll-Desrosiers AR, Mo S et al. (2016). Peripartum neuroactive steroid and γ-aminobutyric acid profiles in women at-risk for postpartum depression. Psychoneuroendocrinology 70: 98–107.

Dennis CL, Falah-Hassani K, Brown HK et al. (2016). Identifying women at risk for postpartum anxiety: a prospective population-based study. Acta Psychiatr Scand 134: 485–493.

Di Florio A, Forty L, Gordon-Smith K et al. (2013). Perinatal episodes across the mood disorder spectrum. JAMA Psychiat 70: 168–175.

Di Florio A, Jones L, Forty L et al. (2014). Mood disorders and parity—a clue to the aetiology of the postpartum trigger. J Affect Disord 152: 334–339.

Dick LP, Gallagher-Thompson D, Thompson LW (1996). Cognitive-behavioural therapy. In: Handbook of the clinical psychology of ageing, John Wiley & Sons 509–544.

Dinsdale NL, Crespi BJ (2017). Revisiting the wandering womb: oxytocin in endometriosis and bipolar disorder. Horm Behav 96: 69–83.

Donadon MF, Martin-Santos R, Osório FDL (2018). The associations between oxytocin and trauma in humans: a systematic review. Front Pharmacol 9: 154.

Drevets WC (2000). Functional anatomical abnormalities in limbic and prefrontal cortical structures in major depression. In: Progress in brain research, Elsevier.

El Costa H, Casemayou A, Aguerre-Girr M et al. (2008). Critical and differential roles of NKp46-and NKp30-activating receptors expressed by uterine NK cells in early pregnancy. J Immunol 181: 3009–3017.

Faas M, Van Der Schaaf G, Schipper M et al. (2003). Glomerular immunoglobulin deposits induce glomerular inflammation in pregnant but not in non-pregnant rats. Am J Reprod Immunol 49: 57–63.

Fairbrother N, Abramowitz JS (2007). New parenthood as a risk factor for the development of obsessional problems. Behav Res Ther 45: 2155–2163.

Farrant M, Nusser Z (2005). Variations on an inhibitory theme: phasic and tonic activation of GABA(A) receptors. Nat Rev Neurosci 6: 215–229.

Faymonville M-E, Roediger L, Del Fiore G et al. (2003). Increased cerebral functional connectivity underlying the antinociceptive effects of hypnosis. Brain Res Cogn Brain Res 17: 255–262.

Figueira RJ, Peabody MF, Lonstein JS (2008). Oxytocin receptor activity in the ventrocaudal periaqueductal gray modulates anxiety-related behavior in postpartum rats. Behav Neurosci 122: 618–628.

Fofie AE, Fewell JE, Moore SL (2005). Pregnancy influences the plasma cytokine response to intraperitoneal administration of bacterial endotoxin in rats. Exp Physiol 90: 95–101.

Forray A, Focseneanu M, Pittman B et al. (2010). Onset and exacerbation of obsessive-compulsive disorder in pregnancy and the postpartum period. J Clin Psychiatry 71: 1061.

Fossati P (2012). Neural correlates of emotion processing: from emotional to social brain. Eur Neuropsychopharmacol 22: S487–S491.

Fransson E, Dubicke A, Byström B et al. (2012). Negative emotions and cytokines in maternal and cord serum at preterm birth. Am J Reprod Immunol 67: 506–514.

Fusté M, Pauls A, Worker A et al. (2017). Brain structure in women at risk of postpartum psychosis: an MRI study. Transl Psychiatry 7: 1286.

Gaynes BN, Gavin N, Meltzer-Brody S et al. (2005). Perinatal depression: prevalence, screening accuracy, and screening outcomes: Summary. In: AHRQ evidence report summaries, Agency for Healthcare Research and Quality (US).

Gillespie SL, Porter K, Christian LM (2016). Adaptation of the inflammatory immune response across pregnancy and postpartum in Black and White women. J Reprod Immunol 114: 27–31.

Gingnell M, Bannbers E, Moes H et al. (2015). Emotion reactivity is increased 4-6 weeks postpartum in healthy women: a longitudinal fMRI study. PLoS One 10: e0128964.

Glover EM, Jovanovic T, Mercer KB et al. (2012). Estrogen levels are associated with extinction deficits in women with posttraumatic stress disorder. Biol Psychiatry 72: 19–24.

Glover EM, Mercer KB, Norrholm SD et al. (2013). Inhibition of fear is differentially associated with cycling estrogen levels in women. J Psychiatry Neurosci 38: 341.

Grant JE, Kim SW, Odlaug BL (2007). N-acetyl cysteine, a glutamate-modulating agent, in the treatment of pathological gambling: a pilot study. Biol Psychiatry 62: 652–657.

Grattan D, Kokay I (2008). Prolactin: a pleiotropic neuroendocrine hormone. J Neuroendocrinol 20: 752–763.

Greicius MD, Flores BH, Menon V et al. (2007). Resting-state functional connectivity in major depression: abnormally increased contributions from subgenual cingulate cortex and thalamus. Biol Psychiatry 62: 429–437.

Grekin R, O'hara MW (2014). Prevalence and risk factors of postpartum posttraumatic stress disorder: a meta-analysis. Clin Psychol Rev 34: 389–401.

Groer MW, Morgan K (2007). Immune, health and endocrine characteristics of depressed postpartum mothers. Psychoneuroendocrinology 32: 133–139.

Guintivano J, Manuck T, Meltzer-Brody S (2018). Predictors of postpartum depression: a comprehensive review of the last decade of evidence. Clin Obstet Gynecol 61: 591–603.

Guintivano J, Putnam KT, Sullivan PF et al. (2019). The international postpartum depression: action towards causes and treatment (PACT) consortium. Int Rev Psychiatry 31: 229–236.

Haghighi M, Jahangard L, Mohammad-Beigi H et al. (2013). In a double-blind, randomized and placebo-controlled trial, adjuvant memantine improved symptoms in inpatients suffering from refractory obsessive-compulsive disorders (OCD). Psychopharmacology (Berl) 228: 633–640.

Hamilton JP, Siemer M, Gotlib IH (2008). Amygdala volume in major depressive disorder: a meta-analysis of magnetic resonance imaging studies. Mol Psychiatry 13: 993–1000.

Hanna J, Goldman-Wohl D, Hamani Y et al. (2006). Decidual NK cells regulate key developmental processes at the human fetal-maternal interface. Nat Med 12: 1065–1074.

Hellgren C, Akerud H, Skalkidou A et al. (2014). Low serum allopregnanolone is associated with symptoms of depression in late pregnancy. Neuropsychobiology 69: 147–153.

Hillerer KM, Reber SO, Neumann ID et al. (2011). Exposure to chronic pregnancy stress reverses peripartum-associated adaptations: implications for postpartum anxiety and mood disorders. Endocrinology 152: 3930–3940.

Hoekzema E, Barba-Müller E, Pozzobon C et al. (2017). Pregnancy leads to long-lasting changes in human brain structure. Nat Neurosci 20: 287–296.

Hoekzema E, Tamnes CK, Berns P et al. (2020). Becoming a mother entails anatomical changes in the ventral striatum of the human brain that facilitate its responsiveness to offspring cues. Psychoneuroendocrinology 112: 104507.

Hucklebridge FH, Smith MD, Clow A et al. (1994). Dysphoria and immune status in postpartum women. Biol Psychol 37: 199–206.

Hwang MJ, Zsido RG, Song H et al. (2015). Contribution of estradiol levels and hormonal contraceptives to sex differences within the fear network during fear conditioning and extinction. BMC Psychiatry 15: 295.

Insel T (2013). Post by former NIMH director Thomas Insel: transforming diagnosis, Chronology of NIMH Directors [Online] [Accessed April 29, 2013].

Insel TR, Wang PS (2010). Rethinking mental illness. JAMA 303: 1970–1971.

Insel T, Cuthbert B, Garvey M et al. (2010). Research domain criteria (RDoC): toward a new classification framework for research on mental disorders. Am J Psychiatry 167: 748–751.

Janak PH, Tye KM (2015). From circuits to behaviour in the amygdala. Nature 517: 284–292.

Jennings KD, Ross S, Popper S et al. (1999). Thoughts of harming infants in depressed and nondepressed mothers. J Affect Disord 54: 21–28.

Jones I, Craddock N (2007). Searching for the puerperal trigger: molecular genetic studies of bipolar affective puerperal psychosis. Psychopharmacol Bull 40: 115–128.

Kaiser RH, Whitfield-Gabrieli S, Dillon DG et al. (2016). Dynamic resting-state functional connectivity in major depression. Neuropsychopharmacology 41: 1822–1830.

Kellner M (2010). Drug treatment of obsessive-compulsive disorder. Dialogues Clin Neurosci 12: 187–197.

Kendler KS (2005). Toward a philosophical structure for psychiatry. Am J Psychiatry 162: 433–440.

Kendler KS (2019). From many to one to many-the search for causes of psychiatric illness. JAMA Psychiat 76 (10): 1085–1091.

Kim P, Leckman JF, Mayes LC et al. (2010). The plasticity of human maternal brain: longitudinal changes in brain anatomy during the early postpartum period. Behav Neurosci 124: 695–700.

Kim P, Feldman R, Mayes LC et al. (2011). Breastfeeding, brain activation to own infant cry, and maternal sensitivity. J Child Psychol Psychiatry 52: 907–915.

Kim P, Strathearn L, Swain JE (2016). The maternal brain and its plasticity in humans. Horm Behav 77: 113–123.

Kinsley CH, Amory-Meyer E (2011). Why the maternal brain? J Neuroendocrinol 23: 974–983.

Kinsley CH, Madonia L, Gifford GW et al. (1999). Motherhood improves learning and memory. Nature 402: 137–138.

Kirsch P, Esslinger C, Chen Q et al. (2005). Oxytocin modulates neural circuitry for social cognition and fear in humans. J Neurosci 25: 11489–11493.

Klampfl SM, Brunton PJ, Bayerl DS et al. (2014). Hypoactivation of CRF receptors, predominantly type 2, in the medial-posterior BNST is vital for adequate maternal behavior in lactating rats. J Neurosci 34: 9665.

Klampfl SM, Schramm MM, Stinnett GS et al. (2016). Brain CRF-binding protein modulates aspects of maternal behavior under stressful conditions and supports a hypo-anxious state in lactating rats. Horm Behav 84: 136–144.

Koch SBJ, Van Zuiden M, Nawijn L et al. (2016). Intranasal oxytocin normalizes amygdala functional connectivity in posttraumatic stress disorder. Neuropsychopharmacology 41: 2041–2051.

Koga K, Mor G (2010). Toll-like receptors at the maternal-fetal interface in normal pregnancy and pregnancy disorders. Am J Reprod Immunol 63: 587–600.

Kohl J, Autry AE, Dulac CJB (2017). The neurobiology of parenting: a neural circuit perspective. Bioessays 39: 1–11.

Koran LM, Saxena S (2000). Issues and strategies in treating refractory obsessive-compulsive disorder. CNS Spectr 5: 24–31.

Kushner MG, Kim SW, Donahue C et al. (2007). D-cycloserine augmented exposure therapy for obsessive-compulsive disorder. Biol Psychiatry 62: 835–838.

Labad J, Menchón JM, Alonso P et al. (2005). Female reproductive cycle and obsessive-compulsive disorder. J Clin Psychiatry 66: 428–435, quiz 546.

Labonté B, Engmann O, Purushothaman I et al. (2017). Sex-specific transcriptional signatures in human depression. Nat Med 23: 1102–1111.

Lambert K, Berry A, Griffin G et al. (2005). Pup exposure differentially enhances foraging ability in primiparous and nulliparous rats. Physiol Behav 84: 799–806.

Laurent HK, Ablow JC (2011). A cry in the dark: depressed mothers show reduced neural activation to their own infant's cry. Soc Cogn Affect Neurosci 7: 125–134.

Laurent HK, Ablow JC (2013). A face a mother could love: depression-related maternal neural responses to infant emotion faces. Soc Neurosci 8: 228–239.

Leckman JF, Goodman WK, North WG et al. (1994). Elevated cerebrospinal fluid levels of oxytocin in obsessive-compulsive disorder. Comparison with Tourette's syndrome and healthy controls. Arch Gen Psychiatry 51: 782–792.

Leibenluft E, Gobbini MI, Harrison T et al. (2004). Mothers' neural activation in response to pictures of their children and other children. Biol Psychiatry 56: 225–232.

Lien Y-J, Chang HH, Yang YK et al. (2016). PS50. The serum oxytocin levels among major depressive and bipolar II disorder. Int J Neuropsychopharmacol 19: 17.

Lonstein JS, Maguire J, Meinlschmidt G et al. (2014). Emotion and mood adaptations in the peripartum female:complementary contributions of GABA and oxytocin. J Neuroendocrinol 26: 649–664.

Lonstein JS, Lévy F, Fleming AS (2015). Common and divergent psychobiological mechanisms underlying maternal behaviors in non-human and human mammals. Horm Behav 73: 156–185.

Lord C, Steiner M, Soares CN et al. (2012). Stress response in postpartum women with and without obsessive-compulsive symptoms: an fMRI study. J Psychiatry Neurosci 37: 78–86.

Luders E, Gingnell M, Poromaa IS et al. (2018). Potential brain age reversal after pregnancy: younger brains at 4–6 weeks postpartum. Neuroscience 386: 309–314.

Luders E, Kurth F, Gingnell M et al. (2020). From baby brain to mommy brain: widespread gray matter gain after giving birth. Cortex 126: 334–342.

Maes M, Verkerk R, Bonaccorso S et al. (2002). Depressive and anxiety symptoms in the early puerperium are related to increased degradation of tryptophan into kynurenine, a phenomenon which is related to immune activation. Life Sci 71: 1837–1848.

Maestripieri D, Badiani A, Puglisi-Allegra S (1991). Prepartal chronic stress increases anxiety and decreases aggression in lactating female mice. Behav Neurosci 105: 663–668.

Maguire J, Mody I (2008). GABA(A)R plasticity during pregnancy: relevance to postpartum depression. Neuron 59: 207–213.

Maguire J, Mody I (2016). Behavioral deficits in juveniles mediated by maternal stress hormones in mice. Neural Plast 2016: 2762518.

Maina G, Albert U, Bogetto F et al. (1999). Recent life events and obsessive–compulsive disorder (OCD): the role of pregnancy/delivery. Psychiatry Res 89: 49–58.

Marsh R, Maia TV, Peterson BS (2009). Functional disturbances within frontostriatal circuits across multiple childhood psychopathologies. Am J Psychiatry 166: 664–674.

Mathematica (2019). New study uncovers the heavy financial toll of untreated maternal mental health conditions. Available: https://www.mathematica.org/news/new-study-uncovers-the-heavy-financial-toll-of-untreated-maternal-mental-health-conditions.

Mcclure L, O'connor AE, Hayward S et al. (2005). Effects of age and pregnancy on the circulatory activin response of sheep to acute inflammatory challenge by lipopolysaccharide. J Endocrinol 185: 139–149.

Mcgrath J, Saha S, Chant D et al. (2008). Schizophrenia: a concise overview of incidence, prevalence, and mortality. Epidemiol Rev 30: 67–76.

Mckeon J, Roa B, Mann A (1984). Life events and personality traits in obsessive-compulsive neurosis. Br J Psychiatry 144: 185–189.

Meinhard N, Kessing LV, Vinberg M (2014). The role of estrogen in bipolar disorder, a review. Nord J Psychiatry 68: 81–87.

Meltzer-Brody S, Stuebe A, Dole N et al. (2011). Elevated corticotropin releasing hormone (CRH) during pregnancy and risk of postpartum depression (PPD). J Clin Endocrinol Metab 96: E40–E47.

Meltzer-Brody S, Larsen JT, Petersen L et al. (2018). Adverse life events increase risk for postpartum psychiatric episodes: a population-based epidemiologic study. Depress Anxiety 35: 160–167.

Misri S, Abizadeh J, Sanders S et al. (2015). Perinatal generalized anxiety disorder: assessment and treatment. J Womens Health 24: 762–770.

Mody I (2019). GABAAR modulator for postpartum depression. Cell 176: 1.

Moleman N, Van Der Hart O, Van Der Kolk BA (1992). The partus stress reaction: a neglected etiological factor in postpartum psychiatric disorders. J Nerv Ment Dis 180: 271–272.

Mor G, Cardenas I, Abrahams V et al. (2011). Inflammation and pregnancy: the role of the immune system at the implantation site. Ann N Y Acad Sci 1221: 80.

Moses-Kolko EL, Perlman SB, Wisner KL et al. (2010). Abnormally reduced dorsomedial prefrontal cortical activity and effective connectivity with amygdala in response to negative emotional faces in postpartum depression. Am J Psychiatry 167: 1373–1380.

Munk-Olsen T, Laursen TM, Pedersen CB et al. (2006). New parents and mental disorders: a population-based register study. JAMA 296: 2582–2589.

Munk-Olsen T, Laursen TM, Meltzer-Brody S et al. (2012). Psychiatric disorders with postpartum onset: possible early manifestations of bipolar affective disorders. Arch Gen Psychiatry 69: 428–434.

Munk-Olsen T, Jones I, Laursen TM (2014). Birth order and postpartum psychiatric disorders. Bipolar Disord 16: 300–307.

Murgatroyd CA, Pena CJ, Podda G et al. (2015). Early life social stress induced changes in depression and anxiety associated neural pathways which are correlated with impaired maternal care. Neuropeptides 52: 103–111.

Muzik M, Mcginnis EW, Bocknek E et al. (2016). PTSD symptoms across pregnancy and early postpartum among women with lifetime PTSD diagnosis. Depress Anxiety 33: 584–591.

National Institute of Mental Health (2008). Definitions of the RDoC domains and constructs. In: Health, Research Domain Criteria Initiative: National Institute of Mental Health. https://www.nimh.nih.gov/research/research-funded-by-nimh/rdoc/definitions-of-the-rdoc-domains-and-constructs.shtml.

Nemeroff CB, Heim CM, Thase ME et al. (2003). Differential responses to psychotherapy versus pharmacotherapy in patients with chronic forms of major depression and childhood trauma. Proc Natl Acad Sci U S A 100: 14293–14296.

Nephew BC, Bridges RS (2011). Effects of chronic social stress during lactation on maternal behavior and growth in rats. Stress 14: 677–684.

Numan M (2007). Motivational systems and the neural circuitry of maternal behavior in the rat. Dev Psychobiol 49: 12–21.

Oatridge A, Holdcroft A, Saeed N et al. (2002). Change in brain size during and after pregnancy: study in healthy women and women with preeclampsia. Am J Neuroradiol 23: 19–26.

Ochsner KN, Ray RR, Hughes B et al. (2009). Bottom-up and top-down processes in emotion generation: common and distinct neural mechanisms. Psychol Sci 20: 1322–1331.

O'hara MW (1986). Social support, life events, and depression during pregnancy and the puerperium. Arch Gen Psychiatry 43: 569–573.

O'hara MW, Wisner KL (2014). Perinatal mental illness: definition, description and aetiology. Best Pract Res Clin Obstet Gynaecol 28: 3–12.

O'hara MW, Neunaber DJ, Zekoski EM (1984). Prospective study of postpartum depression: prevalence, course, and predictive factors. J Abnorm Psychol 93: 158–171.

O'hara MW, Schlechte JA, Lewis DA et al. (1991). Controlled prospective study of postpartum mood disorders: psychological, environmental, and hormonal variables. J Abnorm Psychol 100: 63–73.

Olff M, Langeland W, Witteveen A et al. (2010). A psychobiological rationale for oxytocin in the treatment of posttraumatic stress disorder. CNS Spectr 15: 522–530.

Osborne LM, Monk C (2013). Perinatal depression—the fourth inflammatory morbidity of pregnancy?: theory and literature review. Psychoneuroendocrinology 38: 1929–1952.

Osborne LM, Gispen F, Sanyal A et al. (2017). Lower allopregnanolone during pregnancy predicts postpartum depression: an exploratory study. Psychoneuroendocrinology 79: 116–121.

Osborne LM, Brar A, Klein SL (2019). The role of Th17 cells in the pathophysiology of pregnancy and perinatal mood and anxiety disorders. Brain Behav Immun 76: 7–16.

Pardon M, Gerardin P, Joubert C et al. (2000). Influence of prepartum chronic ultramild stress on maternal pup care behavior in mice. Biol Psychiatry 47: 858–863.

Pawluski JL (2019). The neurobiology of maternal mental illness: current understanding and future directions. Arch Womens Ment Health 22: 407–408.

Pawluski JL, Galea LAM (2007). Reproductive experience alters hippocampal neurogenesis during the postpartum period in the dam. Neuroscience 149: 53–67.

Pawluski JL, Lambert KG, Kinsley CH (2016). Neuroplasticity in the maternal hippocampus: relation to cognition and effects of repeated stress. Horm Behav 77: 86–97.

Pawluski JL, Lonstein JS, Fleming AS (2017). The neurobiology of postpartum anxiety and depression. Trends Neurosci 40: 106–120.

Paykel ES, Emms EM, Fletcher J et al. (1980). Life events and social support in puerperal depression. Br J Psychiatry 136: 339 346.

Payne JL, Maguire J (2019). Pathophysiological mechanisms implicated in postpartum depression. Front Neuroendocrinol 52: 165–180.

Payne JL, Mackinnon DF, Mondimore FM et al. (2008). Familial aggregation of postpartum mood symptoms in bipolar disorder pedigrees. Bipolar Disord 10: 38–44.

Perani CV, Slattery DA (2014). Using animal models to study post-partum psychiatric disorders. Br J Pharmacol 171: 4539–4555.

Perry A, Gordon-Smith K, Webb I et al. (2019). Postpartum psychosis in bipolar disorder: no evidence of association with personality traits, cognitive style or affective temperaments. BMC Psychiatry 19: 395.

Petraglia F, Imperatore A, Challis JR (2010). Neuroendocrine mechanisms in pregnancy and parturition. Endocr Rev 31: 783–816.

Pittenger C, Bloch MH (2014). Pharmacological treatment of obsessive-compulsive disorder. Psychiatr Clin North Am 37: 375–391.

Prevost M, Zelkowitz P, Tulandi T et al. (2014). Oxytocin in pregnancy and the postpartum: relations to labor and its management. Front Public Health 2: 1.

PRGLAC (2019). Task Force on Reserach Specific to Pregnant Women and Lactating Women (PRGLAC) [Online], Eunice Kennedy Shriver National Institute of Child Health and Human Development Advisory Group: National Institute of Child Health and Human Development, Available: https://www.nichd.nih.gov/about/advisory/PRGLAC [Accessed].

Putnam K, Robertson-Blackmore E, Sharkey K et al. (2015). Heterogeneity of postpartum depression: a latent class analysis. Lancet Psychiatry 2: 59–67.

Putnam KT, Wilcox M, Robertson-Blackmore E et al. (2017). Clinical phenotypes of perinatal depression and time of symptom onset: analysis of data from an international consortium. Lancet Psychiatry 4: 477–485.

Quello SB, Brady KT, Sonne SC (2005). Mood disorders and substance use disorder: a complex comorbidity. Sci Pract Perspect 3: 13.

Ramasubbu R, Konduru N, Cortese F et al. (2014). Reduced intrinsic connectivity of amygdala in adults with major depressive disorder. Front Psych 5: 17.

Righetti-Veltema M, Conne-Perréard E, Bousquet A et al. (1998). Risk factors and predictive signs of postpartum depression. J Affect Disord 49: 167–180.

Robertson E, Grace S, Wallington T et al. (2004). Antenatal risk factors for postpartum depression: a synthesis of recent literature. Gen Hosp Psychiatry 26: 289–295.

Rubinow MJ, Mahajan G, May W et al. (2016). Basolateral amygdala volume and cell numbers in major depressive disorder: a postmortem stereological study. Brain Struct Funct 221: 171–184.

Russell JA, Douglas AJ, Ingram CD (2001). Brain preparations for maternity—adaptive changes in behavioral and neuroendocrine systems during pregnancy and lactation. An overview. In: Progress in brain research, Elsevier.

Rutherford HJ, Wallace NS, Laurent HK et al. (2015). Emotion regulation in parenthood. Dev Rev 36: 1–14.

Sabihi S, Dong SM, Durosko NE et al. (2014). Oxytocin in the medial prefrontal cortex regulates maternal care, maternal aggression and anxiety during the postpartum period. Front Behav Neurosci 8: 258.

Sack M, Spieler D, Wizelman L et al. (2017). Intranasal oxytocin reduces provoked symptoms in female patients with posttraumatic stress disorder despite exerting sympathomimetic and positive chronotropic effects in a randomized controlled trial. BMC Med 15: 40.

Schiller CE, O'hara MW, Rubinow DR et al. (2013). Estradiol modulates anhedonia and behavioral despair in rats and negative affect in a subgroup of women at high risk for postpartum depression. Physiol Behav 119: 137–144.

Schiller CE, Meltzer-Brody S, Rubinow DR (2015). The role of reproductive hormones in postpartum depression. CNS Spectr 20: 48–59.

Schiller CE, Johnson SL, Abate AC et al. (2016). Reproductive steroid regulation of mood and behavior. Compr Physiol 6: 1135–1160.

Schüle C, Nothdurfter C, Rupprecht R (2014). The role of allopregnanolone in depression and anxiety. Prog Neurobiol 113: 79–87.

Scott LJ (2019). Brexanolone: first global approval. Drugs 79: 779–783.

Scull A (2015). Madness in civilisation. Lancet 385: 408.

Sekiyama T, Nakatani Y, Yu X et al. (2013). Increased blood serotonin concentrations are correlated with reduced tension/anxiety in healthy postpartum lactating women. Psychiatry Res 209: 560–565.

Seng JS, Low LK, Sperlich M et al. (2009). Prevalence, trauma history, and risk for posttraumatic stress disorder among nulliparous women in maternity care. Obstet Gynecol 114: 839–847.

Seng JS, Rauch SA, Resnick H et al. (2010). Exploring posttraumatic stress disorder symptom profile among pregnant women. J Psychosom Obstet Gynaecol 31: 176–187.

Shivani R, Goldsmith RJ, Anthenelli RM (2002). Alcoholism and psychiatric disorders: diagnostic challenges. Alcohol Res Health 26: 90.

Sibille E, Wang Y, Joeyen-Waldorf J et al. (2009). A molecular signature of depression in the amygdala. Am J Psychiatry 166: 1011–1024.

Sichel DA, Cohen LS, Robertson LM et al. (1995). Prophylactic estrogen in recurrent postpartum affective disorder. Biol Psychiatry 38: 814–818.

Simon NM, Otto MW, Wisniewski SR et al. (2004). Anxiety disorder comorbidity in bipolar disorder patients: data from the first 500 participants in the Systematic Treatment Enhancement Program for Bipolar Disorder (STEP-BD). Am J Psychiatry 161: 2222–2229.

Smith SS, Gong QH, Li X et al. (1998). Withdrawal from 3alpha-OH-5alpha-pregnan-20-One using a pseudopregnancy model alters the kinetics of hippocampal GABAA-gated current and increases the GABAA receptor alpha4 subunit in association with increased anxiety. J Neurosci 18: 5275–5284.

Smith SS, Ruderman Y, Hua Gong Q et al. (2004). Effects of a low dose of ethanol in an animal model of premenstrual anxiety. Alcohol 33: 41–49.

Smith SS, Ruderman Y, Frye C et al. (2006). Steroid withdrawal in the mouse results in anxiogenic effects of 3alpha,5beta-THP: a possible model of premenstrual dysphoric disorder. Psychopharmacology (Berl) 186: 323–333.

Smith SS, Shen H, Gong QH et al. (2007). Neurosteroid regulation of GABA(A) receptors: focus on the alpha4 and delta subunits. Pharmacol Ther 116: 58–76.

Smith CD, Piasecki CC, Weera M et al. (2013). Noradrenergic alpha-2 receptor modulators in the ventral bed nucleus of the stria terminalis: effects on anxiety behavior in postpartum and virgin female rats. Behav Neurosci 127: 582–597.

Stein DJ, Hollander E, Simeon D et al. (1993). Pregnancy and obsessive-compulsive disorder. Am J Psychiatry 150: 1131–1132.

Stevens JS, Van Rooij SJ, Jovanovic T (2016). Developmental contributors to trauma response: the importance of sensitive periods, early environment, and sex differences. In: Behavioral neurobiology of PTSD, Springer.

Stewart TJ, Traber J, Kroll A et al. (2013). Characterization of extracellular polymeric substances (EPS) from periphyton using liquid chromatography-organic carbon detection-organic nitrogen detection (LC-OCD-OND). Environ Sci Pollut Res Int 20: 3214–3223.

Stoffel EC, Craft RM (2004). Ovarian hormone withdrawal-induced "depression" in female rats. Physiol Behav 83: 505–513.

Stuebe AM, Grewen K, Meltzer-Brody S (2013). Association between maternal mood and oxytocin response to breastfeeding. J Womens Health 22: 352–361.

Swain JE (2011). The human parental brain: in vivo neuroimaging. Prog Neuropsychopharmacol Biol Psychiatry 35 (5): 1242-1254.

Tang S, Li H, Lu L et al. (2019). Anomalous functional connectivity of amygdala subregional networks in major depressive disorder. Depress Anxiety 36: 712–722.

Toufexis DJ, Rochford J, Walker C-D (1999). Lactation-induced reduction in rats' acoustic startle is associated with changes in noradrenergic neurotransmission. Behav Neurosci 113: 176–184.

Turan T, Uysal C, Asdemir A et al. (2013). May oxytocin be a trait marker for bipolar disorder? Psychoneuroendocrinology 38: 2890–2896.

Uguz F, Akman C, Kaya N et al. (2007). Postpartum-onset obsessive-compulsive disorder: incidence, clinical features, and related factors. J Clin Psychiatry 68: 132–138.

Vanderkruik R, Barreix M, Chou D et al. (2017). The global prevalence of postpartum psychosis: a systematic review. BMC Psychiatry 17: 272.

Venning EH (1955). Endocrine changes in normal pregnancy. Am J Med 19: 721–723.

Vesga-Lopez O, Blanco C, Keyes K et al. (2008). Psychiatric disorders in pregnant and postpartum women in the United States. Arch Gen Psychiatry 65: 805–815.

Vizi E, Szelenyi J, Selmeczy Z et al. (2001). Enhanced tumor necrosis factor-alpha-specific and decreased interleukin-10-specific immune responses to LPS during the third trimester of pregnancy in mice. J Endocrinol 171: 355–361.

Weissman MM, Klerman GL (1977). Sex differences and the epidemiology of depression. Arch Gen Psychiatry 34: 98–111.

Welter ML, Burbaud P, Fernandez-Vidal S et al. (2011). Basal ganglia dysfunction in OCD: subthalamic neuronal activity correlates with symptoms severity and predicts high-frequency stimulation efficacy. Transl Psychiatry 1: e5.

Wenzel A, Haugen E, Jackson L et al. (2003). Prevalence of generalized anxiety at eight weeks postpartum. Arch Womens Ment Health 6: 43–49.

Werner EA, Gustafsson HC, Lee S et al. (2015). PREPP: postpartum depression prevention through the mother-infant dyad. Arch Womens Ment Health 19: 229–242.

Wesseloo R, Kamperman AM, Munk-Olsen T et al. (2016). Risk of postpartum relapse in bipolar disorder and postpartum psychosis: a systematic review and meta-analysis. Am J Psychiatry 173: 117–127.

Wieck A, Kumar R, Hirst A et al. (1991). Increased sensitivity of dopamine receptors and recurrence of affective psychosis after childbirth. BMJ 303: 613–616.

Wilhelm S, Buhlmann U, Tolin DF et al. (2008). Augmentation of behavior therapy with D-cycloserine for obsessive-compulsive disorder. Am J Psychiatry 165: 335–409.

Williams KE, Koran LM (1997). Obsessive-compulsive disorder in pregnancy, the puerperium, and the premenstruum. J Clin Psychiatry 58: 330–334, quiz 335–336.

Wisner KL, Gelenberg AJ, Leonard H et al. (1999a). Pharmacologic treatment of depression during pregnancy. JAMA 282: 1264–1269.

Wisner KL, Peindl KS, Gigliotti T et al. (1999b). Obsessions and compulsions in women with postpartum depression. J Clin Psychiatry 60: 176–180.

Wisner KL, Sit DK, Mcshea MC et al. (2013). Onset timing, thoughts of self-harm, and diagnoses in postpartum women with screen-positive depression findings. JAMA Psychiat 70: 490–498.

Wonch KE, De Medeiros CB, Barrett JA et al. (2016). Postpartum depression and brain response to infants: Differential amygdala response and connectivity. Soc Neurosci 11: 600–617.

Workman JL, Brummelte S, Galea LAM (2013). Postpartum corticosterone administration reduces dendritic complexity and increases the density of mushroom spines of hippocampal CA 3 arbours in dams. J Neuroendocrinol 25 (2): 119–130.

Yim IS, Glynn LM, Dunkel-Schetter C et al. (2009). Risk of postpartum depressive symptoms with elevated corticotropin-releasing hormone in human pregnancy. Arch Gen Psychiatry 66: 162–169.

Yonkers KA, Wisner KL, Stewart DE et al. (2009). The management of depression during pregnancy: a report from the American Psychiatric Association and the American College of Obstetricians and Gynecologists. Gen Hosp Psychiatry 31: 403–413.

Yonkers KA, Vigod S, Ross LE (2011). Diagnosis, pathophysiology, and management of mood disorders in pregnant and postpartum women. Obstet Gynecol 117: 961–977.

Young EA, Altemus M, Parkison V et al. (2001). Effects of estrogen antagonists and agonists on the ACTH response to restraint stress in female rats. Neuropsychopharmacology 25: 881–891.

Zambaldi CF, Cantilino A, Montenegro AC et al. (2009). Postpartum obsessive-compulsive disorder: prevalence and clinical characteristics. Compr Psychiatry 50: 503–509.

Section II

The pregnant patient: General issues in neurologic care

Handbook of Clinical Neurology, Vol. 171 (3rd series)
Neurology and Pregnancy: Pathophysiology and Patient Care
E.A.P. Steegers, M.J. Cipolla and E.C. Miller, Editors
https://doi.org/10.1016/B978-0-444-64239-4.00006-0

Chapter 6

Epidemiology of neurologic disease in pregnancy

TAMARA BARGHOUTHI, REGAN LEMLEY, MORGAN FIGURELLE, AND CHERYL BUSHNELL*

Department of Neurology, Wake Forest Baptist Medical Center, Winston-Salem, NC, United States

Abstract

Many neurologic diseases in women are influenced by the physiologic and hormonal changes of pregnancy, and pregnancy itself poses challenges in both treatment and evaluation of these conditions. Some diseases, such as epilepsy and multiple sclerosis, have a high enough prevalence in the young female population to support robust epidemiologic data while many other neurologic diseases, such as specific myopathies and muscular dystrophies, have a low prevalence, with data limited to case reports and small case series. This chapter features epidemiologic information regarding a breadth of neurologic conditions, including stroke, epilepsy, demyelinating disease, peripheral neuropathies, migraine, sleep-disordered breathing, and meningioma, in women in the preconception, pregnancy, and postpartum stages.

CEREBROVASCULAR DISEASE

Ischemic stroke

Stroke is estimated to affect 30 per 100,000 pregnancies, which is a rate three times higher than that in the general young adult population (Swartz et al., 2018). Over time, the rate of antenatal stroke in the United States has increased from 15 per 100,000 deliveries in 1994–95 to 22 per 100,000 deliveries in 2006–07, and the rate of postpartum stroke increased from 12 per 100,000 deliveries in 1994–95 to 22 per 100,000 deliveries in 2006–07 (Kuklina et al., 2011; Sells and Feske, 2017). The antenatal and postpartum admissions for pregnancy-related stroke in the United States increased by 47% and 83%, respectively, from 1994–95 to 2006–07, which may be related to the increased prevalence of hypertensive disorders (Kuklina et al., 2011; Too et al., 2018).

The risk of stroke in pregnancy is highest in the peripartum and postpartum periods (Swartz et al., 2018). In one study of pregnancy-related stroke, 48% of strokes occurred in the postpartum period, 41% at the time of delivery, and 11% during pregnancy (Rantanen and Tatlisumak, 2013). Data collected from the Nationwide Inpatient Sample found that 89% of strokes occurred postpartum or at the time of delivery (Treadwell et al., 2008). One study found a ninefold increased risk of stroke in the peripartum period and a threefold increased risk in the early postpartum period (Sells and Feske, 2017). A Swedish cohort study found that ischemic stroke is 33.8 times more likely to occur in the 3 days surrounding delivery and 8.3 times more likely to occur in the 6 weeks following delivery (Moatti et al., 2014).

The risk of pregnancy-related stroke recurrence is 1.8%–2.7% (Moatti et al., 2014). A multicenter study in France found an ischemic stroke recurrence risk of 1.8% with the overall recurrence risk of stroke in women of childbearing age outside of pregnancy being 1% within 1 year and 2.3% within 5 years. The risk was higher in those with a known cause of stroke (Lamy et al., 2000). Pregnancy-related hypertension is the primary cause of both ischemic and hemorrhage stroke in pregnancy (Bushnell et al., 2014). Factors that increase the risk of pregnancy-related stroke include preeclampsia/eclampsia, physiologic changes in coagulation factors, HELLP syndrome, gestational diabetes, hyperemesis gravidarum leading to hemoconcentration, and changes in cerebral vasculature (Swartz et al., 2018).

*Correspondence to: Cheryl Bushnell, MD, MHS, 1 Medical Center Boulevard, Winston-Salem, NC 27157, United States. Tel: +1-336-716-2983, Fax: +1-336-716-7790, E-mail: cbushnel@wakehealth.edu

The risk of pregnancy-related stroke is higher in African American women and in women aged 35 years and above (James et al., 2005). Other risk factors that lead to a large variation in incidence include geography and the associated socioeconomic status of that country or region globally (Swartz et al., 2017).

Pregnancy-related strokes are associated with high morbidity and mortality rates, with stroke in pregnancy being one of the leading causes of maternal mortality (Bushnell and Chireau, 2011; Caso et al., 2017). The mortality rate in studies of pregnancy-related stroke ranges from 2.7% to 20.4% (Swartz et al., 2017). One 2005 study found the mortality rate due to pregnancy-related stroke to be 1.4 per 100,000 deliveries (James et al., 2005). A 2012 population-based cohort and nested case–control study in the United Kingdom found a case-fatality rate of antenatal stroke of 20%. Of those who survived their stroke, 45% were left with significant disability and 30% required a stay in a rehabilitation facility (Scott et al., 2012).

Hemorrhagic stroke

A US study analyzing data from the Nationwide Inpatient Sample found the incidence of pregnancy-related hemorrhagic stroke to be 6.1 per 100,000 deliveries (Bateman et al., 2006). The relative risk of hemorrhagic stroke in pregnancy and the postpartum period is 5.6 times higher than that in nonpregnant women (Shainker et al., 2015). There is increased risk of hemorrhagic stroke during pregnancy and in the puerperium with the highest risk being in the early postpartum period with an incidence of 10.3 per 100,000 deliveries (Sells and Feske, 2017).

A review of stroke types occurring during pregnancy and postpartum suggested that Taiwanese women were more likely to have hemorrhagic stroke than women in the United States, France, and Canada; however, the sample sizes of total strokes were very small and confidence intervals were not provided (Liang et al., 2006). The etiology of hemorrhagic stroke in one retrospective Taiwanese cohort study was vascular malformation in 29%, preeclampsia/eclampsia in 24%, undetermined in 24%, and coagulopathy in 19% (Liang et al., 2006). Multiparity was associated with increased risk of hemorrhagic stroke and subarachnoid hemorrhage (SAH) in a large multicenter Korean study (Jung et al., 2010).

Prognosis is worse for hemorrhagic stroke than for ischemic stroke (Scott et al., 2012). Pregnancy-related stroke accounts for 13.9% of pregnancy-related maternal deaths in the United Kingdom, with intracranial hemorrhage accounting for most of these deaths (Foo et al., 2013). Pregnancy-related intracerebral hemorrhage accounts for 7.1% of pregnancy-related maternal deaths in the United States (Bateman et al., 2006). A greater risk of mortality was associated with hypertension, older age, African American race, Hispanic race, coagulopathy, and substance use. Of those with pregnancy-related hemorrhagic stroke, 18% were discharged to a skilled nursing facility and 44% were discharged to home (Rantanen and Tatlisumak, 2013).

Preeclampsia/eclampsia

Preeclampsia is defined as new-onset hypertension and proteinuria after the 20th week of gestation, and eclampsia is the development of new-onset tonic–clonic seizures in the setting of preeclampsia (Bushnell and Chireau, 2011; Rantanen and Tatlisumak, 2013). Preeclampsia occurs in 2%–10% of pregnancies and is present in 25%–45% of women with pregnancy-related stroke (Rantanen and Tatlisumak, 2013). Preeclampsia/eclampsia is associated with a fourfold increased risk of stroke in pregnancy (James et al., 2005). In one study examining the risk of stroke in pregnancy, preeclampsia/eclampsia was present in 24% of women (4 of 17) with pregnancy-related ischemic stroke and in 14% of women (2 of 14) with pregnancy-related hemorrhagic stroke (Kittner et al., 1996). Preeclampsia/eclampsia most commonly causes hemorrhagic stroke and most often in the postpartum period. They are also associated with the development of posterior reversible encephalopathy syndrome (PRES) and reversible cerebral vasoconstriction syndrome (RCVS) (Bushnell and Chireau, 2011).

The risk of death is 11% in women with preeclampsia complicated by hemorrhagic stroke and is 1.9% in women with preeclampsia/eclampsia without stroke. Women are 60% more likely to have an ischemic stroke unrelated to pregnancy if they have a history of preeclampsia as compared with women without history of preeclampsia (Moatti et al., 2014).

Vascular malformations and aneurysms

The prevalence of arteriovenous malformations (AVMs) is 0.1% in the general population and that of aneurysms is 10% in the general population (Sells and Feske, 2017). AVM rupture in pregnancy is most common in early pregnancy and at the time of delivery (Rantanen and Tatlisumak, 2013). Aneurysm rupture causing SAH occurs most frequently in the third trimester and in the first 6 weeks postpartum (Shainker et al., 2015). A study of vascular malformations causing pregnancy-related hemorrhagic stroke found the cause to be aneurysm in 77% and AVM in 23% of cases. Hemorrhage due to vascular malformation or aneurysm leads to high maternal and fetal mortality, with maternal mortality rates of 28%–35% and fetal mortality of 14%–17% (Sells and Feske, 2017). One study found that recurrent hemorrhage in women with AVMs occurred at a rate of

31% in the first year following the initial hemorrhage and a rate of 6% in subsequent years (Shainker et al., 2015).

Subarachnoid hemorrhage

The incidence of SAH in pregnancy ranges from 3 to 11 per 100,000 pregnancies (Rantanen and Tatlisumak, 2013). SAH in pregnancy is most commonly caused by ruptured aneurysm and is the third leading cause of non-obstetric maternal mortality. In women under 40 years of age, 50% of cerebral aneurysm ruptures are pregnancy related. Aneurysm rupture is more likely in the later stages of pregnancy and up to 6 weeks postpartum. Nonaneurysmal SAH has been reported in pregnancy in the setting of hypertension and eclampsia (Treadwell et al., 2008). The mortality rate of SAH is higher in pregnant than in nonpregnant women (Rantanen and Tatlisumak, 2013). Surgical treatment of the ruptured aneurysm improves maternal and fetal outcomes (Treadwell et al., 2008).

Cerebral venous sinus thrombosis

The incidence of pregnancy-related cerebral venous sinus thrombosis (CVST) ranges from 1 in 2500 deliveries to 1 in 10,000 deliveries in Western countries. CVST most often occurs in the third trimester and in the first 4 weeks postpartum (Bushnell et al., 2014). Around 2% of pregnancy-related stroke is due to CVST (Saposnik et al., 2011). The risk of peripartum CVST is 11.6 per 100,000 deliveries. Risk factors for CVST include older age, hypertension, excess vomiting, infection, hyperhomocystinemia, and cesarean delivery (Treadwell et al., 2008; Saposnik et al., 2011). Pregnancy-related CVST mortality rate is 0.4%, which is lower than the CVST mortality rate in the general population (2%–10%) (Treadwell et al., 2008; Bushnell et al., 2014).

Posterior reversible encephalopathy syndrome

In PRES, focal neurologic symptoms occur secondary to cerebral vasogenic edema, which most often affects the parietal and occipital lobes. Symptoms include headaches, visual changes, altered mental status, and seizures (McDermott et al., 2018). In a Chinese retrospective series of PRES in pregnancy, 26 of 28 women with eclampsia and 7 of 59 women with preeclampsia had PRES. Most patients had complete recovery with residual deficits present in 4 of 36 women (11%) (Wen et al., 2017).

Reversible cerebral vasoconstriction syndrome

RCVS is marked by multifocal, reversible narrowing of the cerebral arteries and manifests as thunderclap headache, sometimes with stroke, SAH, cerebral edema, and/or seizures (McDermott et al., 2018). It occurs more often in women than in men in a 2.4:1 ratio (Miller et al., 2015). The peak incidence of RCVS is at 20–50 years of age (Sheikh and Mathew, 2014). Postpartum RCVS occurs in the first week following delivery in two-thirds of cases but can occur up to 6 weeks postpartum (Ducros, 2012; Miller et al., 2015).

In a retrospective study including 18 women with postpartum RCVS, 9 (50%) had complete recovery, 5 (28%) had residual deficits, and 4 (22%) had a fulminant disease course resulting in death. The most common symptoms in descending order of frequency were headache, focal deficits, visual disturbances, encephalopathy, and seizures. Brain imaging showed hemorrhage in 7 women (39%), infarct in 6 (35%), and vasogenic edema in 6 (35%) (Fugate et al., 2012).

Cerebral autosomal dominant arteriopathy with subcortical infarcts and leukoencephalopathy

Cerebral autosomal dominant arteriopathy with subcortical infarcts and leukoencephalopathy (CADASIL) is a progressive, inherited arteriopathy caused by a NOTCH3 gene mutation and manifests with early-onset strokes and dementia, migraines, seizures, and mood disorders (Donnini et al., 2017). A 2017 review of CADASIL in pregnancy included data on 93 pregnancies and found that none of the women had cerebral vascular events during pregnancy, though there were 16 miscarriages (17.2%) (Donnini et al., 2017). A 2015 study examining the occurrence of neurologic symptoms during pregnancy in women with CADASIL found that 12 of 25 women reported neurologic symptoms in 17 of 43 pregnancies. Symptoms most often occurred in the puerperium and in women older than 30 years of age. The most common symptoms were hemiparesthesia (76%), aphasia (65%), visual disorders (47%), and hemiparesis (36%) (Roine et al., 2005).

Moyamoya disease

Moyamoya disease is characterized by stenosis or occlusion of the internal carotid artery bifurcations as well as abnormalities in the anterior and middle cerebral arteries that lead to both ischemic and hemorrhagic strokes (Suzuki and Kodama, 1983). A 2018 European study of moyamoya vasculopathy in pregnancy included 31 women with 60 pregnancies. Of these women, 94% had ischemic events and 6% had headaches (Acker et al., 2018). A 2013 Japanese study of moyamoya disease in pregnancy included six women. Four of them were diagnosed prior to pregnancy and had no neurologic events during pregnancy or puerperium.

Two patients were diagnosed during pregnancy, one of whom suffered a progressive stroke at 33 weeks of gestation and the other had a minor stroke on day 3 postpartum (Fujimura et al., 2013).

Dissection

Arterial dissection is rare in pregnancy and the puerperium. The annual incidence of spontaneous cervicocephalic arterial dissection is 0.0026%, with 2.4% of these occurring in the postpartum period. There have been case reports of cervical artery dissection associated with preeclampsia (Shanmugalingam et al., 2016).

EPILEPSY

Epilepsy is a heterogeneous disease that affects over 70 million individuals worldwide (Singh and Trevick, 2016). Studies indicate that the incidence rate is close to 50 per 100,000 in high-income countries, and the rate might be two times higher in lower-middle-income countries due to higher prevalence of CNS infections, traumatic brain injuries, and perinatal injuries (Fiest et al., 2017). Incidence rates may be slightly higher in men than in women, though this may be due to a combination of factors, including underreporting by women in countries where women with epilepsy face social stigmatization and higher rate of acquired epilepsies in males (Bangar, 2016).

In the United States, 1.5 million women with epilepsy (WWE) are of reproductive age (Vélez-Ruiz and Pennell, 2016), and between 0.3% and 0.5% of all pregnancies occur among them (MacDonald et al., 2015). In WWE, half of the pregnancies are unplanned, and they do not appear to have increased fertility problems when compared with women without epilepsy (Voinescu and Pennell, 2017).

WWE who are seizure free for 9 months prior to pregnancy have a high (80%–94%) likelihood of remaining seizure free for the duration of pregnancy (Harden et al., 2009). Over half to three-quarters of women will not experience a change in seizure frequency, but in those who experience worsening frequency several factors may be contributing, including anti-epileptic drug (AED) noncompliance, change in AED pharmacokinetics, and sleep deprivation (Voinescu and Pennell, 2017). For example, AED clearance changes in the first trimester for levetiracetam and in the second trimester for oxcarbazepine and topiramate, and lower ratio to target concentrations were associated with seizures worsening (Voinescu and Pennell, 2017). Focal epilepsy may also worsen during pregnancy compared with generalized epilepsies (Voinescu and Pennell, 2017).

A retrospective population-based cohort study found that over 90% of pregnant WWE have favorable pregnancies (Artama et al., 2017). However, several studies have found that epilepsy is associated with increased risk of adverse outcomes.

A prospective trial from 1997 to 2016 involving infants born to woman enrolled in the North American Antiepileptic Drug Pregnancy Registry found that AED usage during pregnancy, regardless of whether the mother had epilepsy or not, was associated with a risk ratio of 2.0 for diagnosis of small for gestational age and 1.5 for preterm delivery compared with women not taking AEDs (Hernández-Díaz et al., 2017). There may also be an increased risk of mild preeclampsia (OR 1.9), gestational hypertension (OR 1.5), and vaginal bleeding late in pregnancy (OR 1.5) (Harden et al., 2009). Women with epilepsy also tend to undergo cesarean section more often than women without epilepsy (Veiby et al., 2009).

In regards to risk of major congenital malformations (MCMs), several studies have identified that AED polytherapy during pregnancy poses an increased risk over monotherapy, though the malformation rate is dependent on the specific drugs used (Vélez-Ruiz and Pennell, 2016). For example, the risk of levetiracetam and lamotrigine causing an MCM is between 2% and 2.5%, while the risk with valproate is 6%–9% (Vélez-Ruiz and Pennell, 2016). The majority of pregnant women with epilepsy treated at tertiary epilepsy centers in the United States are on monotherapy, and the two most popular AEDs are lamotrigine (42.1%) and levetiracetam (37.5%) (Meador et al., 2018).

In a matched case–control study in Denmark, maternal mortality was defined as deaths in pregnancy within the first 42 days after termination in pregnancy. Mortality associated with epilepsy diagnosis was more than five times higher than without this diagnosis (Christensen et al., 2018), and a similar study in the United States showed that there was a more than 10-fold risk of death (MacDonald et al., 2015).

WWE are less likely to report breastfeeding but more likely to report prenatal vitamin or folic acid daily use (Johnson et al., 2018). According to the Neurodevelopmental Effects of Antiepileptic Drugs (NEAD) study, there were no negative cognitive effects in children who had been exposed to AEDs during breastfeeding (Voinescu and Pennell, 2017).

In pregnant women who do not have a previous history of epilepsy and are diagnosed with status epilepticus during pregnancy, etiologies include eclampsia, PRES, cortical vein thrombosis, or limbic encephalitis (Rajiv and Radhakrishnan, 2017).

Preeclampsia affects 2%–8% of all pregnancies worldwide, and eclampsia (preeclampsia complicated by new-onset seizures in a pregnant patient) is much more common in developing countries compared with high-income countries, with rates of 16–69 cases per

10,000 births and 2–3 cases per 10,000 births, respectively (Duley, 2009). Worldwide, eclampsia is responsible for 12% of maternal deaths (Ghulmiyyah and Sibai, 2012). Factors increasing the risk of eclampsia include black or Hispanic race, young maternal age, and nulliparity (Esakoff et al., 2016). A large retrospective study found that though a future seizure disorder is more likely (4.58 per 10,000 person-years) in a pregnancy with eclampsia than in one without a hypertensive disorder (0.72 per 10,000 person-years), the absolute risk of seizure is still low at 1 per 2000 person-years (Nerenberg et al., 2017).

DEMYELINATING DISEASES

Multiple sclerosis

Multiple sclerosis has an uneven distribution around the world: the greatest prevalence rates are in the northern United States, Canada, southern Australia, and New Zealand (>100 cases per 100,000 people) and the lowest in countries such as Colombia and Ecuador (<5 cases per 100,000 people) (Melcon et al., 2014). Numerous studies have reported an increase in disease incidence and prevalence rates over time, and there appears to be an increasing incidence rate of MS in women and an increasing female-to-male ratio of disease (Koch-Henriksen and Sørensen, 2010). For example, in Denmark, disease incidence in women increased from 5.91 to 12.33 per 100,000 people per year from the 1950s to the early 2000s while in men it increased from 4.52 to 6.08 during the same time period (Koch-Henriksen and Sørensen, 2010). In the United States, the female-to-male ratio has increased from 1.4 in 1955 to 2.3 in 2000 (Alonso and Hernan, 2008). Globally, the incidence rate for females is two to three times that for males and is typically diagnosed between ages 20 and 40 (Koch-Henriksen and Sørensen, 2010).

In a prospective European study in 1988 of 269 pregnant women with MS, the rate of MS relapse frequency declined during pregnancy, particularly in the third trimester, increased in the 3-month postpartum period, and then returned to prepregnancy rates (Fabian, 2016). Subsequent studies have demonstrated similar results, and longer retrospective studies have found that pregnancy does not significantly alter the rate of MS relapses over a 10-year period (Karp et al., 2014).

Women with MS do not seem to have significantly increased obstetric or neonatal complications, though some small studies have shown slight increased risk for infections in the mother and for cesarean delivery (Fabian, 2016; Fong et al., 2018).

Neuromyelitis optica

Neuromyelitis optica (NMO) is uncommon, affecting 0.5–4.4 people per 100,000 (Pandit et al., 2015). In monophasic disease types, there is an equal number of female and male cases, but in relapsing disease, it is estimated to be 3–10 times more common in women than in men (Howard et al., 2016). Unlike MS, NMO has a predilection for nonwhites and is more common in people of African and Hispanic descent (Alvarenga et al., 2017).

Unlike with MS, pregnancy does not appear to be protective against disease relapse, which has been shown to increase in the third trimester, and women are at a higher annualized relapse rate (ARR) in the first 3 months after delivery when compared with the ARR in the year prior to pregnancy (Bourre et al., 2012). Women with NMO may also experience greater rates of miscarriage and preeclampsia (Kaplan, 2019).

Acute disseminated encephalomyelitis

Acute disseminated encephalomyelitis (ADEM) is a monophasic inflammatory demyelinating syndrome of the central nervous system that typically affects the pediatric population and occurs following viral infection or vaccination; however, there have been several cases of this disease in adults (Shah et al., 2018). The effect of pregnancy on ADEM has not been established due to very few case reports, but high-dose steroids and plasmapheresis have been successful treatment options (Shah et al., 2000; Zeb and Alegria, 2014).

Progressive multifocal leukoencephalopathy

Progressive multifocal leukoencephalopathy (PML), a white matter disease caused by the JC virus, is rarely reported during pregnancy. However, due to complex immunologic changes during pregnancy, there is concern that pregnant women with MS may have increased risk of PML (Shah et al., 2000). A prospective case–control study in Finland showed that the JC virus titer level did not change during pregnancy in women while the level significantly declined in women without MS, implying that women with MS may indeed have an increased reaction to the JC virus during pregnancy (Saraste et al., 2018).

Central pontine myelinolysis

Data regarding the epidemiology of central pontine myelinolysis (CPM) in pregnancy are limited to case reports. There have also been case reports of patients developing both CPM and Wernicke's encephalopathy in the setting of hyperemesis gravidarum (Bergin and

Harvey, 1992; Peeters et al., 1993; Olindo et al., 1997; Tonelli et al., 1999; Zara et al., 2012; Baouahi and Doumiri, 2014).

NEUROMUSCULAR DISEASE

Data regarding neuromuscular complications in pregnancy are largely limited to case reports and small case series (Sax and Rosenbaum, 2006).

Neuromuscular junction disorders

Myasthenia gravis

Myasthenia gravis (MG) is an autoimmune, postsynaptic neuromuscular junction disorder that causes fluctuating weakness (Nicolle, 2016). The prevalence of MG in the general population is 1 in 5000 (Waters, 2019). It is more common in women than in men with a ratio of 3:2 and incidence is highest in women during the third decade of life (Massey and De Jesus-Acosta, 2014). MG is diagnosed in around 1 in 20,000 pregnancies (Sax and Rosenbaum, 2006). Among women with MG, during pregnancy, 30%–40% experience improvement, 30%–40% have an unchanged course, and 20%–30% experience worsening symptoms. Exacerbations most often occur in the first trimester, in the last 4 weeks of gestation, and in the first 3–4 weeks postpartum. Around 20% of women with MG will develop a respiratory crisis requiring mechanical ventilation during pregnancy (Massey and De Jesus-Acosta, 2014). Pregnant women with MG have the highest risk for exacerbations if they become pregnant within the first year following diagnosis or if they develop an infection postpartum (Sax and Rosenbaum, 2006). MG exacerbations often improve following spontaneous abortion, but the risk and severity of exacerbations are not changed by termination of a pregnancy (Varner, 2013). The course of MG in one pregnancy does not predict the course in any subsequent pregnancy (Massey and De Jesus-Acosta, 2014).

Maternal mortality risk is 4% with risk factors for mortality including respiratory failure, cholinergic crisis, postpartum hemorrhage, and treatment of preeclampsia with magnesium (Sax and Rosenbaum, 2006). Thymectomy prior to pregnancy has been shown to reduce the chances of MG exacerbation during pregnancy, and thymectomies have been performed successfully in pregnant patients with severe MG (Fox et al., 1990). Perinatal mortality is five times higher than the mortality rate of uncomplicated births. In a 2003 retrospective cohort study of 127 births in 79 women with MG, the rate of cesarean section was two times the rate in the control group though many of the woman elected to have cesarean sections. Neonatal MG develops in 10%–20% of infants born to mothers with MG and usually resolves after 1 month. Another neonatal complication that can occur in pregnancies affected by MG is arthrogryposis multiplex congenita with increased risk of neonatal death secondary to pulmonary hypoplasia and polyhydramnios. If arthrogryposis multiplex congenita occurs, there is a high risk of recurrence in subsequent pregnancies (Sax and Rosenbaum, 2006).

Lambert–Eaton myasthenic syndrome

Lambert–Eaton myasthenic syndrome (LEMS) is an autoimmune or paraneoplastic, presynaptic neuromuscular junction disorder that causes fluctuating weakness (Nicolle, 2016). There have been at least 2 case reports of LEMS that have been published (Bayrak et al., 2010; Schneider-Gold et al., 2006).

Botulism

Botulism occurs following exposure to the *Clostridium botulinum* toxin and results in progressive muscular weakness that can lead to death (Badell et al., 2018). A 2017 review of botulism in pregnancy and the postpartum period reported 17 published cases in seven countries. The source of botulism was food in 10 cases, wound contamination secondary to heroin use in two cases, and unknown in five cases. Of the cases that occurred during pregnancy, 11 were in the third trimester, 3 in the second trimester, and 2 in the first trimester. A total of 11 women had neurologic decline and respiratory failure requiring care in an ICU, 2 died, and 2 went into a persistent vegetative state. Of 12 cases that reported on neonatal outcome, 6 were preterm births with 4 of these being medically induced due to maternal decompensation. Moreover, 11 women received botulinum antitoxin without any adverse effect (Badell et al., 2018).

Motor neuron diseases

Amyotrophic lateral sclerosis

There have been a number of case reports of pregnancy in women with amyotrophic lateral sclerosis (Porto and Berndl, 2019).

Spinal muscular atrophy

Spinal muscular atrophy (SMA) has a prevalence of around 1 in 10,000 in the general population. There have been reports of successful pregnancies in women with SMA. Around 80% of women experience difficulties during pregnancy, including worsening of disease and preterm labor. Of pregnant patients with SMA, two-thirds experience exacerbation of muscle weakness, which persists after delivery in over half of these patients (Frazer et al., 2013). In SMA II, the risk of miscarriage and stillbirth is increased in those with poor lung function. In both SMA II and SMA III, there is increased rate of preterm births (Norwood and Rudnik-Schöneborn, 2012).

Acquired muscle conditions

Inflammatory muscle disease

The prevalence of polymyositis (PM) and dermatomyositis (DM) in the general population is around 5 in 1,000,000 with 14% of cases occurring in the reproductive period. Pregnancy outcome in women with PM or DM is associated with the disease status in the mother. One literature review showed that in women with active disease, 43% of pregnancies resulted in fetal death and 33% were complicated by IUGR. There was little risk to the mother and fetus in those with disease in remission. There has been one reported case of focal myositis in pregnancy; this occurred in the first trimester with spontaneous recovery within weeks (Sax and Rosenbaum, 2006).

Hereditary neuromuscular disorders

Inherited myopathies

Myotonias

Myotonic dystrophy. Myotonic dystrophy, or dystrophia myotonia (DM), is the most common inherited muscular dystrophy occurring in around 1 in 8000 adults; it is also the most common muscular dystrophy seen in pregnant women. Some women experience an increase in weakness during pregnancy or in the postpartum period, which usually resolves in the months following delivery (Sax and Rosenbaum, 2006; Frazer et al., 2013). In DM1, DM2, and nondystrophic myotonic syndromes, increase in myotonia and myalgias during pregnancy with resolution following delivery has been reported (Argov and de Visser, 2009).

Pregnant women with DM1 have an increased risk of ectopic pregnancy, preterm labor, fetal loss, preeclampsia, polyhydramnios, prolonged labor, and postpartum hemorrhage. In women with DM1, the rate of perinatal loss is around 10 times higher than that in the general population (Frazer et al., 2013). Women with DM1 with signs of disease prior to pregnancy have a preterm labor rate of 50%, with 30% of these pregnancies ending at 35–38 weeks of gestation and the other 20% ending before 34 weeks of gestation. Preterm labor also occurs in 50% of pregnancies in women with DM2 with signs of disease prior to pregnancy (Argov and de Visser, 2009).

Nondystrophic myotonias. A 2015 study of pregnancy in women with nondystrophic myotonia (NDM) included 25 women with 63 pregnancies (53 live births including 3 twin gestations, 8 miscarriages, 4 terminations, 1 stillbirth). Included NDM types were Thomsen myotonia congenita, Becker myotonia congenita, paramyotonia congenita, myotonia fluctuans, acetazolamide-responsive myotonia, hyperkalemic periodic paralysis with myotonia, and NDM with exact type unknown. Worsening of NDM symptoms occurred in 62% of the pregnancies, and most patients (76%) had symptom improvement within 3 months following delivery (Snyder et al., 2015). A 2009 review of NDMs in pregnancy found pregnancy worsened myotonic symptoms in 40% with chloride channelopathies and in 80% with sodium channelopathies (Trip et al., 2009). Women with myotonia congenita and paramyotonia congenita have much lower rates of maternal and fetal complications in comparison to DM (Sax and Rosenbaum, 2006).

Muscular dystrophy

Limb girdle muscular dystrophy. Pregnancy can occasionally precipitate the onset of limb girdle muscular dystrophy (LGMD) and progression of disease can be influenced by pregnancy (Angelini and Fanin, 2017). In one study of pregnancy in LGMD, over 50% of women reported increased weakness during pregnancy, which generally did not improve following delivery (Frazer et al., 2013).

Hereditary peripheral neuropathies

Charcot–Marie–Tooth disease

Charcot–Marie–Tooth (CMT) disease is a group of hereditary peripheral neuropathies with variable inheritance pattern that causes distal weakness, foot deformities, and sensory loss (Saporta et al., 2011). CMT affects 1 in 2500 people. Pregnant women with CMT are two times more likely to have fetal presentation abnormalities and require assisted (forceps or vacuum) or operative deliveries. Pregnant women with CMT are also two times as likely to have severe postpartum bleeding. There is a 50% risk of worsening of disease during pregnancy if a woman was diagnosed with CMT in early childhood. Around one-third of women with worsening of disease during pregnancy have improvement following delivery and around two-thirds have progressive worsening in the postpartum period. Women with worsening during one pregnancy have a high risk of worsening in subsequent pregnancies (Frazer et al., 2013). Age at delivery is not correlated with risk of worsening disease during pregnancy (Sax and Rosenbaum, 2006).

Hereditary neuropathy with liability to pressure palsies

Axillary and fibular neuropathies have been reported in pregnancy in patients with hereditary neuropathy with liability to pressure palsies (HNPP) (Massey and Guidon, 2014). Postpartum foot drop can occur in women with HNPP (Sax and Rosenbaum, 2006).

Peripheral neuropathies

Postpartum neuropathy

The incidence of postpartum neuropathies occurring secondary to labor and delivery is 0.25%. The most common postpartum neuropathy due to nerve injury during labor and delivery is lateral femoral neuropathy. When lateral femoral neuropathy occurs due to pregnancy itself, it usually happens after 30 weeks of gestation. Other nerves that can be injured during labor and delivery due to prolonged stretch or pressure include the obturator nerve, femoral nerve, sciatic nerve, common peroneal nerve, and the lumbosacral plexus. Risk factors for postpartum neuropathy include prolonged second stage of labor and nulliparity. Prognosis is good with most postpartum neuropathies resolving by 6–8 weeks after delivery (Hosley and McCullough, 2011).

Postpartum foot drop can be caused by injury during labor and delivery to lumbar nerve roots, lumbosacral plexus, sciatic nerve, or common peroneal nerve. Short stature and primiparity are maternal risk factors for lumbosacral plexus injury. Epidural anesthesia is a risk factor for common peroneal neuropathy. Prognosis is dependent on the duration and severity of nerve injury. In pure demyelinating lesions, full recovery usually takes 3–6 months. In axonal lesions, recovery is slower and frequently full recovery does not occur (Sax and Rosenbaum, 2006).

Femoral neuropathy

Pregnancy-related femoral neuropathy is rare. The incidence of femoral neuropathy in pregnancy has decreased over the past 50 years with risk of development being higher in primiparous women. Prognosis is good with recovery of demyelinating lesions usually taking place within 6 months (Sax and Rosenbaum, 2006).

Obturator neuropathy

Pregnancy-related obturator neuropathy is rare and can occur during delivery with contributing factors, including compression of the nerve by the fetal head, compression by hematomas caused by pudendal nerve block, use of forceps, and lithotomy position (Sax and Rosenbaum, 2006). Prognosis is good with recovery usually taking place within 2–6 months (Sax and Rosenbaum, 2006; Massey and Guidon, 2014).

Lumbosacral plexopathy

The incidence of intrapartum lumbosacral plexopathy is estimated to be between 1 in 2000 and 1 in 6400 deliveries with true incidence thought to be underestimated (Katirji et al., 2002). Lumbosacral plexopathy usually occurs in the third trimester or during delivery (Massey and Guidon, 2014). It can occur during delivery due to compression of the lumbosacral trunk by the fetal head. Recovery usually takes place within a few months (Proisy et al., 2014). The likelihood of recurrence in subsequent pregnancies is unknown (Katirji et al., 2002).

Carpal tunnel syndrome

Carpal tunnel syndrome (CTS) is the most frequent compression neuropathy in pregnancy (Briemberg, 2007). A 2010 systematic review of 214 studies of pregnancy-related CTS found that the incidence of clinically diagnosed pregnancy-related CTS was 31%–62% while the incidence of neurophysiologically confirmed cases was 7%–43% (Padua et al., 2010). A 2002 prospective study of CTS in pregnancy by the Italian CTS Study Group found that in pregnant women diagnosed with CTS, 30% had resolution of symptoms shortly after delivery, 11% during lactation, and 5% after lactation stopped; 54% had continued symptoms at 1 year. Prognosis was better when CTS occurred later in pregnancy (Briemberg, 2007).

CTS most commonly occurs in the third trimester and women are more likely to be older than 30, have generalized or hand edema, and be nulliparous. It can also occur first during lactation rather than during pregnancy, and when it occurs during this time period, women are more likely to be older and to be primiparous and less likely to have generalized edema compared with women who develop CTS during pregnancy (Sax and Rosenbaum, 2006).

Neuralgic amyotrophy

Neuralgic amyotrophy is an inflammatory brachial plexus disorder that causes pain, weakness, and sensory changes in muscles innervated by the brachial plexus (Klein et al., 2002). It can be hereditary or idiopathic, and pregnancy and delivery can trigger symptoms in mothers with hereditary neuralgic amyotrophy (Sax and Rosenbaum, 2006).

Meralgia paresthetica

Meralgia paresthetica is common in pregnancy and typically resolves within several months following delivery (Sax and Rosenbaum, 2006; Massey and Guidon, 2014). The lateral femoral cutaneous nerve can be injured during labor and delivery, and predisposing factors for developing the condition in pregnancy include increased abdominal girth and lumbar lordosis. Meralgia paresthetica is unilateral in 80% of cases (Sax and Rosenbaum, 2006).

Lumbosacral radiculopathy

More than 50% of women experience lumbosacral pain during pregnancy, which typically resolves following delivery, and sciatica is estimated to affect 1% of pregnancies (Sax and Rosenbaum, 2006; Proisy et al., 2014). Lumbar disc displacement causing radiculopathy occurs in 1 in 10,000 pregnancies with the L5 and S1 roots commonly affected (Sax and Rosenbaum, 2006). The risk of disc herniation is not increased in pregnancy (Massey and Guidon, 2014). Epidural anesthesia used for delivery has a $\leq$0.1% risk of causing injury to nerve roots or the spinal cord (Sax and Rosenbaum, 2006).

Other neuropathies

Diabetic neuropathy

Diabetic polyneuropathy affects around 50% of patients with diabetes mellitus. Information regarding the course of diabetic neuropathy in pregnancy is limited largely to case reports. Pregnancy is not known to increase the risk of development or progression of diabetic polyneuropathy or autonomic neuropathy (Massey and Guidon, 2014).

Neuropathy due to nutritional deficiency

Pregnant women with hyperemesis gravidarum are at increased risk for vitamin B12, B1, and B6 deficiencies, which can cause polyneuropathy (Massey and Guidon, 2014).

Autoimmune neuropathies

Acute inflammatory demyelinating polyneuropathy

The incidence of acute inflammatory demyelinating polyneuropathy (AIDP) is 0.75–2 per 100,000 per year in the general population. The incidence of AIDP is not increased in pregnancy but in the first 14 days of the postpartum period. The course and severity of AIDP are not affected by pregnancy (Sax and Rosenbaum, 2006). Nearly all cases of AIDP in pregnancy are in the second trimester, third trimester, or first month postpartum (Massey and Guidon, 2014). In a 2004 review of pregnancy complicated by AIDP, one-third of women required ventilator support. There were no neonatal deaths and 35% of women had preterm delivery (Sax and Rosenbaum, 2006).

Chronic inflammatory demyelinating polyneuropathy

Information regarding chronic inflammatory demyelinating polyneuropathy (CIDP) in pregnancy is limited to case reports and small series as it usually presents in adults aged 40–60 and rarely occurs in women of childbearing age. In pregnant women with CIDP, risk of relapse is increased in the third trimester and immediate postpartum period (Sax and Rosenbaum, 2006; Massey and Guidon, 2014).

Multifocal motor neuropathy

In multifocal motor neuropathy (MMN), an increased risk of relapse occurs in the third trimester and postpartum period. One case of neonatal transmission of MMN has been reported (Massey and Guidon, 2014).

HEADACHE

Primary headache disorder

Migraine

Migraine affects 12% of the population, and epidemiologic data in the United States shows that it is more common in women than in men, occurring in 17.1% and 5.6%, respectively, each year (Lipton et al., 2007). The prevalence does not differ between genders before adolescence (Karli et al., 2012), though at the time of menarche there is a steep rise in the prevalence with a 3:1 female-to-male ratio (Granella et al., 2000). Prevalence continues to rise during childbearing years, reaching a peak of 24.4% followed by a decline near menopause (Lipton et al., 2007).

Research has shown that the hormone fluctuations of menarche, the menstrual cycle, and pregnancy seem to affect the frequency and severity of migraines (Granella et al., 2000). During pregnancy, multiple studies have demonstrated that most women with migraine report an improvement in symptoms. Only 4%–8% of women with migraine report an increase in the frequency and intensity of migraines throughout pregnancy (Aube, 1999). Studies show that the frequency of migraine without aura decreases in the second and third trimesters (Digre, 2013), with remission rates ranging from 77% to 83% (O'Neal, 2017). In women who have migraine with aura, remission rates may be only 43.6% (Granella et al., 2000).

A prospective Japanese study found a similar remission rate of 85% in the third trimester, and in the first month following delivery over 50% had recurrence of migraine (Hoshiyama et al., 2012). This same study found that compared with bottle-feeding mothers, breastfeeding mothers have a lower recurrence rate of migraine up to 6 months postpartum (Hoshiyama et al., 2012).

A systematic review of the relationship between migraine, cardiovascular disease, and stroke reported that migraineurs are at increased risk of preeclampsia (OR range 1.08–3.5) and gestational hypertension (OR range 1.23–1.68) during their pregnancy compared with nonmigraineurs (Wabnitz and Bushnell, 2015).

Tension-type headache

Tension-type headache (TTH) is the most common primary headache type with a lifetime prevalence of 88% in women (Pearce and Hansen, 2012). In pregnancy, TTH represents 26% of headaches (Pearce and Hansen, 2012). Several retrospective studies suggest that most women suffering from a primary headache disorder report an improvement during pregnancy, though this association is seen more significantly in migraine than in TTH (Negro et al., 2017). TTH worsens in 5% of women during pregnancy, and 18% of women with TTH report no change (Negro et al., 2017). It is not associated with adverse pregnancy outcomes (Negro et al., 2017). A large Canadian study of women in the postpartum period found that 40% of women reported a headache in the first week after delivery and the majority of cases were TTH (Goldszmidt et al., 2005).

Trigeminal autonomic cephalalgias

Trigeminal autonomic cephalalgias are a group of disorders that includes cluster headache (CH), paroxysmal hemicrania, short-lasting unilateral neuralgiform headache attacks with conjunctival injection and tearing (SUNCT), and short-lasting unilateral neuralgiform headache attacks with cranial autonomic symptoms (SUNA) (Lambru and Matharu, 2012). There are limited epidemiologic data regarding these disorders with the exception of CH. The estimated prevalence in the general population is less than 1 in 1000 and more men are affected than women in a ratio of 2.5–7.5:1 (Giraud and Chauvet, 2009). Prospective studies regarding CH and pregnancy are lacking, as CH is seen in less than 0.3% of pregnancies (Pearce and Hansen, 2012). In one study of CH in pregnancy, 6 of 8 women reported remission of their headaches during pregnancy (Ekbom and Waldenlind, 1981).

Secondary headache disorders

Spontaneous intracranial hypotension

Spontaneous intracranial hypotension (SIH) is an uncommon condition associated with low cerebrospinal fluid pressure and is rarely described in pregnancy (McGrath et al., 2010). In the literature, there are less than 10 case reports of SIH occurring in pregnancy and 1 case report detailing a recurrence of SIH with subsequent pregnancy (McGrath et al., 2010).

Idiopathic intracranial hypertension

Idiopathic intracranial hypertension (IIH) has a female predominance of 8:1 and occurs most frequently in women who are obese (Durcan et al., 1988). The annual incidence of IIH is 0.9 per 100,000 in the general population, 4 per 100,000 in obese women, and 19 per 100,000 in morbidly obese women. The prevalence of obesity in pregnancy increased from 10% in the 1990s to 16%–19% in 2006 (Radhakrishnan et al., 1993). The incidence of IIH in pregnancy is expected to rise along with the increasing rates of obesity (Thirumalaikumar et al., 2014).

The prevalence of IIH among pregnant women is reported to be 2%–12%, and pregnant women develop IIH at approximately the same rate as the general population (Kesler and Kupferminc, 2013). Digre et al. found that 8.3% of IIH patients were pregnant at the onset of IIH and 14.7% were diagnosed with IIH while pregnant (Kesler and Kupferminc, 2013). When IIH presents during pregnancy, onset is typically during the first half of pregnancy with 61% of cases occurring in the first trimester (Kesler and Kupferminc, 2013). Pregnancy alone is not an independent risk factor for the development of IIH (Digre et al., 1984). Obstetric complications are not increased in women with IIH (Evans and Lee, 2010).

MOVEMENT DISORDERS

Parkinson's disease

Parkinson's disease (PD) occurs before the age of 40 in around 5% of cases, and around 400 women under 50 years of age are diagnosed with PD in the United States each year (Kranick et al., 2010; Seier and Hiller, 2017). A 2017 review of literature regarding pregnancy in PD found worsening of PD symptoms in 31 of 64 pregnancies (48%) and unchanged or improved PD symptoms in the remaining 33 (52%). There is no increased rate of fetal or birth complications in PD (Seier and Hiller, 2017).

Chorea

Chorea gravidarum

Chorea gravidarum is associated with rheumatic fever and connective tissue disorders. In women who have had rheumatic fever, chorea in pregnancy occurs in up to 75%, mostly in the first and second trimesters. It resolves spontaneously before delivery in up to one-third of cases and resolves following delivery in the remaining cases. Recurrence in subsequent pregnancies occurs in one of five women (20%) (Miyasaki and AlDakheel, 2014).

Wilson disease

Wilson disease is an autosomal recessive disorder marked by abnormal copper metabolism and accumulation leading to tremors, chorea, psychiatric disease, and liver disease (Miyasaki and AlDakheel, 2014). Pregnancy does not change the course of disease in this condition (Miyasaki and AlDakheel, 2014). A 2004 retrospective analysis of pregnancy in Wilson disease included 16 patients with 59 pregnancies resulting in 30 live births. No women reported a change in their Wilson disease symptoms during pregnancy, but recurrent miscarriages were frequent. There were 24 spontaneous abortions, 2 medical terminations, and 3 stillbirths (Sinha et al., 2004).

Sydenham chorea

Sydenham chorea is a clinical manifestation of acute rheumatic fever and is characterized by involuntary movements of the face and limbs, hypotonia, and behavioral changes (Williams and Swedo, 2015). A 2012 review of pregnancy in those with Sydenham chorea included 20 women and found that 15 of them (75%) developed chorea gravidarum during pregnancy with 80% of cases occurring in the first 6 months of gestation. Of the women who developed chorea, 67% had generalized chorea, 20% had focal or multifocal chorea, and 13% had hemichorea. Recovery following delivery occurred in 11 women, and 4 women had continued chorea in the first 12 months following delivery (Maia et al., 2012).

Myoclonus

There have been two case reports of pregnant women developing spinal myoclonus following the administration of epidural anesthesia or combined spinal–epidural anesthesia for cesarean section (Menezes and Venkat, 2006; Hisano et al., 2010).

Ataxia

Wernicke encephalopathy secondary to hyperemesis gravidarum is the most common cause of ataxia in pregnancy once structural causes are ruled out. A 2010 study of pregnancy in Friedreich ataxia including 31 women with 65 pregnancies resulted in 56 live births. There were equal rates of symptom worsening, improving, and remaining unchanged (Miyasaki and AlDakheel, 2014).

Tourette syndrome

Tourette syndrome is two to three times more common in men than in women (Baker and Hung, 2017). In a 2009 retrospective series of pregnancy in Tourette syndrome including 8 women with 11 pregnancies, 5 women reported improvement of symptoms, 3 reported worsening of symptoms, and another 3 women reported unchanged symptoms. One woman in each category had 2 pregnancies, and in all 3 women the disease course was the same for both pregnancies (Stern et al., 2009).

Dystonia

Dystonia is the third most common movement disorder in the general population with mean age of onset of 27 years. The most common causes of dystonia in pregnancy are Wilson disease, genetic forms of dystonia such as DYT1, and dopa-responsive dystonia. Drug-induced dystonia secondary to neuroleptic use for nausea can occur in pregnancy (Miyasaki and AlDakheel, 2014).

In one study including 27 women who had pregnancies since onset of their dystonia, 4 reported changes in their symptoms during pregnancy. In another study of 10 women who had pregnancies since the onset of their dystonia, 5 reported no changes in symptoms, 3 had partial or complete resolution of symptoms, and 2 had worsening of symptoms (Miyasaki and AlDakheel, 2014; Ciafaloni et al., 2018, 174).

Stiff person syndrome

The prevalence of stiff person syndrome is 1 in 1,000,000 with age of onset peaking at 45 years and two-thirds of cases occurring in females. There have been three reports of pregnancy in women with stiff person syndrome (Miyasaki and AlDakheel, 2014).

CRANIAL NEUROPATHIES

Data regarding the epidemiology of cranial nerve palsies in pregnancy are lacking and largely limited to individual case reports and systematic reviews of case reports. A 2017 review of 21 cases of abducens nerve palsy developing during pregnancy or in the postpartum period found it was most commonly related to hypertensive disorders of pregnancy (38%), followed by epidural blocks (29%), and less commonly, infectious diseases, CNS injury, or an unknown cause. The cases secondary to hypertensive disorders of pregnancy resolved within days to months (Nieto-Calvache et al., 2017).

Cranial nerve palsies rarely develop as a consequence of central neuraxial blocks (epidural, spinal, or combined spinal–epidural) used for anesthesia during labor.

A 2017 systematic review of published cases of postpartum cranial nerve palsy following central neuraxial blocks found 43 cases of cranial nerve palsy involving various cranial nerves as a result of neuraxial blocks. While 8 patients had permanent nerve palsy, 35 patients (81%) recovered completely with the majority recovering over weeks to months. There were no case reports involving the olfactory, oculomotor, glossopharyngeal, accessory, or hypoglossal nerves. Intracranial hypotension was the most common cause (Chambers and Bhatia, 2017).

Horner's syndrome can rarely develop as a consequence of central neuraxial blocks used for anesthesia during labor. A 2018 systematic review of published cases of Horner's syndrome following central neuraxial blocks found 78 cases, 9 of which also had trigeminal nerve palsy and 1 had hypoglossal nerve palsy (Chambers and Bhatia, 2018).

Bell's palsy

The incidence of Bell's palsy is 20 per 100,000, with the incidence in pregnant women being higher at 45 per 100,000. The risk of developing Bell's palsy is six times higher in pregnant women in the third trimester and the first 2 weeks postpartum compared with nonpregnant women (Sax and Rosenbaum, 2006; Massey and Guidon, 2014). Diabetes mellitus and hypertensive disorders of pregnancy, including preeclampsia, are associated with pregnancy-related Bell's palsy (Massey and Guidon, 2014).

A 2011 retrospective study assessing whether Bell's palsy in pregnancy is associated with adverse perinatal outcome found no significant association (Katz et al., 2011). In pregnancy-related Bell's palsy with partial facial paralysis, complete recovery usually takes places within 3–6 months (Sax and Rosenbaum, 2006). In a 2002 study of facial nerve recovery in pregnancy-related Bell's palsy, it was found that the prognosis for good recovery in those with complete facial paralysis was worse in pregnant women (52%) when compared with the general population (77%–88%) (Gillman et al., 2002). The risk of recurrence in subsequent pregnancies is low (Massey and Guidon, 2014).

OPHTHALMOLOGIC DISORDERS

Causes of diplopia in pregnancy include cranial nerve palsies, intracranial hypotension, Wernicke encephalopathy, ocular myasthenia, pituitary tumors, and pituitary apoplexy. In pregnancy, CN VII palsy occurs most frequently followed by CN VI and then CN IV. Wernicke encephalopathy is most often seen in the first trimester whereas exacerbation of ocular myasthenia is most often seen in the first trimester or following delivery. Causes of decreased visual acuity in pregnancy include diabetic retinopathy, ocular vascular events, cerebral vascular events, and preeclampsia/eclampsia. In diabetic women, progression of diabetic retinopathy can occur during pregnancy with regression following delivery (Digre and Kinard, 2014).

Preeclampsia and eclampsia can lead to retinal vein and artery occlusions, central retinal artery occlusion, optic neuritis, ischemic optic neuropathy, and optic atrophy as well as hypertensive changes in the eye, including vascular narrowing, cotton wool spots, hemorrhages, and optic disc edema (Omoti et al., 2008; Digre and Kinard, 2014). Retinal detachment in preeclampsia is typically bilateral and caused by arteriolar vasospasm leading to choroidal ischemia. Most women recover vision completely. Cortical blindness in preeclampsia is most often due to vasogenic edema and is usually reversible (Omoti et al., 2008). Vascular syndromes such as Balint syndrome or homonymous hemianopsias can occur if there is ischemic or hemorrhagic damage to the visual pathways. Cortical blindness occurs in 15% of patients with eclampsia and usually resolves spontaneously but can be permanent if it is secondary to stroke. Blurred vision and vision loss can occur secondary to PRES (Digre and Kinard, 2014).

Thyroid abnormalities occur in 5% of pregnant women with thyroid eye disease being more common in hyperthyroid states. Symptoms of thyroid eye disease include proptosis, diplopia, and photophobia. Graves' disease tends to improve in late pregnancy and relapse in the postpartum state (Omoti et al., 2008; Digre and Kinard, 2014).

NEURO-ONCOLOGY

There is poor understanding of the global incidence and prevalence of primary brain tumors worldwide, and previous systematic reviews have found the global incidence rate to be 4.3–18.6 per 100,000 per year (de Robles et al., 2015). Data is limited regarding primary brain tumors during pregnancy, but the incidence rate is estimated to be in the range of 2.6–15 per 100,000 pregnancies (van Westrhenen et al., 2018). Presenting symptoms of pregnant women with intracranial tumors include seizures, headache, and persistent nausea and vomiting that may be mistaken for hyperemesis gravidarum (Terry et al., 2012).

In patients aged 15–39 the most common nonmalignant tumors are pituitary tumors (3.11 per 100,000 population), meningioma (1.74 per 100,000 population), and nerve sheath tumors (0.91 per 100,000) (Ostrom et al., 2015). Glioma is the most common malignant brain tumor in this age group, accounting for 29% of all CNS tumors and 82% of malignant tumors (Ostrom et al., 2015). Data from the Central Brain Tumor Registry of the United States from years 2008–12 has shown that the incidence of meningioma is significantly higher in females than in males (10.87 compared with 4.98 per 100,000 population) and that tumors of the pituitary are also more common in females (Ostrom et al., 2015).

Management and pregnancy outcomes are diverse and vary depending on tumor location, size, rate of growth, and stage of pregnancy. If possible, it is recommended that surgery and chemotherapy be postponed until after the first trimester (Terry et al., 2012). In a stable patient at term, there does not seem to be any benefit in cesarean section delivery over vaginal delivery (van Westrhenen et al., 2018).

Pituitary tumors

Prolactinomas comprise the majority of pituitary adenomas, and the incidence is four times higher in women of reproductive age than in men of the same age (Cocks Eschler et al., 2018). During a normal pregnancy the pituitary gland will enlarge and increase by 136% (Cocks Eschler et al., 2018), though the majority of pituitary tumors will remain stable and enlarge in only 5%–15% of pregnancies (Stevenson and Thompson, 2005). If tumor growth does occur, dopamine agonists are considered safe and decompression surgery is safe in the second trimester (Cocks Eschler et al., 2018).

Pituitary apoplexy, a medical emergency defined as hemorrhage or infarction of pituitary gland tissue, most often occurs due to growth of an underlying pituitary tumor. It has been estimated to occur in 1 per 10,000 pregnancies (Grand'Maison et al., 2015). The most common presentation of this condition is severe headache with visual field defects, impaired visual acuity, and ophthalmoplegia (Piantanida et al., 2014). Surgical or medical management has been successful with healthy fetal outcomes (Grand'Maison et al., 2015).

Meningioma

Meningiomas represent 15% of intracranial tumors in women of childbearing age (Laviv et al., 2018). Several studies have shown that meningiomas increase in size during pregnancy and can reduce in size after delivery, though the mechanism by which this occurs is poorly understood (Chakravarthy et al., 2018). Proposed explanations include tumor response to circulating hormones due to presence of tumor estrogen and progesterone receptors, increased tumor edema possibly due to an increase in aquaporin 4 during pregnancy, and an increase in prolactin (Chakravarthy et al., 2018). Women who underwent surgical resection during pregnancy had similar overall outcomes compared with women who delayed surgery until the postpartum period (Laviv et al., 2018).

Glioma

A retrospective cohort study in Norway found that pregnancy did not have an effect on survivorship in women diagnosed with low-grade gliomas (WHO grade I) (Rønning et al., 2016). A prospective case–control comparing pregnant and nonpregnant women with higher-grade gliomas (WHO grade II–IV) found that pregnancy also did not alter overall disease course and survivorship (Forster et al., 2019). The majority of glioma tumors do increase in size during pregnancy, though this does not necessarily cause new symptoms or clinical decline (Peeters et al., 2018).

Choriocarcinoma

Gestational choriocarcinoma occurs in 2–7 of every 100,000 pregnancies in the United States, and though most cases occur after abnormal pregnancies (such as ectopic pregnancies), 25% occur after a full-term pregnancy (Tarney et al., 2018). It is a rare but highly metastatic tumor with up to a 20% rate of brain involvement (Hopkins et al., 2014). Data from the SEER program of the National Cancer Institute found that non-Hispanic black women and women over 20 years of age have a higher risk of death with adjusted hazard ratio of 3.87 and 7.18, respectively (Tarney et al., 2018).

Hemangiomas

CNS hemangioblastoma is rare; when age matched with nonpregnant female patients with hemangioblastoma, pregnant women with hemangioblastomas had the same progression (Ma et al., 2018).

Paraneoplastic disorders

Paraneoplastic disorders are rare in pregnancy. There have been at least 12 case reports of NMDA encephalitis during pregnancy, and in the majority of cases both mother and baby had favorable outcomes (Xiao et al., 2017). There have been five case reports of LEMS in pregnancy, and treatment and outcomes in each case varied (Bayrak et al., 2010). Paraneoplastic cerebellar degeneration syndromes such as those involving anti-Yo and anti-Ri antibodies primarily affect women, but there have been no confirmed case reports during pregnancy (Lancaster, 2015).

SLEEP DISORDERS

Sleep-disordered breathing

Obstructive sleep apnea

The estimated prevalence of moderate to severe sleep-disordered breathing (SDB) in women of reproductive age in North America is 3%–6.5%. Some estimates of prevalence exceed 10% in populations composed of more overweight individuals (Abdullah et al., 2016).

Physiologic changes of pregnancy that promote the development of obstructive sleep apnea (OSA) include nasal mucosal congestion, soft tissue deposition in the neck, gestational weight gain, and abdominal mass loading with resulting tracheal shortening (Abdullah et al., 2016). Additionally, pregnancy-induced elevations in progesterone have been shown to increase ventilatory drive at the central chemoreceptor. This increase in ventilatory drive leads to increased diaphragmatic effort and greater negative inspiratory pressures at the level of the upper airway. Ultimately, these changes may increase the chance of upper airway collapse during sleep (Jensen et al., 2008).

Prospective observational studies assessing OSA in pregnancy have shown a prevalence ranging from approximately 17% to 45% in the third trimester (Pamidi and Kimoff, 2018). One study found the prevalence of OSA to be 10.5% and 26.7% in the first and third trimester, respectively (Pien et al., 2014). Another study using portable in-home sleep studies evaluating OSA in normal to low-risk pregnant patients with normal BMI found a prevalence of 3.6% in early pregnancy (6–15 weeks of gestation) and 8.3% in mid-pregnancy (22–31 weeks of gestation) (Facco et al., 2017). In patients with BMI > 30, prevalence rates were found to be much higher with 30% meeting the criteria for sleep apnea in early pregnancy (6–20 weeks of gestation) and 47% in late pregnancy (28–37 weeks of gestation) (Facco et al., 2014). The incidence of new-onset SDB was found to be 20% (Edwards et al., 2005).

Complications of pregnancy in OSA patients include preeclampsia (adjusted OR 2.22), eclampsia (adjusted OR 2.95), and gestational diabetes (adjusted OR 1.51). Women with OSA also have a 2.5–3.5-fold increase in severe complications such as cardiomyopathy, congestive heart failure, and hysterectomy (Abbott et al., 2014).

In the postpartum period the course of SDB is largely unknown (Pamidi and Kimoff, 2018). One small longitudinal case–control study by Edwards et al. found that women treated for OSA in the third trimester demonstrated marked improvement in the apnea-hyponea index 3 months following delivery suggesting improvement in SBD postnatally (Edwards et al., 2005).

Central sleep apnea

Central sleep apnea has not been well studied in pregnant women. Numerous studies on SDB (including both central and OSAs) are suggestive of an overall prevalence of 10%–35% during pregnancy and report that obstructive, rather than central, respiratory events predominate (Pien et al., 2005; Bourjeily et al., 2015). One small case–control study found a total of 2 central apnea events in one pregnant patient and a total of 98 central apnea events in 11 nonpregnant patients. This study reported that overall central sleep apnea index was very low with a median of 0 (Bourjeily et al., 2015).

When comparing the postpartum period with the third trimester of pregnancy, there is an increase in apneic and hypopneic events and in mean apnea and hypopnea duration, with all of the apnea episodes being central in nature (Brownell et al., 1986; Trakada et al., 2003).

Narcolepsy

In the general population the prevalence of narcolepsy with cataplexy is reported to be 25–50 per 100,000 individuals, with an annual incidence of 0.3–0.6 per 100,000 (Calvo-Ferrandiz and Peraita-Adrados, 2018). In most cases the age of onset is between 15 and 40 years, thereby affecting many women in their childbearing years (Maurovich-Horvat et al., 2013). It should be noted that only a small percentage of patients decide to medically treat their narcolepsy during pregnancy (Maurovich-Horvat et al., 2010). Among women who had discontinued their medications in the first trimester of pregnancy, 18% reported an improvement in symptoms, 40% reported no change, another 40% reported a worsening of their symptoms, and 2% were unknown (Maurovich-Horvat et al., 2013). Overall, it is suggested that women with narcolepsy have comparable pregnancy outcomes to the general population, though they may require more psychologic and child-care support in the postpartum period (Maurovich-Horvat et al., 2013). Experts believe that the perceived risks of continuing medication such as modafinil during pregnancy are overestimated and risk to fetus could be essentially nonexistent (Calvo-Ferrandiz and Peraita-Adrados, 2018).

Restless leg syndrome

Restless leg syndrome (RLS) is present in 5% of the US population (Allen et al., 2011). According to the World Health Organization's regional groupings, the prevalence of RLS is 22% in the European region, 14% in the Western Pacific region, 30% in the Eastern Mediterranean region, and 20% in the region of the Americas (Chen et al., 2018).

The prevalence of RLS in pregnancy is 20%–25%, which is threefold higher than in the general population (Bazalakova, 2017). When RLS occurs for the first time in pregnancy, it usually occurs in the second and third trimesters and typically resolves following delivery (Manconi et al., 2004). The association of gestational RLS may be explained by relative iron deficiency during pregnancy (Bazalakova, 2017).

In the postpartum period, women who developed gestational RLS had a fourfold increase in the presence of RLS at 7-year follow-up. Additionally, gestational RLS appears to reemerge in 60% of women in subsequent pregnancies (Abbott et al., 2014).

Insomnia

Disruption in sleep, decreased total sleep time, and decreased sleep quality are common in pregnant women, especially in the third trimester (Abbott et al., 2014). General sleep disturbance rates change across gestation ranging from 13% in the first trimester, 19% in the second, and 66% in the third (Reichner, 2015). The incidence of insomnia also demonstrates a similar trend, with 13% of women reporting symptoms at the beginning of pregnancy and up to 75% reporting mild symptoms by 39 weeks of gestation (Reichner, 2015). Sleep duration decreases in the third trimester, whereas frequent snoring, restless legs, and nightly awakenings all increase significantly (Hedman et al., 2002a).

Parasomnias

Parasomnias can be divided into rapid eye movement (REM) parasomnias and nonrapid eye movement (NREM) parasomnias. REM parasomnias include REM sleep behavior disorder, recurrent isolated sleep paralysis, and nightmare disorder. NREM parasomnias include sleep walking/talking, confusional arousals, and sleep terrors (Bjorvatn et al., 2010).

One study evaluating parasomnias and pregnancy found that many of these disorders decrease during pregnancy and up to 3 months postpartum, which seems to be more pronounced in primiparous mothers (Hedman et al., 2002a). Sleep paralysis was found to decrease during the first trimester and increase in the second and third trimesters (Hedman et al., 2002a). Nightmares affect 40%–50% of pregnant women (Schredl et al., 2016). Reportedly, nightmares decline antenatally and into the postpartum period. In the prepregnancy period, 55.7% of patients reported nightmares compared with 41.2% and 40.3% in the third trimester and 3 months postpartum, respectively (Hedman et al., 2002a). Sleep walking/talking decreased from 23% prepregnancy to 13% in the second trimester (Hedman et al., 2002a). Night terrors occur infrequently in adults with a reported prevalence of <1% of women (Hedman et al., 2002a).

NEUROLOGIC INFECTIOUS DISEASE

Bacterial meningitis

The overall incidence of bacterial meningitis in adults living in the United States is estimated to be 2.6–6 per 100,000 each year (Thigpen et al., 2011).

Streptococcus pneumoniae accounts for 58% of community-acquired acute bacterial meningitis cases (Davis, 2018). The case-fatality rate of pneumococcal meningitis is 10%–20% in developed countries and up to 30%–40% in developing countries (Davis, 2018). There have been five isolated case reports of women diagnosed with pneumococcal meningitis during pregnancy or in the postpartum period and one small case series in Nigeria of 26 cases (15 during pregnancy and 11 during the immediate postpartum period) (Lucas, 1964). The maternal mortality rate was 27% (Lucas, 1964).

Neisseria meningitidis is a major cause of meningitis despite the availability of effective vaccines (Davis, 2018). In sub-Saharan African meningococcal outbreaks of serotype A, attack rates can reach more than 1000 cases per 100,000 population each year (Davis, 2018). Meningococcal meningitis during pregnancy is very rare with less than 5 case reports described in both pregnancy and the postpartum period (Osler, 1942; Brandeberry and Vergon, 1951; Finan et al., 1992).

Listeria monocytogenes causes 9% of acute bacterial meningitis cases worldwide (Davis, 2018). Pregnant women have an increased susceptibility to this infection, with rates 20 times more common than the general population (Olson-Chen, 2018). Estimated incidence rates of pregnancy-related listeriosis is 3 cases per 100,000 births, accounting for up to 35% of all Listeria infections (Madjunkov et al., 2017). Listeriosis can occur anytime during pregnancy, but infections most often occur in the third trimester (66%) (Wadhwa Desai and Smith, 2017). Hispanic patients have incidence rates 1.8 times higher than non-Hispanic patients (Voetsch et al., 2007). Listeria rarely leads to meningoencephalitis; in a case series of 248 mothers diagnosed with listeriosis during pregnancy, only one developed meningoencephalitis as a complication of infection (McLauchlin, 1990).

Haemophilus influenzae meningitis accounts for less than 7% of cases in developed countries (Davis, 2018). One case report could be found of *H. influenzae* meningitis occurring during pregnancy (Sandberg and Dahlberg, 1981).

Staphylococcus aureus composes 5% of meningitis cases and is fatal in 30% (Davis, 2018). A total of two case reports were found in the literature reporting *S. aureus* meningitis during pregnancy and one occurring during the postpartum period (Ali and Kabins, 1977; McComb et al., 1982; Marsaudon and Berthier, 1996). One report was found to be due to complication from obstetric epidural anesthesia (Marsaudon and Berthier, 1996).

Fungal meningitis

Cryptococcus infections are typically reported in patients with immunodeficiency (Costa et al., 2009). The severity of this infection may be greater in pregnant women with mortality rates as high as 25% (Olson-Chen, 2018).

Coccidioides species are found endemically in the southwest region of the United States (Galgiani et al., 2005). The incidence of infection in endemic areas has increased from 5.3 per 100,000 population in 1998 to 42.6 per 100,000 in 2011 (Tsang et al., 2013). There are less than 100 reported cases of coccidioidomycosis occurring in pregnancy (Crum and Ballon-Landa, 2006). African American women had a 13-fold increased risk of developing disseminated disease compared with white women (Crum and Ballon-Landa, 2006).

Toxoplasmosis

Toxoplasmosis is typically asymptomatic in pregnant mothers but has deleterious effects on neonates via congenital toxoplasmosis (Olson-Chen, 2018, 95–104). South America and Africa have the highest rates of seroprevalence while the United States, northern Europe, and Southeast Asia have very low incidence. Seroprevalence has been decreasing over the past several decades, especially in industrialized countries such as France, where there was a decrease from 54% in 1995 to 44% in 2003 (Berghold et al., 2016).

Viral encephalitis

Herpes Simplex Virus type 1 (HSV-1) and Herpes Simplex Virus type 2 (HSV-2).

In the United States, HSV accounts for up to 20% of all identified viral encephalidities (Baldwin and Cummings, 2018). HSV-1 is responsible for 90% of HSV cases (Dodd et al., 2015). There are 17 reported cases of HSV encephalitis presenting during pregnancy (Dodd et al., 2015). Maternal mortality rate of disseminated HSV is as high as 70% without treatment, but it fell to less than 20%–30% with early diagnosis and treatment (Dodd et al., 2015).

Varicella-zoster virus

Varicella-zoster virus (VZV) manifests as cerebellitis in ~1 per 4000 patients with chickenpox (Baldwin and Cummings, 2018). VZV infection rarely occurs during pregnancy in immunocompetent women (Olson-Chen, 2018). However, mortality and morbidity are high in pregnancy with death being five times more likely when compared with the general population and up to 20% of survivors developing epilepsy (Baldwin and Cummings, 2018; Olson-Chen, 2018).

Cytomegalovirus

In the United States the seroprevalence of cytomegalovirus (CMV) is 60% with prior infections acquired mostly during childhood (Olson-Chen, 2018). Primary maternal infection by CMV carries a 30%–40% risk of vertical transmission to the fetus with a higher risk of transmission in immunosuppressed patients (Carlson et al., 2010). CMV is the most common cause of congenital infection complicating ~1% of all live births (Carlson et al., 2010).

West nile virus

West nile virus (WNV) is transmitted to humans via infected mosquitos (Stewart et al., 2013). Prevalence of WNV in North America is increasing (Olson-Chen, 2018). It is estimated that 80%–99% of the infected humans are asymptomatic (Ornoy and Tenenbaum, 2006; Lyons, 2018). Reports of neuroinvasive disease by WNV have ranged from a few hundred to a few thousand each year over the past 10 years (Lyons, 2018), and less than 1% of infected individuals develop neuroinvasive disease (Stewart et al., 2013). There are two reported case studies of WNV encephalitis occurring during pregnancy (Chapa et al., 2003; Bruno et al., 2004).

Brain abscess

There are less than 15 case reports of cerebral abscess complicating pregnancy, sometimes associated with otitis media and sinusitis (Olson-Chen, 2018).

Infectious myelopathy

The incidence of HTLV-1-associated myelopathy among carriers is estimated to be 3 cases per 100,000 infected patients annually (Mizokami et al., 1994). There are two case reports of HTLV-1-associated myelopathy presenting during pregnancy and one case report of acute exacerbation of HTLV-1-associated myelopathy occurring during the postpartum period (Mizokami et al., 1994; Ando et al., 2003). The two pregnant women went on to have normal vaginal deliveries without complications (Ando et al., 2003).

Zika virus

In an epidemiologic study from Brazil, authors estimated that between 500,000 and 1.5 million people have been infected by Zika virus (ZIKV) (Coelho and Crovella, 2017). Maternal infection that is vertically transmitted to the fetus can cause a devastating encephalitis of the developing brain resulting in microcephaly. It is estimated that microcephaly occurs in approximately 3% of live births from infected mothers (Lyons, 2018).

NEUROLOGIC INTENSIVE CARE

Irreversible brain injury

There have been 11 reported cases of irreversible maternal brain injury leading to persistent vegetative state or death by neurologic criteria in pregnancy. In 10 of the cases, pregnancy was prolonged to preserve the life of the fetus. All of the deliveries were preterm and all but one was delivered via cesarean section (Feldman et al., 2000).

Traumatic brain injury

One 2011 study examining outcomes of traumatic brain injury (TBI) in pregnant women found similar mortality rates in pregnant and nonpregnant women with moderate to severe TBI (9.9% vs 9.3%) (Berry et al., 2011).

Spinal cord injury

There is overall good maternal and fetal outcome in women with spinal cord injury (SCI) who become pregnant, though there is an increased incidence of fetal anomalies and miscarriage in women who acquire SCI during pregnancy compared with those who become pregnant after SCI (Baker and Cardenas, 1996; Dawood et al., 2014). One study reported a 12% incidence of increased spasticity in pregnancy in women with SCI (Dawood et al., 2014). Autonomic hyperreflexia, which can be fatal if not properly treated, occurs in up to 85% of patients with SCI above the T5–6 level, and it has been reported in spontaneous and induced labors. Rates of cesarean delivery in women with SCI range from 18% to 67% while the current rate of cesarean delivery in the United States is around 25% (Baker and Cardenas, 1996). Forceps delivery is often required (Fox et al., 1990).

NEUROLOGIC COMPLICATIONS OF SYSTEMIC DISEASE

Sarcoidosis

Neurologic manifestations are rare and occur in 5% of patients with sarcoidosis that manifests in numerous ways, including cranial neuropathy, aseptic meningitis, mass lesions, encephalopathy, vasculopathy, seizures, hydrocephalus, myelopathy, peripheral neuropathy, and myopathy (Krumholz and Stern, 2014). In the literature, there are two reported cases of neurosarcoidosis complicating pregnancy (Maisel and Lynam, 1996; Cardonick et al., 2000).

Systemic lupus erythematosus

Approximately 30% of patients with systemic lupus erythematosus (SLE) develop neurologic complications of the disease. The most common manifestations are headache (39%–61%), seizures (8%–18%), cerebrovascular disease (2%–8%), psychosis (3%–5%), cranial neuropathy (1.5%–2.1%), and movement disorders (1%) (Muscal and Brey, 2010). In pregnant patients with SLE, one small prospective study in Mexico reported that lupus exacerbations were seen in 59.7% of patients, with only 8 out of 75 patients experiencing neurologic disturbance (Mintz et al., 1986). Another small retrospective case series of 39 patients reported no neuropsychiatric complications during pregnancy (Fine et al., 1981). There is one case report detailing a patient with SLE who developed spastic paraparesis thought to be secondary to transverse myelitis during her second pregnancy (Marabani et al., 1989).

PSYCHIATRIC DISEASE

Pregnancy and the postpartum period are thought to be phases of increased susceptibility to psychiatric diseases, including depression and other mood disorders. Risk factors include history of psychiatric disease, exposure to traumatic events, and lack of a partner (Vesga-Lopez et al., 2008). A 2008 study assessed the prevalence of mood, anxiety, and substance-use disorders in pregnant and postpartum women (those pregnant at the time of interview or had been pregnant within the prior 12 months) using interviews conducted in the 2001–02 National Epidemiologic Survey on Alcohol and Related Conditions. It found that the 12-month prevalence of psychiatric diseases was similar in pregnant and postpartum women (0.4%–14.6%) compared with nonpregnant women (0.3%–19.9%). There were lower rates of alcohol and tobacco use in the pregnant and postpartum population and a greater risk of major depressive disorder in the postpartum period (Vesga-Lopez et al., 2008). The prevalence of postpartum depression is estimated at 13%–19% (O'Hara and McCabe, 2013). Postpartum psychosis is the occurrence of an acute psychotic episode that affects 1–2 per 1000 childbearing women in the 2–4 weeks following delivery (Sit et al., 2006).

CONCLUSION

Nearly every neurologic condition occurring in the childbearing age group has been described during pregnancy. However, some are clearly more common, such as migraine, epilepsy, and stroke, compared with others, such as myasthenia or neuromuscular disorders, as summarized in this chapter. It is important to recognize neurologic abnormalities when they occur and refer to a neurologist, as this would aid in determining the diagnosis and possible treatment, if available. In addition the more these conditions are recognized and coded, the more it will improve the understanding of the epidemiology and the underlying influence pregnancy might have on these conditions.

References

Abbott SM, Attarian H, Zee PC (2014). Sleep disorders in perinatal women. Best Pract Res Clin Obstet Gynaecol 28: 159–168.

Abdullah HR, Nagappa M, Siddiqui N et al. (2016). Diagnosis and treatment of obstructive sleep apnea during pregnancy. Curr Opin Anaesthesiol 29: 317–324.

Acker G, Czabanka M, Schmiedek P et al. (2018). Pregnancy and delivery in moyamoya vasculopathy: experience of a single European institution. Neurosurg Rev 41: 615–619.

Ali MA, Kabins SA (1977). Subacute staphylococcal meningitis secondary to postpartum endometritis. South Med J 70: 368–369.

Allen RP, Bharmal M, Calloway M (2011). Prevalence and disease burden of primary restless legs syndrome: results of a general population survey in the United States. Mov Disord 26: 114–120.

Alonso A, Hernan MA (2008). Temporal trends in the incidence of multiple sclerosis: a systematic review. Neurology 71: 129–135.

Alvarenga M, Schimidt S, Alvarenga RP (2017). Epidemiology of neuromyelitis optica in Latin America. Mult Scler J Exp Transl Clin 3: 2055217317730098.

Ando Y, Matsumoto Y, Kakimoto K et al. (2003). Pregnancy complicated by HTLV-I associated myelopathy: two cases. Arch Gynecol Obstet 268: 61–64.

Angelini C, Fanin M (2017). Limb girdle muscular dystrophies: clinical-genetical diagnostic update and prospects for therapy. Expert Opin Orphan Drugs 5: 769–784. https://doi.org/10.1080/21678707.2017.1367283.

Argov Z, de Visser M (2009). What we do not know about pregnancy in hereditary neuromuscular disorders. Neuromuscul Disord 19: 675–679.

Artama M, Braumann J, Raitanen J et al. (2017). Women treated for epilepsy during pregnancy: outcomes from a nationwide population-based cohort study. Acta Obstet Gynecol Scand 96: 812–820.

Aube M (1999). Migraine in pregnancy. Neurology 53: S26–S28.

Badell ML, Rimawi BH, Rao AK et al. (2018). Botulism during pregnancy and the postpartum period: a systematic review. Clin Infect Dis 66: S30–S37.

Baker ER, Cardenas DD (1996). Pregnancy in spinal cord injured women. Arch Phys Med Rehabil 77: 501–507.

Baker JM, Hung AY (2017). Movement disorders in women. Semin Neurol 37: 653–660.

Baldwin KJ, Cummings CL (2018). Herpesvirus infections of the nervous system. Continuum (Minneap Minn) 24: 1349–1369.

Bangar S (2016). Women with epilepsy: clinically relevant issues. Funct Neurol 31: 127–134.

Baouahi H, Doumiri M (2014). Wernicke encephalopathy complicating hyperemesis gravidarum and associated with pontine myelinolysis. Pan Afr Med J 19: 340.

Bateman BT, Schumacher HC, Bushnell CD et al. (2006). Intracerebral hemorrhage in pregnancy: frequency, risk factors, and outcome. Neurology 67: 424–429.

Bayrak AO, Sandikci U, Yon SK et al. (2010). Manifestation of Lambert–Eaton myasthenic syndrome during last trimester of pregnancy. Int J Neurosci 120: 439–441.

Bazalakova M (2017). Sleep disorders in pregnancy. Semin Neurol 37: 661–668.

Berghold C, Herzog SA, Jakse H et al. (2016). Prevalence and incidence of toxoplasmosis: a retrospective analysis of mother–child examinations, Styria, Austria, 1995 to 2012. Euro Surveill 21: 30317.

Bergin PS, Harvey P (1992). Wernicke's encephalopathy and central pontine myelinolysis associated with hyperemesis gravidarum. BMJ 305: 517–518.

Berry C, Ley EJ, Mirocha J et al. (2011). Do pregnant women have improved outcomes after traumatic brain injury? Am J Surg 201: 429–432.

Bjorvatn B, Gronli J, Pallesen S (2010). Prevalence of different parasomnias in the general population. Sleep Med 11: 1031–1034.

Bourjeily G, Sharkey KM, Mazer J et al. (2015). Central sleep apnea in pregnant women with sleep disordered breathing. Sleep Breath 19: 835–840.

Bourre B, Marignier R, Zephir H et al. (2012). Neuromyelitis optica and pregnancy. Neurology 78: 875–879.

Brandeberry KR, Vergon JM (1951). Meningococcic meningitis in term pregnancy. Am J Obstet Gynecol 61: 699–700.

Briemberg HR (2007). Neuromuscular diseases in pregnancy. Semin Neurol 27: 460–466.

Brownell LG, West P, Kryger MH (1986). Breathing during sleep in normal pregnant women. Am Rev Respir Dis 133: 38–41.

Bruno J, Rabito Jr FJ, Dildy 3rd GA (2004). West nile virus meningoencephalitis during pregnancy. J La State Med Soc 156: 204–205.

Bushnell C, Chireau M (2011). Preeclampsia and stroke: risks during and after pregnancy. Stroke Res Treat 2011: 858134.

Bushnell C, McCullough LD, Awad IA et al. (2014). Guidelines for the prevention of stroke in women: a statement for healthcare professionals from the American Heart Association/American Stroke Association. Stroke 45: 1545–1588.

Calvo-Ferrandiz E, Peraita-Adrados R (2018). Narcolepsy with cataplexy and pregnancy: a case–control study. J Sleep Res 27: 268–272.

Cardonick EH, Naktin J, Berghella V (2000). Neurosarcoidosis diagnosed during pregnancy by thoracoscopic lymph node biopsy. a case report. J Reprod Med 45: 585–587.

Carlson A, Norwitz ER, Stiller RJ (2010). Cytomegalovirus infection in pregnancy: should all women be screened? Rev Obstet Gynecol 3: 172–179.

Caso V, Falorni A, Bushnell CD et al. (2017). Pregnancy, hormonal treatments for infertility, contraception, and menopause in women after ischemic stroke: a consensus document. Stroke 48: 501–506.

Chakravarthy V, Kaplan B, Gospodarev V et al. (2018). Houdini tumor: case report and literature review of pregnancy-associated meningioma. World Neurosurg 114: e1261–e1265.

Chambers DJ, Bhatia K (2017). Cranial nerve palsy following central neuraxial block in obstetrics—a review of the literature and analysis of 43 case reports. Int J Obstet Anesth 31: 13–26.

Chambers DJ, Bhatia K (2018). Horner's syndrome following obstetric neuraxial blockade—a systematic review of the literature. Int J Obstet Anesth 35: 75–87.

Chapa JB, Ahn JT, DiGiovanni LM et al. (2003). West Nile virus encephalitis during pregnancy. Obstet Gynecol 102: 229–231.

Chen SJ, Shi L, Bao YP et al. (2018). Prevalence of restless legs syndrome during pregnancy: a systematic review and meta-analysis. Sleep Med Rev 40: 43–54.

Christensen J, Vestergaard C, Hammer Bech B (2018). Maternal death in women with epilepsy: smaller scope studies. Neurology 91: e1716–e1720.

Ciafaloni C, Thornburg L, Bushnell C (2018). Neurological diseases and pregnancy, Oxford University Press, New York, NY.

Cocks Eschler D, Javanmard P, Cox K et al. (2018). Prolactinoma through the female life cycle. Endocrine 59: 16–29.

Coelho AVC, Crovella S (2017). Microcephaly prevalence in infants born to Zika virus-infected women: a systematic review and meta-analysis. Int J Mol Sci 18.

Costa ML, Souza JP, Oliveira Neto AF et al. (2009). Cryptococcal meningitis in HIV negative pregnant women: case report and review of literature. Rev Inst Med Trop Sao Paulo 51: 289–294.

Crum NF, Ballon-Landa G (2006). Coccidioidomycosis in pregnancy: case report and review of the literature. Am J Med 119: 993.e11–17.

Davis LE (2018). Acute bacterial meningitis. Continuum (Minneap Minn) 24: 1264–1283.

Dawood R, Altanis E, Ribes-Pastor P et al. (2014). Pregnancy and spinal cord injury. Obstet Gynaecol 16: 99–107.

de Robles P, Fiest KM, Frolkis AD et al. (2015). The worldwide incidence and prevalence of primary brain tumors: a systematic review and meta-analysis. Neuro Oncol 17: 776–783.

Digre KB (2013). Headaches during pregnancy. Clin Obstet Gynecol 56: 317–329.

Digre KB, Kinard K (2014). Neuro-ophthalmic disorders in pregnancy. Continuum: Lifelong Learn Neurol 20: 162–176.

Digre KB, Varner MW, Corbett JJ (1984). Pseudotumor cerebri and pregnancy. Neurology 34: 721–729.

Dodd KC, Michael BD, Ziso B et al. (2015). Herpes simplex virus encephalitis in pregnancy—a case report and review of reported patients in the literature. BMC Res Notes 8: 118.

Donnini I, Rinnoci V, Nannucci S et al. (2017). Pregnancy in CADASIL. Acta Neurol Scand 136: 668–671.

Ducros A (2012). Reversible cerebral vasoconstriction syndrome. Lancet Neurol 11: 906–917.

Duley L (2009). The global impact of pre-eclampsia and eclampsia. Semin Perinatol 33: 130–137.

Durcan FJ, Corbett JJ, Wall M (1988). The incidence of pseudotumor cerebri. Population studies in Iowa and Louisiana. Arch Neurol 45: 875–877.

Edwards N, Blyton DM, Hennessy A et al. (2005). Severity of sleep-disordered breathing improves following parturition. Sleep 28: 737–741.

Ekbom K, Waldenlind E (1981). Cluster headache in women: evidence of hypofertility(?) Headaches in relation to menstruation and pregnancy. Cephalalgia 1: 167–174.

Esakoff TF, Rad S, Burwick RM et al. (2016). Predictors of eclampsia in California. J Matern Fetal Neonatal Med 29: 1531–1535.

Evans RW, Lee AG (2010). Idiopathic intracranial hypertension in pregnancy. Headache 50: 1513–1515.

Fabian M (2016). Pregnancy in the setting of multiple sclerosis. Continuum: Lifelong Learn Neurol 22: 837–850.

Facco FL, Ouyang DW, Zee PC et al. (2014). Sleep disordered breathing in a high-risk cohort prevalence and severity across pregnancy. Am J Perinatol 31: 899–904.

Facco FL, Parker CB, Reddy UM et al. (2017). Association between sleep-disordered breathing and hypertensive disorders of pregnancy and gestational diabetes mellitus. Obstet Gynecol 129: 31–41.

Feldman DM, Borgida AF, Rodis JF et al. (2000). Irreversible maternal brain injury during pregnancy: a case report and review of the literature. Obstet Gynecol Surv 55: 708–714.

Fiest KM, Sauro KM, Wiebe S et al. (2017). Prevalence and incidence of epilepsy. Neurology 88: 296–303.

Finan MA, Smith SG, Sinnott JT et al. (1992). An interesting case presentation: peripartum meningococcal meningitis. J Perinatol 12: 78–80.

Fine LG, Barnett EV, Danovitch GM et al. (1981). Systemic lupus erythematosus in pregnancy. Ann Intern Med 94: 667–677.

Fong A, Chau CT, Quant C et al. (2018). Multiple sclerosis in pregnancy: prevalence, sociodemographic features, and obstetrical outcomes. J Matern Fetal Neonatal Med 31: 382–387.

Foo L, Bewley S, Rudd A (2013). Maternal death from stroke: a thirty year national retrospective review. Eur J Obstet Gynecol Reprod Biol 171: 266–270.

Forster M-T, Baumgarten P, Gessler F et al. (2019). Influence of pregnancy on glioma patients. Acta Neurochir .

Fox MW, Harms RW, Davis DH (1990). Selected neurologic complications of pregnancy. Mayo Clin Proc 65: 1595–1618.

Frazer KL, Porter S, Goss C (2013). The genetics and implications of neuromuscular diseases in pregnancy. J Perinat Neonatal Nurs 27: 205–214.

Fugate JE, Ameriso SF, Ortiz G et al. (2012). Variable presentations of postpartum angiopathy. Stroke 43: 670–676.

Fujimura M, Akagi K, Uenohara H et al. (2013). Moyamoya disease in pregnancy: a single institute experience. Neurol Med Chir (Tokyo) 53: 561–564.

Galgiani JN, Ampel NM, Blair JE et al. (2005). Coccidioidomycosis. Clin Infect Dis 41: 1217–1223.

Ghulmiyyah L, Sibai B (2012). Maternal mortality from preeclampsia/eclampsia. Semin Perinatol 36: 56–59.

Gillman GS, Schaitkin BM, May M et al. (2002). Bell's palsy in pregnancy: a study of recovery outcomes. Otolaryngol Head Neck Surg 126: 26–30.

Giraud P, Chauvet S (2009). Cluster headache during pregnancy: case report and literature review. Headache 49: 136–139.

Goldszmidt E, Kern R, Chaput A et al. (2005). The incidence and etiology of postpartum headaches: a prospective cohort study. Can J Anaesth 52: 971–977.

Grand'Maison S, Weber F, Bédard M-J et al. (2015). Pituitary apoplexy in pregnancy: a case series and literature review. Obstet Med 8: 177–183.

Granella F, Sances G, Pucci E et al. (2000). Migraine with aura and reproductive life events: a case control study. Cephalalgia 20: 701–707.

Harden CL, Hopp J, Ting TY et al. (2009). Practice parameter update: management issues for women with epilepsy—focus on pregnancy (an evidence-based review): obstetrical complications and change in seizure frequency. Neurology 73: 126–132.

Hedman C, Pohjasvaara T, Tolonen U et al. (2002a). Parasomnias decline during pregnancy. Acta Neurol Scand 105: 209–214.

Hernández-Díaz S, McElrath TF, Pennell PB et al. (2017). Fetal growth and premature delivery in pregnant women on antiepileptic drugs. Ann Neurol 82: 457–465.

Hisano Y, Nakamura K, Kitamura R et al. (2010). Myoclonus following spinal and epidural anesthesia—a case report. Masui 59: 770–772.

Hopkins AN, Alshaeri T, Akst SA et al. (2014). Neurologic disease with pregnancy and considerations for the obstetric anesthesiologist. Semin Perinatol 38: 359–369.

Hoshiyama E, Tatsumoto M, Iwanami H et al. (2012). Postpartum migraines: a long-term prospective study. Intern Med 51: 3119–3123.

Hosley CM, McCullough LD (2011). Acute neurological issues in pregnancy and the peripartum. Neurohospitalist 1: 104–116.

Howard J, Trevick S, Younger DS (2016). Epidemiology of multiple sclerosis. Neurol Clin 34: 919–939.

James AH, Bushnell CD, Jamison MG et al. (2005). Incidence and risk factors for stroke in pregnancy and the puerperium. Obstet Gynecol 106: 509–516.

Jensen D, Duffin J, Lam YM et al. (2008). Physiological mechanisms of hyperventilation during human pregnancy. Respir Physiol Neurobiol 161: 76–86.

Johnson EL, Burke AE, Wang A et al. (2018). Unintended pregnancy, prenatal care, newborn outcomes, and breastfeeding in women with epilepsy. Neurology 91: e1031–e1039.

Jung SY, Bae HJ, Park BJ et al. (2010). Parity and risk of hemorrhagic strokes. Neurology 74: 1424–1429.

Kaplan TB (2019). Management of demyelinating disorders in pregnancy. Neurol Clin 37: 17–30.

Karli N, Baykan B, Ertas M et al. (2012). Impact of sex hormonal changes on tension-type headache and migraine: a cross-sectional population-based survey in 2,600 women. J Headache Pain 13: 557–565.

Karp I, Manganas A, Sylvestre M-P et al. (2014). Does pregnancy alter the long-term course of multiple sclerosis? Ann Epidemiol 24: 504–508.e2.

Katirji B, Wilbourn AJ, Scarberry SL et al. (2002). Intrapartum maternal lumbosacral plexopathy. Muscle Nerve 26: 340–347.

Katz A, Sergienko R, Dior U et al. (2011). Bell's palsy during pregnancy: is it associated with adverse perinatal outcome? Laryngoscope 121: 1395–1398.

Kesler A, Kupferminc M (2013). Idiopathic intracranial hypertension and pregnancy. Clin Obstet Gynecol 56: 389–396.

Kittner SJ, Stern BJ, Feeser BR et al. (1996). Pregnancy and the risk of stroke. N Engl J Med 335: 768–774.

Klein C, Dyck P, Friedenberg S et al. (2002). Inflammation and neuropathic attacks in hereditary brachial plexus neuropathy. J Neurol Neurosurg Psychiatry 73: 45–50.

Koch-Henriksen N, Sørensen PS (2010). The changing demographic pattern of multiple sclerosis epidemiology. Lancet Neurol 9: 520–532.

Kranick SM, Mowry EM, Colcher A et al. (2010). Movement disorders and pregnancy: a review of the literature. Mov Disord 25: 665–671.

Krumholz A, Stern BJ (2014). Neurologic manifestations of sarcoidosis. Handb Clin Neurol 119: 305–333.

Kuklina EV, Tong X, Bansil P et al. (2011). Trends in pregnancy hospitalizations that included a stroke in the United States from 1994 to 2007: reasons for concern? Stroke 42: 2564–2570.

Lambru G, Matharu MS (2012). Trigeminal autonomic cephalalgias: a review of recent diagnostic, therapeutic and pathophysiological developments. Ann Indian Acad Neurol 15: S51–S61.

Lamy C, Hamon JB, Coste J et al. (2000). Ischemic stroke in young women: risk of recurrence during subsequent pregnancies. French Study Group on Stroke in Pregnancy. Neurology 55: 269–274.

Lancaster E (2015). Paraneoplastic disorders. Continuum: Lifelong Learn Neurol 21: 452.

Laviv Y, Bayoumi A, Mahadevan A et al. (2018). Meningiomas in pregnancy: timing of surgery and clinical outcomes as observed in 104 cases and establishment of a best management strategy. Acta Neurochir 160: 1521–1529.

Liang C-C, Chang S-D, Lai S-L et al. (2006). Stroke complicating pregnancy and the puerperium. Eur J Neurol 13: 1256–1260.

Lipton RB, Bigal ME, Diamond M et al. (2007). Migraine prevalence, disease burden, and the need for preventive therapy. Neurology 68: 343–349.

Lucas AO (1964). Pneumococcal meningitis in pregnancy and the puerperium. Br Med J 1: 92–95.

Lyons JL (2018). Viral meningitis and encephalitis. Continuum (Minneap Minn) 24: 1284–1297.

Ma X, Zhang G, Guo T et al. (2018). Management and outcomes of pregnant patients with central nervous system hemangioblastoma. J Clin Neurosci 57: 126–130.

MacDonald SC, Bateman BT, McElrath TF et al. (2015). Mortality and morbidity during delivery hospitalization among pregnant women with epilepsy in the United States. JAMA Neurol 72: 981–988.

Madjunkov M, Chaudhry S, Ito S (2017). Listeriosis during pregnancy. Arch Gynecol Obstet 296: 143–152.

Maia DP, Fonseca PG, Camargos ST et al. (2012). Pregnancy in patients with Sydenham's chorea. Parkinsonism Relat Disord 18: 458–461.

Maisel JA, Lynam T (1996). Unexpected sudden death in a young pregnant woman: unusual presentation of neurosarcoidosis. Ann Emerg Med 28: 94–97.

Manconi M, Govoni V, De Vito A et al. (2004). Restless legs syndrome and pregnancy. Neurology 63: 1065–1069.

Marabani M, Zoma A, Hadley D et al. (1989). Transverse myelitis occurring during pregnancy in a patient with systemic lupus erythematosus. Ann Rheum Dis 48: 160–162.

Marsaudon E, Berthier F (1996). Staphylococcus aureus meningitis after obstetrical epidural anesthesia. Ann Fr Anesth Reanim 15: 389–390.

Massey JM, De Jesus-Acosta C (2014). Pregnancy and myasthenia gravis. Continuum: Lifelong Learn Neurol 20: 115–127.

Massey EW, Guidon AC (2014). Peripheral neuropathies in pregnancy. Continuum: Lifelong Learn Neurol 20: 100.

Maurovich-Horvat E, Tormasiova M, Slonkova J et al. (2010). Assessment of pregnancy outcomes in Czech and Slovak women with narcolepsy. Med Sci Monit 16: SR35–40.

Maurovich-Horvat E, Kemlink D, Hogl B et al. (2013). Narcolepsy and pregnancy: a retrospective European evaluation of 249 pregnancies. J Sleep Res 22: 496–512.

McComb JM, McName PT, Sinnamon DG et al. (1982). Staphylococcal endocarditis presenting as meningitis in pregnancy. Int J Cardiol 1: 325–327.

McDermott M, Miller EC, Rundek T et al. (2018). Preeclampsia: association with posterior reversible encephalopathy syndrome and stroke. Stroke 49: 524–530.

McGrath E, Monaghan TS, Alexander M et al. (2010). Recurrent spontaneous intracranial hypotension in early pregnancy. BMJ Case Rep 2010: bcr0520103040. https://doi.org/10.1136/bcr.05.2010.3040.

McLauchlin J (1990). Human listeriosis in Britain, 1967–85, a summary of 722 cases. 1. Listeriosis during pregnancy and in the newborn. Epidemiol Infect 104: 181–189.

Meador KJ, Pennell PB, May RC et al. (2018). Changes in antiepileptic drug-prescribing patterns in pregnant women with epilepsy. Epilepsy Behav 84: 10–14.

Melcon MO, Correale J, Melcon CM (2014). Is it time for a new global classification of multiple sclerosis? J Neurol Sci 344: 171–181.

Menezes FV, Venkat N (2006). Spinal myoclonus following combined spinal–epidural anaesthesia for caesarean section. Anaesthesia 61: 597–600.

Miller TR, Shivashankar R, Mossa-Basha M et al. (2015). Reversible cerebral vasoconstriction syndrome, part 1: epidemiology, pathogenesis, and clinical course. Am J Neuroradiol 36: 1392–1399.

Mintz G, Niz J, Gutierrez G et al. (1986). Prospective study of pregnancy in systemic lupus erythematosus. Results of a multidisciplinary approach. J Rheumatol 13: 732–739.

Miyasaki JM, AlDakheel A (2014). Movement disorders in pregnancy. Continuum: Lifelong Learn Neurol 20: 148.

Mizokami T, Okamura K, Sato K et al. (1994). Postpartum exacerbation of HTLV-I associated myelopathy/tropical spastic paraparesis followed by an episode of painless thyroiditis. Intern Med 33 (11): 703–705. https://doi.org/10.2169/internalmedicine.33.703.

Moatti Z, Gupta M, Yadava R et al. (2014). A review of stroke and pregnancy: incidence, management and prevention. Eur J Obstet Gynecol Reprod Biol 181: 20–27.

Muscal E, Brey RL (2010). Neurologic manifestations of systemic lupus erythematosus in children and adults. Neurol Clin 28: 61–73.

Negro A, Delaruelle Z, Ivanova TA et al. (2017). Headache and pregnancy: a systematic review. J Headache Pain 18: 106.

Nerenberg KA, Park AL, Vigod SN et al. (2017). Long-term risk of a seizure disorder after eclampsia. Obstet Gynecol 130: 1327–1333.

Nicolle MW (2016). Myasthenia gravis and Lambert–Eaton myasthenic syndrome. Continuum: Lifelong Learn Neurol 22: 1978–2005.

Nieto-Calvache AJ, Loaiza-Osorio S, Casallas-Carrillo J et al. (2017). Abducens nerve palsy in gestational hypertension: a case report and review of the literature. J Obstet Gynaecol Can 39: 890–893.

Norwood F, Rudnik-Schöneborn S (2012). 179th ENMC international workshop: pregnancy in women with neuromuscular disorders: 5–7 November 2010, Naarden, the Netherlands. Neuromuscul Disord 22: 183–190.

O'Hara MW, McCabe JE (2013). Postpartum depression: current status and future directions. Annu Rev Clin Psychol 9: 379–407.

Olindo S, Smadja D, Cabre P et al. (1997). Gayet–Wernicke encephalopathy and centropontine myelionolysis induced by hyperemesis gravidarum. Rev Neurol (Paris) 153: 427–429.

Olson-Chen C (2018). Neurological infections in pregnancy. In: E Ciafaloni, LL Thornburg, CD Bushnell (Eds.), Neurological diseases and pregnancy. Oxford University Press, New York, NY, pp. 95–104.

Omoti AE, Waziri-Erameh JM, Okeigbemen VW (2008). A review of the changes in the ophthalmic and visual system in pregnancy. Afr J Reprod Health 12: 185–196.

O'Neal MA (2017). Headaches complicating pregnancy and the postpartum period. Pract Neurol 17: 191–202.

Ornoy A, Tenenbaum A (2006). Pregnancy outcome following infections by coxsackie, echo, measles, mumps, hepatitis, polio and encephalitis viruses. Reprod Toxicol 21: 446–457.

Osler T (1942). Meningococcal meningitis in advanced pregnancy. Br Med J 1: 13.

Ostrom QT, Gittleman H, Fulop J et al. (2015). CBTRUS statistical report: primary brain and central nervous system tumors diagnosed in the United States in 2008–2012. Neuro Oncol 17: iv1–iv62.

Padua L, Pasquale AD, Pazzaglia C et al. (2010). Systematic review of pregnancy-related carpal tunnel syndrome. Muscle Nerve 42: 697–702.

Pamidi S, Kimoff RJ (2018). Maternal sleep-disordered breathing. Chest 153: 1052–1066.

Pandit L, Asgari N, Apiwattanakul M et al. (2015). Demographic and clinical features of neuromyelitis optica: a review. Mult Scler J 21: 845–853.

Pearce CF, Hansen WF (2012). Headache and neurological disease in pregnancy. Clin Obstet Gynecol 55: 810–828.

Peeters A, Van de Wyngaert F, Van Lierde M et al. (1993). Wernicke's encephalopathy and central pontine myelinolysis induced by hyperemesis gravidarum. Acta Neurol Belg 93: 276–282.

Peeters S, Pagès M, Gauchotte G et al. (2018). Interactions between glioma and pregnancy: insight from a 52-case multicenter series. J Neurosurg 128: 3–13.

Piantanida E, Gallo D, Lombardi V et al. (2014). Pituitary apoplexy during pregnancy: a rare, but dangerous headache. J Endocrinol Invest 37: 789–797.

Pien GW, Fife D, Pack AI et al. (2005). Changes in symptoms of sleep-disordered breathing during pregnancy. Sleep 28: 1299–1305.

Pien GW, Pack AI, Jackson N et al. (2014). Risk factors for sleep-disordered breathing in pregnancy. Thorax 69: 371–377.

Porto LB, Berndl AML (2019). Pregnancy 5 years after onset of amyotrophic lateral sclerosis symptoms: a case report and review of the literature. J Obstet Gynaecol Can 41: 974–980.

Proisy M, Rouil A, Raoult H et al. (2014). Imaging of musculoskeletal disorders related to pregnancy. Am J Roentgenol 202: 828–838.

Radhakrishnan K, Ahlskog JE, Cross SA et al. (1993). Idiopathic intracranial hypertension (pseudotumor cerebri). Descriptive epidemiology in Rochester, Minn, 1976 to 1990. Arch Neurol 50: 78–80.

Rajiv KR, Radhakrishnan A (2017). Status epilepticus in pregnancy: etiology, management, and clinical outcomes. Epilepsy Behav 76: 114–119.

Rantanen K, Tatlisumak T (2013). Stroke in women—oral contraception, pregnancy, and hormone replacement therapy. Curr Vasc Pharmacol 11: 58–73.

Reichner CA (2015). Insomnia and sleep deficiency in pregnancy. Obstet Med 8: 168–171.

Roine S, Pöyhönen M, Timonen S et al. (2005). Neurologic symptoms are common during gestation and puerperium in CADASIL. Neurology 64: 1441–1443.

Rønning PA, Helseth E, Meling TR et al. (2016). The effect of pregnancy on survival in a low-grade glioma cohort. J Neurosurg 125: 393–400.

Sandberg T, Dahlberg T (1981). Meningitis and septicaemia due to haemophilus influenzae type b in pregnancy. Br Med J (Clin Res Ed) 282: 946.

Saporta ASD, Sottile SL, Miller LJ et al. (2011). Charcot Marie tooth (CMT) subtypes and genetic testing strategies. Ann Neurol 69: 22–33.

Saposnik G, Barinagarrementeria F, Brown RD et al. (2011). Diagnosis and management of cerebral venous thrombosis: a statement for healthcare professionals from the American Heart Association/American Stroke Association. Stroke 42: 1158–1192.

Saraste M, Atula S, Hedman K et al. (2018). Humoral response to John Cunningham virus during pregnancy in multiple sclerosis. Mult Scler Relat Disord 21: 11–18.

Sax TW, Rosenbaum RB (2006). Neuromuscular disorders in pregnancy. Muscle Nerve 34: 559–571.

Schneider-Gold C, Wessig C, Höpker M et al. (2006). Pregnancy and delivery of a healthy baby in autoimmune Lambert-Eaton myasthenic syndrome. J Neurol 253 (9): 1236–1237. Epub 2006 Apr 5.

Schredl M, Gilles M, Wolf I et al. (2016). Nightmare frequency in last trimester of pregnancy. BMC Pregnancy Childbirth 16: 346.

Scott CA, Bewley S, Rudd A et al. (2012). Incidence, risk factors, management, and outcomes of stroke in pregnancy. Obstet Gynecol 120: 318–324.

Seier M, Hiller A (2017). Parkinson's disease and pregnancy: an updated review. Parkinsonism Relat Disord 40: 11–17.

Sells CM, Feske SK (2017). Stroke in pregnancy. Semin Neurol 37: 669–678.

Shah A, Tselis A, Mason B (2000). Acute disseminated encephalomyelitis in a pregnant woman successfully treated with plasmapheresis. J Neurol Sci 174: 147–151.

Shah S, Shah D, Skeen MB (2018). Recurrent disseminated encephalomyelitis: a case report and literature review. Mult Scler Relat Disord 26: 86–89.

Shainker SA, Edlow JA, O'Brien K (2015). Cerebrovascular emergencies in pregnancy. Best Pract Res Clin Obstet Gynaecol 29: 721–731.

Shanmugalingam R, Reza Pour N, Chuah SC et al. (2016). Vertebral artery dissection in hypertensive disorders of pregnancy: a case series and literature review. BMC Pregnancy Childbirth 16: 164.

Sheikh HU, Mathew PG (2014). Reversible cerebral vasoconstriction syndrome: updates and new perspectives. Curr Pain Headache Rep 18: 414.

Singh A, Trevick S (2016). The epidemiology of global epilepsy. Neurol Clin 34: 837–847.

Sinha S, Taly AB, Prashanth LK et al. (2004). Successful pregnancies and abortions in symptomatic and asymptomatic Wilson's disease. J Neurol Sci 217: 37–40.

Sit D, Rothschild AJ, Wisner KL (2006). A review of postpartum psychosis. J Womens Health 15: 352–368 2002.

Snyder Y, Donlin-Smith C, Snyder E et al. (2015). The course and outcome of pregnancy in women with nondystrophic myotonias. Muscle Nerve 52: 1013–1015.

Stern JS, Orth M, Robertson MM (2009). Gilles de la Tourette syndrome in pregnancy: a retrospective series. Obstet Med 2: 128–129.

Stevenson CB, Thompson RC (2005). The clinical management of intracranial neoplasms in pregnancy. Clin Obstet Gynaecol 48: 24–37.

Stewart RD, Bryant SN, Sheffield JS (2013). West nile virus infection in pregnancy. Case Rep Infect Dis 2013: 351872.

Suzuki J, Kodama N (1983). Moyamoya disease—a review. Stroke 14: 104–109.

Swartz RH, Cayley ML, Foley N et al. (2017). The incidence of pregnancy-related stroke: a systematic review and meta-analysis. Int J Stroke 12: 687–697.

Swartz RH, Ladhani NNN, Foley N et al. (2018). Canadian stroke best practice consensus statement: secondary stroke prevention during pregnancy. Int J Stroke 13: 406–419.

Tarney CM, Tian C, Craig ER et al. (2018). Relative effects of age, race, and stage on mortality in gestational choriocarcinoma. Int J Gynecol Cancer 28: 338–345.

Terry AR, Barker FG, Leffert L et al. (2012). Outcomes of hospitalization in pregnant women with CNS neoplasms: a population-based study. Neuro Oncol 14: 768–776.

Thigpen MC, Whitney CG, Messonnier NE et al. (2011). Bacterial meningitis in the United States, 1998–2007. N Engl J Med 364: 2016–2025.

Thirumalaikumar L, Ramalingam K, Heafield T (2014). Idiopathic intracranial hypertension in pregnancy. Obstet Gynaecol 16: 93–97.

Tonelli J, Zurrú MC, Castillo J et al. (1999). Central pontine myelinolysis induced by hyperemesis gravidarum. Medicina (B Aires) 59: 176–178.

Too G, Wen T, Boehme AK et al. (2018). Timing and risk factors of postpartum stroke. Obstet Gynecol 131: 70–78.

Trakada G, Tsapanos V, Spiropoulos K (2003). Normal pregnancy and oxygenation during sleep. Eur J Obstet Gynecol Reprod Biol 109: 128–132.

Treadwell SD, Thanvi B, Robinson TG (2008). Stroke in pregnancy and the puerperium. Postgrad Med J 84: 238–245.

Trip J, Drost G, Ginjaar HB et al. (2009). Redefining the clinical phenotypes of non-dystrophic myotonic syndromes. J Neurol Neurosurg Psychiatry 80: 647–652.

Tsang CA, Tabnak F, Vugia DJ et al. (2013). Increase in reported coccidioidomycosis—United States, 1998–2011. Morb Mortal Wkly Rep 62: 217.

van Westrhenen A, Senders JT, Martin E et al. (2018). Clinical challenges of glioma and pregnancy: a systematic review. J Neurooncol 139: 1–11.

Varner M (2013). Myasthenia gravis and pregnancy. Clin Obstet Gynecol 56: 372–381.

Veiby G, Daltveit AK, Engelsen BA et al. (2009). Pregnancy, delivery, and outcome for the child in maternal epilepsy. Epilepsia 50: 2130–2139.

Vélez-Ruiz NJ, Pennell PB (2016). Issues for women with epilepsy. Neurol Clin 34: 411–425.

Vesga-Lopez O, Blanco C, Keyes K et al. (2008). Psychiatric disorders in pregnant and postpartum women in the United States. Arch Gen Psychiatry 65: 805–815.

Voetsch AC, Angulo FJ, Jones TF et al. (2007). Reduction in the incidence of invasive listeriosis in foodborne diseases active surveillance network sites, 1996–2003. Clin Infect Dis 44: 513–520.

Voinescu P, Pennell P (2017). Delivery of a personalized treatment approach to women with epilepsy. Semin Neurol 37: 611–623.

Wabnitz A, Bushnell C (2015). Migraine, cardiovascular disease, and stroke during pregnancy: systematic review of the literature. Cephalalgia 35: 132–139.

Wadhwa Desai R, Smith MA (2017). Pregnancy-related listeriosis. Birth Defects Res 109: 324–335.

Waters J (2019). Management of myasthenia gravis in pregnancy. Neurol Clin 37: 113–120. Neurology of Pregnancy.

Wen Y, Yang B, Huang Q et al. (2017). Posterior reversible encephalopathy syndrome in pregnancy: a retrospective series of 36 patients from mainland China. Ir J Med Sci 186: 699–705.

Williams KA, Swedo SE (2015). Post-infectious autoimmune disorders: Sydenham's chorea, PANDAS and beyond. Brain Res 1617: 144–154.

Xiao X, Gui S, Bai P et al. (2017). Anti-NMDA-receptor encephalitis during pregnancy: a case report and literature review. J Obstet Gynaecol Res 43: 768–774.

Zara G, Codemo V, Palmieri A et al. (2012). Neurological complications in hyperemesis gravidarum. Neurol Sci 33: 133–135.

Zeb Q, Alegria A (2014). Acute disseminated encephalomyelitis (ADEM) following a H3N3 parainfluenza virus infection in a pregnant asthmatic woman with respiratory failure. BMJ Case Rep 2014: bcr2013201072.

Further reading

Landrum LM, Hawkins A, Goodman JR (2009). Pneumococcal meningitis during pregnancy: a case report and review of literature. Infect Dis Obstet Gynecol 2009: 63624.

Leffert LR, Clancy CR, Bateman BT et al. (2016). Treatment patterns and short-term outcomes in ischemic stroke in pregnancy or postpartum period. Am J Obstet Gynecol 214: 723.e1–723.e11.

Tate J, Bushnell C (2011). Pregnancy and stroke risk in women. Womens Health (Lond Engl) 7: 363–374.

Verheecke M, Halaska MJ, Lok CA et al. (2014). Primary brain tumours, meningiomas and brain metastases in pregnancy: report on 27 cases and review of literature. Eur J Cancer 50: 1462–1471.

Handbook of Clinical Neurology, Vol. 171 (3rd series)
Neurology and Pregnancy: Pathophysiology and Patient Care
E.A.P. Steegers, M.J. Cipolla and E.C. Miller, Editors
https://doi.org/10.1016/B978-0-444-64239-4.00007-2

Chapter 7

Neuro-obstetrics: A multidisciplinary approach to care of women with neurologic disease

INGRID A. BRUSSÉ[1]*, ANNA C.M. KLUIVERS[1], DANIELA M. ZAMBRANO ESPINOZA[2], KARA SHETLER[2], AND ELIZA C. MILLER[2,3]

[1]*Department of Obstetrics and Gynecology, Division of Obstetrics and Fetal Medicine, Erasmus University Medical Center, Rotterdam, The Netherlands*

[2]*Department of Neurology, Columbia University Vagelos College of Physicians and Surgeons, New York, NY, United States*

[3]*NewYork-Presbyterian Hospital, Columbia University Irving Medical Center, New York, NY, United States*

Abstract

The term "neuro-obstetrics" refers to a multidisciplinary approach to the care of pregnant women with neurologic comorbidities, both preconceptionally and throughout pregnancy. General preconception care should be offered to all women, including women with neurologic disease. Women with neurologic comorbidities should also be offered specialist preconception care by an obstetrician who consults with a neurologist, anesthesiologist, and if indicated clinical geneticist and/or other specialists.

In women with neurologic comorbidities, neurologic sequelae may influence the course of the pregnancy and delivery. Also, pregnancy may influence the severity of the neurologic condition, depending on the type of disease. Physiologic adaptations during pregnancy and altered pharmacokinetics may cause altered blood serum levels of drugs, leading to decreased or increased drug effects. When administering drugs to a woman who wishes to conceive, it is important to consider possible teratogenic effects and possible secretion in breast milk. Tailoring medication regimens should be considered, preferably preconceptionally. In this chapter, we review general principles of neuro-obstetric care, as well as some specific considerations for neurologists, obstetricians, and anesthesiologists caring for pregnant women with common neurologic conditions.

INTRODUCTION

The overlap of obstetrics and neurology may not be immediately apparent, and practitioners in both fields may question the need for two entire volumes devoted to the topic of what we have termed "neuro-obstetrics." On the one hand, the field of obstetrics is defined by the care concerning pregnancy. The majority of women will experience pregnancy at some point in their lives; it is a highly complex and unique physiologic state that carries the potential for myriad complications. On the other hand, the field of neurology is defined anatomically, encompasses a broad array of conditions, and provides care that a much smaller percentage of women will ever seek. But the very high lifetime incidence of pregnancy ensures that obstetricians will treat numerous patients with chronic neurologic conditions and that neurologists will treat women who are pregnant or wish to become pregnant. A thorough understanding of the interdisciplinary interface is crucial for practitioners in both specialties, for reasons outlined here and explained in detail in Section II of Volume 172.

First, certain neurologic conditions commonly present in and preferentially affect women of childbearing age.

*Correspondence to: Ingrid A. Brussé, M.D., Ph.D., Department of Obstetrics and Gynecology, Division of Obstetrics and Prenatal Medicine Erasmus Medical Centre, Rotterdam, Wytemaweg 80, Rotterdam 3000CA, The Netherlands. Tel: +31-10-703-61-09, Fax: +31-10-703-34-73, E-mail: i.brusse@erasmusmc.nl

Examples include multiple sclerosis (MS), migraine, and certain types of stroke. The diagnosing neurologist must consider reproductive planning when initiating treatment for new diagnoses in this patient demographic. Further, birth control, fertility treatments, and pregnancy itself can all lower the threshold for development of neurologic phenomena. Hormonal therapies can increase risk of ischemic stroke, the classic example being combined oral contraceptive therapy in patients who have migraine with aura. Cardiac output increases in pregnancy, and cerebral autoregulation must occur to ensure that cerebral perfusion is maintained within a narrow window. At the end of pregnancy, cerebral arteries must adapt to counter arterial vasoconstriction (Cipolla, 2013). Preeclampsia/eclampsia, posterior reversible encephalopathy syndrome (PRES), and reversible cerebral vascular syndrome (RCVS) are perhaps the most dramatic manifestations of impaired cerebral autoregulation, the exact causes and mechanisms of which remain unclear. Expression of inhibitory gamma-aminobutyric acid type A ($GABA_A$) receptor subunits is decreased in the cerebral cortex in normal pregnancy, resulting in a lowering of the seizure threshold (Johnson et al., 2015). In these and other ways, women who are pregnant or wish to become pregnant are at unique risk for developing neurologic conditions.

Second, the many physiologic changes that occur in pregnancy can dramatically affect the course of preexisting neurologic diseases and the safety and efficacy of therapies used to treat them. Autoimmune/autoinflammatory conditions such as MS commonly become less active in pregnancy and can flare up in the postpartum period. Most of the medications used for migraine prophylaxis have not been shown to be safe in pregnancy. Many antiepileptic medications carry some risk of teratogenicity, while maternal seizures during pregnancy can have similarly devastating effects on the fetus. Changes in the profile of serum-binding proteins during and after pregnancy can cause wide fluctuations in the bioavailability of medications such as lamotrigine, requiring careful monitoring and dose titration. The neurologist should therefore prioritize counseling regarding birth control and conception planning for these patients in coordination with the obstetrician.

Finally, both neurologists and obstetricians frequently serve as the de facto primary care physicians for their patients, and thus are tasked with the responsibility of ensuring continuity and coordination of care at all stages of the reproductive process. Preconception planning for neurologic patients is crucial for the reasons described earlier; obstetricians providing preconception care should be alert to the indications for referral to a neurologist, including lapse of neurologic care for patients with neurologic conditions and history of stroke or other hypercoagulable disorders. Similar principles apply during pregnancy; while all pregnant patients should be under the care of an obstetrician without referral by a neurologist, this obstetrician should be alert to development of signs and symptoms requiring neurologic referral, such as a new headache type or new neurologic deficits. Coordinated delivery planning may be warranted in patients with elevated intracranial pressure in whom Valsalva may be dangerous or in those with a neuromuscular disorder, motor neuron disorder, or spinal cord injury who may not tolerate vaginal delivery. Postpartum interdisciplinary care is equally critical, for example, to ensure prioritization of ongoing blood pressure monitoring and management for the preeclamptic patient, and may require communication between the obstetrician and the neurologist. Finally, both specialists should be aware of the need to involve other specialists, such as anesthesiologists, ophthalmologists, clinical geneticists, neonatologists, and pediatricians, at various stages of the reproductive process.

This chapter will provide an overview of the many ways in which neurology and obstetrics interface clinically, organized by categories of neurologic disease and with an emphasis on the most important considerations to guide practitioners in providing interdisciplinary care.

The following points should be part of every specialist preconception consultation:

1. Effects of the neurologic disease on pregnancy and pregnancy outcome.
2. Effects of pregnancy on the course of preexisting neurologic disease.
3. Effects of pregnancy on the risk of developing new neurologic disease.
4. Considerations regarding use of medications required to treat neurologic diseases during pregnancy (possible teratogenicity, changes in pharmacokinetics and pharmacodynamics, etc.).
5. Considerations regarding delivery and postpartum care.

PRECONCEPTION CARE: SPECIAL CONSIDERATIONS

All women should have access to general preconception care that includes lifestyle advice and the assessment of possible risk factors in the medical and family history. Neurologic disease may signal increased and/or unique risks for both mother and fetus. For example, prior stroke may indicate metabolic disease or a hypercoagulable disorder that can worsen in pregnancy and cause harm to mother and fetus. History of a genetic epilepsy syndrome or neurodegenerative condition may have implications for heritability. Women taking certain antiepileptics or immunomodulating therapies may be at increased risk of folate deficiency, which is detrimental to fetal neural tube development. In addition, a number of the routine considerations for all patients (with or without a diagnosed neurologic condition) concern either the use of neuroactive substances (caffeine, tobacco, alcohol, illicit drugs, etc.), the effects of various factors on fetal neurologic development (congenital infections, general nutritional supplementation), or both. While factors related to preexisting neurologic conditions will be discussed in greater detail further on, we here present some of the neurologic implications of general risk factor management.

General lifestyle advice for women prior to and during pregnancy

From an obstetric point of view, general preconception recommendations are of great value for all pregnant women, but especially for those with chronic neurologic morbidity. General advice concerning additional supplements to be taken preconceptionally and during pregnancy is presented in Table 7.1.

Alcohol

All alcohol consumption during pregnancy causes fetal alcohol exposure, which is associated with low birth weight, birth abnormalities such as fetal alcohol syndrome, and fetal alcohol spectrum disorders as well as fetal and infant death. When alcohol intake persists during pregnancy, risks increase (Witt et al., 2015; Denny et al., 2017).

Alcohol consumption in both women and men influences fertility and may pose challenges in conception (Bouchery et al., 2011). Also, exposure to alcohol may induce epigenetic alterations in both male and female fetuses (Shukla and Aroor, 2006).

All risks concerning conception, pregnancy, and pregnancy outcome associated with alcohol consumption are dose-dependent, i.e., increase with increased alcohol intake (Bouchery et al., 2011).

Since preconception counseling leads to a significant decrease in alcohol consumption during the first trimester, such counseling concerning alcohol use both before and during pregnancy is desired in all women of reproductive potential (Lassi et al., 2014; Witt et al., 2015).

Tobacco

Cigarette smoking during pregnancy is the most important etiologic determinant of fetal growth restriction in developed countries (Bergmann et al., 2008). Other abnormalities associated with cigarette smoking are gastroschisis, placenta previa, placental abruption, stillbirth, and preterm delivery. Risks remain high even after cessation of smoking in the first trimester (Gould et al., 2017; Perry et al., 2019). However, risks do decrease the earlier cessation occurs, and therefore counseling and appropriate support to motivate cessation are recommended (Blatt et al., 2015).

Caffeine

Caffeine is a widely consumed substance among people of all ages, including women of reproductive age. It is usually found in beverages such as coffee or energy drinks. There is evidence that consumption of up to 300 mg caffeine per day in healthy pregnant women is generally not associated with adverse reproductive and developmental effects (Wikoff et al., 2017).

Table 7.1

Nutritional supplements in pregnancy (World Health Organization, 2013; Youngblood et al., 2013; Hofmeyr et al., 2014; Khaskheli et al., 2016; Mills, 2017; Looman et al., 2018)

Nutritional supplement	Preventative effect	Indication	Special dosages and indications
Folic acid, 400–800 mcg daily	Neural tube defects	For all women, 4 weeks preconception until 10 weeks of pregnancy to reduce the risk of neural tube defects	4–5 mg daily in women with a previous child with a neural tube defect
Vitamin D, 10 mcg daily	Poor bone growth, preeclampsia and gestational diabetes, low birth weight, recurrent wheeze, asthma	For all women preconception	20 mcg daily for women with dark skin and low sunlight exposure
Calcium, 1.52 g daily	Preeclampsia	For women in areas with low dietary calcium intake	
Iron, 200 mg daily	Iron-deficiency anemia	For women with iron deficiency	

Illicit drugs

The use of illicit drugs such as cocaine, morphine, and marijuana during pregnancy is associated with maternal complications and adverse outcomes for infants and children. Different illicit drugs are associated with different complications; however, due to poly drug use (also combined with/including alcohol and tobacco), exact associations are difficult to make in evaluating the literature. Table 7.2 shows problems related to the consumption of commonly used illicit drugs during pregnancy. Disease-specific drug use before and during pregnancy and during lactation will be discussed per disease in section "Specialist Individual Preconception Care".

Preconception risk factor assessment

To develop an appropriate preconception plan, the first preconception consultation should consist of thorough history taking of both partners. Elements that should be discussed are set out in Table 7.3. Depending on the findings, consultation with a specialist should be considered. In patients with existing chronic neurologic morbidity, it is important to consider additional risk factors and comorbidities that should be taken into account.

SPECIALIST INDIVIDUAL PRECONCEPTION CARE

A multidisciplinary individual approach to manage pregnancy and delivery is required to optimize outcome. All women with comorbidities, including neurologic disease, are at increased risk of complications before or during pregnancy as well as during the puerperium. Cooperation and alignment between involved healthcare providers is of utmost importance.

Neurologists are often the first to encounter fertile women with a neurologic morbidity and should stay alert to the possibility that these patients may wish to become pregnant. The key question "would you like to become pregnant in the next year" should always be considered. All women who do should be referred for specialist preconception counseling. Those with neurologic comorbidities should have preconception and/or prenatal consultation with an anesthesiologist, regardless of the expected mode of delivery or wish for analgesia, to prepare for unforeseen emergency situations. For women with a congenital or genetic neurologic disease, questions concerning heredity should be discussed, preferably preconceptionally or in early pregnancy. Consultation with a clinical geneticist should also be considered. Women with neurologic diseases or traumatic brain or spinal cord injury in their history may have serious neurologic sequelae. Resulting disability may require need for extra support or interventions during pregnancy and caring for the infant.

Coordination of care

Various specialists may be included in the care of women with a neurologic comorbidity when they wish to conceive or are pregnant. These specialists include neurologists, obstetricians, anesthesiologists, clinical geneticists, neonatologists, pediatricians, and others depending on the underlying morbidity. As cases might include rare diseases and extraordinary presentation, multidisciplinary consultation is required to allow for optimal care.

The organization for multidisciplinary care may be a challenge due to busy schedules and irregular working schemes that are usually not synchronized. Of course, digital platforms may offer solutions to help facilitate collaboration. Multidisciplinary care in preconception can be organized as presented in Fig. 7.1.

Table 7.2

Adverse effects of illicit drugs in pregnancy

Illicit drug	Adverse effects
Opiates (e.g., morphine, fentanyl)	Neonatal abstinence syndrome (NAS), neural tube defects, congenital heart defects, gastroschisis, miscarriage, stillbirth, preterm delivery, low birth weight (Minozzi et al., 2013)
Cocaine	Preterm birth, premature rupture of membranes, placental abruption and infarction, preeclampsia, growth restriction (Cressman et al., 2014), gastroschisis (Draper et al., 2008)
Amphetamines (e.g., crystal meth)	Gastroschisis (Draper et al., 2008), clefting, cardiac anomalies, fetal growth reduction deficits (Plessinger, 1998)
Cannabis (e.g., marijuana)	Conflicting evidence, possible effect on intellectual development in young children (Draper et al., 2008; Albright and Rayburn, 2009); seems to enhance the teratogenic effects of ethanol (Hansen et al., 2008)

Table 7.3

Risk factor assessment

Medical history
Anemia (e.g., sickle cell anemia, hemoglobinopathy)
Autoimmune diseases (e.g., systemic lupus erythematosus; Kesler and Kupferminc, 2013)
Cardiovascular diseases (e.g., hypertension, congenital heart disease, deep venous thrombosis)
Diabetes (type I, type II, gestational diabetes)
Gastrointestinal diseases (e.g., Crohn's disease, ulcerative colitis)
Gynecological diseases (e.g., myomas, pelvic inflammatory disease)
Infectious diseases (and risk behavior) (e.g., human immunodeficiency virus (HIV), syphilis, hepatitis A/B/C/D/E, human papilloma virus (HPV), group-B streptococcus, group-A streptococcus, zika virus, gonorrhea)
Neurological disorders (e.g., epilepsy, stroke, multiple sclerosis (MS))
Oncological diseases (e.g., leukemia, breast cancer)
Orthopedic diseases (e.g., hip dysplasia, rickets)
Psychiatric diseases (e.g., pica, schizophrenia, depression)
Pulmonary diseases (e.g., asthma, pulmonary embolism, pulmonary hypertension, cystic fibrosis)
Thyroid disorders (e.g., Hashimoto, Graves, thyroiditis)
Urogenital diseases (e.g., hypospadias, obstructive renal disease, dysplastic renal disease)
Previous surgical interventions (abdominal: e.g., appendectomy, gastric bypass, cesarean section; urogenital: e.g., conization, myomectomy)
Medication use (prescribed and "over-the-counter")
Reproductive history
Number of pregnancies
Number of deliveries
Course of pregnancies (e.g., previous preeclampsia), deliveries, and postpartum period
Neonatal course postpartum
Uterine or cervical defects
Miscarriages
Abortions
Premature birth
Intrauterine fetal death (IUFD)
Low birth weight (<2500 g)
Children with congenital anomalies
Family history
Congenital disorders
Consanguinity
Ethnicity
Social history
Home (accommodation, physical and mental abuse, involvement of authorities)
Family and friends (relationship, social safety net)
Education
Employment and income

Continued

Table 7.3

Continued

Lifestyle
Alcohol
Tobacco
Drugs (e.g., opiates, cocaine, amphetamines, cannabis, hallucinogens)
Exposure to chemicals and radiation
Exercise
BMI
Nutrition
Diet (e.g., vegetarian, vegan)
Bulimia or anorexia nervosa
Vitamin supplementation (folate, zinc)
Lactose intolerance

Depending on the population of a hospital or region, multidisciplinary meetings could be planned on a regular basis. Another option could be a multidisciplinary outpatient clinic. Both options allow for improved patient-centered care. In multidisciplinary outpatient clinics, the direct involvement of the patient allows for improved shared decision-making.

In multidisciplinary meetings, cases with neurologic comorbidity, preconceptionally or while pregnant, should be evaluated based on risks and treatment options. To make such multidisciplinary meetings work, it is vital that a representative of all involved specialties is present so that all perspectives on the morbidity and treatment plan are accounted for. For continuity of care, a case manager is recommended.

Based on the most frequently occurring neurologic diseases among women in their reproductive years, the rest of this section will discuss these groups and present an overview of management considerations for obstetricians, neurologists, and anesthesiologists.

Cerebrovascular disease

The incidence of maternal stroke has risen in the past 2 decades (Kuklina et al., 2011), and this has been attributed in part to a higher percentage of patients initiating their pregnancies with one or more comorbidities (Tate and Bushnell, 2011). The complexity of these cases in many instances necessitates the participation of the cardiologist, neurologist, and anesthesiologist early in the course of the pregnancy.

Management considerations for obstetricians

A multidisciplinary approach is crucial in the prevention and management of cerebrovascular diseases.

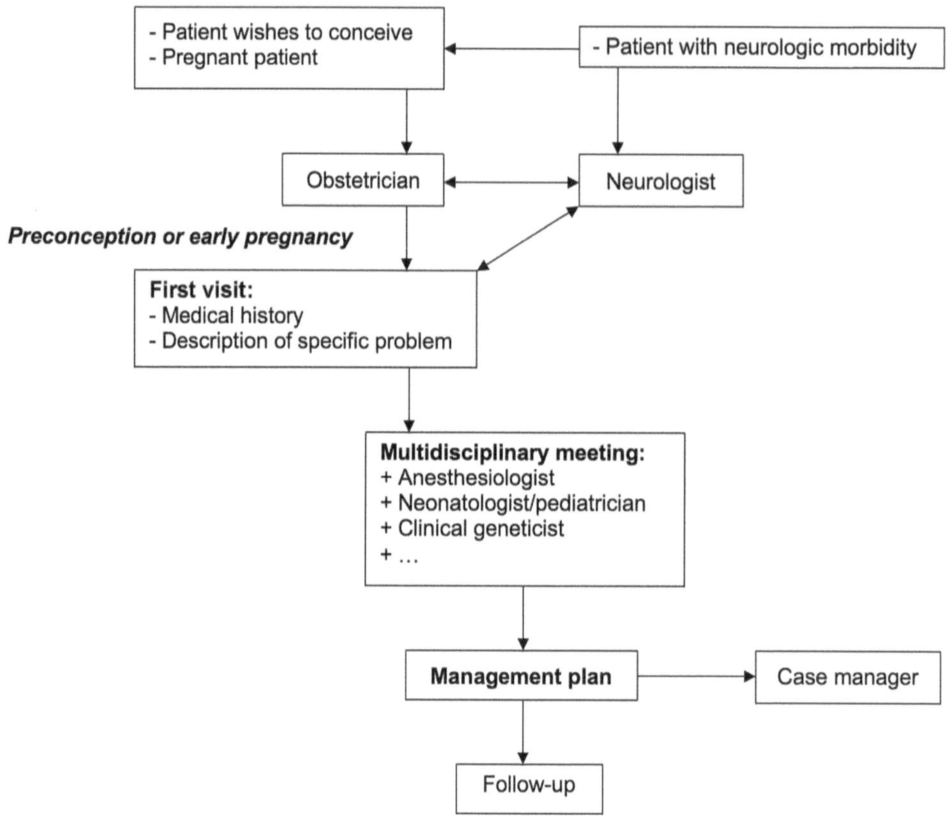

Fig. 7.1. Organization of multidisciplinary preconception care.

Obstetricians have a fundamental role in the prevention of neurologic complications as they are usually the first point of contact for the pregnant woman. They should identify high-risk patients and form an integrated healthcare team based on the patient's specific needs, to adequately monitor and treat conditions that could increase the risk of cerebrovascular complications. Pregnant women are at higher risk for ischemic strokes even in the absence of other risk factors; this is primarily due to a hypercoagulable state that results from the hemodynamic and hormonal changes occurring during gestation reaching their peak during the third trimester and continuing postpartum. Headache and visual changes need to be carefully investigated, as the risk of secondary headaches is significantly higher during pregnancy. Preeclampsia, reversible vasoconstriction syndrome, and venous sinus thrombosis should always be on the differential.

Pregnancy is not contraindicated in women with a history of ischemic and/or hemorrhagic strokes or cerebral venous sinus thrombosis (CVST). However, women with a history of stroke or thrombosis are at an increased risk during pregnancy. It is important to be familiar with the origin of the previous cerebrovascular event to assess for increased risk. If not previously analyzed, screening for inherited and acquired thrombophilia, coagulopathies (i.e., antiphospholipid syndrome, protein C or S deficiency), or cardiac origin (embolisms) is recommended (Marjot et al., 2011).

Prophylactic anticoagulant therapy should be considered, particularly in patients with a personal or familial history of thrombophilia (Scheres et al., 2019). The risk of recurrence of cerebral venous thrombosis is low, but prophylactic antithrombotic medication should be administered and continued during pregnancy and postpartum (Ciron et al., 2013). Consultation with a hematologist may be warranted. In case of a history of CVST, the preferred treatment during pregnancy is low molecular weight heparin (LMWH) given its superior safety profile (Podymow and August, 2008). The neurologist should always involve the obstetrician in the decision to initiate anticoagulation therapy. Warfarin is considered safe during breastfeeding (Warfarin, 2006).

In women with a history of ischemic stroke, aspirin is recommended for secondary stroke prophylaxis (Oza et al., 2017). Although rare but serious complications due to low-dose aspirin cannot be excluded, it appears to be well tolerated (Ahrens et al., 2016).

For patients with a history of thrombophilia, unfractionated and LMWH treatment in both prophylactic and therapeutic dosages can be safely prescribed throughout the entire pregnancy and postpartum. Vitamin K antagonists and direct oral anticoagulant are absolutely contraindicated in the first trimester of pregnancy.

In patients with a history of stroke, a cesarean section should only be performed based on obstetric indications. Induction of labor is sometimes preferred in cases wherein the patient is receiving anticoagulation, to allow for better control. Unfractionated heparin and LMWH should be stopped 24 h before labor induction and continued 24 h after birth, if no contraindications have developed (Katsuragi et al., 2018).

Women with known vascular lesions, including aneurysms and AVMs, have higher rates of cesarean deliveries. In retrospective studies, women with unknown arteriovenous malformations who had vaginal deliveries were not found to have increased numbers of AVM ruptures. Therefore the increased rate of cesarean deliveries may not be justified (Kim et al., 2013). In the case of lesions with a high risk for rupture, pushing may be contraindicated. Instrumental second stage of labor should be considered before deciding to perform a cesarean section.

Hypertension

One out of 10 pregnancies in the United States is complicated by hypertension (Report of the National High Blood Pressure Education Program, 2000; von Dadelszen and Magee, 2005), which is the greatest risk factor for maternal stroke. Patients with a history of chronic hypertension are more likely to develop pregnancy-induced hypertensive disorders. Deciding what the blood pressure target should be during pregnancy and when to treat have been the biggest challenges. Especially concerning milder forms of hypertension, either chronic or pregnancy-induced, there is insufficient evidence to accurately determine risks and benefits of treatment (Abalos et al., 2001).

Hypertension can be managed with a wide variety of antihypertensive drugs. The agent of first choice is methyldopa. Other agents to be considered include labetalol, nifedipine, hydralazine, b-receptor blockers, hydrochlorothiazide, and diazoxide.

The use of ACE inhibitors and angiotensin type 1-receptor antagonists is absolutely contraindicated during pregnancy due to associations with fetal loss, cardiac defects, oligohydramnios, growth restriction, renal agenesis, and neonatal anuric renal failure. If women of reproductive age use these agents, preconception switches to other agents should be considered (Podymow and August, 2008).

Management considerations for neurologists

Patients with a known history of vascular lesions, including arteriovenous malformations and intracranial aneurysms, should optimally get a preconceptional evaluation by a neurologist. In the case of high-risk lesions, a surgical consultation is warranted and can offer women the opportunity to opt for surgical treatment before conception with the objective of avoiding potential complications derived from the hyperdynamic state characteristic of pregnancy and risks related to surgery during pregnancy (Karlsson et al., 2018).

In the case of known vascular lesions without complications, current recommendations favor delaying any interventions until after delivery (Kim et al., 2013; Lv and Li, 2015). In intracranial or subarachnoid hemorrhage secondary to these lesions, all appropriate treatment interventions should be given. Computed tomography and even direct cerebral subtraction angiography use ionizing radiation; however, in emergent situations, these modalities are often preferred. The radiation exposure with proper shielding for both procedures is considerably below teratogenic levels. Surgical or endovascular interventions should be performed to secure these lesions if required, as this directly correlates with maternal and fetal outcomes (Agarwal et al., 2014).

Management considerations for anesthesiologists

Contraindications for neuraxial analgesia as pain management during vaginal delivery or cesarean section have not been published in patients with a history of ischemic stroke.

Anticoagulant therapy should be carefully evaluated according to anesthesiology guidelines before proceeding with epidural or spinal analgesia (Leffert and Schwamm, 2013; Miller and Leffert, 2020).

Epilepsy

Management considerations for obstetricians

Studies have shown conflicting results with trends toward both worsening and improvement of seizure frequency during pregnancy in women with epilepsy (Sabers et al., 1998; Harden et al., 2009; Thomas et al., 2017; Cagnetti et al., 2014). The main predictor of breakthrough seizures during pregnancy is a history of severe or uncontrolled epilepsy before gestation (EURAP Study Group, 2006; Thomas et al., 2017).

Often, breakthrough seizures during pregnancy are attributed to modifiable risk factors, including medication noncompliance or subtherapeutic serum levels of antiepileptic drugs (AEDs). Noncompliance in this group is often associated with fears of congenital malformations, making it fundamental to educate mothers-to-be about potential consequences of seizures during gestation, including direct trauma, hypoxia, and metabolic changes, which often result in significant morbidity and mortality (MacDonald et al., 2015; Christensen et al., 2018). However, it is also important to consider the risks of continuing AED administration during pregnancy and involve the pregnant woman in the decision-making to improve compliance. An overview of various AEDs is presented in Table 7.4, where pregnancy advices and teratogenic risks are portrayed.

Breastfeeding should be encouraged in all women with epilepsy. The levels of AEDs in breast milk are lower than what the fetus was exposed to in utero. In addition, lactation prevents withdrawal symptoms in the newborn, although this condition is rare with newer AEDs.

The risk of having a major seizure while caring for the newborn should be minimized. Safety strategies, including bathing the baby in shallow water and setting a nursery station on the bedroom floor, should be discussed prior to discharge.

Sleep deprivation may pose an extra risk for seizures in women with epilepsy (Schmidt et al., 1983). Close monitoring of blood levels of AEDs is indicated to maintain optimal seizure control and avoid toxicity during the puerperium.

Management considerations for neurologists

The management of epilepsy during pregnancy begins before conception. All women of childbearing age need to receive clear information about contraception, drug interactions, and teratogenesis during their first visit to a neurologist. It is important to repeat this conversation and consider initiating this conversation when patients transfer from different clinics or from pediatric neurology to adult neurology.

Approximately 55% of all pregnancies in women with epilepsy are unintended (Johnson et al., 2018). Patients who are not looking to become pregnant need to be advised that some AEDs may interact with hormonal oral contraceptives, decreasing their efficacy. The rate of pregnancy in women with epilepsy on oral contraceptives is 3.1 per 100 woman-years, which is higher than that in women with no history of epilepsy (0.7 per 100 woman-years). Alternative measures for contraception, such as devices (IUD) and implants, are preferred and should be offered in all cases (Reimers, 2016; Herzog et al., 2017).

In women who prefer oral contraceptives, special considerations need to be observed. Medications that act as inducers of cytochrome P450 (carbamazepine, phenytoin, and phenobarbital) will cause increased metabolism of both estrogen and progesterone, which can result in contraceptive failure. Alternative agents such as lamotrigine, which has minimal effects on oral contraceptives (Curtis et al., 2016; Gooneratne et al., 2017), should be favored.

Newer AEDs, including oxcarbazepine, parampanel, and felbamate, are also known to decrease the hormone serum levels due to increased metabolism. In addition, topiramate decreases ethinylestradiol levels by 30% by a different mechanism (Rosenfeld et al., 1997). For these patients, combined oral contraceptives with an estrogen dose of 0.50 mcg or higher is recommended. Folic acid supplementation has been a topic of debate; however, the current recommendation by the American Academy of Neurology is to prescribe at least 0.4 mg daily prior to conception and through pregnancy (class C) (Harden et al., 2009; Stephen et al., 2019).

In women with epilepsy the neurologist should emphasize the importance of family planning. Young women with optimal seizure control may lose contact with their epilepsy service (Leach et al., 2017). Thus all women should receive education during their first visit about possible complications derived from inadequate seizure control and fetal malformations linked to their medications.

If the patient is planning a pregnancy, individual circumstances, including epilepsy severity, current seizure control, and family history of congenital malformations, need to be considered to develop a therapeutic plan. The aim will always be achieving optimal seizure control while minimizing the exposure to AEDs. Monotherapy at the lowest effective dose is preferred. Drug levels need to be followed carefully through pregnancy due to changes in pharmacodynamics to ensure an adequate plasma concentration of the drug. If changes in the medication are necessary, the patient should take an active role in the decision-making process (Lagana et al., 2016).

Of all patients with epilepsy, 30% experience seizures despite optimal medical management. In patients with inadequate seizure control, it is recommended to counsel them to delay pregnancy until at least a year after acceptable seizure control has been achieved (EURAP Study Group, 2006). Careful analysis is needed to minimize the seizure frequency and severity while controlling for AED exposure and associated teratogenic risks (Leach et al., 2017). Status epilepticus is rare; however, it can lead to serious complications such as aspiration pneumonia, respiratory insufficiency, and metabolic acidosis (Sveberg et al., 2015) resulting in poor outcomes for the mother and the fetus.

Management considerations for anesthesiologists

There is evidence that the risk of seizures is higher peripartum. This has been attributed to inconsistent administration of AEDs, sleep deprivation, and stressors associated with labor and delivery (Schmidt et al., 1983). It is recommended that the patient continue with her

Table 7.4

Antiepileptic drugs (Tomson et al., 2011a, b; Bromley et al., 2013; Gooneratne et al., 2017; Stadelmaier et al., 2017; Koc et al., 2018)

Drug	Pregnancy advice	Risk	Frequent teratogenic effects	Malformation risk (%)	Neurobehavioral effects
Sodium valproate	Absolute contraindication, dose-dependent teratogenic effect; if inevitable: lowest effective dose and monotherapy	Teratogenic effects, risk for fetal valproate syndrome	Neural tube defects, heart defects, oral cleft, hypospadias, craniosynostosis, dysmorphic facial features (e.g., epicanthic folds, broad nasal bridge, small nose with anteverted nostrils)	4.7–10	Up to 66.7%[a]
Phenytoin	10–20 mg vitamin K supplementation after 36 weeks of gestation[b]	Teratogenic effects	Dysmorphic craniofacial features (e.g., wide philtrum, long upper lip)	2.9–6.7	Yes
Phenobarbital	10–20 mg vitamin K supplementation after 36 weeks of gestation[b]	Teratogenic effects	Dysmorphic craniofacial features (e.g., wide philtrum, long upper lip), heart defects, hypospadias, neural tube defects	5.5–7.4	Yes
Carbamazepine	Dose-dependent teratogenic effect; if inevitable: lowest effective dose and monotherapy; 10–20 mg vitamin K supplementation after 36 weeks of gestation[b]	Teratogenic effects	Dysmorphic craniofacial features, heart defects, hypospadias	2.6–5.6	Yes
Lamotrigine	Dose-dependent teratogenic effect; daily doses up to 300 mg not associated with increased malformation risk	Teratogenic effects when used in second and third trimesters	Heart defects, hypospadias, orofacial clefts	2.0–3.4	Unknown
Topiramate	10–20 mg vitamin K supplementation after 36 weeks of gestation[b]	Teratogenic effects	Orofacial clefts	4.2–7.7	Unknown
Levetiracetam	Relatively safe as monotherapy	Teratogenic effects	Heart defects, neural tube defects	0–2.4	Unknown
Oxcarbazepine	10–20 mg vitamin K supplementation after 36 weeks of gestation[b]	Teratogenic effects when used in second and third trimesters	Orofacial clefts	1.8–3.3	Unknown

[a]Permanent disorders in behavior and learning.
[b]Due to reduced vitamin K dependent clotting factors in the newborn and the related increased risk of hemorrhagic disease in the newborn.
In all described AEDs, the risk for occurrence of malformations increases with increased doses.

home medication; in some cases, epidural anesthesia has been suggested to decrease the degree of physical stress. Intravenous lorazepam needs to be available in the labor and delivery rooms.

Demyelinating diseases

Demyelinating diseases, including neuromyelitis optica, acute disseminated encephalomyelitis, progressive multifocal leukoencephalopathy, and MS, are frequently seen among young individuals. MS is the most common form of demyelinating disease, most often first presenting in women at the age group of 20–40, where pregnancy is more common; therefore this section will focus on the approach to MS during pregnancy.

Management considerations for obstetricians

Urinary tract infections (UTIs) are common in patients with MS and frequently attributed to detrusor-sphincter dyssynergia (Mahadeva et al., 2014). In addition, patients are at a higher risk of all infections due to the immunosuppression caused by their therapy. UTIs in this population tend to be asymptomatic and may result in premature labor or, in some cases, pseudoexacerbations, which manifest with focal neurologic deficits. Timely diagnosis and appropriate treatment of UTIs are required.

MS is no contraindication for vaginal delivery. Cesarean section should be performed based only on obstetric indications.

A protective effect against postpartum relapses has been described with breastfeeding. These findings need to be carefully interpreted due to the potential bias resulting from the fact that patients with a higher rate of relapses are more likely to restart their disease-modifying therapy (DMT) immediately after pregnancy, and the findings could potentially reflect patients with milder forms who chose to breastfeed (Langer-Gould et al., 2017; Bove et al., 2018).

Management considerations for neurologists

Most women with MS are diagnosed before pregnancy, which allows time for appropriate counseling by their neurologist before conception. During her first visit to the MS clinic, a patient usually receives education about the disease, including the available pharmacologic and nonpharmacologic interventions. It is important to talk about the significance of family planning early on, particularly as teratogenic effects have been linked to the majority of MS medications (DMT) and the risk of miscarriage and malformations does not stop as soon as the medication is discontinued, requiring a washout period that may last months (Corp. G, 2016; Coyle, 2016; Corp. NP, 2018; Coyle et al., 2019). There is limited experience with DMTs during pregnancy. Therefore it is important to analyze the patient's disease status and evaluate the need for therapy. Recommendations regarding the use of DMTs before and during pregnancy are summarized in Table 7.5. Due to long half-life of most DMTs and their teratogenic origin, stopping these therapies before conception is often recommended. Adequate contraceptives should be considered and advised during washout periods to prevent pregnancy and possible exposure to teratogenic agents.

Pulses of glucocorticoids should be considered in patients with a high risk of relapses who choose to discontinue their DMT before pregnancy (Bove et al., 2014). This treatment strategy can be continued during

Table 7.5

DMTs: advice for usage preconception and during pregnancy (Corp. G, 2016; Coyle, 2016; Corp. NP, 2018; Coyle et al., 2019)

Drug	Preconception	Pregnancy
Fingolimod	Stop, long half-life	Stop
Teriflunomide	Stop, long half-life	Stop
Cladribine	Stop, washout until 6 months after last dose	Stop
Rituximab	Stop, washout until 12 months	Stop
Ocrelizumab	Continue	May be continued, check B-cell counts in newborns before administering live or live-attenuated vaccines (risk of B-cell depletion)
Natalizumab	Continue	May be continued
Glatiramer acetate (GA)	Continue	No negative effects reported
Interferon beta (IFNβ)	Continue	No negative effects reported
Alemtuzumab	Continue	May be continued, thyroid function monitoring due to risk of hypothyroidism and associated risks

pregnancy, with the consideration that glucocorticoid use has been associated with fetal growth restriction (Busada and Cidlowski, 2017).

As a general rule, contraceptive methods should be offered to all patients. If patients already use contraceptive methods, refilling of prescriptions if required is recommended. If patients wish to start on contraception, they need to be informed about the different options with regard to their preferences. Consultation with general practitioner or gynecologists may be warranted. In women with limited mobility, hormonal contraceptives are strongly discouraged due to the potential risk of thromboembolism along with decreased bone mineral density (Curtis et al., 2016).

Patients who are planning a pregnancy frequently ask their neurologist about the potential risk of their children developing the disease. In general, the risk of inheriting MS is low as the mode of inheritance is polygenic with at least 200 genetic susceptibility loci identified (Bove et al., 2018; Coyle et al., 2019). In addition, other factors, including HLA-associated variants, have been associated with the condition. The risk of MS for a child with one affected parent is 2%–5%, whereas if both parents are affected, the risk increases to 20%. Unfortunately, at the moment, there are no tests available to predict the disease. Furthermore, the influence of environmental exposures has an important role in the development of MS.

Pregnancy by itself does not appear to increase the risk of exacerbations; on the contrary, the rate of relapses decreases as pregnancy advances with the lowest rates during the third trimester (Coyle et al., 2019). However, women with active disease prior to pregnancy have a relatively higher risk of exacerbations during pregnancy and the postpartum period (Coyle et al., 2019). Special considerations apply to the first three months postpartum as an increased risk of relapses has been observed in this period.

Patients should be encouraged to exercise, as it has been shown to slow the progression of the disease (Wesselink et al., 2017).

Management considerations for anesthesiologists

There are no restrictions concerning obstetric anesthesia. Nonetheless, a careful history of preexisting neurologic deficits needs to be recorded in the chart of patients with MS by the anesthesiologist, as neurologic deficits may be observed after the use of various anesthetics in the delivery room, which can be mistakenly interpreted as an MS exacerbation or a stroke; this can result in unnecessary diagnostic tests or potentially harmful interventions.

Neuromuscular diseases

Neuromuscular diseases include conditions that affect nerves, the neuromuscular junction, and muscle. This section will provide a broad overview of management considerations in patients with these diseases, focusing on myasthenia gravis (MG) and myotonic dystrophy (MD) as two representative diagnoses.

Management considerations for obstetricians

The main considerations for obstetricians treating women with neuromuscular disease involve counseling about the heritability of diseases and potential adverse effects of pharmacotherapies on mother and fetus.

Myasthenia gravis

Autoimmune MG is a multifactorial neuromuscular disease characterized by fluctuating weakness of extraocular and proximal limb muscles. As women typically present with the condition in their 20s and 30s, many with undiagnosed disease may become pregnant. Notably, severe exacerbations most frequently occur in the first year after presentation. Any woman complaining of intermittent double vision, difficulty swallowing food, or weakness of arms or legs with sustained use (e.g., difficulty lifting the arms after styling hair) should therefore be referred to a neurologist for diagnostic evaluation.

Those with the diagnosis need to be reassured that MG does not affect fertility (Waters, 2019). While the disease is markedly influenced by genetic factors, it exhibits limited heritability. Because of the frequent co-occurrence of MG and autoimmune thyroid disease, the provider should test thyroid antibody status and thyroid function prior to conception. Patients receiving mycophenolate mofetil or methotrexate should be referred to a neurologist for adjustment of therapy prior to conception. If a woman becomes pregnant while using methotrexate, she should receive folic acid and be immediately taken off the drug.

The obstetrician should be aware that prednisolone, azathioprine, cyclosporine, and tacrolimus can contribute to gestational diabetes, leukopenia, liver function abnormalities, and decreased renal function and should monitor accordingly. Transplacental passage of antibodies to the fetal y-subunit of the AchR subunit may cause a severe and often fatal form of fetal arthrogryposis that may be detected on ultrasound as poor fetal movement and polyhydramnios. Currently, there is no in utero treatment for fetal MG. Transient neonatal MG occurs at an incidence rate of 9%–30%. Onset may be somewhere between immediate to several months after birth. There is a high recurrence rate of transient neonatal MG in subsequent pregnancies (Chaudhry et al., 2012). It is good practice to monitor closely infants born to mothers with MG for at least 72 h postpartum.

Importantly, a number of common treatments for complications of pregnancy can have undesired effects in patients with MG. Magnesium sulfate is relatively contraindicated due to muscle relaxation. It has been studied that the use of levetiracetam in preeclampsia/eclampsia as anticonvulsant is effective in most cases, when magnesium sulfate is contraindicated or ineffective. If levetiracetam or other anticonvulsants are ineffective, the use of magnesium should be balanced against the risks (Satia et al., 2016). In addition, not all antihypertensive drugs are suitable for MG patients. β-Adrenoceptor blockers and calcium-channel blockers both have the potential to worsen weakness. Therefore in this patient group, methyldopa or oral hydralazine may be the initial drugs of choice in nonsevere hypertension (Ozcan et al., 2015).

There is no contraindication for vaginal term delivery in patients with MG, but this should always be hospital based with neurologic and neonatal backup available. The course of the first stage of labor in MG patients is not altered because the condition does not affect smooth muscle. The second stage of labor could be complicated by weakened expulsive efforts. Instrumental vaginal delivery may be indicated.

Myotonic dystrophy

MD is a degenerative neuromuscular and neuroendocrine disease. It is associated with cardiac dystrophy, presenting as conduction defects, arrhythmias, or congestive heart failure. There are two types of the disease (MD types 1 and 2); though they share many of the same phenotypic features, they vary in clinical severity and genetic basis. MD type 1 is an autosomal dominant trinucleotide repeat inherited disorder, with the affected gene located on chromosome 19. MD type 2 is autosomal dominant, with an expansion of the CNBP gene (Dalton et al., 2013).

The obstetrician should obtain baseline echocardiography and pulmonary function tests before conception in patients with either form of the disease and should alert the patient to symptoms of arrhythmia. In severely affected women, pregnancy is rare. Referral to a geneticist for preconception counseling may be warranted for consideration of in vitro fertilization and preimplantation genetic diagnosis to avoid giving birth to an affected infant (particularly for patients with MD type 1). Infants born to these mothers may be affected with congenital MD, characterized by severe generalized hypotonia and weakness, respiratory issues, difficulty sucking and swallowing, talipes, arthrogryposis, and learning difficulties. Some of these features may be detectable on prenatal ultrasound. Prenatal diagnosis is possible with DNA analysis from amniocentesis or chorionic villus sampling. MD type 1 is associated with insulin resistance (Takeshima et al., 2018); the obstetrician should monitor closely for gestational diabetes.

Patients with MD may have weakened abdominal and uterine muscles, requiring instrumental vaginal delivery techniques. In case of uterine inertia, oxytocin is a safe pharmacotherapy. Due to possible failure of contraction of the uterus, there is a higher risk for postpartum hemorrhage in these patients.

Management considerations for neurologists

Major considerations for neurologists treating women with neuromuscular disease include fluctuations in disease activity during and after pregnancy and the safety and efficacy of the medications used to treat these conditions during pregnancy.

Myasthenia gravis

Patients taking mycophenolate mofetil and methotrexate should change therapies before conception as these medications are teratogenic and contraindicated in pregnancy. Drugs that have been proven safe during pregnancy are pyridostigmine, prednisolone, azathioprine, cyclosporine/tacrolimus, and intravenous immunoglobulin. While safe, these drugs do carry potential side effects (gestational diabetes, leukopenia, and liver function abnormalities), warranting use of the lowest effective dose. Drug pharmacokinetics may change during pregnancy, causing subtherapeutic serum levels of medication and increased disease activity. This may be particularly relevant late in pregnancy when diaphragm elevation can further compromise respiratory function. The course of MG may alternate in different pregnancies in the same patient; close monitoring of disease activity in each successive pregnancy is therefore crucial. Close postpartum monitoring is also important given that 30% of patients experience exacerbations of MG after delivery.

Myotonic dystrophy

Mexiletine is used in some patients with MD as antimyotonic drug. There is very limited data available concerning mexiletine use during pregnancy; careful monitoring during pregnancy and postpartum to maintain therapeutic drug levels is advised (Gregg and Tomich, 1988). Significant worsening of the symptoms of MD (myotonia and weakness) can occur at any point during pregnancy but is most common in the third trimester. These exacerbations typically abate after delivery.

Management considerations for anesthesiologists

The main consideration for anesthesiologists in delivering infants of mothers with neuromuscular disease is management of potential weakness, particularly of respiratory muscles.

Myasthenia gravis

Anesthetic challenges may arise when patients with MG require general anesthesia. There is always an increased postsurgical risk of respiratory failure (Blichfeldt-Lauridsen and Hansen, 2012). Due to lower levels of nAChRs in patients with MG and the use of cholinesterase inhibitors (pyridostigmine), altered reactions to neuromuscular blocking agents (succinylcholine) may occur. MG patients are at risk for prolonged blockade, and reversion of residual block may be unsuccessful. Regional anesthesia is preferred, as epidural or spinal analgesia may reduce the need for muscle relaxants and offer pain control during delivery or postcesarean section (Blichfeldt-Lauridsen and Hansen, 2012; Norwood et al., 2014).

Myotonic dystrophy

General anesthesia and respiratory depressants should be avoided due to the risk of exacerbation of pulmonary hypoventilation. Nondepolarizing neuromuscular blocking agents should also be avoided.

Headache

Neurologists classify headaches as either primary, due to primary pathology in the nerves in the scalp and other cranial tissues, or secondary, due to a separate pathology that affects these nerves. The first of these categories includes migraine and the trigeminal autonomic cephalalgias, among others. The second includes headaches due to elevated intracranial pressure, infection, vasculitis, and other vascular pathologies. Primary headache disorders (specifically, migraines) are the most common headache types overall and tend to cause minimal complications in pregnancy, though their treatments may require modification (Allais et al., 2019). Secondary headache syndromes, however, may reflect serious illnesses for which pregnant and postpartum women are particularly at risk. It is important to recognize the features of these headaches to minimize morbidity.

Management considerations for obstetricians

Migraine

There is an association of migraine with both preeclampsia and maternal stroke. However, it is not clear whether this is a causal effect (Bushnell and Chireau, 2011). Patients with rare migraines, especially those who are not on daily prophylactic pharmacotherapy, may not require referral to a neurologist. Many prophylactic medications (beta blockers, tricyclic antidepressants, antiepileptics, and abortives such as the triptans) are either contraindicated or poorly studied in pregnancy; patients taking these should confer with their neurologist regarding therapy modification prior to conception (Allais et al., 2019).

Idiopathic intracranial hypertension (IIH)

IIH is a secondary headache syndrome caused by elevated intracranial pressure. Weight gain can adversely affect disease activity, and thus pregnant women should aim not to exceed maximum weight gain recommendations. The primary risk of inadequately treated IIH is visual loss; any patient reporting change in visual acuity or visual fields should be promptly referred to a neurologist or ophthalmologist. First-line pharmacotherapies for IIH include acetazolamide and topiramate. Acetazolamide may cause adverse fetal effects, and topiramate use during pregnancy is contraindicated (Tomson et al., 2011a). Alternative therapies include the use of glucocorticoids or lumbar puncture (Martin and Foley, 2005).

Posterior reversible encephalopathy syndrome (PRES) and reversible cerebral vascular syndrome (RCVS)

These conditions comprise a spectrum of diseases resulting from endothelial dysfunction and subsequent vascular dilatation and leakage (PRES) or vasospasm (RCVS). There is considerable overlap between the two syndromes: the risk of both conditions increases significantly during pregnancy and in the postpartum period, both are associated with hypertension and preeclampsia/eclampsia, and both can cause ischemic stroke or intracranial hemorrhage. The obstetrician should be alerted to the following characteristic symptoms and refer patients promptly to a neurologist or the emergency room if they develop any of the following: visual field deficits, blurred vision, episodes of seizure, focal neurologic deficits, and so-called thunderclap headaches that reach maximal intensity within seconds of onset. If these symptoms develop in the setting of hypertension, it is crucial to lower blood pressure and obtain an MRI and MRA of the brain.

Cerebral venous sinus thrombosis (CVST)

Thrombosis of any of the intracranial veins or venous sinuses can cause headache due to elevated intracranial pressure. If untreated, increased venous pressure can eventually cause intracranial hemorrhage. Women are at increased risk of venous thromboembolic events during pregnancy and in the postpartum period due to hormone-mediated alterations in coagulation pathways. The obstetrician should be alerted to the following characteristic symptoms and refer patients promptly to a neurologist or the emergency room if they develop any of the following: headache that is worse when supine, maximal in the morning, or wakes the patient from sleep; significant nausea/vomiting associated with headaches; and focal neurologic deficits. MR venography can assess for the presence of CVST, which requires treatment using anticoagulation therapy.

For further information concerning CVST, please find this in section "Cerebrovascular disease".

Management considerations for neurologists

Migraine

Observational studies have shown that migraine tends to improve during pregnancy; however, worsening of symptoms and frequency of episodes has also been described (Sacco and Ripa, 2015; Tepper, 2015; Petrovski et al., 2018). Data regarding the safety of many migraine prophylactics and abortives in pregnancy is scarce; see Chapter 11 of Volume 172 for a detailed review of management options.

Idiopathic intracranial hypertension (IIH)

As mentioned earlier, acetazolamide is of unclear safety in pregnancy and topiramate is contraindicated due to teratogenicity. Symptoms and visual fields should be monitored closely during pregnancy, as weight gain can cause worsening of disease. Lumbar puncture with large-volume CSF removal should be considered for management during pregnancy.

Posterior reversible encephalopathy syndrome (PRES) and reversible cerebral vascular syndrome (RCVS)

It is crucial for the neurologist to remember that women are at increased risk for these conditions in the postpartum period as well as during pregnancy. In the setting of preeclampsia/eclampsia, the choice of antihypertensive agent may be different during pregnancy than in the general population due to the safety of these drugs for the fetus; preferred medications include nifedipine, magnesium (also for seizure prophylaxis), nicardipine, and hydralazine. Transcranial Doppler offers a noninvasive method for monitoring the response to therapy in women with RCVS (Marsh et al., 2016). Verapamil has been used successfully to treat RCVS-related headaches and is considered safe when breastfeeding (Verapamil, 2006).

Neuro-oncology

Brain tumors are rare, and the overall incidence of primary brain tumors is the same in pregnant and non-pregnant women of childbearing age. They may be either primary (e.g., gliomas, glioblastomas, meningiomas, pituitary adenomas) or secondary metastases of a primary extracranial malignancy. The systemic cancers that most commonly metastasize to the brain are breast, lung, colon, kidney, and melanoma. All patients with a primary brain neoplasm should be under the care of a neurologist with subspecialty training in neuro-oncology; those with metastatic disease from an extracranial primary should be cared for by a medical oncologist as well as either a neuro-oncologist or a general neurologist. Important factors guiding management of pregnant women with brain tumors are the type of neoplasm, its rate of growth, its location in the brain, and the degree to which it causes symptoms.

Many treatments of brain tumors (radiation, surgical resection, and certain chemotherapeutics such as methotrexate) can be harmful to the fetus and are either contraindicated or high risk in pregnancy. In addition, certain chemotherapeutic agents, especially alkylating chemotherapies, such as cyclophosphamide, busulfan, melphalan procarbazine, and chemotherapeutic combinations that include alkylating chemotherapies, can affect a woman's future fertility (Waimey et al., 2015). For these reasons, interdisciplinary planning is of the utmost importance, starting before conception if at all possible and continuing throughout pregnancy. Other specialists that may need to be involved in these discussions include radiation oncologists, neurosurgeons, anesthesiologists, and, in the case of brain metastases of extracranial primary malignancy, medical oncologists.

Management considerations for obstetricians

Brain tumors can cause secondary neurologic processes that are potentially dangerous; the obstetrician should be aware of these complications and their common symptoms. The tumor may expand and associated edema may increase, causing mass effect and, if severe, herniation and compression of part of the brain. As noted earlier, certain hormone-responsive tumors (prolactinomas and some meningiomas) may grow at an increased rate during pregnancy. Symptoms of increased mass effect include headache, nausea, vomiting, decreased alertness, confusion, and focal neurologic deficits (e.g., weakness, language impairment, and visual changes). Headaches due to increased intracranial pressure are typically worse when the patient is supine and may be present when she wakes in the morning. Hemorrhage into a tumor may cause symptoms similar to those described earlier. Glioblastoma, renal cell carcinoma, and melanoma are particularly prone to hemorrhage. Brain tumors can cause seizure due to cortical irritation; the patient may report intermittent sensory changes or rhythmic movements in one part of the body, periods of language disturbance, or loss of awareness. If a patient reports any of these symptoms during a visit, she should be referred promptly to her neurologist for evaluation.

In case of increased intracranial pressure, vaginal delivery may be assisted with ventouse or forceps to minimize pushing, which can further raise pressure, or cesarean section may be performed. However, there is a lack of empirical data to guide these decisions.

Management considerations for neurologists

The safety of chemotherapeutic agents commonly used to treat brain cancers will be discussed in more detail in a following section of this volume; initiation and continuation of chemotherapy during pregnancy should be pursued only in conversation with the obstetrician. Use of glucocorticoids to treat tumor-associated vasogenic edema also requires careful consideration and discussion with the obstetrician, as glucocorticoid use during pregnancy is associated with fetal growth restriction (Busada and Cidlowski, 2017). If prolonged high-dose steroid treatment is required, prednisone is the first agent of choice. When considering how frequently to monitor tumor growth during pregnancy, the neurologist should keep in mind that certain hormone-responsive tumors, such as prolactinomas and some meningiomas, may grow at an accelerated rate due to hormonal shifts. If there is significant mass effect or obstructive hydrocephalus from tumor, the risk of dural puncture leading to herniation may be too high to safely administer intrapartum neuraxial anesthesia. An alternative plan should be discussed with the obstetrician and anesthesiologist prior to delivery.

Management considerations for anesthesiologists

Any plan for neuraxial anesthesia during delivery should first be discussed with the neurologist to identify any risk of herniation from increased intracranial pressure or tumor location.

CONCLUSION

The importance of coordinated, multidisciplinary care for pregnant women with neurologic comorbidities is increasingly recognized. We propose the term "neuro-obstetrics" to describe the complex care required for these women. All neurologists and obstetricians should be familiar with the basic principles of neuro-obstetric care, including the need for preconception counseling and close collaboration between specialties to optimize maternal and fetal outcomes for women with chronic or acute neurologic conditions.

References

Abalos E, Duley L, Steyn DW et al. (2001). Antihypertensive drug therapy for mild to moderate hypertension during pregnancy. Cochrane Database Syst Rev CD002252.

Agarwal N, Guerra JC, Gala NB et al. (2014). Current treatment options for cerebral arteriovenous malformations in pregnancy: a review of the literature. World Neurosurg 81: 83–90.

Ahrens KA, Silver RM, Mumford SL et al. (2016). Complications and safety of preconception low-dose aspirin among women with prior pregnancy losses. Obstet Gynecol 127: 689–698.

Albright BB, Rayburn WF (2009). Substance abuse among reproductive age women. Obstet Gynecol Clin North Am 36: 891–906, xi–xii.

Allais G, Chiarle G, Sinigaglia S et al. (2019). Migraine during pregnancy and in the puerperium. Neurol Sci 40: 81–91.

Bergmann RL, Bergmann KE, Dudenhausen JW (2008). Undernutrition and growth restriction in pregnancy. Nestle Nutr Workshop Ser Pediatr Program 61: 103–121.

Blatt K, Moore E, Chen A et al. (2015). Association of reported trimester-specific smoking cessation with fetal growth restriction. Obstet Gynecol 125: 1452–1459.

Blichfeldt-Lauridsen L, Hansen BD (2012). Anesthesia and myasthenia gravis. Acta Anaesthesiol Scand 56: 17–22.

Bouchery EE, Harwood HJ, Sacks JJ et al. (2011). Economic costs of excessive alcohol consumption in the U.S., 2006. Am J Prev Med 41: 516–524.

Bove R, Alwan S, Friedman JM et al. (2014). Management of multiple sclerosis during pregnancy and the reproductive years: a systematic review. Obstet Gynecol 124 (6): 1157–1168. https://doi.org/10.1097/AOG.0000000000000541.

Bove R, Rankin K, Chua AS et al. (2018). Oral contraceptives and MS disease activity in a contemporary real-world cohort. Mult Scler 24 (2): 227–230. https://doi.org/10.1177/1352458517692420.

Bromley RL, Mawer GE, Briggs M et al. (2013). The prevalence of neurodevelopmental disorders in children prenatally exposed to antiepileptic drugs. J Neurol Neurosurg Psychiatry 84: 637–643.

Busada JT, Cidlowski JA (2017). Mechanisms of glucocorticoid action during development. Curr Top Dev Biol 125: 147–170.

Bushnell C, Chireau M (2011). Preeclampsia and stroke: risks during and after pregnancy. Stroke Res Treat 2011: 858134.

Cagnetti C, Lattanzi S, Foschi N et al. (2014). Seizure course during pregnancy in catamenial epilepsy. Neurology 83 (4): 339–344. https://doi.org/10.1212/WNL.0000000000000619. Epub 2014 Jun 18.

Chaudhry SA, Vignarajah B, Koren G (2012). Myasthenia gravis during pregnancy. Can Fam Physician 58: 1346–1349.

Christensen J, Vestergaard C, Hammer Bech B (2018). Maternal death in women with epilepsy: smaller scope studies. Neurology 91: e1716–e1720.

Cipolla MJ (2013). The adaptation of the cerebral circulation to pregnancy: mechanisms and consequences. J Cereb Blood Flow Metab 33: 465–478.

Ciron J, Godeneche G, Vandamme X et al. (2013). Obstetrical outcome of young women with a past history of cerebral venous thrombosis. Cerebrovasc Dis 36: 55–61.

Corp. G (2016). Available: http://products.sanofi.us/aubagio/aubagio.html [Accessed 24-05-2019].

Corp. NP (2018). Available: https://www.pharma.us.novartis.com/sites/www.pharma.us.novartis.com/files/gilenya.pdf [Accessed 24-05-2019].

Coyle PK (2016). Management of women with multiple sclerosis through pregnancy and after childbirth. Ther Adv Neurol Disord 9: 198–210.

Coyle PK, Oh J, Magyari M et al. (2019). Management strategies for female patients of reproductive potential with multiple sclerosis: an evidence-based review. Mult Scler Relat Disord 32: 54–63.

Cressman AM, Natekar A, Kim E et al. (2014). Cocaine abuse during pregnancy. J Obstet Gynaecol Can 36: 628–631.

Curtis KM, Tepper NK, Jatlaoui TC et al. (2016). U.S. medical eligibility criteria for contraceptive use, 2016. MMWR Recomm Rep 65: 1–103.

Dalton JC, Ranum LPW, Day JW (2013). Myotonic dystrophy type 2, University of Washington, Seattle, Seattle (WA), PMID: 20301639.

Denny L, Coles S, Blitz R (2017). Fetal alcohol syndrome and fetal alcohol spectrum disorders. Am Fam Physician 96: 515–522.

Draper ES, Rankin J, Tonks AM et al. (2008). Recreational drug use: a major risk factor for gastroschisis? Am J Epidemiol 167: 485–491.

EURAP Study Group (2006). Seizure control and treatment in pregnancy: observations from the EURAP epilepsy pregnancy registry. Neurology 66: 354–360.

Gooneratne IK, Wimalaratna M, Ranaweera AKP et al. (2017). Contraception advice for women with epilepsy. BMJ 357: j2010.

Gould GS, Lim LL, Mattes J (2017). Prevention and treatment of smoking and tobacco use during pregnancy in selected indigenous communities in high-income countries of the United States, Canada, Australia, and New Zealand: an evidence-based review. Chest 152: 853–866.

Gregg AR, Tomich PG (1988). Mexilitene use in pregnancy. J Perinatol 8: 33–35.

Hansen HH, Krutz B, Sifringer M et al. (2008). Cannabinoids enhance susceptibility of immature brain to ethanol neurotoxicity. Ann Neurol 64: 42–52.

Harden CL, Hopp J, Ting TY et al. (2009). Practice parameter update: management issues for women with epilepsy—focus on pregnancy (an evidence-based review): obstetrical complications and change in seizure frequency: report of the Quality Standards Subcommittee and Therapeutics and Technology Assessment Subcommittee of the American Academy of Neurology and American Epilepsy Society. Neurology 73: 126–132.

Herzog AG, Mandle HB, Cahill KE et al. (2017). Predictors of unintended pregnancy in women with epilepsy. Neurology 88: 728–733.

Hofmeyr GJ, Belizan JM, von Dadelszen P et al. (2014). Low-dose calcium supplementation for preventing pre-eclampsia: a systematic review and commentary. BJOG 121: 951–957.

Johnson AC, Nagle KJ, Tremble SM et al. (2015). The contribution of normal pregnancy to eclampsia. PLoS One 10: e0133953.

Johnson EL, Burke AE, Wang A et al. (2018). Unintended pregnancy, prenatal care, newborn outcomes, and breastfeeding in women with epilepsy. Neurology 91: e1031–e1039.

Karlsson B, Johansson AV, Yang HC et al. (2018). A novel method to determine the natural course of unruptured brain arteriovenous malformations without the need for follow-up information. J Neurosurg 129: 10–16.

Katsuragi S, Yoshimatsu J, Tanaka H et al. (2018). Management of pregnancy complicated with intracranial arteriovenous malformation. J Obstet Gynaecol Res 44: 673–680.

Kesler A, Kupferminc M (2013). Idiopathic intracranial hypertension and pregnancy. Clin Obstet Gynecol 56: 389–396.

Khaskheli MN, Baloch S, Sheeba A et al. (2016). Iron deficiency anaemia is still a major killer of pregnant women. Pak J Med Sci 32: 630–634.

Kim YW, Neal D, Hoh BL (2013). Cerebral aneurysms in pregnancy and delivery: pregnancy and delivery do not increase the risk of aneurysm rupture. Neurosurgery 72: 143–149; discussion 150.

Koc G, Keskin Guler S, Karadas O et al. (2018). Fetal safety of levetiracetam use during pregnancy. Acta Neurol Belg 118: 503–508.

Kuklina EV, Tong X, Bansil P et al. (2011). Trends in pregnancy hospitalizations that included a stroke in the United States from 1994 to 2007: reasons for concern? Stroke 42: 2564–2570.

Laganà AS, Triolo O, D'Amico V et al. (2016). Management of women with epilepsy: from preconception to postpartum. Arch Gynecol Obstet 293 (3): 493–503. https://doi.org/10.1007/s00404-015-3968-7. Epub 2015 Nov 30.

Langer-Gould A, Smith JB, Hellwig K et al. (2017). Breastfeeding, ovulatory years, and risk of multiple sclerosis. Neurology 89 (6): 563–569. https://doi.org/10.1212/WNL.0000000000004207. Epub 2017 Jul 12.

Lassi ZS, Imam AM, Dean SV et al. (2014). Preconception care: caffeine, smoking, alcohol, drugs and other environmental chemical/radiation exposure. Reprod Health 11: S6.

Leach JP, Smith PE, Craig J et al. (2017). Epilepsy and pregnancy: for healthy pregnancies and happy outcomes. Suggestions for service improvements from the Multispecialty UK Epilepsy Mortality Group. Seizure 50: 67–72.

Leffert LR, Schwamm LH (2013). Neuraxial anesthesia in parturients with intracranial pathology: a comprehensive review and reassessment of risk. Anesthesiology 119: 703–718.

Looman M, van den Berg C, Geelen A et al. (2018). Supplement use and dietary sources of folate, vitamin D, and n-3 fatty acids during preconception: the GLIMP2 study. Nutrients 10: 962.

Lv X, Li Y (2015). The clinical characteristics and treatment of cerebral AVM in pregnancy. Neuroradiol J 28: 385–388.

MacDonald SC, Bateman BT, McElrath TF et al. (2015). Mortality and morbidity during delivery hospitalization among pregnant women with epilepsy in the United States. JAMA Neurol 72: 981–988.

Mahadeva A, Tanasescu R, Gran B (2014). Urinary tract infections in multiple sclerosis: under-diagnosed and under-treated? A clinical audit at a large university hospital. Am J Clin Exp Immunol 3: 57–67.

Marjot T, Yadav S, Hasan N et al. (2011). Genes associated with adult cerebral venous thrombosis. Stroke 42: 913–918.

Marsh EB, Ziai WC, Llinas RH (2016). The need for a rational approach to vasoconstrictive syndromes: transcranial Doppler and calcium channel blockade in reversible cerebral vasoconstriction syndrome. Case Rep Neurol 8: 161–171.

Martin SR, Foley MR (2005). Approach to the pregnant patient with headache. Clin Obstet Gynecol 48: 2–11.

Miller EC, Leffert L (2020). Stroke in pregnancy: a focused update. Anesth Analg 130 (4): 1085–1096.

Mills JL (2017). Strategies for preventing folate-related neural tube defects: supplements, fortified foods, or both? JAMA 317: 144–145.

Minozzi S, Amato L, Bellisario C et al. (2013). Maintenance agonist treatments for opiate-dependent pregnant women. Cochrane Database Syst Rev CD006318.

Norwood F, Dhanjal M, Hill M et al. (2014). Myasthenia in pregnancy: best practice guidelines from a U.K. multispecialty working group. J Neurol Neurosurg Psychiatry 85: 538–543.

Oza R, Rundell K, Garcellano M (2017). Recurrent ischemic stroke: strategies for prevention. Am Fam Physician 96: 436–440.

Ozcan J, Balson IF, Dennis AT (2015). New diagnosis myasthenia gravis and preeclampsia in late pregnancy. BMJ Case Rep 2015. https://doi.org/10.1136/bcr-2014-208323.

Perry MF, Mulcahy H, DeFranco EA (2019). Influence of periconception smoking behavior on birth defect risk. Am J Obstet Gynecol 220 (6): 588.e1–588.e7.

Petrovski BE, Vetvik KG, Lundqvist C et al. (2018). Characteristics of menstrual versus non-menstrual migraine during pregnancy: a longitudinal population-based study. J Headache Pain 19: 27.

Plessinger MA (1998). Prenatal exposure to amphetamines. Risks and adverse outcomes in pregnancy. Obstet Gynecol Clin North Am 25: 119–138.

Podymow T, August P (2008). Update on the use of antihypertensive drugs in pregnancy. Hypertension 51: 960–969.

Reimers A (2016). Contraception for women with epilepsy: counseling, choices, and concerns. Open Access J Contracept 7: 69–76.

Report of the National High Blood Pressure Education Program (2000). Report of the national high blood pressure education program working group on high blood pressure in pregnancy. Am J Obstet Gynecol 183 (1): S1–S22.

Rosenfeld WE, Doose DR, Walker SA et al. (1997). Effect of topiramate on the pharmacokinetics of an oral contraceptive containing norethindrone and ethinyl estradiol in patients with epilepsy. Epilepsia 38: 317–323.

Sabers A, aRogvi-Hansen B, Dam M et al. (1998). Pregnancy and epilepsy: a retrospective study of 151 pregnancies. Acta Neurol Scand 97 (3): 164–170. https://doi.org/10.1111/j.1600-0404.1998.tb00631.x.

Sacco S, Ripa P (2015). Migraine in pregnancy. J Headache Pain 16: A24.

Satia MN et al. (2016). A study of the obstetric and perinatal outcomes of eclampsia and the use of levetiracetam in its management. Int J Reprod Contracept Obstet Gynecol 5 (12): 4266–4270. https://www.ijrcog.org/index.php/ijrcog/article/view/34/34.

Scheres LJJ, Bistervels IM, Middeldorp S (2019). Everything the clinician needs to know about evidence-based anticoagulation in pregnancy. Blood Rev 33: 82–97.

Schmidt D, Canger R, Avanzini G et al. (1983). Change of seizure frequency in pregnant epileptic women. J Neurol Neurosurg Psychiatry 46: 751–755.

Shukla SD, Aroor AR (2006). Epigenetic effects of ethanol on liver and gastrointestinal injury. World J Gastroenterol 12: 5265–5271.

Stadelmaier R, Nasri H, Deutsch CK et al. (2017). Exposure to sodium valproate during pregnancy: facial features and signs of autism. Birth Defects Res 109: 1134–1143.

Stephen LJ, Harden C, Tomson T et al. (2019). Management of epilepsy in women. Lancet Neurol 18: 481–491.

Sveberg L, Svalheim S, Taubøll E (2015). The impact of seizures on pregnancy and delivery. Seizure 28: 35–38. https://doi.org/10.1016/j.seizure.2015.02.020. Epub 2015 Feb 21.

Takeshima K, Ariyasu H, Ishibashi T et al. (2018). Myotonic dystrophy type 1 with diabetes mellitus, mixed hypogonadism and adrenal insufficiency. Endocrinol Diabetes Metab Case Rep 2018: pii: 17-0143. https://doi.org/10.1530/EDM-17-0143. eCollection 2018.

Tate J, Bushnell C (2011). Pregnancy and stroke risk in women. Womens Health (Lond) 7: 363–374.

Tepper D (2015). Pregnancy and lactation—migraine management. Headache 55: 607–608.

Thomas S, Hovinga ME, Rai D et al. (2017). Brief report: prevalence of co-occurring epilepsy and autism spectrum disorder: The U.S. National Survey of Children's Health 2011–2012. J Autism Dev Disord 47 (1): 224–229. https://doi.org/10.1007/s10803-016-2938-7.

Tomson T, Battino D, Bonizzoni E et al. (2011a). Dose-dependent risk of malformations with anti-epileptic drugs: an analysis of data from the EURAP epilepsy and pregnancy registry. Lancet Neurol 10: 609–617.

Tomson T, Battino D, Perucca E (2011b). Major birth defects after exposure to newer-generation antiepileptic drugs. JAMA 306: 826–827, author reply 827.

Verapamil (2006). SourceDrugs and Lactation Database (LactMed) [Internet]. Bethesda (MD): National Library of Medicine (US).

von Dadelszen P, Magee LA (2005). Antihypertensive medications in management of gestational hypertension-preeclampsia. Clin Obstet Gynecol 48: 441–459.

Waimey KE, Smith BM, Confino R et al. (2015). Understanding fertility in young female cancer patients. J Womens Health (Larchmt) 24: 812–818.

Warfarin (2006), SourceDrugs and Lactation Database (LactMed) [Internet]. Bethesda (MD): National Library of Medicine (US).

Waters J (2019). Management of myasthenia gravis in pregnancy. Neurol Clin 37: 113–120.

Wesselink AK, Rothman KJ, Hatch EE et al. (2017). Age and fecundability in a North American preconception cohort study. Am J Obstet Gynecol 217: 667, e661–667 e668. https://www.ncbi.nlm.nih.gov/pubmed/28917614.

Wikoff D, Welsh BT, Henderson R et al. (2017). Systematic review of the potential adverse effects of caffeine consumption in healthy adults, pregnant women, adolescents, and children. Food Chem Toxicol 109: 585–648.

Witt WP, Mandell KC, Wisk LE et al. (2015). Predictors of alcohol and tobacco use prior to and during pregnancy in the US: the role of maternal stressors. Arch Womens Ment Health 18: 523–537.

World Health Organization (2013). Guideline: calcium supplementation in pregnant woman, WHO.

Youngblood ME, Williamson R, Bell KN et al. (2013). 2012 Update on global prevention of folic acid-preventable spina bifida and anencephaly. Birth Defects Res A Clin Mol Teratol 97: 658–663.

Handbook of Clinical Neurology, Vol. 171 (3rd series)
Neurology and Pregnancy: Pathophysiology and Patient Care
E.A.P. Steegers, M.J. Cipolla and E.C. Miller, Editors
https://doi.org/10.1016/B978-0-444-64239-4.00008-4

Chapter 8

Perinatal pharmacology and safety profiles

KAREL ALLEGAERT[1,2,3]*

[1]*Department of Development and Regeneration, KU Leuven, Leuven, Belgium*

[2]*Department of Pharmaceutical and Pharmacological Sciences, KU Leuven, Leuven, Belgium*

[3]*Department of Clinical Pharmacy, Erasmus University Medical Center, Rotterdam, The Netherlands*

Abstract

Most medications are not adequately evaluated for use during pregnancy, delivery, or the postpartum period, and package inserts fail to provide clear instructions for use in these contexts, despite major concerns among health-care providers and the community as to how to practice evidence-based pharmacotherapy. Valproate fetopathy hereby serves as one of the more recent illustrations of the scope of the problem. At its best, evidence-based pharmacotherapy is driven by a balanced decision between disease-related risks (natural course of the disease) and any risks related to exposure to medications for mother, fetus, or infant. This chapter aims to describe the general patterns of changes in pharmacokinetics (absorption, distribution, metabolism, elimination) in pregnant women and postpartum, with specific emphasis on placental drug transport and additional focus on lactation. The relevance of these changes is illustrated by discussing medications commonly prescribed to treat neurologic conditions.

INTRODUCTION: PERINATAL PHARMACOLOGY REFLECTS PERINATAL PHYSIOLOGY

Clinical pharmacology intends to predict drug-related effects of medication and population-specific pharmacokinetics (PK, concentration time), and pharmacodynamics (PD, concentration effect), including aspects of intra- and interindividual variability. These basic principles of clinical pharmacology obviously also apply to young women of childbearing age, pregnant women, and the postpartum period, including during lactation (Dallmann et al., 2019). The main drivers in clinical pharmacology (PK, PD) are due to differences in physiology: *clinical pharmacology is applied physiology*, so these changes in physiology should be the main drivers of our prescription practices (Ito, 2016).

Human pregnancy induces profound anatomic and physiologic changes, affecting virtually all organ systems of the body. Many of these changes help to maintain and meet the complex demands of fetal development, and—related to this—the needed maternal physiologic adaptations, as the extensive metabolic demands of pregnancy require specific, reversible physiologic and anatomic adaptations. Examples include increases in tidal volume, increases in cardiac output and renal clearance capacity, changes in plasma volume and composition, and a reduction in uterine artery resistance. Along the same lines, a relevant weight gain occurs with a mirrored increase over time in BSA (Dallmann et al., 2017; Westrin et al., 2018b). However, pregnancy is not a dichotomous covariate and its related changes are not uniform: progressive change occurs over the consecutive trimesters of pregnancy, as shown in Fig. 8.1A–C. Both the overall extent in alterations and the trend over the three trimesters of pregnancy are hereby illustrated for (a) body composition and renal function, (b) blood components, and (c) cardiac output,

*Correspondence to: Karel Allegaert, M.D., Ph.D., Department of Development and Regeneration, Neonatal Intensive Care Unit, KU Leuven, Herestraat 49, 3000 Leuven, Belgium. Tel: +32-16-343210; +32-16-230432, Fax: +32-16-343211, E-mail: karel.allegaert@uzleuven.be

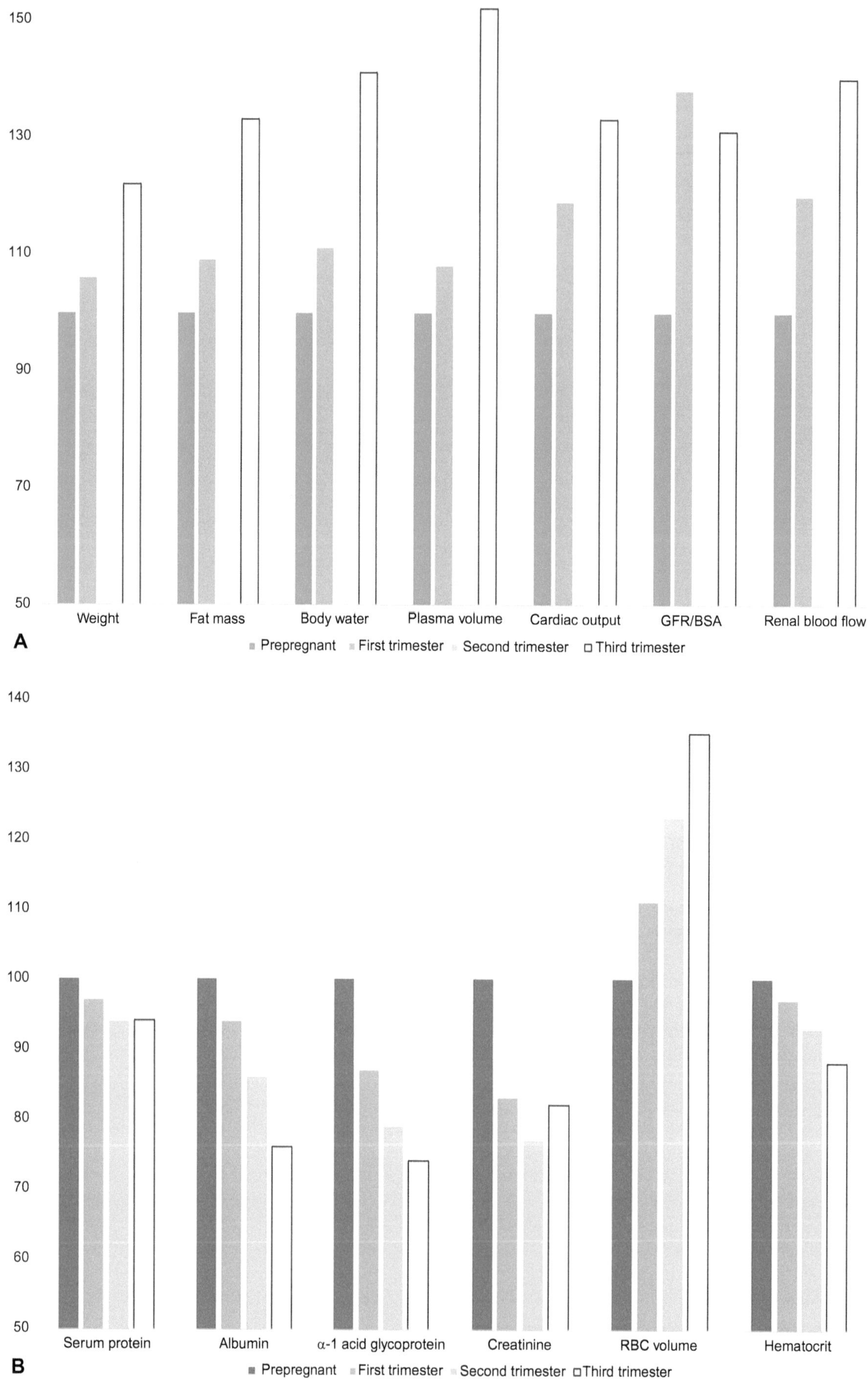
150
130
110
90
70
50
Weight
Fat mass
Body water
Plasma volume
Cardiac output
GFR/BSA
Renal blood flow
A
Prepregnant
First trimester
Second trimester
Third trimester
140
130
120
110
100
90
80
70
60
50
Serum protein
Albumin
α-1 acid glycoprotein
Creatinine
RBC volume
Hematocrit
B
Prepregnant
First trimester
Second trimester
Third trimester

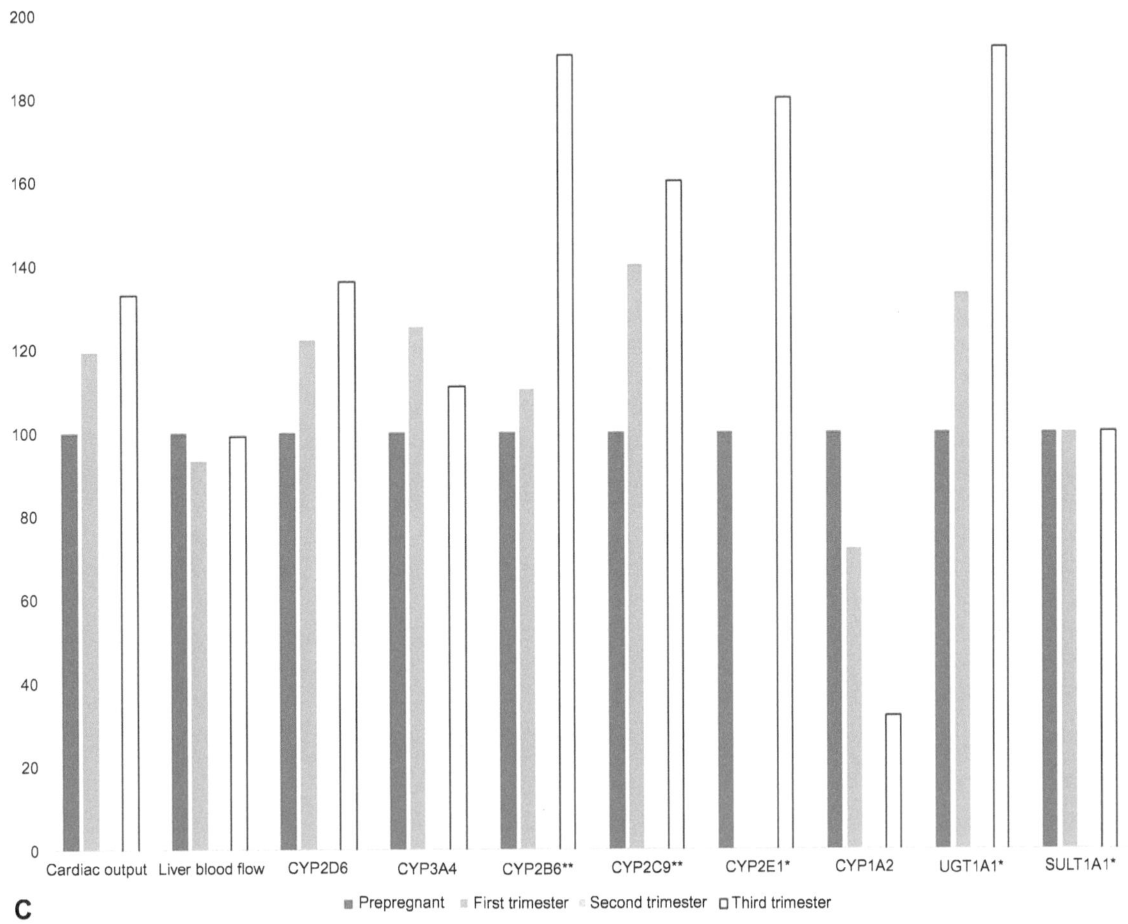

Fig. 8.1. Percentage changes in physiologic parameters across the consecutive trimesters of pregnancy compared to the prepregnancy setting (100%). Percent changes have been calculated at 12, 24, and 36 weeks, based on the individual trend lines as described by Abduljalil et al. to reflect the first, second, and third trimester (Abduljalil et al., 2012). (A) While changes in body weight, fat mass, or body water are rather progressive over the duration of pregnancy, there is an additional increase in plasma volume in the last trimester, while the increases in cardiac output and glomerular filtration rate (GFR, corrected for body surface area, BSA) are most pronounced in the first trimester of pregnancy. (B) Trends in concentrations of serum protein, albumin, alpha-1 acid glycoprotein, creatinine, and hematocrit all display a progressive decrease over the pregnancy, somewhat compensated for by the increases in plasma volume, already mentioned in (A), and red blood cell (RBC) volume. (C) Trends in liver blood flow do not mirror trends in cardiac output; most of the drug-relevant metabolizing enzymes (CYP = cytochrome P450; UGT = UDP-glucuronosyltransferases; SULT = sulfotransferases) display increased activity (CYP2D6, CYP3A4, CYP2B6, CYP2C9, CYP2E1, UGT1A1), while others remain stable (SULT1A1) throughout pregnancy, or even decrease (CYP1A2). The estimated changes in CYP2E1, SULT1A1, and UGT1A1 are taken from the work by Mian et al. (2020); those for CYP2B6 and CYP2C9 are from the research of Ke et al. (2014).

hepatic perfusion, and liver enzyme activities. These data were extracted as proportional changes and calculated at 12, 24, and 36 weeks using the trend lines described by Abduljalil et al. to reflect the first, second, and third trimester (Abduljalil et al., 2012). While the changes in body weight, fat mass, or body water are progressive over the duration of pregnancy, an additional increase in plasma volume occurs in the last trimester. Increases in cardiac output and glomerular filtration rate (GFR, corrected for BSA) are most pronounced in the first trimester of pregnancy. Concentrations of serum protein, albumin, alpha-1 acid glycoprotein, creatinine, and hematocrit all display a progressive decrease over pregnancy, somewhat compensated for by an increase in plasma volume (Fig. 8.1A) and red blood cell (RBC) volume. These trends (lower protein, higher volume) can result in lower concentrations during pregnancy simply by increased dilution and a higher distribution volume. Trends in liver blood flow do not mirror trends in cardiac output, as a relevant portion of the increased cardiac output serves the placental circulation. Most of the drug-relevant metabolizing enzymes (CYP = cytochrome P450; UGT = UDP-glucuronosyltransferases; SULT = sulfotransferases) display increased activity (CYP2D6, CYP3A4, CYP2B6, CYP2C9, CYP2E1, UGT1A1), but others remain stable (SULT1A1) throughout pregnancy, or even decrease (CYP1A2) (Abduljalil et al., 2012; Dallmann et al., 2017).

All these events contribute to the establishment of an efficient and rich transport of oxygen and nutrients to the placenta, with a subsequent transplacental clearance of elimination compounds such as CO_2, creatinine, ureum, bilirubin, or other endogenous or exogenous compounds via the blood flow from the fetus back to the mother, with subsequent maternal elimination. Other changes contribute to fetal protection, e.g., the establishment of maternal immune tolerance toward the fetus, the expression of xenobiotic efflux transporters in the placenta, or the production of amniotic fluid, serving, among other functions, to cushion external shocks against the abdomen of the mother. These changes may affect the pharmacodynamics of, e.g., immune modulating drugs. Furthermore, additional adaptations serve to prepare the mother for parturition, such as the increase in hormone receptors and hormones coordinating the onset and continuation of labor (estriol, prostaglandins, relaxin, and oxytocin), or for subsequent lactation, such as the rise in blood flow and volume of the breast tissue.

Fig. 8.1A–C reflects normal physiologic changes over pregnancy. However, pregnancies may be complicated, with their specific (patho)physiology: e.g., preeclampsia results in water retention and an additional increase in body weight (Gyselaers et al., 2018), impaired GFR (about −15%) when compared with the normal physiologic increase in GFR (Lopes van Balen et al., 2019), and an additional decrease in albumin (Kreepala et al., 2018). Furthermore, the "changes during pregnancy" should normalize somewhere in the postpartum period. Many of these physiologic changes return to normal, and for some this return can be gradual. For example, blood volume decreases rapidly in the first few days under the effect of diuresis and continues to decrease gradually over a period of 8 weeks (Bhatia and Chhabra, 2018). Labor increases cardiac output, along with its components (heart rate and stroke volume), to 80% above prelabor values, and cardiac output returns to normal within 6–8 weeks of pregnancy, with a temporary increase in the first postpartum days (Bhatia and Chhabra, 2018). The phenotypic UGT enzyme activity is largely driven by estradiol. To illustrate this, urine paracetamol glucuronide elimination in young women is affected by pregnancy (higher), the early postpartum period (lower), and exposure to oral contraceptives (higher), compared to healthy female volunteers not on oral contraceptives. This results in at least a twofold variability in paracetamol clearance in young women (Allegaert et al., 2015). Postpartum GFR normalizes to prepregnancy estimates within 4 weeks after delivery (Lopes van Balen et al., 2019). As uterine size decreases, the functional residual capacity of the lung also returns to normal within 1–2 weeks.

Many of the stated anatomic and physiologic changes during pregnancy lead to marked alteration in drug absorption, distribution, metabolism, and excretion (ADME), collectively referred to as pharmacokinetics (PK). Consequently, dosing should be based on the physiologic characteristics of the individual patient. Placental drug transport and potential fetal drug exposure add complexity to this situation (Bouazza et al., 2019). Various methods can be used to study whether, and to what extent, a drug crosses the placenta, such as cell cultures, organ-on-a-chip techniques, ex vivo experiments, or in silico models (Bouazza et al., 2019). The physiologic changes described make it challenging for clinicians to assess symptoms and signs, or to interpret laboratory and radiologic tests. Furthermore, these physiologic alterations can affect the PK and PD of medications, so perinatal pharmacotherapy and safety profiles relate to accurate *selection* and *dosing* of the best medications to treat the medical diagnosis of the mother or fetus with the aim of improving outcome, yet avoiding disproportional side effects for mother, fetus, or infant (Allegaert and van den Anker, 2019).

MATERNAL PHARMACOKINETICS

Many of the anatomic and physiologic changes observed during pregnancy result in marked ADME alterations, collectively referred to as PK. These changes in maternal physiology of relevance for maternal PK are further illustrated in Table 8.1.

Absorption can be altered as a result of pregnancy-induced changes in gastric emptying (possibly progesterone driven), gastric pH and acidity, patterns of intestinal motility, or absorptive surface area (although at present it is not clear whether this is caused by hypertrophy or hyperplasia of intestinal villi) (Dallmann et al., 2019). Although only a limited number of observations during pregnancy are available, it seems reasonable to state that the bioavailability of medications not undergoing first-pass metabolism is generally not significantly different in pregnant women (Koren and Pariente, 2018; Dallmann et al., 2019).

Distribution of medications during pregnancy can be influenced by changes in body and organ weight, regional blood flows, tissue composition, plasma composition and volume, tissue expression of drug influx and efflux transporters, and alterations in the unbound

Table 8.1

Summary of changes in the physiology affecting pharmacokinetics in the maternal and in the placental-fetal compartment, respectively

Maternal compartment		Effects of the placental-fetal compartment	
Absorption	- Decreased gastrointestinal motility → slower, but more complete absorption - Increased gastric pH → affects ionization (absorption) of weak acids and bases - Ventilatory changes: hyperventilation and increased pulmonary blood flow → favors alveolar uptake	Placental barrier function	P-glycoprotein, multidrug resistance proteins, breast cancer resistance protein are active transporters that work against a concentration gradient and protect the fetus against potential toxic effects of environmental toxins and other xenobiotics. Tacrolimus concentrations in the fetal compartment are 70% of the maternal concentrations because of this active transport mechanism
Distribution	- Plasma volume expansion by 50% → decrease in C_{max} - Mean increase in total body water is 8 L - More fat stores - Amniotic fluid may behave as a third space (deep compartment, swallowed by the fetus)	Acid–base equilibrium effect	- Nonionized highly lipid soluble molecules penetrate the placenta more quickly than less lipid soluble ionized molecules - Fetal plasma pH is slightly more acidic than the maternal → *Ion trapping*: weak bases in the maternal circulation will be nonionized and easily cross the placenta; in the fetal blood they become ionized, resulting in a net movement from maternal to fetal compartment
Protein binding	- Dilutional hypoalbuminemia → increased unbound drug fraction - Steroid and placental hormones occupy protein binding sites	Protein binding	- Maternal plasma albumin gradually decreases during pregnancy, while fetal albumin progressively increases → Since only unbound drugs cross the placenta, protein-bound drugs reach higher total concentrations in the fetus, without similar changes in the free concentration
Metabolism	- Hepatic drug metabolism alterations by increased secretion of estrogen and progesterone (induction of microsomal enzyme activity by progesterone, or competitive inhibition of microsomal oxidases by progesterone and oestradiol) → Drug metabolism can be increased, decreased, or remain stable	Feto-placental drug metabolism	- First-pass metabolic effect by placenta - Fetal liver metabolism: from 7 to 8 weeks postconception phase I (oxidation, dehydrogenation, reduction, hydrolysis) and phase II (glucuronidation, methylation and acetylation) enzymatic processes have been documented in the fetal liver; however, because of immaturity their contribution in drug elimination is marginal and most of the umbilical venous blood flow will bypass the liver via the ductus venosus to enter the fetal heart. The main route of elimination is by placental transfer back to the mother, as illustrated for paracetamol
Excretion	- Increased renal blood flow and glomerular filtration rate (by 50%) - Cholestatic effect of estrogen may interfere with clearance of drugs through the biliary system	Feto-placental elimination	- Elimination of drugs from the fetus primarily occurs by diffusion back to the maternal compartment - As most metabolites are more polar than their parent compounds, they are less likely to cross the placental barrier. This may result in metabolite accumulation in various fetal tissues - As pregnancy progresses, higher amounts of drugs are excreted into the amniotic fluid, reflecting maturation of the fetal kidney

fraction of a given medication: e.g., the increase in plasma volume (Fig. 8.1A) during pregnancy may result in an elevated distribution volume of hydrophilic medications, as has been observed for antibiotics such as beta-lactams or aminoglycosides, levetiracetam (Berlin et al., 2019), and metformin (Eyal et al., 2010). Additionally, the growth of tissues and organs uniquely present during pregnancy, such as the placenta, fetus, and amniotic fluid, may further increase the distribution volume for some medications. Secondary to declining plasma concentrations of albumin and α1-acidic glycoprotein (Fig. 8.1B), the unbound fraction of a given medication may increase, with possible critical implications, especially if a medication, such as phenytoin, displays high protein binding and has a narrow therapeutic index (Koren and Pariente, 2018; Westrin et al., 2018b). These compound-specific illustrations stress the relevance of measuring unbound concentrations instead of total concentrations in samples collected in pregnant women. Alternatively, one can aim to correct the total drug plasma concentration for changes in unbound fraction. However, one should cautiously consider both physiologic (pregnancy-related) and pathophysiologic (disease-driven, such as preeclampsia) changes (Lopes van Balen et al., 2019) in, e.g., albumin (Fig. 8.1B). The binding capacity may be further altered by competitive binding with free fatty acids, steroids, or placental hormones. The (free) concentration may also be altered due to the lower hematocrit levels during pregnancy (Fig. 8.1B). Here, again, at least the potential relevance of unbound drug concentrations in pregnant women must be considered before clinical decisions can be made. For example, tacrolimus whole blood concentrations were lower in a small cohort of pregnant patients, while unbound concentrations were not significantly different (Hebert et al., 2013; Hiramatsu et al., 2018). As recently reported, the trough (i.e., usual concentration reached just before the next dose is administered, but in this study 12 h after intake despite single daily dose) tacrolimus concentrations in mothers with systemic lupus erythematosus (SLE) decreased significantly in the second trimester compared with before-pregnancy values (paired data, median level 4–2.2 mg/L). The extent of the decrease in trough concentrations did not in itself lead to SLE deterioration, but tacrolimus *doses* were significantly lower in the second trimester and postpartum (both 3–3.5 vs 2–3 mg/day) in the group that subsequently deteriorated than those of the nondeteriorating group (Hiramatsu et al., 2018).

Clearance is driven by *metabolism* and *elimination* of metabolites or the mother compound. Clinically relevant changes in drug-metabolizing activity of the different isoenzymes have been reported (Fig. 8.1C). It is likely that some of these changes (either increased or decreased activity) are largely caused by altered levels of hormones during pregnancy. As mentioned earlier, UGT enzyme activity is largely driven by estradiol, as reflected in the variability in paracetamol glucuronide elimination in young women, related to pregnancy (higher), early postpartum (lower), or oral contraceptive use (higher) (Allegaert et al., 2015). Along the same lines, 17β-estradiol suppresses CYP1A1/2 gene expression via the aryl hydrocarbon receptor pathway (Quattrochi et al., 1994; Lai et al., 2004). Driven by the >250-fold increase in 17β-estradiol plasma concentrations during pregnancy, metabolism of CYP1A1/2-substrates, such as caffeine and theophylline, is considerably lower than that of nonpregnant subjects (Quattrochi et al., 1994; Lai et al., 2004). The subsequent sharp drop in 17β-estradiol serum concentrations shortly after delivery may be a further mechanistic link to the finding that caffeine clearance on the fourth day postpartum is already comparable to the reference population. Other drugs metabolized by this route are clozapine, olanzapine, fluvoxamine, ondansetron, and haloperidol. If metabolites are of pharmacodynamic relevance, decreased metabolism may also result in decreased efficacy, as described for dacarbazine (a prodrug) when used to treat Hodgkin's lymphoma during pregnancy (Kantrowitz-Gordon et al., 2018).

The expression and activity of many other CYP enzymes, such as CYP2A6, 3A4, 2B6, 2C9, or 2E1, are upregulated during pregnancy (Fig. 8.1C). As an obvious consequence, increases in the oral clearance of various CYP probe substrates are observed in pregnant women, including the CYP2A6 substrates (e.g., nicotine, valproate, efavirenz), CYP2D6 substrates (tramadol, metoclopramide, ondansetron, metoprolol, codeine, paroxetine, dextromethorphan), CYP3A substrates (clorazepate, midazolam), and CYP2B6 (cyclophosphamide, bupropion, methadone). Although a lot of information is available on the effect of pregnancy on many phase I enzymes, less information is available on phase II enzymes, UGT, SULT, and *N*-acetyltransferase (NAT) isoenzymes. The available data suggest that these enzymes remain stable or increase during pregnancy. Relevant drugs that undergo glucuronidation and display altered PK during pregnancy are, e.g., paracetamol, lamotrigine, and propofol. Lamotrigine clearance increases significantly from 5 weeks of gestational age onwards, driven by estradiol and gestational age. As changes occur even before women become aware of their pregnancy, planning and implementation of therapeutic drug monitoring (TDM) and early detection of pregnancy are important (Karanam et al., 2018, 30168175; Westrin et al., 2018b). Propofol clearance is significantly increased (39 vs 29 mLg/kg min, +34%) in young women undergoing cesarean delivery when compared to young women undergoing laparoscopy (Gin et al., 1990).

Although pregnancy is known to cause PK changes, little is known about its impact on serum levels of antipsychotics. Using clinical data (201 TDM samples in 103 women in 110 pregnancies and 512 samples in the same women before or after pregnancy), serum concentrations in the third trimester were significantly lower than baseline for quetiapine (−76%, CYP3A4) and aripiprazole (−52%, CYP2D6), but not for olanzapine (−9%, CYP2D6 and CYP1A2). For the remaining antipsychotics (perphenazine, haloperidol, ziprasidone, risperidone, and clozapine), the dataset suggests a decline for perphenazine (Westrin et al., 2018a). Using a similar study design with clinical data from 2 TDM services for antidepressants, 367 analyses obtained from a total of 290 pregnancies in 281 women and 420 baseline observations from the same women were retrieved. Serum (dose corrected) concentrations in the third trimester were significantly lower for paroxetine (−51%, CYP2D6 > CYP3A4), fluvoxamine (−56%, CYP1A2, CYP2C, CYP3A4), and citalopram (−24%, CYP2C19, 2D6 and 3A4), and higher for sertraline (+68%, CYPs, UGT, and monoamine oxidases), when compared to observations before or after pregnancy. In contrast, escitalopram (CYP2C19 and CYP3A4), fluoxetine (CYP2C19, CYP3A4/5, CYP2D6), and venlafaxine (CYP3A4, CYP2C19, CYP2D6) concentrations did not change significantly (Westrin et al., 2017).

Drug excretion may also be changed in pregnant women. For example, the GFR increases significantly during pregnancy, predominantly as a consequence of increased renal blood flow and ultimately single nephron glomerular hyperfiltration (Fig. 8.1A) (Abduljalil et al., 2012; Lopes van Balen et al., 2019). Furthermore, the activity of some transporters involved in tubular or biliary excretion of medications, such as P-glycoprotein (P-gp) and organic cationic transporter 2, is upregulated (Isoherranen and Thummel, 2013). Consequently, the clearance of drugs excreted unchanged by the renal route is typically higher in pregnant vs nonpregnant women, as for beta-lactam antibiotics, digoxin, and metformin. Levetiracetam and lithium are both almost exclusively driven by GFR elimination. Levetiracetam clearance displays a similar pattern to GFR changes, with the increase in clearance being most pronounced in the first trimester, with only minor additional changes in the second and third trimester of pregnancy (Berlin et al., 2019). In the 2017 study by Wesseloo et al., lithium blood levels (reflecting changes in clearance) decreased in the first trimester (−24%) to reach a nadir in the second trimester (−36%), and then increased somewhat in the third trimester (−21%). In contrast, there was a slight increase postpartum (+9%), while delivery itself was not associated with an acute change in lithium (Wesseloo et al., 2017). Table 8.2 provides some specific examples of PK changes during pregnancy and their practical consequences, relevant to medical conditions discussed in this volume.

Table 8.2

Specific examples of pharmacokinetic (PK) changes during pregnancy and their practical consequences, relevant to medical conditions discussed in this volume

Medicine	Medical condition	Pharmacokinetic changes	Clinical consequence
Metoprolol	Migraine	CYP2D6 activity, significant increase during pregnancy (+80%–180%)	Higher dose needed for similar effect, consider TDM
Methadone	Opioid dependence, pain	CYP3A4 activity, significant increase during pregnancy (57%, 76%, and 94% in the first, second, and third trimester)	Dose requirements may increase, maternal withdrawal possible with the same dose, consider TDM
Lamotrigine	Epilepsy, psychiatry	Increased glucuronidation and renal elimination, increased clearance (82%, 203%, and 197%) in the first, second, and third trimester	Complex adaptations, so that TDM is highly recommended to guide individual dosing throughout pregnancy
Lithium	Bipolar disorder	Raised glomerular filtration	To attain the same exposure needs to be tailored (+25%–50%), guided by TDM
Escitalopram	Depression	CYP2C19 decreased, CYP3A4 increased	No dose adjustment anticipated, clinical follow-up
Levetiracetam	Epilepsy	Raised glomerular filtration, +22%, 18%, and 31% in the first, second, and third trimester	Higher clearance is anticipated, adaptations can be made using TDM

TDM, therapeutic drug monitoring.

PLACENTAL TRANSFER AND FETAL SAFETY

The major function of the placenta is to *transfer* nutrients and oxygen from mother to fetus and to assist in the subsequent removal of waste products back to the mother. Consequently, the wording "placental barrier" holds a false notion, since the placenta is not a true barrier but instead an entry through which the fetus is exposed to compounds, as well as an exit to eliminate compounds. In addition, it plays an important role in the synthesis of hormones (progesterone, estrogens) and peptides (human chorionic gonadotropic, placental growth hormone, insulin growth factor) vital for a successful pregnancy, while its dysfunction has been associated with preeclampsia (Wu et al., 2019). Furthermore, the placenta has metabolic functions, including metabolic processes related to medications, as for some steroids. The "barrier" or "filter" function of the placenta includes passive diffusion, facilitated diffusion, active transport, and pinocytosis.

Passive diffusion: The amount and rate of transfer are primarily determined by the concentration gradient of the medication between the maternal and fetal circulation and placental blood flow. In addition, the physicochemical properties of drugs, such as lipid solubility, ionization constant (pK_a), molecular weight, and protein binding, are critical for placental transfer. Uncharged, low-molecular weight (<500 Da), lipid soluble and unbound compounds, such as paracetamol, can easily cross the human placenta. Additional factors are plasma protein binding (e.g., cefazolin) with lower albumin in the fetal to the maternal compartment and similar free concentrations in both compartments. For weak acid or basic drugs with a pK_a close to the plasma pH, differences in the fetal to the maternal compartment (fetal compartment slightly more acidic) may result in "ion trapping": weak bases are mainly nonionized in the mother and cross the placenta to become ionized in the acidic fetal compartment (local anesthetics) (Table 8.1).

Facilitated diffusion: A diffusion process that does not need energy but is driven by the presence of a carrier substance (such as sodium) in the placenta (ganciclovir, cephalosporins, some glucocorticoids).

Active transport: Substrate transporting proteins P-gp, multidrug resistance proteins, and breast cancer resistance protein are expressed in the fetus-derived epithelial cells making up the exchange border between the fetal and maternal blood compartment. These protein pumps are postulated to have evolved as a protective mechanism against the potential toxic effects of xenobiotics or toxins. To illustrate this, tacrolimus crosses the placenta to result in in utero exposure of approximately 71% of the maternal blood concentrations. The lower fetal blood concentrations are likely due to active efflux P-gp transport of tacrolimus from the fetus back to the maternal compartment (Zheng et al., 2013). Another illustration is the observation that P-gp polymorphisms in part explain the risk of developing neonatal withdrawal syndrome following maternal opioid use, since a less active pump mechanism results in more extensive fetal exposure and more pronounced withdrawal (Lewis et al., 2015).

Pinocytosis: An endocytosis-driven process, facilitating large molecule transfer, such as (endogenous) immunoglobulins or exogenous monoclonal antibodies. PEGylation of monoclonals, like certolizumab, reduces placental transfer (Tun and Lobo, 2015).

In essence, the majority of medications cross the placenta. The subsequent fetal exposure depends on the maternal dose, blood flow patterns, and the previously mentioned placental barrier patterns. The subsequent intended effects or unintended side effects (teratogenicity) relate to this exposure and the relevant fetal PD targets. However, medication-related fetal effects may be secondary to placental PD effects (such as placental prostaglandin synthesis inhibition). Intended fetal exposure (transplacental fetal therapy) may be used effectively for the purpose of lung maturation (dexamethasone, betamethasone), to treat fetal arrhythmia (flecainide, lanoxin) (Alsaied et al., 2017), but the majority of interest is focused on the prevention of side effects (teratogenicity).

Within the scope of this chapter, we mention aspects of teratogenicity related to antiepileptic drugs (AEDs), opioids, and antidepressants. In a recent meta-analysis on AEDs ($n = 96$ studies, 58,461 cases), the higher risk (odds ratio) for major congenital malformations for ethosuximide (3.04), valproate (2.93), topiramate (1.9), phenobarbital (1.83), phenytoin (1.67), and carbamazepine (1.37) was quantified, while lamotrigine and levetiracetam were not associated with an increased risk (Veroniki et al., 2017a). Using a similar methodology ($n = 11$ studies, 933 children), valproate as monopharmacy, or part of polypharmacy, was associated with the highest odds (7.4) for adverse neurodevelopmental outcome, while oxcarbazepine and lamotrigine were associated with an increased risk (albeit with very broad estimates) for autism (Veroniki et al., 2017b). Prenatal opioid exposure has been associated with neonatal withdrawal syndrome, but also with impaired neurocognitive and behavioral outcome (IQ, language, aggression, fear, anxiety, executive functions, visual-motor functions) during childhood (Conradt et al., 2019). For serotonin selective reuptake inhibitors (SSRIs), a recent analysis of more than 9 million births suggested a statistically significant, but small, increased risk of congenital malformations of SSRIs (major congenital malformations 1.11; congenital heart defects 1.24). For all these classes of medications, it is very difficult to fully disentangle the exposure to a given medication from impact related to the indication and its associated environmental or genetic factors, which likely further affect the phenotype and outcome. Fetal exposure can be predicted and modeled, using advanced PK tools, as discussed in the last section of the chapter (Bouazza et al., 2019).

LACTATION AND NEWBORN SAFETY

Breastfeeding is the reference nutrition for newborns and infants. Any health-care provider should be aware of the health impact of lactation on both mother and infant and the subsequent general strategies applied to improve initiation and continuation rates of lactation, commonly referred to as the Baby-Friendly Hospital Initiative (BFHI) (Saadeh, 2012). The 10 steps for successful lactation support women in achieving their lactation intentions and provide guidance in instructing health-care providers to support lactation (Table 8.3) (Saadeh, 2012; Gubler et al., 2013).

However, lactation commonly cooccurs with maternal medication intake. Consequently, the main goal of maternal medication use during lactation is a dual one. The first aim is to provide effective and safe therapy for the maternal condition(s) for which the medication has been prescribed (e.g., postcesarean analgesia, maternal comorbidities, pregnancy-related diseases). The second, simultaneous, aim is the protection of the nursing infant from any (relevant) adverse event related to the maternal treatment. Only rarely do these goals overlap (e.g., maternal intake of galactagogues such as domperidone or metoclopramide, to stimulate or increase milk production) (Forinash et al., 2012; Berlin Jr and van den Anker, 2013).

Table 8.3

The 10 steps to successful lactation (Baby-Friendly Hospital Initiative, BHFI)

- Have a written lactation policy that is routinely communicated to all health-care staff
- Train all health-care staff in skills necessary to implement this policy
- Inform all pregnant women of the benefits and management of lactation
- Help mothers initiate lactation within a half-hour of birth
- Show mothers how to breastfeed and how to maintain lactation, even if they are separated from their infants
- Give newborn infants no food or drink other than human milk unless medically indicated
- Practice rooming-in—allow mothers and infants to remain together 24 h a day
- Encourage lactation on demand
- Give no artificial pacifiers or teats (also called dummies or soothers) to breastfeeding infants
- Foster the establishment of breastfeeding support groups and refer mothers to them on discharge from the clinic or hospital

http://www.tensteps.org/breastfeeding-bfhi-training-materials.shtml.

Despite the fact that lactating women are regular users of medications and that women are often advised to discontinue or stop lactating while taking a medication, only a limited number of medications have been identified as potentially harmful to the newborn. In a cohort of 838 lactating infants with mothers taking medications, Ito et al. quantified the adverse reaction incidence at 11.2% (Ito et al., 1993). The reported adverse reactions were diarrhea (antibiotics), drowsiness (analgesics, sedatives, antidepressants, antiepileptics), and irritability (antihistamines), and all were classified as minor (Ito et al., 1993). Antibiotics (19.3%), analgesics and narcotics (11.2%), antihistamines (9.4%), sedatives, antidepressants, and AEDs (7.1%) were most commonly associated with adverse reactions (Ito et al., 1993).

Health-care providers need to be aware that interruption of lactation is not without its own side effects and risks ("nocebo" effect), as lactation provides many health benefits for mother and infants. It is a misconception that the "*when in doubt, do not provide breastfeeding*" approach has no negative effects (Saadeh, 2012; Veiby et al., 2015; Ito, 2019). The overarching intention of maternal intake of medications during nursing should fulfill two criteria: (i) provide effective and safe medications for the mother while still (ii) assuring safety of nursing newborns from adverse events related to these medications following lactation-related exposure, as well as adverse events related to lactation interruption (Saadeh, 2012; Veiby et al., 2015; Ito, 2019).

Evidence specific to medication exposure to opioids and AEDs from lactation is available. Overall, a significant reduction was seen in the incidence (number needed to treat 5–6) and severity of neonatal abstinence syndrome during lactation (Lefevere and Allegaert, 2015). Along the same lines, the impact of lactation on neurocognitive outcome following AED exposure during pregnancy and lactation was documented by the Neurodevelopmental Effects of AEDs (NEAD) group. Children of mothers on AEDs had a dose-dependent reduction in IQ (−6 to −9 IQ) following fetal valproate exposure when compared to other AEDs, and a higher (11.5% vs 4.8%) risk of impaired fine motor skills compared with controls at 6 years. Subsequent lactation in infants of women using AEDs was associated with *improved* neurodevelopmental outcome compared with those who were fed with formula, and this was "dose" dependent (breastfeeding >6 months, or not) (Meador et al., 2014). Overall, the adjusted IQ was higher (+4 points) in children at the age of 6 years who were breastfed as infants, compared to those who were not. Interestingly, the IQ catch-up was most pronounced in valproate-exposed children (+12, range 1–24) (Meador et al., 2014). As valproate levels in human milk are low and infant serum levels range from undetectable to low, valproate intake is not necessarily a reason for the mother to discontinue lactation. Breastfeeding during valproate monotherapy does not appear to adversely affect infant growth or development, and breastfed infants had higher IQs and better verbal abilities than formula-fed infants at 6 years of age (Meador et al., 2014; Veiby et al., 2015).

A theoretical framework for exploring the extent of neonatal medication exposure through lactation following maternal medication intake is provided in Fig. 8.2. In essence, exposure of a mother providing lactation to any dose of a medication (D_{mother})—even when very limited—results in transfer of the medication into the human milk (D_{infant}). However, concentrations reached in human milk are usually quite low and the oral bioavailability to the infant should also be considered before any relevant pharmacologic exposure (RID = relative infant dose, $D_{mother}/D_{infant} \times$ absorption, fraction) and subsequent effect on the newborn or infant can be anticipated. The difference in exposure in the infant's plasma (* vs **) can be explained by preexisting exposure in the setting of fetal exposure, followed by neonatal exposure (the ** scenario).

The available data suggest that breastfeeding rarely needs to be discouraged, discontinued, or interrupted when the mother needs medications, but some caution may be warranted with analgo-sedatives (opioids, benzodiazepines) (Van den Anker, 2012; Berlin Jr and van den Anker, 2013). In contrast, radioactive-labeled diagnostics, lithium, iodine, gold, and ergotamine alkaloids are high-risk drugs, considered incompatible with lactation. Data on opioids, benzodiazepines, and cannabinoids are discussed here because of their topical relevance. Biologic therapies, including monoclonal antibodies, are more commonly prescribed to treat autoimmune and neoplastic diseases. Although these drugs have a high molecular weight and—related to this—likely a low RID and poor oral bioavailability, absorption by an Fc-receptor expressed on intestinal cells may occur.

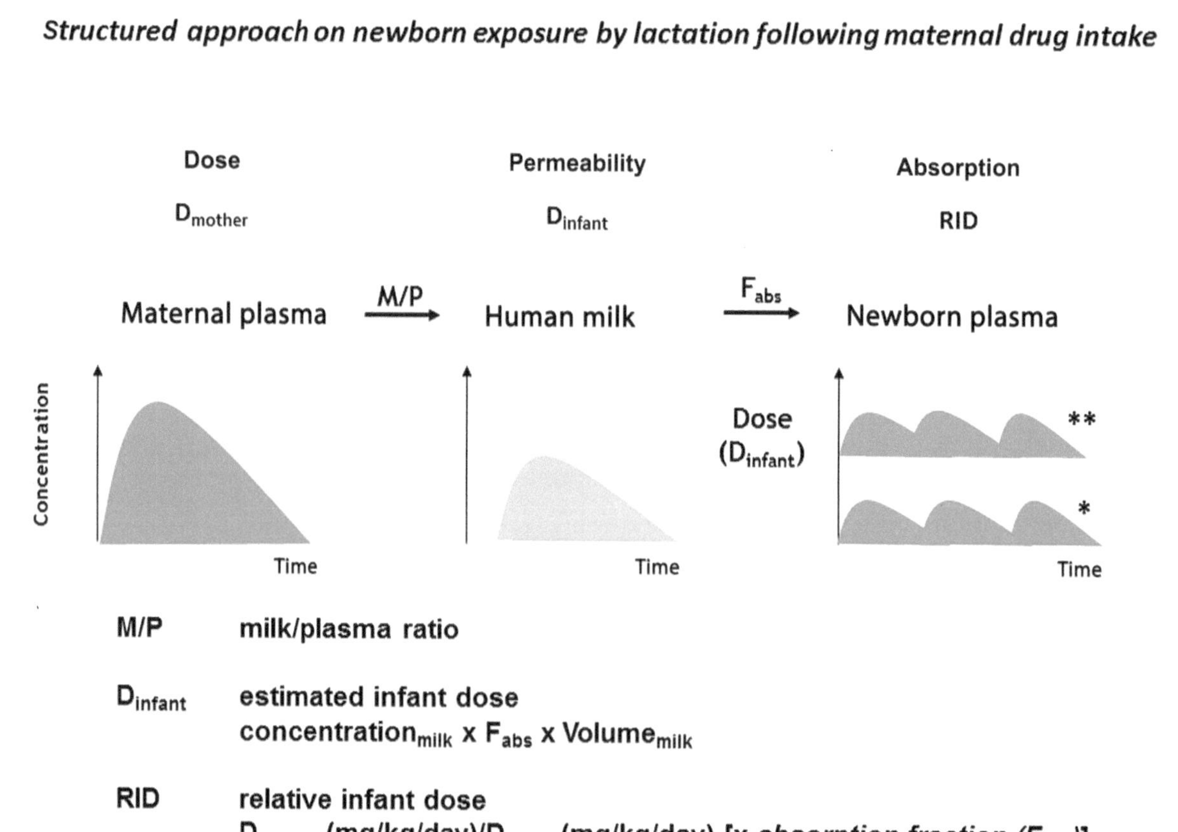

Fig. 8.2. A theoretical framework to explore the extent of neonatal medication exposure through lactation following maternal medication intake.

The potential relevance is discussed, using disease-modifying drugs for multiple sclerosis, later in this section.

Using the framework of Fig. 8.2, lithium during lactation remains a relevant risk factor for the newborn. In a recent systematic review of clinical lactation studies, the mean maternal lithium dose was 904 (293) mg/day, maternal lithium plasma/serum concentration was 0.73 (0.26) mEq/L, and breast milk concentration was 0.84 (0.14) mEq/L. Mean infant lithium plasma/serum concentration was 0.23 (0.26) mEq/L, and 20% of exposed infants showed a transient adverse event (i.e., acute toxicity or thyroid alterations), and all were also prenatally exposed to lithium monotherapy or polytherapy (Imaz et al., 2019). Similar issues exist with opioids, especially when CYP2D6 is involved in their metabolism (codeine, oxycodone, hydromorphone > tramadol), since this isoenzyme is well known for its polymorphisms. Oral absorption of opioids in neonates should be anticipated, while the extent of exposure through mother's milk depends on maternal ingestion (dose) and metabolism. As mentioned earlier, a significant reduction is seen in the incidence (number needed to treat 5–6) and severity of neonatal abstinence syndrome during lactation (Lefevere and Allegaert, 2015). This also provides evidence that there is indeed a relevant transfer with possible effects. The incidence of central nervous system depression in neonates on lactation following maternal exposure to oxycodone, codeine, or paracetamol was retrospectively compared in 533 mother–infant pairs. Lam et al. hereby clearly showed a 20.1% rate of depression in infants of nursing mothers on oxycodone, compared with 16.7% and 0.5% when exposed to codeine or paracetamol by lactation (Lam et al., 2012). Finally, using a sparse sampling study design to assess transfer of tramadol and *O*-desmethyl tramadol in human milk, the relative infant dose of 2%–3% remained very limited. From these observations, it seems reasonable to conclude that short-term maternal use of tramadol is compatible with lactation (Salman et al., 2011; Palmer et al., 2018).

It remains difficult to translate the available reports and data to clinical guidelines, but some guidance is provided here. First, it is reasonable to screen for sedation following maternal exposure to codeine, oxycodone, methadone, or morphine. Clinically quite useful, a high concordance is shown between maternal and neonatal somnolence. When the mother exhibits somnolence, the baby should be examined. This is likely due to the impact of genetic polymorphisms and dosage on the individual maternal–infant-related effects and side effects of these drugs. Severe somnolence commonly emerges after several days of exposure (claimed to be pivotal at 4 days). It is therefore suggested that maternal opioid exposure beyond 72 h warrants specific clinical evaluation, with emphasis on signs of sedation in newborn and mother (Lam et al., 2012; Van den Anker, 2012; Palmer et al., 2018).

Regarding benzodiazepines, lorazepam, midazolam, and diazepam are commonly administered as anxiolytics or sedatives. The mother compounds and some metabolites have been quantified in human milk, and concentrations were low (Cole and Hailey, 1975; Nitsun et al., 2006). In 24 h human milk collection after a single dose in a preanesthesia setting, only 0.005% of the maternal midazolam dose was retrieved. Taking the subsequent oral bioavailability (50%–60%) into account, it is reasonable to assume the exposure will be limited. In contrast, plasma diazepam and its active metabolite (desmethyldiazepam) was measured in neonates up to 7–10 days of postnatal age after single dose administration to the mother during labor (Cole and Hailey, 1975; Nitsun et al., 2006). This is due to its much longer elimination half-life, and therefore short-acting benzodiazepines such as midazolam are more appropriate, when needed. Finally, and supported by the previously mentioned NEAD group, most AEDs are considered safe or moderately safe during breastfeeding, with some medications (barbiturates, benzodiazepines, lamotrigine, ethosuximе) having the potential to reach significant serum levels in breastfed infants. Current available evidence suggests mothers with epilepsy should be supported and encouraged to breastfeed, provided the infant is carefully monitored. For gabapentin (neuropathic pain, restless leg syndrome), the available information suggests that maternal daily doses of gabapentin up to 2.1 g produce relatively low levels in human milk (range 1.2–8.7 mg/L, 1.3%–3.8% of the maternal weight-adjusted dosage) and infant serum (7.7% of the maternal plasma concentration). The infant should be monitored for drowsiness, adequate weight gain, and developmental milestones, especially in younger, exclusively breastfed infants and when using combinations of anticonvulsant or psychotropic drugs (Ohman et al., 2005).

Using a case–control design (to antibiotics) to assess the safety of chronic use of psychotropics in 280 mothers who breastfed, Kronenfeld et al. observed that chronic use of psychotropic monotherapy during lactation was associated with normal growth and gross motor development, by milestone achievements reported by parents. Sleepiness was reported, but it seemed self-limited with no developmental effect (Kronenfeld et al., 2018). The most common diagnoses for psychotropic use were depression (59.6%) and anxiety (33.9%). The majority (68.9%) of women were exposed to SSRIs, and smaller groups to benzodiazepines (13.2%), serotonin–norepinephrine reuptake inhibitors (8.2%), tricyclic antidepressants (3.9%), first-generation antipsychotics (2.1%), second-generation antipsychotics (1.8%), other antipsychotics (0.7%), and norepinephrine reuptake inhibitors (0.7%) (Kronenfeld et al., 2018). Using the currently available evidence, Fischer Fumeaux et al. recently concluded that depressed SSRI-treated pregnant women and their infants likely have a minor increased risk (small absolute risk) for several complications. The interpretation of these findings remains quite difficult, as the available methodology does not allow disentangling of the potential effects of SSRIs from those of the disease itself, or its unmeasured associated risk factors. Thus in lactating women requiring a pharmacologic treatment, SSRIs can still be considered appropriate when effective, as abundant data support their relative safety (Fischer Fumeaux et al., 2019).

Because of the important increase in cannabis and marijuana consumption, as recreational drug, dependent substance, or medication for specific indications in a diversity of formulations and practices, the lactation-related issues must be considered. Given the paucity of evidence, it is currently recommended to avoid using marijuana while pregnant or when breastfeeding (Thompson et al., 2019). Maternal outcomes related to cannabis consumption included panic attacks, delayed response time, increased heart rate, reduced short-term memory, dizziness, and impaired motor performance. Infant outcomes associated with maternal cannabis use and lactation were reduced muscular tone, poor sucking, and growth delay and restriction (Brown et al., 2018). Likely because of the physicochemical properties of the cannabinoid metabolites, such as Δ-9-tetrahydrocannabinol (△9-THC), 11-hydroxy-Δ-9-tetrahydrocannabinol, cannabidiol, or cannabinol, △9-THC was measurable in the majority (63%) of milk samples up to 6 days after maternal marijuana use. This means that exposure by lactation can be prolonged and delayed, so symptoms or events in the infants can be disconnected in time from maternal use (Bertrand et al., 2018).

Multiple sclerosis and lactation also warrants focused assessment, since postpartum is a time of risk for increased disease activity. The best predictor of postpartum relapse is a higher annualized relapse rate in the 2 years prior to pregnancy (Tisovic and Amezcua, 2019). A lower risk of relapse in the 6 months postpartum was seen in women who breastfeed, but no clear evidence exists that lactation provides sufficient protection from relapses to warrant not restarting disease-modifying drugs shortly after delivery in women with active relapsing disease prior to pregnancy (Voskuhl and Momtazee, 2017). Glatiramer acetate is generally considered safe and appears to be one of the preferred disease-modifying agents for treating multiple sclerosis during lactation. The levels of interferon beta-1a in breastmilk are minuscule, and it is considered to be "moderately safe" to use during lactation. Natalizumab is excreted into human milk (RID 2%–5%) in at least some women. The time of the peak level concentrations in human milk varies, but they can be delayed to as long as 6 months. It is therefore recommended that natalizumab be used with caution during breastfeeding, especially while nursing a newborn or preterm infant. Fingolimod and its active metabolite are highly bound in maternal plasma and unlikely to reach the breastmilk in large amounts. However, it is potentially toxic to the breastfed infant because, since this medication is provided by the oral route, subsequent absorption in the infant is possible. Rituximab (molecular weight of 143,860) is probably destroyed in the infant's gastrointestinal tract. If rituximab is required by the mother, it is not a reason to discontinue lactation. There is no clear guidance yet on teriflunomide or dimethyl fumarate. The same holds true for daclizumab or alemtuzumab, but both are large proteins (144 and 145 kDa, respectively), so absorption by the infant is unlikely (Voskuhl and Momtazee, 2017; Tisovic and Amezcua, 2019).

HOW TO IMPROVE THE CURRENT SETTING

It has been reported that more than 80% of pregnant women in Western countries take at least one medication (either prescribed or over-the-counter) (Ceulemans et al., 2019). Yet, little is known about possible and anticipated changes in the PK and PD of medications, because pregnant women are usually excluded from participation in clinical trials, while drug development programs specific to pregnant women or fetuses are very limited (Van Calsteren et al., 2016; Ceulemans et al., 2019). This knowledge gap has been clearly illustrated in a systematic review of PK studies as available in pregnant cohorts compared to nonpregnant cohorts. From 198 studies retrieved, data on 121 different medications were described, including studies for medications relevant to the theme of this book: AEDs (27 studies), cardiovascular medications (17 studies), anesthetics and analgesics (14 studies), antithrombotic agents (8 studies), antidepressants (8 studies), supplements and antiemetics (7 studies), addiction management (5 studies), immunomodulators + antivirals (4 studies), and antipsychotics (2 studies). When considering quality aspects, a number of inconsistencies were seen for several medications, perhaps because of the relatively small sample sizes or uncertainties related to trimester-specific observations. Furthermore, the links between PK changes and clinical outcome were only rarely reported, except in the case of antiretrovirals, antimalarials, and AEDs (Koren and Pariente, 2018).

A similar systematic review on the quality of lactation studies specific to AEDs was also reported (Van der Meer et al., 2015). Quality assessment was based on the guidelines of the Food and Drug Administration (FDA) and the International Lactation Consultant Association. Except for one study on lamotrigine and one case report on gabapentin, no study on AEDs correctly assessed both the absolute infant dose (AID) and milk to plasma (M/P) ratio (Fig. 8.2). Only one study each was found for carbamazepine, phenytoin, and vigabatrin that correctly assessed the AID. The main cause of this low number was the lack of essential details in the published studies, since 25 of 62 studies were case reports, letters, or abstracts. Other major shortcomings included the lack of information on sampling methods (validity of the analysis), the number of samples in a particular dose interval, and the low number of study participants (Van der Meer et al., 2015).

In the absence of robust knowledge and guidance, dosing regimens are often kept the same between pregnant and nonpregnant women in the clinical setting; guidance on lactation is also very heterogeneous and inconsistent. While such an extrapolation may be adequate for some medications with broad target and safety ranges, for other medications the risk exists of incorrect dosing, with potentially subtherapeutic or toxic drug effects for the mother, the fetus, or, during lactation, the infant (Allegaert and van den Anker, 2019).

Knowledge implementation: Translation of knowledge into practice and guidance

Knowledge must be translated and converted into practice to positively impact and improve the health-care of these patients. Along the same lines as the BHFI initiative, specific guidelines can be developed for specific settings. We retrieved guidelines or recommendations covering the broad spectrum of family planning and fetal and maternal prevention during pregnancy and/or lactation for, e.g., multiple sclerosis (Borisow et al., 2012) and epilepsy (Harden et al., 2009). Guidelines for lactation after anesthesia are provided as an illustration, since these guidelines nicely reflect the need for a structured, organized approach and open discussion with the patient (Table 8.4). Besides general aspects related to an overall positive approach to lactation, aiming to limit family disruption and improve the informed consent process and respect for patient preferences, some aspects are also provided related to specific analgesics, mainly the type of opioids metabolized by CYP2D6 (oxycodone, hydromorphone > tramadol) and the duration of exposure (<72 h) (Montgomery et al., 2012; Cobb et al., 2015).

Other initiatives related to knowledge translation and access are the LactMed database and the recently updated labeling process by the FDA, United States. LactMed is a free, easily accessible online database with information on medications and lactation and is one of the newest additions to the National Library of Medicine's TOXNET structure. Such an online tool facilitates updates

Table 8.4

Guidelines on analgesia and anesthesia for the lactating mother (Montgomery et al., 2012; Cobb et al., 2015)

1. Women may breastfeed as normal following surgery
2. Clinicians should be familiar with the relevant pharmacology of the drugs they use. Anesthetic and analgesic drugs are transferred to breast milk in only small amounts. A small number of drugs should be used with caution, or avoided completely, because of possible cumulative effects of high doses, lack of compatibility evidence, or idiosyncratic reactions
3. Women should be advised that expressing and discarding (pumping and dumping) of breastmilk after anesthesia is not necessary. If the mother prefers that her child not be exposed to any medication, prior expression and storage of milk may be possible. If a woman chooses to express and discard after drug exposure, evidence-based material should be provided on the elimination time of the drugs, so that she can resume normal breastfeeding as soon as possible
4. When surgery is unplanned, any mother with a child <2 years should be assumed to be breastfeeding unless there is information to the contrary
5. Local and regional anesthesia interfere the least with the mother's ability to care for her child. Sedation for surgical procedures involves less exposure to drugs than general anesthesia
6. Where possible, day surgery is preferable to avoid disrupting normal routines
7. General anesthesia is associated with sleep disruption. Women having day surgery must have a responsible adult stay with them for the first 24 h. She should be cautious with cosleeping and sleeping while feeding the child in a chair [27940804]
8. Codeine should not be given to breastfeeding women due to genetically mediated differences in metabolism. If codeine is taken, observe the child for drowsiness. It takes 15 h for codeine to leave the maternal system, and therefore the breast milk, completely
9. Breastfeeding support should be available for lactating mothers who are inpatients
10. Patient information leaflets and additional resources should be produced on the compatibility of anesthetic agents, antibiotics, and analgesics during lactation. These should be made available in preoperative assessment clinics and also on wards, as required

when new information becomes available. The date of the latest update is provided, and documents are freely available as open access. Along the same lines, the FDA's Pregnancy and Lactation Labeling Final Rule now requires a narrative description of the available evidence on medication use during pregnancy. This replaces the former FDA medication classification system, which applied a category A, B, C, D, X, or N for every medication and required a large amount of high-quality data before a medication could be defined as pregnancy category A (i.e., safe for the fetus). In the revised Final Rule, narrative descriptions of the evidence available for use of the medication in pregnancy are provided in three consecutive sections: "Pregnancy," "Lactation," and "Females and Males of Reproductive Potential." The "Pregnancy" section provides information on the use of the drug in pregnant women, including dosing and potential fetal risks. The "Lactation" section provides information on the use of the drug during breastfeeding, such as the amount of the drug in human milk and any possible effects on the nursing infant. This Pregnancy and Lactation Labeling Final Rule became effective on June 30, 2015 (Roca et al., 2019).

Another example of knowledge implementation is provided by the recommendations from the Pharmacovigilance Risk Assessment Committee (PRAC), the Safety Committee of the European Medicines Agency, and implemented by the national authorities, on valproate use in young women of childbearing age (Table 8.5). The intention of these recommendations is to avoid any unnecessary fetal exposure by using a stepwise approach (Is this drug really needed?) and by interaction with all stakeholders involved (patient, prescriber, pharmacist, manufacturer, and authorities).

Table 8.5

Summary of the recommendations of the Pharmacovigilance Risk Assessment Committee (PRAC, Safety Committee of the European Medicines Agency) for women with epilepsy, with emphasis on valproate

- In pregnancy—valproate must not be used. However, it is recognized that for some women with epilepsy it may not be possible to stop valproate and they may have to continue treatment (with appropriate specialist care) in pregnancy
- In female patients from the time they become able to have children, valproate must not be used unless the conditions of the new pregnancy prevention program are met
- The Pharmacovigilance Risk Assessment Committee has also recommended that the outer packaging of all valproate medicines must include a visual warning about the risks in pregnancy. In addition to boxed text, this may include a symbol/pictogram, with the details to be adapted at the national level
- A patient reminder card will also be attached to the outer package for pharmacists to discuss with the patient each time the medicine is dispensed
- Companies that market valproate should also provide updated educational materials in the form of guides for health-care professionals and patients

Knowledge generation: Structured data collection and integration

As mentioned earlier, the amount of available data is limited, and quality assessment has raised issues with both maternal PK and lactation studies (Van der Meer et al., 2015; Koren and Pariente, 2018; Ito, 2019). The sharp contrast between extensive drug use and the scarce knowledge on drug safety and efficacy in pregnant women is remarkable. In an attempt to improve this situation, several initiatives, consortia, and collaborations have been initiated to structure and generate knowledge on medication use during pregnancy and postpartum, including lactation. As illustration (and not a complete list), some of these initiatives focus on specific diseases or conditions (infectious diseases, posttransplant, antidepressants), while others focus on, e.g., interventions during pregnancy (prenatal lung maturation) on lactation ("Drugs in Lactation" Analysis Consortium), or aim to develop a research structure needed to generate data (Obstetric Pharmacology Research Unit Network) or research tools (IMI Conception) (Table 8.6). As an additional reflection of this commitment, the FDA recently published a draft guidance titled *Pregnant Women: Scientific and Ethical Considerations for Inclusion in Clinical Trials* (https://www.fda.gov/media/112195/download) that discusses both ethical and scientific issues to be addressed when pregnant or lactating women are being considered for inclusion in drug development clinical trials (Roca et al., 2019). Taken together, these efforts clearly illustrate the intention to promote maternal/fetal health and better inform prescribing decisions during pregnancy and lactation.

Pharmacometric tools are crucial to exploring changes in drug PK during pregnancy and lactation, and they have become instrumental in improving perinatal pharmacology (Bouazza et al., 2019; Dallmann et al., 2019). These tools encompass different mathematical models used to gain a better understanding of PK. Both population PK and physiology PK (PB-PK) modeling and simulation are now established tools, at the level of acceptance by regulatory authorities. Initiatives to stimulate drug development in pregnant women should therefore be accompanied and mirrored by development of valid research tools. However, to realize the field's full potential, data and model sharing, multidisciplinary

Table 8.6

Examples of projects on medicines during pregnancy and the postpartum period, including lactation (alphabetical)

Project name	Project description
ConcePTION: Continuum of Evidence from Pregnancy Exposures, Reproductive Toxicology and Breastfeeding to Improve Outcomes Now	Project within the EU-funded Innovative Medicines Initiative (IMI), which addresses the unmet need for a science- and data-driven approach to define standards for generating data on medicines used during pregnancy and breastfeeding. It aims to change current practices by providing improved tools and methods to generate more valuable, reliable, and timely information to health-care professionals and pregnant and lactating women to enhance optimal care. Further information available at https://www.imi.europa.eu/sites/default/files/uploads/documents/apply-for-funding/call-documents/imi1/DraftTopic2017_PregnancyBreastfeeding.pdf
Co-OPT: The Consortium for the study of Pregnancy Treatments project	Much of the research on the effects of antenatal corticosteroids (ACSs) has focused on outcomes in babies born preterm, while outcomes in babies born at term have been largely ignored. Co-OPT plans to address this. The aim is to describe how antenatal corticosteroid treatment is used across a variety of settings and to determine the short- and long-term outcomes of ACSs, particularly focusing on outcomes when ACSs were given unnecessarily (did not deliver preterm) and/or were inappropriately timed
DLAC: "Drugs in Lactation" Analysis Consortium	Monitoring network designed as a platform for efficient collection of patient milk samples in a real-world setting to generate population predictions of drug excretion levels into human milk. It aims to establish a postmarketing "drug-in-milk" monitoring system for nursing women who are prescribed and using medications. Registered at ClinicalTrials.gov (NCT02755194)
IMPAACT: International Maternal Pediatric Adolescent AIDS Clinical Trials network	Global collaboration of investigators, institutions, community representatives, and other partners organized for the purpose of evaluating interventions to treat and prevent HIV infection and its consequences in infants, children, adolescents, and pregnant/postpartum women through the conduct of high-quality clinical trials. https://www.impaactnetwork.org/index.htm
MONEAD: The Maternal Outcomes and Neurodevelopmental Effects of Antiepileptic Drugs study	Ongoing initiative funded by the National Institute of Neurological Disorders and Stroke (NINDS) and the Eunice Kennedy Shriver National Institute of Child Health and Human Development (NICHD). One of the key questions explored is whether the medicines the mothers-to-be must take to control seizures have a negative, lasting impact on their babies' developing brains. Further information available at https://web.emmes.com/study/monead/neadfind/neadfind.htm
NPRAD: National Pregnancy Registry for Antidepressants	Postmarketing pregnancy registry dedicated to evaluating the safety of antidepressant medications that may be taken by women during pregnancy to treat a wide range of mood, anxiety, or psychiatric disorders. The goal of this Registry is to gather information on the safety of these medications during pregnancy for both mother and child. https://womensmentalhealth.org/research/pregnancyregistry/antidepressants/

Continued

Table 8.6

Continued

Project name	Project description
Obstetric Pharmacology Research Unit Network: Eunice Kennedy Shriver National Institute of Child Health and Human Development (NICHD)	Network that aims to support specialized research to improve safety and efficacy of medication use during pregnancy and while breastfeeding. These projects include basic/translational research involving cells and/or animals as well as clinical studies involving humans, and will better enable clinicians to protect the health of pregnant women, while improving birth outcomes and reducing harm to infants. Further information available at www.utmb.edu/nichd-oprc
PANNA: Pharmacokinetics of newly developed ANtiretroviral Agents in HIV-infected pregNAnt women	Network of European centers for collecting pharmacokinetic data of newly developed antiviral compounds in European pregnant women. http://www.pannastudy.com/network
PROTECT: Pregnancy Study	Completed study designed to pilot direct-to-patient data collection methods for use in postmarketing surveillance of the fetal effects of maternal medication use in pregnancy. Further information available at http://www.imi-protect.eu/documents/Dreyeretel.PROTECTPregnancyStudy-AnExploratoryStudyofSelf-ReportedMedicationUseinPregnantWom.pdf
TRP: Transplant Pregnancy Registry	Ongoing research study focused on the effects of pregnancy on transplant recipients and the effects of immunosuppressive medications on fertility and pregnancy outcomes. https://www.transplantpregnancyregistry.org/

collaboration, and structured efforts are needed to generate more information from datasets already available, to share data and models, and to stimulate communication between pharmacometricians and clinicians to generate specific observations (pathophysiology during pregnancy, breastfeeding) needed to further develop the field.

PB-PK modeling has moved to the front line as a robust approach to predict PK in pregnant women *prior* to initializing clinical trials. PB-PK models are mechanistic models based on a set of differential equations that deterministically simulate or predict drug movements throughout the body and its compartments, within a physiologically realistic structure. Consequently, this necessitates physiology data. To further illustrate this in the setting of perinatal pharmacology, in vitro and in vivo data on placental transfer are needed, to simulate and validate estimated fetal exposure; and to quantify and predict differences in PK in preeclampsia women as compared to healthy pregnant women; accurate descriptive patterns of (patho)physiology (renal, body composition, hepatic, blood composition) are also needed. The interested reader is referred to recent reviews on this topic. In essence, the use of these PK models can generate additional information from already-available datasets, whether concentration-time data or data on perinatal physiology. Subsequent data or model sharing can then facilitate and boost further research. Examples of this practice include the repository of the Drug Disease Model Resources (DDMoRe) consortium, which makes various published models freely available (accessible via http://www.ddmore.eu/product/model-repository), and the Open Systems Pharmacology (OSP) software suite. The OSP Suite makes the formerly commercial software tools PK-Sim® and MoBi® available as freeware (accessible via www.open-systems-pharmacology.org).

ACKNOWLEDGMENTS AND CONFLICT OF INTEREST

K.A. is involved as coinvestigator in the Innovative Medicines Initiative (IMI).

ConcePTION and the Co-Opt project are two projects related to the topic discussed in this chapter.

References

Abduljalil K, Furness P, Johnson TN et al. (2012). Anatomical, physiological and metabolic changes with gestational age during normal pregnancy: a database for parameters required in physiologically based pharmacokinetic modelling. Clin Pharmacokinet 51: 365–396.

Allegaert K, van den Anker JN (2019). Perinatal pharmacology: optimizing pharmacotherapy for pregnant women, their fetuses and infants. Curr Pharm Des 25: 467–468.

Allegaert K, Peeters MY, Beleyn B et al. (2015). Paracetamol pharmacokinetics and metabolism in young women. BMC Anesthesiol 15: 163.

Alsaied T, Baskar S, Fares M et al. (2017). First-line antiarrhythmic transplacental treatment for fetal tachyarrhythmia: a systematic review and meta-analysis. J Am Heart Assoc 6: e007164.

Berlin Jr CM, van den Anker JN (2013). Safety during breastfeeding: drugs, foods, environmental chemicals, and maternal infections. Semin Fetal Neonatal Med 18: 13–18.

Berlin M, Barchel D, Gandelman-Marton R et al. (2019). Therapeutic levetiracetam monitoring during pregnancy: "mind the gap". Ther Adv Chronic Dis 10: 20406223 19851652.

Bertrand KA, Hanan NJ, Honerkamp-Smith G et al. (2018). Marijuana use by breastfeeding mothers and cannabinoid concentrations in breast milk. Pediatrics 142: e20181076.

Bhatia P, Chhabra S (2018). Physiological and anatomical changes of pregnancy: implications for anaesthesia. Indian J Anaesth 62: 651–657.

Borisow N, Döring A, Pfueller CF et al. (2012). Expert recommendations to personalization of medical approaches in treatment of multiple sclerosis: an overview of family planning and pregnancy. EPMA J 3: 9.

Bouazza N, Foissas F, Hirt D et al. (2019). Methodological approaches to evaluate fetal drug exposure. Curr Pharm Des 25: 496–504.

Brown RA, Dakkak H, Seabrook JA (2018). Is breast best? Examining the effects of alcohol and cannabis use during lactation. J Neonatal-Perinatal Med 11: 345–356.

Ceulemans M, Lupattelli A, Nordeng H et al. (2019). Women's beliefs about medicines and adherence to pharmacotherapy in pregnancy: opportunities for community pharmacists. Curr Pharm Des 25: 469–482.

Cobb B, Liu R, Valentine R et al. (2015). Breastfeeding after anesthesia: a review for anesthesia providers regarding the transfer of medications into breast milk. Transl Perioper Pain Med 1: 1–7.

Cole AP, Hailey DM (1975). Diazepam and active metabolite in breast milk and their transfer to the neonate. Arch Dis Child 50: 741–742.

Conradt E, Flannery T, Aschner JL et al. (2019). Prenatal opioid exposure: neurodevelopmental consequences and future research priorities. Pediatrics 144: e20190128.

Dallmann A, Ince I, Meyer M et al. (2017). Gestation-specific changes in the anatomy and physiology of healthy pregnant women: an extended repository of model parameters for physiologically based pharmacokinetic modeling in pregnancy. Clin Pharmacokinet 56: 1303–1330.

Dallmann A, Mian P, van den Anker J et al. (2019). Clinical pharmacokinetic studies in pregnant women and the relevance of pharmacometric tools. Curr Pharm Des 25: 483–495.

Eyal S, Easterling TR, Carr D et al. (2010). Pharmacokinetics of metformin during pregnancy. Drug Metab Dispos 38: 833–840.

Fischer Fumeaux CJ, Morisod Harari M, Weisskopf E et al. (2019). Risk-benefit balance assessment of SSRI antidepressant use during pregnancy and lactation based on best available evidence—an update. Expert Opin Drug Saf 18: 949–963.

Forinash AB, Yancey AM, Barnes KN et al. (2012). The use of galactogogues in the breastfeeding mother. Ann Pharmacother 46: 1392–1404.

Gin T, Gregory MA, Chan K et al. (1990). Pharmacokinetics of propofol in women undergoing elective caesarean section. Br J Anaesth 64: 148–153.

Gubler T, Krähenmann RM et al. (2013). Determinants of successful breastfeeding initiation in healthy term singletons: a Swiss university hospital observational study. J Perinat Med 41: 331–339.

Gyselaers W, Vonck S, Staelens AS et al. (2018). Body fluid volume homeostasis is abnormal in pregnancies complicated with hypertension and/or poor fetal growth. PLoS One 13: e0206257.

Harden CL, Meador KJ, Pennell PB et al. (2009). Practice parameter update: management issues for women with epilepsy—focus on pregnancy (an evidence-based review): teratogenesis and perinatal outcomes: report of the Quality Standards Subcommittee and Therapeutics and Technology Assessment Subcommittee of the American Academy of Neurology and American Epilepsy Society. Neurology 73: 133–141.

Hebert MF, Zheng S, Hays S et al. (2013). Interpreting tacrolimus concentrations during pregnancy and postpartum. Transplantation 95: 908–915.

Hiramatsu Y, Yoshida S, Kotani T et al. (2018). Changes in the blood level, efficacy, and safety of tacrolimus in pregnancy and the lactation period in patients with systemic lupus erythematosus. Lupus 27: 2245–2252.

Imaz ML, Torra M, Soy D et al. (2019). Clinical lactation studies of lithium: a systematic review. Front Pharmacol 10: 1005. https://doi.org/10.3389/fphar.2019.01005.

Isoherranen N, Thummel KE (2013). Drug metabolism and transport during pregnancy: how does drug disposition change during pregnancy and what are the mechanisms that cause such changes? Drug Metab Dispos 41: 256–262.

Ito S (2016). Mother and child: medication use in pregnancy and lactation. Clin Pharmacol Ther 100: 8–11.

Ito S (2019). Emerging research paradigm for infant drug exposure through breast milk. Curr Pharm Des 25: 528–533.

Ito S, Blajchman A, Stephenson M et al. (1993). Prospective follow up of adverse reactions in breast-fed infants exposed to maternal medication. Am J Obstet Gynecol 168: 1393–1399.

Kantrowitz-Gordon I, Hays K, Kayode O et al. (2018). Pharmacokinetics of dacarbazine (DTIC) in pregnancy. Cancer Chemother Pharmacol 81: 455–460.

Karanam A, Pennell PB, French JA et al. (2018). Lamotrigine clearance increases by 5 weeks gestational age: relationship to estradiol concentrations and gestational age. Ann Neurol 84: 556–563.

Ke AB, Nallani SC, Zhao P et al. (2014). Expansion of a PBPK model to predict disposition in pregnant women of drugs cleared via multiple enzymes, including CYP2B6, CYP2C9 and CYP2C19. Br J Clin Pharmacol 77: 554–570.

Koren G, Pariente G (2018). Pregnancy-associated changes in pharmacokinetics and their clinical implications. Pharm Res 35: 61.

Kreepala C, Kitporntheranunt M, Sangwipasnapaporn W et al. (2018). Assessment of preeclampsia risk by use of serum ionized magnesium-based equation. Ren Fail 40: 99–106.

Kronenfeld N, Ziv Baran T, Berlin M et al. (2018). Chronic use of psychotropic medications in breastfeeding women: is it safe? PLoS One 13: e0197196.

Lai KP, Wong MH, Wong CK (2004). Modulation of AhR-mediated CYP1A1 mRNA and EROD activities by 17beta-estradiol and dexamethasone in TCDD-induced H411E cells. Toxicol Sci 78: 41–49.

Lam J, Kelly L, Ciszkowski C et al. (2012). Central nervous system depression of neonates breastfed by mothers receiving oxycodone for postpartum analgesia. J Pediatr 160: 33–37.e2.

Lefevere J, Allegaert K (2015). Question: is breastfeeding useful in the management of neonatal abstinence syndrome? Arch Dis Child 100: 414–415.

Lewis T, Dinh J, Leeder JS (2015). Genetic determinants of fetal opiate exposure and risk of neonatal abstinence syndrome: knowledge deficits and prospects for future research. Clin Pharmacol Ther 98: 309–320.

Lopes van Balen VA, van Gansewinkel TAG, de Haas S et al. (2019). Maternal kidney function during pregnancy: systematic review and meta-analysis. Ultrasound Obstet Gynecol 54: 297–307.

Meador KJ, Baker GA, Browning N et al. (2014). Breastfeeding in children of women taking antiepileptic drugs: cognitive outcomes at age 6 years. JAMA Pediatr 168: 729–736.

Mian P, van den Anker JN, van Calsteren K et al. (2020). Physiologically based pharmacokinetic modeling to characterize acetaminophen pharmacokinetics and N-acetyl-p-benzoquinone imine (NAPQI) formation in non-pregnant and pregnant women. Clin Pharmacokinet 59 (1): 97–110. https://doi.org/10.1007/s40262-019-00799-5.

Montgomery A, Hale TW, Academy of breastfeeding medicine (2012). ABM clinical protocol #15: analgesia and anesthesia for the breastfeeding mother, revised 2012. Breastfeed Med 7: 547–553.

Nitsun M, Szokol JW, Saleh HJ et al. (2006). Pharmacokinetics of midazolam, propofol, and fentanyl transfer to human breast milk. Clin Pharmacol Ther 79: 549–557.

Ohman I, Vitols S, Tomson T (2005). Pharmacokinetics of gabapentin during delivery, in the neonatal period, and lactation: does a fetal accumulation occur during pregnancy? Epilepsia 46: 1621–1624.

Palmer GM, Anderson BJ, Linscott DK et al. (2018). Tramadol, breast feeding and safety in the newborn. Arch Dis Child 103: 1110–1113.

Quattrochi LC, Vu T, Tukey RH (1994). The human CYP1A2 gene and induction by 3-methylcholanthrene. A region of DNA that supports AH-receptor binding and promoter-specific induction. J Biol Chem 269: 6949–6954.

Roca C, Sahin L, Yao L (2019). Collaboration in regulatory science to facilitate therapeutic development for pregnant women. Curr Pharm Des 25: 609–614.

Saadeh RJ (2012). The baby-friendly hospital initiative 20 years on: facts, progress, and the way forward. J Hum Lact 28: 272–275.

Salman S, Sy SK, Ilett KF et al. (2011). Population pharmacokinetic modeling of tramadol and its O-desmethyl metabolite in plasma and breast milk. Eur J Clin Pharmacol 67: 899–908.

Thompson R, DeJong K, Lo J (2019). Marijuana use in pregnancy: a review. Obstet Gynecol Surv 74: 415–428.

Tisovic K, Amezcua L (2019). Women's health: contemporary management of MS in pregnancy and postpartum. Biomedicine 7: E32.

Tun GS, Lobo AJ (2015). Evaluation of pharmacokinetics and pharmacodynamics and clinical efficacy of certolizumab pegol for Crohn's disease. Expert Opin Drug Metab Toxicol 11: 317–327.

Van Calsteren K, Gersak K, Sundseth H et al. (2016). Position statement from the European Board and College of Obstetrics and Gynaecology (EBCOG): the use of medicines during pregnancy—call for action. Eur J Obstet Gynecol Reprod Biol 201: 189–191.

Van den Anker JN (2012). Is it safe to use opioids for obstetric pain while breastfeeding? J Pediatr 160: 4–6.

Van der Meer DH, Wieringa A, Wegner I et al. (2015). Lactation studies of anticonvulsants: a quality review. Br J Clin Pharmacol 79: 558–565.

Veiby G, Bjørk M, Engelsen BA et al. (2015). Epilepsy and recommendations for breastfeeding. Seizure 28: 57–65.

Veroniki AA, Cogo E, Rios P et al. (2017a). Comparative safety of anti-epileptic drugs during pregnancy: a systematic review and network meta-analysis of congenital malformations and prenatal outcomes. BMC Med 15: 95.

Veroniki AA, Rios P, Cogo E et al. (2017b). Comparative safety of antiepileptic drugs for neurological development in children exposed during pregnancy and breast feeding: a systematic review and network meta-analysis. BMJ Open 7: e017248.

Voskuhl R, Momtazee C (2017). Pregnancy: effect on multiple sclerosis, treatment considerations, and breastfeeding. Neurotherapeutics 14: 974–984.

Wesseloo R, Wierdsma AI, van Kamp IL et al. (2017). Lithium dosing strategies during pregnancy and the postpartum period. Br J Psychiatry 211: 31–36.

Westrin AA, Brekke M, Molden E et al. (2017). Selective serotonin reuptake inhibitors and venlafaxine in pregnancy: changes in drug disposition. PLoS One 12: e0181082.

Westrin AA, Brekke M, Molden E et al. (2018a). Treatment with antipsychotics in pregnancy: changes in drug disposition. Clin Pharmacol Ther 103: 477–484.

Westrin AA, Reimers A, Spigset O (2018b). Should pregnant women receive lower or higher medication doses? Tidsskr Nor Laegeforen 138: 17.

Wu JL, Jia L, He MZ et al. (2019). Placental origins of preeclampsia: potential therapeutic targets. Curr Med Sci 39: 190–195.

Zheng S, Easterling TR, Hays K et al. (2013). Tacrolimus placental transfer at delivery and neonatal exposure through breast milk. Br J Clin Pharmacol 76: 988–996.

Handbook of Clinical Neurology, Vol. 171 (3rd series)
Neurology and Pregnancy: Pathophysiology and Patient Care
E.A.P. Steegers, M.J. Cipolla and E.C. Miller, Editors
https://doi.org/10.1016/B978-0-444-64239-4.00009-6

Chapter 9

Neuroimaging and radiation exposure in pregnancy

ANNIKA ANDERSON[1], JESSICA SINGH[1], AND RILEY BOVE[1,2]*

[1]*Department of Neurology, Weill Institute for Neurosciences, University of California San Francisco, San Francisco, CA, United States*

[2]*Department of Neurology, School of Medicine, University of California San Francisco, San Francisco, CA, United States*

Abstract

Physiologic changes occurring in pregnancy and postpartum can have secondary effects on the maternal nervous system. While most alterations to neurologic function during pregnancy are transient, there is an elevated risk for more serious complication in the peripartum period, such as cerebrovascular events or exacerbation of preexisting neurologic conditions. Due to the morbidity and mortality associated with these neurologic manifestations in some cases, timely diagnostic evaluation is essential. In the pregnant population, the use of diagnostic techniques such as computed tomography (CT) and magnetic resonance imaging (MRI), commonly employed to evaluate emergent neurologic abnormalities, requires special consideration of the potential risks associated with prenatal exposure. This review discusses several neurologic conditions affecting women during pregnancy for which diagnostic imaging may be warranted. Concerns relating to CT and MRI procedures, radiation exposure in utero, and exposure to intravenous contrast by placental transfer and breastfeeding are also reviewed.

INTRODUCTION

During pregnancy, a sequence of changes occurs to the maternal physiology, including alterations in the hemodynamic, hematologic, endocrine, immune, metabolic, and musculoskeletal systems, to accommodate implantation and fetal development (Torgersen and Curran, 2006; Lee, 2007; Laway and Mir, 2013). As a result of these changes, women may experience a range of common neurologic manifestations, such as headache (Digre, 2013) and lower back pain (Lee, 2007). However, these physiologic changes also predispose women to more life-threatening pregnancy-related complications such as eclampsia, stroke, and pituitary disorders. Additionally, pregnancy can modify disease activity in preexisting neurologic conditions, such as neuromyelitis optica spectrum disorders (NMOSD). While some of these pregnancy-related complications confer a high mortality risk, early diagnosis and clinical intervention can vastly improve patient outcomes. Expedient diagnosis in this population, however, is uniquely challenging due to overlaps in symptom presentation between various pregnancy-related neurologic conditions. In such cases, neuroimaging may be necessary to inform timely intervention.

During pregnancy, the use of both computed tomography (CT) and magnetic resonance imaging (MRI), which are routinely employed to evaluate neurologic abnormalities, requires careful consideration of potential risks to the fetus as well as diagnostic benefits to the mother. Where possible, elective imaging should be deferred to the postpartum period to minimize prenatal exposure. When neuroimaging prior to delivery is necessary, comprehensive review and discussion of the risks associated with imaging procedures should occur between the treating neurologist and the pregnant patient.

*Correspondence to: Riley Bove, Department of Neurology, Weill Institute for Neurosciences, University of California, San Francisco, 675 Nelson Rising Lane, San Francisco, CA, United States. Tel: +1-415-353-2069, Fax: +1-415-353-2633, E-mail: riley.bove@ucsf.edu

SAFETY OF NEUROIMAGING IN PREGNANCY AND LACTATION

When there is a need to evaluate new or worsening neurologic symptoms in the pregnant population, selection of the optimal imaging modality should seek to maximize diagnostic power while minimizing the potential for harm to the fetus. Several considerations should inform this selection, including prenatal radiation exposure (ionizing vs nonionizing), gestational age, and the utility of contrast-enhanced imaging. Numerous independent consensus panels provide guidelines for the appropriate use of diagnostic imaging and intravenous contrast in the pregnant population: the American College of Radiology (ACR) in collaboration with the Society of Pediatric Radiology (SPR) (ACR, 2018, 2020), the American Congress of Obstetricians and Gynecologists (ACOG, 2017), and the European Society of Urogenital Radiology (ESUR, n.d.).

Computed tomography (CT)

CT uses specialized X-ray technology to visualize irregularities in both soft tissue and bone. The rapidity of CT relative to other imaging techniques makes it the preferred imaging modality in the setting of trauma and acute neurologic emergencies. Head, neck, and spinal CT are routinely used within the general population to identify the location and severity of neurologic complications, including hemorrhage, blood clots, ischemia, and select neoplasms. In pregnant women, CT should be used in a cautious manner because exposure to ionizing radiation (IR) in utero has the potential to be teratogenic, mutagenic, and/or carcinogenic (De Santis et al., 2005; Patel et al., 2007). Practically, shielding of the fetus is recommended during CT imaging.

The effects of prenatal exposure to IR have been evaluated through several distinct lines of study: animal experiments, observations of prenatal irradiation from occupational exposure, observation of prenatal exposure after the 1945 atomic bombings in Japan, and observation of patients treated with therapeutic radiation prior to the recognition of prenatal risks (Yamazaki and Schull, 1990; Osei and Faulkner, 1999; Wakeford and Little, 2003; De Santis et al., 2005; Patel et al., 2007). These effects are typically categorized as either "deterministic" or "stochastic."

Deterministic effects

Deterministic effects result in teratogenesis due to cellular injury to the fetus during gestation. These effects are associated with a specific threshold dose and are dependent on gestational age (Table 9.1; ACR, 2018). Below a

Table 9.1

Summary of suspected in utero induced deterministic radiation effects*

Menstrual or gestational age	Conception age	<50 mGy (<5 rad)	50–100 mGy (5–10 rad)	>100 mGy (>10 rad)
0–2 weeks (0–14 days)	Prior to conception	None	None	None
3rd and 4th weeks (15–28 days)	1st–2nd weeks (1–14 days)	None	Probably none	Possible spontaneous abortion
5th–10th weeks (29–70 days)	3rd–8th weeks (15–56 days)	None	Potential effects are scientifically uncertain and probably too subtle to be clinically detectable	Possible malformations increasing in likelihood as dose increases
11th–17th weeks (71–119 days)	9th–15th weeks (57–105 days)	None	Potential effects are scientifically uncertain and probably too subtle to be clinically detectable	Risk of diminished IQ or mental retardation, increasing in frequency and severity with dose
18th–27th weeks (120–189 days)	16th–25th weeks (106–175 days)	None	None	IQ deficits not detectable at diagnostic doses
>27 weeks (>189 days)	>25 weeks (>175 days)	None	None	None applicable to diagnostic medicine

*Stochastic risks are suspected, but data are not consistent [6]. For exposure to a newborn child, the lifetime attributable risk of developing cancer is estimated to be 0.4% per 10 mGy (1 rad) dose to the baby. The potential risks in utero for the second and third trimesters and part of the first trimester may be comparable, but the uncertainties in this estimate are considerable.

Reproduced with permission of the American College of Radiology from ACR, SPR, 2018. ACR-SPR practice parameter for imaging pregnant or potentially pregnant adolescents and women with ionizing radiation (resolution 39). American College of Radiology, Reston, VA.

given threshold, there is relatively negligible risk of these effects (De Santis et al., 2005). Between conception and implantation (0–2 weeks of gestation), IR exposure has an "all or none" effect (Miller, 1990; Adam, 2012); if the embryo survives, other teratogenic effects are unlikely. During organogenesis (weeks 2–15), greater fetal sensitivity to IR can result in mental retardation, growth restriction, and organ malformation (Yamazaki and Schull, 1990; Osei and Faulkner, 1999; De Santis et al., 2005; Patel et al., 2007). Risk for IQ deficits may still occur later in pregnancy (15–25 weeks), but at much higher threshold doses (Yamazaki and Schull, 1990; Osei and Faulkner, 1999; Patel et al., 2007). Beyond a gestational age of 25 weeks, teratogenicity is not observed at exposures that significantly exceed those encountered through diagnostic imaging (>10 rad) (Osei and Faulkner, 1999; ACR, 2018).

Stochastic effects

In contrast, stochastic effects are mutagenic and increase the risk for later development of malignancy. Unlike deterministic effects, stochastic effects are not associated with any specific threshold dose and, in theory, can occur with any exposure to IR. Even at low doses, prenatal IR exposure is associated with a risk of childhood cancer, especially leukemia (Wakeford and Little, 2003). The risk for childhood cancer has been linked to increasing dose (approximately 6% per 100 rad) and earlier exposure in pregnancy (Wakeford and Little, 2003). However, IR exposure from diagnostic imaging is thought to confer a low risk of childhood cancers, given that the effective fetal exposures are orders of magnitude less than those studied to evaluate risk (Osei and Darko, 2013).

As low as reasonably achievable (ALARA) principle

In practice, the effective fetal IR doses encountered in diagnostic imaging are significantly lower than the thresholds for teratogenesis. While the effective fetal IR dose associated with CT cannot be measured directly, it is calculated based on background radiation (estimated at 0.23 rad; Osei and Faulkner, 1999), maternal size, examination parameters, and whether the fetus is in the direct field of irradiation. Of note, in imaging studies where the fetus is outside the field of view, such as occurs with head and cervical CT, radiation is significantly attenuated—with an estimated dose <0.01 rad (Osei and Faulkner, 1999). Even in imaging studies in which the fetus is within the field of view (such as lumbar spine or abdominal/pelvic CT), effective fetal dose falls within 0.28–3.0 rad (Osei and Faulkner, 1999).

Given the uncertainty surrounding carcinogenesis, an effort to minimize prenatal IR exposure should be guided by the principle of ALARA (ACOG, 2017; ACR, 2018). When possible, alternative imaging techniques, such as MRI, or deferral of CT to the postpartum period should be considered.

Magnetic resonance imaging (MRI)

MRI is often the preferred imaging approach during pregnancy as it does not entail prenatal exposure to IR. It uses a form of nonionizing radiation and low-energy radiofrequencies and has a wider range of soft tissue contrast, providing better sensitivity to visualize brain and spinal cord irregularities.

There is no conclusive data to date indicating that MRI exposure causes harm to the fetus (ACR, 2008). Studies have determined that MRI performed with up to a 1.5-T magnet is safe throughout pregnancy, including in the first trimester, when temperature effects on organogenesis would be a primary concern (De Wilde et al., 2005). Elective imaging should still be postponed to the postpartum period where possible to avoid theoretical risks (fetal exposure to strong magnetic fields, elevated temperatures due to energy deposition by radio waves, and acoustic noise).

The ACR recommends that MRI can be obtained during any trimester of pregnancy, provided three conditions are met: "1. The information requested from the MR study cannot be acquired by means of nonionizing means (e.g., ultrasonography). 2. The data is needed to potentially affect the care of the patient or fetus during the pregnancy. 3. The referring physician believes that it is not prudent to wait until the patient is no longer pregnant to obtain this data" (Kanal et al., 2013).

Safety of contrast-enhanced imaging

Intravenous contrast agents are commonly used in neuroimaging to provide better structural distinction and enhance the visibility of abnormalities. The contrast agents used in CT and MRI are iodinated and gadolinium-based, respectively. During pregnancy and lactation, the decision to administer intravenous contrast requires additional consideration of fetal and neonatal exposure.

During pregnancy

Both iodinated and gadolinium-based contrast have been shown to cross the placenta (Novak et al., 1993; Marcos et al., 1997; Moon et al., 2000; Prola-Netto et al., 2018); however, longitudinal evaluation of prenatal exposure to these agents is limited in humans.

In the case of iodinated contrast, the Food and Drug Administration (FDA) evaluated these as category B under previous categorization schemata. Despite evidence that iodinated contrast crosses the placenta and is present in small concentrations in fetal tissue, there is a lack of data in pregnant women to inform recommendations or drug-associated risks (FDA, 2018). Reassuringly, in animal studies, there have been no observable teratogenic effects of prenatal exposure (Nelson et al., 1982). Given the potential for fetal thyroid exposure to iodine, the possibility of neonatal hypothyroidism has been considered. Moreover, a study evaluating thyroid function in a small cohort of newborns exposed in utero found negligible evidence of thyroid suppression at delivery (Bourjeily et al., 2010). According to the ACR, given absence of documented cases of neonatal hypothyroidism from maternal intravascular water-soluble iodinated contrast agents to date (ACR, 2018, 2020), and given that newborns are routinely evaluated for congenital hypothyroidism through TSH level measurement at birth (De Santis et al., 2005; ACOG, 2017; ESUR, n.d.), no additional thyroid evaluation is recommended in the newborn. The ACR-SPR, ACOG, and ESUR guidelines advise that iodinated contrast be administered during pregnancy (including the first trimester) when necessary to maternal care, and recommend the standard-of-care follow-up with neonatal thyroid function testing in the first week of life (ACOG, 2017; ACR, 2018, 2020; ESUR, n.d.).

Animal studies of prenatal exposure to gadolinium-based contrast, albeit at concentrations much higher than clinically used, have identified risk for early termination and developmental irregularities (Okuda et al., 1999). Additionally, development of nephrogenic system fibrosis from exposure to gadolinium-based contrast in adults with renal impairment supports the need for special attention to exposed neonate kidney function (Abraham and Thakral, 2008); however, a small prospective study of renal function in premature infants exposed in utero did not indicate any immediate adverse effects (Amin et al., 2017). The recommended protocol for use of gadolinium-based contrast agents in pregnancy per the ACOG, ESUR, and ACR, and according to the FDA, is to administer the smallest possible dose required for imaging and only when necessary to maternal care (ACOG, 2017; ACR, 2018, 2020; ESUR, n.d.; FDA, 2011).

During lactation

Based on strong evidence for maternal and neonatal benefits from breastfeeding, both the World Health Organization and American Academy of Pediatrics (AAP) recommend exclusive breastfeeding for at least the first 6 months of an infant's life (Johnston et al., 2012). Historically, temporary discontinuation of breastfeeding for 24 h after the administration of contrast media in lactating women has been advised; however, both iodinated and gadolinium-based contrast have limited lipid solubility. Thus minimal concentrations of these agents have been observed to transfer to breastmilk: less than 1% of the maternal dose, of which less than 1% is absorbed by the infant gastrointestinal tract during digestion (Nielsen et al., 1987; Kubik-Huch et al., 2000; Webb et al., 2005). Given the low risk of transmission and the potential for premature weaning with temporary cessation, the AAP, ACR, ACR-SPR, ACOG, and ESUR guidelines advise that breastfeeding can proceed uninterrupted postadministration of iodinated and gadolinium-based contrast in lactating women (ACOG, 2017; ACR, 2018, 2020; ESUR, n.d.; Thomas, n.d.).

RADIOGRAPHIC CHARACTERISTICS OF NEUROLOGIC CONDITIONS ASSOCIATED WITH PREGNANCY

Cerebrovascular complications

During pregnancy, the cerebrovascular system is affected by hemodynamic and hematologic changes occurring to accommodate fetal development and perfusion of the placenta. Increased blood volume and cardiac output may cause the blood pressure to rise (Torgersen and Curran, 2006; Lee, 2007). Simultaneously, the increase in circulating estrogen stimulates production of clotting factors (I, II, VII, VIII, IX, and XII) and reduction of endogenous anticoagulants (antithrombin III and protein S), leading to a procoagulant state (Torgersen and Curran, 2006; Lee, 2007; Treadwell et al., 2008). Vascular homeostasis is also modified via pregnancy-related endocrine shifts that alter the permeability and integrity of vessel walls (Leo et al., 2017). This peripartum physiologic environment may elevate risk of cerebrovascular complications such as stroke and thrombosis.

Ischemic stroke

While stroke risk in pregnant women remains comparable with that in the general population, it is notably elevated in the first 6 weeks postpartum (Kubik-Huch et al., 2000; Treadwell et al., 2008).

Additional contributing risk factors include advanced maternal age and hypertension (Treadwell et al., 2008). Women presenting with ischemic stroke may demonstrate impaired speech production or comprehension, facial or limb weakness, visual deficits, vertigo, or impaired sensation.

Ischemic stroke is best evaluated by diffusion-weighted imaging (DWI) sequences on MRI. When imaged early, infarcting tissue will appear hyperintense on DWI, with restricted diffusion due to cytotoxic edema. Perfusion CT and contrast-enhanced MRI can be performed when stroke is suspected postpartum to evaluate for ischemic changes, surrounding edema, and vessel blockages.

Amniotic fluid embolism

Amniotic fluid embolism (AFE) is an uncommon, but life-threatening condition that occurs when amniotic fluid, fetal cells, hair, or other debris cross the placenta into the maternal circulation. AFE has a reported incidence within the range of 2–6 cases per 100,000 deliveries (Ito et al., 2014). The maternal inflammatory response to an AFE can trigger respiratory distress, hypotension, cardiac arrest, and coagulopathy (Ito et al., 2014). Brain MRI performed without contrast in AFE may reveal reduced diffusion and T2 hypointensities indicative of ischemic change.

Cerebral venous thrombosis (CVT)

Cerebral venous thrombosis (CVT) is a rare form of thromboembolism that accounts for up to 6% of maternal deaths (Nie et al., 2015). While CVT can occur at any time during pregnancy, the risk is greatest 2–4 weeks postpartum and is more frequently observed in young mothers and mothers who underwent cesarean section (Saposnik et al., 2011). Additional contributing risk factors include dehydration, intracranial hypotension, and traumatic delivery. The clinical presentation of CVT includes headache, seizures, encephalopathy, papilledema, and focal neurologic deficits.

MRI (Fig. 9.1; Bonneville, 2014) is the preferred imaging modality to detect a CVT. Time-of-flight (TOF) MRI allows for rapid identification of obstructed flow without the need for contrast enhancement. Infarcted tissue appears hyperintense on DWI. Evaluation for hemorrhage should include gradient echo (GRE) and susceptibility-weighted sequences. Despite limited diagnostic utility, hyperdense regions on noncontrast CT can indicate blood products from hemorrhage or edema.

Subarachnoid hemorrhage (SAH)

Aneurysmal subarachnoid hemorrhage (aSAH) occurring during pregnancy may arise as a result of increased cardiac output during gestation, as well as hormonal factors that could favor intracranial aneurysm formation, enlargement, and/or rupture (Treadwell et al., 2008; Guida et al., 2012). The common neurologic manifestations of SAH include acute onset of a "thunderclap" headache, followed by rapid deterioration of neurologic status.

In acute settings where expedient diagnosis is needed, noncontrast head CT can be used to rapidly identify an intracranial hemorrhage; MRI will have higher sensitivity in identifying nonacute hemorrhage, as well as distinguishing other acute conditions such as preeclampsia/eclampsia, CVT, and vascular malformations (Acciarresi et al., 2019). To identify aneurysms, CT angiography, digital subtraction angiography, and TOF MRI can be used.

Preeclampsia, eclampsia, and HELLP syndrome

Preeclampsia affects 2%–8% of women after 20 weeks of gestation and up to 8 weeks postpartum (Ghulmiyyah and Sibai, 2012). The clinical signs of preeclampsia include hypertension, proteinuria, and, in some cases, peripheral edema. Symptoms include headache, visual deficits, and altered mental status. Women with severe preeclampsia may develop HELLP (hemolysis, elevated liver enzymes, and low platelet count) syndrome. Between 1% and 2% of women with preeclampsia will develop tonic–clonic seizures, indicating progression to eclampsia (Ghulmiyyah and Sibai, 2012).

Establishing a diagnosis of preeclampsia or eclampsia does not typically require neuroimaging. However, imaging may be indicated if eclampsia occurs postpartum and/or if resolution of visual and focal neurologic deficits has not occurred, to evaluate for hypertensive encephalopathy (vide infra).

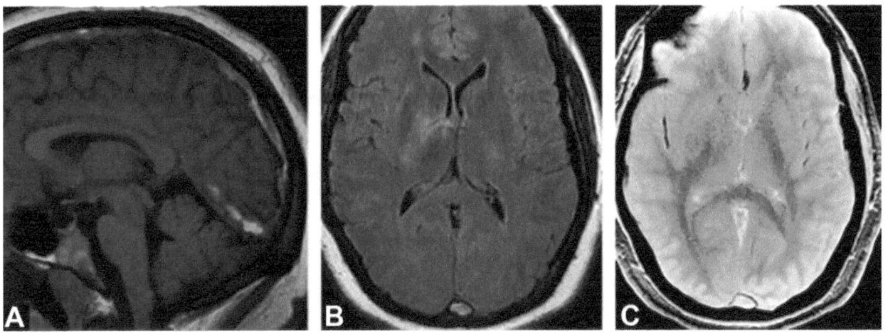

Fig. 9.1. Cerebral venous thrombosis. Visible hyperintense thrombus on MRI sequences using (A) sagittal T1, (B) axial T2, and (C) T2 in gradient echo. Reproduced with permission of Elsevier from Bonneville F (2014). Imaging of cerebral venous thrombosis. Diagn Interv Imaging 95: 1145–1150.

Posterior reversible encephalopathy syndrome (PRES)

Posterior reversible encephalopathy syndrome (PRES) is classically characterized by vasogenic edema in the posterior lobes. Along with pregnancy, it can be triggered by hypertension, but also infection, uremia, malignancy, autoimmune disorders, thrombotic thrombocytopenic purpura (TTP), and use of immunosuppressant medications (Fugate et al., 2010). Although it classically involves the posterior (occipital and parietal) lobes, other regions (frontal) may also be affected (Kastrup et al., 2015). PRES is characterized by clinical manifestations that include headache, visual deficits, seizures, diplopia, and hallucinations.

Typically, MRI reveals a nonuniform pattern of hyperintensity on T2/fluid-attenuated inversion recovery images and apparent diffusion coefficient mapping, and isointensity on DWI, in affected cortical and subcortical regions (McKinney et al., 2007; Tetsuka and Ogawa, 2019). Reversibility of neuroimaging abnormalities is a critical aspect of the syndrome. The occipital and parietal lobes are classically affected, although other regions, including frontal lobes and cerebellum, may be involved as well (Kastrup et al., 2015). When contrast is administered, leptomeningeal enhancement with/without cortical enhancement may be seen, possibly related to dysfunction of the blood–brain barrier (Karia et al., 2016). DWI may reveal vasogenic or cytotoxic edema. The neuroimaging features in preeclampsia/eclampsia, TTP, and PRES may be indistinguishable (McKinney et al., 2007).

Postpartum angiopathy (PPA)/reversible cerebral vasoconstriction syndrome (RCVS)

Reversible cerebral vasoconstriction syndrome (RCVS), commonly referred to as "Call-Fleming syndrome," is characterized by multifocal vasoconstriction of the cerebral arteries. When occurring in the postpartum period, it is referred to as postpartum angiopathy (PPA). Risk for PPA is greatest in the first 4 weeks postpartum, particularly in women with hypertension, with abnormal endocrine function, or who use vasoactive medications (such as serotonin reuptake inhibitors) (Singhal and Bernstein, 2005; Burton and Bushnell, 2019). Women with PPA may present with sudden onset of thunderclap headache, nausea, photophobia, encephalopathy, or seizures (Singhal and Bernstein, 2005; Burton and Bushnell, 2019). Given the symptomatic overlap, PPA is often difficult to distinguish from PRES and eclampsia. While usually self-limiting, severe cases of PPA can trigger more serious hemorrhagic and ischemic changes.

MR and CT angiography, both of which require contrast administration, can be used to visualize the multifocal constriction of intracranial arteries in PPA. However, these may appear normal in early stages (Singhal and Bernstein, 2005). Noncontrast MRI in PPA is recommended to evaluate for T2 hyperintense white matter lesions indicating regions affected by restricted blood flow, hemorrhage, or infarction. Both CT and MRI of the brain in PPA reveal regions of edema commonly affecting the occipital and parietal lobes (Fig. 9.2; Chandrashekaran et al., 2007). On CT, the affected regions will appear hypointense. Transcranial Doppler ultrasound is a noninvasive alternative that can be used to monitor the velocity of intracranial blood flow and resolution of PPA (Singhal and Bernstein, 2005).

Demyelinating diseases of the central nervous system

Demyelinating conditions such as multiple sclerosis (MS) and NMOSD are most commonly diagnosed in women of reproductive age (Bove and Chitnis, 2013; Borisow et al., 2017; Gold et al., 2019). During pregnancy, endocrine-mediated immunologic changes, as well as placental factors, have different modulatory effects on disease activity (Qiu et al., 2019).

Multiple sclerosis (MS)

In MS, the immunotolerant state of pregnancy has been linked to reduced MS disease activity, or fewer relapses (Qiu et al., 2019), followed by an increased risk of relapses in the first 3 months postpartum (Confavreux et al., 1998; Vukusic et al., 2004; Hellwig et al., 2012). Exclusive breastfeeding postpartum may provide beneficial hormonal cues in nursing women that reduce the risk for postpartum MS relapse (Langer-Gould et al., 2009; Hellwig et al., 2015). Some women may present with a first demyelinating event during this postpartum period. Unfortunately, in the current disease-modifying therapy (DMT) era, women discontinuing lymphocyte trafficking DMTs (fingolimod and natalizumab) in anticipation of pregnancy have been found to experience greater risk of relapse even during pregnancy, due to "rebound" disease activity (Alroughani et al., 2018).

It is typically advised to obtain a "baseline" MRI of the brain and cervical and thoracic spine in women planning conception, so that any new or worsening symptoms consistent with demyelination that occurs during pregnancy can be evaluated by comparing a new, unenhanced MRI with this baseline. Foci of demyelination typically appear as hyperintense on T2/FLAIR-weighted images. Brain lesions are distributed to the periventricular, intracallosal, cerebral, cerebellar, optic nerve, and brainstem regions (Tillema and Pirko, 2013). Lesions in the spinal

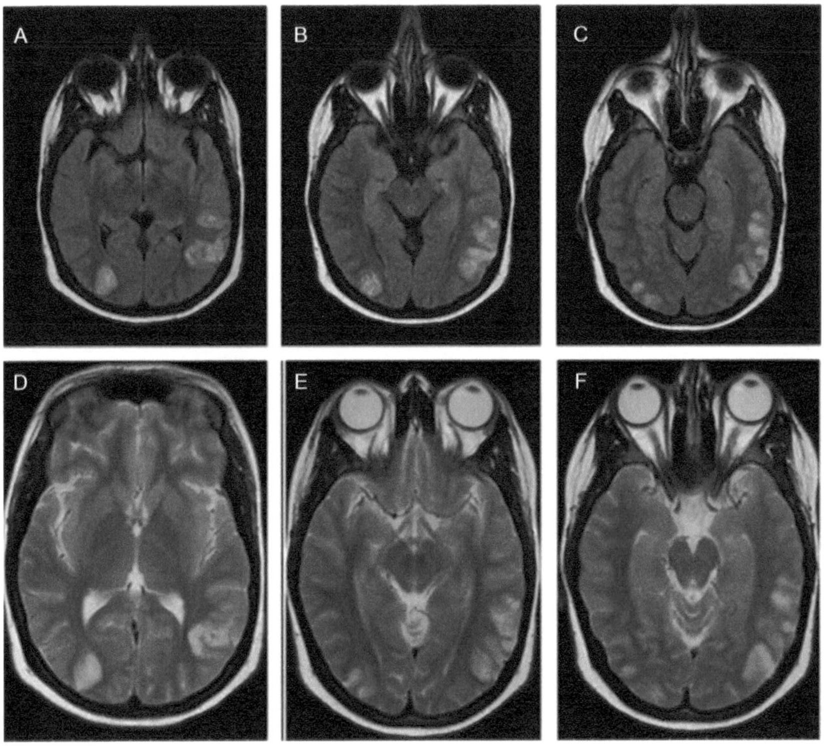

Fig. 9.2. Postpartum angiopathy. Ischemic changes secondary to postpartum angiopathy are visible as hyperintense lesions in the right occipital, left temporal, and left parietal lobes on MRI (A–C) axial FLAIR and (D–F) axial T2-weighted sequences. Reproduced with permission of Elsevier from Chandrashekaran S, Parikh S, Kapoor P et al. (2007). Postpartum reversible cerebral vasoconstriction syndrome. Am J Med Sci 334: 222–224.

cord are peripherally located (Tillema and Pirko, 2013). Given the risk of postpartum relapse, early contrast-enhanced MRI following delivery should be used to monitor for active lesions and inform treatment planning.

Neuromyelitis optica spectrum disorders (NMOSD)

In contrast to MS, women with NMOSD experience an increase in disease activity both during pregnancy as well as postpartum (Borisow et al., 2017). This trend has been linked to hormonal and immunologic shifts during pregnancy, predominately Th2-mediated immunity, which is associated with increased production of aquaporin4-immunoglobin, a diagnostic and pathogenic marker of NMOSD (Qiu et al., 2019).

Given the risk of exacerbation, unenhanced MRI can be used to monitor for interval disease activity during pregnancy (Fig. 9.3; Mpateni et al., 2018). On T2-weighted imaging, hyperintense lesions are localized primarily in the optic nerve and spinal cord. In contrast to MS, there are notably sparse or no white matter lesions on brain MRI in NMOSD patients (Tillema and Pirko, 2013); spinal cord lesions are more centrally located and typically longitudinal, spanning three or more

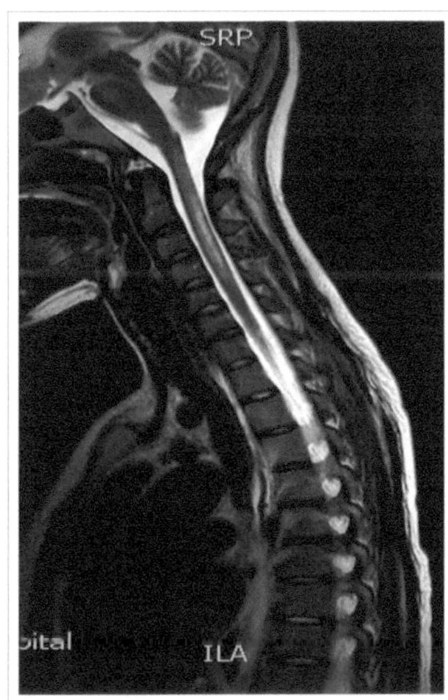

Fig. 9.3. Neuromyelitis optica spectrum disorders. Extensive hyperintense lesion spanning the cervicomedullary junction to C4 spinal segment on T2-weighted MRI of the cervical spine in a patient with NMOSD. Reproduced under the terms of the Creative Commons Attribution License from Mpateni SS, Sihlali NC, Gardiner EC et al. (2018). Magnetic resonance imaging findings in a patient with seropositive neuromyelitis optica. SA J Radiol 22: 1306.

vertebra (Tillema and Pirko, 2013). As in MS, early-postpartum, contrast-enhanced MRI is recommended to evaluate for disease activity.

Neoplasms

Fluctuating estrogen, progesterone, adrenocorticotropic hormone, and prolactin levels can increase the rate of neoplasm growth in the central nervous system during pregnancy (Doyle et al., 2009; Molitch, 2017). As a result, women may experience headache from increased intracranial pressure and other neurologic manifestations due to compression of adjacent structures.

Meningiomas

Meningiomas, the most common intracranial tumors, originate from arachnoid cap cells (Marosi et al., 2008). Although typically benign and slow growing, meningiomas are hormone responsive and have been observed to undergo accelerated growth during pregnancy in response to increased blood estrogen levels (Marosi et al., 2008). Growth of meningiomas during pregnancy can lead to increased intracranial pressure and compression of adjacent brain tissue; depending on location, this could result in visual impairment and seizures (Watts et al., 2014).

MRI is the preferred technique for evaluating meningiomas, and if monitoring is performed during pregnancy, this should involve an MRI with contrast. Meningiomas typically present as well-defined masses within the subdural space extending to form a "dural tail" in some cases (Fig. 9.4; Watts et al., 2014). On T2-weighted images, meningiomas are hyperintense relative to gray matter. Homogenous postcontrast enhancement is observed (Watts et al., 2014).

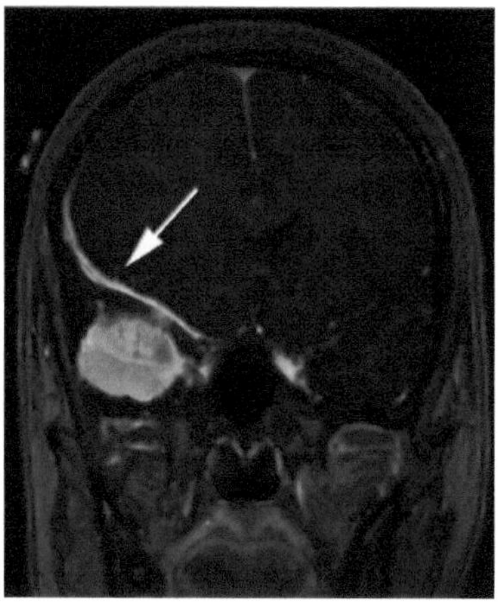

Fig. 9.4. Meningioma. Coronal postcontrast T1-weighted MRI sequence demonstrating homogeneous enhancement within the mass, occupying the right middle cranial fossa, and accompanied by a broad dural tail (*arrow*). Reproduced under the terms of the Creative Commons Attribution License from Watts J, Box G, Galvin A et al. (2014). Magnetic resonance imaging of meningiomas: a pictorial review. Insights Imaging 5: 113–122.

Gestational trophoblastic disease (Choriocarcinoma)

Choriocarcinoma is a rare malignancy that arises from placental trophoblastic tissue. It has a high rate of metastasis, including to the brain in 20% of cases (Shaaban et al., 2017). These metastases are highly vascularized and thereby prone to hemorrhage. Common clinical presentations in women with brain metastases include headache, focal neurologic deficits, seizure, and elevated β-human choriogonadotropin (Gavanier et al., 2019).

As with other intracranial tumors, choriocarcinomas are best evaluated by MRI. However, CT imaging of choriocarcinoma metastases will show hemorrhagic masses with notable surrounding edema (Shaaban et al., 2017). On MRI, choriocarcinoma presents as mass-like lesions with different signal intensity depending on the size of the tumor, surrounding edema, and hemorrhage (Shaaban et al., 2017). Peripheral enhancement is typical when contrast is administered (Shaaban et al., 2017). If a lesion is noted, GRE and DWI can evaluate for blood products indicative of additional metastases (Fig. 9.5, Mercado et al., 2017).

Pituitary disorders

Hormonal fluctuations during pregnancy, particularly rising estradiol, stimulate prolactin-producing cells in the pituitary gland (Laway and Mir, 2013). As a result, the pituitary gland has been observed to grow up to 136% in size during gestation (Karaca et al., 2010). Although the gland typically returns to normal size within 6 months of delivery (Dinc et al., 1998), lactotroph hypertrophy in combination with pregnancy-induced hemodynamic and hematologic conditions can increase the risk of ischemia, hemorrhage, and prolactinoma growth.

Pituitary adenomas

Pituitary adenomas are usually benign tumors, classified by size as either microadenomas (<10 mm) or macroadenomas (>10 mm) (Molitch, 2017). Typically, they are managed with dopamine agonists (DAs), but surgical resection is warranted in cases of severe neurologic deficits (Varlamov et al., 2020). Roughly 50% of pituitary adenomas are prolactinomas (Molitch, 2017). Thus estrogen stimulation of lactotrophs and discontinuation

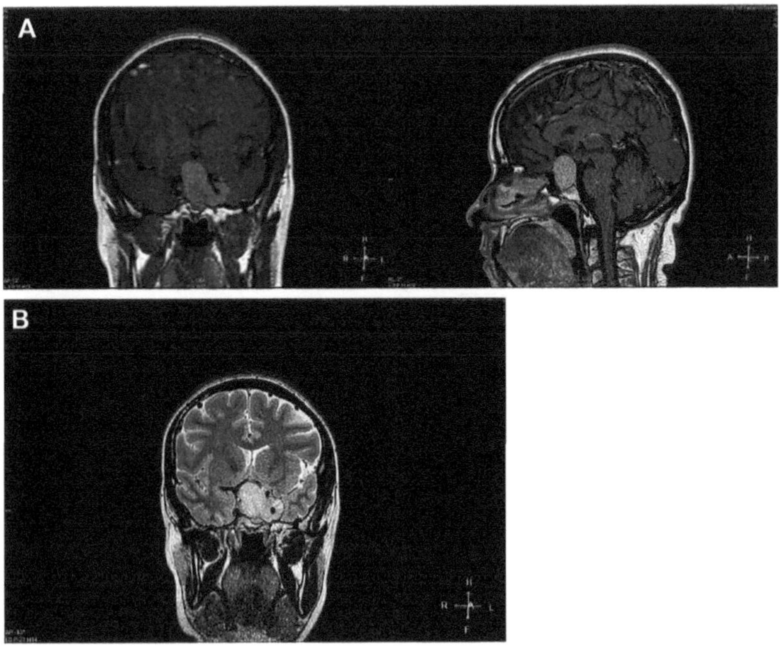

Fig. 9.5. Pituitary macroadenoma. MRI of pituitary macroadenoma visible on (A) coronal and sagittal postcontrast T1-weighted and (B) coronal T2-weighted sequences. Reproduced with permission of Elsevier from Mercado M, Melgar V, Salame L et al. (2017). Clinically nonfunctioning pituitary adenomas\pathogenic, diagnostic and therapeutic aspects. Endocrinol Diabetes Nutr 64: 384–395.

of DAs, if done prior to conception, can have trophic effects on micro- and macroadenomas. This growth can cause compression of adjacent tissue and manifest in clinical symptoms such as headache, visual deficits, and hypopituitarism.

MRI of both micro- and macroadenomas will reveal asymmetric enlargement of the pituitary gland. When contrast is administered, adenomas enhance less intensely than normal tissue, resulting in nonuniform enhancement throughout the pituitary gland.

Pituitary apoplexy (PA)

As the pituitary gland enlarges during pregnancy, it can outgrow its vascular supply. When compounded by additional risk factors, such as gestational hypertension or preexisting pituitary adenoma, hemorrhage or infarction can result. PA initially presents with headache, nausea, and emesis; however, bitemporal hemianopsia and ophthalmoplegia can arise from compression of adjacent structures, the optic chiasm and cranial nerves (II, IV, and VI) (Ranabir and Baruah, 2011; Laway and Mir, 2013).

If acute hemorrhage is suspected, CT can be used to evaluate for an enlarged pituitary gland and hyperdense regions in the sella. Noncontrast MRI provides better temporal resolution of PA, with sagittal images demonstrating expansion of the gland and varying T1 iso- and T2 hyperintensity, depending on the stage of hemorrhage (Boellis et al., 2014). DWI should be used to evaluate for early ischemic changes, characterized by increased signal intensity relative to normal, noninfarcting gray or white matter.

Sheehan syndrome (SS)

Hemorrhage and hypotensive shock in the peripartum period can result in Sheehan syndrome (SS), which is caused by the restriction of blood flow to and subsequent necrosis of the pituitary gland (Laway and Mir, 2013). This can occur secondary to PA. In SS, hypopituitarism is most commonly indicated by lack of breastmilk production and prolonged postpartum amenorrhea (Laway and Mir, 2013).

Unenhanced MRI is the imaging modality of choice in SS. However, both MRI and CT will reveal an initial enlargement of the pituitary gland, which over time resolves to a partial or completely empty sella, typically occupied by cerebrospinal fluid (CSF) (Laway and Mir, 2013). This empty space presents as a hypointense region on T1-weighted images and a hyperintense region on T2-weighted images by MRI.

Lymphocytic hypophysitis (LH)

Lymphocytic hypophysitis (LH), while not exclusive to the pregnant population, can arise peripartum. It is an autoimmune condition characterized by lymphocytic infiltration of the pituitary gland and infundibulum,

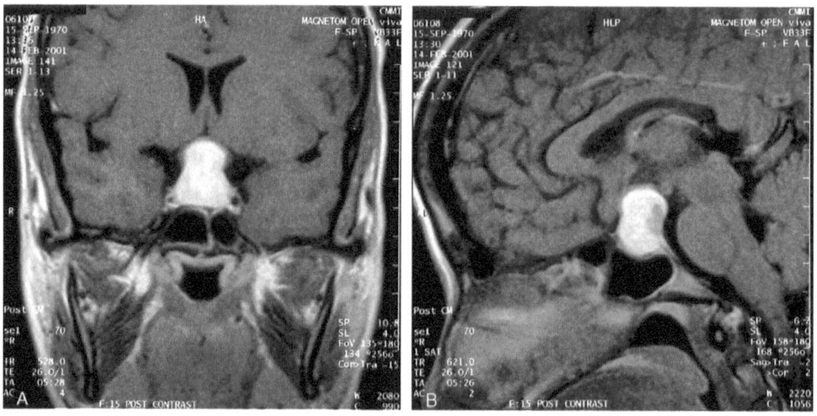

Fig. 9.6. Lymphocytic hypophysitis. Postcontrast T1-weighted MRI sequences demonstrating nonuniform enhancement and extension of the pituitary into the suprasellar region from (A) coronal and (B) sagittal perspectives. Reproduced with permission of Elsevier from Ng WH, Gonzales M, Kaye AH (2003). Lymphocytic hypophysitis. J Clin Neurosci 10: 409–413.

resulting in enlargement of the gland. Subsequent compression of surrounding tissues can manifest symptomatically through headache and visual deficits and can result in hypopituitarism, making LH difficult to distinguish from pituitary adenoma (Karaca et al., 2010; Laway and Mir, 2013).

On MRI, the pituitary gland and infundibular stalk appear enlarged and T1 hyperintense in both LH and pituitary adenoma (Fig. 9.6; Ng et al., 2003). However, on dynamic contrast-enhanced studies, LH is distinguished by homogenous enhancement in the early phase, in comparison with adenomas that exhibit diminished enhancement relative to pituitary tissue. In later phases, enhancement will persist within the adenoma but not in normal tissue (Powrie et al., 1995; Unluhizarci et al., 2001).

Back pain

Back pain is a common neurologic manifestation of pregnancy. Laxity of spinal ligaments induced by hormonal changes, as well as uterine pressure on the lumbar plexus, can result in transient pain that typically resolves after delivery (Lee, 2007). However, atypical back pain, especially when accompanied by motor or sensory deficits or bowel and bladder dysfunction, requires more thorough diagnostic evaluation, typically by MRI.

Epidural processes

Epidural abscesses and hematomas can arise spontaneously, possibly due to intra-abdominal pressure during delivery, or as a complication of epidural and spinal anesthesia (Arko et al., 2014). They can present through fever, pain, and weakness. On MRI, epidural abscesses and phlegmons demonstrate T2 hyperintensity. On contrast-enhanced imaging, epidural abscesses can be

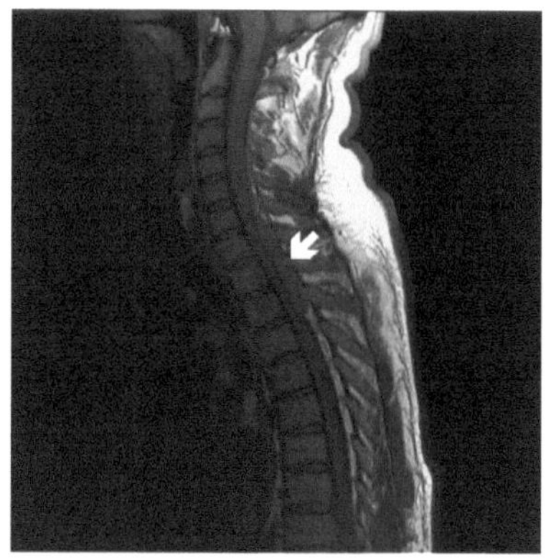

Fig. 9.7. Epidural hematoma. T1-weighted MRI of the cervical and thoracic spine demonstrates an isointense region (*arrow*) between the C7 and T4 vertebrae. NB: This hematoma was not related to a pregnancy. Reproduced with permission of Elsevier from Hussenbocus SM, Wilby MJ, Cain C (2012). Spontaneous spinal epidural hematoma: a case report and literature review. J Emerg Med 42: e31–34.

distinguished from phlegmon based on the pattern of enhancement: there is typically peripheral enhancement with abscesses and homogeneous enhancement with phlegmon (Sandhu and Dillon, 1991). Hematomas are best seen on sagittal MRI images, which reveal well-defined regions of T1-weighted hypo- or isointensity depending on the age of hematoma (Fig. 9.7; Hussenbocus et al., 2012).

Others

Epilepsy

During pregnancy, women with preexisting seizure disorders require careful monitoring to ensure control of seizure activity. Metabolic changes, weight gain, and decreased albumin can alter the effects of antiepileptic drugs and result in mild increase in seizure activity during pregnancy (Torgersen and Curran, 2006).

Typically, seizures in a woman with known preexisting epilepsy do not require imaging during pregnancy. However, the new onset of seizures can be indicative of several pregnancy-related neurologic complications discussed, including: eclampsia, PRES, RCVS, SAH, CVT, or neoplasm growth. Given the overlaps in presentation of these conditions, neuroimaging is necessary for diagnostic evaluation and treatment.

Intracranial hypotension

When performed during pregnancy, lumbar puncture or spinal and epidural anesthesia can introduce perforations in the dura that cause leakage of the CSF. As a result, women may experience acute onset of headache, neck stiffness, nausea, cognitive dysfunction, tinnitus, and vertigo (Gordon, 2009). While typically self-resolving, severe cases can escalate to subdural hematoma, brainstem compression, cranial nerve palsies, and herniation of the cerebellar tonsils (Gordon, 2009). MRI imaging may reveal these sequelae, brain sag, or dural tears.

Wernicke's encephalopathy

Wernicke's encephalopathy (WE) occurs in the setting of thiamine deficiency. It is often precipitated by chronic alcoholism but can be attributed to metabolic changes and malnutrition caused by hyperemesis gravidarum in the context of pregnancy (Talib et al., 2018). Acute encephalopathy, ataxia, and ophthalmoplegia comprise the symptomatic triad of WE (Talib et al., 2018). In severe cases, coma can occur (Jung et al., 2012). While early intervention with high-dose thiamine can resolve this condition, more permanent damage occurs in untreated, chronic cases known as Korsakoff's syndrome.

MRI is the preferred imaging technique to visualize T2 hyperintensities around the third and fourth ventricles, mammillary bodies, paramedian thalamic nuclei, and cerebral aqueduct, typical of WE (Zuccoli and Pipitone, 2009). Postpartum imaging with contrast administration may demonstrate enhancement in these regions. DWI is recommended to assess for cytotoxic and vasogenic edema in the periventricular space (Chung et al., 2003; Zuccoli and Pipitone, 2009; Jung et al., 2012).

SUMMARY

While a number of neurologic conditions may present or worsen during pregnancy, judicious use of imaging can play a major role in early diagnosis and treatment with the goal of reducing morbidity and mortality. According to updated guidelines, no specific interruption of breastfeeding is recommended with contrast administration postpartum.

References

Abraham JL, Thakral C (2008). Tissue distribution and kinetics of gadolinium and nephrogenic systemic fibrosis. Eur J Radiol 66: 200–207.

Acciarresi M, Altavilla R, Mosconi MG et al. (2019). Management of intracranial haemorrhage, unruptured aneurysms and arteriovenous malformations during and after pregnancy. Curr Opin Neurol 32: 36–42.

ACOG (2017). ACOG comittee opinion no. 723, October 2017 (replaces committee opinion no. 656, February 2016). Guidlines for diagnostic imaging during pregnancy. Obstet Gynecol 130: e210–e216.

ACR (2008). ACR guidelines and technical standards. Practice guideline for imaging pregnant or potentially pregnant adolescents and women with ionizing radiation, first edn American College of Radiology, Reston, VA.

ACR SPR (2018). ACR-SPR practice parameter for imaging pregnant or potentially pregnant adolescents and women with ionizing radiation (resolution 39), American College of Radiology, Reston, VA.

ACR (2020). ACR manual on contrast media. V10.3 [online]. Available at: https://www.acr.org/-/media/ACR/Files/Clinical-Resources/Contrast_Media.pdf. Accessed December 12th.

Adam MP (2012). The all-or-none phenomenon revisited. Birth Defects Res A Clin Mol Teratol 94: 664–669.

Alroughani R, Alowayesh MS, Ahmed SF et al. (2018). Relapse occurrence in women with multiple sclerosis during pregnancy in the new treatment era. Neurology 90: e840–e846.

Amin R, Darrah T, Wang H et al. (2017). Editor's highlight: in utero exposure to gadolinium and adverse neonatal outcomes in premature infants. Toxicol Sci 156: 520–526.

Arko L, Quach E, Nguyen V et al. (2014). Medical and surgical management of spinal epidural abscess: a systematic review. Neurosurg Focus 37: E4.

Boellis A, di Napoli A, Romano A et al. (2014). Pituitary apoplexy: an update on clinical and imaging features. Insights Imaging 5: 753–762.

Bonneville F (2014). Imaging of cerebral venous thrombosis. Diagn Interv Imaging 95: 1145–1150.

Borisow N, Kleiter I, Gahlen A et al. (2017). Influence of female sex and fertile age on neuromyelitis optica spectrum disorders. Mult Scler 23: 1092–1103.

Bourjeily G, Chalhoub M, Phornphutkul C et al. (2010). Neonatal thyroid function: effect of a single exposure to iodinated contrast medium in utero. Radiology 256: 744–750.

Bove R, Chitnis T (2013). Sexual disparities in the incidence and course of MS. Clin Immunol 149: 201–210.

Burton TM, Bushnell CD (2019). Reversible cerebral vasoconstriction syndrome. Stroke 50: 2253–2258.

Chandrashekaran S, Parikh S, Kapoor P et al. (2007). Postpartum reversible cerebral vasoconstriction syndrome. Am J Med Sci 334: 222–224.

Chung SP, Kim SW, Yoo IS et al. (2003). Magnetic resonance imaging as a diagnostic adjunct to Wernicke encephalopathy in the ED. Am J Emerg Med 21: 497–502.

Confavreux C, Hutchinson M, Hours MM et al. (1998). Rate of pregnancy-related relapse in multiple sclerosis. Pregnancy in multiple sclerosis group. N Engl J Med 339: 285–291.

De Santis M, Di Gianantonio E, Straface G et al. (2005). Ionizing radiations in pregnancy and teratogenesis: a review of literature. Reprod Toxicol 20: 323–329.

De Wilde JP, Rivers AW, Price DL (2005). A review of the current use of magnetic resonance imaging in pregnancy and safety implications for the fetus. Prog Biophys Mol Biol 87: 335–353.

Digre KB (2013). Headaches during pregnancy. Clin Obstet Gynecol 56: 317–329.

Dinc H, Esen F, Demirci A et al. (1998). Pituitary dimensions and volume measurements in pregnancy and post partum. MR assessment. Acta Radiol 39: 64–69.

Doyle S, Messiou C, Rutherford JM et al. (2009). Cancer presenting during pregnancy: radiological perspectives. Clin Radiol 64: 857–871.

ESUR. (n.d.). Guidelines on Contrast Agents 10.0. [online]. Available at http://www.esur.org/fileadmin/content/2019/ESUR_Guidelines_10.0_Final_Version.pdf. Accessed December 12.

FDA (2018). Omnipaque (iohexol): highlights of prescribing information. [online]. Available at: https://www.accessdata.fda.gov/drugsatfda_docs/label/2018/018956s101lbl.pdf. Accessed December 12th.

FDA (2011). Gadavist (gadobutrol): highlights for prescribing information. [online]. Available at: https://www.accessdata.fda.gov/drugsatfda_docs/label/2019/201277s017lbl.pdf#page=42. Accessed 12 December 2011.

Fugate JE, Claassen DO, Cloft HJ et al. (2010). Posterior reversible encephalopathy syndrome: associated clinical and radiologic findings. Mayo Clin Proc 85: 427–432.

Gavanier D, Leport H, Massardier J et al. (2019). Gestational trophoblastic neoplasia with brain metastasis at initial presentation: a retrospective study. Int J Clin Oncol 24: 153–160.

Ghulmiyyah L, Sibai B (2012). Maternal mortality from preeclampsia/eclampsia. Semin Perinatol 36: 56–59.

Gold SM, Willing A, Leypoldt F et al. (2019). Sex differences in autoimmune disorders of the central nervous system. Semin Immunopathol 41: 177–188.

Gordon N (2009). Spontaneous intracranial hypotension. Dev Med Child Neurol 51: 932–935.

Guida M, Altieri R, Palatucci V et al. (2012). Aneurysmal subarachnoid haemorrhage in pregnancy: a case series. Transl Med UniSa 2: 59–63.

Hellwig K, Haghikia A, Rockhoff M et al. (2012). Multiple sclerosis and pregnancy: experience from a nationwide database in Germany. Ther Adv Neurol Disord 5: 247–253.

Hellwig K, Rockhoff M, Herbstritt S et al. (2015). Exclusive breastfeeding and the effect on postpartum multiple sclerosis relapses. JAMA Neurol 72: 1132–1138.

Hussenbocus SM, Wilby MJ, Cain C et al. (2012). Spontaneous spinal epidural hematoma: a case report and literature review. J Emerg Med 42: e31–e34.

Ito F, Akasaka J, Koike N et al. (2014). Incidence, diagnosis and pathophysiology of amniotic fluid embolism. J Obstet Gynaecol 34: 580–584.

Johnston M, Landers S, Noble L et al. (2012). Breastfeeding and the use of human milk. Pediatrics 129: e827–e841.

Jung YC, Chanraud S, Sullivan EV (2012). Neuroimaging of Wernicke's encephalopathy and Korsakoff's syndrome. Neuropsychol Rev 22: 170–180.

Kanal E, Barkovich AJ, Bell C et al. (2013). ACR guidance document on MR safe practices: 2013. J Magn Reson Imaging 37: 501–530.

Karaca Z, Tanriverdi F, Unluhizarci K et al. (2010). Pregnancy and pituitary disorders. Eur J Endocrinol 162: 453–475.

Karia SJ, Rykken JB, McKinney ZJ et al. (2016). Utility and significance of gadolinium-based contrast enhancement in posterior reversible encephalopathy syndrome. Am J Neuroradiol 37: 415–422.

Kastrup O, Schlamann M, Moenninghoff C et al. (2015). Posterior reversible encephalopathy syndrome: the spectrum of MR imaging patterns. Clin Neuroradiol 25: 161–171.

Kubik-Huch RAG-AN, Frenzel T, Seifert B et al. (2000). Gadopentetate dimeglumine excretion into human breast milk during lactation. Radiology 216: 4.

Langer-Gould A, Huang SM, Gupta R et al. (2009). Exclusive breastfeeding and the risk of postpartum relapses in women with multiple sclerosis. Arch Neurol 66: 958–963.

Laway BA, Mir SA (2013). Pregnancy and pituitary disorders: challenges in diagnosis and management. Indian J Endocrinol Metab 17: 996–1004.

Lee LK (2007). Physiological adaptations of pregnancy affecting the nervous system. Semin Neurol 27: 405–410.

Leo CH, Jelinic M, Ng HH et al. (2017). Vascular actions of relaxin: nitric oxide and beyond. Br J Pharmacol 174: 1002–1014.

Marcos HB, Semelka RC, Worawattanakul S (1997). Normal placenta: gadolinium-enhanced dynamic MR imaging. Radiology 205: 493–496.

Marosi C, Hassler M, Roessler K et al. (2008). Meningioma. Crit Rev Oncol Hematol 67: 153–171.

McKinney AM, Short J, Truwit CL et al. (2007). Posterior reversible encephalopathy syndrome: incidence of atypical regions of involvement and imaging findings. Am J Roentgenol 189: 904–912.

Mercado M, Melgar V, Salame L et al. (2017). Clinically nonfunctioning pituitary adenomas\pathogenic, diagnostic and therapeutic aspects. Endocrinol Diabetes Nutr 64: 384–395.

Miller RW (1990). Effects of prenatal exposure to ionizing radiation. Health Phys 59: 57–61.

Molitch ME (2017). Diagnosis and treatment of pituitary adenomas: a review. JAMA 317: 516–524.

Moon AJ, Katzberg RW, Sherman MP (2000). Transplacental passage of iohexol. J Pediatr 136: 548–549.

Mpateni SS, Sihlali NC, Gardiner EC et al. (2018). Magnetic resonance imaging findings in a patient with seropositive neuromyelitis optica. SA J Radiol 22: 1306.

Nelson JA, Livingston GK, Moon RG (1982). Mutagenic evaluation of radiographic contrast media. Invest Radiol 17: 183–185.

Ng WH, Gonzales M, Kaye AH (2003). Lymphocytic hypophysitis. J Clin Neurosci 10: 409–413.

Nie Q, Guo P, Ge J et al. (2015). Cerebral venous sinus thrombosis with cerebral hemorrhage during early pregnancy. Neurosciences (Riyadh) 20: 48–51.

Nielsen ST, Matheson I, Rasmussen JN et al. (1987). Excretion of iohexol and metrizoate in human breast milk. Acta Radiol 28: 523–526.

Novak Z, Thurmond AS, Ross PL et al. (1993). Gadolinium-DTPA transplacental transfer and distribution in fetal tissue in rabbits. Invest Radiol 28: 828–830.

Okuda Y, Sagami F, Tirone P et al. (1999). Reproductive and developmental toxicity study of gadobenate dimeglumine formulation (E7155) (3)—study of embryo-fetal toxicity in rabbits by intravenous administration. J Toxicol Sci 24: 79–87.

Osei EK, Darko J (2013). Foetal radiation dose and risk from diagnostic radiology procedures: a multinational study. ISRN Radiol 2013: 318425.

Osei EK, Faulkner K (1999). Fetal doses from radiological examinations. Br J Radiol 72: 773–780.

Patel SJ, Reede DL, Katz DS et al. (2007). Imaging the pregnant patient for nonobstetric conditions: algorithms and radiation dose considerations. Radiographics 27: 1705–1722.

Powrie JK, Powell M, Ayers AB et al. (1995). Lymphocytic adenohypophysitis: magnetic resonance imaging features of two new cases and a review of the literature. Clin Endocrinol (Oxf) 42: 315–322.

Prola-Netto J, Woods M, Roberts VHJ et al. (2018). Gadolinium chelate safety in pregnancy: barely detectable gadolinium levels in the juvenile nonhuman primate after in utero exposure. Radiology 286: 122–128.

Qiu K, He Q, Chen X et al. (2019). Pregnancy-related immune changes and demyelinating diseases of the central nervous system. Front Neurol 10: 1070.

Ranabir S, Baruah MP (2011). Pituitary apoplexy. Indian J Endocrinol Metab 15: S188–S196.

Sandhu FS, Dillon WP (1991). Spinal epidural abscess: evaluation with contrast-enhanced MR imaging. Am J Neuroradiol 12: 1087–1093.

Saposnik G, Barinagarrementeria F, Brown Jr RD et al. (2011). Diagnosis and management of cerebral venous thrombosis: a statement for healthcare professionals from the American Heart Association/American Stroke Association. Stroke 42: 1158–1192.

Shaaban AM, Rezvani M, Haroun RR et al. (2017). Gestational trophoblastic disease: clinical and imaging features. Radiographics 37: 681–700.

Singhal AB, Bernstein RA (2005). Postpartum angiopathy and other cerebral vasoconstriction syndromes. Neurocrit Care 3: 91–97.

Talib V, Sultana S, Hamad A et al. (2018). Wernicke encephalopathy due to hyperemesis gravidarum in pregnancy: a case report. Cureus 10: e2991.

Tetsuka S, Ogawa T (2019). Posterior reversible encephalopathy syndrome: a review with emphasis on neuroimaging characteristics. J Neurol Sci 404: 72–79.

Thomas J. (n.d.). Medications and breastfeeding: tips for giving accurate information to mothers. [online]. Available at: https://www.aap.org/en-us/advocacy-and-policy/aap-health-initiatives/Breastfeeding/Pages/Medications-and-Breast feeding.aspx. Accessed February 14.

Tillema JM, Pirko I (2013). Neuroradiological evaluation of demyelinating disease. Ther Adv Neurol Disord 6: 249–268.

Torgersen KL, Curran CA (2006). A systematic approach to the physiologic adaptations of pregnancy. Crit Care Nurs Q 29: 2–19.

Treadwell SD, Thanvi B, Robinson TG (2008). Stroke in pregnancy and the puerperium. Postgrad Med J 84: 238–245.

Unluhizarci K, Bayram F, Colak R et al. (2001). Distinct radiological and clinical appearance of lymphocytic hypophysitis. J Clin Endocrinol Metab 86: 1861–1864.

Varlamov EV, Hinojosa-Amaya JM, Fleseriu M (2020). Magnetic resonance imaging inthe management of prolactinomas; a review of the evidence. Pituitary 23: 16–26.

Vukusic S, Hutchinson M, Hours M et al. (2004). Pregnancy and multiple sclerosis (the PRIMS study): clinical predictors of post-partum relapse. Brain 127: 1353–1360.

Wakeford R, Little MP (2003). Risk coefficients for childhood cancer after intrauterine irradiation: a review. Int J Radiat Biol 79: 293–309.

Watts J, Box G, Galvin A et al. (2014). Magnetic resonance imaging of meningiomas: a pictorial review. Insights Imaging 5: 113–122.

Webb JA, Thomsen HS, Morcos SK (2005). The use of iodinated and gadolinium contrast media during pregnancy and lactation. Eur Radiol 15: 1234–1240.

Yamazaki JN, Schull WJ (1990). Perinatal loss and neurological abnormalities among children of the atomic bomb. Nagasaki and Hiroshima revisited, 1949 to 1989. JAMA 264: 605–609.

Zuccoli G, Pipitone N (2009). Neuroimaging findings in acute Wernicke's encephalopathy: review of the literature. Am J Roentgenol 192: 501–508.

Handbook of Clinical Neurology, Vol. 171 (3rd series)
Neurology and Pregnancy: Pathophysiology and Patient Care
E.A.P. Steegers, M.J. Cipolla and E.C. Miller, Editors
https://doi.org/10.1016/B978-0-444-64239-4.00010-2

Chapter 10

Neuro-anesthesiology in pregnancy

KYRA BERNSTEIN, HANNA HUSSEY, PATRICK HUSSEY, KAILA GORDO, AND RUTH LANDAU*

Department of Anesthesiology, Columbia University Irving Medical Center, New York, NY, United States

Abstract

Management of the pregnant patient requiring neurosurgery poses multiple challenges, juxtaposing pregnancy-specific considerations with that accompanying the safe provision of intracranial or spine surgery. There are no specific evidence-based recommendations, and case-by-case interdisciplinary discussions will guide informed decision-making about the timing of delivery vis-à-vis neurosurgery, the performance of cesarean delivery immediately before neurosurgery, consequences of neurosurgery on subsequent delivery, or even the optimal anesthetic modality for neurosurgery and/or cesarean delivery.

In general, identifying whether increased intracranial pressure poses a risk for herniation is crucial before allowing neuraxial procedures. Modified rapid sequence induction with advanced airway approaches (videolaryngoscopic or fiberoptic) allows improved airway manipulation with reduced risks associated with endotracheal intubation of the obstetric airway. Currently, very few anesthetic drugs are avoided in the neurosurgical pregnant patient; however, ensuring access to critical care units for prolonged monitoring and assistance of the respiratory-compromised patient is necessary to ensure safe outcomes.

INTRODUCTION

It is estimated that up to 2% of pregnant women will undergo nonobstetric surgery during pregnancy, although most commonly, these procedures are abdominal emergencies, such as acute appendicitis, cholecystitis, or intestinal obstructions, with preterm delivery, preterm labor without delivery, and intrauterine fetal demise after surgery being the most common adverse outcomes after these surgical procedures (Sachs et al., 2017).

Indications for neurosurgery in the pregnant patient include intracranial and spinal surgeries, along with interventional neuroradiology procedures, for management of one of the following: subarachnoid hemorrhage (SAH), intracranial hemorrhage (ICH), arteriovenous malformation (AVM), ischemic stroke, symptomatic intracranial tumor, cerebral abscesses, spinal cord tumors and lesions, and traumatic brain injury (Wang and Paech, 2008). There are no large series describing outcomes of neurosurgery during pregnancy, but tight control of maternal blood pressure, avoidance of maternal hypotension, and continued fetal heart rate monitoring during the surgical procedure are recommended to ensure adequate fetoplacental perfusion and optimized obstetric outcomes (Po et al., 2019).

BASIC CONSIDERATIONS RELATED TO THE PROVISION OF ANESTHESIA IN THE OBSTETRIC PATIENT

With thorough knowledge of all pregnancy-related physiologic changes, obstetric anesthesiologists balance the risks associated with the provision of anesthetic and analgesic medications to the mother, with those involving the fetus. Depending on the surgery, there may be choices, and a multidisciplinary approach allows the weighing of risks and complications to enable decision-making.

*Correspondence to: Ruth Landau, M.D., Virginia Apgar Professor of Anesthesiology, Director of Obstetric Anesthesia, Columbia University Irving Medical Center, Director of the Center for Precision Medicine, Department of Anesthesiology, CHONY North, CHN-1123, 3959 Broadway, New York, NY 10032, United States. Tel: +1-212-305-7474, Fax: +1-212-342-2028, E-mail: rl262@cumc.columbia.edu

Neuraxial anesthesia

For cesarean delivery, and whenever possible for nonobstetric surgery during pregnancy, neuraxial anesthesia, which includes spinal, epidural, or combined spinal–epidural (CSE) anesthesia, is the preferred anesthetic modality (Fig. 10.1).

Neuraxial anesthesia will not only reduce anesthesia-related morbidity and mortality (Guglielminotti et al., 2019), it also allows (1) avoidance of volatile agent-induced uterine relaxation and uterine atony, (2) prevention of intraoperative recall also referred to as "awareness during anesthesia," (3) enhanced postoperative pain management via long-acting lipophilic opioids, (4) decreased fetal exposure to the potentially toxic effects of general anesthesia, and (5) overall avoidance of airway manipulation, which may be challenging in the obstetric patient (Kinsella et al., 2015).

Spinal anesthesia, typically performed with small-gage pencil-point needles (25–27G), is the preferred and most common technique for cesarean delivery offering rapid onset and reliable surgical block, with overall low doses of local anesthetics and opioids (Fig. 10.2).

Epidural anesthesia is used in women in labor who had received neuraxial labor analgesia, and subsequently will require a cesarean delivery. Titration of the indwelling epidural catheter (19–20G) will achieve a surgical block that will be sufficient for cesarean delivery and may also be used for postcesarean pain relief. The advantage of epidural anesthesia is that it may be titrated slowly, when sudden sympathectomy is undesirable (e.g., women with pulmonary hypertension). CSE anesthesia, typically performed with a large epidural needle (17–18G) through which a small-gage pencil-point needle (25–27G) will be inserted before threading of the epidural catheter (19–20G), allows for a quick, dense block, as well as titration of the local anesthetic over a longer period of time, and may be favored over spinal anesthesia if de novo anesthesia is provided in the operating room.

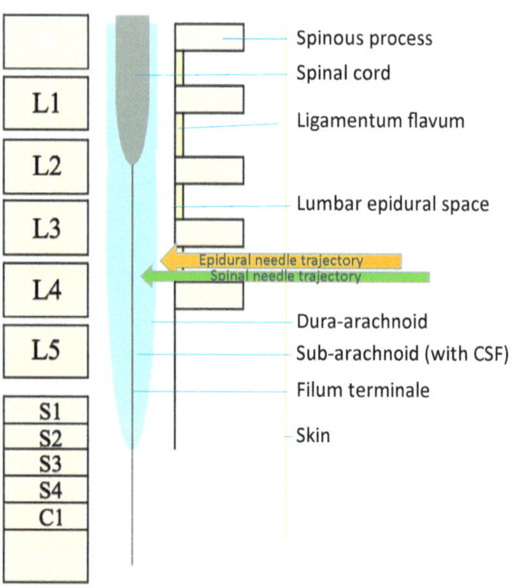

Fig. 10.1. Simplified anatomy of lumbar spinal, epidural, and subarachnoid spaces. The *orange arrow* illustrates the epidural needle trajectory. The *green arrow* depicts dural/arachnoid passage into subarachnoid space (with cerebrospinal fluid).

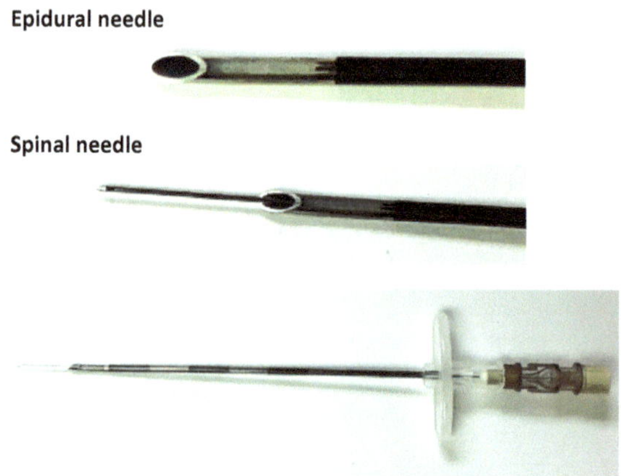

Fig. 10.2. Needles used for combined spinal–epidural (CSE) applying the needle through needle technique. The 27G spinal needle protrudes through the 17G epidural needle and reaches the subarachnoid space (dural/arachnoid puncture). Epidural needle: Tuohy or Weiss, usually 17G (cutting edge). Spinal needle: Whitacre or Sprotte, usually 25 or 27G (pencil-point edge). Shown here, a 27G Whitacre needle protruding 11 mm through a 17G epidural needle.

Maternal hypotension is the most common side effect of neuraxial blockade, although use of vasopressor infusions (typically phenylephrine) to prevent and manage spinal-induced hypotension is now standard (Kinsella et al., 2018).

Pregnant patients are more sensitive to local anesthetics, likely due to hormonal and anatomical changes. As pregnancy progresses, there is dilation of epidural veins and increased abdominal pressure, which decrease the size of the epidural space and cerebrospinal fluid (CSF) volume in the subarachnoid space. This ultimately improves the spread of local anesthetics in the space.

However, one must consider the complications of high intracranial pressure (ICP) and dural puncture, as a rapid change in CSF pressure may precipitate intracranial hemorrhage and herniation (Leffert and Schwamm, 2013). Epidural injection of medications can also cause an increase in ICP due to compression of the dural sac; therefore slow and incremental injection is recommended if increased ICP is suspected.

Patients with intracranial lesions are often assumed to have high ICP, and the risk of herniation is frequently cited as a contraindication to neuraxial anesthesia. However, women with space-occupying lesions that have no mass effect, no hydrocephalus, and no clinical or imaging findings suggestive of increased ICP are not at increased risk of herniation from a dural puncture (Leffert and Schwamm, 2013). Those at high risk of herniation from a dural puncture have lesions that compress normal brain tissue and cause it to shift across the midline or downward, with or without obstruction to the flow of CSF (Leffert and Schwamm, 2013).

Neuraxial medications at clinically relevant doses are thought not to cross the placenta, and women are informed that medication given epidurally or intrathecally will have minimal to no effect on the fetus/neonate. Contraindications to neuraxial procedures include patient refusal, infection at site, significant coagulopathy or recent administration of thromboprophylaxis (Leffert et al., 2018), hypovolemic shock, and increased ICP. In contrast, common neurological disorders such as multiple sclerosis (MS), epilepsy, and neuromuscular disorders, or unruptured aneurysm or AVMs are not contraindications to neuraxial procedures (see later under specific sections).

General anesthesia

Airway management in pregnancy is more challenging than in the nonobstetric population due to several factors, leading to a higher incidence of failed intubation for general anesthesia (Kinsella et al., 2015). Failed tracheal intubation in the pregnant patient is a dramatic situation as the presence of fetus(es) means that more than one life could potentially be compromised if severe hypoxia occurred during difficult airway management.

The incidence of failed intubation in the obstetric population is eight times higher than that in the general population. It has remained unchanged over the past 4 decades at 1:390 for general anesthesia in the obstetric population and 1:443 for cesarean deliveries (Kinsella et al., 2015). In busy tertiary obstetric centers with higher general anesthesia rates for cesarean deliveries and 24/7 specialist coverage the failed intubation incidence may actually be lower (1:462) (Teoh et al., 2012).

Factors contributing to the challenges of the pregnant airway are airway edema, respiratory and metabolic changes, weight gain and obesity, breast enlargement, gastroesophageal changes, and increased aspiration risk. Increased maternal blood volume and higher estrogen levels result in mucosal edema, capillary engorgement, and increased tissue friability. Hence intubation, insertion of nasal airways, use of orogastric or nasogastric tubes, are each associated with increased bleeding tendency. Epistaxis and soft palate hematoma can occur even after little or minimal unprovoked trauma (Teoh et al., 2013). Edema distorts laryngeal anatomy narrowing apertures and mandates intubation with tracheal tubes having smaller diameters.

Pregnancy causes several changes in the pulmonary system that ultimately lead to decreased oxygen reserve and more rapid desaturation compared with a nonpregnant patient (Guglielminotti et al., 2019). This makes supplemental oxygen and preoxygenation prior to induction of anesthesia a pivotal part of airway management in a pregnant patient. The enlarging gravid uterus pushes the diaphragm cephalad causing 15%–30% reduced expiratory reserve volumes and decreased functional residual capacity. Early airway closure can occur at normal tidal volume breathing exacerbated in supine, Trendelenburg, and pregnant women with high body mass index. Increased oxygen consumption, and potentially the pain and/or stress of labor, potentiates rapid hypoxemia necessitating denitrogenation (administration of a maximal fraction of inspired oxygen (FiO_2) with tight-fitting mask) prior to rapid sequence induction (RSI) to achieve the longest apneic duration before desaturation. This is best achieved by elevating the head of the bed by 25 degrees (Dixon et al., 2005). The standard technique for preoxygenation is to breathe 100% oxygen for 3–5 min of tidal volume (Tanoubi et al., 2009); however, given the emergent nature of general anesthesia in obstetrics, eight deep breaths over 60 s have been shown to provide adequate denitrogenation as measured by end-tidal fractional oxygen concentration (F_{ETO2}) (Chiron et al., 2004).

While RSI has historically implied succinylcholine administration to achieve rapid onset muscle paralysis before endotracheal intubation, succinylcholine is absolutely contraindicated in patients with known or suspected elevated ICP. Alternatives to RSI with succinylcholine in the obstetric population include use of rocuronium (Sharp and Levy, 2009), which may now be reversed at any time point if needed with sugammadex. The Society for Obstetric Anesthesia and Perinatology (SOAP) has provided recommendations on the use of sugammadex in the obstetric patient (https://soap.org/wp-content/uploads/2019/06/SOAP_Statement_Sugammadex_During_Pregnancy_Lactation_APPROVED.pdf).

Mammomegaly in supine pregnant women often impedes the insertion and manipulation of the laryngoscope to achieve good glottic visualization. The "head-ramped" position (where the external auditory meatus is aligned horizontally with the sternal notch) is advocated (El-Orbany et al., 2011).

Although maternal mortality from pulmonary aspiration of gastric contents has declined to negligible rates in the past 3 decades, pregnant women remain at risk for regurgitation and aspiration of gastric contents due to hormonal changes (increased gastrin, decreased motilin, and progesterone-induced relaxation of gastrointestinal smooth muscle decreasing lower esophageal sphincter tone). Historical data identified a threefold higher aspiration risk in women undergoing cesarean delivery, with an overall incidence of aspiration of 1:2131 in the general population undergoing anesthesia vs 1:661 in obstetric patients (Olsson et al., 1986). Pregnant women are therefore considered to have a "full stomach" by second trimester, and prophylaxis with inhibitors of gastric acid secretion or sodium citrate and RSI remain the standard of care.

Preserving hemodynamics is an important goal during surgery to maintain maternal and uteroplacental perfusion. The main cardiovascular changes in pregnancy include an increase in cardiac output, increase in intravascular volume, decrease in systemic vascular resistance, and supine aortocaval compression. Cardiac output increases to about 50% above baseline by the third trimester due to increases in both heart rate and stroke volume. Strategies to maintain perfusion involve maternal positioning with left displacement of gravid uterus, fluids, and vasopressors.

Another important consideration in pregnancy is that anesthetic requirement is reduced by approximately 25% by the first trimester. The volatile anesthetic agents most commonly used to maintain anesthesia in pregnancy are isoflurane and sevoflurane, as they have been shown to reduce cerebral metabolic rate and have the least effect on ICP. However, they also cause uterine relaxation and are often decreased or turned off (switched to intravenous propofol infusion) once the baby is delivered to avoid uterine atony and excessive maternal blood loss.

Overall, general anesthesia for cesarean delivery or nonobstetric procedures is used only when neuraxial technique is contraindicated or when it cannot be avoided. A comprehensive understanding of maternal physiology, its implications regarding anesthesia, and the difficult airway algorithm is crucial in providing a safe general anesthetic to a pregnant patient.

CHRONIC NEUROLOGIC CONDITIONS AND ANESTHETIC MANAGEMENT OF THE PREGNANT PATIENT REQUIRING SURGERY

Numerous neurologic conditions may constitute maternal comorbidities, and some of the more common or more critical conditions that may complicate management of the obstetric patient during pregnancy or at the time of delivery are reviewed here.

Epilepsy

The prevalence of epilepsy in pregnancy is 0.3%–0.7%, making it one of the most common neurologic conditions in pregnancy (Stephen et al., 2019). There is a higher incidence of status epilepticus in pregnant compared with nonpregnant women and increased mortality in epileptic pregnant patients (80 per 100,000 patients) compared with nonepileptic pregnant patients (6 per 100,000) (MacDonald et al., 2015). There are no contraindications to general or neuraxial anesthesia in epileptic parturients, and no routine changes in medications used for anesthetic care are made. Antiepileptic drugs require careful titration during pregnancy, with care taken to minimize teratogenicity for the fetus, although most antiepileptic drugs are teratogenic.

Most cases of status epilepticus in pregnancy are not due to underlying epilepsy, and differential diagnosis should include eclampsia, posterior reversible encephalopathy syndrome, reversible cerebral vasoconstriction syndrome, cortical venous sinus thrombosis, and autoimmune encephalitis. Initial treatment involves stabilizing airway, breathing and circulation, and differentiating eclamptic from noneclamptic seizures (Rajiv and Radhakrishnan, 2019). Eclamptic seizures require management with magnesium sulfate and control of hypertension. Noneclamptic seizures will be managed in the same way as in nonpregnant patients (e.g., benzodiazepines, including lorazepam and midazolam). Midazolam has the added benefit of being titratable, short acting, and available to be administered intramuscularly as well as intravenously.

Fetal care throughout the initial phase of treatment necessitates fetal monitoring, delivery near term after the maternal condition has been stabilized, anticipation of potential premature delivery, administration of betamethasone when indicated, and neonatal ICU involvement and counseling.

Multiple sclerosis

Women with MS who become pregnant usually encounter a "remission" of symptoms particularly during the third trimester, with a risk for relapse in the immediate postpartum period. The mode of delivery, presence or absence of anesthetic, and anesthetic technique do not influence the rate of postpartum relapse (Dorotta and Schubert, 2002; Mankowitz, 2018; Lavie et al., 2019).

Historical concern with neuraxial procedures and specifically spinal anesthesia centered on the potential neurotoxic effect of local anesthetic on demyelinated nerves, with concern that lidocaine may unmask MS symptoms (Stoelting et al., 2008). However, spinal, epidural, and CSE anesthesia have all been administered safely in patients with MS and are not contraindicated (Pasto et al., 2012; Lu et al., 2013; Harazim et al., 2018).

A study of 423 pregnancies in 415 patients with MS at 21 centers in Italy demonstrated no correlation between type of anesthetic care and disease progression at 1 year (Pasto et al., 2012). However, increased disability (defined by higher EDSS score) at conception was correlated with disease progression. This suggests that patients with more disability may benefit from aggressive MS treatment in the postpartum period.

Current obstetric anesthesia recommendations entail no change in practice in women with MS, and CSE for neuraxial labor analgesia and spinal anesthesia for cesarean delivery are routinely offered to pregnant women with MS.

General anesthesia may be necessary during any delivery for failed block, massive hemorrhage and resuscitation, or emergent obstetric indications such as cord prolapse and fetal bradycardia. Preoperative documentation of neurologic examination and symptoms is critical on arrival to the obstetric unit to risk-stratify patients and identify potential new symptoms. Particular attention should be paid to symptoms of chronic aspiration, respiratory muscle weakness, and baseline lower extremity exam. Interactions with medications should be reviewed for potential effects on anesthetic agents (e.g., resistance to nondepolarizing neuromuscular blockade in patients on anticonvulsants, sensitivity with baclofen use) and side effects (e.g., pulmonary fibrosis from cyclophosphamide) (Dorotta and Schubert, 2002). In general, there are no contraindications to intravenous hypnotics or inhalational anesthetics. Succinylcholine is avoided due to potential for hyperkalemia and i.v. lidocaine due to concern for potential disease progression. Stressors such as hyperthermia and anxiety in the surgical period may also lead to disease progression and should be managed in an active manner.

Women with MS warrant particular attention postcesarean delivery due to the potential for hypoventilation as a result of baseline disease and use of epidural or spinal morphine for pain control, and the SOAP guidelines on monitoring for respiratory depression should be followed (Bauchat et al., 2019).

In the event of a postdural puncture headache, epidural blood patches have been safely administered in women with MS, although some advocate for slow injection of blood and somatosensory evoked potential monitoring (Makris et al., 2014).

Myasthenia gravis

While pregnancy may cause myasthenia gravis (MG) flares in 50% of patients, it does not alter the overall course of the disease (Toscano and Thornburg, 2019), which is unpredictable during this period (Hopkins et al., 2014).

Based on the 2016 international consensus statement on management of MG, treatment goals include optimization of disease state with oral pyridostigmine and prednisone as the immunosuppressant of choice, with azathioprine and cyclosporine as back-up options (Sanders et al., 2016).

Vaginal delivery is preferred because surgery itself is a risk factor in MG crisis, although patients with MG have an increased risk of operative vaginal delivery and a cesarean delivery rate of up to 60% due to muscular fatigue (Ducci et al., 2017). Magnesium sulfate is contraindicated in the event of preeclampsia in a woman with MG due to exacerbation of muscle weakness and respiratory depression. Barbiturates or phenytoin may be used instead, and if magnesium sulfate use is required, intubation and ventilatory support must be readily available (Toscano and Thornburg, 2019). Reversal of neuromuscular blockade with sugammadex has been reported in a woman with MG. Neonates born to myasthenic mothers should be carefully observed for signs of weakness due to transplacental passage of antibodies (Mankowitz, 2018).

Anesthetic preparation for gravid myasthenic patients includes a dedicated anesthetic consultation prior to patient presentation on the labor and delivery unit (Hopkins et al., 2014).

MG does not constitute a contraindication for neuraxial anesthesia, and early neuraxial labor analgesia is encouraged to mitigate the potentially crisis-inducing stress of labor and will allow for an assisted second stage. (Varner, 2013) Amide local anesthetic agents should be used in neuraxial preparations as ester local anesthetics are metabolized by pseudocholinesterases, which are rendered less effective by pyridostigmine (e.g., avoid chloroprocaine).

General anesthesia should be avoided if possible, although it may be required for emergent obstetric indications. Myasthenic patients may demonstrate an amplified and prolonged response to nondepolarizing neuromuscular blockers, whereas they may have a diminished response to succinylcholine (Stoelting et al., 2008). Respiratory depression in the postoperative period may be profound and prolonged intubation and monitoring in an ICU may be required.

ACUTE NEUROLOGIC DISEASES THAT MAY REQUIRE ANESTHETIC CARE IN PREGNANT WOMEN

A variety of acute neurologic conditions may affect women during pregnancy, which may either require surgery as a therapeutic intervention or interact with the provision of anesthesia at the time of delivery, whether it be neuraxial labor analgesia or anesthesia for cesarean delivery.

The most common conditions are reviewed here.

Lumbar disc herniation

Lumbar disc herniation (LDH) has been reported to affect 1 in 10,000 pregnant women. Low back pain affects up to 50%–75% of women during pregnancy and may be due to the anterior shift of the center of gravity and subsequent hyperlordosis in response to the growing fetus(es), placenta(es), and uterus.

Failure to treat significant herniation can lead to long-term deficits, and 2% of patients with LDH may develop cauda equina syndrome. Worsening back pain deserves a thorough neurologic exam. Magnetic resonance imaging scan is recommended in pregnant women to confirm a herniated disc or evaluate cauda equina symptoms (Ahern et al., 2019).

Spine surgery

With failure of conservative management, a decompression laminectomy (e.g., microdiscectomy) in the prone or lateral position may be indicated if pain is worsening, with suspected spinal cord compression or with bowel/bladder dysfunction (Di Martino et al., 2017; Hayakawa et al., 2017; Kapetanakis et al., 2017; Ahern et al., 2018; Kovari and Horvath, 2018). General anesthesia will typically be provided, although microdiscectomy in the right lateral position with spinal anesthesia has also been reported (Kovari and Horvath, 2018).

Neuraxial labor analgesia and cesarean delivery anesthesia

With regards to neuraxial labor analgesia or cesarean delivery anesthesia, disc herniation is not an absolute contraindication in the former. However, patients with symptomatic disc disease may have an increased risk for neurologic complications. Relevant for future pregnancies and anticipated neuraxial procedures, prior lumbar discectomy surgery does not alter the efficacy of subsequent neuraxial labor analgesia (Bauchat et al., 2012).

Guillain–Barre syndrome

In the general population, the incidence of Guillain–Barre syndrome (GBS) is about 2/100,000 per year, and there were 2 cases of GBS during pregnancy in the Vaccine Safety Datalink cohort of 2.5 million women of reproductive age in the United States (Myers et al., 2019).

There does not appear to be any significant protection or predisposition of developing GBS in pregnancy, although vaccination (e.g., Zika, flu, H1N1) is associated with GBS (Tomimatsu et al., 2016; Myers et al., 2019).

Treatment during pregnancy should follow the same principles as for nonpregnant individuals (Pacheco et al., 2016). Effective treatment options include plasmapheresis and intravenous immunoglobulin therapy, both of which are considered safe in pregnancy. Respiratory muscle weakness may require mechanical ventilation support, and delivery in the ICU has been reported in such circumstances (Jain et al., 2019). Neuraxial anesthesia is not contraindicated; however, deficits should be documented and rare cases of relapses have been reported (Wiertlewski et al., 2004; Meenakshi-Sundaram et al., 2014). The benefits of neuraxial anesthesia appear to outweigh the risks, bearing in mind that women with GBS may be more sensitive to local anesthetics and may experience an exaggerated sympathetic block, although this has not always been reported to be an issue (Brooks et al., 2000; Kocabas et al., 2007; Kim et al., 2013; Volquind et al., 2013; Bouslama et al., 2017).

There is no contraindication for general anesthesia if required; however, autonomic dysfunction may be present and anesthesia should be carried out carefully depending on the extent of dysfunction. Succinylcholine should be avoided (Feldman, 1990), and nondepolarizing neuromuscular blockers should be used with caution as there is increased sensitivity.

Intracranial mass lesions

Of the intracranial tumors due to neurologic symptoms that have been seen in pregnancy, gliomas represent the majority, followed by meningioma and acoustic neuroma. Tumors such as meningiomas and pituitary adenomas can be hormone responsive and therefore may enlarge during pregnancy. Metastatic intracranial lesions are more common than primary tumors in pregnancy. Symptoms may be confused with other diseases in pregnancy, and careful history and physical exam are imperative.

Major concerns with intracranial tumors include ICP and increased risk for tumor-induced seizures. The decision to proceed with the surgical resection of an intracranial tumor during pregnancy vs after delivery should be based on presenting symptoms and the gestational age (Verheecke et al., 2014; Laviv et al., 2018a, b).

Neurosurgery (craniotomy)

There are scarce reports on the management of pregnant women undergoing tumor resections, most of which took place with general anesthesia for cesarean delivery immediately followed by the tumor resection (Sahu et al., 2010; Kazemi et al., 2014; Khurana et al., 2014). Hyperventilation and osmotic diuresis with mannitol are recommended strategies. Awake craniotomy, usually with dexmedetomidine for sedation, has also been reported in the obstetric population (Abd-Elsayed et al., 2013; Handlogten et al., 2015; Meng et al., 2016; Hedayat et al., 2017; Kamata et al., 2017; Al Mashani et al., 2018).

Neuraxial labor analgesia and cesarean delivery anesthesia

Neuraxial anesthesia has generally been contraindicated in patients with known increased ICP, and inadvertent dural puncture has been thought to increase the risk of cerebral herniation. If a primary tumor or metastasis is located remotely from CSF pathways, it will likely cause no or inconsequential ventricular compression and have no impact on CSF flow. In contrast, if the lesion partially or completely obstructs CSF flow, then the risk of brain herniation due to either intentional (spinal) or unintentional dural puncture will be increased. Space-occupying lesions that narrow the foramen magnum or are situated in the posterior fossa can place a parturient at significant risk of herniation. This can occur from benign or malignant tumors anywhere in the posterior fossa or at the opening to the foramen magnum. It can also occur from low-lying cerebellar tonsils, due to either a preexisting Arnold–Chiari malformation or intracranial hypotension from a persistent CSF leak.

As previously indicated, injection of medication into the lumbar epidural space (bolus) may also increase ICP due to the upward displacement of CSF into the intracranial compartment (Hilt et al., 1986). However, without evidence of increased ICP or obstructive CSF flow, neuraxial anesthesia can be offered.

The question about whether an epidural procedure might be preferable over a spinal one is valid, although inadvertent dural puncture can occur during epidural placement, even in expert hands, and this would occur with large-gage needle (17 or 18G). Therefore a small-gage pencil-point spinal needle (25 or 27G) might be preferable, and it has been shown to result in shorter duration and minimal CSF leak.

Planned cesarean delivery is preferred to avoid the high ICP associated with Valsalva maneuvers during childbirth; however, assisted vaginal delivery may be feasible with a carefully titrated neuraxial technique.

If general anesthesia is indicated, precautions for increased ICP should be in place. This often requires placement of an intraarterial line to maintain cerebral perfusion pressure. To minimize the hemodynamic responses to laryngoscopy, intubation and extubation are warranted.

Intracranial vascular pathology

Implications for anesthesia management of the pregnant patient with intracranial arterial or vascular pathology requires a thorough understanding of the type of vascular abnormality and the risk of rupture in the peripartum period. For pregnant women with an SAH due to aneurysmal rupture, neurosurgical management should be the same as if she were not pregnant (Roth and Deck, 2019).

Advanced maternal age, African American race, Hispanic ethnicity, hypertensive disorders, coagulopathy, tobacco, drug or alcohol abuse, intracranial venous thrombosis, sickle cell disease, and hypercoagulability are risk factors for pregnancy-related SAH (Bateman et al., 2012).

Neurosurgery (craniotomy/embolization)

There are anecdotal reports of the management of endovascular emergencies during pregnancy, which have reported clipping or coiling (Tarnaris et al., 2012; Kataoka et al., 2013; Kim et al., 2014; Fritzsche et al., 2017).

For patients requiring combined procedures (craniotomy/embolization and cesarean delivery), general anesthesia might be preferable although neuraxial anesthesia for cesarean delivery followed by general anesthesia for the craniotomy may be recommended and offers the advantages of allowing the patient to experience the delivery of her child and have neuraxial analgesic medication for postcesarean pain management.

Neuraxial labor analgesia and cesarean delivery anesthesia

Avoiding neuraxial anesthesia in parturients with intracranial hemorrhage or in the setting of altered intracranial vasculature out of concern for precipitating neurologic complications has historically been advocated. However, risks with general anesthesia in this population and information about the vascular abnormality and its risk of rupture are crucial to making informed decisions about anesthetic management.

Unruptured brain aneurysms have historically been considered an absolute indication for cesarean delivery to avoid prolonged Valsalva maneuvers and expulsive efforts-associated increases in cerebral blood and CSF volume and increases in ICP. However, the risk of aneurysmal SAH was not shown to be increased during pregnancy, labor, or postpartum compared with that in the general population, probably because Valsalva maneuvers with substantial pushing during the second stage of labor may be avoided with effective epidural analgesia and followed by instrumental vaginal delivery, suggesting that vaginal deliveries with "maternal sparing" may be offered to women with unruptured aneurysms (Tiel Groenestege et al., 2009). In addition, a significant proportion of peripartum SAH is likely nonaneurysmal in etiology (Bateman et al., 2012).

For patients with a recent intracranial hemorrhage from an aneurysm or AVM, evaluating the risk of mass effect, ICP, or CSF obstruction is paramount before proceeding with a neuraxial procedure.

For patients with a stable aneurysm who do not require simultaneous aneurysm repair, neuraxial anesthesia is appropriate if hemodynamic stability is ensured.

Stroke

The incidence of maternal strokes, ischemic (IS) or hemorrhagic (HS), was estimated to be 30 in 100,000 pregnancies (Swartz et al., 2017) and includes nontraumatic intracerebral hemorrhage (ICH) and SAH, dural sinus thrombosis, and cerebral venous thrombosis (CVT). Recent data suggests that strokes account for 7.4% of maternal deaths in the United States (Miller and Leffert, 2019). Pregnancy-specific risk factors for stroke include hypertensive disorders of pregnancy via endothelial dysfunction and impaired cerebral autoregulation, gestational diabetes, severe postpartum hemorrhage, and cesarean delivery (Miller and Leffert, 2019). Stroke during pregnancy most often occurs close to the time of delivery although up to 50% occurs in the immediate postpartum period (Swartz et al., 2017; Too et al., 2018).

Pregnant women suspected of having IS or HS should be evaluated for the same therapies as nonpregnant women (Ladhani et al., 2018; Kozberg and Camargo, 2019; Miller and Leffert, 2019; Elgendy et al., 2020), and immediate multidisciplinary care is warranted to avoid any delays in managing the maternal emergency.

Recombinant tissue plasminogen activator (r-tPA) has emerged as a safe treatment option for ischemic stroke in the pregnant patient (Ladhani et al., 2018). If fibrinolysis is contraindicated, catheter-based thrombolysis or thrombectomy are alternative treatment options. Pregnancy should not be considered a contraindication to angiography and endovascular thrombectomy for proximal large vessel occlusions causing acute disabling stroke (Ladhani et al., 2018; Limaye et al., 2019; Szuchy Kristiansen et al., 2019; Watanabe et al., 2019; Wiącek et al., 2019).

Neuraxial labor analgesia or cesarean delivery anesthesia

Anesthetic management will largely depend on the recent administration of anticoagulation/thrombolysis and if ICP is increased.

In the setting of unruptured AVMs and aneurysm, assisted second stage of labor is sometimes advised, and in that case, neuraxial labor analgesia will be beneficial. Occasionally, cesarean delivery is recommended with immediate neurosurgical intervention to follow.

Traumatic brain injury

Domestic/intimate partner violence and motor vehicle accidents (MVAs) are the predominant causes of trauma during pregnancy (Einav et al., 2013; Rossignol, 2016). Pregnant trauma victims have a twofold increased mortality rate compared with their nonpregnant counterparts, whether trauma was violent (homicide, assault) or nonviolent (MVA or accidental fall) (Deshpande et al., 2017).

The Advanced Trauma Life Support® (ATLS) guidelines provide a framework for rapid assessment and management of the injured patient and have been demonstrated to improve patient outcomes. Modifications to ATLS guidelines that may be considered for the pregnant casualty are provision of supplemental oxygen because of maternal susceptibility to hypoxia and desaturation, preference for establishment of intravenous access above the diaphragm, and left lateral positioning of the patient as soon as possible, because of the possibility of reduced venous return secondary to uterine pressure on the vena cava, aortocaval syndrome, or supine hypotensive syndrome (Einav et al., 2013). However, some updates have deemphasized spinal immobilization in favor of restriction of spinal motion in case of spinal cord injury (Galvagno Jr. et al., 2019). Although there is no consensus on the best method of intubation in patients with cervical-spine injury, fiberoptic techniques may be preferable in pregnant patients because of the additional difficulty that may arise from pregnancy and an unstable neck.

FETAL AND NEONATAL CONSIDERATIONS IN THE PREGNANT PATIENT UNDERGOING NEUROSURGERY

Neurosurgical procedures may require anticonvulsant therapy in the perioperative and postoperative periods. All anticonvulsants cross the placenta. The fetal congenital anomaly rate in pregnant women with epilepsy on anticonvulsant drugs is 4%–8% compared with the general population (Chestnut, 2009). Such medications have even coined the broad term of fetal anticonvulsant syndrome consisting of orofacial, cardiovascular, digital malformation, and neural tube defects. These drugs include the following: phenytoin, carbamazepine, phenobarbital, and valproic acid. There is unequivocal data on newer anticonvulsant therapies. For example, felbamate was approved for monotherapy treatment of epilepsy but was later restricted secondary to its association with aplastic anemia and liver failure (Chestnut, 2009). Historically, lamotrigine has been utilized in pregnancy given its relative safety during this period, but it requires complex titration schedules.

Levetiracetam has demonstrated similar safety profiles in the pregnant population. The rate of major congenital malformation associated with exposure to levetiracetam monotherapy is comparable to that in the nonepilepsy population (Koubeissi, 2013). Levetiracetam is very easy to titrate and can be given both orally and intravenously. Frequently, it is used in the perioperative setting for seizure prophylaxis. Regardless, neurologic guidance should be sought as pregnancy-induced changes can alter the clearance, unbound fractions, and half-lives of many anticonvulsant drugs.

Other induction medications appear safe in pregnancy and include short-acting opioids, local anesthetics, and intravenous anesthetic agents. Prior to the advent of propofol, thiopental was frequently used and considered safe in pregnancy. In several countries, propofol was stated to be contraindicated in pregnancy. However, both agents have demonstrated acceptable safety profiles during their use in pregnancy. Of note, intravenous magnesium sulfate (30–60 mg/kg bolus) given after induction is effective in patients with SAH or preeclampsia/eclampsia.

Volatile anesthesia, such as sevoflurane and isoflurane, involves inhalation agents of choice in neuroanesthesia as they decrease the cerebral metabolic rate and provide cerebral protection. During pregnancy, it is important to note that the minimum alveolar concentration of volatile agents is reduced by approximately 25%. Nitrous oxide should be avoided as it increases ICP and cerebral blood flow and cerebral oxygen metabolic rate, impairs autoregulation, and expands air bubbles.

Either total intravenous anesthesia or balanced intravenous and volatile anesthesia and/or solely volatile anesthesia are acceptable maintenance options for surgery during pregnancy. Intraoperative blood pressure maintenance is crucial and often intraarterial blood pressure monitoring is recommended for induction. This is crucial to preserve both cerebral and uteroplacental perfusion.

Intraoperatively, a variety of measures are applied to avoid elevations in ICP. Severe hyperventilation ($PaCO_2 < 25$ mmHg) may cause uterine artery vasoconstriction and leftward shift of maternal oxyhemoglobin dissociation curve (Lars Peter Wang, 2008). Thus it is recommended to keep $PaCO_2$ between 25 and 30 mmHg. Mannitol given to pregnant women can accumulate in the fetus leading to fetal hyperosmolality, reduced fetal lung fluid production and urinary blood flow, and increased plasma sodium concentration. However, mannitol in doses of 0.25–0.5 mg/kg appears safe in pregnancy. There is no data to support the use of hypertonic saline (23.4%) for refractory elevated ICP in pregnant women; in fact, historical studies report on the use of hypertonic saline in medical abortions. A single dose of steroid is not teratogenic and accelerates fetal lung maturity, and antiemetic drugs appear safe to use in pregnancy. Calcium channel blockers have not been shown to increase teratogenic risk and are regarded as safe in pregnancy (Alabdulrazzaq, 2012).

Nimodipine is commonly used to reduce the incidence of intracranial vasospasm, and it has been used in pregnant patients without apparent adverse events in fetal or neonatal outcomes. Esmolol 0.5–1 mg/kg may cause fetal bradycardia and cautious use in pregnancy is advised.

It should be assumed that all postpartum woman will be breastfeeding. Only few drugs are absolutely contraindicated during breastfeeding, and they include some cytotoxic drugs, some immunosuppressive drugs, and radioactive compounds that are rarely implemented during a neurosurgical procedure. Current recommendations no longer involve the "pump and dump" dogma (Dodd and Sharpe, 2018), and women are encouraged to breastfeed after anesthesia (Reece-Stremtan et al., 2017; Martin et al., 2018).

CONCLUSION

Management of pregnant women with acute or chronic neurological conditions may potentially include providing anesthesia for two distinct events: (1) the anesthetic required for a neurosurgical procedure during pregnancy, which might entail, depending on the gestational age, continuous intraoperative monitoring of the fetal heart rate and (2) neuraxial labor analgesia or anesthesia for cesarean delivery at the time of delivery.

Factors to be considered before proceeding with a neuraxial procedure in women with neurological conditions involve the risk of herniation with increased ICP; however, in most instances, if a primary tumor or metastasis is located remotely from CSF pathways, it will likely cause no or inconsequential ventricular compression and have no impact on CSF flow; therefore a multidisciplinary evaluation is recommended to weigh the risks and benefits of providing general vs neuraxial anesthesia, whenever possible.

With general anesthesia, manipulation of the "pregnant" airway carries its own risks, and general anesthesia for cesarean delivery or nonobstetric procedures is used only when neuraxial technique is contraindicated or when it cannot be avoided.

Overall, there are very few anesthetic drugs that should be avoided in the neurosurgical pregnant patient; however, ensuring access to critical care units for prolonged monitoring and assistance of the respiratory-compromised patient is necessary for safe outcomes.

References

Abd-Elsayed AA, Diaz-Gomez J, Barnett GH et al. (2013). A case series discussing the anaesthetic management of pregnant patients with brain tumours. F1000Res 2: 92.

Ahern DP, Gibbons D, Dodds M et al. (2018). Operative management of perinatal lumbar disc herniation and cauda equina syndrome: a case series. Ir Med J 111: 843.

Ahern DP, Gibbons D, Johnson GP et al. (2019). Management of herniated lumbar disk disease and cauda equina syndrome in pregnancy. Clin Spine Surg 32: 412–416.

Al Mashani AM, Ali A, Chatterjee N et al. (2018). Awake craniotomy during pregnancy. J Neurosurg Anesthesiol 30: 372–373.

Alabdulrazzaq F (2012). Fetal safety of calcium channel blockers. Can Fam Physician 58: 746–747.

Bateman BT, Olbrecht VA, Berman MF et al. (2012). Peripartum subarachnoid hemorrhage: nationwide data and institutional experience. Anesthesiology 116: 324–333.

Bauchat JR, McCarthy RJ, Koski TR et al. (2012). Prior lumbar discectomy surgery does not alter the efficacy of neuraxial labor analgesia. Anesth Analg 115: 348–353.

Bauchat JR, Weiniger CF, Sultan P et al. (2019). Society for obstetric anesthesia and perinatology consensus statement: monitoring recommendations for prevention and detection of respiratory depression associated with administration of neuraxial morphine for cesarean delivery analgesia. Anesth Analg 129: 458–474.

Bouslama MA, Brahim A, Kaabia O et al. (2017). Anesthesia for cesarean section in parturient with Guillain Barre syndrome: a case report. Tunis Med 95: 229–232.

Brooks H, Christian AS, May AE (2000). Pregnancy, anaesthesia and Guillain Barre syndrome. Anaesthesia 55: 894–898.

Chestnut D (2009). Chestnut's obstetric anesthesia: principles and practive, fourth edn Mosby/Elsevier, Philadelphia.

Chiron B, Laffon M, Ferrandiere M et al. (2004). Standard preoxygenation technique versus two rapid techniques in pregnant patients. Int J Obstet Anesth 13: 11–14.

Deshpande NA, Kucirka LM, Smith RN et al. (2017). Pregnant trauma victims experience nearly 2-fold higher mortality compared to their nonpregnant counterparts. Am J Obstet Gynecol 217: 590 e591–590 e599.

Di Martino A, Russo F, Denaro L et al. (2017). How to treat lumbar disc herniation in pregnancy? A systematic review on current standards. Eur Spine J 26: 496–504.

Dixon BJ, Dixon JB, Carden JR et al. (2005). Preoxygenation is more effective in the 25 degrees head-up position than in the supine position in severely obese patients: a randomized controlled study. Anesthesiology 102: 1110–1115. discussion 1115A.

Dodd SE, Sharpe EE (2018). Pump and dump; anesthesiologists lead the feed. Anesthesiology 128: 1046–1047.

Dorotta IR, Schubert A (2002). Multiple sclerosis and anesthetic implications. Curr Opin Anaesthesiol 15: 365–370.

Ducci RD, Lorenzoni PJ, Kay CS et al. (2017). Clinical follow-up of pregnancy in myasthenia gravis patients. Neuromuscul Disord 27: 352–357.

Einav S, Sela HY, Weiniger CF (2013). Management and outcomes of trauma during pregnancy. Anesthesiol Clin 31: 141–156.

Elgendy IY, Gad MM, Mahmoud AN et al. (2020). Acute stroke during pregnancy and puerperium. J Am Coll Cardiol 75: 180–190.

El-Orbany M, Woehlck H, Salem MR (2011). Head and neck position for direct laryngoscopy. Anesth Analg 113: 103–109.

Feldman JM (1990). Cardiac arrest after succinylcholine administration in a pregnant patient recovered from Guillain–Barre syndrome. Anesthesiology 72: 942–944.

Fritzsche FS, Regelsberger J, Schmidt NO et al. (2017). Maternal aneurysmal subarachnoid hemorrhage during pregnancy as an interdisciplinary task. Z Geburtshilfe Neonatol 221: 276–282.

Galvagno Jr. SM, Nahmias JT, Young DA (2019). Advanced trauma life support((R)) update 2019: management and applications for adults and special populations. Anesthesiol Clin 37: 13–32.

Guglielminotti J, Landau R, Li G (2019). Adverse events and factors associated with potentially avoidable use of general anesthesia in cesarean deliveries. Anesthesiology 130: 912–922.

Handlogten KS, Sharpe EE, Brost BC et al. (2015). Dexmedetomidine and mannitol for awake craniotomy in a pregnant patient. Anesth Analg 120: 1099–1103.

Harazim H, Stourac P, Janku P et al. (2018). Obstetric anesthesia/analgesia does not affect disease course in multiple sclerosis: 10-year retrospective cohort study. Brain Behav 8: e01082.

Hayakawa K, Mizutani J, Suzuki N et al. (2017). Surgical management of the pregnant patient with lumbar disc herniation in the latter stage of the second trimester. Spine (Phila Pa 1976) 42: E186–E189.

Hedayat H, Felbaum DR, Reynolds JE et al. (2017). Awake craniotomy for the treatment of a cortical pseudoaneurysm in a pregnant patient. Cureus 9: e1921.

Hilt H, Gramm HJ, Link J (1986). Changes in intracranial pressure associated with extradural anaesthesia. Br J Anaesth 58: 676–680.

Hopkins AN, Alshaeri T, Akst SA et al. (2014). Neurologic disease with pregnancy and considerations for the obstetric anesthesiologist. Semin Perinatol 38: 359–369.

Jain R, Rathi PS, Telang K et al. (2019). A case of Guillain–Barre syndrome with pregnancy who delivered in ICU: a rare outcome of rare co-occurrence. BMJ Case Rep 12 (11): e230650. https://doi.org/10.1136/bcr-2019-230650.

Kamata K, Fukushima R, Nomura M et al. (2017). A case of left frontal high-grade glioma diagnosed during pregnancy. JA Clin Rep 3: 18.

Kapetanakis S, Giovannopoulou E, Blontzos N et al. (2017). Surgical management for lumbar disc herniation in pregnancy. J Gynecol Obstet Hum Reprod 46: 753–759.

Kataoka H, Miyoshi T, Neki R et al. (2013). Subarachnoid hemorrhage from intracranial aneurysms during pregnancy and the puerperium. Neurol Med Chir (Tokyo) 53: 549–554.

Kazemi P, Villar G, Flexman AM (2014). Anesthetic management of neurosurgical procedures during pregnancy: a case series. J Neurosurg Anesthesiol 26: 234–240.

Khurana T, Taneja B, Saxena KN (2014). Anesthetic management of a parturient with glioma brain for cesarean section immediately followed by craniotomy. J Anaesthesiol Clin Pharmacol 30: 397–399.

Kim H, Ryu J, Hwang JW et al. (2013). Anesthetic management for cesarean delivery in a Guillain–Barre syndrome patient—a case report. Korean J Anesthesiol 64: 268–271.

Kim KD, Chang CH, Choi BY et al. (2014). Endovascular treatment of a ruptured posterior inferior cerebellar artery aneurysm during pregnancy. J Korean Neurosurg Soc 55: 273–276.

Kinsella SM, Winton AL, Mushambi MC et al. (2015). Failed tracheal intubation during obstetric general anaesthesia: a literature review. Int J Obstet Anesth 24: 356–374.

Kinsella SM, Carvalho B, Dyer RA et al. (2018). International consensus statement on the management of hypotension with vasopressors during caesarean section under spinal anaesthesia. Anaesthesia 73: 71–92.

Kocabas S, Karaman S, Firat V et al. (2007). Anesthetic management of Guillain–Barre syndrome in pregnancy. J Clin Anesth 19: 299–302.

Koubeissi M (2013). Levetiracetam: more evidence of safety in pregnancy. Epilepsy Curr 13: 279–281.

Kovari VZ, Horvath L (2018). Surgical management of cauda syndrome in third trimester of pregnancy focusing on spinal anesthesia and right lateral positioning during surgery as possible practices. Eur Spine J 27: 483–488.

Kozberg MG, Camargo EC (2019). Management of maternal stroke and mitigating risk. Curr Treat Options Cardiovasc Med 21: 72.

Ladhani NNN, Swartz RH, Foley N et al. (2018). Canadian stroke best practice consensus statement: acute stroke management during pregnancy. Int J Stroke 13: 743–758.

Lars Peter Wang M (2008). Neuroanesthesia for the pregnant woman. Anesth Anal 107: 193–200.

Lavie C, Rollot F, Durand-Dubief F et al. (2019). Neuraxial analgesia is not associated with an increased risk of postpartum relapses in MS. Mult Scler 25: 591–600.

Laviv Y, Bayoumi A, Mahadevan A et al. (2018a). Meningiomas in pregnancy: timing of surgery and clinical outcomes as observed in 104 cases and establishment of a best management strategy. Acta Neurochir 160: 1521–1529.

Laviv Y, Ohla V, Kasper EM (2018b). Unique features of pregnancy-related meningiomas: lessons learned from 148 reported cases and theoretical implications of a prolactin modulated pathogenesis. Neurosurg Rev 41: 95–108.

Leffert LR, Schwamm LH (2013). Neuraxial anesthesia in parturients with intracranial pathology: a comprehensive review and reassessment of risk. Anesthesiology 119: 703–718.

Leffert L, Butwick A, Carvalho B et al. (2018). The society for obstetric anesthesia and perinatology consensus statement on the anesthetic management of pregnant and postpartum women receiving thromboprophylaxis or higher dose anticoagulants. Anesth Analg 126: 928–944.

Limaye K, Van de Walle Jones A, Shaban A et al. (2019). Endovascular management of acute large vessel occlusion stroke in pregnancy is safe and feasible. J Neurointerv Surg. https://doi.org/10.1136/neurintsurg-2019-015578. Epub ahead of print.

Lu E, Zhao Y, Dahlgren L et al. (2013). Obstetrical epidural and spinal anesthesia in multiple sclerosis. J Neurol 260: 2620–2628.

MacDonald SC, Bateman BT, McElrath TF et al. (2015). Mortality and morbidity during delivery hospitalization among pregnant women with epilepsy in the United States. JAMA Neurol 72: 981–988.

Makris A, Piperopoulos A, Karmaniolou I (2014). Multiple sclerosis: basic knowledge and new insights in perioperative management. J Anesth 28: 267–278.

Mankowitz SKW (2018). Consults in obstetric anesthesiology, first edn Springer Science+Business Media, New York, NY.

Martin E, Vickers B, Landau R et al. (2018). ABM clinical protocol #28, peripartum analgesia and anesthesia for the breastfeeding mother. Breastfeed Med 13: 164–171.

Meenakshi-Sundaram S, Swaminathan K, Karthik SN et al. (2014). Relapsing Guillain–Barre syndrome in pregnancy and postpartum. Ann Indian Acad Neurol 17: 352–354.

Meng L, Han SJ, Rollins MD et al. (2016). Awake brain tumor resection during pregnancy: decision making and technical nuances. J Clin Neurosci 24: 160–162.

Miller EC, Leffert L (2019). Stroke in pregnancy: a focused update. Anesth Analg 130: 1085–1096.

Myers TR, McCarthy NL, Panagiotakopoulos L et al. (2019). Estimation of the incidence of Guillain–Barre syndrome during pregnancy in the United States. Open Forum Infect Dis 6: ofz071.

Olsson GL, Hallen B, Hambraeus-Jonzon K (1986). Aspiration during anaesthesia: a computer-aided study of 185,358 anaesthetics. Acta Anaesthesiol Scand 30: 84–92.

Pacheco LD, Saad AF, Hankins GD et al. (2016). Guillain–Barre syndrome in pregnancy. Obstet Gynecol 128: 1105–1110.

Pasto L, Portaccio E, Ghezzi A et al. (2012). Epidural analgesia and cesarean delivery in multiple sclerosis post-partum relapses: the Italian cohort study. BMC Neurol 12: 165.

Po G, Olivieri C, Rose CH et al. (2019). Intraoperative fetal heart monitoring for non-obstetric surgery: a systematic review. Eur J Obstet Gynecol Reprod Biol 238: 12–19.

Rajiv KR, Radhakrishnan A (2019). Status epilepticus in pregnancy—can we frame a uniform treatment protocol? Epilepsy Behav 101 (Pt B): 106376. https://doi.org/10.1016/j.yebeh.2019.06.020.

Reece-Stremtan S, Campos M, Kokajko L et al. (2017). ABM clinical protocol #15: analgesia and anesthesia for the breastfeeding mother, revised 2017. Breastfeed Med 12: 500–506.

Rossignol M (2016). Trauma and pregnancy: what anesthesiologist should know. Anaesth Crit Care Pain Med 35: S27–S34.

Roth J, Deck G (2019). Neurovascular disorders in pregnancy: a review. Obstet Med 12: 164–167.

Sachs A, Guglielminotti J, Miller R et al. (2017). Risk factors and risk stratification for adverse obstetrical outcomes after appendectomy or cholecystectomy during pregnancy. JAMA Surg 152: 436–441.

Sahu S, Lata I, Gupta D (2010). Management of pregnant female with meningioma for craniotomy. J Neurosci Rural Pract 1: 35–37.

Sanders DB, Wolfe GI, Benatar M et al. (2016). International consensus guidance for management of myasthenia gravis: executive summary. Neurology 87: 419–425.

Sharp LM, Levy DM (2009). Rapid sequence induction in obstetrics revisited. Curr Opin Anaesthesiol 22: 357–361.

Stephen LJ, Harden C, Tomson T et al. (2019). Management of epilepsy in women. Lancet Neurol 18: 481–491.

Stoelting RK, Hines RL, Marschall KE (2008). Stoelting's anesthesia and co-existing disease, fifth edn Churchill Livingstone/Elsevier, Philadelphia.

Swartz RH, Cayley ML, Foley N et al. (2017). The incidence of pregnancy-related stroke: a systematic review and meta-analysis. Int J Stroke 12: 687–697.

Szuchy Kristiansen E, Holm Vestergaard H, Modrau B et al. (2019). Acute ischemic stroke in late pregnancy treated with intravenous thrombolysis and endovascular therapy. Case Rep Neurol 11: 41–46.

Tanoubi I, Drolet P, Donati F (2009). Optimizing preoxygenation in adults. Can J Anaesth 56: 449–466.

Tarnaris A, Haliasos N, Watkins LD (2012). Endovascular treatment of ruptured intracranial aneurysms during pregnancy: is this the best way forward? Case report and review of the literature. Clin Neurol Neurosurg 114: 703–706.

Teoh W, Yeoh S, Ithnin F et al. (2012). Incidence of difficult & failed intubations during obstetric general anaesthesia in a tertiary referral center. Br J Anaesth 108: ii212.

Teoh WH, Yeoh SB, Tan HK (2013). Airway management of an expanding soft palate haematoma in a parturient. Anaesth Intensive Care 41: 680–681.

Tiel Groenestege AT, Rinkel GJ, van der Bom JG et al. (2009). The risk of aneurysmal subarachnoid hemorrhage during pregnancy, delivery, and the puerperium in the Utrecht population: case-crossover study and standardized incidence ratio estimation. Stroke 40: 1148–1151.

Tomimatsu T, Sugihara M, Nagai T et al. (2016). Guillain–Barre syndrome after trivalent influenza vaccination during pregnancy. Eur J Obstet Gynecol Reprod Biol 201: 225–226.

Too G, Wen T, Boehme AK et al. (2018). Timing and risk factors of postpartum stroke. Obstet Gynecol 131: 70–78.

Toscano M, Thornburg LL (2019). Neurologic diseases in pregnancy. Curr Opin Obstet Gynecol 31: 97–109.

Varner M (2013). Myasthenia gravis and pregnancy. Clin Obstet Gynecol 56: 372–381.

Verheecke M, Halaska MJ, Lok CA et al. (2014). Primary brain tumours, meningiomas and brain metastases in pregnancy: report on 27 cases and review of literature. Eur J Cancer 50: 1462–1471.

Volquind D, Fellini RT, Rose GL et al. (2013). Anesthesia for cesarean section in a patient with Guillain–Barre syndrome: case report. Braz J Anesthesiol 63: 369–371.

Wang LP, Paech MJ (2008). Neuroanesthesia for the pregnant woman. Anesth Analg 107: 193–200.

Watanabe TT, Ichijo M, Kamata T (2019). Uneventful pregnancy and delivery after thrombolysis plus thrombectomy for acute ischemic stroke: case study and literature review. J Stroke Cerebrovasc Dis 28: 70–75.

Wiącek M, Kaczorowski R, Oboz-Adaś A et al. (2019). Acute ischemic stroke in a third trimester of pregnancy—cesarean section followed by mechanical thrombectomy. Int J Neurosci 1–4. https://doi.org/10.1080/00207454.2019.1702539. Epub ahead of print.

Wiertlewski S, Magot A, Drapier S et al. (2004). Worsening of neurologic symptoms after epidural anesthesia for labor in a Guillain–Barre patient. Anesth Analg 98: 825–827. table of contents.

Handbook of Clinical Neurology, Vol. 171 (3rd series)
Neurology and Pregnancy: Pathophysiology and Patient Care
E.A.P. Steegers, M.J. Cipolla and E.C. Miller, Editors
https://doi.org/10.1016/B978-0-444-64239-4.00011-4

Chapter 11

Neurocritical care of the pregnant patient

MARTINA S. BURN[1], SANGINI S. SHETH[1], AND KEVIN N. SHETH[2]*

[1]*Department of Obstetrics, Gynecology & Reproductive Sciences, Yale School of Medicine, New Haven, CT, United States*
[2]*Department of Neurology, Yale School of Medicine, New Haven, CT, United States*

Abstract

Introduction: An estimated 0.1%–0.8% of obstetric patients require admission to an intensive care unit (ICU) during pregnancy or the puerperium. When neurologic emergencies occur in pregnancy, collaboration between the neurointensivist, obstetric anesthesiologist, and obstetrician is key in minimizing morbidity and mortality.

Principles: Care of the critically ill pregnant patient mirrors that of the critically ill nonpregnant patient with some minor exceptions. Special care must be taken to consider the normal physiologic changes of pregnancy as well as possible fetal exposure to medical interventions. Timing and method of delivery must be carefully considered when caring for patients with neurologic emergencies. Common neurologic emergencies in pregnancy include hypertensive disorders of pregnancy, intracranial neoplasms, noneclamptic seizures, cerebrovascular disorders, and ventriculoperitoneal shunt malfunctions.

Conclusion: While neurologic emergencies in pregnancy are overall rare, when they do occur, they can be devastating. As in the nonpregnant population, prompt recognition and rapid intervention are crucial in optimizing patient outcomes. When neurologic emergencies occur in pregnancy, maternal and fetal care is optimized through a multidisciplinary care team.

INTRODUCTION

An estimated 0.1%–0.8% of obstetric patients require admission to an intensive care unit (ICU) during pregnancy or the puerperium (American College of Obstetricians and Gynecologists, 2016). Neurologic emergencies in pregnancy are overall rare, but pregnant women are at increased risk for several conditions that may require ICU-level care. Given the pregnant patient's unique physiology and pathophysiology, critically ill obstetric patients present a particular challenge to both the intensivist and the obstetrician. Care of the pregnant patient is further complicated by consideration of the fetus as a second patient. When neurologic emergencies occur in pregnancy, collaboration between the neurointensivist, obstetric anesthesiologist, and obstetrician is key in minimizing morbidity and mortality in this unique patient population (Soubra and Guntupalli, 2005; American College of Obstetricians and Gynecologists, 2016).

The first part of this chapter will focus on general principles important in caring for pregnant patients who requires neurocritical care, while the second part will discuss specific conditions that most commonly lead to neurologic emergencies in pregnancy.

GENERAL CONSIDERATION FOR NEUROCRITICAL CARE IN THE PREGNANT PATIENT

Surgical considerations

TIMING

Urgent surgery should be performed as indicated, regardless of the trimester (Committee on Obstetric Practice

*Correspondence to: Kevin N. Sheth, M.D., Chief, Neurocriitical Care & Emergency Neurology, Neurology, Yale University School of Medicine, 15 York Street, Bldg LLCI, Suite 1003, New Haven, CT 06520, United States. Tel: +1-443-615-4729, E-mail: kevin.sheth@yale.edu

and the American Society of Anesthesiologists, 2017). In cases that cannot be deferred until after delivery, nonurgent surgery is generally performed in the second trimester as organogenesis is complete by then and risk of preterm delivery is relatively low. Since miscarriage rates are highest early in pregnancy, this time period is avoided whenever possible. There is, however, no strong or consistent evidence that anesthesia or surgery causes miscarriages in the first trimester. The third trimester is avoided whenever possible simply due to the size of the gravid uterus (Committee on Obstetric Practice and the American Society of Anesthesiologists, 2017).

Positioning

After 20 weeks gestation, patients should be placed in left lateral tilt to prevent inferior vena cava (IVC) compression by the gravid uterus.

Diagnostic imaging

Radiation exposure in pregnancy is a common and often real concern, but a missed or delayed diagnosis can be catastrophic. Radiation risk is dose dependent, and the risk begins to increase as radiation increases. There have been no reports of fetal anomalies, growth restriction, or abortion with radiation exposure of less than 50 mGy (Brent, 1989; Committee on Obstetric Practice, 2017). Radiation from a noncontrast head CT is orders of magnitude below this threshold at 0.001–0.01 mGy (Committee on Obstetric Practice, 2017). Iodinated contrast can also be used when necessary for diagnosis.

MRI does not appear to have any adverse fetal effects and can be used preferentially for diagnostic imaging in a pregnant patient. The use of gadolinium, however, is not recommended in pregnancy as it crosses the placenta and also due to its long half-life and unknown fetal effects (Kanal et al., 2007).

Fetal monitoring

Acute episodes

Given the risk of hemodynamic collapse during acute neurologic emergencies, continuous fetal monitoring should be initiated as soon as possible for viable pregnancies (typically after 23 weeks of gestation depending on the hospital). While maternal hemodynamics are generally prioritized over fetal hemodynamics, rapid identification of fetal distress and resuscitation can be critical (Sheth and Sheth, 2012). When the maternal condition is stable, continuous fetal monitoring can be safely discontinued.

Surgical management

For previable pregnancies, fetal monitoring should consist of fetal heart rate assessment before and after the procedure.

For viable pregnancies, a discussion between the obstetrician and patient should precede operative management. In cases where full fetal resuscitation is desired, continuous fetal monitoring should be performed throughout the duration of the procedure with the understanding that emergency cesarean delivery will be performed in the event of fetal distress.

Antenatal steroids

For viable pregnancies, antenatal corticosteroids should be administered as soon as preterm delivery becomes a possibility. In the preterm infant, corticosteroids help promote lung maturity and prevent neonatal morbidity and mortality (Liggins and Howie, 1972). Maximum efficacy is observed when delivery occurs 2–7 days following administration (Roberts and Dalziel, 2006). Two doses of betamethasone 12 mg are administered by IM injections that are 24 h apart.

Ventilation considerations

Pregnancy is a state of respiratory alkalosis characterized by an arterial pH of 7.40–7.47 compared with a nonpregnant arterial pH of 7.35–7.45. Mechanical ventilation of the pregnant patient should mimic this normal physiology by adjusting minute ventilation to maintain a $PaCO_2$ between 30 and 32 mmHg (Lapinsky et al., 1995).

Patient location

Outcomes extrapolated from medical and surgical ICU literature suggest that obstetric patients fare better in a closed ICU under the direct care of an intensivist (Berghella, 2017). The preferred location of a patient should be determined on a case-by-case basis. When a long-term stay is anticipated, patients are generally transferred to the ICU to facilitate care for the duration of the critical illness. In select cases, when an emergency presents during active labor, the labor and delivery unit may be the preferred location. A multidisciplinary approach is key, and the obstetrician should be closely involved in providing daily input regarding patient care regardless of patient location (Berghella, 2017).

Delivery planning

Whenever a patient with a viable fetus is admitted to the ICU, a delivery plan should be established.

This discussion should include preferred location, mode, and timing of delivery, as well as plans for an emergency delivery. Intubated patients who cannot close the glottis to push and patients who cannot safely valsalva may require an assisted second stage. Cesarean section should preferably not be performed in an ICU setting given the rarity of this occurrence, lack of preparedness for such an event, limited space, and high rates of nosocomial infections (Berghella, 2017). An emergency delivery cart, complete with operative equipment and neonatal resuscitation equipment, should be at the ICU bedside at all times in the event of an unanticipated, emergent need for delivery.

HYPERTENSIVE DISORDERS OF PREGNANCY: PREECLAMPSIA, ECLAMPSIA, HELLP Syndrome

Epidemiology

According to a 2010 systematic review, hypertensive disorders of pregnancy are the most prevalent indications for ICU admission in the obstetric population and account for 16% of maternal mortality (Pollock et al., 2010; Steegers et al., 2010). Preeclampsia, defined as new-onset hypertension in the setting of either proteinuria or end organ dysfunction, affects 2%–8% of pregnancies worldwide (Steegers et al., 2010; Sheth and Sheth, 2012; Gestational Hypertension and Preeclampsia, 2019). Eclampsia, a severe variant of preeclampsia associated with new-onset seizure, accounts for 5000 maternal deaths annually (Pollock et al., 2010; Steegers et al., 2010; Sheth and Sheth, 2012).

The exact pathophysiology of preeclampsia is poorly understood and is likely multifactorial, involving abnormalities in placental function. Preeclampsia and eclampsia, if left untreated, can have profound neurologic consequences, including seizure, reversible encephalopathy, ischemic stroke, intracerebral hemorrhage, and HELLP (hemolysis, elevated liver enzymes, low platelet count) syndrome, and thus require early identification and treatment (Karnad et al., 2004; Frontera and Ahmed, 2014).

Diagnosis

Preeclampsia is defined as new-onset hypertension with associated proteinuria or new-onset hypertension with associated end organ dysfunction. Classic symptoms include general feeling of unease, sense of impending doom, headache, vision changes, scotomata, and right upper quadrant pain (Sibai, 2003; Sibai, 2005). The specific diagnostic criteria for preeclampsia are summarized in Table 11.1.

Table 11.1

Diagnostic criteria for preeclampsia

Hypertension + Proteinuria	• Systolic blood pressure > 140 mmHg or diastolic blood pressure > 90 mmHg on two occasions >4 h apart after 20 weeks of gestation • >300 mg protein on 24-h urine specimen or Pr/Cr > 0.3 mg/dL on random urine specimen
	or
Hypertension (in the absence of proteinuria) + End organ dysfunction	• Systolic blood pressure > 140 mmHg or diastolic blood pressure > 90 mmHg on two occasions >4 h apart after 20 weeks of gestation • Thrombocytopenia, defined as platelet count <100,000 × 10^9/L • Renal insufficiency, defined as Cr > 1.1 mg/dL or twice the baseline creatinine • Impaired liver function, defined as transaminitis twice the upper limit of normal • Pulmonary edema (in the absence of other etiology) • Severe headache/visual disturbance (refractory to medical management and without alternate diagnosis)

Table adapted from ACOG, "Hypertension in pregnancy," table 2-1 (Gestational Hypertension and Preeclampsia, 2019. ACOG practice bulletin no. 202. American College of Obstetricians and Gynecologists. Obstet Gynecol 133, e1–e25).
mmHg, millimeters of mercury; h, hours; mg, milligram; Pr/Cr, protein to creatinine ratio; mg/dL, milligrams per deciliter; L, liter.

Eclampsia, a severe variant of preeclampsia, is defined as preeclampsia with new-onset tonic–clonic, focal, or multifocal seizures in the absence of other causative conditions such as epilepsy, cerebral arterial ischemia and infarction, intracranial hemorrhage, or drug use (Gestational Hypertension and Preeclampsia, 2019). While preeclampsia does not always progress to eclampsia, antecedent symptoms such as severe and persistent occipital or frontal headaches, blurred vision, photophobia, and altered mental status are present 78%–83% of the time (Sibai, 2003, 2005; Cooray et al., 2011).

HELLP syndrome is thought to be a severe form of preeclampsia. The diagnosis is made based on laboratory evidence of hemolysis, transaminitis twice the upper limit of normal, and a platelet count <100,000 × 10^9/L

(Gestational Hypertension and Preeclampsia, 2019). Among other sequelae, HELLP syndrome can lead to disseminated intravascular coagulopathy and catastrophic hemorrhage.

Management

Stabilization of blood pressure

Stabilization of blood pressure is one of the key components of preeclampsia management. Systolic blood pressures >160 mmHg and diastolic blood pressures >110 mmHg should be treated within 30–60 min to prevent complications, including left ventricular failure and intracranial hemorrhage (Gestational Hypertension and Preeclampsia, 2019). The degree of systolic hypertension along with the rate of increase of mean arterial pressure may be the best predictors of cerebral injury or infarction (Sibai, 2003, 2005; Karnad et al., 2004; Frontera and Ahmed, 2014; Gestational Hypertension and Preeclampsia, 2019).

First-line antihypertensive agents include IV hydralazine (5–10 mg for a maximum dose of 20 mg), IV labetalol (20–40 mg every 15 min for a maximum dose of 300 mg), and oral nifedipine (10–20 mg every 30 min for a maximum dose of 180 mg) (Anon, 2000; Sibai, 2003; Duley et al., 2006; Gestational Hypertension and Preeclampsia, 2019). Care should be taken so that there is no drop of systolic and diastolic blood pressures to levels <140 mmHg and <90 mmHg, respectively, to ensure uteroplacental blood flow and prevent fetal hemodynamic compromise (Soubra and Guntupalli, 2005; Steegers et al., 2010; Gestational Hypertension and Preeclampsia, 2019). During an acute hypertensive episode, continuous fetal monitoring should be initiated as soon as possible to ensure adequate uteroplacental perfusion.

Management/prevention of eclampsia

In a patient with preeclampsia with signs of end organ dysfunction or eclampsia, an IV loading dose of 4–6 g of magnesium sulfate in 100 mL of fluid should be administered over 15–20 min followed by a continuous infusion of IV magnesium sulfate at a rate of 2 g/h until 24 h after delivery (Aagard-Tillery and Belfort, 2005; Sibai, 2005; Sheth and Sheth, 2012; Gestational Hypertension and Preeclampsia, 2019). Alternatively, in a patient without IV access, IM magnesium sulfate may be administered using a 10-g loading dose of 50% magnesium sulfate given as a 5-g dose in the upper outer quadrant of each buttock, followed by 5-mg IM injection every 4 h until 24 h after delivery (Sheth and Sheth, 2012; Gestational Hypertension and Preeclampsia, 2019).

A serum magnesium level of 4.8–9.6 mg/dL (4–8 mEq/L) is considered the therapeutic range at which eclamptic seizures are typically prevented (Gestational Hypertension and Preeclampsia, 2019). It is important to remember that magnesium is renally cleared, and thus care should be taken to avoid magnesium toxicity, especially in a patient with renal impairment. Patients should be monitored for signs of magnesium toxicity, which begin at 8.5–12 mg/dL (7–10 mEq/L) with loss of deep tendon reflexes (Lu and Nightingale, 2000). Life-threatening side effects include respiratory paralysis at 12–16 mg/dL (10–13 mEq/L), altered cardiac conduction at >18 mg/dL (>15 mEq/L), and cardiac arrest at >30 mg/dL (>25 mEq/L) (Lu and Nightingale, 2000). If needed, calcium gluconate (1 g IV) can be administered to counteract magnesium toxicity.

During an eclamptic seizure, supportive care should be provided to prevent maternal injury and hypoxia. After an eclamptic seizure, a period of fetal bradycardia is often observed. Fetal status generally recovers within 10 min and with standard resuscitative efforts, including administration of intravenous fluids and supplemental oxygen and maternal left lateral tilt. When fetal bradycardia or persistent late decelerations continue despite aggressive resuscitation efforts beyond 15 min, delivery may need to be expedited as long as maternal condition has stabilized as fetal reserve is limited and placental abruption may have occurred (Sibai, 2005; Gestational Hypertension and Preeclampsia, 2019).

Delivery considerations

The definitive treatment for preeclampsia and eclampsia is delivery, although it is important to remember that a subset of hypertensive disorders of pregnancy first present in the postpartum period. Recommended delivery timing, provided maternal and fetal stability, is between 34 and 37 weeks depending on presence or absence of signs of end organ dysfunction (Gestational Hypertension and Preeclampsia, 2019).

If the decision to move toward delivery is made, vaginal delivery is the preferred method. Induction of labor can be initiated following maternal stabilization. Cesarean delivery should be reserved for standard obstetric indications even in the setting of eclampsia or HELLP syndrome.

INTRACRANIAL NEOPLASMS

Epidemiology

The incidence of intracranial neoplasms in the United States is 30 per 100,000 (Ostrom et al., 2018). Tumors are predominantly gliomas and meningiomas, each representing approximately one-third of cases

(DeAngelis, 1994). While the incidence of intracranial neoplasms is not affected by pregnancy, physiologic changes can lead to tumor expansion and initial symptom manifestation during pregnancy. Tumor expansion in pregnancy is commonly attributed to hormonal regulation and expansion of maternal fluid compartments. Depending on tumor location, even very small increases in tumor size can have profound neurologic consequences.

Growth of certain tumors may be accelerated through hormonal regulation. Meningiomas can express estrogen or progesterone receptors, acoustic neuromas can express estrogen receptors (Carroll et al., 1999; Ostrom et al., 2018), and pituitary adenomas can expand through estrogen-driven lactotroph hyperplasia.

Volume expansion has also been identified as a mechanism of tumor expansion in pregnancy. Intravascular volume expands by 50% in pregnancy with a maximal increase by 32 weeks of gestation (Berghella, 2017). This physiologic volume expansion and fluid retention of pregnancy can increase tumor edema and enlarge vascular tumors such as meningiomas and acoustic neuromas (Michelsen and New, 1969). Given these physiologic changes, preexisting intracranial neoplasms may present themselves for the first time during pregnancy and symptoms may be exacerbated by tumor growth and volume expansion.

Diagnosis

Intracranial neoplasms present similarly in pregnancy as outside of pregnancy and include new-onset headache or visual field changes, nausea, vomiting, and new-onset seizure. These symptoms can be missed when they are incorrectly managed as normal signs and symptoms of pregnancy.

Nausea and vomiting in the first trimester is a common manifestation of hormonal changes in pregnancy; however, nausea and vomiting worsening in late gestation is atypical and suggests another etiology. Additionally, new-onset seizures are often attributed to eclampsia, but intracranial pathology should remain on the differential. Headaches are another common complaint during this period, with migraines affecting up to 18% of pregnancies (Berghella, 2017). Pregnant women with new-onset headaches during pregnancy or atypical headaches should be evaluated carefully.

As in the nonpregnant population, MRI is the imaging modality of choice for evaluation of intracranial neoplasms and is safe for use in pregnancy. The use of gadolinium, however, is not recommended as it crosses the placenta and also due to its long half-life and unknown fetal effects (Kanal et al., 2007).

Management

Medical management

Whenever possible, symptom control should be the mainstay of treatment in pregnancy. For benign tumors such as meningiomas or acoustic neuromas or for slow-growing tumors with low malignant potential such as gliomas, excision can be deferred until after delivery to optimize fetal outcomes without compromise of maternal outcomes (Pennel, 2002; Karnad and Guntupalli, 2005; Stevenson and Thompson, 2005).

For seizure prophylaxis, anticonvulsants should be considered as in the general population. A single agent should be initiated at the lowest effective dose. No changes should be made to a woman's medications if she is well maintained on her current anticonvulsant regimen (Pennel, 2002). To minimize tumor edema, steroid therapy should also be considered (Stevenson and Thompson, 2005). Both dexamethasone and prednisone cross the placenta. Dexamethasone is pregnancy category C and prednisone is pregnancy category D.

Surgical management

Surgical excision of intracranial neoplasms should be reserved for high-grade tumors and those causing severe symptoms that cannot be delayed (Raps et al., 1994). In the late preterm period, risks and benefits of antenatal surgery can be weighed against preterm delivery on a case-by-case basis.

Delivery considerations

Intracranial pressure can increase to as high as 70 cm H_2O (normal, 20 cm H_2O) during the first and second stages of labor. Patients with intracranial neoplasms can have baseline increases in intracranial pressures, placing them at risk of cerebral herniation during valsalva or placement of neuraxial anesthesia (Stevenson and Thompson, 2005). An assisted second stage of labor with vacuum or forceps may be indicated in instances when valsalva is contraindicated. Vaginal delivery is often still preferred over cesarean delivery to minimize fluid shifts associated with the perioperative period and increases in intracranial pressure associated with intubation.

NONECLAMPTIC SEIZURES

Epidemiology

The majority of seizures in pregnancy are epileptic rather than eclamptic (Noe and Pack, 2010). Epilepsy is not a contraindication to pregnancy, and 90% of women who attempt pregnancy will have positive outcomes

(Anon, 1998). Based on a literature review, 54%–80% of women noted no change in seizure frequency while 14%–32% noted an increase (Harden et al., 2009). This expert literature review concluded that there is insufficient evidence at present to determine if there is a change in seizure frequency in women with epilepsy (Harden et al., 2009). There is good data, however, showing that women who had a seizure in the prepregnancy months had a higher risk of seizure during pregnancy (OR 15, 95% CI 9–25.1) (Thomas et al., 2012).

A seizure triggers a chain reaction that decreases uteroplacental blood flow through lactic acid build up, increased cardiac output, increased intra-abdominal pressures, and finally shunting of blood flow to the brain and muscles away from the uterus (Gestational Hypertension and Preeclampsia, 2019). The prevention and quick treatment of seizures is key to preventing maternal and fetal morbidity and mortality.

Diagnosis

Standard imaging/workup should be initiated for new-onset seizure, including MRI and EEG. Both of these modalities are safe for use in pregnancy. The first priority is securing the airway preventing injury, and providing supportive care. As is routine with seizure, reversible causes such as infection, electrolyte disturbances, medications, and use of recreational drugs should be considered and reversed, if applicable.

Management

Supportive care

Management of noneclamptic seizures in pregnant women is generally the same as in nonpregnant women with a few minor exceptions. The patient should be placed in the left lateral position to decrease IVC compression by the gravid uterus in an attempt to increase systemic blood flow and prevent aspiration (Anon, 1997; Atallah, 2003; Sibai, 2005). Care should be taken to avoid prolonged hypotension or hypoxemia to ensure both maternal and fetal well-being. As previously discussed, maternal seizure is often followed by a period of fetal bradycardia. Bradycardia is usually transient; however, when it persists past 15 min, fetal condition becomes critical and delivery must be expedited (Sibai, 2005; Gestational Hypertension and Preeclampsia, 2019).

Treatment

An IV dose of 2-mg lorazepam should be administered and repeated every 5 min until cessation of tonic–clonic seizures. Alternatively, IV diazepam boluses of 5–10 mg may be used (Pollock et al., 2010; Steegers et al., 2010; Sheth and Sheth, 2012; Gestational Hypertension and Preeclampsia, 2019). If the patient is not already on oral phenytoin, an IV phenytoin loading dose should be given at the same time at 18 mg/kg (rate not to exceed 50 mg/min) (Sheth and Sheth, 2012; Frontera and Ahmed, 2014; Gestational Hypertension and Preeclampsia, 2019). If the seizure is not controlled at this time, the patient should be intubated and midazolam infusion initiated (Sheth and Sheth, 2012; Frontera and Ahmed, 2014; Gestational Hypertension and Preeclampsia, 2019).

An important consideration is that antiepileptic drug levels are affected by pregnancy fluid shifts and metabolism and have the risk of falling to subtherapeutic levels (Anon, 1998). Therefore, the levels should be checked frequently with increases made to maintain levels within therapeutic range.

CEREBROVASCULAR DISORDERS

Epidemiology

Rates of cerebrovascular accidents are much higher in the pregnancy and postpartum periods at 34 per 100,000 than outside of pregnancy (James et al., 2005). The highest risk period appears to be intrapartum and postpartum, accounting for 40% and 50% of incidents, respectively (James et al., 2005). Hemorrhagic and ischemic stroke occur at similar frequencies in pregnancy.

Hemorrhagic strokes arise in the setting of preeclampsia, ruptured aneurysms, and arteriovenous malformations (Marder et al., 2014). One rare type of intracranial hemorrhage seen in pregnancy is a convexal subarachnoid hemorrhage, a type of nonaneurysmal subarachnoid hemorrhage with several etiologies, including reversible cerebral vasoconstriction syndrome and posterior reversible encephalopathy syndrome. In the setting of pregnancy, convexal subarachnoid hemorrhages are usually related to hypertensive disorders of pregnancy (Marder et al., 2014).

Ischemic strokes arise in the setting of cerebral venous sinus thrombosis (CVT), preeclampsia, and cardioembolic stroke (Feske, 2007). In addition to standard risk factors, several physiologic changes of pregnancy predispose women to cerebrovascular disorders. Pregnancy significantly changes the body's hemodynamics with a 50% increase in total body water and increase in cardiac output, stroke volume, and heart rate. Additionally, venous stasis occurs through decreased systemic vascular resistance and compression of the IVC by the gravid uterus. Furthermore, pregnancy induces a hypercoagulable state most pronounced in the third trimester with increased production of clotting factors (Grear and Bushnell, 2013). It is not clear exactly how long after delivery this heightened risk of thrombosis persists, but data shows that thrombosis rates remain elevated at least 12 weeks postpartum (Kamel et al., 2014).

Diagnosis

Pregnancy should not cause any delay in diagnostic imaging. Whenever immediately available, a noncontrast MRI is the imaging modality of choice, given its safety profile in pregnancy. However, in cases where MRI will cause any diagnostic delay, a noncontrast CT can be obtained to expedite diagnosis and subsequent therapy.

Management

General management

Management of stroke in pregnancy generally follows the same guidelines as management of stroke in the general population with some important modifications. Most therapy is centered around protecting salvageable brain tissue. Care should be taken to prevent medical complications, including aspiration, seizure, and trauma, and to optimize maternal factors that may worsen prognosis, including blood pressure, blood glucose, and infection.

Ischemic stroke

Specific guidelines for management of ischemic stroke revolve around preventing progression and facilitating reperfusion. Specifically, aspirin administration within 48 h of symptom onset is recommended in pregnant patients not receiving thrombolysis, IV heparin, or oral anticoagulation (Lansberg et al., 2012; Powers et al., 2018). Thrombolysis is an option in the first 4.5 h of stroke onset as brain tissue is ischemic but not yet hypoxic. There is mixed data on the use of IV alteplase in pregnancy. Alteplase does not cross the placenta and is pregnancy category C. According to the 2018 American Heart Association/American Stroke Association guidelines, alteplase may be considered in pregnancy in cases where the anticipated benefits of treating moderate or severe stroke outweigh the risks of hemorrhage (Powers et al., 2018).

Cerebral venous thrombosis

CVT is the presence of a blood clot in the dural venous sinus, preventing drainage of blood from the brain. The annual incidence is 0.22–1.57 per 100,000 and occurs disproportionally in pregnant women (Coutinho et al., 2012). The most common signs of CVT include headache, visual problems, focal neurologic deficits, seizures, and encephalopathy (Bousser et al., 1985). The mainstay of treatment is anticoagulation therapy with IV heparin or low molecular weight heparin (LMWH) and when indicated are considered safe in pregnancy (Powers et al., 2018). Neither medication crosses the placenta and LMWH is pregnancy category B and IV heparin is pregnancy category C. Patients maintained on LMWH should be transitioned to UFH at 36 weeks of gestation, given its shorter half-life, in attempts to prevent intrapartum hemorrhage and allow for regional anesthesia. Warfarin crosses the placenta and is pregnancy category D and thus should not be used in pregnancy.

In addition to anticoagulation, treatment revolves around prevention of intracranial hypertension and seizures. Thrombosis can lead to significant edema and rapid deterioration of maternal status. In such cases, decompressive craniectomy can be lifesaving and should not be delayed on account of pregnancy (Ebke et al., 2011).

Hemorrhagic stroke

Management of hemorrhagic stroke also parallels management in the nonpregnant population. It is important to note that 14%–55% of hemorrhagic strokes in pregnancy can be attributed to hypertensive disorders of pregnancy (McDermott et al., 2018). Hemorrhage caused by AVMs or aneurysms can be managed by surgical clipping or embolization of the lesion. Fetal and maternal outcomes favored early surgical clipping for aneurysms over conservative management in one study (Dias and Sekhar, 1990). When an embolization procedure is favored, CT angiography can be safely performed with abdominal shielding. While no clear guidelines exist, women who have not had definitive treatment of ruptured aneurysms or AVMs may benefit from an assisted second stage or cesarean delivery to minimize valsalva and consequent fluctuations in intracranial pressures (Raps et al., 1994).

VENTRICULOPERITONEAL SHUNT MALFUNCTION

Epidemiology

A number of neurologic conditions require alleviation of intracranial pressures with a ventriculoperitoneal (VP) shunt. The presence of a VP shunt is not a contraindicated in pregnancy, but as up to 50% of shunts malfunction during pregnancy, close monitoring is required (Maheut-Lourmière and Si, 2000). Disruptions in peritoneal catheter flow can occur as the pregnancy progresses and intra-abdominal pressures rise, leading to elevations in intracranial pressure.

Diagnosis

Signs of increased intracranial pressure include confusion, lethargy, nausea, vomiting, headache, and nystagmus. If shunt malfunction is suspected, CT or MRI should be obtained and compared with baseline imaging whenever possible.

Management

Shunt management

If there is a change in imaging findings from preconception, a shunt revision is necessary. This can be performed per routine procedure in the first and second trimesters when the gravid uterus is nonobstructive. In the third trimester, a ventriculoatrial shunt or third ventriculostomy can be considered to avoid preterm labor onset and uterine trauma (Ng and Kitchen, 2008).

Delivery considerations

While data is limited, vaginal delivery is not contraindicated in patients with a functioning VP shunt (Ng and Kitchen, 2008). Efforts can be made to shorten the second stage, and an operative vaginal delivery can be considered to minimize increases in intracranial pressure. In cases where there is shunt malfunction, cesarean delivery is the recommended route of delivery (Ng and Kitchen, 2008), and in cases of cesarean section, prophylactic antibiotics should be administered to prevent shunt infection.

CONCLUSIONS

While neurologic emergencies in pregnancy are overall rare, when they do occur, they can be devastating. As in nonpregnant populations, prompt recognition and rapid intervention are crucial in optimizing patient outcomes. Management of neurologic emergencies in pregnancy is similar to management outside of pregnancy; however, care must be taken to consider both the maternal and the fetal condition. When neurologic emergencies occur in pregnancy, collaboration between the neurointensivist, obstetric anesthesiologist, and the obstetrician is key in minimizing morbidity and mortality in this unique patient population (Soubra and Guntupalli, 2005; American College of Obstetricians and Gynecologists, 2016).

References

Aagard-Tillery KM, Belfort MA (2005). Eclampsia: morbidity, mortality and management. Clin Obstet Gynecol 48: 12–33.

American College of Obstetricians and Gynecologists (2016). Committee on practice bulletins—obstetrics. practice bulletin no. 170: critical care in pregnancy. Obstet Gynecol 128: e147. Reaffirmed 2017.

Anon (1997). Committee on educational bulletins of the American college of obstetricians and gynecologists. ACOG educational bulletin. Seizure disorders in pregnancy. No. 231, December 1996. Int J Gynaecol Obstet 56: 279–286.

Anon (1998). Practice parameter: management issues for women with epilepsy (summary statement). Report of the Quality Standards Subcommittee of the American Academy of Neurology. Neurology 51: 944.

Anon (2000). Report of the national high blood pressure education program working group on high blood pressure in pregnancy. Am J Obstet Gynecol 183: S1–S22.

Atallah AN (2003). Anticonvulsant therapy for eclampsia: RHL practical aspects, The WHO Reproductive Health Library. World Health Organization.

Berghella V (2017). Maternal fetal evidence based guidelines, Taylor & Francis Group, Boca Raton, FL.

Bousser MG, Chiras J, Bories J et al. (1985). Cerebral venous thrombosis—a review of 38 cases. Stroke 16: 199.

Brent RL (1989). The effect of embryonic and fetal exposure to x-ray, microwaves, and ultrasound: counseling the pregnant and nonpregnant patient about these risks. Semin Oncol 16: 347.

Carroll RS, Zhang J, Black PM (1999). Expression of estrogen receptors alpha and beta in human meningiomas. J Neurooncol 42: 109.

Committee on Obstetric Practice (2017). Committee opinion no. 723: guidelines for diagnostic imaging during pregnancy and lactation. Obstet Gynecol 130: e210.

Committee on Obstetric Practice and the American Society of Anesthesiologists (2017). Committee opinion no. 696: nonobstetric surgery during pregnancy. Obstet Gynecol 129: 777.

Cooray SD, Edmonds SM, Tong S et al. (2011). Characterization of symptoms immediately preceding eclampsia. Obstet Gynecol 118: 995–999.

Coutinho JM, Zuurbier SM, Aramideh M et al. (2012). The incidence of cerebral venous thrombosis: a cross-sectional study. Stroke 43: 3375.

DeAngelis LM (1994). Central nervous system neoplasms in pregnancy. Adv Neurol 64: 139.

Dias MS, Sekhar LN (1990). Intracranial hemorrhage from aneurysms and arteriovenous malformations during pregnancy and puerperium. Neurosurgery 27: 855–866.

Duley L, Henderson-Smart DJ, Meher S (2006). Drugs for rapid treatment of very high blood pressure during pregnancy. Cochrane Database Syst Rev 3: CD001449.

Ebke M, Jurgens KU, Tomandl B et al. (2011). Surgical treatment of space occupying edema and hemorrhage due to cerebral venous thrombosis during pregnancy. Neurocrit Care 15: 166–169.

Feske SK (2007). Stroke in pregnancy. Semin Neurol 27: 442.

Frontera JA, Ahmed W (2014). Neurocritical care complications of pregnancy and puerperium. J Crit Care 29: 1069–1081.

Gestational Hypertension and Preeclampsia (2019). ACOG practice bulletin no. 202. American College of Obstetricians and Gynecologists. Obstet Gynecol 133: 211–214.

Grear KE, Bushnell CD (2013). Stroke and pregnancy: clinical presentation, evaluation, treatment, and epidemiology. Clin Obstet Gynecol 56: 350–359. https://doi.org/10.1097/GRF.0b013e31828f25fa.

Harden CL, Hopp J, Ting TY et al. (2009). Practice parameter update: management issues for women with epilepsy—focus on pregnancy (an evidence-based review): obstetrical complications and change in seizure frequency: report of the quality standards subcommittee and therapeutics and technology assessment Subcommittee of the American Academy of neurology and American Epilepsy Society. Neurology 73: 126.

James AH, Bushnell CD, Jamison MG et al. (2005). Incidence and risk factors for stroke in pregnancy and the puerperium. Obstet Gynecol 106: 509.

Kamel H, Navi BB, Sriram N et al. (2014). Risk of thrombotic event after the 6-week postpartum period. N Engl J Med 370: 1307–1315.

Kanal E, Barkovich AJ, Bell C et al. (2007). ACR guidance document for safe MR practices: 2007. Am J Roentgenol 188: 1447.

Karnad DR, Guntupalli KK (2005). Neurologic disorders in pregnancy. Crit Care Med 33: S362–S371.

Karnad DR, Lapsia V, Krishnan A et al. (2004). Prognostic factors in obstetric patients admitted to an Indian intensive care unit. Crit Care Med 32: 1294–1299.

Lansberg MG, O'Donnell MJ, Khatri P et al. (2012). Antithrombotic and thrombolytic therapy for ischemic stroke: antithrombotic therapy and prevention of thrombosis, 9th ed: American College of Chest Physicians Evidence-Based Clinical Practice Guidelines. Chest 141: e601S.

Lapinsky SE, Kruczynski K, Slutsky AS (1995). Critical care in the pregnant patient. Am J Respir Crit Care Med 152: 427.

Liggins GC, Howie RN (1972). A controlled trial of antepartum glucocorticoid treatment for prevention of the respiratory distress syndrome in premature infants. Pediatrics 50: 515.

Lu JF, Nightingale CH (2000). Magnesium sulfate in eclampsia and pre-eclampsia: pharmacokinetic principles. Clin Pharmacokinet 38: 305.

Maheut-Lourmière J, Si CT (2000). Hydrocephalus during pregnancy with or withoutneurosurgical history in childhood. Practical advice for management. Neurochirurgie 46: 117.

Marder CP, Narla V, Fink J et al. (2014). Subarachnoid hemorrhage: beyond aneurysms. Am J Roentgenol 202: 25–37.

McDermott M, Miller EC, Rundek T et al. (2018). Preeclampsia: association with posterior reversible encephalopathy syndrome and stroke. Stroke 49: 524.

Michelsen JJ, New PF (1969). Brain tumour and pregnancy. J Neurol Neurosurg Psychiatry 32: 305.

Ng J, Kitchen N (2008). Neurosurgery and pregnancy. J Neurol Neurosurg Psychiatry 79: 745.

Noe KH, Pack AM (2010). Women issues and epilepsy. Continuum 16: 159–178.

Ostrom QT, Gittleman H, Truitt G et al. (2018). CBTRUS statistical report: primary brain and other central nervous system tumors diagnosed in the United States in 2011–2015. Neuro Oncol 20: iv1.

Pennel PB (2002). Pregnancy in the woman with epilepsy: maternal and fetal outcomes. Semin Neurol 22: 299–307.

Pollock W, Rose L, Dennis CL (2010). Pregnant and postpartum admissions to the intensive care unit: a systematic review. Intensive Care Med 36: 1465. Https://doi.org/10.1007.s00134-010-1951-0.

Powers WJ, Rabinstein AA, Ackerson T et al. (2018). 2018 guidelines for the early Management of Patients with Acute Ischemic Stroke: a guideline for healthcare professionals from the American Heart Association/American Stroke Association. Stroke 49: e46.

Raps EC, Galetta SL, Flamm ES (1994). Neuro-intensive care of the pregnant woman. Neurol Clin 12: 601–611.

Roberts D, Dalziel S (2006). Antenatal corticosteroids for accelerating fetal lung maturation for women at risk of preterm birth. Cochrane Database Syst Rev 3: CD004454.

Sheth SS, Sheth KN (2012). Treatment of neurocritical care emergencies in pregnancy. Curr Treat Options Neurol 14: 197. https://doi.org/10.1007/s11940-011-0161-6.

Sibai BM (2003). Diagnosis and management of gestational hypertension and preeclampsia. Obstet Gynecol 102: 181–192.

Sibai BM (2005). Diagnosis, prevention, and management of eclampsia. Obstet Gynecol 105: 402–410.

Soubra SH, Guntupalli KK (2005). Critical illness in pregnancy: an overview. Crit Care Med 33: S248–S255.

Steegers EA, von Dadelszen P, Duvekot JJ et al. (2010). Pre-eclampsia. Lancet 376: 631–644.

Stevenson CB, Thompson RC (2005). The clinical management of intracranial neoplasms in pregnancy. Clin Obstet Gynecol 48: 24–37.

Thomas SV, Syam U, Devi JS (2012). Predictors of seizures during pregnancy in women with epilepsy. Epilepsia 53: e85.

Handbook of Clinical Neurology, Vol. 171 (3rd series)
Neurology and Pregnancy: Pathophysiology and Patient Care
E.A.P. Steegers, M.J. Cipolla and E.C. Miller, Editors
https://doi.org/10.1016/B978-0-444-64239-4.00012-6

Chapter 12

Neurologic complications of medical conditions in pregnancy

MARY A. O'NEAL*

Department of Neurology, Brigham and Women's Hospital, Boston, MA, United States

Abstract

This chapter discusses some of the neurologic complications of medical diseases that may occur in pregnancy. It reviews both the effects of pregnancy on the underlying disorder and how the medical condition may influence pregnancy outcomes. The most up-to-date information about risk stratification and disease management is presented. The specific disorders reviewed include sickle cell disease, antiphospholipid antibody syndrome, cardiac valve diseases, HIV infection, systemic lupus erythematosus, and fibromuscular dysplasia.

SICKLE CELL DISEASE

Neurologic manifestations

Sickle cell disease (SCD) is a group of genetic disorders with the most severe forms related to homozygous sickle cell anemia and the beta thalassemia mutations. Patients with SCD are at risk for both ischemic and hemorrhagic strokes (Sarnaik and Lusher, 1982). Vaso-occlusion can result in ischemic infarction. In addition, some patients may develop bilateral stenosis of the arteries around the circle of Willis, leading to a moyamoya pattern of vasculopathy (Webb and Kwiatkowski, 2013). Further, these patients are at risk for hemorrhage, both intracranial and subarachnoid, cerebral venous thrombosis (CVT), and fat embolism (Ciurea et al., 2006; Brandão et al., 2009; Adekile, 2015; Dhakal et al., 2015; Boon et al., 2016).

Preconception counseling

For patients with SCD who want to conceive, preimplantation genetic testing should be offered, although many women with SCD become pregnant without such testing. Nonaffected embryos can then be implanted. Other options for people affected with the disease include a sperm donor for men and surrogate carriers for women to avoid pregnancy complications.

Effects of pregnancy on maternal SCD

Available evidence has found that SCD confers increased risks for both maternal and fetal complications during pregnancy (Oteng-Ntim et al., 2015). Anemia, painful crisis, and infection are the most common pregnancy complications (Al Jama et al., 2009). One large study showed that of all neurologic complications, pregnant women with SCD were at greater risk for CVT (OR 4.9; 95% CI 2.2–10.9) but not ischemic stroke (Villers et al., 2008).

Effects of SCD on pregnancy outcomes

Women with SCD are at higher risk of developing preeclampsia (PE) and eclampsia, particularly those with hypertension (Lari et al., 2017). A large retrospective cohort study of 2,027,323 women with singleton pregnancies looked at outcomes of those with SCD compared with controls. Babies of mothers with SCD were more likely to have fetal anomalies. In addition, pregnancies

*Correspondence to: Mary A. O'Neal, M.D., Department of Neurology, Brigham and Women's Hospital, 44 Walter St. Roslindale, Boston, MA 02131, United States. Tel: +1-617-251-7114, Fax: +1-617-732-7432, E-mail: maoneal@bwh.harvard.edu

of these women were more likely to be complicated by preterm delivery, severe preeclampsia, small-for-gestational-age infants, and a higher cesarean section rate (Kuo and Caughey, 2016).

Management of SCD preconception and during pregnancy

Hydroxyurea is stopped when pregnancy is achieved, given the lack of safety data concerning malformation rates in animal studies. High-dose folate at 5 mg daily is used to prevent neural tube defects (RCOG, 2011). Most women with SCD have high iron levels and should not receive additional supplementation unless there is documented iron deficiency. Women with SCD are hyposplenic, which puts them at risk for infection from encapsulated organisms. Vaccination status should be determined and updated before pregnancy. Women with SCD have a higher risk of thrombotic events and should get prophylactic antithrombotic therapy (usually low molecular weight heparin—LMWH) during hospitalizations. Those at high risk may need prophylactic anticoagulation throughout their pregnancy. A study to look at prophylactic transfusion during pregnancy showed a decrease in the number of painful crisis, but no difference in other measures of maternal and neonatal outcomes (Koshy et al., 1988). Some experts do transfuse women with hemoglobin SS beginning in the third trimester (Howard et al., 1995). However, current evidence suggests that there is not enough data to support any significant difference in morbidity and mortality outcomes based on transfusion strategy (Jackson et al., 2018). Transfusion is indicated during pregnancy for several complications, including acute stroke, multiorgan failure, acute chest syndrome, and symptomatic anemia (NIH, 2004). In children with sickle cell anemia showing abnormal results on transcranial Doppler ultrasonography, prophylactic transfusion greatly reduces the risk of a first stroke (Adams et al., 1998). Analysis of the available trials show no clear benefit of prophylactic transfusions during pregnancy, but expert opinion supports this practice in women at high risk for complications (Okusanya and Oladapo, 2016).

Women with SCD are at high risk of developing gestational hypertension, PE, and eclampsia and should be closely monitored for these conditions. In addition, low-dose aspirin should be prescribed beginning in the second trimester to decrease the risk of PE and eclampsia (Askie et al., 2007).

ANTIPHOSPHOLIPID ANTIBODY SYNDROME

Antiphospholipid antibody syndrome (APLS) is an autoimmune disorder that causes both arterial and venous thrombosis and pregnancy loss. It is associated with antiphospholipid antibodies (aPL) and may be seen with systemic lupus erythematosus (SLE) and other autoimmune conditions, or it can occur as a primary disorder. The updated Sapporo criteria for APLS are listed in Table 12.1 (Miyakis et al., 2006).

Table 12.1

Sapporo criteria for APLS: at least one clinical and one laboratory criteria are required

Clinical criteria

1. Vascular thrombosis

One or more clinical arterial or venous thrombosis in any tissue or organ. Thrombosis must be confirmed by objective criteria. No evidence of an inflammatory etiology

2. Pregnancy morbidity

a. One or more unexplained deaths of morphologically normal fetuses at or beyond the 10th week of gestation documented by ultrasound or by direct examination of the fetus, or

b. One or more premature births of morphologically normal neonates before the 34th week of gestation because of eclampsia or severe preeclampsia

c. Three or more unexplained consecutive spontaneous abortions before the 10th week of gestation, with maternal anatomic or hormonal abnormalities and paternal and maternal chromosomal causes excluded

Laboratory criteria

1. Lupus anticoagulant present in plasma on two or more occasions, at least 12 weeks apart
2. Anticardiolipin antibody of IgG and/or IgM isotype in serum or plasma, present in medium or high titer on two or more occasions, at least 12 weeks apart
3. Anti-b2 glycoprotein-I antibody of IgG and/or IgM isotype in serum or plasma, present in high titer on two or more occasions, at least 12 weeks apart

Etiology of the obstetric complications

The etiology of early pregnancy loss is believed to be associated with thrombotic complications related to APLS. However, there is mounting evidence that many of the other obstetric complications may be related to inflammatory mediated endothelial dysfunction causing abnormal angiogenesis and an ischemic placental vascular unit (Abrahams et al., 2017).

Effect of APLS on pregnancy outcomes

In a large multicenter prospective trial, pregnant women with APLS were followed for adverse fetal outcomes.

Adverse outcomes were defined as fetal death after 12 weeks of gestation, neonatal death, delivery prior to 36 weeks of gestation due to PE and eclampsia, placental insufficiency, or small-for-gestational-age infants. The study found that lupus anticoagulant was the only aPL associated with adverse fetal outcomes after the first trimester. In addition, a clinical history of thrombosis associated with APLS was a strong risk factor for an adverse outcome in pregnancy (Yelnik et al., 2016). Another retrospective study looking at pregnancy outcomes in women with APLS showed that a higher number of antibodies is strongly correlated with poor outcomes. Women with APLS with a single aPL had higher obstetric complications. However, for triple-positive women (presence of lupus anticoagulant, anticardiolipin antibody, and anti-β2) even treated with low-dose aspirin and prophylactic LMWH, the chance of having a liveborn neonate was only 30% (Saccone et al., 2017).

Effects of pregnancy on neurologic complications

Ischemic stroke is a common neurologic manifestation (Mikdashi et al., 2007). The hypercoagulability of pregnancy enhances the risk associated with aPL.

Management in pregnancy

The management is divided into those with APLS with thrombotic complications versus those with pregnancy-related complications. Patients with prior thrombotic complications when pregnancy is achieved are treated with low-dose aspirin and therapeutic LMWH. Those women with obstetric complications either early <10 weeks and more than 3 miscarriages are treated with prophylactic LMWH and low dose aspirin. Pregnant women who have APLS alone are treated with low-dose aspirin (Uthman et al., 2019).

CARDIAC VALVE DISEASE

Neurologic manifestations

Bioprosthetic valves confer a low risk of embolism (6.6%–7.5%), and aspirin is an appropriate therapy for these women during pregnancy (Rafiq et al., 2017). In contrast, patients with a mechanical heart valve (MHV) have a high risk of thromboembolic events, including stroke (Lawley et al., 2015). A comparison of surgically placed mechanical valves versus bioprosthetic aortic valves in patients over 70 showed that the rate of intracranial bleeding was 8.8% vs 6.0% (HR 2.12; CI 1.09–4.15; $P=0.028$); ischemic stroke, 18.9% vs 16.1% ($P=0.341$); and infective endocarditis, 3.7% vs 2.8% ($P=0.242$) (Kytö et al., 2019). Data around incidence of prosthetic valve versus mechanical valves in the mitral position endocarditis appears to be similar with both hemorrhage and ischemic stroke being more common in the mechanical valve group (Yao et al., 2003). Pregnancy is a hypercoagulable state and places women with MHV at even higher risk for thrombotic complications.

Medications

Warfarin is the best studied and the preferred agent to prevent thrombotic events. However, in the first trimester, it is associated with major congenital malformations (MCMs), and in the second and third trimesters, it increases the risk of fetal loss and neonatal adverse advents (D'Souza et al., 2017).

LMWH does not cross the placenta and is safe for the fetus. This benefit must be weighed against the higher maternal risks of thromboembolism (8.7%) and mortality (2.9%) (D'Souza et al., 2017). Further, careful monitoring of the level of anticoagulation is needed throughout pregnancy.

The application of unfractionated heparin (UFH) is limited to IV use prior to a planned delivery. Subcutaneous UFH during pregnancy is not recommended as it has been associated with significant maternal risks, including death related to valve thrombosis (McLintock, 2011; D'Souza et al., 2017).

Thus anticoagulation in pregnancy in women with an MHV is a balancing act between maternal and fetal safety. In 2014, the American Heart Association/American College of Cardiology issued guidelines for management of anticoagulation in pregnant women with an MHV. The risks of warfarin-related MCMs is related to dose and is lower in regimens where a therapeutic INR can be achieved on doses less than or equal to 5 mg daily. Therefore in these women, continuation of warfarin, even in the first trimester, is reasonable (Nishimura et al., 2014). The guidelines recommend using dose-adjusted LMWH during the first trimester for pregnant patients with an MHV if the dose of warfarin is greater than 5 mg per day and to switch to warfarin in the second and third trimesters with maintenance of a therapeutic INR. Other guidelines differ in that they also support the use of dose-adjusted LMWH in the second and third trimesters. Low-dose aspirin is recommended during the second and third trimesters for high-risk women (Bates et al., 2012; Nishimura et al., 2014).

Preconception management of women with an MHV

Prior to conception, there needs to be an in-depth discussion regarding the risks of pregnancy. At that time, a detailed anticoagulation plan should be drafted to include management before pregnancy, during each trimester, delivery, and postpartum. The pregnancy anticoagulation plan will vary according to the patient's thrombotic risk profile as well as her desire to favor maternal versus fetal risks during the pregnancy. Risk factors that increase valve thrombosis and embolic complications include: atrial fibrillation or flutter, prior embolic complications, multiple mechanical valves, and type of valve (Cannegieter et al., 1994).

A detailed plan for anticoagulation for both a planned and urgent delivery should be made with input from cardiology, maternal fetal medicine, and obstetric anesthesia. In general, the indication for a caesarian section is decided on obstetric grounds; exceptions would be a woman with advanced heart failure or who is not hemodynamically stable (Regitz-Zagrosek et al., 2011). In addition, urgent cesarean section is indicated for women who go into labor while on therapeutic anticoagulation as risks of fetal hemorrhage are high (Vahanian et al., 2007). Anticoagulation would need to be reversed prior to delivery and the safety of neuraxial anesthesia versus general anesthesia assessed. Table 12.2 shows the recommended timing of stopping anticoagulants prior to neuraxial anesthesia (Horlocker et al., 2018).

HIV INFECTION

General principles

The goals of antiretroviral therapy (ART) in pregnancy are twofold: to decrease maternal complications of the disease and reduce perinatal transmission. The efficacy of ART in decreasing disease transmission has been clearly demonstrated. In a study of over 8000 mother–infant pairs, perinatal HIV transmission was almost zero in mothers who began ART before conception and had a sustained suppression of viral load (Mandelbrot et al., 2015). Neurologic complications related to opportunistic infections were directly related to viral load and CD4 count.

Treatment-naïve women

For women with newly diagnosed HIV disease, earlier initiation of ART was strongly associated with diminished rates of perinatal disease transmission. Perinatal infection rates were lower in women who became pregnant on ART than those initiating ART during pregnancy. In treatment-naïve women, each additional week of treatment reduced the odds of transmission by 8% (Hoffman et al., 2010). Therefore treatment should be initiated as soon as the diagnosis is made.

The ART regimen is based first on the resistance pattern of the woman's HIV and then on choosing drugs with a known safety profile in pregnancy. In general, dual nucleoside reverse transcriptase inhibitors with either a protease inhibitor or integrase inhibitor as the third drug are chosen (see Table 12.3).

Women on antiretroviral treatment

The recommendation is for women on successful ART with good viral load suppression to remain on their current regimen. However, alternations in drug dosing may be needed to accommodate the pregnancy-induced

Table 12.2

Timing of stopping anticoagulants prior to neuraxial anesthesia as recommended by the American Society of Regional Anesthesia and Pain Medicine

Anticoagulant/ antiplateletagent	Time required off the agent to safely place neuraxial anesthesia; testing required	Pregnancy category rating
Warfarin	4–5 days; normal INR	X in the first trimester
Heparin-IV or SQ	4–6 h; normal PTT	B
LMWH	>24 h	B
Aspirin	May continue	C

B: Animal reproduction studies have failed to demonstrate a risk to the fetus, and there are no adequate and well-controlled studies in pregnant women. Or animal studies have shown an adverse effect, but adequate and well-controlled studies in pregnant women have failed to demonstrate a risk to the fetus in any trimester.
C: Animal reproduction studies have shown an adverse effect on the fetus, but there are no adequate and well-controlled studies in humans; however, potential benefits may warrant use of the drug in pregnant women despite potential risks.
X: Studies in animals or humans have demonstrated fetal abnormalities, and/or there is positive evidence of human fetal risk based on adverse reaction data from investigational or marketing experience; moreover, the risks involved in use of the drug in pregnant women clearly outweigh potential benefits.

Table 12.3

Recommendations for use of antiretroviral drugs in pregnant HIV-1-infected women for maternal health and interventions to reduce perinatal HIV transmission in the United States

Dual NRT1 backbone	Facts
Tenofovir disoproxil fumarate/emtrictabine or Tenofovir disoproxil fumarate plus lamivudine	Daily administration
	Potential renal toxicity
Abacavir/lamivudine	Daily administration
	HLA-B5701 testing is required before use due to the risk of hypersensitivity reaction
	Not recommended if pretreatment HIV RNA level is >100,000 copies/mL
Third drug: integrase inhibitor	
Dolutegravir	Daily administration
	Not recommended for initiation during the first trimester
	Rapid viral load reduction
	Useful when drug interactions with protease inhibitor regimens are a concern
	Well tolerated
Raltegravir	Twice-daily administration
	Rapid viral load reduction
	Useful when drug interactions with protease inhibitor regimens are a concern
	Expanding experience with use in pregnancy
Third drug: protease inhibitor	
Atazanavir plus ritonavir	Daily administration
	Extensive experience in pregnancy
	Can cause maternal hyperbilirubinemia
Darunavir plus ritonavir	Twice-daily administration
	Well tolerated
	Increasing experience with use in pregnancy

From the Panel on Treatment of HIV-Infected Pregnant Women and Prevention of Perinatal Transmission.

changes in drug metabolism. If the viral load suppression is not good, the reasons for this, including drug resistance and poor adherence, need to be explored.

Mode of delivery and infant prophylaxis

For women with a viral load <1000 copies/mL the mode of delivery is based on obstetric concerns as the risk of HIV fetal transmission is low. For those with a viral load >1000 copies/mL, unknown viral load, or poor adherence to their ART regimen, a cesarean section at 38 weeks is recommended. Zidovudine is given as a pre-exposure prophylaxis to the fetus. In a trial, combined antepartum zidovudine, intravenous zidovudine during labor, and 6 weeks of infant zidovudine prophylaxis reduced perinatal transmission by 67% (Connor et al., 1994). For infants of mothers whose viral load is suppressed the regimen is 4–6 weeks of zidovudine whereas combination prophylaxis with either a two- or three-drug regimen is needed for infants at high risk (Nielsen-Saines et al., 2012).

Neurologic complications

The likely etiology of neurologic complications in women infected with HIV is best predicted by their CD4 count. For example, in patients with CD4 cell counts ranging from 200 to 500/microliters, HIV-associated cognitive and motor disorders are common. In severely immunosuppressed patients with CD4 cell counts <200/microliters, the most important diagnostic considerations include opportunistic infections and AIDS-associated tumors, such as primary central nervous system lymphoma. The leading diagnostic considerations in a patient with advanced immunosuppression are toxoplasma encephalitis, primary CNS lymphoma, progressive multifocal leukoencephalopathy, HIV encephalopathy, and CMV encephalitis (Skiest, 2002). In these cases, the clinical presentation, imaging findings (such as the presence of a mass and whether there is enhancement), as well as cerebrospinal fluid analysis are important determinants to establish a correct diagnosis. In those patients recently started on ART, immune reconstitution inflammatory syndrome can occur. This syndrome describes a collection of inflammatory disorders associated with paradoxical worsening of preexisting infectious processes following the initiation of ART (French, 2009).

Pregnancy outcomes

In study from a rural population in South Africa involving untreated pregnant women with HIV, there were significant adverse pregnancy outcomes, including maternal antenatal deaths, spontaneous abortions, and stillbirths. In addition, these women had a significantly increased risk of having a low-birth-weight baby (Rollins et al., 2007). There are conflicting studies about the effects of ART on pregnancy outcomes with suggestions of increased risk of preterm birth, intrauterine growth restriction, and anemia (Dadhwal et al., 2017). Study results are limited due to confounders such as access to prenatal care and nutritional state.

CONNECTIVE TISSUE DISORDERS

Connective tissue disorders refer to a group to diseases that affect the fascia. They are caused by either autoimmune or genetic mechanisms. Autoimmune disorders are more prevalent in women, who account for 78% of those afflicted (Fairweather and Rose, 2004). Two connective tissue disorders that are more prevalent in women and can affect pregnancy outcomes will be discussed: SLE and fibromuscular dysplasia (FMD).

SYSTEMIC LUPUS ERYTHEMATOSUS

Diagnosis and sex-specific concerns

SLE is an inflammatory disorder with heterogeneous manifestations. The American College of Rheumatology formulated the classification criteria for SLE (see Table 12.4). It affects women seven times more frequently than men (Chakravarty et al., 2007). The disease usually starts in women during their reproductive years. In pregnancy, SLE can cause significant maternal and fetal morbidity. For example, in a prospective study undertaken from 1987 to 2002, the impact of lupus disease activity on pregnancy outcomes was observed in 267 pregnancies. High-activity lupus in the first and second trimesters led to a threefold increase in pregnancy loss (Clowse et al., 2005). Thus patients are recommended to postpone pregnancy until their disease is stable for 6 months.

Preconception management

A preconception consultation with both a rheumatologist and a high-risk obstetrician is ideal to assess disease activity, medication safety during pregnancy, and risks of pregnancy-related complications. In addition, other associated medical conditions such as

Table 12.4

ACR classification criteria for SLE need 4/11 criteria

Symptom	Definition
Malar rash	Erythema over malar regions of the face
Discoid rash	Raised erythematous patches with keratotic scaling and follicular plugging
Photosensitivity	Skin sensitivity to sunlight
Oral ulcers	Usually painless
Arthitis	Nonerosive with tenderness or swelling
Pericarditis, pleuritis	Can be by history
Renal dysfunction	Proteinuria of 0.5 g daily or cellular casts
Neurologic disorder	Seizures or psychosis without provocative factors
Hematologic abnormalities	Hemolytic anemia or leukopenia or lymphopenia or thrombocytopenia
Positive serology	Positive ANA-generally high titer, Positive antidouble-stranded DNA, anti-Sm nuclear antigen, or positive antiphospholipid antibodies.
Immunologic disorder	Anti-DNA—Antibody to native DNA in abnormal titer **OR** Anti-Sm—Presence of antibody to Sm nuclear antigen **OR** Positive antiphospholipid antibody on: 1. An abnormal serum level of IgG or IgM anticardiolipin antibodies **OR** 2. A positive test result for lupus anticoagulant using a standard method **OR** A false-positive serologic test for syphilis known to be positive for at least 6 months and confirmed by *Treponema pallidum* immobilization orfluorescent treponemal antibody absorption test

hypercoagulability and prior obstetric outcomes need to be thoroughly explored. During this visit, a plan should be devised about which medications should be continued, how a flare will be treated, and baseline laboratories obtained. Women with SLE at high risk during pregnancy include those with active disease.

Neurologic and psychiatric manifestations

The neurologic manifestations of SLE are now referred to as neuropsychiatric lupus (NPSLE) and not lupus vasculitis as there is no evidence for active vasculitis leading to the CNS involvement (Kivity et al., 2015). Estimates as to the prevalence vary given type of study, definition, and population evaluated. A meta-analysis of 10 prospective studies showed the prevalence of NPSLE to be over 50% (Unterman et al., 2011). In this study, headache, mood disorders, and cognitive disturbance were the most frequent manifestations of NPSLE.

Psychosis

The prevalence of psychosis is estimated at 17% (Appenzeller et al., 2008). It appears that active SLE likely via an antibody-mediated mechanism causes psychosis (Bonfa et al., 1987; Hanly et al., 2011). Lupus-mediated psychosis needs to be distinguished from that related to steroids. Usually, other disease markers demonstrate increased inflammation and active disease.

Stroke

Overall the risk of ischemic stroke in lupus in 2.5 times higher than in the general population (Holmqvist et al., 2015). It is also higher in younger cohorts. The etiology of stroke is heterogeneous, including small vessel vasculopathy from hypertension as well as a direct disease-associated vasculopathy, embolic stroke from Libman Sacks endocarditis, or related to the APLS. Patients with SLE also have an increased risk for both subarachnoid and intracranial hemorrhage. In a study looking at stroke in pregnancy the presence of SLE increased the risk of stroke by an OR of 15.2 (James et al., 2005). Appropriate treatment is dependent on stroke mechanism.

Seizures

Seizures occur in 12%–14% of patients with lupus with over half occurring in the first year of diagnosis (Mikdashi et al., 2005; González-Duarte et al., 2008). They can be focal or generalized. The seizure etiology is varied, including metabolic causes such as uremia, electrolyte disturbances, hypertension, medication toxicity, and stroke. The risk of seizures appears correlated with higher disease activity. Particularly, focal seizures with impaired awareness are associated with psychosis. A large prospective cohort study of 1631 patients with lupus followed for 3.5 years found that seizures did not require antiepileptic drugs (AEDs) in 76% of patients (Hanly et al., 2012). As most seizures are provoked, AEDs should be reserved for patients with two or more unprovoked seizures, evidence of structural brain injury, and an epileptiform electroencephalogram (Bertsias et al., 2010). In those women who develop epilepsy during their reproductive years, an antiepileptic medication with a known safety profile in pregnancy should be chosen.

Headache

Headache is reported in many patients with lupus, with over a third meeting the criteria for migraine (Tjensvoll et al., 2011), although studies have not proven any increased prevalence of headache. Caution needs to be exercised to exclude secondary causes of headache, including PE and eclampsia, meningitis, CVT, and hemorrhage. In the absence of red flags such as fever, presence of APLS, and focal neurologic findings, concerning medications such as anticoagulants, isolated headache can be treated similarly to a primary benign headache.

Other neurologic manifestations

Chorea is rare occurring in only 1% of patients with lupus (Sanna et al., 2003). It may be the first manifestation of NPSLE and is often associated with aPL (Baizabal-Carvallo et al., 2013). Focality to the movement disorder should raise a concern about infarction. Myelopathy and cranial neuropathies and particularly optic neuritis can also occur in SLE (Sanna et al., 2003). Their treatment depends on whether the underlying cause is inflammatory, infectious, or thrombotic.

Effects of pregnancy on maternal SLE

Pregnancy increases the likelihood of having lupus flares with frequencies ranging from 42% to 60% (Petri et al., 1991; Carvalheiras et al., 2010). Risk factors associated with flares include history of lupus nephritis, stopping hydroxychloroquine, and active disease in the preceding 6 months (Saavedra et al., 2002; Clowse et al., 2005, 2006).

Effects of SLE on pregnancy outcomes

Women with SLE are at enhanced risk of pregnancy loss, affecting 15%–25% of all pregnancies (Smyth et al., 2010). A review of the Nationwide Inpatient Sample from 2000 to 2003 showed that maternal mortality was 20-fold higher among women with lupus (Clowse et al., 2008). Complications include need for cesarean section, premature birth, fetal growth restriction, gestational hypertension, and PE and eclampsia.

Predictors for pregnancy complications included disease activity the year prior to conception and presence of hypertension and APLS (Borella et al., 2014).

Preconception planning

The patient's medications should be reviewed prior to conception and adjusted to appropriately balance control of disease and minimize fetal risk by the use of medications with the best pregnancy safety profile. In general, hydroxychloroquine is continued in pregnancy as it has been shown to reduce lupus flares without increasing miscarriages, stillbirth, pregnancy loss, or congenital abnormalities (Clowse et al., 2006). Low-dose aspirin is initiated at 12 weeks of gestation to decrease the risk of PE and eclampsia. Prednisone is tapered to the lowest dose known to control disease. Table 12.5 shows the common medications used in lupus treatment and their pregnancy categories.

Fibromuscular dysplasia (FMD)

FMD is a noninflammatory disorder that involves medium-sized vessels usually affecting the renal and cervical arteries. The prevalence of FMD is unknown as many patients are asymptomatic. In a 25-year study of consecutive autopsies done at the Mayo clinic, FMD was seen in 0.02% of the carotids (Schievink and Björnsson, 1996). It is more common in women with a prevalence rate of 91% (Olin et al., 2012). The reasons for this sex difference in prevalence are unclear.

Neurologic manifestations

The most common neurologic manifestations of FMD are headaches, with a significant proportion of these being migraine and pulsatile tinnitus (Olin et al., 2012). It is also associated with aneurysm, which occurred in 21.7% of patients with dissection in 25.7% (Kadian-Dodov et al., 2016). Women with renal artery involvement have been reported to have a higher risk for PE and eclampsia (Vance et al., 2015). There are no guidelines for management of FMD in patients with stroke, so in general it is managed similarly to other patients with large- or medium-vessel stroke without a surgical indication with the use of low-dose aspirin. In pregnancy, the time for concern is at delivery. There are no studies to guide appropriate management. Limiting pushing in patients with known cervical FMD seems prudent; therefore an assisted second stage labor should be considered (Cunningham et al., 2019).

Known aneurysms are best evaluated prior to pregnancy onset. A study looking at inpatient data over 2 decades showed a risk of intracranial aneurysm rupture during pregnancy of 1.4% and during delivery of 0.05%. This data suggests that the risk of aneurysm rupture is similar in pregnant women compared with nonpregnant women of the same age (Kim et al., 2013). Another study looking at the risk of aneurysm rupture found a several-fold increase on the day of delivery (Weir and Macdonald, 1996). Given these conflicting data, no satisfactory general recommendation is supported by evidence. It is likely safe for women with small, uncomplicated, unruptured aneurysms to delivery vaginally (Kim et al., 2013). Cesarean section would be justified when the aneurysm characteristics such as size, location, morphology, and hemodynamic factors suggest high risk (Juvela et al., 2013). Given the predilection for dissection/hemorrhage, neurologic consultation is prudent for women with known FMD and an unexplained headache or neck pain in the postpartum period.

Table 12.5

Pregnancy and breastfeeding recommendations for medications used to treat SLE

Drug	Pregnancy recommendations	Breastfeeding recommendations
Nonsteroidal antiinflammatories	Unclear risk in first trimester, avoid after 30 weeks due to risk of premature closure of the patent ductus arteriosus	Compatible
Glucocorticoids	Lowest possible dose	Compatible
Azathioprine	Preferred immunosuppressant for active disease	Probably compatible
Cyclosporine	Unknown risk	Need to monitor for toxicity
Tacrolimus	Reasonable alternative to other more potent immunosuppressants	Not recommended
Rituximab	Unknown risk	Unknown risk
Cyclophosphamide	Avoid	Avoid
Hydroxychloroquine	Continue	Compatible
Methotrexate	Avoid	Avoid

SUMMARY

Pregnancy in women with medical diseases needs to be carefully planned. Ideally, there should be an individualized preconception, pregnancy, and delivery plan based on the woman's known manifestations of their disease and involve the pertinent subspecialists, preferably neurologists. Pregnancies of these high-risk women need to be carefully monitored for early detection of complications by their care teams. This includes neurologic complications in the postpartum period.

References

Abrahams VM, Chamley LW, Salmon JE (2017). Antiphospholipid syndrome and pregnancy: pathogenesis to translation. Arthritis Rheumatol 69: 1710–1721.

Adams RJ, McKie VC, Hsu L et al. (1998). Prevention of a first stroke by transfusions in children with sickle cell anemia and abnormal results on transcranial Doppler ultrasonography. N Engl J Med 339: 5–11.

Adekile A (2015). Risk of cerebral haemorrhage in children with sickle cell disease. Dev Med Child Neurol 57: 116–117.

Al Jama FE, Gasem T, Burshaid S et al. (2009). Pregnancy outcome in patients with homozygous sickle cell disease in a university hospital, Eastern Saudi Arabia. Arch Gynecol Obstet 280: 793–797.

Appenzeller S, Cendes F, Costallat LT (2008). Acute psychosis in systemic lupus erythematosus. Rheumatol Int 28: 237–243.

Askie LM, Duley L, Henderson-Smart DJ et al. (2007). Antiplatelet agents for prevention of pre-eclampsia: a meta-analysis of individual patient data. Lancet 369: 1791–1798.

Baizabal-Carvallo JF, Bonnet C, Jankovic J (2013). Movement disorders in systemic lupus erythematosus and the antiphospholipid syndrome. J Neural Transm 120: 1579–1589.

Bates SM, Greer IA, Middeldorp S et al. (2012). VTE, thrombophilia, antithrombotic therapy, and pregnancy: antithrombotic therapy and prevention of thrombosis, 9th ed: American College of Chest Physicians Evidence-Based Clinical Practice Guidelines. Chest 141: e691S–e736S.

Bertsias GK, Ioannidis JP, Aringer M et al. (2010). EULAR recommendations for the management of systemic lupus erythematosus with neuropsychiatric manifestations: report of a task force of the EULAR Standing Committee for Clinical Affairs. Ann Rheum Dis 69: 2074–2082.

Bonfa E, Golombek SJ, Kaufman LD et al. (1987). Association between lupus psychosis and anti-ribosomal P protein antibodies. N Engl J Med 317: 265–271.

Boon IS, Starkey KJ, Samsonova O et al. (2016). In the thick of it: cerebral venous sinus thrombosis precipitated by iron-deficiency anaemia and sickle cell trait. BMJ Case Rep 17: 2016.

Borella E, Lojacono A, Gatto M et al. (2014). Predictors of maternal and fetal complications in SLE patients: a prospective study. Immunol Res 60: 170–176.

Brandão RA, de Carvalho GT, Reis BL et al. (2009). Intracranial aneurysms in sickle cell patients: report of 2 cases and review of the literature. Surg Neurol 72: 296–299.

Cannegieter SC, Rosendaal FR, Briët E (1994). Thromboembolic and bleeding complications in patients with mechanical heart valve prostheses. Circulation 89: 635–641.

Carvalheiras G, Vita P, Marta S et al. (2010). Pregnancy and systemic lupus erythematosus: review of clinical features and outcome of 51 pregnancies at a single institution. Clin Rev Allergy Immunol 38: 302–306.

Chakravarty EF, Bush TM, Manzi S et al. (2007). Prevalence of adult systemic lupus erythematosus in California and Pennsylvania in 2000: estimates obtained using hospitalization data. Arthritis Rheum 56: 2092–2094.

Ciurea SO, Thulborn KR, Gowhari M (2006). Dural venous sinus thrombosis in a patient with sickle cell disease: case report and literature review. Am J Hematol 81: 290–293.

Clowse ME, Magder LS, Witter FR et al. (2005). The impact of increased lupus activity on obstetric outcomes. Arthritis Rheumatol 52: 514–521.

Clowse ME, Magder L, Witter F et al. (2006). Hydroxychloroquine in lupus pregnancy. Arthritis Rheum 54: 3640–3647.

Clowse ME, Jamison M, Myers E et al. (2008). A national study of the complications of lupus in pregnancy. Am J Obstet Gynecol 199: 127.e1-6.

Connor EM, Sperling RS, Gelber R et al. (1994). Reduction of maternal-infant transmission of human immunodeficiency virus type 1 with zidovudine treatment. Pediatric AIDS Clinical Trials Group Protocol 076 Study Group. N Engl J Med 331: 1173–1180.

Cunningham TK, Draper H, Rajesh U (2019). Management of a pregnancy with underlying fibromuscular dysplasia with a history of stroke and carotid artery dissection. J Obstet Gynaecol 39: 417–419.

Dadhwal V, Sharma A, Khoiwal K et al. (2017). Pregnancy outcomes in HIV-infected women: experience from a tertiary care center in India. Int J MCH AIDS 6: 75–81.

Dhakal LP, Bourgeois K, Barrett KM et al. (2015). The "starfield" pattern of cerebral fat embolism from bone marrow necrosis in sickle cell crisis. Neurohospitalist 5: 74–76.

D'Souza R, Ostro J, Shah PS et al. (2017). Anticoagulation for pregnant women with mechanical heart valves: a systematic review and meta-analysis. Eur Heart J 38: 1509–1516.

Fairweather D, Rose NR (2004). Women and autoimmune diseases. Emerg Infect Dis 10: 2005–2011.

French MA (2009). HIV/AIDS: immune reconstitution inflammatory syndrome: a reappraisal. Clin Infect Dis 48: 101–107.

González-Duarte A, Cantú-Brito CG, Ruano-Calderón L et al. (2008). Clinical description of seizures in patients with systemic lupus erythematosus. Eur Neurol 59: 320–323.

Hanly JG, Urowitz MB, Su L et al. (2011). Autoantibodies as biomarkers for the prediction of neuropsychiatric events in systemic lupus erythematosus. Ann Rheum Dis 70: 1726–1732.

Hanly JG, Urowitz MB, Su L et al. (2012). Seizure disorders in systemic lupus erythematosus results from an international, prospective, inception cohort study. Ann Rheum Dis 71: 1502–1509.

Hoffman RM, Black V, Technau K et al. (2010). Effects of highly active antiretroviral therapy duration and regimen on risk for mother-to-child transmission of HIV in Johannesburg, South Africa. J Acquir Immune Defic Syndr 54: 35–41.

Holmqvist M, Simard JF, Asplund K et al. (2015). Stroke in systemic lupus erythematosus: a meta-analysis of population-based cohort studies. RMD Open 1: e000168.

Horlocker TT, Vandermeulen E, Kopp SL et al. (2018). Regional anesthesia in the patient receiving antithrombotic or thrombolytic therapy: American Society of Regional Anesthesia and Pain Medicine Evidence-Based Guidelines (fourth edition). Reg Anesth Pain Med 43: 263–330.

Howard RJ, Tuck SM, Pearson TC (1995). Pregnancy in sickle cell disease in the UK: results of a multicentre survey of the effect of prophylactic blood transfusion on maternal and fetal outcome. Br J Obstet Gynaecol 102: 947–951.

Jackson B, Fasano R, Roback J (2018). Current evidence for the use of prophylactic transfusion to treat sickle cell disease during pregnancy. Transfus Med Rev 32: 220–224.

James AH, Bushnell CD, Jamison MG et al. (2005). Incidence and risk factors for stroke in pregnancy and the puerperium. Obstet Gynecol 106: 509–516.

Juvela S, Poussa K, Lehto H et al. (2013). Natural history of unruptured intracranial aneurysms: a long-term follow-up study. Stroke 44: 2414–2421.

Kadian-Dodov D, Gornik HL, Gu X et al. (2016). Dissection and aneurysm in patients with fibromuscular dysplasia: findings from the U.S. registry for FMD. J Am Coll Cardiol 68: 176–185.

Kim YW, Deal D, Hoh BL (2013). Cerebral aneurysms in pregnancy and delivery: pregnancy and delivery do not increase the risk of aneurysm rupture. Neurosurgery 72: 143–150.

Kivity S, Agmon-Levin N, Zandman-Goddard G et al. (2015). Neuropsychiatric lupus: a mosaic of clinical presentations. BMC Med 13: 43.

Koshy M, Burd L, Wallace D et al. (1988). Prophylactic red-cell transfusions in pregnant patients with sickle cell disease. A randomized cooperative study. N Engl J Med 319: 1447–1452.

Kuo K, Caughey AB (2016). Contemporary outcomes of sickle cell disease in pregnancy. Am J Obstet Gynecol 215: 505.e1-5.

Kytö V, Myllykangas ME, Sipilä J et al. (2019). Ann Thorac Surg. 108 (5): 1354–1360.

Lari NF, DeBaun MR, Oppong SA (2017). The emerging challenge of optimal blood pressure management and hypertensive syndromes in pregnant women with sickle cell disease: a review. Expert Rev Hematol 10: 987–994.

Lawley CM, Lain SJ, Algert CS et al. (2015). Prosthetic heart valves in pregnancy, outcomes for women and their babies: a systemic review and meta-analysis. BJOG 122: 1446–1455.

Mandelbrot L, Tubiana R, Le Chenadec J et al. (2015). No perinatal HIV-1 transmission from women with effective antiretroviral therapy starting before conception. Clin Infect Dis 61: 1715–1725.

McLintock C (2011). Anticoagulant therapy in pregnant women with mechanical prosthetic heart valves: no easy option. Thromb Res 127 (Suppl 3): S56–S60.

Mikdashi J, Krumholz A, Handwerger B (2005). Factors at diagnosis predict subsequent occurrence of seizures in systemic lupus erythematosus. Neurology 64: 2102–2107.

Mikdashi J, Handwerger B, Langenberg P et al. (2007). Baseline disease activity, hyperlipidemia, and hypertension are predictive factors for ischemic stroke and stroke severity in systemic lupus erythematosus. Stroke 38: 281–285.

Miyakis S, Lockshin AT, Branch DW et al. (2006). International consensus statement on an update of the classification criteria for definite antiphospholipid syndrome (APS). J Thromb Haemost 4: 295–306.

Nielsen-Saines K, Watts DH, Veloso VG et al. (2012). Three postpartum antiretroviral regimens to prevent intrapartum HIV infection. N Engl J Med 366: 2368–2379.

NIH (2004). The management of sickle cell disease, National Institutes of Health; National Heart, Lung, and Blood Institute, Division of Blood Diseases and Resources NIH publication 04-2117, revised 2004.

Nishimura RA, Otto CM, Bonow RO et al. (2014). AHA/ACC guideline for the management of patients with valvular heart disease: a report of the American College of Cardiology/American Heart Association Task Force on Practice Guidelines. J Am Coll Cardiol 63: e57.

Okusanya BO, Oladapo OT (2016). Prophylactic versus selective blood transfusion for sickle cell disease in pregnancy. Cochrane Database Syst Rev12: CD010378. https://doi.org/10.1002/14651858.

Olin JW, Froehlich J, Gu X et al. (2012). The United States registry for fibromuscular dysplasia: results in the first 447 patients. Circulation 125: 3182–3190.

Oteng-Ntim E, Meeks D, Seed PT et al. (2015). Adverse maternal and perinatal outcomes in pregnant women with sickle cell disease: systematic review and meta-analysis. Blood 125: 3316–3325.

Petri M, Howard D, Repke J (1991). Frequency of lupus flare in pregnancy. The Hopkins Lupus Pregnancy Center experience. Arthritis Rheum 34: 1538–1545.

Rafiq S, Steinbrüchel DA, Lilleør NB et al. (2017). Antithrombotic therapy after bioprosthetic aortic valve implantation: warfarin versus aspirin, a randomized controlled trial. Thromb Res 150: 104–110.

RCOG (2011). Management of sickle cell disease in pregnancy. Green–top guideline no. 61 1–20.

Regitz-Zagrosek V, Lundqvist CB, Borghi C et al. (2011). ESC guidelines on the management of cardiovascular diseases during pregnancy: the task force on the Management of Cardiovascular Diseases during Pregnancy of the European Society of Cardiology (ESC). Eur Heart J 32: 3147–3197.

Rollins NC, Coovadia H, Bland R et al. (2007). Pregnancy outcomes in HIV-infected and uninfected women in rural and urban South Africa. J Acquir Immune Defic Syndr 44: 321–328.

Saavedra MA, Cruz-Reyes C, Vera-Lastra O et al. (2002). Impact of previous lupus nephritis on maternal and fetal outcomes during pregnancy. Clin Rheumat 31: 813–819.

Saccone G, Berghella V, Maruotti GM et al. (2017). Antiphospholipid antibody profile based obstetric outcomes of primary antiphospholipid syndrome: the PREGNANTS study. Am J Obstet Gynecol 216: 525.e1.

Sanna G, Bertolaccini ML, Cuadrado MJ et al. (2003). Neuropsychiatric manifestations in systemic lupus erythematosus: prevalence and association with antiphospholipid antibodies. J Rheumatol 30: 985–992.

Sarnaik SA, Lusher JM (1982). Neurological complications of sickle cell anemia. Am J Pediatr Hematol Oncol 4: 386–394.

Schievink WI, Björnsson J (1996). Fibromuscular dysplasia of the internal carotid artery: a clinicopathological study. Clin Neuropathol 15: 2–6.

Skiest DJ (2002). Focal neurological disease in patients with acquired immunodeficiency syndrome. Clin Infect Dis 34: 103–115.

Smyth A, Oliveira GH, Lahr BD et al. (2010). A systematic review and meta-analysis of pregnancy outcomes in patients with systemic lupus erythematosus and lupus nephritis. Clin J Am Soc Nephrol 5: 2060–2068.

Tjensvoll AB, Harboe E, Gøransson LG et al. (2011). Migraine is frequent in patients with systemic lupus erythematosus: a case-control study. Cephalalgia 31: 401–408.

Unterman A, Nolte JE, Boaz M et al. (2011). Neuropsychiatric syndromes in systemic lupus erythematosus: a meta-analysis. Semin Arthritis Rheum 41: 1–11.

Uthman I, Noureldine MHA, Ruiz-Irastorza G et al. (2019). Management of antiphospholipid syndrome. Ann Rheum Dis 78: 155–161.

Vahanian A, Baumgartner H, Bax J et al. (2007). Guidelines on the management of valvular heart disease: the task force on the management of valvular heart disease of the European society of cardiology. Eur Heart J 28: 230–268.

Vance CJ, Taylor RN, Craven TE et al. (2015). Increased prevalence of preeclampsia among women undergoing procedural intervention for renal artery fibromuscular dysplasia. Ann Vasc Surg 29: 1105–1110.

Villers MS, Jamison MG, De Castro LM et al. (2008). Morbidity associated with sickle cell disease in pregnancy. Am J Obstet Gynecol 199: 125.e1.

Webb J, Kwiatkowski JL (2013). Stroke in patients with sickle cell disease. Expert Rev Hematol 6: 301–316.

Weir B, Macdonald RL (1996). Management of intracranial aneurysms and arteriovenous malformations during pregnancy. In: RH Wilkins, SS Rengachary (Eds.), Neurosurgery. McGraw-Hill, New York, pp. 2421–2427.

Yao H, Miyamoto T, Mukai S et al. (2003). Long-term results of mitral valve replacement: biological xenograft versus mechanical valves. J Artif Organs 6: 30–36.

Yelnik CM, Laskin CA, Porter TF et al. (2016). Lupus anticoagulant is the main predictor of adverse pregnancy outcomes in aPL-positive patients: validation of PROMISSE study results. Lupus Sci Med 3: e000131.

Handbook of Clinical Neurology, Vol. 171 (3rd series)
Neurology and Pregnancy: Pathophysiology and Patient Care
E.A.P. Steegers, M.J. Cipolla and E.C. Miller, Editors
https://doi.org/10.1016/B978-0-444-64239-4.00013-8

Chapter 13

The ethics of neurologically complicated pregnancies

DOROTHY SMOK[1]* AND KENNETH M. PRAGER[2]

[1]*Department of Obstetrics Gynecology, Columbia University Vagelos College of Physicians and Surgeons, New York, NY, United States*

[2]*Department of Medicine, Columbia University Irving Medical Center, New York, NY, United States*

Abstract

Bioethical conflicts in pregnancy are distinguished from those in other areas of medicine due to competing interests between mother and fetus because of their shared biology. Historically, prior to the advent of fetal therapy and advances in medical technology, the maternal–fetal complex was considered to be a single entity. With advances in medicine, treatment options can now be directed at both the mother and the fetus, and a duality has evolved in the maternal–fetal unit. Thus at some point during pregnancy, two individuals rather than just one are the responsibility of the physician.

In determining how to properly care for the pregnant woman with a neurologic condition, therapeutic choices must take into consideration the impact a treatment will have on both the mother and the fetus. Since what benefits one may harm the other, tension results from the need to choose.

This chapter will highlight ethical conflicts arising at the interface of obstetrics and neurology. We will delve into situations where difficult reproductive and therapeutic decisions must be made in pregnant women with intellectual disabilities, stroke, brain tumors, and epilepsy. The complexity of brain death in pregnancy will be analyzed, acknowledging the influence of politics, law, and religion that bears on ethical decision-making.

In approaching ethical dilemmas encountered in pregnancies complicated by neurologic conditions, frameworks based on principles, virtues, care, and feminist ethics, and case precedents will be applied to facilitate ethically appropriate shared decision-making. We hope that this chapter will provide valuable guidance for providers caring for this complex obstetric population.

INTRODUCTION

The term “ethics” refers to well-grounded standards of right and wrong that inform how we should act in a range of situations. They make us question what our behavior should be and what virtues should be cultivated in our moral lives (Chervenak and McCullough, 1992; Beauchamp and Childress, 2013; Steinbock et al., 2013; Kuhse et al., 2015). In medicine, a structured approach to examine the ethical issues that arise in clinical practice is provided by the discipline of clinical ethics, which tends to the question of a moral choice for the clinician. Ethical dimensions of clinical care become more relevant when value-laden choices lead to conflicts between patients, family, providers, and the medical establishment at large. Importantly, the nuances of ethical choices will be realized only if their existence is appreciated. Thus a mindful approach toward every patient and every clinical situation is required to ensure that care is delivered in the manner in which it should be delivered and not merely in the way in which it could be delivered.

Ethical conflicts in clinical practice may be addressed through the application of systematic processes that are rooted in philosophical theory. Such conflicts are examined from two perspectives, that of the physician and that

*Correspondence to: Dorothy Smok, M.D., Atlantic Maternal Fetal Medicine, 435 South Street, Suite 380, Morristown, NJ 07960. Tel: +1-201-919-4954, E-mail: dp2073@columbia.edu

of the patient, under the assumed axiomatic truth that the physician's obligation is to promote the interests of the patient (Beauchamp and McCullough, 1984).

Several types of frameworks are utilized to recognize and manage ethical issues in medicine. A practical approach may be pursued through applied ethics, which refers to the use of philosophical methods to solve moral problems, practices, and policies in the professions, technology, and government (Beauchamp, 2003). Originally developed by Beauchamp and Childress (2013), the four principles, where ethical principles of respect for autonomy, beneficence, nonmaleficence, and justice are applied, represent an applied ethics approach. In contrast, normative approaches such as virtue ethics, among others, involve abstract debates that occur about the best or most appropriate approach (Pellegrino, 1995).

Distinct worldviews and different moral values raise the question whether one ethical framework can be applied internationally. Since dominant ethical frameworks evolve within the context of religion, culture, politics, and economics, each has limits in terms of general applicability. As an example, individual-centered models of moral reasoning emphasizing autonomy, which are highlighted in Western bioethics, may be less applicable in societies where duties and obligations dominate behavior. In certain cultures, the rights of others are valued over those of the individual and require the application of communitarian models of bioethics, where duty-based principles and beneficence-based obligations form the core framework (Farmer and Campos, 2004; Fox and Swazey, 2005, 2008; ten Have and Gordijn, 2014). Social inequalities and poverty, which are pervasive around the globe, lead to a human rights based approach that emphasizes justice and dignity (Feuillet-Liger and Orfali, 2018). Thus across cultures and in a pluralistic society, the clinician grappling with value-laden conflicts will often find that there is no one best approach and more than one ethical framework is required to attain appropriate resolutions to difficult clinical problems.

The role of spirituality and religion in bioethics deserves special mention. Bioethics has become secularized over the past 50 years and has come to rely heavily on the law as a source of morality (Callahan, 1990). While the degree of importance that religion plays varies between individuals and among cultures, a substantial portion of the population considers religion to have value in their daily lives (Silver, 2019). In certain cultures where religion and spirituality are paramount, knowledge of the patient's beliefs is essential to deliver culturally competent medical care. Thus a holistic approach that strives to understand the wholeness of a person will attend to underlying physical, emotional, and spiritual concerns and incorporate faith into the framework where applicable (Pew Research accessed 2019).

What follows is a description of the most common ethical frameworks used in clinical problem solving.

ETHICAL THEORIES

The four principles

The four principles, introduced by Tom Beauchamp and James Childress, are among the most commonly applied and practical ethical frameworks in medicine. Using this approach, four guiding principles of bioethical analysis are grouped into four general categories (Beauchamp, 2007). The four prima facie principles are respect for autonomy, beneficence, nonmaleficence, and justice. Each principle is to be followed unless it conflicts with one or more of the others, and no principle is ranked higher than another (nonhierarchical). A criticism of this approach is that is does not provide for a method of choosing between the principles when they are in conflict (Gillon, 1994).

Autonomy

Respect for patient autonomy acknowledges an individual's right to make decisions based on her own personal values and beliefs (President's Commission for the Study of Ethical Problems in Medicine, Biomedical and Behavioral Research (US), 1982). This principle upholds the ethical doctrine of informed consent, where a patient is enabled to make informed and voluntary decisions about her medical care and what is done to her body.

The principle of autonomy applies to individuals with medical decision-making capacity who are capable of providing informed consent and choice, and to those who are incapable of making autonomous decisions, including children, unconscious patients, and those with cognitive disability. Medical decision-making capacity is a higher-order functional capacity that refers to an individual's cognitive and emotional ability, or capacity, to accept, decline, or choose among treatment options (Okonkwo et al., 2008). Individuals who are unable to exercise autonomous decisions require protection of their autonomy through either a living will, consent from a surrogate, or, in their absence, a hospital ethics committee or legal action.

Beneficence and nonmaleficence

Beneficence, from the Latin word *beneficentia*, means "kindness, generosity," and this principle refers to the moral obligation to act in a manner that will benefit others. However, in trying to exert a positive effect, a risk of harm may exist, and therefore the principle of nonmaleficence must be taken into consideration as a net benefit over harm. Thus these principles consider the balance of risks versus benefits, benefits over burdens. Many medical treatments involve some harm, even if minimal, but the harm should not be disproportionate to the benefits of the treatment. Though most clinical scenarios involve a clear-cut preponderance of beneficence over maleficence, or the opposite, allowing clinicians to easily decide on a plan of treatment, there are situations where these two principles are roughly equally potent making it potentially very difficult to make a clinically and ethically sound recommendation. Chervenak and McCullough propose that beneficence supersedes nonmaleficence in obstetric ethics (Chervenak and McCullough, 1992). A common application of beneficence and nonmaleficence to obstetrics involves considerations of the two modes of delivery. Both vaginal and cesarean deliveries have their own complications. In general, spontaneous vaginal delivery is expected to result in fewer complications than a cesarean and is therefore the default mode of delivery if no contraindications exist.

Justice

Justice is synonymous with fairness and is viewed as the moral obligation to justly settle competing claims and accord others what is due to them. This principle may be applied at different levels, including at the local and societal levels. Obligations of justice fall into different categories, including matters that address the fair distribution of scarce resources (distributive justice), those that respect people's rights (rights based justice), and those that ensure laws are morally acceptable (legal justice). This principle requires equal treatment of those who are alike, unless otherwise indicated by scientific and clinical evidence.

Casuistry

Casuistry refers to case-based reasoning, a method that uses precedent cases to guide behavior rather than rules or theories. It is analogous to case law wherein interpretation of similar cases provides ethical guidance (Jonsen and Toulmin, 1988; Arras, 1991; Jonsen, 1995).

Feminist ethics and ethics of care

Feminist ethics brings to light society's androcentric values and the impact that gendered thinking has had in the fields of ethics and bioethics. This framework strives to address health disparities and bias against women and marginalized groups (based on race, class, sexual orientation, etc.) by applying the principles autonomy and justice. Feminist theory values emotions and argues that it is indispensable to moral reasoning in the ethical practice of clinical medicine. Moral theories surrounding women's lived experiences are created with this framework, which strives to make sense of the experiences of women as moral agents (Brennan, 1999) and fills a gap that philosophy has omitted. In doing so, feminist ethics affirms the need for social inclusivity in ethical discourse to eliminate oppression of women and marginalized groups in philosophical discussion and in the greater world (Tuana, 2011).

Ethics of care is related to feminist ethics in emphasizing values of relationships, commitment, empathy, compassion, caring, and love. These qualities have also been emphasized in women's traditional activities and experiences. This theory affirms the moral significance of the fundamental elements of relationships in life and builds on the motivation to care for those who are dependent and vulnerable. To do so, the practitioner must have a well-developed sense of empathy that can lead to a benevolent response (Kuhse and Singer, 1998). This perspective overlaps with virtue ethics and emphasizes the caregiver's qualities and characteristics, rather than duties or rules. Ethics of care centers on "care," the action that guides the relationship between medical professionals and patients seeking their care, and is fundamental to health-care overall.

Virtue ethics

Virtue ethics emphasizes the role of character and virtue in decision-making, rather than a desired outcome or a duty-based approach. The qualities of character, including trustworthiness, prudence, justice, fortitude, temperance, and integrity, are those that are possessed by morally virtuous individuals and in medicine reflect how a morally virtuous physician would act. Virtue ethics correlates with principlism and guides behavior according to this principle.

Four-quadrant approach

The four-quadrant approach utilizes the metaphor of a toolbox to organize the various frameworks in a process that facilitates resolution of ethical conflicts. Within this framework, ethical problems are analyzed in the context of four areas, including medical indications, patient preferences, quality of life, and contextual features—social, economic, administrative, and legal (Jonsen et al., 1982).

Bioethics of pregnancy

Ethical concerns in pregnancy speak to issues that relate to a developing human life, which matures within a woman's body. Although this chapter is concerned with the ethics of pregnancy in neurologically complicated pregnancies, significant ethical questions exist for these patients prior to conception. Neurologic disorders have been linked with adverse pregnancy outcomes for women and the developing fetus. A preventive approach that incorporates preconception counseling should be adopted in the care of this patient population to identify ethical conflicts before they arise. Preconception care can have a positive impact on future pregnancies of women with neurologic disorders by anticipating, identifying, and addressing issues that may negatively impact the health of the mother and the developing fetus. This can occur in the setting of well-women care or when care is sought for a woman with a chronic neurologic condition.

Many pregnancies are unintended, and therefore a strategy to integrate preconception education into the primary care setting is essential. Women with underlying neurologic conditions should be presented with information about the risk of pregnancy complications potentially associated with their medical condition, as this may influence their decision about whether or not to pursue pregnancy or defer it until their health can be optimized. Information addressing maternal morbidity and mortality, disease prognosis, potential conflict involving treatment of the mother and its impact on fetal health, and the serious question of pregnancy's impact on the woman's life expectancy may inform a woman's reproductive decision-making. In the ideal setting, the patient should have time to consider these often complex issues regarding medical risks associated with her pregnancy. If, after considering these issues, the woman opts for pregnancy, her physician must make plans to optimize both her health and that of the fetus. If the patient elects to delay or prevent pregnancy, methods of contraception should be addressed (Dunlop et al., 2008; Ciafaloni et al., 2018; Garthwaite and Wilkes, 2019).

The most contentious topics in bioethics pertain to the creation and destruction of life, starting from conception with assisted reproductive technology, to abortion, and everything in between, including fetal therapy. These issues call attention to women's health, their autonomy, and right to privacy and also bring awareness to the embryo, fetus, and future child. Issues of maternal autonomy, women's rights, fetal personhood and rights, the sanctity of life, and religion, all converge in the abortion debate and lead to unresolvable disagreements among secular and religious people. Furthermore, it is not within the province of science to define when a cluster of cells becomes a living human being. Thus these unanswered questions emerge in pregnancy and linger throughout gestation, requiring thoughtful discourse when addressing ethical conflicts that need to consider a person's value system and call upon philosophy, religion, and the law to inform the discussion.

Viewing ethical issues in pregnancy as maternal–fetal conflicts overly simplifies the complexity of circumstances that require pregnant women to weigh their best interests against those of the fetus. Rather than strictly focusing on any one theory, different ethical frameworks are applied to guide decision-making in obstetrics that draw on rights and principles, virtues, feminist theories, and case-based analysis. The professional responsibility model of obstetric ethics was proposed by Chervenak and associates, which emphasizes medical science and compassionate clinical care of both the pregnant woman and fetal patient through autonomy and beneficence-based obligations toward the pregnant woman and beneficence-based obligations toward the fetus as guided by the pregnant woman (Chervenak et al., 2011). These decisions occur in a landscape influenced by policy, law, and religion, which significantly impacts how the care of pregnant women is rendered.

The fetus intrinsically depends on the pregnant woman, and this relationship distinguishes ethical issues in obstetrics from other areas of medicine. As technology has introduced the ability to perform fetal interventions in utero for the benefit of the fetus, this action has affirmed the idea of viewing the fetus as a patient. The ethical significance of conceptualizing the fetus as a patient is that it depends on the fluid relationship between the fetus in utero and the future child after birth who achieves independent moral status. Thus a graded moral status for the fetus has been introduced as one way to frame this issue. When, and whether, the fetus has moral status is a controversial matter that cannot be reconciled (Dunstan, 1984; Chervenak and McCullough, 1993; Fleischman et al., 1998), but when a woman decides to continue a pregnancy, the fetus may be considered a "potential child," whose interests evolve over time, and the interests of which will ultimately be attained at the time of delivery when the child is born (Minkoff et al., 2014). Many women consider decision-making in pregnancy to be similar to that of a parent toward a child and apply the best interest standard to the fetus.

APPLICATION OF ETHICAL FRAMEWORKS AT THE INTERFACE OF OBSTETRICS AND NEUROLOGY

Deciding for others

Determination of capacity and competence

Valid informed consent for medical treatment is based on common law that requires the uncoerced patient with decisional capacity to be provided with adequate information regarding the nature and purpose of the proposed treatment, along with the risks, benefits, and available treatment alternatives (Appelbaum and Grisso, 1995).

When proposing medical treatment to patients, the physician must be sure that they are able to understand the information presented to them and the associated consequences of the available choices. This refers to an individual's capacity, or ability, to make informed decisions and provide informed consent. It speaks to the patient's ability to engage in the process of understanding the risks and benefits of a proposed intervention. Capacity is not an all-or-none phenomenon but is present in degrees. An individual may have the capacity to make certain decisions but not others, and capacity can change over time. Capacity standards are generally based on four abilities. Patients must have the ability to understand their illness, assess the risks and benefits of the proposed diagnostic tests or treatments, manipulate information rationally, and communicate their choice (Appelbaum and Grisso, 1988). Patients with capacity have the right to refuse even interventions that physicians feel are in their best interests.

In contrast, competence is a legal state decided by the court and not physicians. It refers to the degree of mental soundness required to make decisions about a specific issue. It is a process that determines the capability of a person to act on his or her own behalf and whether he or she has the mental capacity to participate in legal proceedings or transactions. A person's functional deficits (mental illness, intellectual disability, etc.) can render him or her incompetent, due to deficits judged to be so great that they prevent the person from being able to meet the demands of a specific decision-making situation (Berg et al., 1995). For practical purposes, distinction will not be made between these two terms in this chapter; however, the reader should be mindful of the important differences between the two in the context of law and that criteria for valid consent to medical treatment and capacity standards may vary from state to state and presumably from country to country.

Surrogate decision-makers and advance directives

Deciding for patients is simpler when surrogates are familiar with their values. However, in neurology the underlying nature of a person's cognitive disability or illness may prevent the patient's value system from ever being communicated, appreciated, or understood. Some patients may be so impaired as to have no substantive value system at all. If the patient is deemed to lack capacity for purposes of informed consent, a surrogate decision-maker should be identified and the hospital ethics committee and legal department should be consulted. The term "surrogate" generally refers to a substitute decision-maker, for example, a health-care agent/proxy, next-of-kin, or guardian. Differences between these correspond to the ways through which surrogates get their decision-making authority. When a competent individual appoints a person to decide on health-care matters in the event they become unable to decide for themselves, the surrogate is referred to as a "proxy," "agent," or "attorney-in-fact." A court-appointed surrogate is referred to as a "guardian" or "conservator" (Pope, 2012). If neither of these exist, then the next-of-kin becomes the surrogate decision-maker, although the laws governing who is the surrogate vary from state to state. A typical order or priority starts with a spouse (or domestic partner if this status is recognized), adult child, parent, siblings, and then other relatives or a close friend.

Surrogates are obligated to follow the expressed wishes of the patient and to act in the patient's best interests, taking into account the patient's personal values, goals of care, and wishes, to the extent known.

If the patient has a court-appointed guardian with authority to make health-care decisions, he or she is the authorized surrogate who executes decisions to the extent described in the guardianship order. Of note, a court-appointed guardian will have priority over other surrogates (Devettere, 2009). If disputes occur between surrogates or if a surrogate is felt not to be acting in good faith, a court proceeding may be required.

Living wills, medical directives, or advance directives are legal documents that allow people to express their preferences about what medical treatment they wish to receive in the event that they lack capacity and are therefore unable to provide informed consent. For purposes of this discussion, we will refer to them collectively as advance directives (Kinzbrunner and Policzer, 2011).

A health-care power of attorney gives one or more persons the power to act as a decision-making proxy or agent. A power of attorney is accepted in all states, but the rules and requirements differ from state to state (Table 13.1).

Standards for making proxy decisions

To protect an incapacitated patient's previously expressed autonomous preferences and current best interests, surrogate decision-making should be guided in three different ways: a patient's "precedent autonomy," i.e., expressed wishes of the previously competent patient; "substituted judgment" if precedent autonomy is not available; "best interests" standard if neither of the preceding is available (American Medical Association, 2016).

Surrogate decision-making according to the expressed wishes of a previously competent patient follow health-care decisions that the patient provided prior to losing her decision-making capacity. This standard allows surrogates to uphold the patient's autonomy. The surrogate makes decisions in accordance with the patient's written advance directive or prior verbal statements.

Table 13.1

Elements of advance directives

Standards in surrogate decision making	
	Best interests
	Substituted judgment
Durable power of attorney	Document appoints an individual(s) as the decision-maker(s) on the patient's behalf if they are incapacitated and unable to express their desires for treatment
Living will	This document specifies the healthcare desires of the individual. These choices most commonly include cardiopulmonary resuscitation (CPR), and endotracheal intubation/mechanical ventilation
	Additional options include percutaneous endoscopic gastrostomy (PEG) tube placement, intravenous fluid (IVF), antibiotics, among others
Do not resuscitate	This document states the patient's wishes to refuse resuscitative efforts and is usually signed by the patient's healthcare provider. In most states, these written orders are the only advance directives that emergency room (ER) providers are legally obligated to consider. Patient may elect which resuscitative procedures they wish to accept or refuse
Physician's orders for life-sustaining treatment (POLST)	This document serves as an agreement between the doctor and patient, summarizing the patient's wishes and records them as medical orders. In states where POLST forms are recognized, the medical orders carry throughout the healthcare system in all settings, including the prehospital setting

Adapted from House, S.A., Ogilvie, W.A., 2018. Advanced directives. In: StatPearls [Internet]. StatPearls Publishing.

Substituted judgment

Under the standard of substituted judgment, the surrogate defers to the patient's known values, beliefs, and past decisions to make a decision as to what the patient would likely have decided in the given situation. In this manner, decision-making strives to follow the patient's own standards as a guide, in the absence of an advance directive. This standard requires that the surrogate be familiar with the patient to appreciate her wishes. Critics of this approach argue that substituted judgment is often inaccurate, and a false sense of autonomy is provided with this approach unless there have been relevant prior discussions between the patient and the surrogate.

Best interests

When precedent autonomy and substituted judgment are not available, the best interests standard is applied. In this case, the surrogate applies his or her own values to interpret what is in the patient's best interest. In this manner, the surrogate acts beneficently by weighing the risks and benefits of a proposed intervention. If a designated surrogate makes a decision that threatens the patient's best interests, the decision may be challenged. This standard is applied in pediatric decision-making and in incompetent adults (Beauchamp and Childress, 2013).

Advance directives in pregnancy

Physicians may not impose treatment without an individual's informed consent or, in cases where they lack capacity, that of their surrogates, who may decline treatment in accordance with the patient's advance directive or best interests. At times, patients or surrogates may demand treatment against the recommendation of their physicians. Tension arises in such cases if the physician feels that such treatment is medically inappropriate or unethical because it would not benefit the mother or the fetus (Gostin, 2014).

These concepts are applicable to pregnancy, where any treatment of the fetus must be performed through the body of mother and requires her informed consent, or that of a legally appointed surrogate decision-maker. In cases where a pregnant woman is so critically ill that prospects for maternal recovery diminish, or if brain death is determined, the calculus of benefits and burdens involves the developing fetus, who does not possess autonomy but does have interests. When the mother has little prospect for recovery, continuation of the pregnancy will be of value to her only in theory by benefitting the future child that she will deliver.

The right to decline or withdraw life-sustaining treatment was upheld in 1976 by the New Jersey Supreme Court in the case of Karen Quinlan (*In re Quinlan*, 1976). In 1990 the first case to reach the US Supreme

Court addressing the constitutional right to die was *Cruzan v. Director, Missouri Department of Health* (1990); the court upheld that individuals have the right to refuse life-sustaining treatment. Subsequently, in 1991, Congress passed the Patient Self-Determination Act that requires health-care providers who accept Medicare or Medicaid to inform all adult patients in writing of their right to prepare an advance directive, thus increasing public awareness and use of advance directives. Still, pregnant women are unlikely to have created an advance directive because of their young age and health. The population of pregnant women at high risk of complications due to chronic medical conditions is more likely to be encouraged by their obstetric providers to complete advance directives when they enter into care, given their increased risk of morbidity and mortality during pregnancy. However, despite the presence of an advance directive, not all states recognize these legal documents for women during pregnancy because they may impinge on the rights of the fetus and invalidate them even if present.

In emergency situations, ethical medical practice requires physicians to default to providing life-sustaining treatment if a patient's wishes are unknown or if she is unable to consent for medical care. For a pregnant woman, the state's interest in preserving her life, and also that of the fetus, prompts a similar conservative approach. In cases where an incapacitated pregnant woman had previously made known her wish to forego life-sustaining treatment, a conflict is created between the state's interest in preserving fetal life and the autonomous prior directives of the incapacitated woman (Manninen, 2016). In a diverse country like the United States, state statutes governing treatment decisions for incapacitated pregnant women vary between states and over time. The heterogeneity within the United States is representative of how different cultures and religions assign different rights to the fetus.

As an illustration, as of this writing, in the United States, 30 states either invalidate a woman's advance directive during pregnancy or restrict decisions of their surrogate decision-makers. DeMartino et al. (2019) reviewed official state advance directive documents from each state's legislature, administrative agency, and bar association. They observed considerable heterogeneity among state laws in restricting health-care options available to decisionally incapacitated women during pregnancy as well as a lack of transparency in disclosing the restrictions (of the 28 states with statutory pregnancy restrictions, the restrictions were not disclosed in 68% of official advance directive forms) (Table 13.2). If such variability exists within one country, from state to state, how much do ethical frameworks vary between countries? No one ethical principle can be applied to all.

Table 13.2

Restrictions on choices for life-sustaining therapies in pregnancy in the United States by state

States with pregnancy restriction on choices for life-sustaining therapy	Alabama, Alaska, Colorado, Delaware, Florida, Georgia,[a] Idaho, Illinois, Indiana, Iowa, Kansas, Kentucky, Louisiana, Michigan, Missouri, Montana, Nebraska, Nevada, New Hampshire, North Dakota, Ohio, Oklahoma,[a] Pennsylvania, Rhode Island, South Carolina, South Dakota, Texas, Utah, Washington,[b] Wisconsin
States without pregnancy restriction	Arkansas, Arizona, California, Connecticut, District of Columbia, Hawaii, Maine, Massachusetts, Minnesota, Mississippi, Maryland, New Mexico, New Jersey, New York, North Carolina, Oregon, Tennessee, Vermont, Virginia, West Virginia, Wyoming

Adapted from DeMartino, E.S., Sperry, B.P., Doyle, C.K., et al., 2019. US state regulation of decisions for pregnant women without decisional capacity. JAMA 321, 1629–1631.
[a]State restricts withdrawal of life-sustaining therapy in absence of clear and convincing evidence of patient's wishes.
[b]State restricts decisions for pregnant women through official advance directive documents but does not have statutory pregnancy restriction.

The case of Marlise Muñoz, a 33-year-old pregnant woman in Texas who suffered brain death at 14 weeks of gestation due to a pulmonary embolism, highlights the complexities of such a situation (*Munoz v. John Peter Smith Hospital*, n.d.). The patient's family requested removal of life-sustaining treatment, upholding what she would have wanted. The hospital refused, citing the Texas Health and Safety Code that was said to proscribe the removal of life-sustaining treatment from a pregnant patient. However, in actuality, the statute banned withdrawal of life-sustaining treatment from patients in a persistent vegetative state (PVS), not from patients who were brain dead. After 2 months a state district judge ruled against the hospital and ordered it to remove Muñoz from organ-sustaining equipment.

The availability of advance directives sheds light on what a woman's core beliefs are and guide surrogates to make decisions consistent with her priorities. Advance directives offer a level of protection ensuring that interests of others do not override those of the incapacitated person. Women have the legal right to terminate a pregnancy prior to fetal viability. If a pregnant woman's directive expresses that life-sustaining treatment be removed in the event of brain death, for example, the embryonic or fetal demise that would occur creates a controversy that ties into the national abortion debate. On the one hand, breaching the bodily integrity against the wishes of one person to save the life of another is considered a human rights violation. For example, the state cannot force someone to donate an organ even if it would save someone else's life. Accordingly, forcing women to undergo interventions in pregnancy to save the life of the fetus is condemned by professional organizations. On the other hand, some courts have prioritized the life of the fetus and forced women to undergo cesarean deliveries or other procedures to protect the fetus. When appealed, these cases have usually been overturned in favor of the women. *In re A.C.* (n.d.) serves as a landmark case that established the rights of informed consent and bodily integrity in pregnancy. In this case, a pregnant woman dying of cancer was forced to undergo a cesarean delivery with the intent of saving the fetus. The federal appeals court vacated a lower court ruling that authorized a forced cesarean delivery, stating that the lower court had "erred in subordinating [the woman's] right to bodily integrity in favor of the state's interest in potential life." Both the mother and neonate died (Annas, 1988).

Medicine must be practiced within the scope of the law. However, what some physicians may consider ethical may in fact be illegal, and conversely, some physicians may consider certain legal medical options to be unethical. Walking this fine line may be difficult at times, and the law may place physicians in a morally precarious situation that forces them to choose between violating the law and what they view as their ethical duty to their patient. In these cases, a multidisciplinary approach may be very helpful. Members of the primary clinical team, ethics committee, legal department, family members, and stakeholders need to be brought together to proceed through the layers of complexity in which these cases are buried (Gostin, 2014).

REPRODUCTIVE DECISIONS IN INTELLECTUALLY DISABLED PREGNANT WOMEN

Background

Matters pertaining to reproductive choice in intellectually disabled women will be addressed in this section; these include decisions regarding pregnancy termination, pregnancy continuation, and sterilization.

Prevalence of intellectual disability in the population is estimated to be 1%–2%, affecting approximately 745,000 women of reproductive age (McKenzie et al., 2016). The reproductive needs of this group of women are rarely addressed, and the fact that these individuals have sexual desires, needs, and expectations is often ignored. For example, in women with autism spectrum disorders, sexuality has been identified as an important issue but is often avoided by clinicians. A majority of individuals with autism spectrum disorders, with and without intellectual disability, express interest in dating and physical intimacy and are engaged in sexual behavior but lack basic factual knowledge about anatomy and sexuality (Stokes and Kaur, 2005).

Women with intellectual disabilities include those with mild-to-severe mental retardation, Down syndrome and other genetic conditions, autism spectrum disorders, acquired disorders of cognition, congenital abnormalities, and acquired conditions (such as traumatic brain injury). These women are at increased risk of having more lifetime sex partners, low contraceptive usage, higher rates of unwanted pregnancies, sexual assault, and a higher risk of acquiring sexually transmitted infections. Strategies to mitigate the damaging effects of these situations are required (Greenwood and Wilkinson, 2013).

Individuals with intellectual disability represent a marginalized group with higher rates of poverty, psychiatric conditions, and chronic health conditions compared with individuals of normal intelligence. To address disparities in health outcomes in individuals with intellectual disability, high-quality culturally sensitive health-care is essential. To achieve this goal, strenuous efforts on the part of surrogates, the community, and public health programs are required. Ethical policies regarding matters of reproduction are fundamental to an individual's health and to the medical ethics of a community and are an important goal of public health. Therefore respect for a woman's autonomy and her privacy are essential when she is unable to advocate for herself due to cognitive impairment.

Women with intellectual disability need education on matters of reproduction to develop their own personal values and beliefs. If the patient's intellectual disability precludes her from forming well-thought-out values, her surrogate will need to be engaged. If this is accomplished, shared decision-making between the provider and patient (and surrogate decision-maker, as required) can proceed according to the values of the patient. Reproductive concerns that may arise include those surrounding sexuality, risks of sexual abuse, menstrual hygiene, the impact of comorbid medical conditions exacerbating the menstrual cycle, medication interactions with contraception, and behavioral issues. To address these situations, physicians must attend to requests surrounding medical manipulation of the menstrual cycle, contraception, sterilization, and abortion.

Abuses of power in reproductive decision-making are well recognized in the history of bioethics. In 1924, during the heyday of American belief in the pseudoscience of eugenics, Dr. Albert Sidney Priddy, the superintendent of the Virginia State Colony for Epileptics and Feebleminded, petitioned his Board of Directors to sterilize Carrie Buck, an 18-year-old patient described as having a mental age of 9, because of her threat to society of having children with intellectual disability. The Board of Directors ordered the sterilization. Carrie's guardians then appealed the case to the Circuit Court of Amherst County, which sustained the board's decision, and the case then progressed to the US Supreme Court. Buck and her guardian contended that her right to procreation based on the Due Process and Equal Protection Clause of the 14th Amendment was violated. However, 3 years later, the court ruled in an 8-1 decision that it was in the State's best interest to have Carrie Buck sterilized. Consequently, *Buck v Bell* (1927) established the constitutionality of state-authorized sterilization for eugenic purposes and it has never been overturned. Historian Paul Lombardo wrote, "*Buck* represents a milestone in the affirmation of governmental power over individual rights, but more specifically, *Buck* is a landmark in the endorsement of intrusive medical procedures as tools to be used for state ends" (Lombardo, 1985).

In the aftermath of this decision, federal regulations were created to provide protection for individuals when sterilization was requested through federally funded programs. A standardized consent form is now required that attests that the individual to be sterilized "appears mentally competent and knowingly and voluntarily consents to be sterilized" (https://www.law.cornell.edu/cfr/text/42/441.258).

The intent of the policy is to prevent involuntary sterilization of women who either do not consent to the procedure or lack capacity to consent. The policy protects the individual's autonomy by ensuring that the woman is competent and that informed consent has been obtained. Although protecting women who cannot give informed consent or who refuse sterilization, the policy also has the unfortunate effect of preventing sterilization of women who might assent to the procedure and who would be best served by sterilization when their surrogates request the procedure to prevent unwanted pregnancies in this population of women who are vulnerable to sexual exploitation.

Sterilization and abortion for women with cognitive impairment

The legacy of involuntary sterilization programs in the United States has resulted in restrictions of procedures that impact women's reproductive health, specifically sterilization, contraception, and abortion. These are accentuated when cognitively disabled women seek sterilization for legitimate reasons. For example, an intellectually disabled woman may not understand what pregnancy is or that a child is created as a result of sexual activity; she may not be able to care for herself let alone a newborn infant. For these and other reasons, caretakers may request sterilization out of what they perceive to be the best interest of the patient. Statutes authorizing and governing sterilization of mentally disabled individuals vary among states and are grounded in their *parens patriae* authority, permitting sterilization in various situations if certain criteria are met.

The court will apply various standards to determine whether to proceed with sterilization or termination of pregnancy (McIntyre, 2007): (1) substituted judgment, (2) the mandatory criteria rule, and (3) the best interest standard. The substituted judgment criterion allows the woman's surrogates to opine as to what the woman would want if she had full capacity to make the decision in question. This standard assumes that family members and/or close friends are acting in good faith when they state what they feel the woman would want in a given situation.

The mandatory criterion rule places a high burden of proof on the petitioning party. This rule requires the judge to find by clear, cogent, and convincing evidence that the individual lacks competence required to make her own decision regarding sterilization, that there is a need for contraception, and that no alternatives to sterilization exist. While this standard implements safeguards to ensure that the individual's health is prioritized, the burden of proof may be so high as to be unachievable.

Through the "best interest" standard, the judiciary considers specific criteria to determine whether sterilization is in the cognitively disabled individual's best interest, as in the mandatory criteria rule. However, this standard allows for more flexibility and judicial discretion.

To ensure that decisions are made in the patient's best interest and not solely on perceived beliefs about a person with disability, a hybrid approach has been proposed, which combines the aforementioned standards to maximize the reproductive rights of mentally ill women (Kundnani, 2013).

Clinicians should be aware of their state laws regarding abortion and sterilization. Women with intellectual disability rely on federally funded government health services, and consequently, this population is directly affected by federal legislation pertaining to access to sterilization and abortion.

Case study

An 18-year-old intellectually disabled woman was found to be pregnant (first trimester) when hospitalized for status epilepticus. The patient's history was significant for a generalized seizure disorder and cognitive impairment in the setting of a genetic syndrome. She was nonadherent with taking her antiseizure medications and relied on her mother, with whom she lived, to oversee her care. It was determined that the patient lacked capacity in matters regarding the pregnancy, based on her lack of understanding of how she became pregnant and what pregnancy and childbirth entailed, although she was determined to have capacity to designate a health-care proxy. When asked about the pregnancy and how she felt about it, the patient offered contradictory statements saying that "a baby would be nice" but that the baby was making her feel sick and that she wanted the baby "out of her." The patient told the gynecologist that she wanted a termination and had scheduled a visit to further discuss abortion, but then changed her mind at a follow-up visit.

The patient's mother felt that termination was in her daughter's best interest for several reasons. First, she said that her daughter was essentially a child, given the fact that her mental age was equivalent to that of a 5-year-old. Second, she was concerned about the negative impact that uncontrolled seizures would have on the developing fetus, and the effect of the medications on fetal development. Third, she was aware of the fact that pregnancy lowered her daughter's seizure threshold, would make the seizures more difficult to control, and that recurrent seizures would have a very negative impact on her daughter's overall health and specifically might cause brain damage.

The clinician had the task of ascertaining how to render optimal medical care for the patient and fetus, as well as arriving at an ethically sound decision of how to uphold the patient's autonomy while taking into consideration the wishes of her surrogate. To achieve this goal, knowledge of the social context and accurate medical facts of the case was essential. A multidisciplinary meeting was convened, including neurology, obstetrics, social work, the patient's family, and the ethics committee. The ethical question centered on the request to terminate a pregnancy based on the wishes of someone other than the patient herself, who did not seem to understand what pregnancy was or what termination would entail.

The patient was determined to lack capacity and named her mother as her surrogate. The caregivers collectively agreed that the mother had the patient's best interests in mind with no evidence of secondary gain influencing her decision. Beneficence and nonmaleficence informed ethical decision-making in this case. The mother's decision to terminate the pregnancy was based on the best interests of the patient.

The clinicians weighed the potential risks associated with termination and compared them with the desired good that it might offer the patient. In discussing the abortion, it became apparent to the clinicians that the patient did not understand what termination entailed. They also noted that the patient did not understand that if she continued with the pregnancy, she would go through labor to deliver a child. The good for the patient that would result from termination was optimizing seizure control, preventing possible brain damage from more uncontrolled seizures, enabling her to return home instead of spending the remainder of her pregnancy hospitalized to maximize seizure control, and preventing the rigors of labor and delivery in a woman with the mental age of a 5-year-old who would have no conception of what was happening to her.

The patient did not understand the concept of caring for a child, and so the responsibility of raising the baby would either be left to her mother or result in adoption. The gestational age of the pregnancy was within the appropriate time frame for termination of pregnancy according to state law, and the procedure was performed per the surrogate's request with the patient's assent.

In this case, care of a patient who lacked capacity was thoughtfully provided through a multidisciplinary approach involving clinicians, social workers, and the ethics committee. Appreciation of the patient's social and medical circumstances allowed the clinical team to support the surrogate in deciding what was in the patient's best interest. Assent of the patient eliminated conflict that might have otherwise occurred.

TREATMENT OF NEUROLOGIC CONDITIONS IN PREGNANCY: SHARED DECISION-MAKING

Balancing benefits and burdens of treatment options in neurologic conditions impacting maternal and fetal health

Medical conditions affecting the mother will affect the health of her fetus. Therefore optimal care of medical conditions in pregnancy is essential for both maternal and fetal well-being. Unique challenges arise for the clinician when therapeutic interventions for the mother may be harmful for the fetus. Conversely, intrauterine fetal therapy may be associated with harm for the mother. Thus the benefits and risks to both the mother and fetus must be balanced when striving to determine best ethical treatment options in pregnancy.

Care of pregnant women with neurologic conditions requires consideration of three things: (1) the impact of the pregnancy on the illness, (2) the impact of the illness on the pregnancy, and (3) the risks of maternal treatment on fetal development. When striving to elucidate risks associated with any therapy in pregnancy, the maternal and fetal risks of alternative treatments should be evaluated as well as the risks of not treating the illness, since abstaining from treatment is an option.

Neurologic conditions encountered in pregnancy include epilepsy, eclampsia, headaches, aneurysms, arteriovenous malformations, brain tumors, cerebral venous thrombosis, central nervous system infections, myasthenia gravis, Guillain-Barre syndrome, spinal cord disease, and multiple sclerosis, among others (Karnad and Guntupalli, 2005).

To inform the discussion, decisions regarding the general care of epilepsy in pregnancy will be reviewed to illustrate the tension created between beneficence-based decisions for the mother that may possibly harm the fetus. The least harmful course of treatment is appropriate to uphold the principle of nonmaleficence.

Background

Epilepsy is a common neurologic condition affecting pregnant women. These pregnancies are associated with higher rates of morbidity and mortality. Therefore treatment of epilepsy with antiepileptic drugs (AEDs) is required, and is generally not withheld, to optimize the health of the mother and fetus (MacDonald et al., 2015). However, treatment with AEDs in pregnancy has been associated with increased risks of congenital malformations in children (Tomson and Battino, 2012). For example, major congenital malformations are reported in 6.1% of children whose mothers took AEDs during pregnancy compared with 2.8% in the children of untreated women with epilepsy and 2.2% in the general population (Tomson and Battino, 2012). Certain medications are associated with greater fetal risks than others, including valproate sodium, a drug commonly used in women of reproductive age for management of seizures, migraines, and bipolar disorder (Tomson et al., 2018). In the United States, about 1.5 million outpatients received valproate, and roughly 22% of these were women of reproductive age (FDA, accessed 2019). An increased risk of anatomical malformations and impaired behavioral and cognitive function has been observed in children exposed to this medication during fetal development (Meador et al., 2009). As a result, several regulatory bodies across the world have banned the use of valproate during pregnancy and in women of childbearing age, unless no satisfactory alternatives exist or a pregnancy prevention program is implemented (much like for the well-established teratogen accutane) (Tkachenko et al., 2019). Informed and shared decision-making is essential if women considering pregnancy choose to continue valproate in pregnancy.

Shared decision-making

Ethical principles provide a framework to guide decision-making when counseling on medication use in pregnancy. Respect for patient autonomy holds that a pregnant woman has a right to weigh the risks and benefits of medical interventions related to her pregnancy based on her personal values and beliefs. Beneficence and maleficence are often in conflict when considering medications in pregnancy, since what helps one member of the maternal–fetal dyad may hurt the other.

Shared decision-making is a process where health professionals and patients work together to make health-care choices. This form of clinical decision-making supports patient autonomy in reviewing details of the benefits and burdens of possible treatments. The values of the pregnant woman are elicited as they impact on maternal and fetal health. Barriers to shared decision-making may include low health literacy or cultural differences where the physician might give inadequate weight to the woman's wishes.

Features of a shared decision-making model include the introduction of choice, description of options, and helping patients express their preferences while supporting them in arriving at a decision (Elwyn et al., 2012). In the case of a woman taking valproate who is considering pregnancy, the patient will first assess different management strategies (to continue the medication, to continue at a lower dose, discontinue the medication, or change to a different medication; see Table 13.3). Discussion of the safety profile and efficacy of the considered AEDs for the mother and risks to the fetus will follow. With the aid of the physician, the patient can then decide whether to proceed with the pregnancy and which AED to use. If valproate is the only medication that is effective in controlling the patient's seizures and they are so severe that her morbidity or mortality rate is increased, the patient and physician may agree that continuation of the medication will be required, despite the increased risks to the fetus. Knowing this, a patient may choose to avoid pregnancy or, if pregnant, may choose to terminate. She may explore the option of a gestational carrier or adoption. Strategies to mitigate the fetal risk should be pursued, with folic acid supplementation and treatment with one agent using the lowest effective dose.

BRAIN DEATH AND DISORDERS OF CONSCIOUSNESS IN PREGNANCY

Catastrophic brain injuries are among the most devastating situations that can befall patients and their loved ones. The complexity of ethical and legal decision-making surrounding the determination of brain death and discontinuation of organ support increases substantially when this occurs in pregnancy, given the presence of a fetus. Although women have the right to refuse medical treatment, or opt for abortion, based on the principle of autonomy, when they lack capacity their interests may come into conflict with those of other stakeholders, including their surrogates and the law.

Care of a pregnant woman with a catastrophic brain injury involves four major issues: (1) the medical facts, including prognosis for meaningful neurologic recovery, the impact of the insult on the fetus, and the gestational age at which it occurred; (2) maternal advance directives;

Table 13.3

Shared decision-making process (sodium valproate in pregnancy)

	Discussion				
	Risks		Benefits		
Option	Maternal	Fetal	Maternal	Fetal	Decision
Valproate continued	Typical medication side effects	Teratogenesis; neurodevelopmental delay; fetal effects of uncontrolled seizures	Seizure control	Maternal seizure control	Medication continued if seizures are not controlled on other medications or on lower doses
Valproate decreased	Seizure activity increased consequences of seizures (injury, etc.)	Teratogenesis; neurodevelopmental delay; fetal effects of uncontrolled seizures	Seizure control	Reduced risk of congenital malformations and neurodevelopmental delay compared to full dose of medication	Patient accepts risks of seizure exacerbation. Medication is continued if seizures are not controlled on other medications
Valproate discontinued	Seizure activity increased consequences of seizures (injury, etc.)	Fetal effects of uncontrolled seizures	Unnecessary exposure to valproate avoided	Elimination of valproate-associated teratogenicity and neurodevelopmental delay	Women with epilepsy in remission who wish to taper and avoid other medications
Medication changed to other	Seizure activity increased consequences of seizures (injury, etc.)	Adverse effects of substituted drug. including potential teratogenesis	Seizure control	Maternal seizure control. Less risk of teratogenicity	Seizures controlled with alternative antiepileptic medications

Based on Tomson, T., Marson, A., Boon, P., et al., 2015. Valproate in the treatment of epilepsy in girls and women of childbearing potential. Epilepsia 56, 1006–1019; MacFarlane, A., Greenhalgh, T., 2018. Sodium valproate in pregnancy: what are the risks and should we use a shared decision-making approach? BMC Pregnancy Childbirth 18, 200; Elwyn, G., Frosch, D., Thomson, R., et al., 2012. Shared decision making: a model for clinical practice. J Gen Intern Med 27, 1361–1367.

(3) involvement of surrogate decision-makers; and (4) state laws where the clinical care is rendered.

Medical facts

Clarifying the medical facts: Disorder of consciousness and brain death

Terms describing different states of consciousness after neurologic injury vary and include coma, the minimally conscious state (MCS), and the PVS. Because precise diagnosis of these states impacts on the ethics of decision-making, a brief overview of the disorders of consciousness is important. Distinction of patients with MCS from those with coma and PVS is important since there are meaningful differences in clinical outcome (Giacino et al., 2002). Catastrophic brain injury may lead to coma, and surviving patients will either recover, progress to brain death, or experience prolonged disorders of consciousness that exist between these two extremes. Coma is a condition where there is complete failure of the arousal system with no spontaneous eye opening or response to vigorous sensory stimulation (Plum and Posner, 1982). Coma does not persist. It either leads to recovery or develops into PVS, MCS, or brain death. PVS is described as a state of "wakeful unresponsiveness" with a complete absence of awareness of self or others and a preserved capacity for spontaneous or stimulus-induced arousal (Jennett and Plum, 1972). The MCS is distinguished from the PVS by partial preservation of conscious awareness "in which minimal but definite behavioral evidence of self or environmental awareness is demonstrated" (Giacino et al., 2002). These clinical situations evoke heated medical, political, and legal debates that surround brain injury, as was seen in the case of Terry Schiavo, a 41-year-old woman in PVS as a result of cardiac arrest and anoxic brain injury. Schiavo remained in a PVS for 15 years before her death on March 31, 2005, after her feeding tube was removed (Weijer, 2005).

Criteria for brain death were first introduced by an Ad Hoc Committee of the Harvard Medical School to Examine the Definition of Brain Death (1968) after Mollaret and Goulon (1959) presented the concept of *coma dépassé* in 1959 (Wijdicks, 2001). Subsequently, the President's Commission for the Study of Ethical Problems in Medicine and Biomedical and Behavioral Research concluded in 1981 that brain death should be endorsed as legal death, resulting in the Uniform Determination of Death Act (UDDA), which states that "an individual who has sustained either (1) irreversible cessation of circulatory and respiratory functions, or (2) irreversible cessation of all functions of the entire brain, including the brain stem, is dead." The diagnosis of brain death is primarily clinical with three essential findings: coma, absence of brainstem reflexes, and apnea.

Disorders of consciousness and brain death are rare events in pregnancy. Decisions regarding care of the pregnant patient will center on optimizing chances for maternal recovery, while taking stock of the impact the insult may have had on fetal health, and the effects on the fetus of the interventions required for maternal support (i.e., impact of ongoing medication and radiation exposure on fetal development, risk of maternal infections, and risks of preterm labor and delivery at extreme prematurity).

How should decision-making proceed?

For obvious reasons, defining the maternal condition will drive decision-making. Pregnant women in a coma, MCS, or PVS are alive with clear interests in shaping the course of their treatment. Women with documented brain death are legally dead. Once a pregnant woman is brain dead, discussion about the appropriate form of medical treatment for her is no longer valid. However, depending on the goals of treatment regarding the fetus, continued organ support may be appropriate. In these rare situations, the ethical analysis must consider the following issues: the values of the mother, if known; the welfare of the fetus as reflected by fetal gestational age and in relation to viability; and what is mandated by state law. The position statement of the American Academy of Neurology states that "a lawful advance directive expressing the premortal wishes of the brain dead patient regarding the care of her unborn child should be reviewed and considered subject to legal interpretation." In addition, requests by lawful surrogates to withdraw organ-sustaining treatment if a fetus has not reached a viable gestational age should be honored, as should requests for the continued use of organ-sustaining treatment in a brain-dead pregnant woman with a viable fetus (Russell et al., 2019). Although successful maternal organ support following brain death is possible, it is fraught with potential complications.

Gestational age and fetal condition

Preterm birth is generally defined as one that occurs after 20 weeks of gestation and prior to 37 completed menstrual weeks of gestation. Viability refers to the lower limit of gestational age when newborn survival is possible after delivery. It evolves along a continuum, between $20^{0/7}$ weeks and $25^{6/7}$ weeks of gestation, and delivery at this time period is defined as a periviable birth (American College of Obstetricians and Gynecologists and Society for Maternal-Fetal Medicine, 2017). It should be appreciated that "viability" is a changing standard that is dependent on technological advances, and there is no firm line that separates viable from nonviable gestations. It is known that as gestational age at delivery decreases, neonatal morbidity and mortality rates increase and complications of prematurity account for more infant deaths than any other cause (Raju et al., 2014; Gabbe et al., 2016). Surviving infants have increased risks of impairment in vision and hearing, chronic lung disease, cerebral palsy, and childhood developmental disorders. Given the real threat that prematurity places on a child's health, the gestational age at the time of maternal injury and delivery are significant factors to consider if continuing life-sustaining treatment for purposes of reaching viability is planned.

In addition to gestational age, the nature of the maternal injury on fetal development will impact fetal health. Regardless of the fetal evaluation performed with imaging and other assessments, the information obtained will be indirect, and the fetal outcome will be uncertain, with a potential for fetal demise or severe fetal injury that may have occurred at the time maternal brain injury occurred. Fetal effects include those caused by altered maternal physiology at the time of the insult with potential fetal hypoxemia, risk of intrauterine infection, risk of periviable delivery and high probability of preterm birth, and the uncertain outcome of allowing fetal development with ongoing maternal somatic support. Thus if a fetus has not reached viability or if a fetus of any gestational age has sustained brain injury of substantial magnitude, requests by surrogates to withdraw organ support should be honored (American Academy of Neurology—AAN). In the case of a fetus without overt brain injury who has reached viability, decision-making regarding potential removal of maternal life support is, depending on State law, in the hands of the lawful surrogate.

THE ETHICS COMMITTEE

Pregnancy is a normal biological process that may occur in the setting of neurologic disease and may be associated with serious health complications. An ethics consultation may help clarify questions that may arise regarding the physician's obligation to the pregnant woman, her family, and fetus. Since 1992 the Joint Commission has required that every health-care institution have a standing mechanism to address ethical issues and resolve disputes. This task may be achieved with a clinical ethics consultation service. An institutional ethics committee is another mechanism that provides a formalized forum for ethical deliberation where clinical cases are presented to committees that consist of representatives from the medical, legal, and public spheres. The need for hospital ethics committees was realized after the Karen Ann Quinlan case in 1976 due to legal controversies regarding refusal of life-sustaining treatment (Annas and Grodin, 2016).

Ethics committees serve several purposes, including providing clinical ethics consultation and education, assisting with policy development, and mediating conflicts such as those involving brain death. Ethical issues unique to pregnancy that may prompt an ethics consultation include questions surrounding pregnancy termination, advance directives in pregnant patients, conflicts between treatments having opposite effects on the well-being of mother and fetus, refusal of blood products during labor and delivery, and fetal surgery, among others.

When an ethical problem arises, resolution will rely on a multitude of factors, including clarifying the medical facts, effectively communicating with the involved parties, knowing the relevant state laws, and organizing meetings with input from a wide variety of professionals. And even then, a single "correct" answer may not be apparent, but rather a range of acceptable moral options may be available. Through this process, compassionate, honest, and respectful clinical care will be rendered.

References

Ad Hoc Committee of the Harvard Medical School to Examine the Definition of Brain Death (1968). A definition of irreversible coma: report of the Ad Hoc Committee of the Harvard Medical School to Examine the Definition of Brain Death. JAMA 205: 337–340.

American College of Obstetricians and Gynecologists and Society for Maternal-Fetal Medicine (2017). Periviable birth. Obstetric care consensus no. 6. Obstet Gynecol 130: e187–e199.

American Medical Association (2016). Opinion 2.1.2 Decisions for adult patients who lack capacity. In: Code of medical ethics, Available at: https://www.ama-assn.org/sites/default/files/media-browser/code-of-medical-ethics-chapter-2.pdf. Published. Accessed November 2019.

Annas GJ (1988). At law: she's going to die: the case of Angela C. Hastings Cent Rep 18: 23–25.

Annas G, Grodin M (2016). Hospital ethics committees, consultants, and courts. AMA J Ethics 18: 554–559.

Appelbaum PS, Grisso T (1988). Assessing patients' capacities to consent to treatment. N Engl J Med 319: 1635–1638.

Appelbaum PS, Grisso T (1995). The MacArthur treatment competence study. I. Law Hum Behav 19: 105–126.

Arras JD (1991). Getting down to cases: the revival of casuistry in bioethics. J Med Philos 16: 29–51.

Beauchamp TL (2003). The nature of applied ethics. In: A companion to applied ethics, Blackwell Publishing, MA, USA. pp. 1–16.

Beauchamp TL (2007). The 'four principles' approach to health care ethics. In: Principles of health care ethics, 29: 3–10.

Beauchamp TL, Childress JF (2013). Principles of biomedical ethics, seventh edn AUFL, New York.

Beauchamp TL, McCullough LB (1984). Medical ethics: the moral responsibilities of physicians, Prentice-Hall, Englewood Cliffs, NJ.

Berg JW, Appelbaum PS, Grisso T (1995). Constructing competence: formulating standards of legal competence to make medical decisions. Rutgers Law Rev 48: 345.

Brennan S (1999). Recent work in feminist ethics. Ethics 109: 858–893.

Buck v Bell 274 US 200 (1927) 205 (1927).

Callahan D (1990). Religion and the secularization of bioethics. In: Ethics and medical decision-making, Routledge71–73.

Chervenak FA, McCullough LB (1992). What is obstetric ethics? Clin Obstet Gynecol 35: 709–719.

Chervenak FA, McCullough LB (1993). Ethical issues in recommending and offering fetal therapy. West J Med 159: 396.

Chervenak FA, McCullough LB, Brent RL (2011). The professional responsibility model of obstetrical ethics: avoiding the perils of clashing rights. Am J Obstet Gynecol 205: 315–e1.

Ciafaloni E, Thornburg LL, Bushnell CD (2018). Neurological diseases and pregnancy: a coordinated care model for best management, Oxford University Press.

Cruzan v. Director, Missouri Department of Health. 497 U.S. 261, 110 S. Ct. 2841, 111 L. Ed. 2d 224 (1990)

DeMartino ES, Sperry BP, Doyle CK et al. (2019). US state regulation of decisions for pregnant women without decisional capacity. JAMA 321: 1629–1631.

Devettere RJ (2009). Practical decision making in health care ethics: cases and concepts, Georgetown University Press.

Dunlop AL, Jack BW, Bottalico JN et al. (2008). The clinical content of preconception care: women with chronic medical conditions. Am J Obstet Gynecol 199: S310–S327.

Dunstan GR (1984). The moral status of the human embryo: a tradition recalled. J Med Ethics 10: 38.

Elwyn G, Frosch D, Thomson R et al. (2012). Shared decision making: a model for clinical practice. J Gen Intern Med 27: 1361–1367.

Farmer P, Campos NG (2004). New malaise: bioethics and human rights in the global era. J Law Med Ethics 32: 243–251.

Feuillet-Liger B, Orfali K (2018). The reality of human dignity in law and bioethics, Springer, Cham.

Fleischman AR, Chervenak FA, McCullough LB (1998). The physician's moral obligations to the pregnant woman, the fetus, and the child. In: Seminars in perinatology, vol. 22. WB Saunders184–188.

Fox RC, Swazey JP (2005). Examining American bioethics: its problems and prospects. Camb Q Healthc Ethics 14: 361–373.

Fox RC, Swazey JP (2008). Observing bioethics, Oxford University Press.

Gabbe SG, Niebyl JR, Simpson JL et al. (2016). Obstetrics: normal and problem pregnancies E-book, Elsevier Health Sciences.

Garthwaite H, Wilkes S (2019). Preconception counselling. InnovAiT 12: 85–92.

Giacino JT, Ashwal S, Childs N et al. (2002). The minimally conscious state: definition and diagnostic criteria. Neurology 58: 349–353.

Gillon R (1994). Medical ethics: four principles plus attention to scope. BMJ 309: 184.

Gostin LO (2014). Legal and ethical responsibilities following brain death: the McMath and Muñoz cases. JAMA 311: 903–904.

Greenwood NW, Wilkinson J (2013). Sexual and reproductive health care for women with intellectual disabilities: a primary care perspective. Int J Family Med 2013.

In re A.C *In re* A.C., 573 A.2d 1235 (D.C. App. 1990).

In re Quinlan. *In re Quinlan*, 70 N.J. 10, 355 A.2d 647 (1976), *certdenied*, 429 U.S. 922 (1976).

Jennett B, Plum F (1972). Persistent vegetative state after brain damage: a syndrome in search of a name. Lancet 299: 734–737.

Jonsen AR (1995). Casuistry: an alternative or complement to principles? Kennedy Inst Ethics J 5: 237–251.

Jonsen AR, Toulmin SE (1988). The abuse of casuistry: a history of moral reasoning, University of California Press.

Jonsen AR, Siegler M, Winslade WJ (1982). Clinical ethics a practical approach to ethical decisions in clinical medicine.

Karnad DR, Guntupalli KK (2005). Neurologic disorders in pregnancy. Crit Care Med 33: S362–S371.

Kinzbrunner BM, Policzer JS (2011). End-of-life care: a practical guide, McGraw Hill Professional.

Kuhse H, Singer P (1998). A companion to bioethics malden, Blackwell Publishing, MA, USA, pp. 98–105.

Kuhse H, Schüklenk U, Singer P (2015). Bioethics: an anthology, John Wiley & Sons.

Kundnani R (2013). Protecting the right to procreate for mentally ill women. S Cal Rev L Soc Just 23: 59.

Lombardo PA (1985). Three generations, no imbeciles: new light on Buck v. bell. N Y Univ Law Rev 60: 30.

MacDonald SC, Bateman BT, McElrath TF et al. (2015). Mortality and morbidity during delivery hospitalization among pregnant women with epilepsy in the United States. JAMA Neurol 72: 981–988.

Manninen BA (2016). Sustaining a pregnant cadaver for the purpose of gestating a fetus: a limited defense. Kennedy Inst Ethics J 26: 399–430.

McIntyre M (2007). Buck v. bell and beyond: a revisited standard to evaluate the best interests of the mentally disables in sterilization context. Univ Ill Law Rev 4: 1303–1327.

McKenzie K, Milton M, Smith G et al. (2016). Systematic review of the prevalence and incidence of intellectual disabilities: current trends and issues. Curr Dev Disord Rep 3: 104–115.

Meador KJ, Baker GA, Browning N et al. (2009). Cognitive function at 3 years of age after fetal exposure to antiepileptic drugs. N Engl J Med 360: 1597–1605.

Minkoff H, Marshall MF, Liaschenko J (2014). The fetus, the "potential child," and the ethical obligations of obstetricians. Obstet Gynecol 123: 1100–1103.

Mollaret P, Goulon M (1959). Le coma dépassé ["a state beyond coma"]. Rev Neurol 101: 3–15.

Munoz v. John Peter Smith Hospital, n.d. Tarrant County, Texas District Court Order Cause No. 096-270080-14. Available at: http://thaddeuspope.com/images/MUNOZ_202053415-Judges-Order-on-Munoz-Matter.pdf. Accessed November 2019.

Okonkwo OC, Griffith HR, Copeland JN et al. (2008). Medical decision-making capacity in mild cognitive impairment: a 3-year longitudinal study. Neurology 71: 1474–1480.

Pellegrino ED (1995). Toward a virtue-based normative ethics for the health professions. Kennedy Inst Ethics J 5: 253–277.

Plum F, Posner JB (1982). The diagnosis of stupor and coma, vol. 19. Oxford University Press, USA.

Pope TM (2012). Legal fundamentals of surrogate decision making. Chest 141: 1074–1081.

President's Commission for the Study of Ethical Problems in Medicine, Biomedical and Behavioral Research (US) (1982). Making health care decisions, vol. 1. President's Commission for the Study of Ethical Problems in Medicine and Biomedical and Behavioral Research.

Raju TN, Mercer BM, Burchfield DJ et al. (2014). Periviable birth: executive summary of a joint workshop by the Eunice Kennedy Shriver National Institute of Child Health and Human Development, Society for Maternal-Fetal Medicine, American Academy of Pediatrics, and American College of Obstetricians and Gynecologists. Am J Obstet Gynecol 210: 406–417.

Russell JA, Epstein LG, Greer DM et al. (2019). Brain death, the determination of brain death, and member guidance for brain death accommodation requests: AAN position statement. Neurology 92: 228–232.

Silver L (2019). Where Americans and Europeans agree—and differ—in the values they see as important [Online]. Available at: https://www.pewresearch.org/areas-of-expertise/religion/. Accessed: 17 November 2019.

Steinbock B, London AJ, Arras J (Eds.), (2013). Ethical issues in modern medicine: contemporary readings in bioethics. McGraw-Hill.

Stokes MA, Kaur A (2005). High-functioning autism and sexuality: a parental perspective. Autism 9: 266–289.

ten Have H, Gordijn B (Eds.), (2014). Handbook of global bioethics. vol. 4. Springer, New York.

Tkachenko E, Singer S, Sharma P et al. (2019). US Food and Drug Administration reports of pregnancy and pregnancy-related adverse events associated with isotretinoin. JAMA Dermatol 155: 1175–1179.

Tomson T, Battino D (2012). Teratogenic effects of antiepileptic drugs. Lancet Neurol 11: 803–813.

Tomson T, Battino D, Bonizzoni E et al. (2018). Comparative risk of major congenital malformations with eight different antiepileptic drugs: a prospective cohort study of the EURAP registry. Lancet Neurol 17: 530–538.

Tuana N (2011). In: EN Zalta (Ed.), "Approaches to feminism", The Stanford Encyclopedia of Philosophy, Spring 2011 edn. http://plato.stanford.edu/archives/spr2011/entries/feminism-approaches/.

Weijer C (2005). A death in the family: reflections on the Terri Schiavo case. CMAJ 172: 1197–1198.

Wijdicks EF (2001). The diagnosis of brain death. N Engl J Med 344: 1215–1221.

Section III

Prenatal neurologic diagnoses: Pathophysiology, diagnosis, and management

Handbook of Clinical Neurology, Vol. 171 (3rd series)
Neurology and Pregnancy: Pathophysiology and Patient Care
E.A.P. Steegers, M.J. Cipolla and E.C. Miller, Editors
https://doi.org/10.1016/B978-0-444-64239-4.00014-X

Chapter 14

The developing brain by trimester

PAUL GOVAERT[1,2,3]*, FABIO TRIULZI[4], AND JEROEN DUDINK[5]

[1]*Department of Neonatology, Erasmus University Medical Center, Rotterdam, The Netherlands*
[2]*Department of Neonatology, ZNA Middelheim, Antwerp, Belgium*
[3]*Department of Rehabilitation and Physical Therapy, Gent University Hospital, Gent, Belgium*
[4]*Department of Pediatric Neuroradiology, Università Degli Studi di Milano, Milan, Italy*
[5]*Department of Neonatology, University Medical Center, Utrecht, The Netherlands*

Abstract

Transient anatomical entities play a role in the maturation of brain regions and early functional fetal networks. At the postmenstrual age of 7 weeks, major subdivisions of the brain are visible.

At the end of the embryonic period, the cortical plate covers the neopallium. The choroid plexus develops in concert with it, and the dorsal thalamus covers about half the diencephalic third ventricle surface. In addition to the fourth ventricle neuroepithelium the rhombic lips are an active neuroepithelial production site. Early reciprocal connections between the thalamus and cortex are present. The corticospinal tract has reached the pyramidal decussation, and the arteries forming the mature circle of Willis are seen. Moreover, the superior sagittal sinus has formed, and at the rostral neuropore the massa commissuralis is growing.

At the viable preterm age of around 24 weeks PMA, white matter tracts are in full development. Asymmetric progenitor division permits production of neurons, subventricular zone precursors, and glial cells. Myelin is present in the ventral spinal quadrant, cuneate fascicle, and spinal motor fibers. The neopallial mantle has been separated into transient layers (stratified transitional fields) between the neuroepithelium and the cortical plate. The subplate plays an important role in organizing the structuring of the cortical plate. Commissural tracts have shaped the corpus callosum, early primary gyri are present, and opercularization has started caudally, forming the lateral fissure. Thalamic and striatal nuclei have formed, although GABAergic neurons continue to migrate into the thalamus from the corpus gangliothalamicum.

Near-term PMA cerebral sublobulation is active. Between 24 and 32 weeks, primary sulci develop. Myelin is present in the superior cerebellar peduncle, rubrospinal tract, and inferior olive. Germinal matrix disappears from the telencephalon, except for the GABAergic frontal cortical subventricular neuroepithelium.

THE DEVELOPING BRAIN PER[1] TRIMESTER: DEVELOPMENT TO ACQUIRED LESIONS

In the embryonic period the central nervous system develops substructures. Throughout the second and third fetal trimesters, transient anatomical entities are important in the maturation of brain regions and early functional networks. This review by clinicians (neonatologists and a radiologist) aims to summarize the evolution of these transient structures and studies their relevance in fetal and neonatal brain injury. The focus is not on fetal cranial

[1]Abbreviations used in the chapter are listed at the end of the chapter before References section.

*Correspondence to: Paul Govaert, M.D., Ph.D., Department of Neonatology, ZNA Middelheim, Antwerp, Wilfordkaai 10, bus 9, Temse 9140, Belgium. Tel: +32-465238292, E-mail: govaert@icloud.com

ultrasound. Development of brain vessels is added in anticipation of relevance for future fetal vascular studies. For white matter (WM) and hindbrain development, some molecular genetic information is added. Figs. 14.1–14.4 provide an overview of brain development per trimester.

At 7 weeks postmenstrual age (PMA) (5 weeks after conception), all major subdivisions of the brain are visible (at stage 15 of O'Rahilly and Müller): telencephalon, diencephalon, mesencephalon, metencephalon, and myelencephalon. The future hemispheres become identifiable a week after the telencephalon impar, which contains the preoptic portion of the third ventricle, is recognized. As the hemispheres grow the longitudinal (interhemispheric) fissure between them becomes evident at PMA 7 weeks. In addition the passage between each lateral ventricle and the telencephalon impar narrows to an interventricular foramen. A thickening of the hippocampus begins at 5 weeks, and its arched shape is detectable half a week later. Based on the reference list that follows, an image is compiled of the brain at three different stages of development. These images serve as background for the subsequent descriptions.

The subsequent sections deal with developing and transient structures in such manner that the clinician is prepared to understand lesions of germinal matrix (GM), vascular disorders in the fetal/perinatal period, abnormal gyral development, WM injury, cerebellar lesions, and abnormal fluid collection. For the first figures, an overview of brain development is derived from Retzius (1896), Feess-Higgins and Larroche (1987), Nieuwenhuys et al. (1988), Naidich et al. (1994), Kostovic and Jovanov-Milosevic (2006), O'Rahilly and Müller (2006, 2008), Bystron et al. (2008), Price et al. (2011), Kadic and Predojevic (2012), Altman and Bayer (2015), Vasung et al. (2016), Keunen et al. (2017), and Ortega et al. (2018).

At the end of the embryonic period the cortical plate—first appearing in the insular area lateral to the lateral ganglionic eminence at stage 21—has begun development from the alar telencephalic neuroepithelium and is gradually covering nearly the entire neopallium as a thin sheet (Fig. 14.1). The cortical plate has a triple origin: a primitive plexiform cortex (forming a primary plexiform layer with Cajal–Retzius cells), a subplate from stage 20 (day 50), and the definitive mammalian neocortex (layers II–VI) intercalating between them. The cortical plate develops with a lateral-to-dorsal and a central-to-frontal gradient and the choroid plexus develops in concert with it. The relatively indented area of the insula is visible in the basal center of the curved hemispheres. The diencephalon is separated by mesenchyme (the future transverse fissure) from the

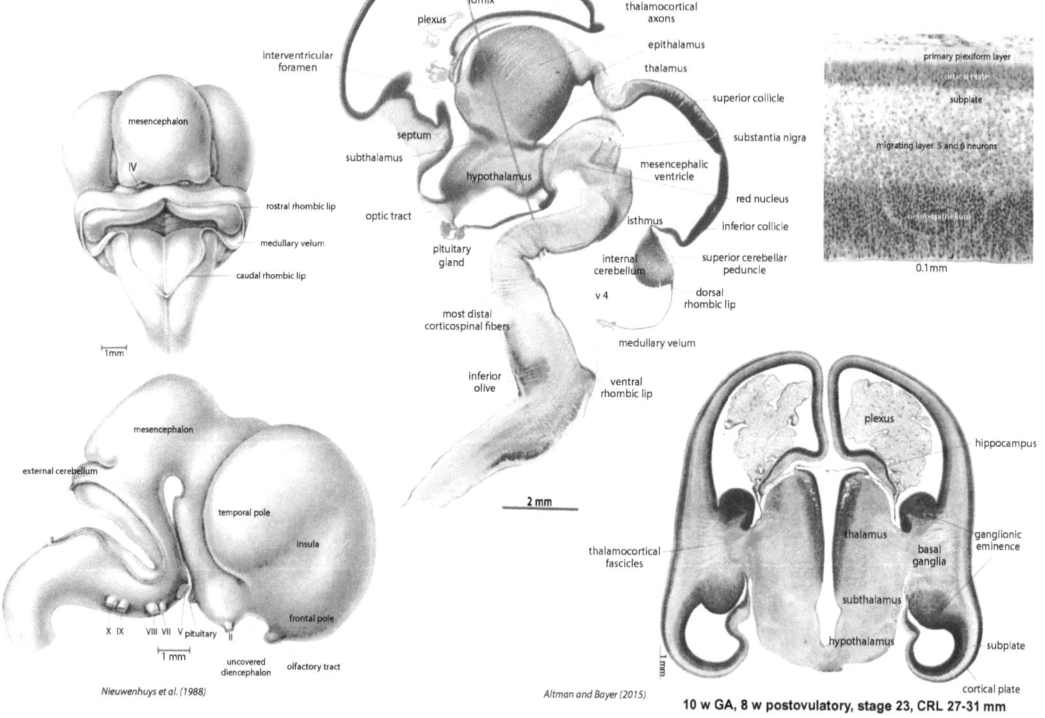

Fig. 14.1. The brain at 10 weeks GA, 8 weeks postovulatory, stage 23, CRL 27–31 mm, end of the embryonic period.

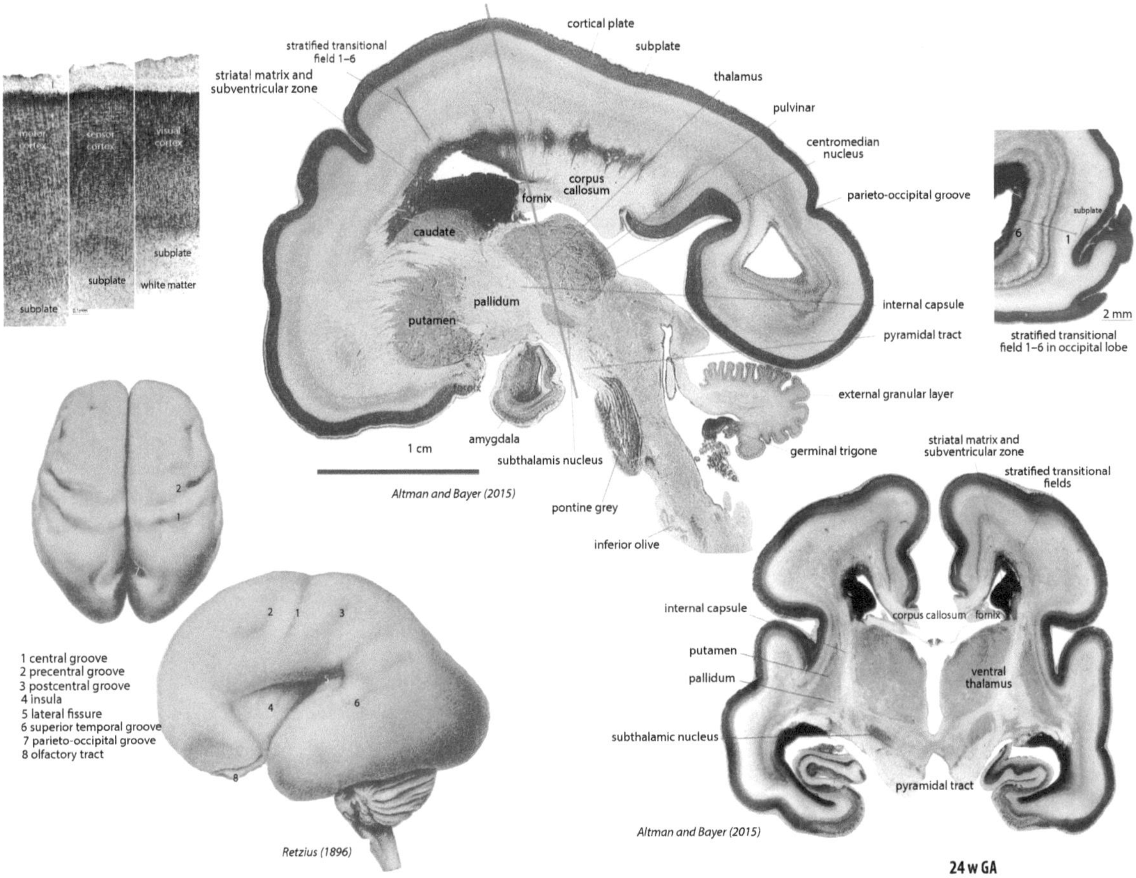

Fig. 14.2. The brain at 24 weeks GA, at viable preterm age.

telencephalon and is still partially uncovered. Olfactory structures and the optic commissure are formed at stage 20, around day 51. The basal ganglia and amygdaloid nuclei are present in the form of their definitive nuclei. The dorsal thalamus covers about half the diencephalic third ventricle surface. In addition to fourth ventricle neuroepithelium the rhombic lips (RLs) have become an active cerebellar neuroepithelial production site, first forming the floccular cortex, inferior olive, and pontine gray nuclei. Early reciprocal connections between the thalamus and the cortex are present in the lateral prosencephalic fasciculus (predecessor of the internal capsule that has just formed), and the pioneering thalamic axons have reached the occipital area. Moreover, the corticospinal tract (CST) has reached the pyramidal decussation. Arteries forming the mature circle of Willis are present, MCA and ACA branches overlie the insular cleavage, and the superior sagittal sinus (SSS) has gradually formed. At the rostral neuropore (closed at stage 11, day 24) the midline massa commissuralis is growing, in preparation of commissural development. Functional synapses ascend from the spinal cord up from the 7th week of gestation. By 10 weeks of GA the fetal motor repertoire contains general movements, some isolated limb movements, head flexion and rotation, yawning, immature breathing, and jaw opening.

The second trimester, the fetal period before current viable preterm age, is characterized by maturation of the cortical plate and development of WM (white matter) tracts (Fig. 14.2). Cortical layering proceeds in a different manner between various regional types of cortex. Asymmetric progenitor division permits production of postmitotic migrating neurons, subventricular zone (SVZ) precursors, and glial cells. At the end of the second trimester, around viable age, myelin is still absent in cerebrum and cerebellum but present in the ventral spinal quadrant, cuneate fascicle, and spinal motor efferent fibers. The neopallial mantle has been separated into transient layers between the neuroepithelium (ventricular and SVZ) and the cortical plate. The subplate is important in organizing the structuring of the cortical plate under the influence of thalamocortical and later corticocortical afferents. The stratified transitional fields reflect different cortical neuronal precursors on their way to the cortex, and these fields grow especially between weeks 14 and 18 of gestation. The matrix neuroepithelium is

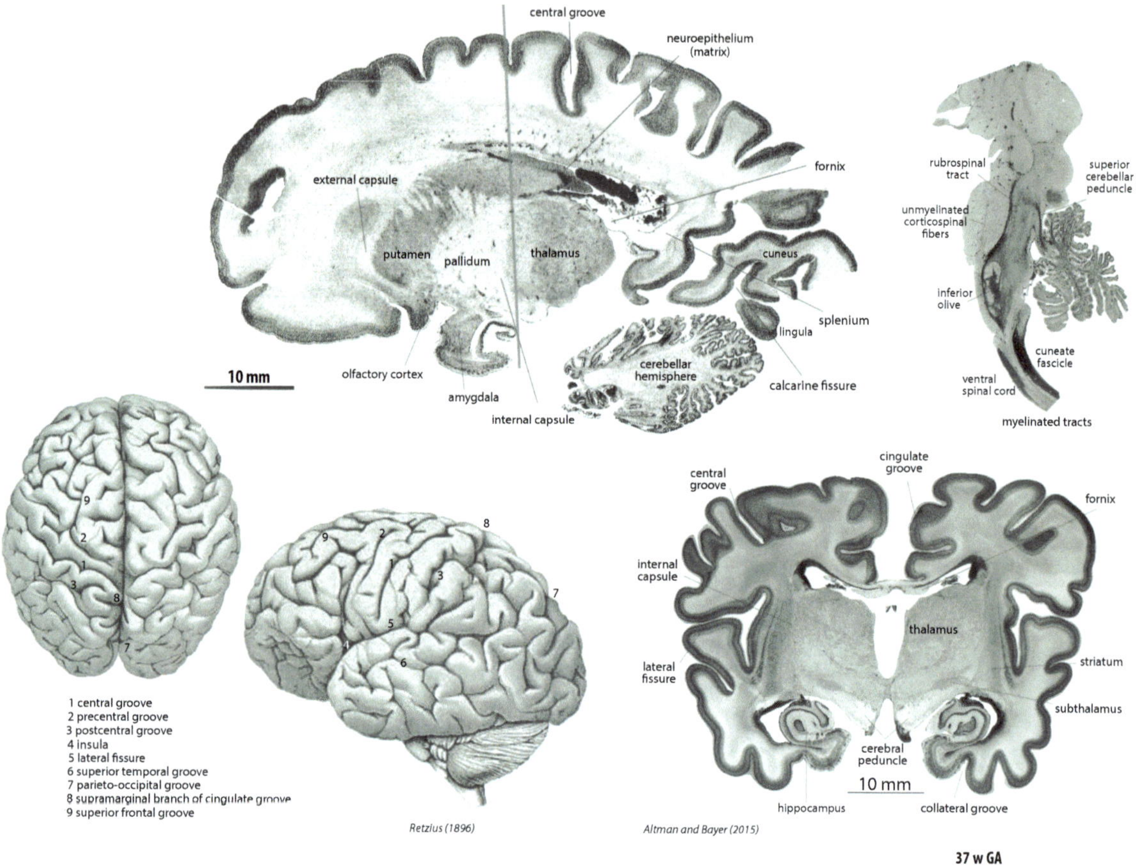

Fig. 14.3. The brain at 37 weeks GA, near term.

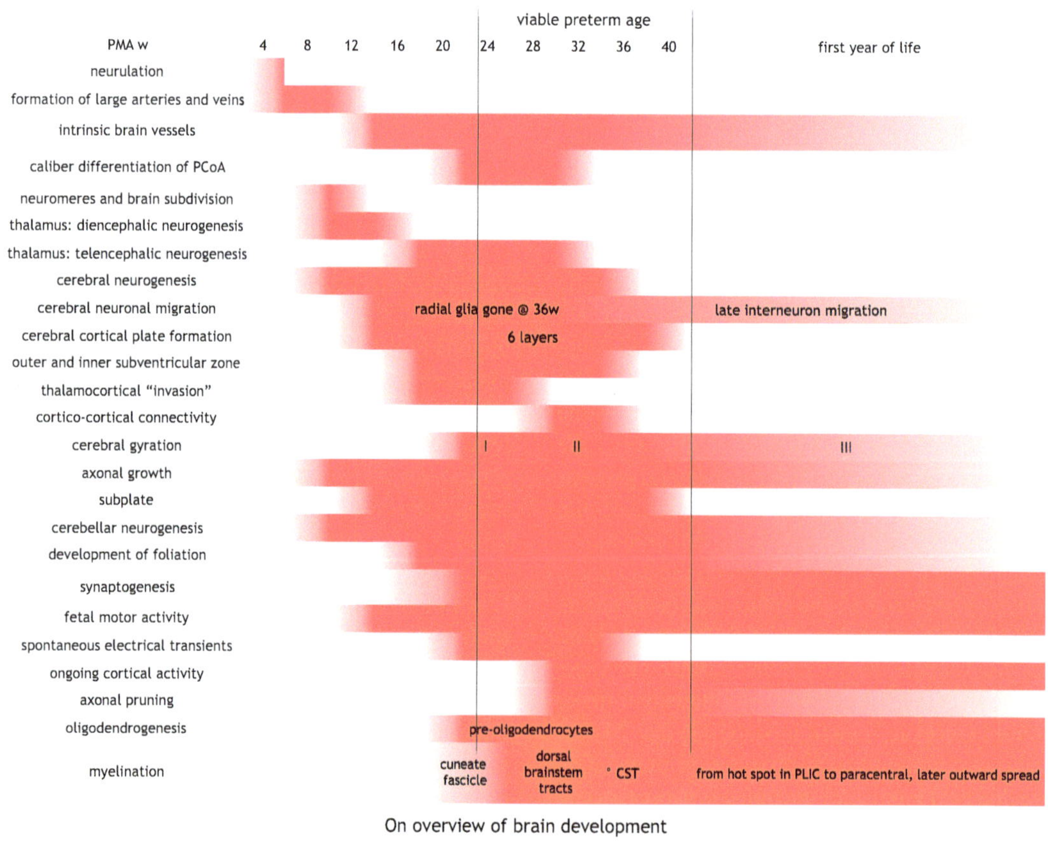

Fig. 14.4. An overview of brain development.

very prominent from the caudothalamic grooves to the dorsal temporal horn. The fourth ventricle neuroepithelium and external granular layer actively produce cerebellar cells. Commissural WM tracts have shaped the corpus callosum (CC) and underneath it the cavum septi pellucidi is formed. Early primary gyri are present (parieto-occipital, calcarine and later central) and opercularization has started caudally, forming the lateral fissure. The calcarine sulcus separates the lingula from the cuneus, for upper and lower visual field processing, respectively. Thalamic and striatal nuclei are formed, although GABAergic neurons will migrate into thalamus from the corpus gangliothalamicum throughout the third trimester. By 24 weeks of GA, goal-oriented movements have appeared. Fetal motor activity is especially high between 14 and 19 weeks, and this is chronologically related to subplate development and intense synaptogenesis. From 20 weeks on, movements come in cycles between resting periods. Electrocortical activity (from subplate) appears just before 20 weeks. Painful stimuli have been shown to alter blood flow from around 16 weeks.

By this stage, sublobulation, the result of secondary gyration, is active and will continue after term birth (Fig. 14.3). Between 24 and 32 weeks, primary sulci and gyri develop as a consequence of gradual intercalation of migrating cells from the SVZ into the cortical plate. Primary sulci develop in a well-defined order, and their appearance at postmortem precedes detection with imaging for at least 2 weeks. By 37 weeks the CST and the rest of the cerebral WM are still mostly unmyelinated, but myelin is present in the superior cerebellar peduncle, the rubrospinal tract, and the inferior olive. CST myelination mainly proceeds in the first 4 months after term birth, except for an early core of myelinated fibers in the posterior limb of the internal capsule (PLIC) directed at postcentral cortex, which is already present before term birth. Cerebral WM myelination is almost entirely postnatal. The GM disappears from the telencephalon around term except for the frontal cortical subventricular neuroepithelium, which continues to produces GABAergic neurons for several months. At term, eye blinking and facial mimicry have developed, and general movements have decreased in number but increased in complexity. Spinothalamic tracts myelinate just before 30 weeks, and so by 26 weeks somatosensory evoked potentials start to make an entry. Auditory and visual evoked responses mature later.

GERMINAL MATRIX (GM)

Germinal matrix development

The name GM is used interchangeably with neuroepithelium. GM is a transient layer at the surface of the ventricles, present in the fetal brain between 8 and 36 weeks of gestation (Kinoshita et al., 2001) (Figs. 14.4 and 14.5). GM volume reaches its maximum around 25 weeks of gestational age (Battin et al., 1998; Kinoshita et al., 2001; Habas et al., 2010; Del Bigio, 2011; Scott et al., 2011). The clear appearance of the alar–basal junction after 8–9 weeks of gestation separates a basal site of neuronal precursor proliferation from a dorsal part: these are referred to as basal (subpallial) neuroepithelium and pallial (or cortical) neuroepithelium, respectively. This separation is a bicameral event following hemisphere formation at 7 weeks of GA. The entire GM is not uniform, but it can be referred to as a protomap with very early regional specification according to the functional specialization of any specific region.

Subpallial GM is considered the "factory" for the production of most cortical brain cells: glutamatergic (pyramidal) projection neurons, some GABAergic interneurons, and macroglial (oligodendroglial as well as astroglial) precursors in the latter part of pregnancy. These different cells stem from multipotent intermediate progenitor cells following a switch from symmetrical to asymmetrical cell division (Kriegstein and Alvarez-Buylla, 2009). Neurons and macroglia originate in the ventricular zone (VZ, area close to the ventricle lumen) but later on also in a secondary germinal area above the VZ, the SVZ. SVZ appears at 8 weeks of GA and persists until birth and in a rudimentary form even into adulthood (the subependymal zone); it is subdivided into the smaller inner (iSVZ) and the expanded outer SVZ (oSVZ), with incoming fibers in between. After 15 weeks the SVZ forms large ganglionic eminences along the infero-lateral walls of the frontal horns of the lateral ventricles (Ulfig, 2002; Del Bigio, 2011; Vasung et al., 2016), and for that reason this richly vascularized area is the major site of fetal and preterm brain hemorrhage. Reduction of GM volume is site specific and continues until the SVZ regresses at 36 weeks. Part of the SVZ persists as a source of progenitor cells, providing interneurons for the cortical plate in infancy and for the olfactory bulb as well as oligodendrocytes through adulthood (Corbin et al., 2008). Precursor cells first migrate laterally and then change direction, engaging in tangential migration within telencephalon. Cells move through the IZ and upon reaching the telencephalon migrate actively in the direction of the cortical VZ; here, they pause for a period obtaining cues from the local environment before resuming their migration radially toward the correct layer of the developing cortex (Nadarajah and Parnavelas, 2002; Ayala et al., 2007). Tangential migration is also known for thalamic pulvinar as its interneurons arise from telencephalic GM and migrate to the diencephalon from the 5th to the 8th month of gestation.

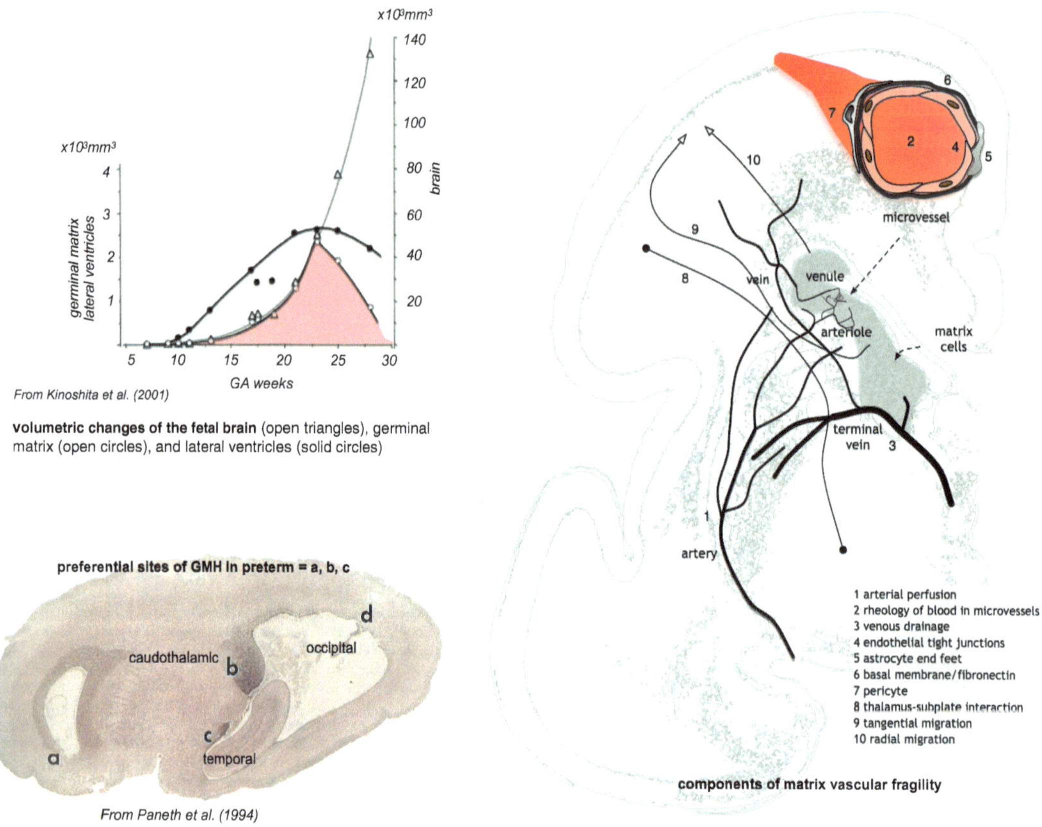

Fig. 14.5. Germinal matrix development and fragility.

Cortical GM parallels the ventricle cavities, lateral and above the subpallial GM. Neuroblasts from this cortical neuroepithelium migrate radially (orthogonal to the ventricular axis) on a glial scaffold to finally form the cortex. The mesial cortical subplate along the ventricle cavity—much smaller than the lateral cortical plate—produces precursors for the limbic system and amygdala. Basal radial glial cells are the neural progenitors in the cortical VZ/SVZ, and their function governs cortical growth and gyrification (Ortega et al., 2018). Glutamatergic neurons for the upper cortical plate, some interneurons, some oligodendrocytes (mainly after birth), and astrocytes are generated in the cortical SVZ. The first cohort of postmitotic neurons leaves the cortical VZ to reach the developing cortex following this radial migration pattern in a process called "somatic translocation" (Nadarajah et al., 2001). Neuronal migration first gives rise to the primordial plexiform layer or preplate (PP) located just underneath the pia mater (Bystron et al., 2008). In radial migration from pallial GM, "glia-guided locomotion" causes a group of cell bodies to move along the leading trailing process, as a unit through the radially oriented glial fibers of the RGCs (Nadarajah et al., 2001). The next wave of migrating cells will form the subplate that later becomes subcortical WM (see later). Neuronal migration proceeds from pericentral and insular areas to the convexity and to the cerebral poles.

Between 15 and 27 weeks of GA, the intensity of neuronal proliferation decreases, and so the relative volume of proliferative compartments declines. In contrast, the subplate—rich in synapses and extracellular matrix—grows during the first two trimesters, occupying up to 45% of telencephalon and reaching maximum volume and thickness around 32 weeks of GA. This maximum volume coincides with intensive growth of long corticocortical fibers. The addition of such associative fibers and the formation of the centrum semiovale separate the sagittal stratum of axons reaching the cortex from the subplate. While the SVZ does not produce electrical activity, the subplate does. Volume, gyrification index (GI), and surface area of the cortical plate grow exponentially during the last gestational weeks (GWs). This cortical expansion correlates with the insertion of cortical columns, dendritic differentiation, and ingrowth of axons. RGCells transform into astrocytes (GFAP and vimentin-positive cells) from early fetal life on until term.

Pool differentiation and proliferation in the SVZ depends on sonic hedgehog (Shh) signaling, WNT signaling, expression of basic fibroblast growth factor, chemokines (CXCL1α), and neurotransmitters. The SVZ can thus be disturbed by hereditary mutations, spontaneous mutations in specific cell lines, and external factors such as hypoxia, ischemia, and hemorrhage. As GM is a major source for oligodendrocytes, their number can be reduced and differentiation can be impaired due to matrix injury such as hemorrhage; this may therefore contribute to WM injury (Ulfig, 2002). The final stages of development of the cortical plate are areal differentiation and gyration (see further) in the third trimester, and after birth, myelination of cerebral tracts, pruning of cells and synapses, and elaboration of dendritic arbors. Especially in the first year of life, dendritic expansion and myelin formation are intense.

Germinal matrix hemorrhage (GMH)

Hemodynamic factors predispose to wall rupture in small GM venules in many (fetuses and) preterm infants (Ghazi-Birry et al., 1997; Ballabh, 2010). Preterm brain hemorrhage consequently occurs primarily in GM (GMH) and not readily in cortex or WM (El-Khoury et al., 2006), in part because vascular density is high compared with that of WM and the cerebral cortex (Ballabh et al., 2004). The vascular network in GM is complex, but regular vascular profiles such as arterioles, capillaries, and veins are not present (Anstrom et al., 2004). Different arterial perforators feed GM: from the anterior cerebral artery, anterior and posterior choroidal arteries, and hence the unproved suggestion of possible interarterial watershed injury of GM. GM vessels drain along subependymal veins, wherein medullary veins from the deep WM coalesce between matrix and caudate head (Bassan, 2009). At 23 weeks of GA, thin-walled subependymal veins are well identified in GM; they mature with increasing GA (Anstrom et al., 2004; Bassan, 2009). Pericyte coverage and density are limited in GM vasculature compared with WM and cortex (Braun et al., 2007). Blood–brain barrier (BBB) components include endothelial tight junctions, astrocyte end-feet coverage, basal lamina, and capillary pericytes, all of which play a role in the stabilization of microvessels. Weakness of any of those components predisposes to GMH (Xu et al., 2008). Changes in BBB permeability correlate with the distribution of tight junction proteins (Ballabh et al., 2005), although their role in the development of GMH is controversial. Astrocytic end-feet perivascular coverage possibly plays a role in the fragility of GM vasculature (El-Khoury et al., 2006). Together with the basal lamina, fibronectin in extracellular matrix also plays a role in providing structural stability of vessels. Xu et al. (2008) found that fibronectin protein levels in GM were significantly lower than those in the WM or cerebral cortex. Fibronectin levels in WM and cortex increase significantly with advancing GA, but not in GM. The expression of type IV collagen chains and perlecan is not different between cortex, WM, or GM (Anstrom et al., 2005), but mutations in type IV collagen are known to cause fetal brain hemorrhage, hydrocephalus, porencephaly, cerebellar destruction, and arterial aneurysms (Verbeek et al., 2012).

Fetal drug treatment can influence the risk of GMH. Prenatal lung ripening with glucocorticoids (GC) has contributed to increased survival and reduced morbidity in preterm infants: the incidence of GMH-IVH is reduced by 50%. In a histopathological study of rabbits and human fetuses, GCs stabilized GM vessels through suppressed angiogenesis, pruned neovasculature, and enhanced pericyte coverage (Vinukonda et al., 2010). In animal studies, indomethacin was found to inhibit prostaglandin synthesis and decrease cerebral blood flow. Ment et al. demonstrated that indomethacin in beagle pups led to increased deposition of collagen V and laminin, and this increase in stability of the basal lamina probably explains the reduction in incidence of IVH (Ment et al., 1992).

Antenatal cystic matrix regression (germinolysis, subependymal pseudocyst)

GM is abundant within ganglionic eminences in the floor of the lateral ventricle during the second trimester, but in the third, it persists mostly around the foramen of Monro—in the caudothalamic groove—and along the temporal horn superolateral wall (Fig. 14.6). Hemorrhage and/or selective cell death in this matrix, both in and ex utero, leads to cyst formation in the affected area in the weeks after the event. Because these cysts have glial walls encircled by germinal cells but not by ependyma, the term "pseudocyst" is appropriate (Larroche, 1972; Shackelford et al., 1983). Glial trabeculae often bridge the cyst walls, invaded by nodules of immature matrix cells. The prevalence of a pseudocyst in the caudothalamic groove in healthy term newborns is around 1% (Shen and Huang, 1985; Heibel et al., 1993). A cyst in choroid plexus (usually one round cavity) behind the foramen of Monro is hard to distinguish from a unilocular germinolytic cyst (Beltinger and Saule, 1988; Mito et al., 1989; Lu et al., 1992). Germinolysis of in utero origin typically presents as one or several pseudocysts in the caudothalamic groove. It has not yet been shown that an echographic distinction can be made between germinolysis following either hemorrhage or any other cause. GM cysts are indeed associated with several factors, including nonbacterial fetopathy

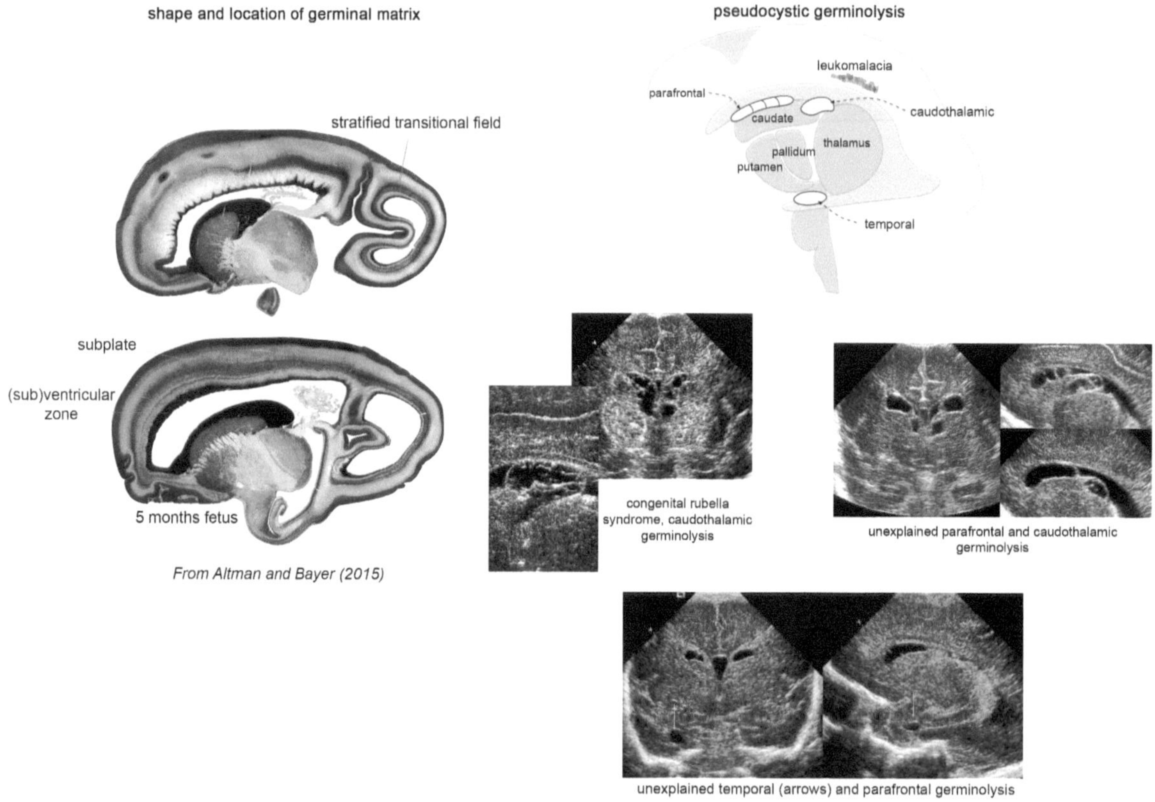

Fig. 14.6. Antenatal cystic germinolysis.

(CMV, rubella), twin-to-twin transfusion, prenatal and perinatal asphyxia, inborn errors of metabolism (peroxisomal disorder, mitochondrial disorder, organic aciduria), and anomalies of the karyotype. Several perforator stroke types may also damage parts of GM.

A special variant, parafrontal germinolysis, has been described as originating around mid-second to early third trimester (Thun-Hohenstein et al., 1983; Rademaker et al., 1993; Pal et al., 2001). Parafrontal germinolysis has been reported in infants with CMV fetopathy. One observes an often symmetrical cyst under the lateral wall of the frontal horn. It is located medial to and below typical (cystic) periventricular leukomalacia and is more anterolateral than the classic germinolytic cyst at the foramen of Monro. It lies in front of these foramina and is separated from the lateral ventricle by a single membrane. On parasagittal section the cavities are elongated and often formed into two or three sections separated by thin septa (like a string of beads). Some cysts are almost 2 cm long. These cysts have erroneously been referred to as coarctation of the frontal horn. At around 4 months after term, nearly all cysts are integrated into the ventricle wall. Isolated paraventricular frontal horn cysts appear benign, with no effect on neurodevelopment at the age of 1 year (Trawber et al., 2010; Cevey-Macherel et al., 2013). MRI in early childhood did not reveal gliosis around such benign parafrontal cysts and ventriculomegaly is as a rule not associated.

There are rare occasions of GM cystic change outside the caudothalamic and parafrontal areas. One such site is along the *temporal* or occipital horn, e.g., in association with congenital CMV (Ramenghi et al., 1997).

Based on sonographic characteristics (number, size, shape, and anatomical location) the following classification can be derived from published descriptions (Fernandez Alvarez et al., 2009; Trawber et al., 2010; Cevey-Macherel et al., 2013; Esteban et al., 2015):

Type 1: Isolated cyst at caudothalamic groove and/or parafrontal horn with oval shape and well-defined margins; largest axis <9 mm.

Type 2: Multiple cysts localized at the parafrontal horn with well-defined margins and typical string of beads pattern.

Type 3: Cyst without well-defined margins with atypical location posterior to caudothalamic notch (e.g., in temporal and occipital horn) and/or with great axis ≥ 9 mm and/or with atypical morphology (this type may warrant further diagnostic tests).

The functional consequence of germinolysis needs further research. As an isolated finding, germinolytic changes are not associated with severe developmental problems, although exceptions exist. Bilateral multiple subependymal pseudocysts may offer enough reason to undertake investigations even if the findings are incidental in an otherwise normal patient (Fernandez Alvarez et al., 2009). Parents of otherwise healthy newborns with a single cyst should be reassured without resorting to other diagnostic tools. Nevertheless, there have been indications that lesions of perinatal GM may influence the potential of cortical frontal GABAergic neurons, the significance of which will have to be studied in relation to behavioral and cognitive outcomes (Paredes et al., 2016). This may be especially relevant for postnatal-onset germinolytic changes (Horsch et al., 2010): as postnatal-onset germinolysis starts with a long period of hyperechoic change before cavitation, the term "germinolysis" (of either hyperechoic or cystic type) may thus better describe the wide range of nonhemorrhagic changes in GM.

GM enlargement has been reported with fetal MRI (Righini et al., 2016). In a cohort of eight cases with GM anomalies and associated changes the median gestational age at the time of MR in these infants was 21 weeks (19–29 weeks). Five fetuses had cavitation in the GM and three showed enlargement of the GM region. For each fetal case, associated cranial anomalies were also reported. Enlargement of matrix has been associated with molar tooth malformation and with TUBA1A mutation with microcephaly. It is not clear whether this matrix enlargement in utero bears similarities to postnatal hyperechoic germinolysis.

DEVELOPMENT OF ARTERIES

Vasculogenesis

Vasculogenesis, the formation of new vessels, starts in the connecting stalk and the extra-embryonic membranes (yolk sac or umbilical vesicle) during the third week (Sadler, 2011; Moore et al., 2016). Mesodermal precursors differentiate into endothelial precursors or angioblasts (vessel-forming cells), which aggregate to form angiogenic cell clusters known as blood islands. Cavities appear within the blood islands by the confluence of intercellular clefts. Angioblasts flatten to form endothelial cells (influenced by VEGF) that arrange themselves around these cavities in the blood islands to form the primordial endothelium. The new endothelium-lined cavities soon fuse to form networks of endothelial channels. Mesenchymal cells that surround the primordial endothelial blood vessels differentiate into muscular and connective tissue elements of the vessels, and the BBB develops in parallel (Risau, 1997; Engelhardt, 2003). Vessels sprout by a process called angiogenesis, by endothelial budding into adjacent nonvascularized areas and fusion with other vessels, also regulated by VEGF. Final modeling and stabilization of vessels are accomplished by PDGF and TGF-b (Park and Gerecht, 2014). The heart and great vessels develop from mesenchymal cells in the cardiogenic area. Paired, endothelium-lined channels develop during the third week and fuse to form a primordial heart tube. The tubular heart joins blood vessels in the embryo, connecting stalk, chorion, and umbilical vesicle to form a cardiovascular system. By the end of the third week, blood is flowing, and the heart begins to beat on day 21 or 22. The embryonic heartbeat can be detected by Doppler ultrasonography during the fourth week, approximately 6 weeks after the last normal menstrual period.

Phases of arterial development

The phases of arterial development are listed according to Padget (1948) (Table 14.1) and revisited by Raybaud (2010a) (Figs. 14.7 and 14.8). During weeks 2–4 the open neural plate is fed by diffusion from the amniotic fluid. From weeks 5 to 8 the closed neural tube is enveloped by a dense network of developing connective tissue called the meninx primitiva (from mesoderm and neural crest at di-mesencephalic level), which contains a primitive vascular plexus irrigated from primitive dorsal aorta and drained by cardinal veins. This plexus is at first connected to the yolk sac. Initially the meshwork of vessels is indeterminate in the sense of arteries or veins, and as such referred to as primary head plexus. This vascular plexus develops from ventral forebrain and midbrain, later also from hindbrain, and extends dorsally as it grows. Supply of oxygen and substrate is from diffusion out of meninx primitiva and from the ventricular fluid. Layers in the meninx primitiva develop differently: the internal layer develops a capillary meshwork and the superficial layer contains larger vessels that become major veins and arteries, initially connected to the dorsal aorta and cardinal veins. An intermediate layer will contain branches in between the internal and superficial vessels. In week 6 the primary brain vesicles differentiate and a portion of the meninx primitiva invaginates into the neural tube from its prosencephalic and rhombencephalic roof, depositing primitive choroid plexus and early vessels that evolve into choroidal arteries and veins. This choroidal network will sprout definitive brain arteries. Initial choroidal venous drainage remodels to become the definitive venous system, whereas early arteries remain permanent and grow by branching.

The internal carotid, basilar, and vertebral arteries are formed during the first three embryonic vascular stages of Padget in the branchial period (approximately 4–12 mm). When pharyngeal arches form during the 4th and 5th weeks of development, each arch receives

Table 14.1

Phases of arterial development.

Phases	Time, GA	Arterial stage, Padget	Description
Diffusion	2–4 weeks	0	The open neural plate is fed by diffusion from the amniotic fluid
Primitive vascular plexus	Days 28–29 Early pharyngeal arches; carotido-vertebral connections	1	The primitive carotid artery supplies fore- and hindbrain through carotido-vertebral connections (carotid artery to LNA = longitudinal neural artery): such connections persist for 4–8 days and vanish at stage 3; they may persist as (clinical) variants carotid branches: – anterior: olfactory branch, will become ACA – posterior: forms midbrain plexus, will become mesencephalic artery – midline: intercarotid branch behind Rathke's pouch – trigeminal artery: most rostral presegmental branch to LNA, level of nV ganglion – otic artery: level of otic vesicle – dorsal aortic branches: hypoglossal artery and proatlantal artery (=first segmental C1 artery)
	Days 29–31	2	Caudal branches of the ICA above the mesencephalic plexus form the PCoAs (posterior communicating arteries) by connecting to the rostral LNA; trigeminal and hypoglossal arteries wither from the carotid end; LNAs coalesce into the basilar artery with initially mainly perfusion from the proatlantal artery
Choroidal	Days 32–34 Pharyngeal arches developed	3	All major arteries are formed, mainly as choroidal arteries The MCA appears as a small lateral bud of the ACA, which curves around the telencephalic vesicle rostrally; the primitive AChA courses to the diencephalon; a dorsal ophthalmic artery arises ventrally at the branching area of ACA and MCA; from the caudal PCoA arise a mesencephalic and a posterior choroidal artery; the trigeminal artery is usually interrupted at this stage; the vertebral arteries form from longitudinal anastomoses between cervical segmental arteries; the superior cerebellar artery arises at the rostral basilar artery
	Days 35–39 Early mature arterial pattern	4	Presence of ACA and ACoA of dorsal and ventral ophthalmic arteries; prominent AChA; development of the MCA proceeds, which is still plexiform and not a true artery, but already the major blood source for the hemisphere at this stage
	Days 40–43	5	Choroid tissue forms projections from meninx primitiva into ventricle cavities, selecting branches from ACA, AChA, and posterior choroidal artery (all these arteries are initially primarily choroidal feeders); the mesencephalic artery forms a rich plexus over the quadrigeminal plate; MCA is primarily a striatal feeder and becomes a single artery by fusion of plexus vessels; its branches start to penetrate the cerebral hemisphere; the stem of the ACA gives rise to the olfactory artery
Definitive	Days 44–51	6	Heubner's artery appears near medial striatum; the ACA continues to grow medially toward the contralateral ACA, eventually leading to the formation of the anterior communicating artery, with a branch to corpus callosum
	>Day 51 Postpharyngeal arch period; intrinsic brain vessels formed	7	Dorsal extensions from the mesencephalic artery become PCAs that perfuse inferior and posterior hemispheres at the expense of the AChA; these arteries are perfused from the basilar and not carotid artery; AICA and PICA become more apparent; the definitive PCA connects with the BA as branches from the primitive fetal PCA fuse medially to form the distal end of the BA

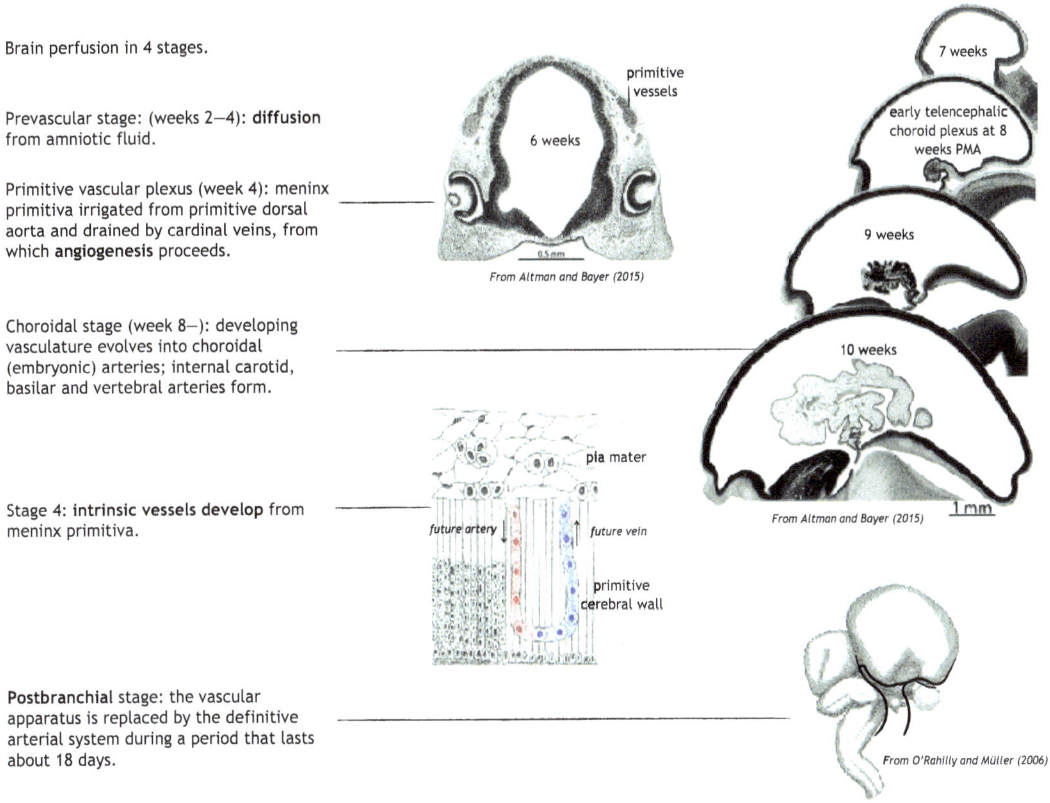

Fig. 14.7. Development of brain perfusion.

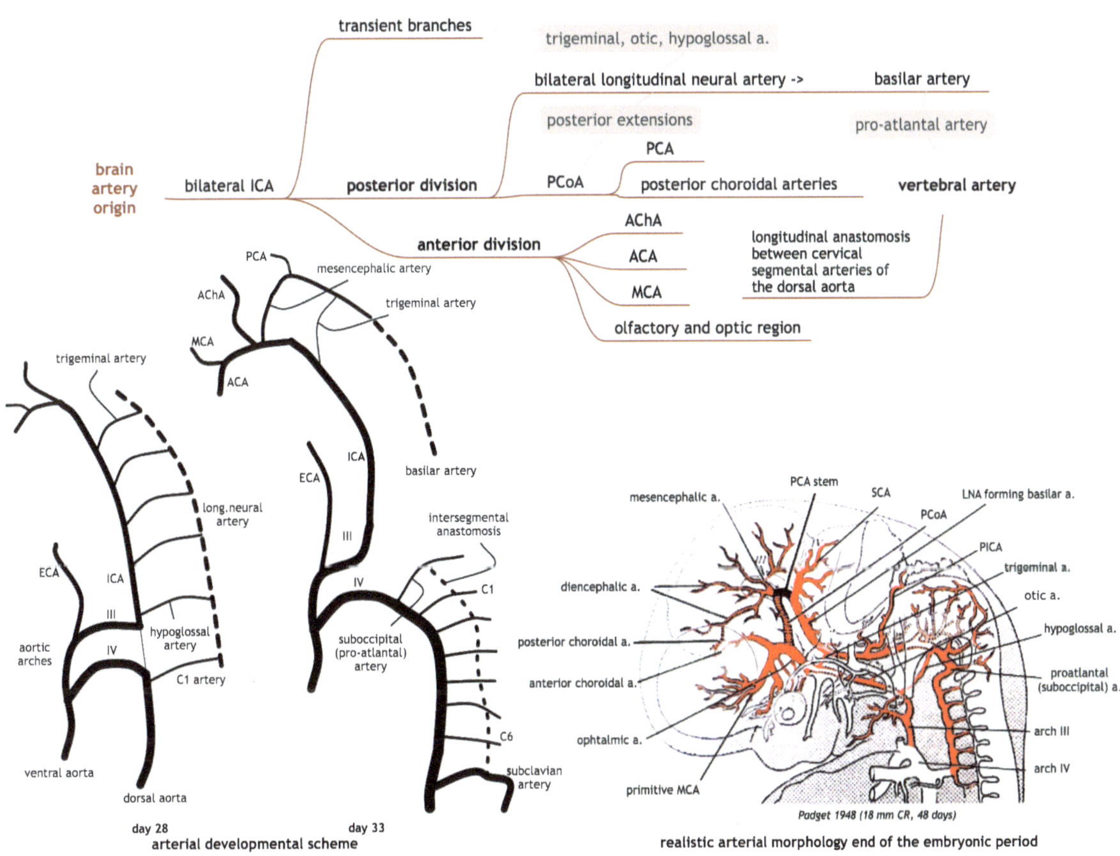

Fig. 14.8. Sources of large brain arteries.

its own cranial nerve and artery. These arteries, the aortic arches, arise from the aortic sac, the most distal part of the truncus arteriosus. The third aortic arch forms the common carotid artery and the first part of the internal carotid artery. The remainder of the internal carotid artery is formed by the cranial portion of the dorsal aorta. Six pairs of aortic arches are formed, coursing around the five branchial arches by the 32nd day. In the postbranchial phase, the vascular apparatus is replaced by the adult arterial system during a period that lasts about 18 days.

Intrinsic brain vessels

During PMA weeks 7 and 8 the neural tube is too thick to be supplied by diffusion alone and intrinsic vessels develop from the vessels of the meninx primitiva. Two episodes of intrinsic angiogenesis are in agreement with parenchymal development: first intrinsic vessels develop in GM between ventricles and subpallium, later they also develop in the cortical plate. At 7 weeks, intrinsic vessels form first in the brainstem with ramifications at the VZ, as well as in rhombencephalic choroid plexus. A dense vascular plexus then forms over the diencephalon and into tela choroidea, starting from the PCoAs. Diencephalon also receives vessels from the midbrain plexus and further along the postero-medial hemispheres (the future PCA) at the same time that perfusion to this area shifts from carotid to basilar. Also at 7 weeks and very dense at 8 weeks is the striatal plexus that develops from the olfactory artery (later ACA), which is the stem of the future MCA. The primitive AChA sprouts vessels into basal diencephalon, lateral ventricle choroid plexus, and uncus temporalis.

At the cerebral cortical level, starting from the pial limiting membrane, new vessels bud from the ones in the meninx primitiva, with a tip cell for progression and behind it stalk cells forming the lumen (Marin-Padilla, 2012). These vessels reach matrix areas and form a rich subventricular plexus between 12 and 15 weeks. Such primitive penetrating vessels make horizontal connections in the matrix area, wherein flow direction will change a ventriculopetal vessel into an artery and outgoing vessels into veins (Raybaud, 2010a,b). In the (sub)ventricular zone, this branching is very dense between 23 weeks and term (Pape and Wigglesworth, 1979; Chang et al., 2013). Matrix capillaries remain sinusoid (have no muscle wall), contrary to trunks in WM that become arteries or veins. Branching near the cortex is not seen before 20 weeks, and the horizontal connections formed there remain capillaries (fewer than two RBCs), whereas related penetrating trunks become arterioles and venules (up to ten RBCs per diameter) (Kuban and Gilles, 1985). A new supply of cortical radial and spiraloid penetrating branches develops at 27 weeks, and later on shorter branches keep on forming between larger branches as the hemispheres grow. A muscularis layer, typical of arterial character, develops in striatal perforators around 20 weeks from their pial end, to reach the caudate head under matrix by about 26 weeks (Nelson et al., 1991). In WM, trunks still lack muscularis until near term, with a small caliber of 10–25 µm at 24 weeks and 20–40 µm near term; venules near the ventricle (collector veins) are about double to triple that size (Anstrom et al., 2002). Smooth muscle cells and pericytes come from the neural crest in vessels at or rostral to the circle of Willis; caudal to this point (rhombencephalon and below), they come from the mesoderm (Etchevers et al., 2001).

The clinical relevance of variations in development of the large arteries of the brain is probably limited in the perinatal period (Fig. 14.9). Carotido-vertebrobasilar anastomoses are patent for a period of 7–10 days. Failure to reshape according to hemodynamic needs may result in persistent primitive carotido-basilar arteries, the most common of which is the persistent trigeminal artery (PTA), observed in 1–2 angiograms/1000 (Salas et al., 1998; Dimmick and Faulder, 2009; Kathuria et al., 2011; ten Donkelaar, 2011). The PTA leaves the cavernous part of the ICA and courses either medially or laterally with respect to the abducens nerve to the basilar artery, usually between the origins of SCA and AICA. Its presence can change caliber and presence of other arteries of the circle of Willis, especially the PCA. The PTA regresses at the 45 mm stage because of the subsequent development of the PCA, which takes over blood supply from the BA (Vasovic et al., 2012; Gregg and Gailloud, 2017). The persistence of the PTA in adults mirrors the 11–14 mm embryonic stage (41–43 days). PTA has been associated with PHACE syndrome, Kippel-Feil syndrome, and aplasia of the inferior vermis.

The PCoA gradually decreases in size (or not) following development of the basilar artery as the fetal carotid contribution via the PCoA gradually changes. In a study of human fetuses from the 13th to the 24th week (Vasovic et al., 2008), six basic types and many subtypes were recognized depending on vessel size: bilateral symmetric transitory type (18.6%, P1 and PCoA equal size), fetal type (9.3%, PCoA larger), and adult type (33.1%, P1 larger), as well bilateral asymmetric type (fetal transitory in 5.8%, adult-transitory in 14.5%, and adult-fetal in 18.6%). This variation in caliber of the PCoA may become relevant in association with progressive arterial diseases any time later in life. Permanent configuration of the posterior part of the circle of Willis is not yet acquired in the period from the 4th to the 6th gestational month, but caliber differentiation starts at this time. In fact the

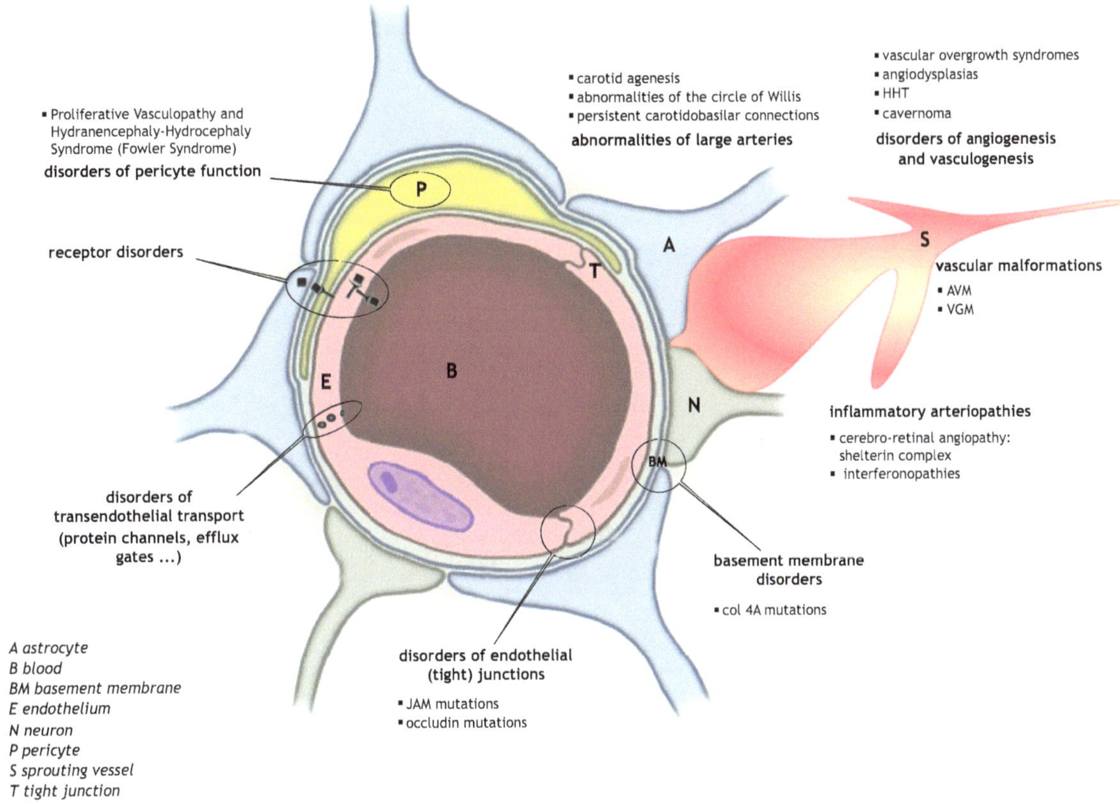

Fig. 14.9. Disorders of vascular development in utero.

formation of the PCA is complex, as the vessel is built up by a caudal branch from the carotid artery and the common trunk of mesencephalic, diencephalic, and posterior choroidal arteries, a further proximal portion of the primitive posterior choroidal artery and one of its branches, and a new vessel developing against the mesial occipital ventricle wall (Moffat, 1961).

Variation in development of hindbrain arteries is common. The cerebellar primordium is vascularized by the metencephalic plexus that will form BA and SCAs. Due to the development of the pontine flexure the rhomboid lips approach and the cerebellum comes into contact with the myelencephalon at 23 mm crown rump length (CRL); thus the myelencephalic plexus represents an acquired source of vascularization for the approaching cerebellum. The VAs, BA, SCAs, AICAs, and primitive lateral vertebrobasilar anastomoses become well recognizable (Gregg and Gailloud, 2017). The PICA is not yet defined due to the persistence of a plexus of thin vessels at the lateral aspect of the myelencephalon, indicating that its origin and course are established at the end of the embryonic period, in relation to the growth of the neocerebellum. The different patterns of PICA reflect the variable retention of the primitive lateral vertebrobasilar anastomosis in the trunk of the definitive PICA (Macchi et al., 2005).

DEVELOPMENT OF VEINS

Initial venous development (Streeter, 1915; Padget, 1956)

Initially cephalic blood drainage from the primary head plexus consists of one main bilateral venous channel, the primary head vein, starting in the region of the midbrain to run caudally along the embryonic brain (Fig. 14.10 and Table 14.2). The primary head vein is composite in origin: rostral to the vagus nerve it is an intrinsic vein of the head; more caudally it is a neck vein named the anterior cardinal vein, later to become the internal jugular vein. Three anastomosing loops form initially: the anterior, the middle, and the posterior dural plexus. A large rostral tributary to the eye region is modified into the ophthalmic vein. Metamorphosis of these primitive plexuses is based on adjustment to the environment: passive migration induced by mass change, active migration by flow remodeling with dwindling of a previous channel, either gradually or abruptly. As the three plexuses spread upward, the outlines of dura mater and arachnoid space appear, first in ventral parts. This results in cleavage of the more superficial primary head vein and its plexuses (dural system) from the subjacent vessels that arise from and drain the intrinsic brain veins (arachnoid–pial system). This deeper system continues to drain into the former at

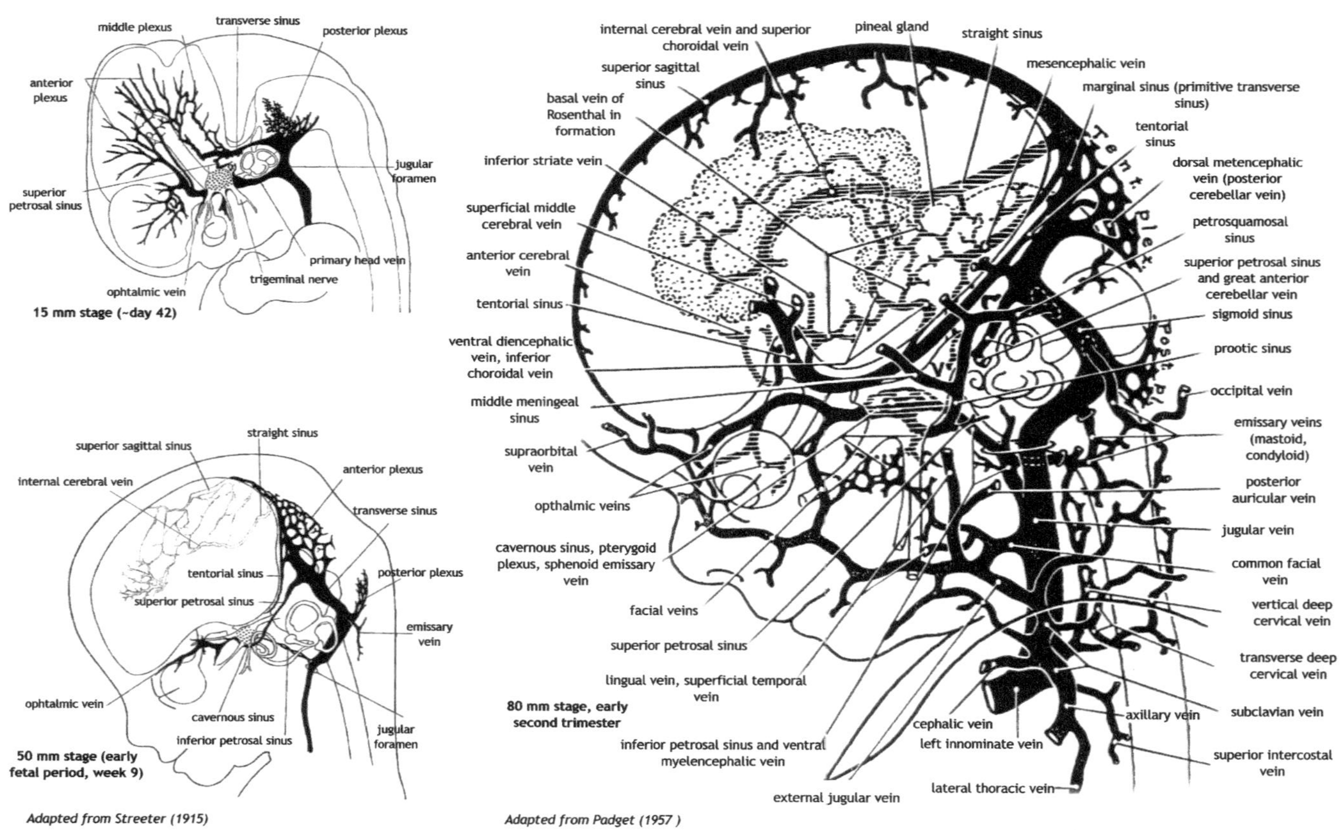

Fig. 14.10. Development of brain veins.

Table 14.2

Phases of venous development.

Stage	Venous changes
28 days, CRL (crown rump length) 4 mm	Primary head plexus drains into anterior cardinal veins (later internal jugular veins)
32 days, CRL 5–6 mm	Superficial anastomosing loops from head plexus (anterior: forebrain and midbrain; middle: cerebellum) form paired paramedian sinus
37 days, CRL 7–12 mm	Paramedian sinus forms marginal sinus (anterior plexus); marginal sinus will become SSS and TS; emergence of paired primitive telencephalic (later superficial middle cerebral vein) and diencephalic veins
41 days, CRL 12–14 mm	Meninges separate into dura and pia/arachnoid; pial veins traverse arachnoid space to dural sinus and run perpendicular to overlying pial arteries; diencephalic veins form dorsal and ventral branches
	Mesencephalic veins emerge from anterior dural plexus
44–48 days, CRL 16–18 mm	Middle and posterior dural plexuses form sigmoid sinus; residual stem is pro-otic sinus
51 days, CRL 20–24 mm	Anterior and middle plexuses become tentorial plexus
56 days, CRL 40 mm	Medial marginal sinuses fuse to one SSS; pro-otic sinus forms cavernous sinus; inferior petrosal sinus forms caverno-jugular connection
	Choroid plexus drains into single median prosencephalic vein
10–12 weeks, CRL 60–80 mm	TS lateralizes; superior petrosal sinus forms
	Paired internal cerebral veins replace median prosencephalic vein and drain growing basal ganglia; basal vein of Rosenthal forms by variable anastomosis of telencephalic, diencephalic and mesencephalic veins

certain places, notably in more dorsal parts. It is the dural system that is involved in the formation of the dural sinuses. Owing to the expansion of the labyrinth the course of the primary head vein, ventrolateral to the otic capsule, becomes unfavorable. A new channel is established dorsal of the otic capsule. In this way a trunk for drainage of the head is established dorsal to the primary head vein and as far as the jugular foramen. Of the primary head vein thus remain the internal jugular vein and the part around the trigeminal nerve, referred to as cavernous sinus, into which will drain veins from the base of the brain, the orbital and maxillary regions. The newly formed dorsal channel becomes the transverse sinus (TS), of which the part between the superior petrosal sinus and the jugular foramen becomes its sigmoid portion. Thus in the 21 mm advanced embryo the dural channels in the region of the temporal bone have acquired their permanent connections, with the exception of the inferior petrosal sinus, which appears later.

A subdivision of the anterior plexus extends forward to the midline as the plexus sagittalis, being interposed as a vertical curtain between the hemispheres. From its dorsal meshes develops a unique longitudinal channel which we refer to as the SSS. The SSS is formed first in its anterior portions by the selection and enlargement of the most favorable vein with a corresponding disappearance of the others; in the posterior part, there is coalescence of adjacent veins later on (Kopuz et al., 2010). As the hemispheres extend backward the sinus correspondingly elongates. The sagittal plexus very early exhibits a tendency to drain more to one side of the head than to the other, very often toward the right (see later). The straight sinus is formed in the ventral part of the sagittal plexus, and its caudal adjustment is essentially like that of the SSS. In the fetal period the TS bends backward until it comes to lie at an angle of 90 degrees with the internal jugular vein. As the sinus becomes more definitely established the tentorial plexus becomes smaller and it undergoes traction by its changing environment. The confluens sinuum is usually plexiform in character and represents the last trace of the embryonic tentorial plexus. Fetal torcular thrombosis no doubt relates to the complicated mode of development of this venous crossing.

The transverse-to-sigmoid connection is a developmental, perinatal venous point of vulnerability in the context of thrombosis and hemorrhage, for the following reasons (Padget, 1956; Raets et al., 2013, 2015).

First, the tentorial sinus dwindles in the perinatal period: The fetal superficial middle cerebral vein (SMCV) drains through the tentorial sinus into the transverse rather than the cavernous sinus. In adults the SMCV is usually connected to the cavernous sinus by the intermediary sphenoparietal sinus (SPS) and there is no tentorial sinus any more. Variations of paracavernous sinus veins are therefore common (Suzuki and Matsumoto, 2000; Tanoue et al., 2006). In term infants, it is possible to visualize with ultrasound a large tentorial venous vessel that enters the TS through a keyhole with high and turbulent velocities. The natural involution of this vessel still has to be studied in fetuses and preterm infants, as also its relation to TS thrombosis or temporal lobe hematoma. Hemorrhage into the temporal lobe (basal and temporal parts) or around the tentorium and in the cerebellum may or may not be associated with TS thrombosis involving temporal bridging veins (Baram et al., 1988). In adults, several patterns of TS bridging veins exist: between torcular and the curve via transverse into sigmoid sinus, bridging veins may form multiple entries and gather into a candelabra or tentorial lakes. Mesial tentorial lakes mainly drain straight into the sinus or the torcular and mesial/middle TS and lateral lakes into the lateral TS and transverse-petrosal junction (Muthukumar and Palaniappan, 1998).

Second, there exists relative jugular obstruction in the fetal period: From a high and straight early fetal position, the straight sinus is flattened and displaced downward by enlarging cerebral hemispheres. Several intrafalcine veins connect the primitive axis of internal cerebral vein, great cerebral vein, and straight sinus below with the SSS. If such vein persists after birth, it is called a persistent falcine sinus. Around term the straight sinus typically drains into the left limb of the SSS. The TS on each side is small and plexiform in its medial part at the fetal age of 3 months; it begins to balloon from its lateral end later in the fourth fetal month, gradually extending medially to reach the confluence and becoming evenly enlarged by the 5th month (Okudera et al., 1994). This may be due to the narrow caliber of the sigmoid sinuses and jugular bulbs at that time. Sigmoid sinuses are slender at the 80 mm stage. It is only after birth that the pressure waves up along the jugular veins entrain progressive dilatation of the superior jugular vein into a bulb, which then becomes a reservoir and suction pump, gently draining blood from the brain. During dilatation of the TS the occipital and marginal sinuses also dilate, emissary veins develop in condylar and mastoid areas, septae and fenestrae may be formed in unevenly dilated sinuses, and lacunae form near the TS (later tentorial lakes).

Third, TS develops asymmetrically: All adult angiographic studies confirmed drainage mainly to the right side in about 40%, equal drainage in a similar 40%, drainage mainly to the left in slightly less than 20%; 2% have drainage only to the right and 0.53% only to the left (Durgun et al., 1993; Fukusumi et al., 2010). Persistent occipital sinuses (OSs) are recognized in about 60% and persistent falcial sinuses in 2.5%. In the abnormal strict unilateral Type I TS, typically the right one is minute; in such instances the possibility of an anomalous configuration of the great veins above the heart should be considered. The relative straight course of the venous system down to the superior caval vein on the right is probably the reason why the growing hemisphere preferentially drains via the SSS to the right TS: venous drainage on the left is more circuitous via the innominate vein. Accordingly, asymmetry of the internal jugular veins is noted with ultrasound in 62.5% adults (Lichtenstein et al., 2001). This functional venous asymmetry has been confirmed in term neonates using MR venography. In a neonatal cadaver study (Kopuz et al., 2010), termination patterns were reclassified into six types, taking into account the presence of OSs: the SSS showed continuity with right TS (with OS) (30.3%, Type I) or multiple OSs (21.2%, Type II); with left TS (with OS) (12.1%, Type III) or with multiple OSs (6.1%, Type IV); with both TSs (9.1%, Type V); symmetrically bifurcated with a large OS (21.2%, Type VI). Prior knowledge of venous asymmetry may have an influence on the decision to use deep jugular venous lines for neonatal venous access.

Other developmental findings have clinical relevance

It may be of relevance in understanding hemorrhagic lesions to know about ongoing remodeling just before and around viable preterm age (Padget, 1956). By the early fetal period (vascular stage 7a; 80 mm CRL), many definitive adult veins have formed, but some embryonic veins are still present (Padget, 1956). The primitive TS receives an elongated tentorial sinus, itself continuous with the SMCV. The primitive internal cerebral vein drains the large superior choroidal vein into the straight sinus and has only one feeder, a thalamic tributary. The basal vein of Rosenthal (BVR) is formed by coalescence of the deep middle cerebral, inferior striate and diencephalic veins, a mesencephalic vein and a tributary of the StS (Chung and Weon, 2005). The lateral mesencephalic vein, connected to the primitive metencephalic or great anterior cerebellar vein, may drain the BVR or just merely become a collateral when the BVR drains into the great cerebral vein. The stem of the great anterior cerebellar vein will become the superior petrosal sinus, outlet of the cavernous into the sigmoid sinus. The dwindling pro-otic sinus is continuous with the petrosquamosal sinus. Both become diploic in adult life. After the third fetal month, superior striate and medullary veins, draining into anterior and posterior terminal veins, form around the ICV. Anastomosing veins of Trolard and Labbé on the cerebral convexity—connecting SMCVs, superior sagittal and TSs—are in fact shaped only in the fetal period.

Neck rotation can influence venous flow. By compression of the jugular veins in the lower neck a drop in flow velocities in the SSS can be registered in many neonates; asymmetrical findings are compatible with the earlier reported asymmetrical adult TS anatomy (Cowan and Thoresen, 1983). Escape along nonjugular veins thus seems common and variable. Rotation occludes the jugular vein on the side to which the face is rotated; it is therefore logical (although difficult to prove) to avoid excessive neck rotation especially in very low birth weight infants with fluid restriction and/or venous congestion.

The range of perinatal disorders where brain veins are implicated is broad (Raets et al., 2015). Best known are sinus thrombosis and deep WM venous infarction associated with GMH. This presents as hemorrhage in nearly all cases. To understand the mechanism behind the bleeding, fetal brain specialists must be able to correlate extent and location of suspected hemorrhage with the vein(s) affected. Neonatal brain injury implicating veins can be summarized as follows: (1) *preterm brain lesions*—medullary hemorrhagic venous infarction (associated with GMH/IVH (intraventricular hemorrhage); primary (prothrombotic conditions, collagen A4 mutation) cerebellar hematoma); (2) *term brain lesions*—thrombosis (sinovenous: general, internal cerebral vein, basal vein, TS), lobar hematoma (major lobar hematoma, extracorporeal membrane oxygenation-related hematoma, contusion), open vessel ischemia (polycythemia, dehydration, severe congestion); and (3) *others*—vascular anomaly (arteriovenous malformation, vein of Galen malformation, developmental venous anomaly, sinus pericranii, cerebral angiodysplasia, cerebellar vascular dysplasia, Sturge–Weber syndrome, hereditary hemorrhagic telangiectasia), infantile subdural hematoma, infectious thrombophlebitis, plexus bleeding with IVH. Young infants are more vulnerable to subdural hematoma because of relatively large venous lakes in the dura around the posterior fossa (Browder et al., 1975). That ultrasound, including color Doppler, can depict sinovenous thrombosis at an early stage presents therapeutic options prior to parenchymal injury.

PRIMARY GYRATION

Normal development

Development of cerebral sulci and gyri is a sequential process, to be observed with fetal imaging or by serial postnatal imaging following preterm birth (Fig. 14.11). The process of primary sulcation/gyration (folding) occurs mainly between 19 and 32 weeks of gestation (review Nishikuni and Carvalhal Ribas, 2013), driven by complex forces in cortex, underlying WM and GM (Sawada et al., 2012; Lewitus et al., 2013; Zilles et al., 2013; Altman and Bayer, 2015; Striedter et al., 2015). Development of the six cortical layers occurs during the first half of this period, coinciding with increasing thickness of subplate, in itself driven not only by thalamocortical but also by commissural and corticocortical contacts (Vasung et al., 2016). The thickest subplate is formed in perisylvian associative cortical areas. Further, postprimary sulcation, beyond 32 weeks of gestation until a peak of gyrification in the latter part of the first year of life, is driven by specialization of regions that generate mechanical forces in and immediately below the cortex. After 32 weeks, most secondary (around 33 weeks) and tertiary (around term) sulci are formed, and primary sulci have further matured by becoming deeper and tighter. As gyration proceeds the volume of WM increases, the relative volume of the ventricles decreases, the STFields disappear and the matrix areas shrink.

Based on cerebral growth and the developmental pattern in cynomolgus monkeys, the gyrification process can be divided into four stages (Sawada et al., 2012): *stage 1*—demarcation of cerebral lobes and limbic gyri and emergence of corticocortical long associative fibers; *stage 2*—demarcation of primary neocortical gyri and expansion of cerebrum; *stage 3*—emergence of secondary and tertiary sulci; and *stage 4*—growth of sulcal length and depth and cortical maturation. Primary sulci mature in a particular sequence before the 32nd week of gestation (Chi et al., 1977; Feess-Higgins and Larroche, 1987). The surface first indents, later deepens into a groove with banks, and finally bifurcates at the ends of the groove, starting around 30 weeks of gestation. Secondary and tertiary sulci develop after the 31st week (Slagle et al., 1989). Bending (tortuosity) of primary sulci starts around 30 weeks (Worthen et al., 1986). All over the brain (submerged or surfaced), annectant gyri (French term "plis de passage") connect adjacent brain areas often perpendicular to the sulci (Broca, 1888; Cunningham, 1892; Retzius, 1896, with references to contemporary and preceding authors). Further research is needed to understand whether these submerged connections predict later functional specialization (see later).

Detailed attention to the brain surface in imaging is the only way to study gyration in life. The lag between development at postmortem and detection with fetal MRI is well documented (Levine and Barnes, 1999; Girard and Gambarelli, 2001). A fetal Total Maturation Score scale with MRI has been proposed to assess morphologic brain maturity of fetuses between 25 and 35 weeks of GA in a clinical setting: sulcation of frontal, occipital, and insular cortex, as well as presence and depth of superior and inferior temporal sulcus, can be used as a part of the score (Vossough et al., 2013). On ultrasound scans an early indication of a developing sulcus is a broad band of echospeckle upon tangential insonation. Fetal sonography shows the interhemispheric

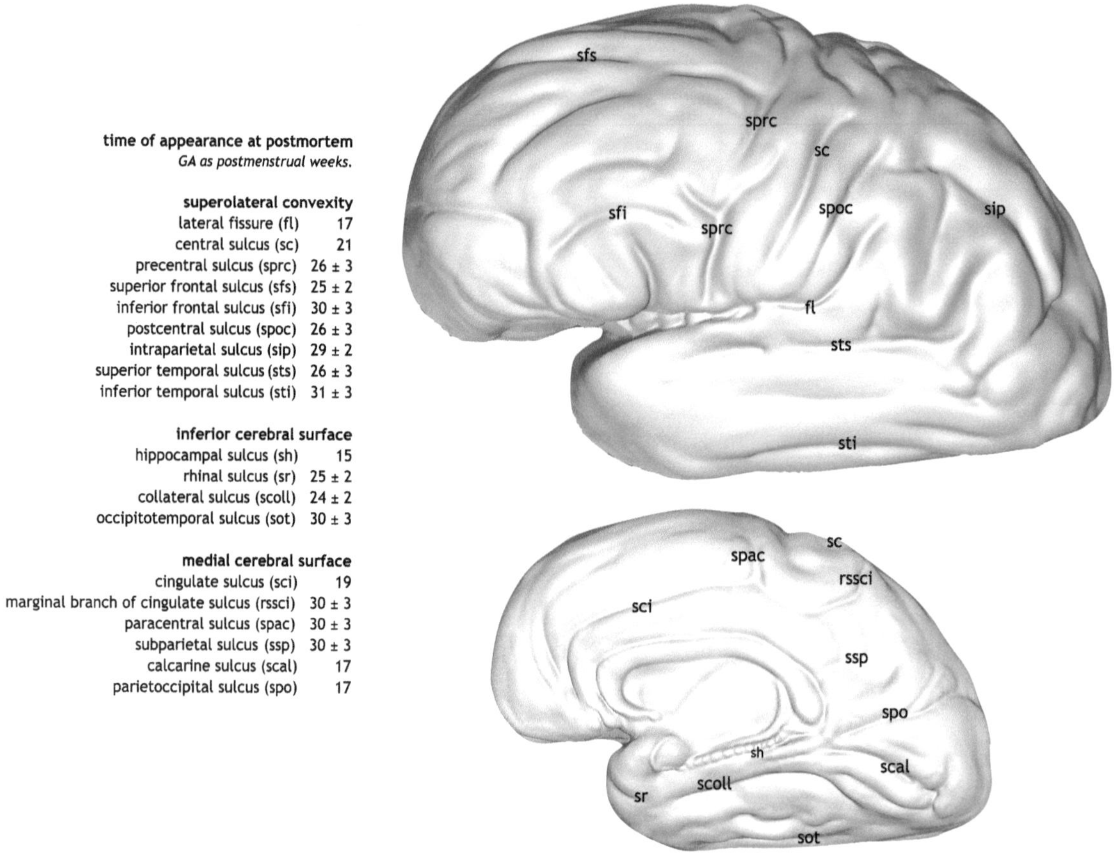

Fig. 14.11. Timelines of primary sulci at postmortem.

fissure at the 8th and the lateral fissure at the 19th week. The lateral fissure, in fact not a parenchymal sulcus but the result of opercularization of the developing insula, develops after 15 weeks. By 22 weeks, calcarine and parieto-occipital sulci should be visible with ultrasound. The frontal part of the cingulate sulcus, the central and calcarine sulci appear on cranial ultrasound before the 24th week. Toi et al. (2004) studied sulcation with transabdominal ultrasound in 50 normal fetuses, postmenstrual ages ranging from 15.6 to 29.6 weeks: the earliest PMA at which specific sulci could be seen in any fetus were as follows—parieto-occipital fissure at 18.5 weeks, calcarine sulcus at 18.5 weeks, cingulate sulcus at 23.2 weeks, and convexity sulci at 23.2 weeks; the PMA at which these sulci were always visible were as follows—parieto-occipital fissure >20.5 weeks, calcarine sulcus >21.9 weeks, cingulate sulcus >24.3 weeks, and convexity sulci >27.9 weeks; the circular sulcus at the margin of the insula was initially smooth but became angular after about 17 weeks by parieto-temporal opercularization, which means that initially the insula/operculum angle in an axial plane is obtuse and that an acute angle is first evident at 23.2 weeks and in all fetuses older than 24.5 weeks. Pistorius et al. (2010) described feasibility of mapping folding in fetuses, using practical scores per different sectional plane. In routine clinical practice, evaluation of sulcus parieto-occipitalis, calcarinus, and cinguli is possible, in addition to measurements at the insula of Reill and lateral fissure. In preterm infants a composite ultrasound score has also been proposed but not validated in prospective research (Murphy et al., 1989).

Sulcus cinguli develops on neonatal ultrasound from 24 weeks onward as single or multiple but discontinuous lines; it becomes one line at 28–33 weeks first in its anterior part and develops multiple branches after 33 weeks. The cingulate sulcus begins below the rostrum, and it sweeps around the genu paralleling the CC. It separates the medial frontal gyrus from the cingulate gyrus below. It ends with a constant supramarginal branch cutting in on the convexity of the parietal lobe, separating precuneus from the paracentral lobule. The marginal ramus has a constant relationship with the central sulcus, being formed between 24 weeks and term. The cingulate sulcus appears duplicated in about one-third of the hemispheres, mainly in its anterior segment, by the presence of a paracingulate sulcus (Paus et al., 1996; Spasojevic et al., 2010). This doubling of the anterior cingulate sulcus occurs twice as often on the left versus the right. Interruptions are frequently (around one-third) noted along its course. These interruptions lead to transitional gyri between mesial frontal cortex and cingulate gyrus, the "plis de passage fronto-limbiques of Broca." A small posterior branch points in the direction of the subparietal sulcus with which it may anastomose. A branch (paracentral sulcus) extending up from the cingulate sulcus in the region in front of the paracentral lobule is present in about half of the hemispheres.

The *pericentral sulci* mainly develop between 20 and 24 weeks (Afif et al., 2014), and the sensorimotor cortex develops in stages (Table 14.3). The central cerebral sulcus is first identified (at postmortem) at 18 weeks, the right appearing before the left; it is first identifiable in its inferior part but develops contemporary to but independent from the central insular sulcus. In early stages the projection from the inferior extremity of the central sulcus is located anterior to that of the superior extremity of the central insular sulcus. The precentral cerebral sulcus is first observed at 20–22 weeks, usually in two separate parts, and the postcentral sulcus at 22–24 weeks. On CUS, they make their appearance 1 or 2 weeks later. According to Cunningham (1892) the sulcus centralis usually cuts the upper convexity in adults; in two-fifths cases, it ends near or just at the convexity. On the mesial hemisphere, when present, the sulcus ends with a backward curve, never joining the ramus supramarginalis of the sulcus cinguli, the large sulcus that ends behind it and invariably cuts the convexity to reach the lateral hemisphere. The sulcus centralis nears the convexity at around 30 weeks of gestation. At the lower end the sulcus centralis ends just above the lateral fissure in the majority; in one-fifth cases, it is connected to the lateral fissure. In most fetuses the lower two-thirds of the sulcus form from a large early shallow fissure, later followed by a deep pit or depression for the upper third. This deeper depression near the convexity joins the lower larger fissure soon after appearance. The elevated intervening part has been referred to as the "pli de passage frontoparietal moyen" by Broca. It is this elevation that may remain relatively large and high in some brains, leading to division of the central groove (De Bisschop et al., 2019). The elevated portion corresponds to an area of motor development for the contralateral hand. Although a comparable double origin exists for the precentral sulcus, the two precentral sulci in general remain separate throughout development.

The *insula of Reil*, initially exposed, develops at the cleavage between the frontal and temporal lobes in which branches from the carotid artery reach the brain surface in early fetal life. Later on, opercularization, the covering of the insula by rapidly growing cerebral parenchyma, renders the insula at term no longer visible from the surface. Insular sulci are visible by 34 weeks on ultrasound, but develop between 28 and 34 weeks, usually two short sulci anterior to the central sulcus. At 24 weeks PMA the surrounding cortex overlaps the posterior part of the sylvian region, with only partial covering of the anterior insula and the anterior peri-insular sulcus. The superior part of the anterior insula is gradually hidden by development of the fronto-orbital opercula. The central cerebral artery arises from the central insular artery (branch of the superior division of the MCA).

Clinically relevant variations: Although it is well acknowledged that cortical folding varies between individuals (Turner, 1948; Chi et al., 1977; Naidich et al., 1994), and that the right hemisphere usually matures before the left with respect to gyration (Chi et al. (1977) on the basis of postmortem findings, Kasprian et al. (2011) using fetal MRI, Afif et al. (2014) on the basis of postmortem findings, Zhang et al. (2015) based on MR findings), there are few perinatal in vivo reports on the importance and cause of such individual variation outside of the context of disorders of migration (see later). The exception is the study of cingular groove development with CUS by Slagle et al. (1989). The discussion of cerebral sulcal asymmetries is beyond our

Table 14.3

Stages of development of central and insular sulci.

Afif et al. (2007, 2014)	Central sulci	Insular sulci
13–17 weeks GA		Stage 1: Appearance of the first sulcus
18–19 weeks GA	Stage 1: Appearance of the inferior part of the central cerebral sulcus	Stage 2: Development of the peri-insular sulci
20–22 weeks GA	Stage 2: Development of the pericentral lateral regions and beginning of opercularization	Stage 3: Central sulci and opercularization of the insula
24–26 weeks GA	Stage 3: Development of parietal and temporal cortices and covering of the postcentral insular region	Stage 4: Covering of the posterior insula
27–28 weeks GA	Stage 4: Maturation of the central cerebral regions	Stage 5: Closure of the posterior sylvian fissure

scope. The study of clinically relevant variation of primary sulci is, even with MRI, still in a very early stage.

Abnormal development

Changes induced by lesions: Focal cortical brain injury can directly alter final gyral morphology during either primary or postprimary gyration. Any significant lesion of the cortex itself, be it an arterial or venous infarct, watershed injury, inflammation, or other, changes gross cortical morphology. In postmortem descriptions, this is referred to as ulegyria, i.e., atrophy and alteration in shape of gyri (Friede, 1989; Paneth et al., 1994). When lesions to the cortex occur before the end of the brunt of neuroblast migration (around 25 weeks), they lead to focal interruption of cortical formation, histologically often referred to as polymicrogyria, the existence of far too many erratic small gyri in a specific region (Friede, 1989).

Deep lesions may, however, also alter the cortical plate above it. Perinatal human clastic deep brain lesions, below subplate and cortex, like a medullary venous infarct in a preterm infant at 23–25 weeks, can cause macroscopically recognizable cortical dysplasia in the form of polymicrogyria (Govaert et al., 2004). At postmortem, changes in the cerebral cortex have been reported above bilateral deep WM lesions referred to as leukomalacia (Marin-Padilla, 1996, 1997), now lumped under the umbrella "punctate white matter lesions." A maturational delay but not an alteration in shape (compared with the normal detailed description from a discontinuous sulcus around 26 weeks to continuity by 30 weeks and branching in the following 4 weeks) was documented by postnatal sonographic study of the developing cingular groove following unilateral medullary venous infarction in preterm infants with GA below 32 weeks (Slagle et al., 1989).

Changes by preterm delivery: MR has described diffuse altered gyrification in preterm infants. Using MRI with dedicated postprocessing (mainly using the inner cortex to WM border), it has been demonstrated that twins have a delayed but harmonious maturation, with reduced surface and sulcation index compared with singletons, whereas the gyrification of IUGR newborns is discordant to normal development, with a more pronounced reduction of surface in relation to a given sulcation index (Dubois et al., 2008). Using global surface measurements, an effect of preterm birth on cortical folding can be shown (Lefèvre et al., 2016). MR studies start to focus on specific regional variations (Kersbergen et al., 2016), confirming the higher pace of right versus left sulcation of postmortem studies. Since long, postmortem studies have demonstrated that hemispheres in the third trimester are already asymmetric at the level of Heschl gyri, planum temporale, and superior temporal sulcus (STS). In a cross-sectional neonatal MRI study, Dubois et al. (2010) aimed to blindly and automatically map early asymmetries over the immature cortex in 25 newborns from 26 to 36 weeks of GA. Asymmetries were highlighted in three specific cortical regions: a deeper sulcus temporalis superior on the right side, a larger posterior region of the lateral fissure on the left side, and ipsilateral to a larger left planum temporale. Such a study confirms in vivo that perisylvian regions are asymmetric from early on, suggesting their anatomical specificity for the emergence of functional lateralization in language processing prior to language exposure.

Gyrification theories and relation to the perinatal period

Mechanistic explanations for cortical folding tend to focus either on forces external to the developing cortex or on intrinsic forces (Fig. 14.12). The latter models tend to emphasize axonal tension, differential proliferation, or differential tangential expansion of developing structures that are bonded to one another. Various explanations are not mutually exclusive (Striedter et al., 2015). Cortical folding starts around the 16th week of gestation and increases rapidly during the first trimester; the GI reaches a transient maximum between weeks 66 and 80 postconception. From this time on the GI declines by 18% from a maximal value of 3.03 to the adult level, which is reached at an age of almost 23 years. The decrease in GI continues into adulthood, whereas perinatal overshooting of synaptic density seems to reach a plateau much earlier than the GI. The first cortical folds usually develop into the final deepest parts of sulci and are termed sulcal pits on fetal MRI. A sulcal pit can be identified in a sulcal catchment basin by using structural information of small connecting gyri (the focal elevation of the sulcal bottom, also called an annectant gyrus, submerged or not), as described above for the "pli frontoparietal moyen" in the central groove. The first major folds appear to show greater spatial invariance during development as they deepen and have a stronger spatial covariance with functional areas under closer genetic control than later developing sulci. Invariant sulcal pit distributions across individuals may be due to the stability of the human-specific protomap. The clinical relevance of the relation among annectant gyri, sulcal pits, and definitive sulci constitutes a distinct research field.

Insights into the role of the outer SVZ in gyral development (Lewitus et al., 2013; Zilles et al., 2013; Striedter et al., 2015) explain how deep brain lesions may indeed alter primary gyration. Primary gyration is driven by genetic mechanisms that control proliferation of cells

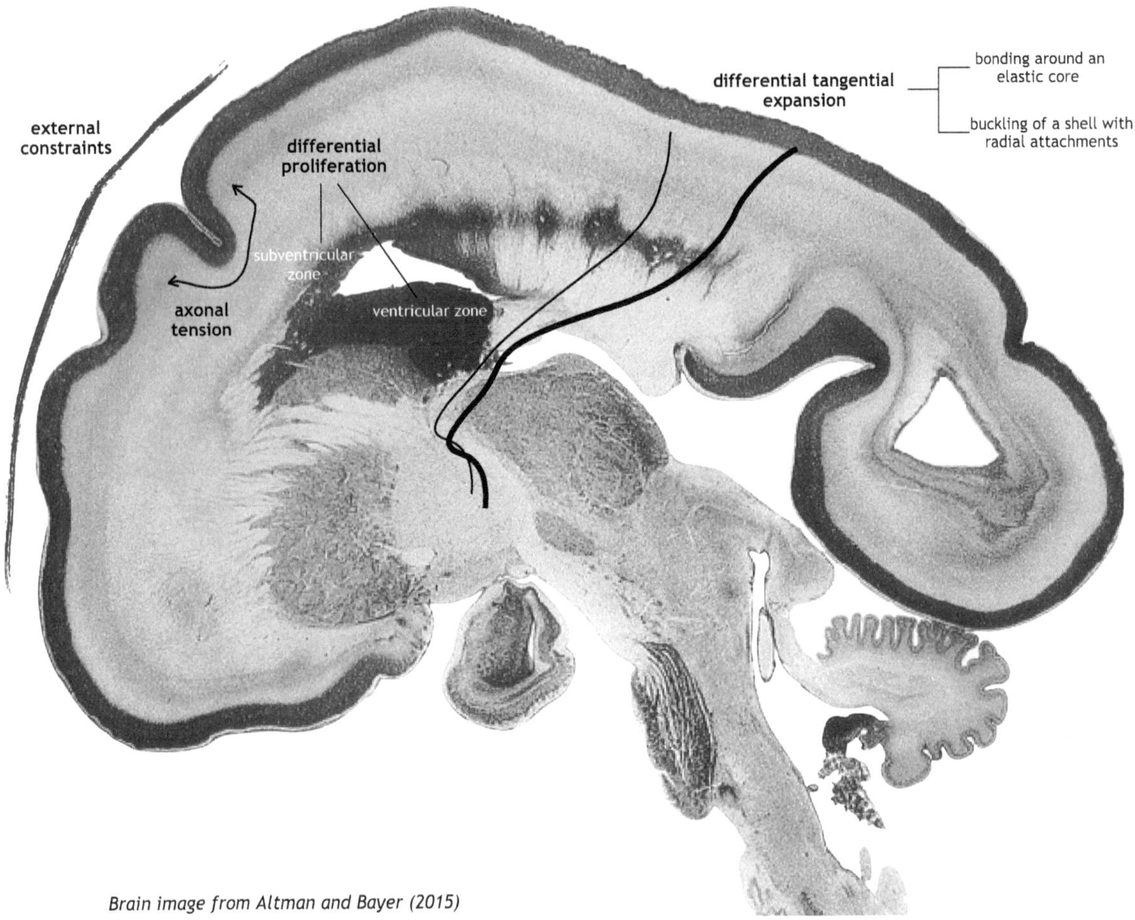

Fig. 14.12. Gyrification mechanisms.

in the *outer SVZ*. A discrepant conical expansion of cells in this zone occurs under future gyral crowns. The number of cells produced and their spreading lead to gyral folding. Basal radial glia are progenitors forming transit-amplifying progenitors in the OSVZ. Beyond 32 weeks, postprimary sulcation continues until around 1 year of age after term birth and is also driven by external, in addition to intrinsic, genetic factors. Sulcal length and width increase in this stage. Tension in immediate subcortical WM then becomes important, whereas such tension is not (or less?) relevant to primary gyration. Primary gyration does co-occur with early development of long associative tracts but is not related to myelination. In evolution, there seems to be an advantageous trend that favors folding over cortical thickening. The stage of likely interference with primary gyration is in the fetal (20–23 weeks) and early preterm (24–32 weeks) phases of cerebral development when thalamocortical, callosal, and cholinergic afferents first connect to outer subplate targets and wait before connecting targets in the deep cortical plate (Kostovic and Jovanov-Milosevic, 2006). This is a period of massive cell production in ventricular and SVZs, during which the subplate is large due to abundant extracellular matrix and early synapse formation and when surface negative large EEG transients emerge from this subplate. It is also a period of vigorous structural plasticity. It precedes corticocortical connectivity at around 33–35 weeks of GA and later reorganization following disappearance of transient structures. The time window in which interference can exist with gyration was determined by experimental study. It was demonstrated by applying frontal cortical resections in primates during gestation that not only local but also distant changes (in the remote occipital areas on both sides, even with unilateral lesions) occur in the formation of primary sulci: new and deep sulci appear, consequent not to abnormal prolongation of neurogenesis but to rerouting of thalamocortical fibers coming from the mediodorsal thalamic nucleus (Goldman-Rakic, 1980). This remote change in primary sulci occurs only when fetuses are lesioned before E124 for rhesus monkeys (equivalent age around PMA 30 weeks in humans), because mediodorsal thalamic nuclei transsynaptically whither when their cortical target is lesioned after that moment. In summary, there is need to focus on the effect of deep brain lesions on primary gyration.

DEVELOPMENT OF WHITE MATTER (WM)

Cerebral mantle layers

Between 24 and 32 weeks of GA, ongoing events comprise growth of axons, maximal growth of the subplate, and entrance of thalamic afferents into the developing cortex (Fig. 14.13).

Layering of the cerebral mantle, including development of WM and its tracts, follows complex cell production in GM (see earlier) (Kriegstein and Alvarez-Buylla, 2009). Specifically, during early brain development, neuroepithelial cells divide in both a symmetric and an asymmetric fashion to regenerate the stem cell compartment of the VZ and generate neurons. Neuronal migration leads to formation of the primordial plexiform

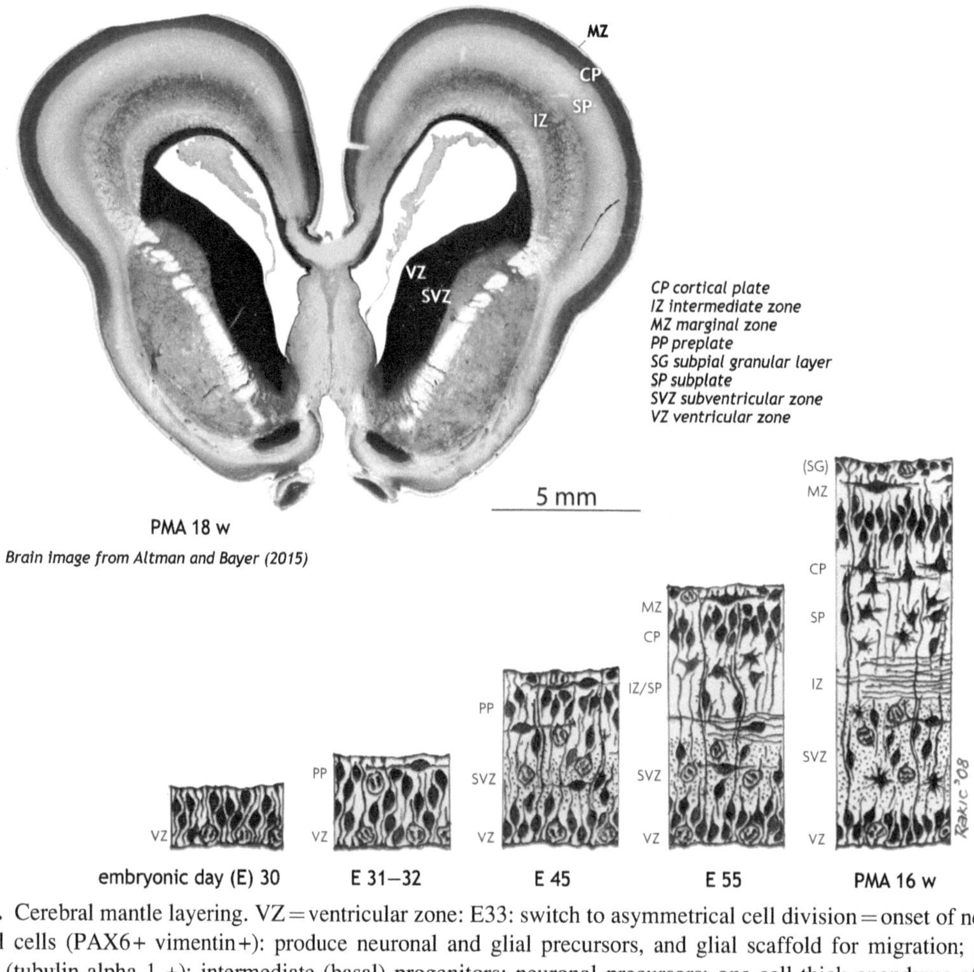

Fig. 14.13. Cerebral mantle layering. VZ = ventricular zone: E33: switch to asymmetrical cell division = onset of neurogenesis; radial glial cells (PAX6+ vimentin+): produce neuronal and glial precursors, and glial scaffold for migration; short neural precursors (tubulin-alpha 1 +); intermediate (basal) progenitors: neuronal precursors; one-cell-thick ependyma after GW 27. PP = preplate: E31: predecursor cells; E40: radial and tangential in-migrating Cajal–Retzius cells (CR); later: GABAergic and glutamatergic interneurons before arrival of first cortical plate cells; CR cells produce reelin, have horizontal axons that synapse with arriving pyramidal neurons; at GW 11: formation of subpial granular layer (SG) to spread all over cortical plate by GW 14, to regress after GW 28 and disappear before term (human specific). SVZ = subventricular zone: multiple symmetrical divisions to produce neurons and glial cells; after GW 20 divided into outer and inner SVZ by tangential incoming fibers. CP = cortical plate: gradient of development from rostrolateral to dorsocaudal; inside-out settlement of migrating cells; first cells in situ at GW 8–9; peak of migration 3–5 months of pregnancy; differentiation > GW 22; six layers by seventh month; projection neurons from VZ, association and other neurons from SVZ; GABAergic neurons from cortical plate (radial) and subpallial GM (tangential). IZ/SP = intermediate/subplate zone: no proper boundary between IZ and SP; contains migrating cells and long-range axons; GW 9 markers of synaptic activity in IZ below early CP = early SP activity; GW 10: descending axons from subplate guide thalamic afferent axons; synaptic contacts at E50 = waiting compartment; considerable thickening of SP at GW 9–12, mainly in association with cortical areas; columnar aspect around GW 12; maximum thickness at around two-thirds of gestation; 6 months after birth interstitial WM cells only remaining; IZ = later white matter. From Bystron, I., Blakemore, C., Rakic, P., 2008. Development of the human cerebral cortex: Boulder Committee revisited. Nat Rev Neurosci 9, 110–122.

layer or preplate (PP) located just underneath the pia mater (Bystron et al., 2008). Subsequent migrating neurons form the cortical plate (layers II–VI) in an "inside-out" manner splitting the preplate in an outer layer, the "marginal zone" (MZ or layer I) containing the Cajal–Retzius cells, and in the subplate (SP). Beneath the SP the intermediate zone (IZ) is mainly formed by afferent and efferent fibers while just underneath it the SVZ is constituted by actively proliferating cells that give rise to neuronal IPCs and glia mainly in late gestation and in postnatal life (Nadarajah et al., 2002; Ayala et al., 2007).

As cells migrate between GW 9 and 20, there are concentric fields that become apparent in future WM: stratified transitional fields (Altman and Bayer, 2015). Early in this period, Cajal–Retzius cells, present in the subpial layer, orchestrate cortical plate formation. This is also the period of formation of fibers in corticofugal, thalamocortical, and commissural networks; diffusible molecules and axonal growth and sprouting are intense during this period of cell migration. Early fibers are the optic radiation and the sensorimotor tracts to the future pericentral area. Cells sojourning and forming early fiber contacts form six transitional fields: STF 1, deep WM; STF 2, large pyramidal neurons for layer V and VI; STF 3, layer IV granular neurons (with columnar "honeycomb" appearance in some areas); STF 4, fibers in different directions; STF 5, small pyramidal cells for layers II and III; and STF 6, callosal fibers. Between 10.5 and 13 weeks of GA, thalamocortical fibers from the IZ are initially dispersed throughout the SVZ. Thalamocortical axonal "invasion" takes place between 12.5 and 15 weeks of GA and is followed by callosal fiber invasion, which separates the inner from the outer SVZ (Isasegi et al., 2018).

Oligodendrocyte precursors, the major myelin forming cells, crucial to WM maturation, undergo changes in morphology and antigenic profile expressing several different superficial antigen overtimes. These cells are subject to stage-specific vulnerability (preoligodendrocyte stage) being selectively affected by different insults based on the maturation stage and with glutamatergic excitotoxic cell death as common pathway (Butts et al., 2008; Murugan et al., 2013; Kuhn et al., 2019).

Major forebrain connections

After appropriate migration the excitatory projection neurons born in the VZ of the dorsal telencephalon exert their function through several types of long-distance projections: corticocortical callosal projections, corticospinal projections, corticothalamic projections, and thalamocortical projections (Leyva-Díaz and López-Bendito, 2013). The bundles of WM are subject to complex trajectories along their travel through the CNS; this implies underlying processes of neuronal specification, axonal guidance, and midline crossing.

Corticocortical callosal projections: The hemispheres communicate through three forebrain commissures, namely CC, anterior commissure (AC), and hippocampal commissure (HC). CC is formed by axons from pyramidal neurons (callosal projection neurons) organized in two layers: (1) the deep layer neurons are characterized by a dual axonal branch projecting to the internal capsule and to the ipsi- or contralateral cortex and (2) superficial layer neurons extend local collaterals to the deep and superficial layer neurons of the CP as well as commissural homotopic axons into the contralateral cortex through CC (Fame et al., 2011). A key regulator of callosal neuron specification is Satb2, a DNA-binding transcription factor that represses Ctip2 transcription factor. As they migrate toward the cortex, these neurons already start to direct their axons away from the cortical plate; this mechanism is in part mediated from Sema3A signals that are expressed in a gradient with progressively lower concentrations toward the midline in the cortex (Zhao et al., 2011). Around 10 weeks in humans, progressive deepening of the interhemispheric fissure forms the sulcus medianus telencephalic medii, causing approximation of cortical plates at the corticoseptal boundaries. At this level, glial cells migrate from the SVZ to form the midline zipper glia (week 12) to function as an interhemispheric glial bridge for pioneering callosal axons. Besides this, two other glial structures are essential to correctly convey the axons toward the midline zipper, namely the induseum griseum glia and the glial wedge (Raybaud, 2010a, b); both release repulsive molecules (Slit2, Draxin, Wnt5a) to correctly help axons cross the midline (Fothergill et al., 2014).

Corticospinal projections: CST fibers derive from large pyramidal neurons (corticospinal motor neurons) located in layer V of the sensorimotor area and mainly project to the spinal cord (direct and crossed pyramidal tracts) and to a lesser extent to striatum and brainstem nuclei. The identification of a large set of specific corticospinal motor neuron genes suggests that fate determination relies on expressed transcription factors and other molecules: *Fezf2* (Molyneaux et al., 2005), *Ctip2* (Arlotta et al., 2005), and *Sox5* (Lai et al., 2008). A small number of pioneering axons form the forefront of the CST. After axonal fate specification, they travel from the cortex into the subcortical regions toward the spinal cord. The trajectory of these axons is modulated by different cues at several specific anatomical choice-points (Sakai and Kaprielian, 2012). Once fibers reach the internal capsule, Slit1 and Slit2 signals (acting on ROBO1 and ROBO2 receptors expressed in developing CST axons) guarantee dorsoventral patterning of the fiber bundles and prevent axonal dispersal across the midline (Lòpez-Bendito et al., 2007). Several important molecules are implicated in the pyramidal decussation such as the Sema6A-PlexinA3/A4 mediated repulsion, Netrin-1 and Unc5 with the cell adhesion molecule L1, and neural cell adhesion molecule (NCAM). Insulin-like growth factor and a gradient of Wnt1-5a regulates the subsequent downward development of the corticospinal axons. A decreasing gradient concentration of EphA4 along the spinal cord is responsible for the appropriate exit of axons from the major axonal bundle toward the spinal gray matter at the level of the cervical and lumbar spinal cord enlargements (Leyva-Díaz and López-Bendito, 2013). The pyramidal tract reaches the level of the pyramidal decussation at the end of the embryonic period, i.e., at 8 weeks of development (Müller and O'Rahilly, 1990). Pyramidal decussation is complete by 17 weeks of gestation, and the rest of the spinal cord is invaded by 19 weeks (lower thoracic cord) and 29 weeks (lumbosacral cord) of gestation (Humphrey, 1960). Owing to this, protracted development disorders of the pyramidal tract may occur over the entire prenatal period: aplasia, hypoplasia, hyperplasia, destructive lesions, anomalies of crossing, and disorders of myelination (ten Donkelaar et al., 2004).

Corticothalamic and thalamocortical projections: Corticothalamic and thalamocortical axons represent an integrated system necessary for the regulation of the thalamic inputs coming from the sensory outer world (Figs. 14.14–14.18). Corticothalamic neurons are mainly located in layers V and VI of the CP, and their fate determination seems to be determined by the expression of several transcription factors (Fezf2, Ctip2, Sox5). Thalamic cells develop from progenitors located in caudal diencephalon where the mid-diencephalic organizer expressing Shh molecule acts as the principal local organizer for thalamic development and cell fate specification (Vue et al., 2009). Thalamic axons terminate in several layers of the cortical plate (layer IV mainly but also layers I–II/III and VI) to relay peripheral sensory inputs to specific areas of the cortex (Leyva-Díaz and López-Bendito, 2013). Specific subsets of molecular repellents and attractants work together as guidance cues for the developing axons while traveling toward their targets, either cortical or thalamic (Leyva-Díaz and López-Bendito, 2013). Corticothalamic axonal reciprocity is largely established by 30 weeks PMA.

DEVELOPMENT OF CEREBELLUM

The cerebellum has a superficial, multifolded, trilaminate cortex overlying a core of WM. Although the size of the cerebellum is only one-ninth that of the cerebrum, it holds almost two-thirds of the central nervous system neurons (Barton and Venditti, 2014). From early on the cerebellum has been associated with motor function: motor learning, motor coordination, posture, and balance (Glickstein et al., 2009). Studies have repeatedly confirmed that cerebellum is also involved in nonmotor functions, including cognition, language, attention, and behavior (Schmahmann, 2019). Both reduced prenatal cerebellar growth and prenatal cerebellar injury are linked to impaired neurodevelopmental outcome (e.g., autism, ADHD, dyslexia) (Wang et al., 2014). In humans, cerebellar development continues during the first year of life as it is one of the latest brain structures to complete its neurogenesis. Given this prolonged development, cerebellum is particularly sensitive to clinical and environmental variables that have an impact on early intra- and extrauterine brain growth (Koning et al., 2017).

Cerebellar structure and function

The cerebellum has three major divisions: the vermis, the paravermis (defined as two narrow intermediate regions adjacent to the vermis), and two hemispheres on either side (Voogd, 2002; Rahimi Balaci et al., 2017) (Fig. 14.19 and Table 14.4). The cerebellar peduncles, located on the ventral surface of the cerebellum, are three pairs of large fiber tracts connecting it to the rest of the brain. The inferior cerebellar peduncles connect cerebral hemispheres, brainstem, and spinal cord to the cerebellum. The middle cerebellar peduncles connect the pons to the cerebellum, and the superior cerebellar peduncles, major output tract, connect the cerebellum to the midbrain and diencephalon (Voogd, 2002). Most of the efferent fibers in the superior cerebellar peduncle decussate in the midbrain and pass via the contralateral red nucleus to reach the contralateral thalamus. In a rostrocaudal direction, fissures divide the cerebellum into three lobes: anterior, posterior, and flocculo-nodular. The first two are divided further into lobules and folia, increasing the cerebellar surface. From a phylogenetic view, another threefold division can be made: the neocerebellum (most of the hemispheres), the paleocerebellum (vermis and a small part of the hemispheres), and the archicerebellum (the flocculo-nodular lobe) (Nieuwenhuys, 1967; Stoodley and Schmahmann, 2018). The cerebellum has a three-layered cortex that surrounds an inner core of WM. There are four pairs of cerebellar nuclei, often improperly named "deep" nuclei (Haines et al., 2003), which are arranged (from lateral to medial) in the following order: dentate (most lateral), emboliform, globose, and fastigial nuclei (most medial). They receive inhibitory input from Purkinje cell axons. They are the "output neurons" of the cerebellar cortex.

In the 18th century, Luigi Rolando reasoned that cerebellum had the right structure to serve as "brain battery,"

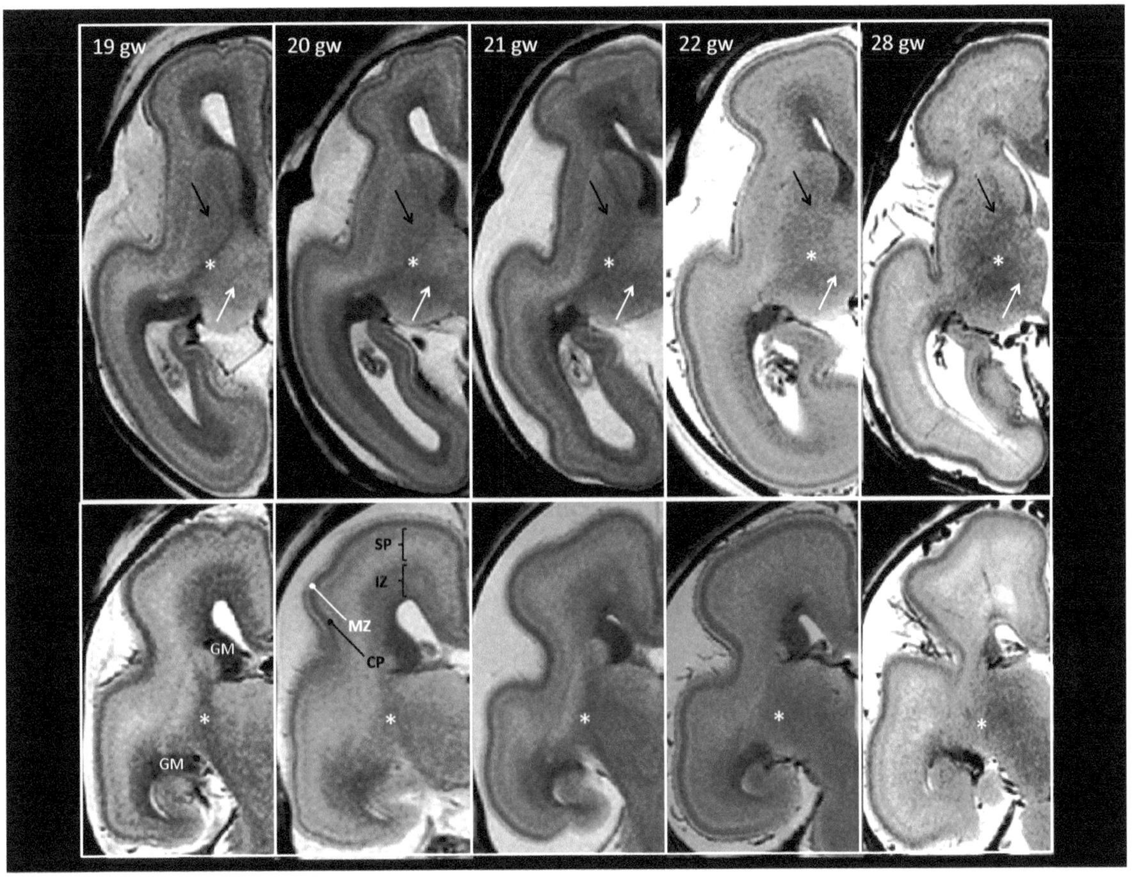

Fig. 14.14. (*Top*) Postmortem magnetic resonance imaging (PMMRI) performed on a 3.0 T scanner of nonfixed in situ fetal brains assessed within 24 h of death. Axial T2-weighted (T2W) fast spin-echo (FSE) images performed on fetuses from 19th to 28th GW studied after therapeutic termination of pregnancy (TOP) due to extracranial abnormalities or after spontaneous intrauterine death. From left to right are depicted 19th, 20th, 21st, 22nd, and 28th GW brain fetuses, respectively. The laminar organization of the cerebral hemispheres and signal changes of the PLIC with respect to the surrounding brain structures at different gestational ages can be seen. Between 19 and 20 GW the PLIC (*) is clearly more hypointense compared with the adjacent thalami (*white arrow*) and the lentiform nuclei (*black arrow*). From 21 to 22 GW, due to the progressively more hypointense signal of the mesial part of the lentiform nuclei and the anterolateral part of the thalami, the contrast between the PLIC and these structures decreases. At 28 GW the PLIC is relatively hyperintense compared with the thalami and nuclei pallidi. (*Bottom*) Coronal T2W-FSE PMMRI images of the same fetuses as in Fig. 14.1, in which the signal changes of the PLIC (*) with respect to the adjacent brain structures between 19 and 28 GW can be seen again. The laminar organization of the cerebral hemispheres at different GW is clearly visible as well. From 19 to 28 GW the cerebral hemispheres show transient layers from the pia to the ventricle, which becomes progressively less separate and more blurred into each other. From 19 to 22 GW the cerebral hemispheres show the following transient layers: marginal zone (MZ), cortical plate (CP), subplate (SP), and IZ. The SP is no longer visible at 28 GW, when a clear demarcation between the SB and the IZ does not exist anymore. The markedly hypointense layer recognizable along the ependymal surface of the caudothalamic groove and at the level of the temporal horn of the lateral ventricle corresponds to the GM and is well evident from 19 to 28 GW although progressively smaller.

given i's overlapping laminae (the so-called arbor vitae) (Coco and Perciavalle, 2015). Since then, cerebellar injury has been associated with motor learning and control. The cerebellum remains essential for its role in controlling purposeful, smooth movements: cerebellar circuits receive information of motor planning from the cerebral cortex, and it integrates this with feedback information from the brain stem and the spinal cord about the sensory consequences of motion. Although some reports described cerebellar injury and consequent nonmotor disabilities, up to the 1990s it was almost universally believed to be primarily involved in movement (Schmahmann, 2019). The "cerebellar cognitive affective syndrome," described by Schmahmann and Sherman (1998), is characterized by language deficits, problems with executive functioning, difficulties in spatial cognition, and even changes in personality. To take part in cognitive functions the cerebellar cortex receives

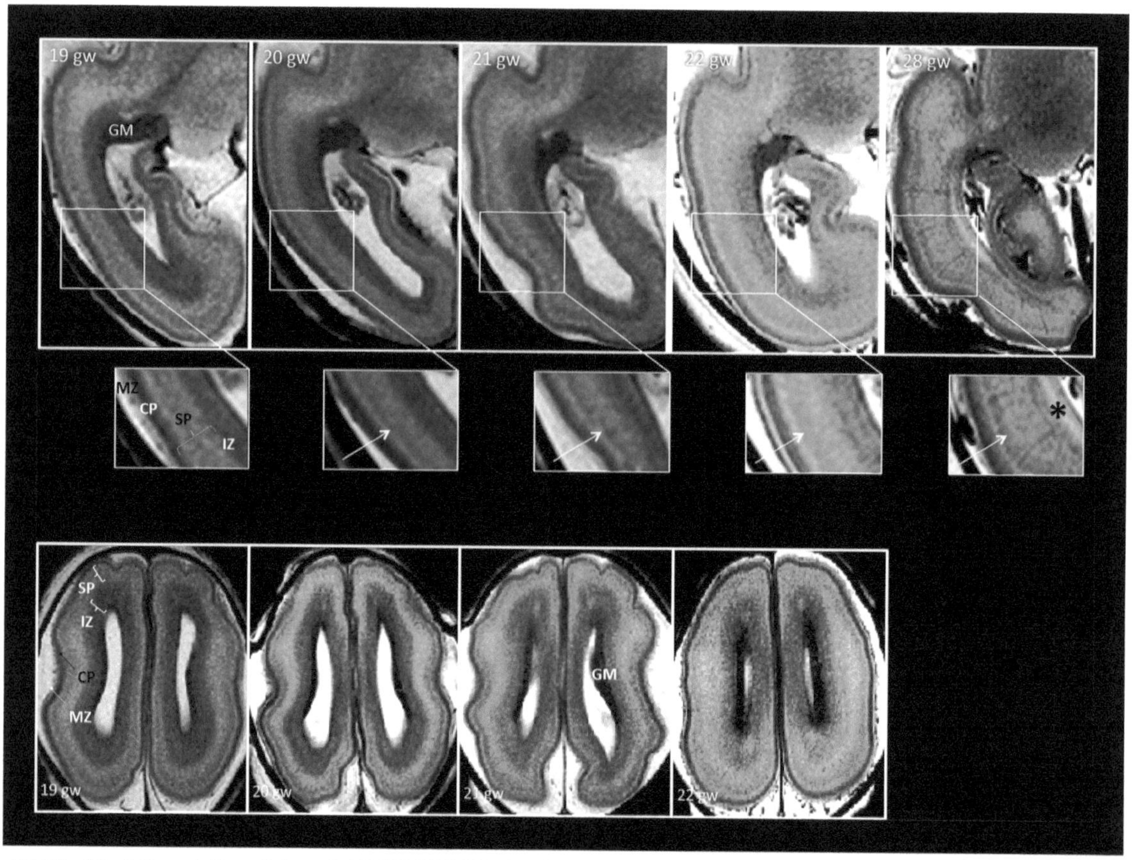

Fig. 14.15. (*Top*) From left to right, axial T2W-FSE PMMRI images of fetuses of 19, 20, 21, 22, and 28 GW, respectively, who underwent TOPs for extracranial abnormalities or intrauterine death. A detail of the cerebral parenchyma is shown in the boxes. The laminar organization of the fetal brain parenchyma is clearly visible. From the pia to the ventricles are shown MZ, CP, SP, and IZ. The GM in well documented as a hypointense layer at the level of the ependymal surface of the temporal horn of the lateral ventricle. The MZ (layer I) is the outermost and thinner layer of the fetal brain parenchyma. It hosts the Cajal–Retzius cell, and its complex organization is due to the presence of a network of cortical inputs from both Cajal–Retzius neurons and afferents to dendrites of CP neurons. The CP is displayed as a T2 low signal intensity rim just below the hyperintense MZ. CP is clearly visible from 19 to 28 GW. It is composed of postmigratory neurons. Underlying the CP is the SP. It is mainly composed of growing cortical afferents migrating toward the CP and abundant extracellular matrix, which is responsible of the high T2W signal. From 19 to 22 GW, it is characterized by a classic three-layer aspect. The most superficial and thin layer borders the CP, while the deepest one borders the IZ; they are both T2 hyperintense. The intermediate layer (*arrow*) is a thin T2 hypointense rim; it represents the thalamocortical axons in the superficial layer of the SP, which are waiting to enter the CP. This hypointense rim is still visible at 28 GW while at this time the SP is not recognizable anymore; the maximum contrast between SB and IZ is reached at 19–20 GW. The IZ is an hypointense structure disposed in the periventricular area. It is composed of growing axons, immature glial cells, and migratory neurons. At 28 GW, SP and IZ are no longer distinguishable; at this time a thin T2 hyperintense layer (*) appears to border the ventricle in the region of the future optic radiation. GM is well evident from 19 to 28 GW although progressively smaller. (*Bottom*) Axial T2W-FSE PMMRI images at the level of the lateral ventricles of fetuses from 19 to 22 GW. The transient laminar organization of the cerebral parenchyma can be seen clearly from 19 to 22 GW with the more conspicuous differentiation between layers at 19–20 GW. The classic three-layer aspect of the SP is still evident at 22 GW although the tissue contrast between layers has become less evident. GM is evident from 19 to 22 GW as a thick hypointense rim bordering the ependymal lining of the lateral ventricles.

input from multiple sites of the central nervous system (Stoodley and Schmahmann, 2018). It processes incoming information in a uniform manner, and the output of the cerebellum is via the cerebellar nuclei to a large number of different locations within the CNS. There are three major histological subdivisions in cerebellum: cortex, WM, and cerebellar nuclei. The adult cerebellar cortex has three histologic layers: the deep granular cell layer, the Purkinje cell layer, and the superficial molecular layer. The cerebellar cortex has a relative invariant histology compared with the cerebral cortex. Cerebellar cortical neurons and deep nuclei are linked in microcircuits

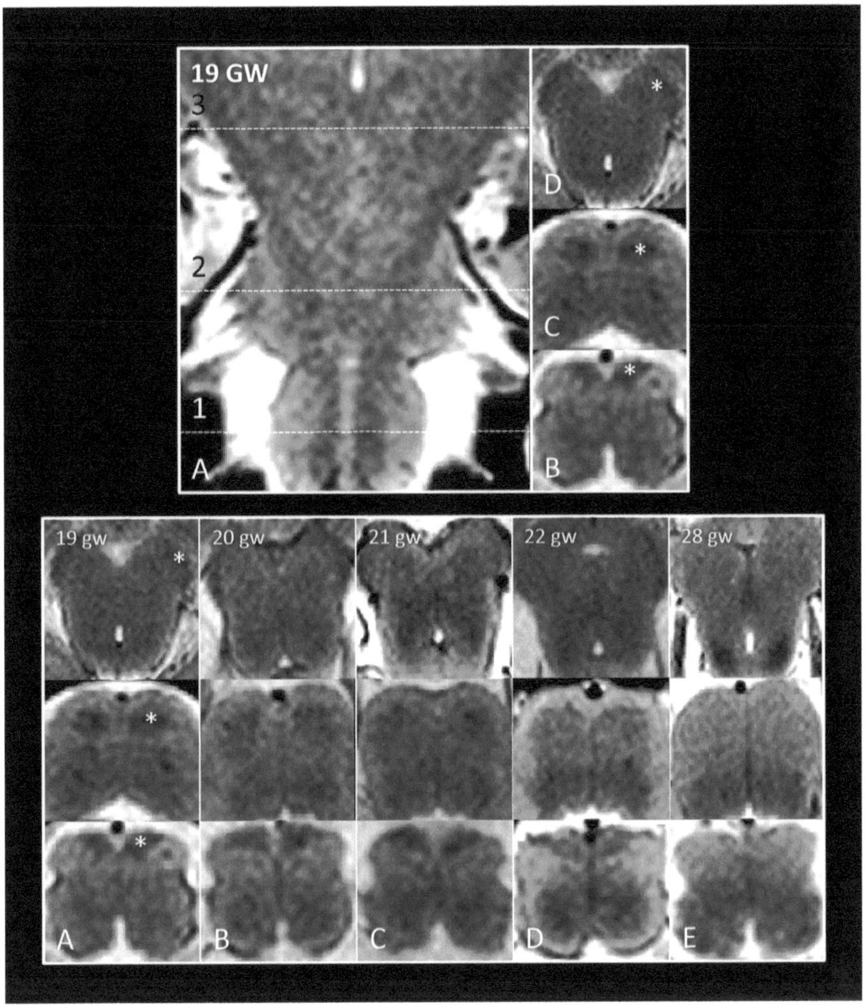

Fig. 14.16. (*Top*) (A) Coronal T2W-FSE PMMRI of a 19 GW fetus and corresponding axial images obtained at the level of lines 1, 2, 3 passing through the medulla (B), pons (C), and midbrain (D), respectively. The CST (*) is well recognizable from the surrounding parenchyma as a hypointense structure at any level of the brainstem. Its dark appearance on T2W imaging is not due to myelin deposition, which begins to develop later during the last phase of gestation, but rather due to immature oligodendrocytes and densely packed WM fibers. (*Bottom*) Axial T2W-FSE PMMRI images obtained at the level of medulla, pons, and midbrain, respectively (from bottom row to top row), of fetuses from 19 to 22 GW (A–D) and 28 GW (E). The CST, in particular at the level of the pons, is progressively less clearly discernible from the surrounding brainstem parenchyma from 19 to 28 GW. At 28 GW the CST is not visible anymore as it becomes isointense with its surroundings. From 19 to 28 GW the dorsal medulla and the dorsal pons appear hypointense.

(corticonuclear microcomplexes) (Ito, 2006). The superficial layer contains inhibitory interneurons (stellate and basket cells) and excitatory climbing fibers (see further). The Purkinje layer contains interneurons (candelabrum cells) and astrocytes (Bergmann glia). The granular layer (deepest layer) contains small excitatory neurons (granule cells, inhibitory Golgi cells, inhibitory Lugaro cells, unipolar brush cells) (Joyner et al., 2017). Scott (1963) and Hawkes and Leclerc (1987) showed that the cerebellar cortex has chemoarchitectonic zones. Zebrin antibodies (in the cerebellum, Zebrin II is expressed only in Purkinje cells) show a pattern of alternating zebrin-positive and -negative bands of Purkinje cells. Stripes are symmetrically distributed, highly reproducible in the same species, and conserved through evolution.

There are two main pathways informing cerebellar cortex (Gilman, 1985; Ito, 2006). Climbing fibers come from the inferior olivary nucleus and have direct synaptic contacts with Purkinje cells dendrites (Rahimi Balaei et al., 2017). Mossy fibers from pons branch extensively when entering the cerebellar cortex, and their fibers excite a group of granule cell dendrites. Both climbing and mossy fibers have collateral branches that excite neurons of the cerebellar nuclei. Axons of the granule cells project toward the cortical surface and when

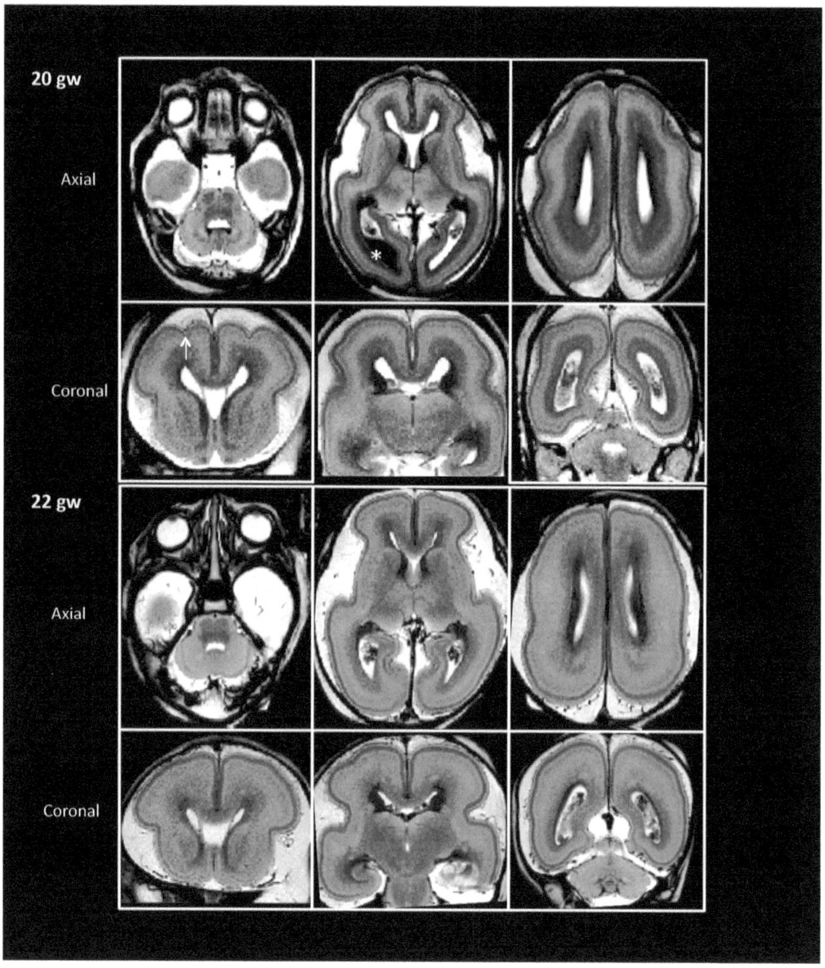

Fig. 14.17. (*Top*) Axial and coronal T2W-FSE MRI sections of a normal 20 GW fetus. The hypointense material in the right ventricular trigone (*) is blood from intracranial hemorrhage during delivery. The frontal cortical indentations are due to post-mortem artifacts (*arrow*). (*Bottom*) Axial and coronal T2W-FSE MRI sections of a normal 22 GW fetus.

reaching the molecular layer divide into a "T." Their axons run along the axis of the cerebellar folia (called parallel fibers), cross the dendritic arbors Purkinje cells, and make excitatory in passing (en-passant) synapses (Rahimi Balaei et al., 2017). The axons of the granule cells also deliver excitatory input to stellate, golgi, and basket cells. Since the input of a single parallel fiber is small, excitatory summation is needed to activate a Purkinje cell. Granule cells are connected to Purkinje cells in an order of approximately 100,000:1; furthermore, 40:1 is the ratio of Purkinje cells to deep nuclear cells. Purkinje cells seem to perform the main computation of the cerebellum (Gilman, 1985). They are arranged in a series of longitudinal parasagittal bands. Each band receives input from a cluster of inferior olive cells via climbing fibers and projects to a specific part of the cerebellar nuclei. Finally, output via the cerebellar deep nuclei is mostly high-frequency tonic excitation, directed toward the midbrain and the thalamus. This output is modulated by the firing pattern of inhibitory Purkinje cells. A distinct long-range inhibitory axon tract allows the cerebellar nuclei to influence the activity of the inferior olive.

Cerebellar development

The developing cerebellum is unique as it has two distinct germinal zones (progenitor zones) (Fig. 14.20). In other parts of the CNS the VZ produces all neurons and macroglia. In cerebellum the fourth ventricle VZ but also the RL produce neurons. The fourth ventricle VZ is the neuroepithelium of the alar plate of the rhombencephalon: it produces neurons that use gamma-aminobutyric acid (GABA) as neurotransmitter and also produces astroglia. The RL generates the excitatory neurons of the cerebellum (they use glutamate as neurotransmitter). The overabundant granule cells are unique in their ontogenesis as they are generated from granule cell

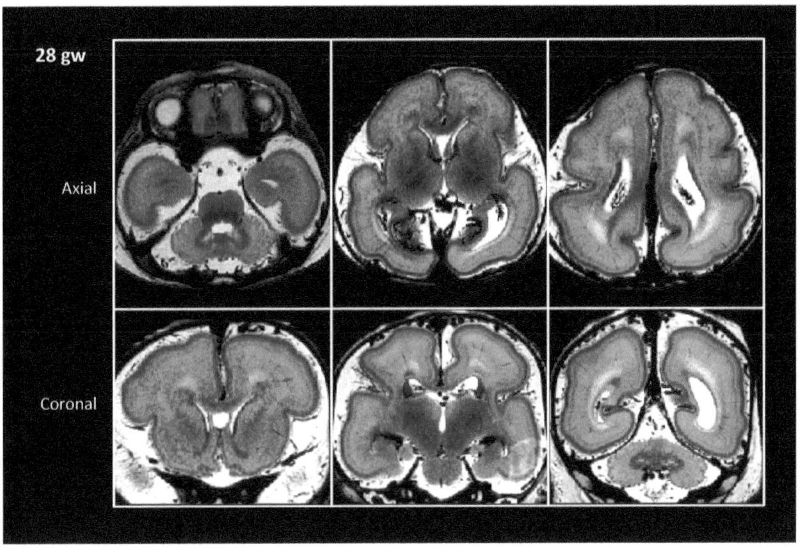

Fig. 14.18. Axial and coronal T2W-FSE MRI sections of a normal 28 GW fetus. The CP is still well recognizable as a hypointense rim. The three-layer aspect of the SP is not visible anymore, and there is no clear boundary between SP and IZ; this is due to the gradual disappearance of the extracellular matrix in the SP. The hypointense layer in the SP corresponding to thalamocortical axons is less delineated likely because of the migration of thalamocortical afferents in the CP. The periventricular areas with the immature periventricular WM become more hyperintense than surrounding parenchyma.

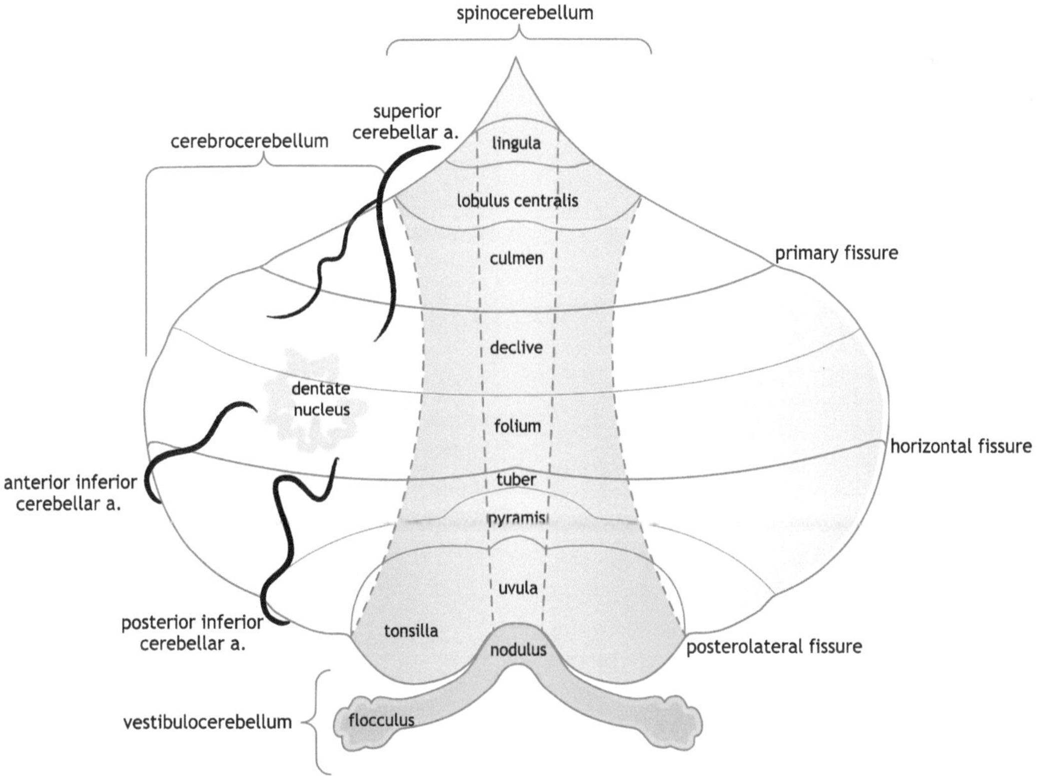

Fig. 14.19. Functional cerebellar anatomy.

Table 14.4

Anatomical divisions of the cerebellum.

Vermis lobule	Fissure	Hemisphere lobule
I, II	Precentral	HI, HII
III	Preculminate	HIII
IV	Intraculminate	HIV
V	Primary	HV
VI	Superior posterior	HVI
VIIAF	Horizontal	Crus I
VIIAT	Ansoparamedian	Crus II
VIIB	Prepyramidal/ prebiventer	HVIIB
VIIIA	Intrabiventer	HVIIIA
VIIIB	Secondary	HVIIIB
IX	Posterolateral	HIX
X		HX

Roman numbers are often used to identify lobules in the vermis and the lobules of the hemispheres are commonly named with the prefix H, followed by the Roman number corresponding numbers of the vermis. Each lobule is connected to the neocortex in its specific pattern and presumed to have its own functional role.

precursors, a secondary progenitor pool. New granule cells migrate inward and form the IGL across the entire cerebellum (Rahimi Balaei et al., 2017). Most oligodendrocytes of the cerebellum seem to originate from the VZ outside of the cerebellum.

Human cerebellar development begins in the embryonic period and continues into the first year after birth (Leto et al., 2016). The cerebellum develops from the dorsal portions (alar plate) of the metencephalon (rhombomere 1) and the neural folds (RLs) (Leto et al., 2016; Voogd, 2017; Haldipur et al., 2018). The rhombencephalon is divided into seven segments along the anterior–posterior axis. These segments are referred to as rhombomeres (rh) (Leto et al., 2016). The expression of a series of transcription factors and signaling molecules defines the cerebellar territory (Sajan et al., 2010). The alar plate of the metencephalon undergoes bilateral expansion in the dorsolateral region and forms rhombomere 1. These (rostral) extensions of the alar plate eventually join in the midline to form the vermis. The RL is a specialized epithelium and as mentioned previously is the origin of cerebellar granule cells, but it also gives rise to the ventrally located pontine nuclei and the inferior olive (Leto et al., 2016). During the third trimester of pregnancy the cerebellar volume increases 5–10 times and the surface area increases an astonishing 50 times due to the formation of folia and lobules (Dobbing and Sands, 1973; Dobbing, 1974). At mid-gestation, myelination of the human cerebellum starts, shown by both in vitro histology and in vivo neuroimaging as myelination of the ICP and the SCP (Yakovlev and Lecours, 1967). After term birth a phase of rapid myelination of around 3–4 months can be seen, explaining a doubling of the cerebellar volume in this time period. Multiple interdependent processes occur simultaneously such that mechanisms driving cerebellar development also influence hindbrain and midbrain development, explaining why the pons is often hypoplastic in patients with cerebellar hypoplasia. Genes critical for cerebellar development are also expressed in other parts of the brain and in the skin (Patel and Barkovich, 2002).

Cerebellar malformations

The incidence of posterior fossa malformations is estimated to be around 1/5000 live births (Fig. 14.21 and Table 14.5). Several classifications were published based on different criteria: genetic diagnosis, pre- and postnatal neuroimaging findings, clinical characteristics, and neurodevelopmental outcome (Niesen, 2002; Patel and Barkovich, 2002; Parisi and Dobyns, 2003; Barkovich et al., 2007; Doherty et al., 2013; Cotes et al., 2015; Aldinger and Doherty, 2016). Due to the immaturity of the fetal brain and limitations of prenatal neuroimaging a correct prenatal diagnosis can be very challenging (Doherty et al., 2013). To support pregnancy decision-making and/or postnatal management for both parents and caregivers, it is important to have as much of information on neurodevelopmental consequences as possible. However, most current outcome studies have several drawbacks (e.g., their retrospective design, different case definitions, sometimes poor imaging quality, often small numbers of patients, a short follow-up program, a lack of standardized neurodevelopmental assessments, etc.). Therefore we still have limited data on long-term outcomes in children diagnosed with cerebellar malformations (Poretti et al., 2014). To classify abnormalities of the fossa posterior based on prenatal (MR and ultrasound) imaging, three categories are in common use: (1) Dandy–Walker malformation (DWM), (2) Dandy–Walker variant (DWV), and (3) mega cisterna magna (MCM). However, their definitions do not always seem to provide guidance in diagnostic workup and prognosis (Doherty et al., 2013). Several authors have proposed using a classification based on early clinical and imaging findings (Doherty et al., 2013; Poretti et al., 2014).

Genetic defects of neurogenesis in cerebellar germinal zones

1. Abnormal neurogenesis of the cerebellar VZ, e.g., mutations in the PTF1A gene, which is a known regulator of the VZ and which can lead to cerebellar agenesis (Hoshino, 2006).

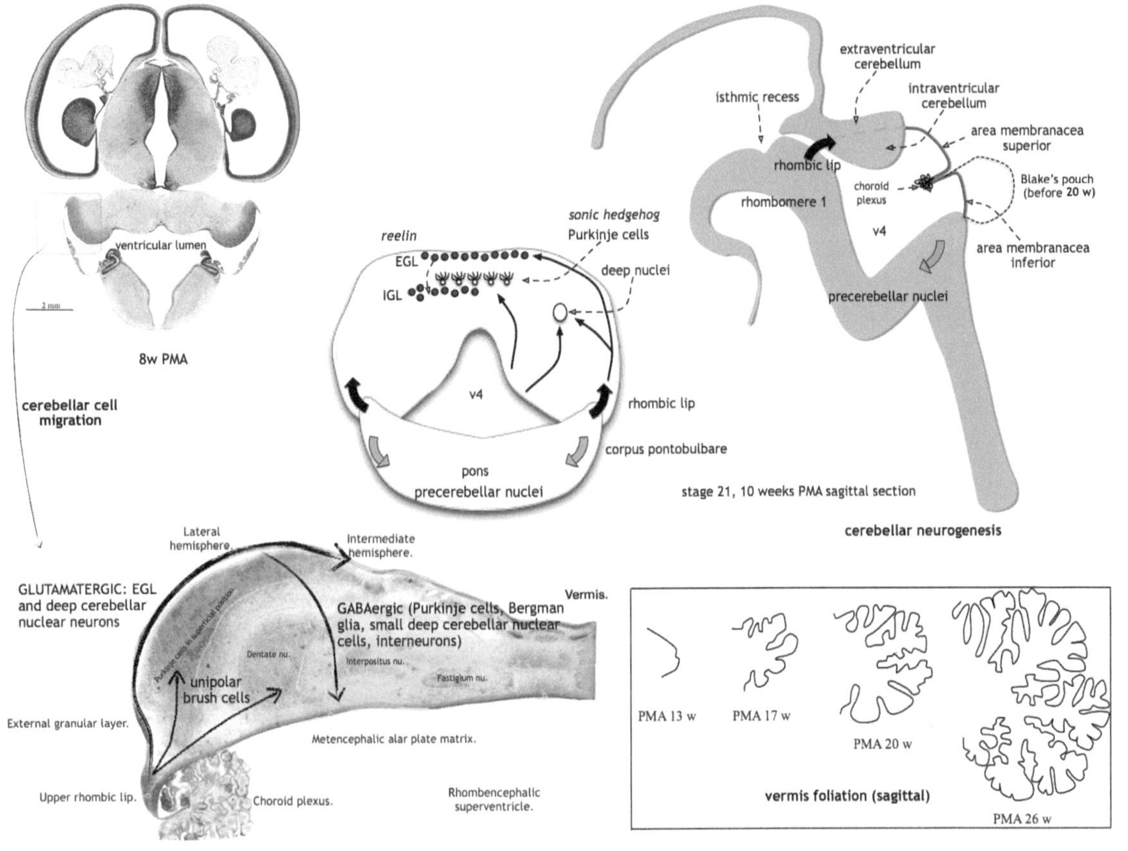

Fig. 14.20. Development of cerebellum.

2. Abnormal neurogenesis of the rhombic limb, e.g., patients with mutations of the FOXC1 gene, a regulator for growth factors and other signaling molecules for neurogenesis of the rhombic limb. Mutations show a phenotype that ranges from isolated vermis hypoplasia to MCM and DWM (Haldipur et al., 2014).
3. Abnormal neurogenesis of the external granule layer, e.g., seen in both Joubert syndrome and Joubert syndrome-related disorders (JSRDs). Genes associated with JSRD are responsible for proteins encoding for normal structure or function of primary cilia (Guo et al., 2019). Primary cilia are cell sensors needed for the reception of Shh, and the Shh pathway is essential for production of granule cells.

Genetic defects in migration, synaptogenesis, and connectivity

Several genes are involved in orderly migration of the neurons in the cerebellum during development. For example, REELIN is one of the genes critical for migration of neurons, including radial migration in the cerebellum. Mutations of the REELIN gene can cause lissencephaly with cerebellar hypoplasia (Miyata et al., 2010; Ranaivoson et al., 2016). Gene mutations encoding for microtubules (such as TUBA1A) can generate a similar phenotype (Arrigoni et al., 2019). There are several mutations of genes, associated with synaptic organization, linked with cerebellar malformations. CASK genes are needed for maintenance of neuronal connectivity and mutations are found in some patients with pontocerebellar hypoplasia (Najm et al., 2008).

Prenatal cerebellar disruption

Disruptions can be defined as morphological defects resulting from extrinsic interference with development. The human cerebellum has a particular susceptibility to prenatal disruption (Dobbing et al., 1970; Poretti et al., 2009b, 2016) as development starts early in embryonic week 5 and is protracted. The cerebellum has a very steep growth during the third trimester of pregnancy. Disruptors include maternal illness (e.g., imbalances in vitamin, nutrition or hormonal status, severe, prolonged periods of mental stress) (Koning et al., 2017), maternal teratogenic exposures (e.g., alcohol, cocaine, nicotine, opioid exposure), ischemic stroke, and trauma (e.g., amnion rupture). Most frequently reported disruptors are hemorrhage (Hayashi et al., 2015; see later) and

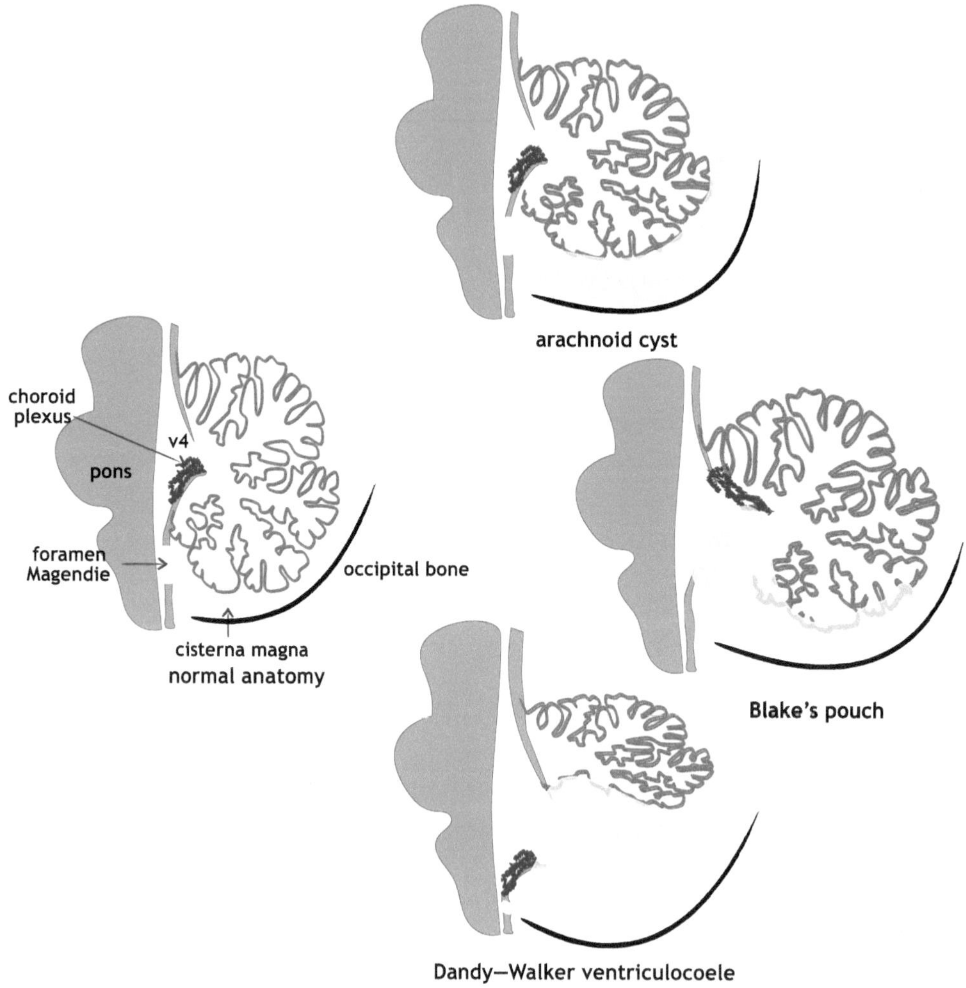

Fig. 14.21. Posterior fossa cavities.

infectious agents (e.g., cytomegalovirus, toxoplasmosis, rubella, herpes, zika) (Volpe, 2009). Cerebellar underdevelopment (decreased cerebellar volume) is often seen with IUGR and/or in infants born preterm. Two mechanisms that might occur simultaneously and seem most likely to explain this disturbance of cerebellar growth are: (1) direct effects on the fast-growing cerebellum in the late second and third trimesters and (2) remote effects via altered transsynaptic trophic interaction (Volpe, 2009). Examples of direct effects on cerebellar development (both intra- and extrauterine) are: inflammation/infection (see later), hemosiderin deposition (iron accumulation, free radical attack), poor nutrition, global hypoxia–ischemia, severe anemia requiring intrauterine transfusions (RhD alloimmunization), and glucocorticoids (Simonazzi et al., 2016). A well-known remote transsynaptic effect is a loss of function in a brain area neurally connected to (but remote from) a separate lesioned brain area (diaschisis). The evidence supporting a role for diaschisis is suggested by the impressive correlation of cerebellar changes in infants with periventricular leukomalacia and cerebral medullary venous infarction (Argyropoulou et al., 2003; Shah et al., 2006; Yoshida et al., 2008). Although it might be that the WM in cerebellum itself is directly affected similar to supratentorial WM (cerebellar leukomalacia) (Tsuru et al., 1995), transsynaptic effects are better described.

Prenatal cerebellar hemorrhage

Cerebellar hemorrhage (CBH) may occur in utero, either in isolation or in combination with telencephalic GMH (Poretti et al., 2009a). Diagnosis of prenatal CBH is often via ultrasonography, where depending on timing, the lesion is either hyper- and/or hypoechoic (Ranzini et al., 1998). MRI (T1, T2, DWI, and SWI sequences) can be used for a confirmation of diagnosis, assessment of timing, and a more detailed assessment of size and

Table 14.5

Malformations in the posterior fossa.

Category	Key findings	Syndromes and associated findings	Etiology (genes involved)
Dandy–Walker malformation (DWM)	Hypoplasia (or absence) of vermis Inferior portion of vermis is elevated and rotated counterclockwise Dilatation of 4th ventricle Hemispheres often hypoplastic Posterior fossa enlarged Elevation of tentorium Enlarged cisterna magna	Isolated or part of complex syndromes Additional malformations in 30–50%: callosal dysgenesis/agenesis, occipital encephalocele, polymicrogyria, heterotopia Hydrocephalus in 90%	ZIC1/ZIC4 FOXC1
Cerebellar vermis hypoplasia/aplasia (former Dandy–Walker variant—DWV)	Small vermis, variably elevated	Joubert and Meckel syndrome Molar tooth sign, polydactyly, encephalocele, kidney malformations	Joubert and Meckel genes, VLDLR, PMM2, OPHN1
Unilateral cerebellar hypoplasia	Vermis often small or dysplastic Vermis angle normal or elevated One small hemisphere	PHACE syndrome In utero infection	In utero vascular insult or infections
Global cerebellar hypoplasia	Small vermis Small hemispheres Normal or depressed tentorium Normal to increased CM	Walker Warburg, Cockayne syndrome Microcephaly, calcification, polymicrogyria, hydrops, lissencephaly	POMT1, REELIN, ERCC8, ERCC6 Prenatal CMV infection
Rhombencephalosynapsis	No vermis Fused hemispheres	Gomez–Lopez–Hernandez syndrome Cardiac, limb, renal, and vertebral abnormalities Absent septum pellucidum Ventriculomegaly Holoprosencephaly	
Pontocerebellar hypoplasia	Small vermis Small hemispheres Normal or depressed tentorium Normal to increased CM	Pontocerebellar hypoplasia	PMM2, VRK1, RARS2, TSEN2, TSEN34, TSEN54 Metabolic disorders
Mega cisterna magna (MCM) (axial diameter > 10 mm)	CM size increased Variable dilatation of pericerebellar subarachnoid spaces		
Cerebellar dysplasia	Small vermis Small hemispheres	Lissencephaly	TUBA1A, REELIN
Cysts: Arachnoid cyst Blake's pouch cyst	Compressed vermis, normal or elevated angle Tentorium normal or elevated Cisterna magna often enlarged Often compressed hemispheres (often asymmetric)	Arachnoid cysts can be beside the midline Blake's pouch cyst should disappear after the 5th month of gestation; when permeabilization fails, Blake's cysts may persist after birth Hydrocephalus often present	
Chiari type I malformation	Vermis in low position Hemispheres often in low position Tentorium normal or low CM size normal or small	Occasional in multiple syndromes	
Chiari type II malformation	Vermis in low position Compressed hemispheres, wrapped around brainstem Often low position tentorium CM size small	Multifactorial, occasional chromosomal abnormalities Waardenburg and other syndromes Open neural tube defect ("banana" and lemon sign)	

location of the hemorrhage. Early diagnosis is relevant because of the consequences on long-term neurodevelopment. With a systematic review, Hortensius et al. (2018) concluded that in children born very preterm (<32 weeks gestation) with isolated CBH, up to a third are severely impaired in cognitive, motor, linguistic, and behavioral domains. Involvement of the vermis and heavy bleeding increase the incidence of severe impairment. Hayashi et al. (2015) reported similar impairment in children with prenatal diagnosis of CBH. Based on neuroimaging studies the majority of prenatal CBH seems to occur between GWs 21 and 25. The underlying mechanisms are insufficiently understood. It has been postulated that CBH often follows bleeding from the GM (both at the subpial external granule cell layer or the subependymal layer at the roof of the fourth ventricle) (Martin et al., 1976; Haines et al., 2013). However, autopsy studies revealed limited evidence to support these locations as the primary source (Williams, 1981; Haines et al., 2013; Fumagalli et al., 2015; Pierson and Al Sufiani, 2016). The cortical region that will become the "internal granule cell layer" starts out as a paucicellular zone, being populated by enormous amounts of granule cells (precursors migrate from the external granule layer). This region is a zone of very rapid angiogenesis with immature, poorly formed capillary beds, which are in a continual process of remodeling. Because supratentorial intraventricular hemorrhage often coexists with CBH, they share overlapping pathogenetic mechanisms (i.e., impaired autoregulation) (Fumagalli et al., 2015). This pathogenetic mechanism would explain timing, location, coexistence with IVH, and multifocal distribution starting near the cerebellar surface. Venous obstruction (e.g., trauma/pressure) might cause blood to pool in well-developed capillary beds in the center of the folia, leading to a central hemorrhagic lesion (venous infarction) involving all layers of the cerebellum (Williams, 1981). Based on morphological and clinical aspects of CBH in preterm infants, six types can be discerned: (1) subarachnoid, (2) folial, (3) lobar, (4) bilateral lobar, (5) giant lobar, including vermis, and (6) contusional (Ecury-Goossen et al., 2010). Maternal risk factors include coagulopathies (e.g., use of anticoagulation medication), thrombophilia, (pre-)eclampsia, sepsis, trauma (leading to increased fetal cerebral venous pressure), seizures, cocaine use, and anticoagulation medication (Sharony et al., 1999; Malinger et al., 2006; Castillo, 2007; Nomura et al., 2009). Antepartum causes of CBH include vascular malformation (e.g., capillary telangiectasia), hematological disorders causing thrombosis of the placenta and/or fetal arteries (leading to secondary hemorrhagic infarction), congenital infection (cytomegalovirus, parvovirus), and congenital tumors (Sharony et al., 1999; Malinger et al., 2006). Postmortem cerebellar examination of infants with CBH documented adjacent cortical changes (reduced thickness due to neuronal loss and gliosis) and an increased incidence of neuronal loss and gliosis in the inferior olivary nuclei and dentate nuclei (Grunnet and Donald Shields, 1976; Castillo, 2007). More research is needed to understand this impact of CBH on nearby structures and on functional neurodevelopment. In contrast to CBH stroke, in utero cerebellar arterial ischemic stroke is rarely reported (Johnsen et al., 2002; Govaert et al., 2009a,b; Ansari et al., 2011).

Congenital cerebellar infection

Congenital infections can be transmitted by transplacental passage (fetal infection) or during birth (parturitional infection). Both timing in relation to pregnancy and infectious agent will determine the effect on the cerebellum by TORCHES: toxoplasmosis, others (e.g., syphilis), rubella, cytomegalovirus (CMV), and herpes simplex virus (HSV). In congenital CMV, hypoplasia of the cerebellum, abnormal foliation (abnormal orientation, altered cortical thickness), and calcifications have been reported (De Vries et al., 2004; Rossi et al., 2016; Sanapo et al., 2017). Congenital toxoplasmosis can induce similar effects on foliation (Rossi et al., 2016). Congenital parvovirus infection has been linked to CBH and hypoplasia (Sanapo et al., 2017). It has been proposed that in addition to microbial effects, hemodynamic changes caused by blood transfusion for severe anemia caused by parvovirus infection might contribute to the lesion (Sanapo et al., 2017).

FLUID PRODUCTION AND VENTRICULAR SIZE

CSF space development

Differentiating neurons leave the neocortical neuroepithelium in large numbers early during the 9th week postconception, which is associated in time with the rapid expansion of choroid plexus in the equally expanded lateral ventricles. Before appearance of plexus tissue, CSF pressure increases in the lumen of the neural tube and the volume of the prosencephalon grows, suggesting initial production of CSF by structures other than the choroid plexus. The cranial meninges originate from several sources such as the prechordal plate, the parachordal mesoderm, and the neural crest (O'Rahilly and Müller, 1986). The subarachnoid spaces appear on the 32nd day at the ventral aspect of the rhombencephalon, then extend caudally and dorsally (Sakkaa et al., 2011). However, the fourth ventricle (foramen of Magendie) is not open before postmenstrual week 16. CSF production starts weeks before that moment (Robinson and Goldstein, 2007), enabling ventricular volumina to increase. Still later opening of the foramina of Luschka in the lateral recesses of the fourth ventricle can be related to prolonged dilatation of the fourth ventricle, referred to as Blake's pouch (Paladini et al., 2012). Formation of the subarachnoid spaces is therefore not exclusively due to CSF pressure, but capillaries appear to play a role in the secretion and absorption of CSF during embryogenesis. A microscopic flow of CSF has been described in brain interstitial spaces (Iliff et al., 2012), the significance of which is not known in the fetus. Fluid circulation in the brain further depends on maturation of arachnoid villi, development of glymphatics, and presence and development of nasal lymphatic fluid drainage (Adeeb et al., 2013; Nakada and Kwee, 2019). Arachnoid villi mature near term, and growth into arachnoid granulations is mainly a postnatal event (Gomez et al., 1982). The human subcommissural organ is a gland with differentiated ependyma located caudal to the pineal gland at the entrance to the aqueduct (Galarza, 2002). It is the first secretory structure of the brain that differentiates in the late embryonic period, to become an active secretory structure at mid-gestation, and to regress during the first year of postnatal life (Rodriguez et al., 2001). Changes in the subcommissural organ have been described in all species developing congenital hydrocephalus and in human fetuses, but the precise role of the glycoproteins secreted here is still unknown in the human fetal context.

Anatomists have defined ventricular morphology early on (Retzius, 1896). The lumen of the spinal cord, the central canal, is continuous with that of the brain vesicles. The cavity of the rhombencephalon is the fourth ventricle, that of the diencephalon is the third ventricle, and those of the cerebral hemispheres are the lateral ventricles. The lumen of the mesencephalon becomes very narrow and is known as the aqueduct of Sylvius. Each lateral ventricle communicates with the third ventricle through the interventricular foramen of Monro.

Choroid plexus development

Growth of choroid plexus may have a role in the extension of the ventricular fluid compartments by increasing ventricular surface area and thus promotion of a widespread proliferation of neocortical precursors (Desmond and Jacobsen, 1977; Altman and Bayer, 2015) (Fig. 14.22). The choroid plexus sits between two barrier systems (blood and cerebrospinal fluid—CSF) and has as primary goal the production and regulation of CSF (Catala, 1998). Choroid plexus first appears in the roof of the fourth ventricle at stage 18 (day 41), in the lateral ventricles at stage 19, and in the third ventricle at stage 21 (Ariëns Kappers, 1958; Bartelmez and Dekaban, 1962). The primordia appear as simple or club-shaped folds protruding into the ventricles, emanating from pial mesenchymal vascular extensions covered by single ependymal epithelium, later named vela interposita. The time at which choroid plexus starts to secrete CSF has not been clearly determined, but carbonic anhydrase (for bicarbonate excretion) is present 2 weeks after appearance of plexus. It may be that an early nutritional role for plexus is present, but this function is lost during development. In week 8, from the choroidal fissure extending out, the lobular character of the plexus gradually changes into an arrangement of wavy folds; capillaries form single loops under the cuboidal epithelium (Kraus and Jirásek, 2002). Bone morphogenetic proteins and tropomyosin are involved in specification of plexus development. Four stages of lateral ventricular plexus development have been defined (Dziegielewska et al., 2001): *stage 1*: 7–9 weeks after conception, no glycogen, ill-defined blood islets, pseudostratified epithelium; *stage 2*: 9–17 weeks after conception, abundant glycogen, primary villi, low columnar epithelium, subepithelial capillaries, very large relative size in relation to ventricle; *stage 3*: fetal period, moderate glycogen, many primary villi, cuboidal epithelium, mature capillaries in villous core; and *stage 4*: after 29 weeks, villi with multiple fronds, disappearing glycogen, cuboidal or squamous epithelium, relatively small plexus in relation to ventricle size. The mechanisms involved in formation of a blood–CSF barrier at choroid plexus combine diffusion restraint (tight junctions between the plexus epithelial cells) and specific exchange. One barrier mechanism unique to the developing brain transfers specific proteins from blood to CSF via tubulocisternal endoplasmic reticulum in plexus epithelial cells. This results in a high concentration of proteins in early CSF. Proteins may function as colloid osmotic agents for ventricle expansion but also as specific carriers and growth factors. Some proteins are not transferred but are locally produced, such as transthyretin and IGF II. Barrier mechanisms in the immature brain reflect developmental specialization rather than immaturity.

Human fetal choroid plexus sometimes contains a large central mass of amorphous material with a few poorly distinguishable blood vessels, surrounded by a single layer of epithelial cells (Voetmann, 1949; Netsky and Shuangshoti, 1975). Its significance is uncertain, but the resemblance with Wharton's jelly

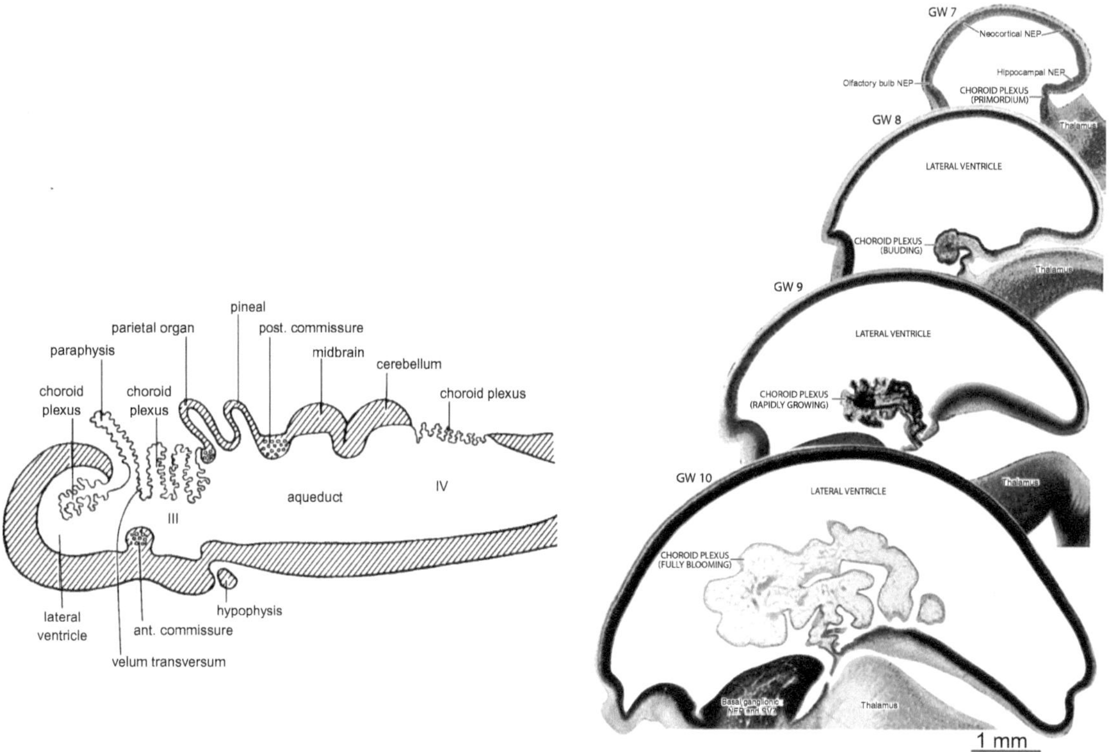

Image from Netsky and Shuangshoti (1975) *Image from Altman and Bayer (2015)*

Fig. 14.22. Embryonic sagittal section with early plexus formation.

suggests it may have a mechanical, expanding function. In many fetuses, plexus cysts appear (and often regress) in which fluid accumulates in the mesenchymal space without epithelial lining (unlike colloid cysts). The cysts have irregular angiomatous capillaries in their walls, unlike the normal capillary loops (Kraus and Jirásek, 2002). As an isolated finding, this is within normal development, often an incidental sonographic finding.

Already at 9 weeks, with transvaginal ultrasound, the plexus can be seen at the level of both lateral ventricles. Between 9 and 11 weeks, plexus covers most of the surface of the standard transventricular fetal ultrasound axial section. As pregnancy progresses the volume ratio between choroid plexus and the ventricles decreases. Moreover, as the ventricles develop the choroid plexus grows to occupy the atrium, covering the posterior thalamus (Timor-Tritsch et al., 2012). The finding of isolated choroid plexus separation by more than 3 mm from the medial ventricle wall is usually temporary in the second trimester, resolving in most cases within 4 weeks of the initial diagnosis in the second trimester. Most infants with this finding have no abnormalities (Bronsteen et al., 2006). During stage 21 the choroid plexuses become vascularized. Choroid plexus vessels are first seen with Doppler ultrasound at 10–11 weeks of GA.

Fetal ventriculomegaly
(see also Chapter 15 of this volume)

Definition

Evaluation with fetal imaging of the width of cerebral lateral ventricles is an essential part of the standard sonographic examination of the fetus. The calipers are placed at the level of the glomus of the choroid plexus, perpendicular to the ventricular atrium, inside the echoes generated by the wall of the ventricles, all this in an axial section through the brain containing the cavum septi pellucidi and the ambient cistern. Normal values: 7.6 ± 0.6 mm in the second trimester of pregnancy. Ventriculomegaly (VM) is defined as the presence of an excess of lateral ventricle width ≥ 10 mm (at least 3 s.d. above the mean) between 15 and 40 weeks of gestation. Unfortunately, most VM cannot be detected before 24 weeks of gestational age. Three degrees exist: mild/borderline ventriculomegaly, atrial width of 10–12 mm; moderate ventriculomegaly, atrial width of 12.1–15 mm; and severe ventriculomegaly, atrial width ≥ 15.1 mm.

Mild and moderate are often lumped into one category (Fig. 14.23). Isolated ventriculomegaly (IVM) is withheld in the absence of sonographic evidence of associated malformations or markers of aneuploidy at the time of the initial presentation. Some degree of asymmetry of the lateral ventricles exists in the human fetal brain,

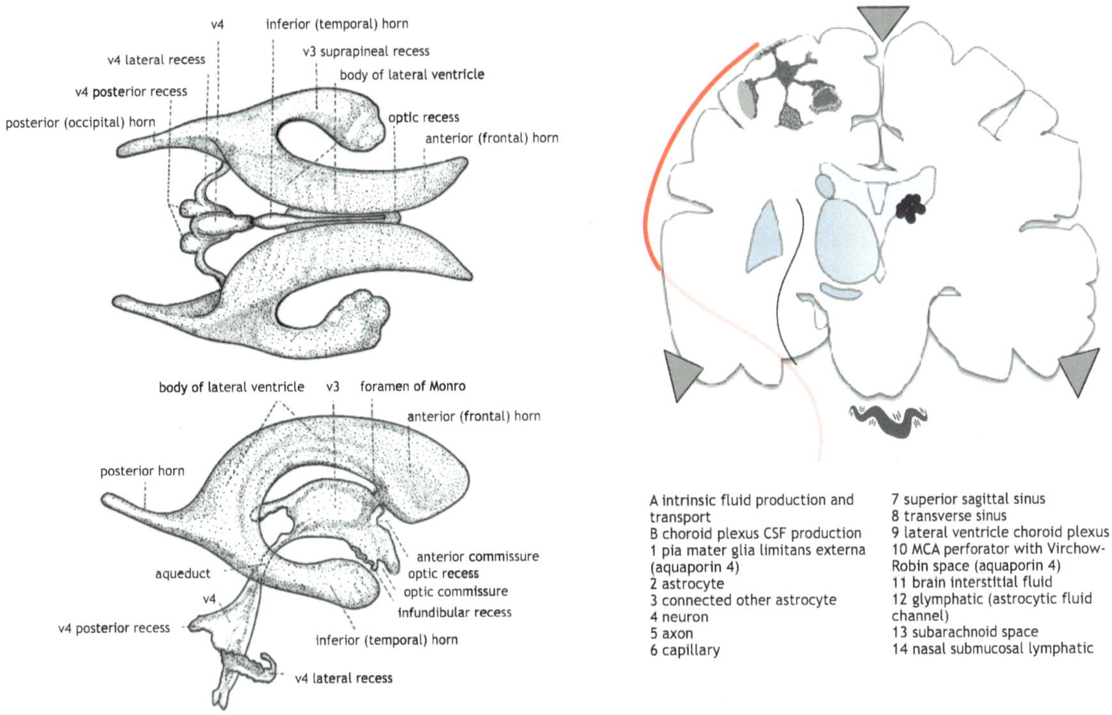

Fig. 14.23. Mature ventricle anatomy; fluid production and circulation scheme.

and occasionally, this can be appreciated prenatally, which is not abnormal when both ventricles are within normal size. However, unilateral pathological ventriculomegaly does occur and is occasionally diagnosed prenatally. The etiology and outcome of unilateral and bilateral ventriculomegaly are similar. Asymmetric bilateral VM is defined by a difference of $\geq$2 mm between both ventricles. VM tends to be unilateral in cases of destruction and bilateral in cases of malformation. The higher incidence of mild VM in male fetuses is related to the fact that their mean atrial size is slightly greater than that of females.

The incidence of ventriculomegaly is 0.15%–0.7% of all pregnancies. Severe ventriculomegaly affects around 1/1000 newborns. Ventriculomegaly is caused by an imbalance between production and absorption of CSF, secondary to:

1. disorders of neuronal proliferation or migration (microcephaly, schizencephaly, lissencephaly type Walker–Warburg, and Miller–Dieker);
2. disorders of organogenesis (agenesis CC, holoprosencephaly, DWM);
3. destructive processes, ex vacuo (hypoxia, infections, tumors, vascular insults); and
4. abnormal turnover of CSF: nonobstructive (overproduction of CSF, such as choroid plexus papiloma) and obstructive (noncommunicating or communicating).

Imaging aspects

Classic sonographic criteria for the diagnosis of hydrocephalus (a problem of excess CSF in the ventricles) in the postnatal period cannot be completely extended to the fetus. Indeed, in utero, hydrocephalus may be present even with a normal head circumference, and the underlying cause of an altered flow of CSF cannot be identified in all cases. However, progressive and severe fetal ventricular enlargement, together with decreased or absent pericerebral space and parenchymal thinning, can be presumed to be indicative of hydrocephalus. It is therefore often not difficult to differentiate between obstructive hydrocephalus and encephaloclastic sequelae with "hydrocephalus *ex vacuo*": (1) in the latter, we deal with microcephalic or low normocephalic children, with normal fontanel tension; (2) occipital as well as frontal ventricular segments are affected compared with prominent occipital dilatation in obstructive hydrocephalus; and (3) with brain atrophy one finds dilated subarachnoid spaces and a dilated interhemispheric fissure (also seen with chronic subdural effusions causing subarachnoid space distension). Hydrocephalus associated with increased pressure and impaired liquor circulation may secondarily induce damage due to microvascular ischemia, impaired axoplasmic transport, oligodendroglial cell death and impaired myelination, accumulation of toxic factors, and decrease of trophic factors in CSF.

VM can be associated with malformation patterns of the CNS. In the Chiari II malformation, VM is associated with a small posterior fossa, small dysmorphic cerebellum, effaced cisterna magna, and a spinal defect with myelomeningocele. The association of VM with the presence of an enlarged posterior fossa and complete or partial agenesis or hypoplasia of the cerebellar vermis is typical of DWM. The "teardrop" appearance of laterally shifted lateral ventricles and upward displacement of the third ventricle suggest callosal agenesis. Aqueduct stenosis/atresia typically causes triventricular supratentorial hydrocephalus. "Communicating" as connotation for VM infers dilatation of all ventricles, including the cisterna magna, but with the inference of a more distal hindrance of liquor circulation. Colpocephaly refers to ventriculomegaly of the occipital horns without frontal dilatation. This may be due to a developmental disturbance (in fetal months 1–4) or destruction of occipital periventricular WM. Colpocephaly can be found in association with lissencephaly and pachygyria, with porencephaly, with agenesis of the CC and heterotopia. Due to its nonspecific character, it is advisable not to consider colpocephaly a disease proper, but rather part of an overlying cerebral anomaly.

VM associated with destructive lesions can be uni- or bilateral. When VM is associated with hyperechoic foci located in the brain and germinolytic cysts, fetal infection should be suspected. If multiple nodular subependymal hyperechoic dots are seen, sometimes grouped in hyperechogenic bands, cytomegalovirus, ZIKA, or rubella infection should be suspected. If free-floating particles are detected within the cerebral ventricles, often dilated because this particulate matter narrows ventricular foramina, a differential diagnosis with intraventricular hemorrhage should be made.

Diagnostic testing

The prevalence of aneuploidy in severe fetal ventriculomegaly is 3%–5%, trisomy 21 being the most common among these. Other cytogenetic syndromes with hydrocephalus include trisomy 13, 18, and 9, and triploidy. In VM with associated other findings the chance of finding aneuploidy rises to around one in four.

Mendelian conditions of significance are hydrolethalus syndrome (CRASH), Meckel–Grüber, Miller–Dieker, Walker–Warburg, Kartagener, Smith–Lemli–Opitz, VACTERL with hydrocephalus, some craniosynostoses, oculoauriculovertebral spectrum, Apert, Seckel, Neu–Laxova, acrocallosal, and Aicardi syndrome.

Aqueduct stenosis, if not secondary to infection, tumor, or hemorrhage, can be genetic: both disturbances in the secretion of sialoglycoproteins from the subcommissural organ and ependymal denudation might induce collapse of the aqueduct (McAllister II, 2012). The X-linked hydrocephalus spectrum (former Bickers–Adams syndrome), prevalence 1/30,000 live births, is associated with mutations in the NCAM L1CAM (located on Xq28). The L1-spectrum also contains MASA (mental retardation, aphasia, shuffling gait, adducted thumbs) and XLSP (X-linked spastic paraplegia), allelic mutations of the same gene. L1 is important for neurite growth, axonal fasciculation, and neuronal migration. Absence of the pyramidal tracts is pathognomonic and linked with thumb adduction. Infants with XLH may have fused thalami and colliculi, and many do not have a CC. Affected males have an IQ in the range of 20–50 and develop lower limb spasticity and aphasia. Adducted thumbs can also occur isolated in the congenital clasped thumb syndrome or be part of the whistling face syndrome. A cortical thumb is also flexed but in addition the fingers are in spastic flexion over it and the arm is spastic.

Structural anomalies in addition to ventriculomegaly are found in 6% with mild VM and in 56% with moderate VM. Associated abnormalities not detected at time of first examination (false negative diagnosis) amount to 10%–13%. The incidence of congenital infection (fetopathy by cytomegalovirus, herpes, toxoplasmosis, rubella, parvo B19) in mild/moderate ventriculomegaly versus severe VM is 1%–5% vs 10%–20%, respectively. The overall incidence of congenital infection is 1.5%. Very rarely does alloimmune thrombocytopenia cause intraventricular hemorrhage with subsequent VM.

There is no consensus on the prognostic significance of ventriculomegaly, understandable given the complexity of mechanisms and causes. The follow-up duration varied in reported studies. Different modalities of neurological follow-up were used: formal assessment of cognitive and motor development, interviews with families, or not specified. There was often a lack of a control group. Children were assessed at different ages, and no distinction was made between mild, moderate, or severe VM. Progressive dilatation carries the worst prognosis. Crudely, abnormal neurodevelopment in VM is around 12%: VM < 12 mm, 6–12%; VM > 12.1 mm, 16.4%; and VM > 15.1 mm, 37%.

Outcome depends on progression of VM and the evolution to asymmetrical bilateral VM. The best prognoses are found in the category due to hemorrhage or cyst and with callosal agenesis. There is a high rate of spontaneous abortion, in utero fetal demise, and neonatal death (up to 60%) if VM is associated with other brain anomalies. The mortality rate is much lower if the lesion is isolated, with a survival rate of at least 80% and over 60% of these having normal neurologic development. Early detection and progression of VM indicate a less favorable prognosis. The rate of neurodevelopmental delay in infants with a prenatal diagnosis of isolated mild VM is about 11%, and it is not certain whether this is increased over that in the general population. Hahner et al. (2017) conducted a careful sonographic study at 26 and 30 weeks of gestation, using a scoring system as well as linear measurements, and suggested that subtle changes in the formation of the insula, sulcus cinguli, sulcus parieto-occipitalis, and sulcus calcarinus may need further study in this context.

ABBREVIATIONS

ACA, anterior cerebral artery; AChA, anterior choroidal artery; ACoA, anterior communicating artery; AICA, anterior inferior cerebellar artery; BA, basilar artery; C, cervical; CM, cisterna magna; CMV, cytomegalovirus; CRL, crown rump length; DWM, Dandy-Walker malformation; DWV, Dandy-Walker variant; GA, gestational age; ICA, internal carotid artery; LNA, longitudinal neural artery; MCA, middle cerebral artery; MCM, mega cisterna magna; PCA, posterior cerebral artery; PCoA, posterior communicating artery; PICA, posterior inferior cerebellar artery; SSS, superior sagittal sinus; TS, transverse sinus.

References

Adeeb N, Deep A, Griessenauer CJ et al. (2013). The intracranial arachnoid mater: a comprehensive review of its history, anatomy, imaging, and pathology. Childs Nerv Syst 29: 17–33.

Afif A, Bouvier R, Buenerd A et al. (2007). Development of the human fetal insular cortex: study of the gyration from 13 to 28 gestational weeks. Brain Struct Funct 212: 335–346.

Afif A, Trouillas J, Mertens P (2014). Development of the sensorimotor cortex in the human fetus: a morphological description. Surg Radiol Anat 37: 153–160.

Aldinger KA, Doherty D (2016). The genetics of cerebellar malformations. Semin Fetal Neonatal Med 21: 321–332.

Altman J, Bayer SA (2015). Development of the human neocortex: a review and interpretation of the histological record. A Free eBook from the Laboratory of Developmental Neurobiology, Inc. www.neurondevelopment.org © 2015. Ocala, FL.

Ansari SA et al. (2011). Bilateral in utero cerebellar infarction. J Child Neurol 26: 895–899.

Anstrom JA, Brown WR, Moody DM et al. (2002). Anatomical analysis of the developing cerebral vasculature in premature neonates: absence of precapillary arteriole-to-venule shunts. Pediatr Res 52: 554–560.

Anstrom JA, Brown WR, Moody DM et al. (2004). Subependymal veins in premature neonates: implications for hemorrhage. Pediatr Neurol 30: 46–53.

Anstrom JA, Thore CR, Moody DM et al. (2005). Morphometric assessment of collagen accumulation in germinal matrix vessels of premature human neonates. Neuropathol Appl Neurobiol 31: 181–190.

Argyropoulou MI et al. (2003). MRI measurements of the pons and cerebellum in children born preterm; associations with the severity of periventricular leukomalacia and perinatal risk factors. Neuroradiology 45: 730–734.

Ariëns Kappers JA (1958). Structural and functional changes in the telencephalic choroid plexus during brain ontogenesis. In: GEW Wolstenholme, CM O'Connor (Eds.), The cerebrospinal fluid. Little, Brown, Boston, MA, pp. 3–25.

Arlotta P, Molyneaux BJ, Chen J et al. (2005). Neuronal subtype-specific genes that control corticospinal motor neuron development in vivo. Neuron 45: 207–221.

Arrigoni F et al. (2019). The spectrum of brainstem malformations associated to mutations of the tubulin genes family: MRI and DTI analysis. Eur Radiol 29: 770–782.

Ayala R, Shu T, Tsai LH (2007). Trekking across the brain: the journey of neuronal migration. Cell 128: 29–43.

Ballabh P (2010). Intraventricular hemorrhage in premature infants: mechanism of disease. Pediatr Res 67: 1–8.

Ballabh P, Braun A, Nedergaard M (2004). Anatomic analysis of blood vessels in germinal matrix, cerebral cortex, and white matter in developing infants. Pediatr Res 56: 117–124.

Ballabh P, Hu F, Kumarasiri M et al. (2005). Development of tight junction molecules in blood vessels of germinal matrix, cerebral cortex, and white matter. Pediatr Res 58: 791–798.

Baram TZ, Butler IJ, Nelson Jr MD et al. (1988). Transverse sinus thrombosis in newborns: clinical and magnetic resonance imaging findings. Ann Neurol 24: 792–794.

Barkovich AJ, Millen KJ, Dobyns WB (2007). A developmental classification of malformations of the brainstem. Ann Neurol 62: 625–639.

Bartelmez GW, Dekaban AS (1962). The early development of the human brain. Contrib Embryol Carnegie Instn 37: 13–32.

Barton RA, Venditti C (2014). Rapid evolution of the cerebellum in humans and other great apes. Curr Biol 24: 2440–2444.

Bassan H (2009). Intracranial hemorrhage in the preterm infant: understanding it, preventing it. Clin Perinatol 36: 737–762.

Battin MR, Maalouf EF, Counsell SJ et al. (1998). Magnetic resonance imaging of the brain in very preterm infants: visualization of the germinal matrix, early myelination, and cortical folding. Pediatrics 101: 957–962.

Beltinger C, Saule H (1988). Sonography of subependymal cysts in congenital rubella syndrome. Eur J Pediatr 148: 206–207.

Braun A, Xu H, Hu F et al. (2007). Paucity of pericytes in germinal matrix vasculature of premature infants. J Neurosci 27: 12012–12024.

Broca P (1888). Descriptions élémentaires des circonvolutions cérébrales de l'homme. In: Mémoires d'Antropologie, Memo Paris: C Reinwald 707–804.

Bronsteen R, Lee W, Vettraino I et al. (2006). Isolated choroid plexus separation on second trimester sonography: natural history and postnatal importance. J Ultrasound Med 25: 343–347.

Browder J, Kaplan HA, Krieger AJ (1975). Venous lakes in the suboccipital dura mater and falx cerebelli of infants: surgical significance. Surg Neurol 4: 53–55.

Butts BD, Houde C, Mehmet H (2008). Maturation-dependent sensitivity of oligodendrocyte lineage cells to apoptosis: implications for normal development and disease. Cell Death Differ 15: 1178–1186.

Bystron I, Blakemore C, Rakic P (2008). Development of the human cerebral cortex: Boulder Committee revisited. Nat Rev Neurosci 9: 110–122.

Castillo M (2007). Selective vulnerability and the cerebellum in neonates. Am J Neuroradiol 28: 20–21. Available at: http://www.ajnr.org/content/28/1/20.abstract.

Catala M (1998). Embryonic and fetal development of structures associated with the cerebrospinal fluid in man and other species. Part I: the ventricular system, meninges and choroid plexuses. Arch Anat Cytol Pathol 46: 153–169.

Cevey-Macherel M, Forcada Guex M, Bickle Graz M et al. (2013). Neurodevelopment outcome of newborns with cerebral subependymal pseucocyst at 18 and 46 months: a prospective study. Arch Dis Child 98: 497–502.

Chang H, Cho KH, Hayashi S et al. (2013). Site- and stage-dependent differences in vascular density of the human fetal brain. Childs Nerv Syst 30: 399–409.

Chi JG, Dooling EC, Gilles FH (1977). Gyral development of the human brain. Ann Neurol 1: 86–93.

Chung JI, Weon YC (2005). Anatomic variations of the deep cerebral veins, tributaries of Basal vein of rosenthal: embryologic aspects of the regressed embryonic tentorial sinus. Interv Neuroradiol 11: 123–130.

Coco M, Perciavalle V (2015). Where did the motor function of the cerebellum come from? Cerebellum Ataxias 2: 1–5.

Corbin JG, Gaiano N, Juliano SL et al. (2008). Regulation of neural progenitor cell development in the nervous system. J Neurochem 106: 2272–2287.

Cotes C et al. (2015). Congenital basis of posterior fossa anomalies. Neuroradiol J 28: 238–253.

Cowan F, Thoresen M (1983). Ultrasound study of the cranial venous system in the human new-born infant and the adult. Acta Physiol Scand 117: 131–137.

Cunningham DJ (1892). Contribution to the surface anatomy of the cerebral hemispheres. In: The fissure of Rolando, 161–192, Extensive reference to older literature.

De Bisschop B, Camfferman F, van Hengel-Jacobs M et al. (2019). Abnormal primary gyration in relation to deep brain injury in preterm infants. Acta Paediatr 109: 204–205.

De Vries LS et al. (2004). The spectrum of cranial ultrasound and magnetic resonance imaging abnormalities in congenital cytomegalovirus infection. Neuropediatrics 35: 113–119.

Del Bigio MR (2011). Cell proliferation in human ganglionic eminence and suppression after prematurity-associated haemorrhage. Brain 134: 1344–1361.

Desmond ME, Jacobsen AG (1977). Embryonic brain enlargement requires cerebrospinal fluid pressure. Dev Biol 57: 188–198.

Dimmick SJ, Faulder KC (2009). Normal variants of the cerebral circulation at multidetector CT angiography. Radiographics 29: 1027–1043.

Dobbing J (1974). The later growth of the brain and its vulnerability. Obstet Gynecol Surv 53: 2–6.

Dobbing J, Sands J (1973). Quantitative growth and development of human brain. Arch Dis Child 48: 757–767.

Dobbing J et al. (1970). Vulnerability of developing brain: I. Some lasting effects of X-irradiation. Exp Neurol 28: 442–449.

Doherty D, Millen KJ, Barkovich AJ (2013). Midbrain and hindbrain malformations: advances in clinical diagnosis, imaging, and genetics. Lancet Neurol 12: 381–393.

Dubois J, Benders M, Borradori-Tolsa C et al. (2008). Primary cortical folding in the human newborn: an early marker of later functional development. Brain 131: 2028–2041.

Dubois J, Benders M, Lazeyras F et al. (2010). Structural asymmetries of perisylvian regions in the preterm newborn. Neuroimage 52: 32–42.

Durgun B, Ilglt ET, Cizmeli MO et al. (1993). Evaluation by angiography of the lateral dominance of the drainage of the dural venous sinuses. Surg Radiol Anat 15: 125–130.

Dziegielewska KM, Ek J, Habgood MD et al. (2001). Development, evolution and ageing: development of the Choroid Plexus. Microsc Res Tech 52: 5–20.

Ecury-Goossen GM et al. (2010). The clinical presentation of preterm cerebellar haemorrhage. Eur J Pediatr 169: 1249–1253.

El-Khoury N, Braun A, Hu F et al. (2006). Astrocyte end-feet in germinal matrix, cerebral cortex, and white matter in developing infants. Pediatr Res 59: 673–679.

Engelhardt B (2003). Development of the blood-brain barrier. Cell Tissue Res 314: 119–129.

Esteban H, Blondiaux E, Audureau E et al. (2015). Prenatal features of isolated subependymal pseudocysts associated with adverse pregnancy outcome. Ultrasound Obstet Gynecol 46: 678–687.

Etchevers HC, Vincent C, Le Douarin NM et al. (2001). The cephalic neural crest provides pericytes and smooth muscle cells to all blood vessels of the face and forebrain. Development 128: 1059–1068.

Fame RM, MacDonald JL, Macklis JD (2011). Development, specification and diversity of callosal projection neurons. Trends Neurosci 34: 41–50.

Feess-Higgins A, Larroche J-C (1987). Le Développement du Cerveau Foetal Humain, Masson, Atlas Anatomique, Paris.

Fernandez Alvarez JR, Amess PN, Gandhi RS et al. (2009). Diagnostic value of subependymal pseudocyst and choroid plexus cyst on neonatal cerebral ultrasound: a meta-analysis. Arch Dis Child Fetal Neonatal Ed 94: F443–F446.

Fothergill T, Donahoo AL, Douglass A et al. (2014). Netrin-DCC signaling regulates corpus callosum formation through attraction of pioneering axons and by modulating Slit2-mediated repulsion. Cereb Cortex 24: 1138–1151.

Friede RL (1989). Developmental neuropathology, Springer-Verlag, Berlin/Heidelberg.

Fukusumi A, Okudera T, Takahashi S et al. (2010). Anatomical evaluation of the dural sinuses in the region of the torcular herophili using three dimensional CT venography. Acad Radiol 17: 1103–1111.

Fumagalli M et al. (2015). From germinal matrix to cerebellar haemorrhage. J Matern Fetal Neonatal Med 28: 2280–2285.

Galarza M (2002). Evidence of the subcommissural organ in humans and its association with hydrocephalus. Neurosurg Rev 25: 205–215.

Ghazi-Birry HS, Brown WR, Moody DM et al. (1997). Human germinal matrix: venous origin of hemorrhage and vascular characteristics. Am J Neuroradiol 18: 219–229.

Gilman S (1985). The cerebellum and neural control. JAMA Neurol 42: 310.

Girard N, Gambarelli D (2001). Normal fetal brain. Magnetic resonance imaging. An atlas with anatomic correlations. In: F Brunelle, D Show (Eds.), Pediatric fetal imaging, Rickmansworth, UK.

Glickstein M, Strata P, Voogd J (2009). Cerebellum: history. Neuroscience 162: 549–559.

Goldman-Rakic PS (1980). Morphological consequences of prenatal injury to the primate brain. Prog Brain Res 53: 3–19.

Gomez DG, DiBenedetto AT, Pavese AM et al. (1982). Development of arachnoid villi and granulations in man. Acta Anat (Basel) 111: 247–258.

Govaert P, Swarte R, De Vos A et al. (2004). Sonographic appearance of the (ab)normal insula of Reil. Dev Med Child Neurol 46: 610–616.

Govaert P, Ramenghi L et al. (2009a). Diagnosis of perinatal stroke II: mechanisms and clinical phenotypes. Acta Paediatr, Int J Paediatrics 98: 1720–1726.

Govaert P, Dudink J et al. (2009b). Top of the basilar artery embolic stroke and neonatal myoclonus. Dev Med Child Neurol 51: 324–327.

Gregg L, Gailloud P (2017). The role of the primitive lateral basilovertebral anastomosis of padget in variations of the vertebrobasilar arterial system. Anat Rec 300: 2025–2038.

Grunnet ML, Donald Shields W (1976). Cerebellar hemorrhage in the premature infant. J Pediatr 88: 605–608.

Guo J et al. (2019). Primary cilia signaling promotes axonal tract development and is disrupted in joubert syndrome-related disorders models. Dev Cell 51: 759–774.e5.

Habas PA, Kim K, Corbett-Detig JM et al. (2010). A spatiotemporal atlas of MR intensity, tissue probability and shape of the fetal brain with application to segmentation. Neuroimage 53: 460–4701.

Hahner N, Puerto B, Perez-Cruz M et al. (2017). Altered cortical development in fetuses with isolated nonsevere venbtriculomegaly assessed by neurosonography. Prenat Diagn 38: 365–375.

Haines DE, Olry R, Dietrichs E (2003). If there are "Deep" cerebellar nuclei, where are the "Superficial" ones? J Hist Neurosci 12: 203–205.

Haines KM, Wang W, Pierson CR (2013). Cerebellar hemorrhagic injury in premature infants occurs during a vulnerable developmental period and is associated with wider neuropathology. Acta Neuropathol Commun 1: 69.

Haldipur P et al. (2014). In: R Krumlauf (Ed.), Foxc1 dependent mesenchymal signalling drives embryonic cerebellar growth. 3: eLife Sciences Publications, Ltd p. e03962.

Haldipur P, Dang D, Millen KJ (2018). Embryology. Handb Clin Neurol 154: 29–44.

Hawkes R, Leclerc N (1987). Antigenic map of the rat cerebellar cortex: the distribution of parasagittal bands as revealed by monoclonal anti-purkinje cell antibody mabQ113. J Comp Neurol 256: 29–41.

Hayashi M et al. (2015). Prenatal cerebellar hemorrhage: fetal and postnatal neuroimaging findings and postnatal outcome. Pediatr Neurol 52: 529–534.

Heibel M, Heber R, Bechinger D et al. (1993). Early diagnosis of perinatal cerebral lesions in apparently normal full-term newborns by ultrasound of the brain. Neuroradiology 35: 85–91.

Horsch S, Kutz P, Roll C (2010). Late germinal matrix hemorrhage-like lesions in very preterm infants. J Child Neurol 809–814.

Hortensius LM et al. (2018). Neurodevelopmental consequences of preterm isolated cerebellar hemorrhage: a systematic review. Pediatrics 142: e20180609.

Hoshino M (2006). Molecular machinery governing GABAergic neuron specification in the cerebellum. Cerebellum 5: 193–198.

Humphrey T (1960). The development of the pyramidal tracts in human fetuses,correlated with cortical differentiation. In: DB Tow-er, JP Schadé (Eds.), Structure and function of the cerebral Cortex. Elsevier, Amsterdam, pp. 93–103.

Iliff JJ, Wang M, Liao Y et al. (2012). A paravascular pathway facilitates CSF flow through the brain parenchyma and the clearance of interstitial solutes, including amyloid beta. Sci Transl Med 4: 147ra111.

Isasegi IZ, Rados M, Krsnik Z et al. (2018). Interactive histogenesis of axonal strata and proliferative zones in the human fetal cerebral wall. Brain Struct Funct 223: 3919–3943.

Ito M (2006). Cerebellar circuitry as a neuronal machine. Prog Neurobiol 78: 272–303.

Johnsen SD et al. (2002). Cerebellar infarction: an unrecognized complication of very low birthweight. J Child Neurol 17: 320–324.

Joyner AL, Willett R, Lawton A (2017). Cellular and genetic programs underlying cerebellum development. In: H Marzban (Ed.), Development of the cerebellum from molecular aspects to diseases. Springer International Publishing, Cham, pp. 45–65.

Kadic AS, Predojevic M (2012). Fetal neurophysiology according to gestational age. Semin Fetal Neonatal Med 17: 256–260.

Kasprian G, Langs G, Brugger PC et al. (2011). The prenatal origin of hemispheric asymmetry: an in utero neuroimaging study. Cereb Cortex 21: 1076–1083.

Kathuria S, Gregg L, Chn J et al. (2011). Normal arterial cerebral development and variations. Semin Ultrasound CT MR 32: 242–251.

Kersbergen KJ, Leroy F, Išgum I et al. (2016). Relation between clinical risk factors, early cortical changes, and neurodevelopmental outcome in preterm infants. Neuroimage 142: 301–310.

Keunen K, Counsell SJ, Benders MJ (2017). The emergence of functional architecture during early brain development. Neuroimage 160: 2–14.

Kinoshita Y, Okudera T, Tsuru E et al. (2001). Volumetric analysis of the germinal matrix and lateral ventricles performed using MR images of postmortem fetuses. Am J Neuroradiol 22: 382–388.

Koning IV et al. (2017). Impacts on prenatal development of the human cerebellum: a systematic review. J Matern Fetal Neonatal Med 30: 2461–2468.

Kopuz C, Aydin ME, Kale A et al. (2010). The termination of superior sagittal sinus and drainage patterns of the lateral, occipital at confluens sinuum in newborns: clinical and embryological implications. Surg Radiol Anat 32: 827–833.

Kostovic I, Jovanov-Milosevic N (2006). The development of cerebral connections during the first 20-45 weeks' gestation. Semin Fetal Neonatal Med 11: 415–422.

Kraus I, Jirásek JE (2002). Some observations of the structure of the choroid plexus and its cysts. Prenat Diagn 22: 1223–1228.

Kriegstein A, Alvarez-Buylla A (2009). The glial nature of embryonic and adult neural stem cells. Annu Rev Neurosci 32: 149–184.

Kuban KCK, Gilles FH (1985). Human telencephalic angiogenesis. Ann Neurol 17: 539–548.

Kuhn S, Gritti L, Crooks D et al. (2019). Oligodendrocytes in development, myelin generation and beyond. Cell 8: 1424.

Lai T, Jabaudon D, Molyneaux BJ et al. (2008). SOX5 controls the sequential generation of distinct corticofugal neuron subtypes. Neuron 57: 232–247.

Larroche J-C (1972). Sub-ependymal pseudo-cysts in the newborn. Biol Neonate 21: 170–183.

Lefèvre J, Germanaud D, Dubois J et al. (2016). Are developmental trajectories of cortical folding comparable between cross-sectional datasets of fetuses and preterm newborns? Cereb Cortex 26: 3023–3035.

Leto K et al. (2016). Consensus paper: cerebellar development. Cerebellum 15: 789–828.

Levine D, Barnes PD (1999). Cortical maturation in normal and abnormal fetuses as assessed with prenatal MR imaging. Radiology 210: 751–758.

Lewitus E, Kelavaand I, Huttner WB (2013). Conical expansion of the outer subventricular zone and the role of neocortical folding in evolution and development. Front Hum Neurosci 7: 1–10, Article 424.

Leyva-Díaz E, López-Bendito G (2013). In and out from the cortex: development of major forebrain connections. Neuroscience 254: 26–44.

Lichtenstein D, Saifi R, Augarde R et al. (2001). The Internal jugular veins are asymmetric. Usefulness of ultrasound before catheterization. Intensive Care Med 27: 301–305.

Lòpez-Bendito G, Flames N, Ma L et al. (2007). Robo1 and Robo2 cooperate to control the guidance of major axonal tracts in the mammalian forebrain. J Neurosci 27: 3395–3407.

Lu JH, Emons D, Kowalewski S (1992). Connatal periventricular pseudocysts in the neonate. Pediatr Radiol 22: 55–58.

Macchi V, Porzionato D, Guidolin D et al. (2005). Morphogenesis of the posterior inferior cerebellar artery with three-dimensional reconstruction of the late embryonic vertebrobasilar system. Surg Radiol Anat 27: 56–60.

Malinger G et al. (2006). Fatal outcome following foetal cerebellar haemorrhage associated with placental thrombosis. Eur J Paediatr Neurol 10: 93–96.

Marín-Padilla M (1996). Developmental neuropathology and impact of perinatal brain damage. I: Hemorrhagic lesions of neocortex. J Neuropathol Exp Neurol 55 (7): 758–773.

Marín-Padilla M (1997). Developmental neuropathology and impact of perinatal brain damage. II: white matter lesions of the neocortex. J Neuropathol Exp Neurol 56 (3): 219–235.

Marin-Padilla M (2012). The human brain intracerebral microvascular system: development and structure. Front Neuroana 6: 38.

Martin R, Roessmann U, Fanaroff A (1976). Massive intracerebellar hemorrhage in low-birth-weight infants. J Pediatr 89: 290–293.

McAllister II JP (2012). Pathophysiology of congenital and neonatal hydrocephalus. Semin Fetal Neonatal Med Fetal Neurol 1: 285–294.

Ment LR, Stewart WB, Ardito TA et al. (1992). Indomethacin promotes germinal matrix microvessel maturation in the newborn beagle pup. Stroke 23: 1132–1137.

Mito T, Ando Y, Takeshita K et al. (1989). Ultrasonographical and morphological examination of subependymal cystic lesions in maturely born infants. Neuropediatrics 20: 211–214.

Miyata T et al. (2010). Migration, early axonogenesis, and Reelin-dependent layer-forming behavior of early/posterior-born Purkinje cells in the developing mouse lateral cerebellum. Neural Dev 5: 23.

Moffat DB (1961). The development of the posterior cerebral artery. J Anat 95: 485–495.

Molyneaux BJ, Arlotta P, Hirata T et al. (2005). Fezl is required for the birth and specification of corticospinal motor neurons. Neuron 47: 817–831.

Moore KL, Persaud TVN, Torchia MG (2016). Before we are born: essentials of embryology and birth defects, ninth edn. Elsevier.

Müller F, O'Rahilly R (1990). The human brain at stages 21–23, with particular reference to the cerebral cortical plate and to the development of the cerebellum. Anat Embryol (Berl) 182: 375–400.

Murphy NP, Rennie J, Cooke RWI (1989). Cranial ultrasound assessment of gestational age in low birthweight infants. Arch Dis Child 64: 569–572.

Murugan M, Ling E-A, Kaur C (2013). Dysregulated glutamate uptake by astrocytes causes oligodendroglia death in hypoxic periventricular white matter damage. Mol Cell Neurosci 56: 342–354.

Muthukumar N, Palaniappan P (1998). Tentorial venous sinuses: an anatomic study. Neurosurgery 42: 363–371.

Nadarajah B, Parnavelas JG (2002). Modes of neuronal migration in the developing cerebral cortex. Nat Rev Neurosci 3: 423–432.

Nadarajah B, Brunstrom JE, Grutzendler J et al. (2001). Two modes of radial migration in early development of the cerebral cortex. Nat Neurosci 4: 143–150.

Nadarajah B, Alifragis P, Wong RO et al. (2002). Ventricle-directed migration in the developing cerebral cortex. Nat Neurosci 5: 218–224.

Naidich TP, Grant JL, Altman N et al. (1994). The developing cerebral surface. Neuroimaging Clin N Am 2: 201–224.

Najm J et al. (2008). Mutations of CASK cause an X-linked brain malformation phenotype with microcephaly and hypoplasia of the brainstem and cerebellum. Nat Genet 40: 1065–1067.

Nakada T, Kwee IL (2019). Fluid dynamics inside the brain barrier: current concept of interstitial flow, glymphatic flow, and cerebrospinal fluid circulation in the brain. Neuroscientist 25 (2): 155–166.

Nelson MD, Gonzalez-Gomez I, Gilles FH (1991). Dyke award. The search for human telencephalic ventriculofugal arteries. Am J Neuroradiol 12: 215–222.

Netsky MG, Shuangshoti S (1975). The choroid plexus in health and disease, Wright, Bristol.

Niesen CE (2002). Malformations of the posterior fossa: current perspectives. Semin Pediatr Neurol 9: 320–334.

Nieuwenhuys R (1967). Comparative anatomy of the cerebellum. In: CA Fox, RS Snider (Eds.), The cerebellum, vol. 25, Pages ii-viii, 1–355, Progress in Brain Research. Elsevier 1–93.

Nieuwenhuys R, Voogd J, van Huijzenz C (1988). The human central nervous system, third revised edn Springer-Verlag.

Nishikuni K, Carvalhal Ribas G (2013). Study of fetal and postnatal morphological development of the brain sulci. J Neurosurg Pediatr 11: 1–11.

Nomura ML et al. (2009). Prenatal diagnosis of isolated fetal cerebellar hemorrhage associated with maternal septic shock. Prenat Diagn 29: 169–171.

O'Rahilly R, Müller F (1986). The meninges in human development. J Neuropathol Exp Neurol 45: 588–608.

O'Rahilly R, Müller F (2006). The embryonic human brain. An atlas of developmental stages, third edn Wiley-Liss, Hoboken, NJ.

O'Rahilly R, Müller F (2008). Significant features in the early prenatal development of the human brain. Ann Anat 190: 105–118.

Okudera T, Huang YP, Ohta T et al. (1994). Development of posterior fossa dural sinuses, emissary veins, and jugular bulb: morphological and radiologic study. Am J Neuroradiol 15: 1871–1883.

Ortega JA, Memi F, Radonjic N et al. (2018). The subventricular zone: a key player in human neocortical development. Neuroscientist 24: 156–170.

Padget DH (1948). The development of the cranial arteries in the human embryo. Contrib Embryol 212: 205–271.

Padget DH (1956). The cranial venous system in man in reference to development, adult configuration, and relation to the arteries. Am J Anat 98: 307–355.

Pal BR, Preston PR, Morgan MEI et al. (2001). Frontal horn thin walled cysts in preterm neonates are benign. Arch Dis Child Fetal Neonatal Ed 85: F187–F193.

Paladini D, Quarantelli M, Pastore G et al. (2012). Abnormal or delayed development of the posterior membranous area of the brain: anatomy, ultrasound diagnosis, natural history and outcome of Blake's pouch cyst in the fetus. Ultrasound Obstet Gynecol 39: 279–287.

Paneth N, Rudelli R, Kazam E et al. (1994). Brain damage in the preterm infant. In: Clinics in developmental medicine no. 131, Mac Keith Press, London.

Pape KE, Wigglesworth JS (1979). Haemorrhage, ischaemia and the perinatal brain. In: Clinics in developmental medicine no. 69/70, Spastics International Medical Publications, London.

Paredes MF, James D, Gil-Perotin S et al. (2016). Extensive migration of young neurons into the infant human frontal lobe. Science 7: 354.

Parisi MA, Dobyns WB (2003). Human malformations of the midbrain and hindbrain: review and proposed classification scheme. Mol Genet Metab 80: 36–53.

Park KM, Gerecht S (2014). Harnessing developmental processes for vascular engineering and regeneration. Development 141: 2760–2769.

Patel S, Barkovich AJ (2002). Analysis and classification of cerebellar malformations. Am J Neuroradiol 23: 1074–1087.

Paus T, Tomaiuolo F, Otky N et al. (1996). Human cingulate and paracingulate sulci: pattern, variability, asymmetry and probabilistic map. Cereb Cortex 6: 207–214.

Pierson CR, Al Sufiani F (2016). Preterm birth and cerebellar neuropathology. Semin Fetal Neonatal Med 21: 305–311.

Pistorius LR, Stoutenbeek P, Groenendaal F et al. (2010). Grade and symmetry of normal fetal cortical development: a longitudinal two- and three-dimensional ultrasound study. Ultrasound Obstet Gynecol 36: 700–708.

Poretti A, Prayer D, Boltshauser E (2009a). Morphological spectrum of prenatal cerebellar disruptions. Eur J Paediatr Neurol 13: 397–407.

Poretti A, Prayer D, Boltshauser E (2009b). Review article. Morphological spectrum of prenatal cerebellar disruptions. Eur J Paediatr Neurol 13: 397–407.

Poretti A, Boltshauser E, Doherty D (2014). Cerebellar hypoplasia: differential diagnosis and diagnostic approach. Am J Med Genet C Semin Med Genet 166: 211–226.

Poretti A, Boltshauser E, Huisman TAGM (2016). Prenatal cerebellar disruptions: neuroimaging spectrum of findings in correlation with likely mechanisms and etiologies of injury. Neuroimaging Clin N Am 26: 359–372.

Price DJ, Jarman AP, Mason JO et al. (2011). Building brains: an introduction to neural development, Wiley-Blackwell.

Rademaker KJ, De Vries LS, Barth PG (1993). Subependymal pseudocysts: ultrasound diagnosis and findings at follow-up. Acta Paediatr Scand 82: 394–399.

Raets MM, Sol JJ, Govaert P et al. (2013). Serial cranial US for detection of cerebral sinovenous thrombosis in preterm infants. Radiology 269: 879–886.

Raets M, Dudink J, Raybaud C et al. (2015). Brain vein disorders in newborn infants. Dev Med Child Neurol 57: 229–240.

Rahimi Balaei M, Ashtari N, Bergen H (2017). The embryology and anatomy of the cerebellum. In: H Marzban (Ed.), Development of the cerebellum from molecular aspects to diseases. Springer International Publishing, Cham, pp. 33–43.

Ramenghi LA, Domizzio S, Quartulli L et al. (1997). Prenatal pseudocysts of the germinal matrix in preterm infants. J Clin Ultrasound 25: 169–173.

Ranaivoson FM, von Daake S, Comoletti D (2016). Structural insights into reelin function: present and future. Front Cell Neurosci 10: 137. Available at: https://www.frontiersin.org/article/10.3389/fncel.2016.00137.

Ranzini AC et al. (1998). Prenatal sonographic appearance of hemorrhagic cerebellar infarction. J Ultrasound Med 17: 725–727.

Raybaud C (2010a). Normal and abnormal embryology and development of the intracranial vascular system. Neurosurg Clin N Am 21: 399–426.

Raybaud C (2010b). The corpus callosum, the other great forebrain commissures, and the septum pellucidum: anatomy, development, and malformation. Neuroradiology 52: 447–477.

Retzius G (1896). Das Menshenhirn: Studien in der Makroskopischen Morphologie, PA Norstedt, Sockholm, pp. 1–167.

Righini A, Cesaretti C, Conte G et al. (2016). Expanding the spectrum of human ganglionic eminence region anomalies on fetal magnetic resonance imaging. Neuroradiology 58: 293–300.

Risau W (1997). Mechanisms of angiogenesis. Nature 386: 671–674.

Robinson AJ, Goldstein R (2007). The cisterna magna septa. J Ultrasound Med 26: 83–95.

Rodriguez EM, Oksche A, Montecinos H (2001). Human subcommissural organ, with particular emphasis on its secretory activity during the fetal life. Microsc Res Tech 52: 573–590.

Rossi A et al. (2016). Neuroimaging of infectious and inflammatory diseases of the pediatric cerebellum and brainstem. Neuroimaging Clin N Am 26: 471–487.

Sadler TW (2011). Langman's medical embryology, sixth edn Lippincott Williams & Wilkins.

Sajan SA, Waimey KE, Millen KJ (2010). Novel approaches to studying the genetic basis of cerebellar development. Cerebellum 9: 272–283.

Sakai N, Kaprielian Z (2012). Guidance of longitudinally projecting axons in the developing central nervous system. Front Mol Neurosci 5: 59.

Sakkaa L, Coll G, Chazala J (2011). Anatomy and physiology of cerebrospinal fluid. Eur Ann Otorhinolaryngol Head Neck Dis 128: 309–316.

Salas E, Ziyal IM, Sekhar LN et al. (1998). Persistent trigeminal artery: an anatomic study. Neurosurgery 43: 557–561.

Sanapo L et al. (2017). Fetal anemia, cerebellar hemorrhage and hypoplasia associated with congenital Parvovirus infection. J Matern Fetal Neonatal Med 30: 1887–1890.

Sawada K, Fukunishi K, Kashima M et al. (2012). Fetal gyrification in cynomolgus monkeys: a concept of developmental stages of gyrification. Anat Rec 295: 1065–1074.

Schmahmann JD (2019). The cerebellum and cognition. Neurosci Lett 688: 62–75.

Schmahmann JD, Sherman JC (1998). The cerebellar cognitive affective syndrome. Brain 121: 561–579.

Scott TG (1963). A unique pattern of localization within the cerebellum. Nature 200: 793.

Scott JA, Habas PA, Kim K et al. (2011). Growth trajectories of the human fetal brain tissues estimated from 3D reconstructed in utero MRI. Int J Dev Neurosci 29: 529–536.

Shackelford GD, Fulling KH, Glasier CM (1983). Cysts of the subependymal germinal matrix: sonographic demonstration with pathologic correlation. Radiology 149: 117–121.

Shah DK et al. (2006). Reduction in cerebellar volumes in preterm infants: relationship to white matter injury and neurodevelopment at two years of age. Pediatr Res 60: 97–102.

Sharony R et al. (1999). Prenatal diagnosis of fetal cerebellar lesions: a case report and review of the literature. Prenat Diagn 19: 1077–1080.

Shen E-Y, Huang F-Y (1985). Subependymal cysts in normal neonates. Arch Dis Child 60: 1072–1074.

Simonazzi G et al. (2016). Fetal cerebellar damage in fetuses with severe anemia undergoing intrauterine transfusions. J Matern Fetal Neonatal Med 29: 389–392.

Slagle TA, Oliphant M, Gross SJ (1989). Cingulate sulcus development in preterm infants. Pediatr Res 26: 598–602.

Spasojevic G, Stojanovic Z, Suscevic D et al. (2010). Morphological variations of the limbic-lobar border cortex on the inner side of the human brain hemisphere. Period Biol 112: 89–95.

Stoodley CJ, Schmahmann JD (2018). Functional topography of the human cerebellum. Handb Clin Neurol 154: 59–70.

Streeter GL (1915). The development of the venous sinuses of the dura mater in the human embryo. Am J Anat 18: 145–178. https://embryology.med.unsw.edu.au/embryology/index.php/Paper_-_The_development_of_the_venous_sinuses_of_the_dura_mater_in_the_human_embryo.

Striedter GF, Srinivasan S, Monuki ES (2015). Cortical folding: when, where, how, and why? Annu Rev Neurosci 38: 291–307.

Suzuki Y, Matsumoto K (2000). Variations of the superficial middle cerebral vein: classification using three-dimensional CT angiography. Am J Neuroradiol 21: 932–938.

Tanoue S, Kiyosue H, Okahara M et al. (2006). Para-cavernous sinus venous structures: anatomic variations and pathologic conditions evaluated on fat-suppressed 3D fast gradient-echo MR images. Am J Neuroradiol 27: 1083–1089.

ten Donkelaar HJ (2011). Clinical neuroanatomy brain circuitry and its disorders ch 2. In: Vascularization of the brain and spinal cord, Springer Verlag.

ten Donkelaar HJ, Lammens M, Wesseling P et al. (2004). Development and malformations of the human pyramidal tract. J Neurol 251: 1429–1442.

Thun-Hohenstein L, Forster I, Kunzle C et al. (1983). Transient bifrontal solitary periventricular cysts in term neonates. Neuroradiology 36: 241–244.

Timor-Tritsch I, Monteagudo A, Pilu G et al. (2012). Ultrasonography of the prenatal brain, third edn McGraw–Hill Professional60–63.

Toi A, Lister WS, Fong KW (2004). How early are fetal cerebral sulci visible at prenatal ultrasound and what is the normal pattern of early fetal sulcal development ? Ultrasound Obstet Gynecol 24: 706–775.

Trawber R, Rao S, Srinivasjois R et al. (2010). Outcomes of preterm neonates with frontal horn cysts: a retrospective study. J Child Neurol 25: 1377–1381.

Tsuru A, Mizuguchi M, Takashima S (1995). Cystic leukomalacia in the cerebellar folia of premature infants. Acta Neuropathol 90: 400–402.

Turner OA (1948). Growth and development of the cerebral cortical pattern in man. Arch Neurol Psychol 59: 1–12.

Ulfig N (2002). Ganglionic eminence of the human fetal brain-new vistas. Anat Rec 267: 191–195.

Vasovic L, Jovanovic ID, Ugrenovic SZ et al. (2008). Normal subtypes of the posterior part of the cerebral arterial circle in human fetuses. Surg Neurol 70: 287–294.

Vasovic L, Jovanovic I, Ugrenovic S et al. (2012). Trigeminal artery: a review of normal and pathological features. Childs Nerv Syst 28: 33–46.

Vasung L, Lepage C, Rados M et al. (2016). Quantitative and qualitative analysis of transient fetal compartments during prenatal human brain development. Front Neuroanat 10: 11.

Verbeek E, Meuwissen ME, Verheijen FW et al. (2012). COL4A2 mutation associated with familial porencephaly and small-vessel disease. Eur J Hum Genet 20: 844–851.

Vinukonda G, Dummula K, Malik S et al. (2010). Effect of prenatal glucocorticoids on cerebral vasculature of the developing brain. Stroke 41: 1766–1773.

Voetmann E (1949). On the structure and surface area of the human choroid plexuses: a quantitative anatomical study. Acta Anat 8: 1–116.

Volpe JJ (2009). Cerebellum of the premature infant: rapidly developing, vulnerable, clinically important. J Child Neurol 24: 1085–1104.

Voogd J (2002). The anatomy of the cerebellum. Trends Neurosci 21: 370–375. https://doi.org/10.1016/s0166-2236(98)01318-6.

Voogd J (2017). The development of the cerebellum: from the beginnings. In: H Marzban (Ed.), Development of the cerebellum from molecular aspects to diseases. Springer International Publishing, Cham, pp. 1–32.

Vossough A, Limperopoulos C, Putt ME et al. (2013). Development and validation of a semiquantitative brain maturation score on fetal MR images: initial results. Radiology 268: 200–207.

Vue TY, Bluske K, Alishahi A et al. (2009). Sonic hedgehog signaling controls thalamic progenitor identity and nuclei specification in mice. J Neurosci 29: 4484–4497.

Wang SSH, Kloth AD, Badura A (2014). The cerebellum, sensitive periods, and autism. Neuron 83: 518–532.

Williams RS (1981). Haemorrhage, ischaemia and the perinatal brain (clinics in developmental medicine, nos. 69/70). eds. Karen E. Pape and J. S. Wigglesworth, with a preface by Gordon B. Avery Spastics International Medical Publications, 1979. Ann Neurol 9: 516.

Worthen NJ, Gilbertson V, Lau C (1986). Cortical sulcal development seen on sonography: relationship to gestational parameters. J Ultrasound Med 5: 153.

Xu H, Hu F, Sado Y et al. (2008). Maturational changes in laminin, fibronectin, collagen IV, and perlecan in germinal matrix, cortex, and white matter and effect of betamethasone. J Neurosci Res 86: 1482–1500.

Yakovlev PL, Lecours AR (1967). The myelogenetic cycles of regional maturation of the brain. In: A Minkowski (Ed.), Resional development of the brain in early life. Blackwell, Oxford, pp. 3–70.

Yoshida S et al. (2008). Pontine hypoplasia in children with periventricular leukomalacia. Am J Neuroradiol 29: 425–430.

Zhang Y, Inder TE, Neil JJ et al. (2015). Cortical structural abnormalities in very preterm children at 7 years of age. Neuroimage 109: 469–479.

Zhao H, Maruyama T, Hattori Y (2011). A molecular mechanism that regulates medially oriented axonal growth of upper layer neurons in the developing neocortex. J Comp Neurol 519: 834–848.

Zilles K, Palomero-Gallagher N, Amunts K (2013). Development of cortical folding during evolution and ontogeny. Trends Neurosci 36: 275–284.

Handbook of Clinical Neurology, Vol. 171 (3rd series)
Neurology and Pregnancy: Pathophysiology and Patient Care
E.A.P. Steegers, M.J. Cipolla and E.C. Miller, Editors
https://doi.org/10.1016/B978-0-444-64239-4.00015-1

Chapter 15

Diagnosis and management of congenital neurologic disease during pregnancy

ARTHUR M. MANDEL*

The Neurological Institute, Columbia University Irving Medical Center, New York, NY, United States

Abstract

Advanced techniques in neuroimaging and genetics, as well as the publication of longer-term prognostic studies, have led to fetal neurology becoming an essential part of prenatal obstetric care. A multidisciplinary approach to providing prenatal counseling is now commonly used in most academic medical centers.

Common conditions seen are ventriculomegaly, agenesis of the corpus callosum and other midline abnormalities, and posterior fossa anomalies. The prognosis frequently depends on the severity of the condition and the presence of other anomalies.

Certain pathologic processes, such as myelomeningocele, can be treated surgically in the prenatal period, and there is ongoing research regarding potential treatments of other conditions such as tuberous sclerosis. Acquired conditions such as hemorrhage, tumor, and ischemic stroke can also be seen prenatally, and their early diagnosis may inform postnatal care.

The development of fetal ultrasonography, fetal MRI, and amniocentesis with advanced genetic techniques has resulted in tremendous progress in the field of fetal neurology. Typically, central nervous system pathology seen on the anatomical ultrasound at 18–20 weeks leads to a fetal MRI and subsequent visits with a pediatric neurologist, neurosurgeon, neonatologist, geneticist, and high-risk obstetrician as indicated. Certain conditions, such as acrania/anencephaly and Walker-Warburg syndrome, may be detected by ultrasonography earlier in the pregnancy (Katorza et al., 2018).

For proper prenatal counseling, it is essential to obtain a family history and obstetric history (including prior miscarriages, travel history, medication history, and history of illness). In addition, since timing may be of utmost importance, the clinician must be familiar with local and national laws regarding termination of pregnancy. Perhaps most importantly, parents of the fetus may be emotionally overwhelmed and may find it difficult to understand technical information in a single visit, and so repeated visits with multiple specialists can add clarity.

This chapter will discuss various common and uncommon conditions seen in a fetal neurology practice and focus on epidemiology, etiology, pathology, clinical associations, and prognosis.

FETAL VENTRICULOMEGALY

Fetal ventriculomegaly is a very common neuroimaging anomaly seen prenatally, with an estimated prevalence of 0.5–2 in 1000 births (Edwards, 1958; Laurence et al., 1968; Garel et al., 2003; Nomura et al., 2010). It should ideally be regarded as a common end result rather than a specific entity, as it has multiple causes with variable prognoses. In addition to etiology, the degree of

*Correspondence to: Arthur M. Mandel, M.D., Ph.D., Assistant Professor of Neurology at CUMC, Director of Fetal Neurology, Division of Pediatric Neurology, The Neurological Institute, Columbia University Irving Medical Center, 180 Fort Washington Ave, 5th floor, New York, NY 10032, United States. Tel: +1-212-342-6867, Fax: +1-212-342-6865, E-mail: amm28@cumc.columbia.edu

ventriculomegaly also determines prognosis. The measurement is made at the atrium of the lateral ventricle, which remains stable between the 15th and 40th week of pregnancy (Cardoza et al., 1988; Pilu et al., 1989). Ventriculomegaly is defined as an atrial width >10mm and commonly is subdivided into mild (10–12mm), moderate (12.1–15mm), and severe (>15mm) (Fox et al., 2018) (Fig. 15.1). The ventricular size is typically measured using ultrasonography, and ventriculomegaly is usually detected during a fetal survey at 18–20 weeks.

It is important to perform a detailed evaluation of the other parts of the ventricular system to determine whether there is an obstructive process. Ventriculomegaly is distinguished from hydrocephalus, which specifically refers to increased cerebrospinal fluid pressure resulting in ventricular dilatation.

An important distinction is whether the ventriculomegaly is isolated or not. A comprehensive ultrasound evaluation of the central nervous system and the extracranial organ systems should be performed by a clinician experienced in the diagnosis of fetal anomalies. Fetal MRI has also been used to assess for other CNS abnormalities, with studies reporting up to a 22% chance of finding additional abnormalities, most commonly agenesis of the corpus callosum (Griffiths et al., 2010; Rossi and Prefumo, 2014). The finding of additional anomalies can be significant for prognosis, and the absence of other abnormalities can provide reassurance regarding a more favorable outcome. However, there is also some evidence that in the hands of an experienced clinician using dedicated neurosonography, no additional information is obtained from MRI in the case of mild isolated ventriculomegaly (Paladini et al., 2019).

Amniocentesis is helpful for diagnosis or exclusion of a genetic or infectious cause. Infection, although seen in only <2% of fetal ventriculomegaly cases, can inform prognosis; the most common infectious agents seen are toxoplasmosis and CMV; other typical findings such as calcification may not be present (Senat et al., 1999; Pasquini et al., 2014). There is some initial indication that prenatal antiviral therapy may be helpful for treatment of CMV infection (Leruez-Ville et al., 2016). Exposure risk for Zika virus infection, another potential cause of ventriculomegaly, should be determined, although a longitudinal study of Zika-exposed fetuses showed only a very small proportion with fetal neuroimaging anomalies (including ventriculomegaly) (Mulkey et al., 2019). There have also been cases of ventriculomegaly associated with other viruses, including parvovirus B19, LCMV, enterovirus 71, and parainfluenza virus type 3 (Katz et al., 1996; Seidman et al., 1996; Chow et al., 2000; Jamieson et al., 2006).

There are specific genetic mutations causing congenital hydrocephalus, most notably in the L1CAM gene on the X chromosome, which also frequently causes adducted thumbs and may also cause other limb abnormalities or agenesis of the corpus callosum (Weller and Gartner, 2001; Vos and Hofstra, 2010). A family history may contribute to the diagnosis. Typically, L1CAM hydrocephalus results in ventriculomegaly in the severe range in males. More generally, ventriculomegaly may be associated with an abnormal karyotype or microarray study (most commonly trisomy 21) (Agathokleous et al., 2013; Bronshtein et al., 2017). Amniocentesis may also be useful to assess levels of alpha-fetoprotein and acetylcholinesterase, probably reflecting an occult neural tube defect (NTD), which can also be associated with ventriculomegaly (Coleman et al., 2015).

Other potential etiologies of ventriculomegaly include intracranial hemorrhage (which can obstruct CSF flow or interfere with the resorption of CSF by the arachnoid villi), posterior fossa malformation such as a Dandy-Walker malformation (DWM) (obstructing flow at the level of the foramina of Luschka and

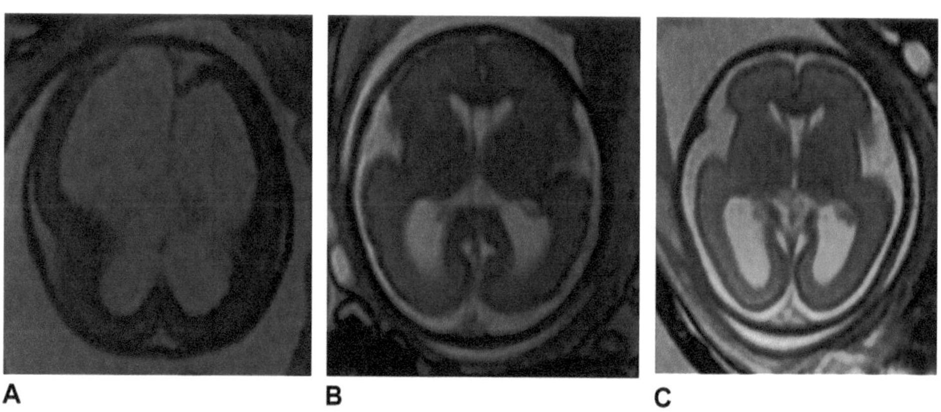

Fig. 15.1. Axial gradient echo MRI showing bilateral ventriculomegaly: (A) 21 weeks of gestation, severe ventriculomegaly (measuring 21mm on right, 24.6mm on left); (B) 28 weeks of gestation, moderate ventriculomegaly (1.4mm on right, 1.5mm on left); and (C) 27 weeks of gestation, mild ventriculomegaly (12mm bilaterally).

Magendie), CNS malformations such as holoprosencephaly or agenesis of the corpus callosum (affecting CSF flow), tumors (obstruction of flow), or the very rare choroid plexus papilloma (excessive CSF production) (McKechnie et al., 2012).

For isolated mild and moderate ventriculomegaly, prognosis is usually good (with the caveat that true isolation cannot be definitively determined prenatally). A 2014 meta-analysis of 20 studies assessing ventriculomegaly of 10–15 mm, postnatally confirmed to be an isolated finding, showed a prevalence of developmental delay of 7.9%, similar to the rate in the general population (Pagani et al., 2014). Subdividing mild and moderate ventriculomegaly results in little, if any, difference (Melchiorre et al., 2009; Bar-Yosef et al., 2017). Severe ventriculomegaly is more likely to cause cognitive disability: a 2018 meta-analysis of 11 studies of isolated severe ventriculomegaly showed mild/moderate disability in 18.6% and severe disability in 39.6%. In addition, the rate of neonatal demise or stillbirth was 12.1% (as compared with 97%–98% survival in mild to moderate cases) (Gaglioti et al., 2005; Carta et al., 2018). If there are other abnormalities present, whether in the central nervous system or elsewhere, the outcome is typically worse, depending on the specific anomaly.

Unilateral ventriculomegaly, which occurs in approximately 50% of cases, usually carries a similar prognosis as bilateral ventriculomegaly (Falip et al., 2007; Griffiths et al., 2017a, b). Ventricular asymmetry, when there is no evidence of ventriculomegaly, has not been associated with future neurodevelopmental disability (Meyer et al., 2018).

Recommended practice is to repeat at least one ultrasound in the third trimester to assess for change. Isolated mild or moderate ventriculomegaly may improve or even resolve later in the pregnancy, typically with a good outcome. In a 2009 report, pooled data analysis of 13 studies of ventriculomegaly between 10 and 15 mm showed resolution in 159/468 (34%), stability in 248/445 (55.7%), and progression in 54/344 (15.7%) (not all studies included in the analysis described all three outcomes) (Melchiorre et al., 2009).

MIDLINE ABNORMALITIES

Abnormalities of the Corpus Callosum

The corpus callosum primarily develops between the 8th and 20th week of gestation and plays a significant role in connecting the two cerebral hemispheres, thereby integrating visual, sensory, motor, and cognitive processes (Hinkley et al., 2012; Palmer and Mowat, 2014). Abnormalities in its development include complete agenesis (ACC), partial agenesis (pACC), hyperplasia, hypoplasia, or dysplasia (Leombroni et al., 2018a). By far, the most common anomalies seen prenatally are agenesis, both partial and complete, with estimated prevalence of 1 in 4000–5000 live births, although with 2%–3% rates in neurodevelopmentally disabled patients (Paul et al., 2007; Glass et al., 2008; Sotiriadis and Makrydimas, 2012) (Fig. 15.2).

ACC or pACC can be isolated, or there may be associated findings such as posterior fossa abnormalities, interhemispheric cysts, and neuronal migration disorders. There may also be findings outside the CNS, most commonly craniofacial (e.g., cleft lip/palate, hypertelorism, macrocephaly), but also involving the genitourinary, cardiac, gastrointestinal, or musculoskeletal systems (Tang et al., 2009; Santo et al., 2012). Diagnosis is typically made with transvaginal ultrasound in the 18th–20th week, with absence of the corpus callosum seen by direct visualization in the coronal and sagittal planes. The pericallosal arteries are typically not visualized. In the axial plane, certain findings raise suspicion of agenesis, including colpocephaly, and laterally displaced, narrow frontal horns of the lateral ventricles. The absence of the cavum septum pellucidum (CSP), which is normally present between 18 and 37 weeks, is

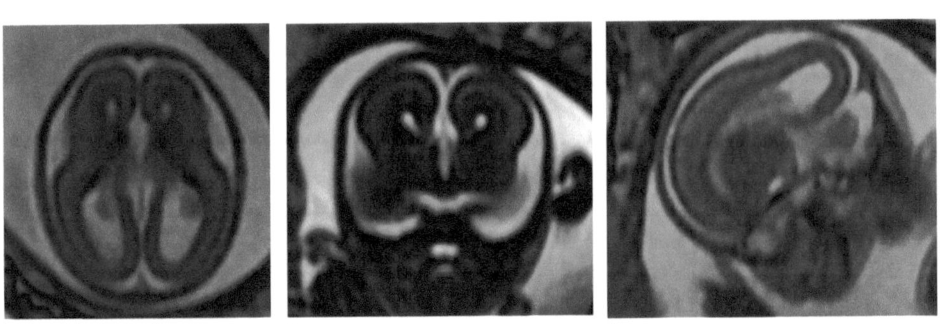

Fig. 15.2. Agenesis of the corpus callosum and mild ventriculomegaly, at 21 weeks of gestation. Gradient echo sequences in the (A) axial, (B) coronal, and (C) sagittal planes. Note the presence of colpocephaly in the axial plane, and the absence of the corpus callosum in the coronal and sagittal planes.

also common. However, this is nonspecific and can also be seen in a variety of other conditions, such as hydrocephalus, holoprosencephaly, schizencephaly, and porencephaly (Leombroni et al., 2018a).

In pACC, the CSP is frequently present, making diagnosis more challenging. Biometric tables of the normal corpus callosum may be helpful; the main finding is the relatively short rostral-caudal length of the corpus callosum (Rizzo et al., 2011; Pashaj et al., 2013). There may also be associated findings such as an abnormally shaped CSP (Shen et al., 2015).

Fetal MRI can be performed for confirmation of findings and also to assess for associated anomalies, both within the CNS and extra-axially. A subgroup analysis of ACC in a multicenter prospective cohort study showed a diagnostic accuracy of 40.0% for ultrasound and 92.7% for MRI, with a change in prognostic information in 45.6% cases after MRI, with 44.3% resulting in a "significant," "major," or "decisive" effect on clinical management (Griffiths et al., 2017a, b). However, it is unclear whether there is such a decisive improvement with MRI when expert neurosonography is used (Paladini et al., 2017).

There is a variety of genetic conditions associated with ACC. Karyotype can reveal chromosomal conditions such as trisomy 13, trisomy 18, and mosaic trisomy 8; an abnormal karyotype has been reported to be associated with 17.8% of cases with other anomalies, and with 4.8% of isolated ACC (7.5% of isolated pACC) (Santo et al., 2012; Leombroni et al., 2018a). Reports of chromosomal microarray analysis have shown a 9.4% rate of pathogenic de novo copy number variants in cases of isolated ACC (Sajan et al., 2013; Edwards et al., 2014).

There have been a multitude of single gene mutations associated with ACC. Some of these syndromes have distinctive phenotypes, such as acrocallosal (polydactyly, intellectual disability, craniofacial anomalies), Anderman (intellectual disability, motor/sensory neuropathy), and MASA (adducted thumbs, intellectual disability, and hydrocephalus) syndromes, associated with the KIF7, SLC12A6, and L1CAM genes, respectively (Larbrisseau et al., 1984; Stumpel and Vos, 1993; Putoux et al., 2011). Whole exome and whole genome sequencing has been helpful in identifying underlying causes of agenesis if classic clinical features of a known syndrome are not present. For example, compound heterozygous variants in the CDK5RAP gene have been linked to ACC with whole exome sequencing (Jouan et al., 2016). In addition, there are genetic syndromes without known gene associations, such as Aicardi syndrome, which manifests as chorioretinal anomalies, seizures, and intellectual disability (Aicardi, 2005).

Less commonly, ACC is attributable to nongenetic etiologies, including infectious, toxic, or vascular/hypoxic conditions (Palmer and Mowat, 2014). Typically, infectious causes such as TORCH infections or Zika virus are seen in nonisolated cases. Toxic associations include fetal alcohol syndrome and maternal phenylketonuria.

The prognosis of ACC primarily depends on whether or not it is isolated. A meta-analysis of 27 studies of isolated cases (by imaging), with normal karyotype, revealed a normal outcome in 76.0% of ACC cases and 71.4% of pACC cases confirmed at birth (D'Antonio et al., 2016c). Disabilities ranged from gross and fine motor control, language, cognitive function, ocular coordination, and epilepsy. For ACC, 16.0% had a borderline/moderate outcome and 8.2% had a severe outcome. For isolated pACC, outcomes were 14.9% borderline/moderate and 12.5% severe. Predicting clinical outcome for ACC and pACC, however, remains a challenging task, because available studies use a wide variability of tools to determine neurodevelopment, vary in the timing of follow-up, and use different outcome measures. There are also potential ascertainment biases in available studies of isolated ACC or pACC, because some genetic and structural findings are not observable in the prenatal or immediate postnatal period.

Diagnosis of other abnormalities of the corpus callosum, such as dysplasia, hypoplasia (thinning of the corpus callosum), or hyperplasia (thickening of the corpus callosum), is typically made postnatally. Biometric measurements as described earlier may be helpful. Large enough studies to assess prognosis have not been published. Hypoplasia is more commonly associated with CNS or extra-axial anomalies, and multiple associations have been described, such as compression from a mass, ischemia, infection, teratogens, alcohol, and metabolic disorders. Hypoplasia of the corpus callosum has also been associated with autism (Casanova et al., 2011). It is also frequently associated with other developmental anomalies, but if isolated, may be transient with a normal outcome (Shinar et al., 2016). Dysplasia is seen in pyridoxine-dependent epilepsy and with other etiologies, including infectious or genetic causes, or structural lesions (Oesch et al., 2018).

Holoprosencephaly

Holoprosencephaly is a set of disorders resulting from failure of ventral induction, a process that begins in the fifth week of gestation. There is subsequent inability of the prosencephalon (forebrain) to undergo cleavage, and midline structures are not formed correctly (Calloni et al., 2019). It is a very common malformation, with incidences in epidemiologic studies of approximately 0.4% prenatally, although birth prevalence is much lower (approximately 1 per 10,000 live births)

(Matsunaga and Shiota, 1977; Olsen et al., 1997; Leoncini et al., 2008). There is frequent association with craniofacial anomalies, ranging from severe (e.g., cyclopia, synophthalmia, and a single nostril in a proboscis) to less severe (hypotelorism, midface hypoplasia, cleft lip/palate, ocular colobomas, a flat nasal bridge, and a single central incisor). There may be microcephaly, which can range from mild to pronounced. Extracranial manifestations such as cardiovascular malformations, clubfoot, myelomeningocele, renal anomalies, and intestinal malformations may also be present.

The general classification scheme is of four types: alobar, semilobar, lobar, and the middle interhemispheric variant. Alobar holoprosencephaly is the most severe form and is characterized by a single ventricle, absence of hemispheric division, the corpus callosum, and olfactory bulbs and tracts, and fusion of the thalamus, basal ganglia, and hypothalami (Fig. 15.3). It can be diagnosed as early as the 9th–10th week, with the absence of the interhemispheric fissure, a single ventricle, and absence of the "butterfly sign," which is the normal appearance of the choroid plexus within two separate cerebral hemispheres.

In semilobar holoprosencephaly, there is absence of the interhemispheric fissure anteriorly, with fusion of the frontal horns of the lateral ventricles. The thalami are partially fused, and the third ventricle is small. The posterior horns of the lateral ventricles are present.

Lobar holoprosencephaly may not be easy to detect until later in the second trimester as it is generally diagnosed from the absence of the CSP, which is formed at 18 weeks. There is fusion of the frontal horns of the lateral ventricles and of the fornix columns. The cingulate gyri may also be fused; the corpus callosum may be absent or hypoplastic, or it may be normal. The olfactory bulbs may be hypoplastic, and the thalami may be fused. The interhemispheric fissure is present.

The middle inhemispheric variant has well-separated occipital lobes and anterior poles of the frontal lobes, but there is failed division of the posterior frontal and parietal lobes. In addition, the corpus callosum may show partial absence, in the body (or it may be absent or hypoplastic), and the thalami/caudate nuclei are not completely separated (Simon et al., 2002). It is usually not diagnosed prenatally, unless suspicion is raised because of absence of the CSP or an anomaly of the corpus callosum.

Holoprosencephaly has multiple genetic and nongenetic causes, frequently occurring together. Maternal diabetes is a strong risk factor, with an estimated 200-fold increase in risk; potential mechanisms for injury include fetal hyperglycemia resulting in oxidative stress, apoptosis, and hypoxia, and/or interference in glycolysis, all affecting neural crest cell development (Barr Jr. et al., 1983; Ornoy et al., 2015; Summers et al., 2018). Twin pregnancy is another risk factor, as are teratogens such as alcohol, retinoic acid, and isotretinoin (Johnson and Rasmussen, 2010; Weber and Sebire, 2010). There is a concern that statins may carry risk, since genetic defects in cholesterol metabolism have been linked (Edison and Muenke, 2005). Sexually transmitted infections have also been associated (Croen et al., 2000; Miller et al., 2010).

Abnormal karytope is seen in 25%–45% of fetuses with holoprosencephaly, most commonly trisomy 13 but also triploidy and, less commonly, trisomy 18 or other trisomies (Olsen et al., 1997; Solomon et al., 2010; Petryk et al., 2015). Genetic syndromes that may manifest with holoprosencephaly include Smith-Lemli-Opitz, Pallister-Hall, Rubinstein-Taybi, and Meckel syndromes. There are also multiple genes associated, such as genes in the Sonic Hedgehog signaling pathway (SHH, GLI2, DISP1, PTCH1) and those influencing other aspects of brain development (SIX3, TGIF, ZIC2, NODAL) (Petryk et al., 2015; Leombroni et al., 2018b).

Alobar holoprosencephaly is usually lethal prenatally, and surviving infants have severe deficits, including profound intellectual disabilities, seizures, feeding difficulties, and profound motor delays. Endocrinopathies and hypothalamic dysfunction are common, resulting in potentially emergent conditions such as temperature instability and diabetes insipidus. The facial anomalies

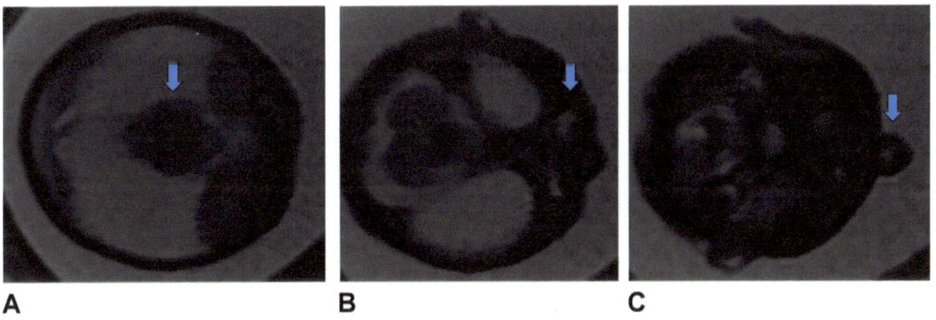

Fig. 15.3. Third trimester gestation, alobar holoprosencephaly, axial gradient echo. *Arrows* indicating (A) fused thalamus, (B) hypotelorism, and (C) proboscis with single nostril.

may impair respiratory and oromotor function, requiring a tracheostomy for survival. Survival is correlated with the severity of facial anomaly (Croen et al., 1996). There are reports of rare cases with survival into adulthood (Raam et al., 2011).

Patients with less severe forms of holoprosencephaly frequently survive beyond the neonatal period and may reach adulthood. Intellectual disability is typical, and there are commonly endocrinopathies, hypothalamic dysfunction, motor impairment, epilepsy, and respiratory disease (Levey et al., 2010).

Septo-Optic Dysplasia

Septo-optic dysplasia (SOD), also known as de Morsier syndrome, is characterized by the triad of agenesis of the corpus callosum/septum pellucidi, optic nerve hypoplasia, and hypopituitarism (De Morsier, 1956; Saranac and Gucev, 2014). It is part of a clinical spectrum, with other potential abnormalities such as schizencephaly (which has been termed "septo-optic dysplasia plus" syndrome) (Gutierrez-Castillo et al., 2018).

The optic nerve hypoplasia and the hypopituitarism may not be evident until after birth; the former may be evident later in the third trimester, especially by MRI, if not seen earlier. It may also be difficult to distinguish SOD from lobar holoprosencephaly prenatally; potential distinguishing characteristics include the shape of the frontal horns (flatter appearance in holoprosencephaly) and the fornices (fused in holoprosencephaly). Maternal urine and serum estriol levels may be low in SOD, and fetal cord blood may show hormonal deficiencies (ACTH, growth hormone, TSH) (Volpe et al., 2009; Leombroni et al., 2018b).

There have been multiple genes implicated in SOD, such as HESX1, SOX2, and SOX3, although a genetic cause is found in <1% of cases. Other causes such as viral infections, maternal cocaine use, and teratogens such as valproic acid have also been associated (McMahon and Braddock, 2001).

Prognosis is quite variable, depending on the level of hypopituitarism and associated anomalies. There may be blindness, nystagmus, strabismus, micro-ophthalmia, or even nearly normal vision. There may also be associated epilepsy, cognitive impairment, or motor deficits, especially if there is associated schizencephaly (Miller et al., 2000).

Agenesis of the Cavum Septum Pellucidum

The CSP is a structure seen between 18 and 37 weeks of gestation, occasionally persisting through infancy, located between the frontal horns of the lateral ventricles, but it does not contain choroid plexus or communicate with the lateral ventricles and is not part of the ventricular system. Its absence may be an isolated finding, but it may also be associated with other anomalies such as agenesis of the corpus callosum, septo-optic dysplasia, and holoprosencephaly. Absence is relatively rare, with an estimated prevalence of 2–3 per 100,000 live births (Barkovich and Norman, 1989) (Fig. 15.4).

Diagnosis of absence of CSP is usually made at the anatomical ultrasound, since typically it is not seen earlier. If there is severe ventriculomegaly with hydrocephalus, the CSP may not be visualized. If it is absent, it is essential to assess the interhemispheric fissure and frontal horns of the lateral ventricles to check for holoprosencephaly or septo-optic dysplasia. Follow-up imaging during the pregnancy may also elucidate whether the CSP is truly isolated, or whether there are other associated neuronal migration disorders, which may appear

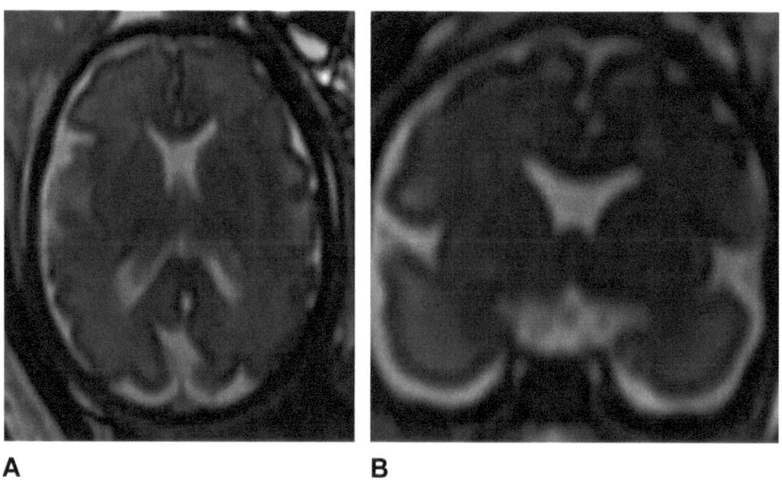
A B

Fig. 15.4. Absence of the cavum septum pellucidum, at 28 weeks of gestation. Gradient echo sequences in the (A) axial and (B) coronal planes.

only in late gestation or postnatally (Barkovich and Norman, 1989; Malinger et al., 2005).

There have only been small studies attempting to address prognosis. If the septal agenesis is not isolated, prognosis relates to the associated anomalies. In the isolated case, confirmed postnatally, studies have indicated a normal prognosis in the short term. For example, in a study of eight patients with prenatally diagnosed CSP agenesis, seven were developmentally normal as assessed at ages ranging from 8 months to 6 years, and the child with abnormal development also had associated severe ventriculomegaly (Vawter-Lee et al., 2018). In another study of 17 patients with isolated CSP diagnosed prenatally, 14 had normal neurologic evaluations, and of 8 undergoing neuropsychologic evaluations, 6 were normal. Of note, three patients were diagnosed with septo-optic dysplasia postnatally, emphasizing the difficulty in assessing the hypothalamic–pituitary axis and the optic tracts prenatally (Damaj et al., 2010). In another study of 16 fetuses with isolated CSP and postnatal follow-up, 14 had normal development, 1 had mild motor and language delay, and 1 had infantile hypotonia but normal developmental milestones (Malinger et al., 2012).

Posterior Fossa Abnormalities

Advances in fetal neuroimaging have enhanced the understanding of a wide spectrum of disorders, ranging from cystic malformations to abnormal cerebellar anatomy to cerebellar hypoplasia (Guibaud and des Portes, 2006; Paladini and Volpe, 2006; Pilu et al., 2006; Massoud and Guibaud, 2018). To properly assess these disorders, it is essential to obtain imaging of the mid-sagittal plane and visualize the position of the tentorium cerebelli and the presence and size of the cerebellar vermis. The vermis is not fully formed until the 16th–20th gestational week, and so it is important not to prematurely diagnose vermian hypoplasia (VH) or absence.

Dandy-Walker Malformation (DWM)

DWM is a complex constellation of findings, including VH/agenesis, cystic dilatation of the fourth ventricle, and an enlarged posterior fossa with upward displacement of the tentorium cerebelli and torcula herophili. Frequently, there are signs of hydrocephalus, with dilatation of the third and lateral ventricles (Kumar et al., 2001). The terms "Dandy-Walker variant" or "Dandy-Walker complex" have been used to describe less severe forms of DWM, usually in cases when the posterior fossa is not enlarged, but the usage of these terms is discouraged because of their lack of specificity (Bosemani et al., 2015).

It may be difficult to distinguish DWM from other fluid collections causing upward rotation of the cerebellum. The measurement of the brainstem–vermian and brainstem–tentorium angles, which range from 4 to 17 and 21 to 44 degrees, respectively, may be helpful. Angles greater than 45 degrees suggest DWM (Volpe et al., 2012).

DWM is the most common congenital anomaly of the posterior fossa, with a large-scale multicenter European study showing a prevalence of 6.79 per 100,000 births, including pregnancy terminations and late fetal deaths (Santoro et al., 2019). Embryologically, the roof of the rhombencephalon is not formed correctly, either because of vermian developmental arrest or compression of the vermis caused by failure of fourth ventricle foramina fenestration and subsequent enlargement of the Blake's pouch. It may be isolated, or there may be associated genetic syndromes, chromosomal abnormalities, fetal alcohol exposure, or congenital infections. There are rare gene mutations associated: FOXC1, FGF17, LAMC1, NID1, ZIC1, and ZIC4 (Zamora and Ahmad, 2019).

A meta-analysis of ultrasonography fetal studies with apparently isolated DWM showed a rate of 60.9% CNS and 42.6% extra-CNS structural anomalies. Chromosomal anomalies were present in 16.3%, most commonly deletions (7.6%). Ventriculomegaly requiring a ventriculoperitoneal shunt was common, occurring in 62.7% of cases (D'Antonio et al., 2016b).

Prognosis of DWM depends on the underlying etiology and whether associated conditions are present. Prenatal counseling has been somewhat constrained by inconsistency of research studies regarding the definition of DWM. In prenatally diagnosed isolated DWM the rate of an abnormal neurodevelopmental outcome in the meta-analysis discussed earlier was 58.2%, but the rate of abnormal outcome varied among the studies analyzed from 0% to 100%, most likely because of differing definitions of the diagnosis (D'Antonio et al., 2016a). The prognostic spectrum includes motor delay, with hypotonia and spasticity, intellectual disability, and seizures. Cerebellar signs are variable and may not be present. Hydrocephalus is very common (up to 90% of cases), frequently requiring shunting (Lerman-Sagie et al., 2018). Involvement of extra-CNS anomalies or genetic/chromosomal abnormalities worsens the prognosis (Klein et al., 2003; Guibaud et al., 2012).

Mega Cisterna Magna (MCM)

MCM refers to a retro- and infracerebellar CSF space greater than 10 mm on mid-sagittal images. It is thought to arise during embryogenesis from a defect of the posterior membranous area of the roof of the fourth ventricle, with mildly deficient fenestration of Blake's pouch

(Robinson and Ederies, 2016). The vermis is intact as are the cerebellar hemispheres, the fourth ventricle is normal, and the torcula is not elevated. It can be associated with other anomalies: in a meta-analysis and systematic review of prenatally diagnosed MCM, there were 12.6% and 16.6% additional CNS (usually ventriculomegaly) and extra-CNS anomalies, respectively. There were no associated chromosomal anomalies, and there were no other anomalies detected postnatally (D'Antonio et al., 2016a). Also, 13.8% had abnormal neurodevelopmental outcome, with some studies showing a significantly lower rate in isolated cases of MCM (D'Antonio et al., 2016a; Liu et al., 2017).

Vermian Hypoplasia (VH)/Dysgenesis

The cerebellar vermis can be abnormally rotated, malformed, or hypoplastic (Fig. 15.5). Abnormalities of the vermis may be difficult to distinguish from Blake's pouch cyst (BPC), which rotates the vermis; if visualized, the position of the choroid plexus may be a distinguishing feature (Robinson, 2014). There are known genetic associations such as CHARGE syndrome (coloboma, heart defects, choanal atresia, growth retardation, genital abnormalities, and ear abnormalities) with VH as well as abnormal cerebellar foliation (Yu et al., 2013). In the largest series published to date, isolated cases of VH had a normal neurodevelopmental outcome at school age (Tarui et al., 2014).

Blake's Pouch Cyst (BPC)

Blake's pouch is a transient protrusion of the posterior membranous area of the fourth ventricular roof, initially existing as an ependymal-lined closed cavity (Fig. 15.6). From the seventh or eighth week until the fourth month of gestation, it communicates with the subarachnoid space to form the foramen of Magendie. When it fails to communicate with the subarachnoid space, it is termed a BPC or persistent Blake's pouch (Azab et al., 2014). There may be associated anomalies, both within the CNS (11.5%) and extra-CNS (25.3%), but if isolated, there is no significant association with abnormal neurodevelopmental delay (D'Antonio et al., 2016a). It may also regress spontaneously (Ramaswamy et al., 2013).

Cerebellar Hypoplasia

There may be decreased volume of the cerebellum in one hemisphere or globally, and there may also be involvement of the pons. Unless severe, cerebellar hypoplasia is not reliably diagnosed until the sixth month of gestation, when there is more profound neuroblast proliferation (Guibaud, 2004).

Unilateral hypoplasia typically results from hemorrhage or ischemia, or it may result from an infection such as CMV. It has been described as a prenatal sign of the PHACES syndrome (posterior fossa brain malformations, hemangioma, arterial lesions, cardiac abnormalities, and eye abnormalities), which has been further associated with deviated vermis and a dilated fourth ventricle connected to an ipsilateral retrocerebellar cyst, the so-called tilted telephone receiver sign (Leibovitz et al., 2018). It can also be seen with a COL4A1 mutation (Vermeulen et al., 2011). If isolated, unilateral cerebellar hypoplasia usually has a normal neurodevelopmental outcome; cognition may be affected if the vermis is involved (Massoud et al., 2014).

Global cerebellar hypoplasia has a variety of etiologies, including chromosomal anomalies (e.g., trisomies 13, 18, 21; trisomy and monosomy 21q; ring chromosome 6), specific genetic disorders (e.g., CHARGE syndrome, Pallister-Killian syndrome, velocardiofacial syndrome), metabolic disorders (e.g., Zellweger syndrome, congenital disorders of glycosylation, Smith-Lemli-Opitz syndrome), hemorrhage, ischemia, infection,

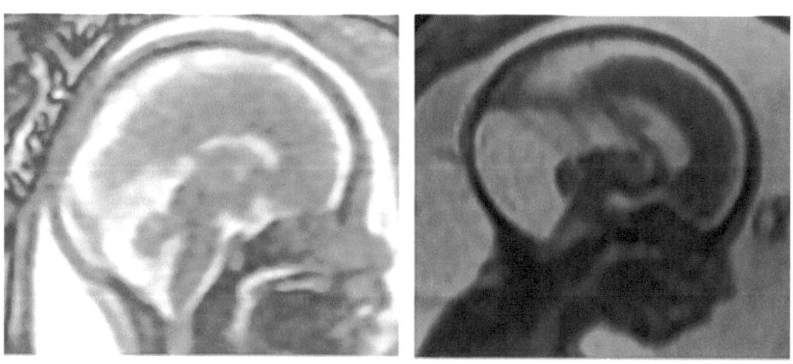

Fig. 15.5. T2 sagittal, at 24 weeks of gestation showing (A) inferior vermian hypoplasia (tegmento-vermian angle 38.4 degrees) and (B) Dandy-Walker malformation with severe global hypoplasia of the cerebellum and torcular lambdoid inversion (tegmento-vermian angle 65 degrees).

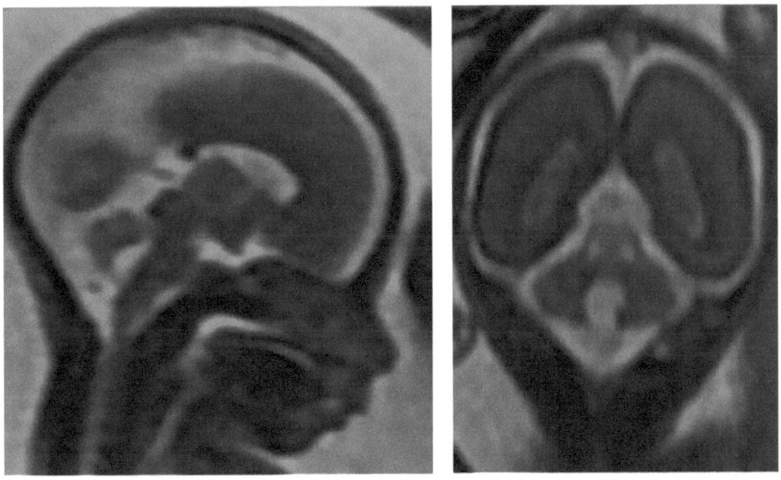

A B

Fig. 15.6. Blake's pouch cyst, at 23 weeks of gestation. Normal cerebellar hemispheres and slight inferior ballooning at the level of the inferior vermis are suggestive of a persistent Blake's pouch cyst. Gradient echo sequences in the (A) sagittal and (B) coronal planes.

and teratogens (phenytoin and valproic acid)/illicit drug/alcohol exposure. Prognosis relates to etiology; in the isolated case, there may be a good outcome (Poretti et al., 2014; Lerman-Sagie et al., 2018; Zou et al., 2018). If there is involvement of the pons, which is difficult to assess on prenatal imaging, there may be an underlying pontocerebellar hypoplasia genetic syndrome (PCH gene), with a more severe prognosis (van Dijk et al., 2018).

Joubert Syndrome (JS)

JS is a caused by a defect in ciliary function, resulting in ataxia, oculomotor apraxia, cognitive impairment, hypotonia, and breathing dysregulation. There may also be systemic involvement of the kidneys, liver, eyes, liver, and skeleton (Poretti et al., 2014). Multiple genes in multiple loci have been identified as causative, and these genes are also involved in other ciliopathies such as Bardet-Biedl and Meckel syndromes (Massoud and Guibaud, 2018).

The classic neuroimaging hallmark is the "molar tooth sign" (MTS), which is characterized by bilateral elongation and thickening of the superior cerebellar peduncles. Rarely has the MTS been identified prenatally (Fluss et al., 2006; Pugash et al., 2011). Analysis of the shape and dimensions of the fourth ventricle has also been used as a diagnostic marker (Quarello et al., 2014).

Rhombencephalosynapsis (RS)

RS is characterized by fusion of the cerebellar hemispheres and absence or partial absence of the cerebellar vermis. There may also be supratentorial anomalies, most commonly ventriculomegaly due to aqueductal stenosis (Arrigo and Grech, 2017). A genetic cause has not yet been identified (Aldinger et al., 2018). Rarely, RS can be identified prenatally, usually in association with ventriculomegaly and VH (Izzo et al., 2015; Poretti and Boltshauser, 2015; Kline-Fath et al., 2018).

RS may be associated with ataxia, hypotonia, and abnormal eye movements; cognitive function is usually impaired but may be normal (Poretti et al., 2009). Development may relate to the degree of ventriculomegaly or other supratentorial findings.

Kinked Brainstem

A kinked brainstem is an anomaly that is usually seen with ventriculomegaly (Fig. 15.7). The differential diagnosis includes alpha-dystroglycanopathies such as Walker-Warburg syndrome, X-linked hydrocephalus (L1CAM mutations), and tubulinopathies (TUBA1A mutations) (Amir et al., 2016). Walker-Warburg syndrome may be distinguished because it typically has associated ocular abnormalities. Tubulinopathies frequently involve the basal ganglia (which may show dysplasia).

NEURAL TUBE DEFECTS (NTDs)

Neurulation, a tightly regulated process by which the central nervous system in vertebrates is formed, occurs between days 17 and 28 of gestation. NTDs occur when the neurulation process is interrupted, and their location and severity depend on the particular closure site affected. They can be divided into open or closed defects depending on whether the neural tissue is exposed and contacts the amniotic fluid (Avagliano et al., 2019). The combination of failed neural tube closure followed by exposure to toxic elements in the amniotic fluid in

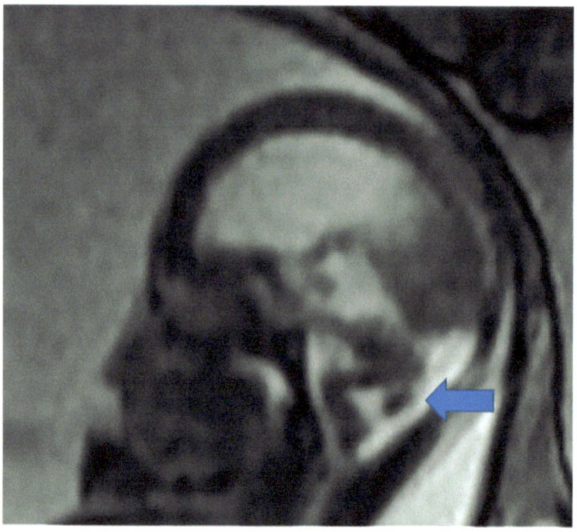

Fig. 15.7. Kinked brainstem, at 20 weeks of gestation, with enlarged third and lateral ventricles, cerebellar and vermian hypoplasia, and kinking of brainstem (*arrow*). Sagittal gradient echo.

open defects is considered to be a "two-hit" process, resulting in neurodegeneration in utero (Sahni and Ohri, 2019).

The prevalence of NTDs has been decreasing with the broader use of supplemental folate during pregnancy. There are variations among regional and ethnic groups, with a prevalence of approximately 1/1000 live births worldwide (Copp et al., 2013). The importance of folate supplementation cannot be overstated: there was a successful decrease in the prevalence from 120/10,000 in 2004 to 31.5/10,000 in northern China following a large-scale folic acid program (Li et al., 2006; Liu et al., 2016).

Most cases of NTDs are sporadic, although there is an increased risk for recurrence in siblings and concordance in monozygotic as compared with dizygotic twins (Windham et al., 1982; Ross et al., 2017). The suggestion is of a polygenic or oligogenic process with contribution from environmental factors, including teratogens, particularly valproic acid, which is strongly associated. There are also specific chromosomal abnormalities and genetic syndromes associated, such as trisomies 13 and 18, Meckel-Gruber syndrome, and VACTERL (vertebral defects, anal atresia, cardiac defects, trachea-esophageal fistula, renal anomalies, and limb abnormalities) (Sepulveda et al., 2004; Sahni and Ohri, 2019). Maternal obesity and diabetes have also been implicated, as has maternal fever or exposure to high temperatures early in pregnancy (Stothard et al., 2009; Sukanya et al., 2012; Dreier et al., 2014). Approximately 20% of NTDs, most commonly encephalocele, are associated with extracranial congenital malformations (Stoll et al., 2011).

Either maternal serum alpha-fetoprotein or ultrasonography is used as a prenatal screening test; generally, ultrasound is a more sensitive test unless there are technical limitations such as maternal obesity or difficult fetal positioning (Norem et al., 2005). Elevated alpha-fetoprotein can only be used to detect an open NTD and is not specific because it may be associated with other conditions such as gastroschisis, omphalocele, congenital nephrosis, and fetal demise. Ultrasound is more sensitive in the second trimester as compared with the first, especially for spina bifida, and so if a first trimester study is normal, it should be repeated in the second semester (Cameron and Moran, 2009).

Anencephaly and Craniorachischisis

Anencephaly is an NTD in which the neural tube does not close in its cranial region, resulting in the destruction of cerebral tissues by the amniotic environment. It is much more common than craniorachischisis, in which both the spinal and cranial portions of the neural tube do not close. These conditions are usually diagnosed early in pregnancy by ultrasound. Almost all cases of either condition do not survive through term, and those that do, die within the first 2 days after birth. The differential diagnosis includes amniotic banding, which can also cause destruction of the head; since amniotic banding is a sporadic process, whereas NTDs carry a recurrence risk for subsequent pregnancies, it can be helpful for prenatal counseling to distinguish. Amniotic bands cause asymmetric cranial defects, and there are typically other abnormalities in the limbs and body wall, unlike NTDs, which arise from symmetric processes (Goldstein and Filly, 1988; Avagliano et al., 2019).

Myelomeningocele, Myelocele, and Meningocele

If the caudal portion of the neural tube fails to close, the vertebral column and spinal cord may be malformed, with herniation of the meningeal sac through the vertebral arch. Protrusion of the cystic mass containing the meninges and neural tissue without overlying skin results in a myelomeningocele. If the spinal cord is exposed without being covered by the meningeal tissue, the defect is termed a myelocele. A meningocele, which is a closed NTD, occurs when the meninges protrudes through the vertebral column, forming a sac without the spinal cord, although there may be some neural elements. Myelomeningocele is far more common than myelocele or meningocele.

Myelomeningocele and myelocele are almost always associated with a Chiari II malformation, likely caused in part by mechanical traction: downward displacement of

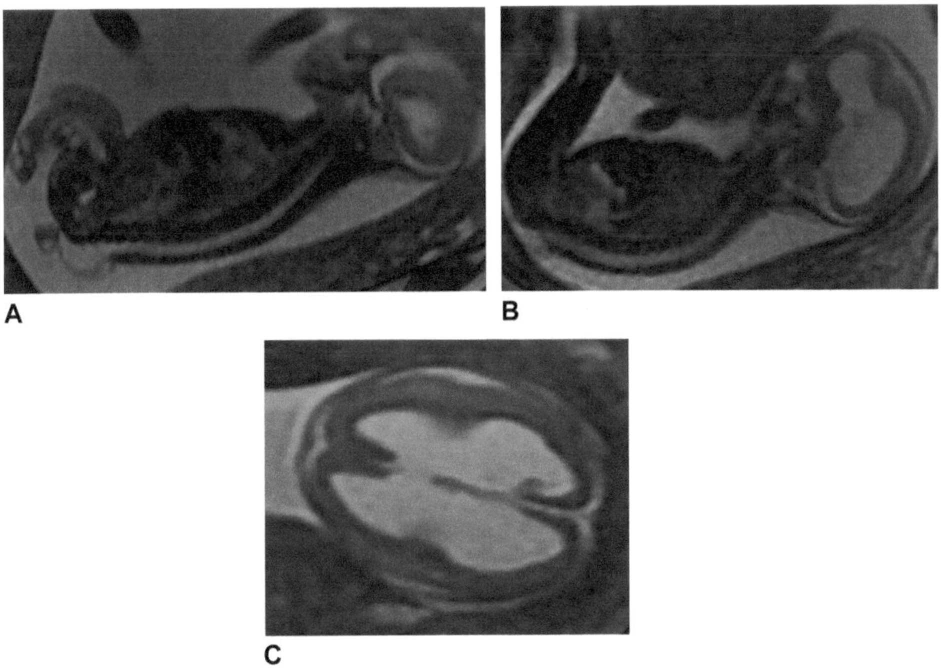

Fig. 15.8. Chiari II malformation, at 20 weeks of gestation. (A) Axial gradient echo, small myelomeningocele sac within lower spine. (B) Axial gradient echo, ventriculomegaly. (C) Sagittal gradient echo, small, crowded posterior fossa, ventriculomegaly, agenesis of corpus callosum.

the cerebellum and brainstem, affecting ventricular flow with subsequent hydrocephalus (Fig. 15.8). Frequently other structures of the fetal brain are also affected, resulting in cognitive impairment typically manifesting as learning disabilities, as well as brainstem signs such as apnea (Dennis and Barnes, 2010; English et al., 2010; Patel et al., 2015; Shellhaas et al., 2018).

Ultrasonography during the second trimester is highly sensitive and specific for myelomeningocele, and MRI can provide supplementary information regarding abnormalities in the brain (Lennon and Gray, 1999; Mirsky et al., 2015). The herniated sac in the spinal column may be directly visualized. In addition, there are classic indirect signs seen on ultrasonography: the “lemon sign,” in which there is scalloping of the frontal bones, and the “banana sign,” in which the cerebellar hemispheres are curved anteriorly and the cisterna magna is obliterated (Nicolaides et al., 1986). Hydrocephalus may also be observed if there is a Chiari II malformation. There may be irregularities of the bony spine, which may be observed in the first trimester (Baxi et al., 1990).

Early detection of myelomeningocele has become particularly important with the advent of prenatal surgical repair. Conceptually, the rationale for fetal surgery is twofold: to decrease exposure of the neural tissue to toxic elements in the amniotic fluid (reducing the “second hit” of the “two-hit” process discussed earlier), and to reduce the mechanical process causing the Chiari II abnormalities in the brain (Dewan and Wellons, 2019). The Management of Myelomeningocele Study (MOMS) was a prospective, randomized clinical trial comparing prenatal (before 26 weeks of gestation) and standard postnatal myelomeningocele closure at three tertiary care medical centers in the United States. There were several risks associated with the prenatal surgery, including prematurity, oligohydramnios, and maternal pulmonary edema, but generally there was significant improvement in ventriculoperitoneal shunt placement rate, abnormal radiolographic signs, and motor skills, and the trial was terminated early because of its efficacy (Adzick et al., 2011).

Prognosis depends on the level of the spinal lesion as well as the extent of cerebral injury. Most commonly, the lumbar and sacral regions are involved, although the defect can occur at any level of the spine. Frequently, there is severe deficit of motor and sensory function, as well as bowel and bladder incontinence. Hydrocephalus usually requires a ventriculoperitoneal shunt, with higher rates for more cephalad lesions (Rintoul et al., 2002). As discussed earlier, brain abnormalities may contribute to learning disabilities or other cognitive impairment. The expectation is that morbidity from myelomeningocele will decrease as prenatal surgery becomes more common. There is some evidence that myelocele has more Chiari II abnormalities than myelomeningocele, possibly from a protective effect of the herniated myelomeningeal sac (Rethmann et al., 2017; Nagaraj et al., 2018). Meningocele usually has a much better prognosis than myelomeningocele, as typically brain abnormalities are not involved, although there is a significant association with tethered cord (Yun-Hai et al., 2015).

Encephalocele

An encephalocele is a closed NTD characterized by a protrusion of the brain and/or meninges through a skull defect. They are usually sporadic, although there is association with some genetic conditions such as Meckel-Gruber, Knobloch, and Walker-Warburg syndromes (Lo et al., 2008). They may contain meningeal tissue only, meningeal and brain tissue, or meningeal, brain, and ventricular tissue, known as meningocele, encephalomeningocele, or encephalomeningocystocele, respectively. They can also be classified according to location: anterior, parietal, and occipital. Anterior encephaloceles are further subdivided into frontal, sincipital, and basal locations, with frontal encephaloceles usually in the glabellar region and associated with craniofacial abnormalities. Basal encephaloceles are internal, within the nose, pharynx, and orbit (Avagliano et al., 2019). Occipital encephaloceles may be associated with a Chiari III malformation, with herniation of the posterior fossa along with hydrocephalus (Ivashchuk et al., 2015). The differing types of encephaloceles have a geographic distribution: for example, anterior encephaloceles are more frequent in Southeast Asia and central Africa, and occipital encephaloceles are more common in North America and Western Europe (Naidich et al., 1992; Mahapatra and Suri, 2002).

Clinical outcome depends most on associated abnormalities such as hydrocephalus, microcephaly, and seizures. Outcome also depends on the presence of cerebral or cerebellar tissue within the herniated sac, which makes surgery more challenging. Prognosis varies from normal neurodevelopment to severe developmental delay (Bui et al., 2007; Da Silva et al., 2015).

TUBEROUS SCLEROSIS COMPLEX (TSC)

TSC is a phenotypically heterogeneous syndrome caused by inactivating mutations in the genes TSC1 or TSC2 (Randle, 2017). It affects multiple organ systems and can be overwhelming for the family, and so early and frequent counseling is important (Ng et al., 2015). Early treatment with mTOR inhibitors (everolimus or sirolimus), with antiepileptic medications such as vigabatrin, and with neurosurgery has been shown to be beneficial in certain cases, and so it is important to make an early diagnosis (Curatolo et al., 2018; Franz and Krueger, 2018; Ostrowsky-Coste et al., 2019).

As fetal imaging techniques develop, TSC is becoming easier to diagnose prenatally. Most often, the diagnosis is suspected with the presence of cardiac rhabdomyomas, which are the most commonly observed fetal cardiac tumor (Chen et al., 2019; Pavlicek et al., 2019). These tumors typically resolve early in life, but may present complications requiring surgery if they are very large causing obstruction or interfere with the cardiac conduction system (Kwiatkowska et al., 2017). There have been reports of successful reduction in rhabdomyoma size via prenatal treatment with sirolimus or everolimus (Patel et al., 2018; Chen et al., 2019; Garg et al., 2019; Park et al., 2019; Vachon-Marceau et al., 2019).

The dominant clinical presenting symptoms of TSC are neurologic, with a broad array of manifestations, including epilepsy, autism, and cognitive impairment. Cortical tubers and subependymal nodules, two hallmarks of TSC and the cerebral lesions most commonly seen, most likely form between 10 and 20 weeks of gestation but are usually detected after 21 weeks (Wortmann et al., 2008; Prabowo et al., 2013). MRI is more sensitive than ultrasonography for detecting lesions (Dragoumi et al., 2018) (Fig. 15.9).

Genetic diagnosis prenatally can be confirmatory, with sequencing of the TSC1 and TSC2 genes. If there is a relative with a known mutation, targeted sequencing can have a high predictive value (Dragoumi et al., 2018). However, negative genetic testing should be viewed with caution, because there are TSC patients who do not carry mutations identified by conventional screening methods (Nellist et al., 2015).

SCHIZENCEPHALY

Schizencephaly is defined as a transmantle column of dysplastic gray matter extending across the cerebrum or cerebellum from pia to ependyma (Naidich et al., 2015). It can be unilateral or bilateral, and unifocal or multifocal. It is divided into two types: type 1, also known as closed-lip schizencephaly, in which there is no communication with the lateral ventricle; and type 2, or open-lip schizencephaly, in which the subpial space is in communication with the lateral ventricle. Open-lip schizencephaly is distinguished from porencephaly, which can look similar but results from a destructive lesion in the white matter, and does not have gray matter lining the CSF cleft (Griffiths, 2018).

The vast majority of cases are sporadic and most often attributed to a destructive lesion, whether vascular, hypoxic–ischemic, or infectious (Nabavizadeh et al., 2014). Earlier reports of associations with the EMX2 gene have not been supported with more data (Tietjen et al., 2007; Merello et al., 2008).

Diagnosis is best made with MRI and may be missed until after the first year of life, which may explain why a report of live birth prevalence of 0.54 per 10,000 was higher than previous estimates (Szabo et al., 2010).

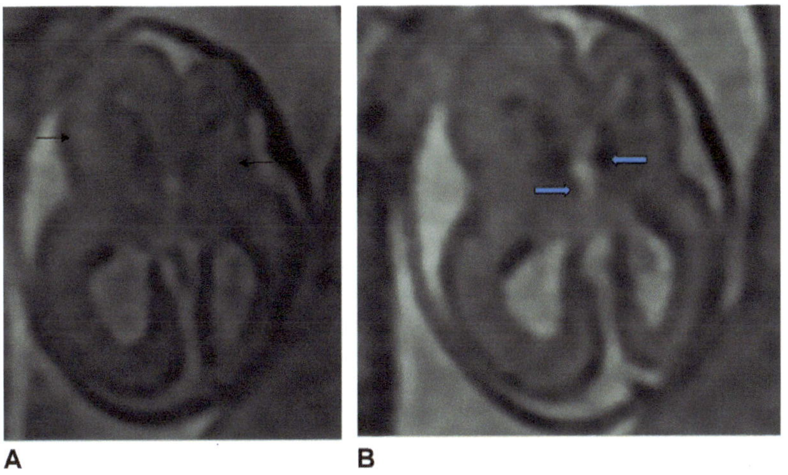

Fig. 15.9. Tuberous sclerosis complex, at 19 weeks of gestation. Axial T2 images showing tubers in (A) subependymal and (B) cortical locations. Also with hemimegaloencephaly (right cerebral hemisphere).

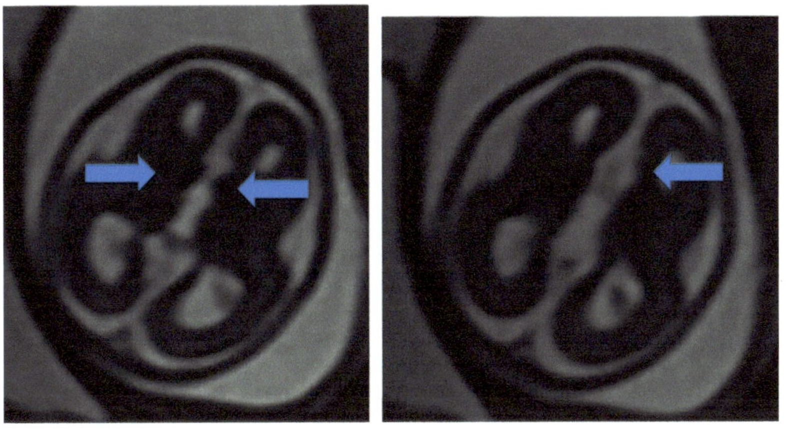

Fig. 15.10. Periventricular nodular heterotopia, at 22 weeks of gestation. Axial gradient echo sequences show irregularity of the cortex along the medial frontal lobes (*arrows*) suspicious for periventricular/cortical nodular heterotopia. Also with wide separation of frontal horns of lateral ventricles, colpocephaly, consistent with agenesis of the corpus callosum.

In a review of the published literature (734 cases in total), bilateral clefts were present in 41.4% and open-lip schizencephaly was seen in 61.3% (Braga et al., 2018). It is most often seen in the frontal and parietal lobes. Associated anomalies include microcephaly, septo-optic dysplasia, and polymicrogyria. Clinical manifestations are variable, depending on the location and extent of the lesion as well as associated anomalies; motor impairment is common as are neurocognitive deficits and epilepsy.

PERIVENTRICULAR NODULAR HETEROTOPIA (PNH)

Neuronal migrational abnormalities can lead to heterotopias, which are collections of neurons of multiple sizes and shapes. The most common form is PNH, in which the neurons line the ventricular surface (Deloison et al., 2018).

PNH can be bilateral of unilateral. There are multiple genetic etiologies, including Filamin A (FLNA) gene mutations, which are X-linked dominant and usually affect females. They typically manifest as bilaterally symmetric diffuse heterotopias, with the majority near the frontal horns of the lateral ventricles, and are frequently associated with cerebellar vermis hypoplasia and cardiovascular abnormalities (Parrini et al., 2006). Other genes that have been associated with PNH are ARFGEF2, C6orf70, and NEDD4L (Conti et al., 2013; Bardon-Cancho et al., 2014; Broix et al., 2016).

Diagnosis is usually made in the third trimester, and MRI is more sensitive than ultrasonography, although there are some features on ultrasound that are suggestive: (1) a square shape of the frontal horns, (2) irregular borders of the lateral ventricles, and (3) nodules protruding into the ventricular lumen (Blondiaux et al., 2013; Deloison et al., 2018) (Fig. 15.10).

Developmental outcome varies; typically, FLNA mutations have normal intelligence or mild cognitive deficits, variable severity of epilepsy, and there may be behavioral or psychiatric comorbidities (Sole et al., 2009; Fry et al., 2013). As discussed earlier, there are also associations with extracerebral symptoms that may affect quality of life.

BRAIN TUMORS

Congenital brain tumors are usually diagnosed in the third trimester; imaging signs are macrocephaly and the intracranial mass. The most common fetal brain tumor is a teratoma, comprised of at least two of the three embryologic layers of ectoderm, mesoderm, and endoderm. Mature teratomas have low mitotic activity and well-differentiated tissue elements and are potentially curable by gross total resection. Immature teratomas have incompletely differentiated components and are usually infiltrative and disseminated, with a poor prognosis (Shekdar and Schwartz, 2017). Other fetal brain tumors include choroid plexus tumors (most commonly choroid plexus papilloma but also choroid plexus carcinoma and atypical choroid plexus papilloma), which may cause hydrocephalus through mechanical obstruction and/or overproduction of CSF (Crawford and Isaacs Jr., 2019) (Fig. 15.11).

FETAL STROKE

Stroke in the fetus can occur as early as 14 weeks of gestation, but it is usually asymptomatic, and there may not be clinical symptoms until as late as 8–10 months postnatally, presenting with hemiparesis or a hand preference. Fetal stroke can be ischemic or hemorrhagic, focal or global, and there are multiple risk factors relating to a broad spectrum of etiologies (Kirkham et al., 2018).

Hemorrhagic strokes are more easily visualized radiographically than ischemic strokes in the fetus, and they are also more likely to occur at a lower gestational age (Fig. 15.12). Epidemiologic studies have estimated an incidence of 0.5–0.9 per 1000 pregnancies (Sanapo et al., 2017). Hemorrhages are more commonly supratentorial and may be periventricular, intraventricular, intraparenchymal, or in the subdural, subarachnoid, and epidural spaces. Intraventricular hemorrhage, the most common form, is likely related to the fragile nature of the germinal matrix vasculature, as seen in prematurely born infants, and can produce ventriculomegaly, hydrocephalus, and in severe cases, cause periventricular venous infarction with porencephaly (Takanashi et al., 2003; Huang et al., 2006; Abdelkader et al., 2017). Intraparenchymal hemorrhages may relate to endothelial dysfunction/thrombocytopenia such as seen in alloimmune thrombocytopenia, or a mutation in the COL4A1

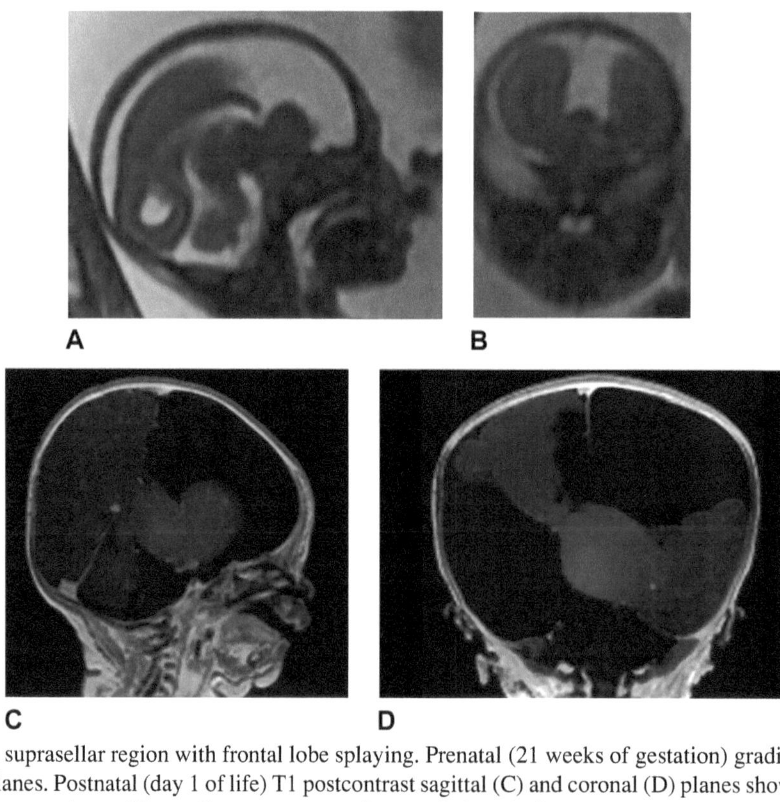

Fig. 15.11. Tumor in suprasellar region with frontal lobe splaying. Prenatal (21 weeks of gestation) gradient echo in the sagittal (A) and coronal (B) planes. Postnatal (day 1 of life) T1 postcontrast sagittal (C) and coronal (D) planes show increase in size, with displacement of brain parenchyma bilaterally, or severe cystic encephalomalacia.

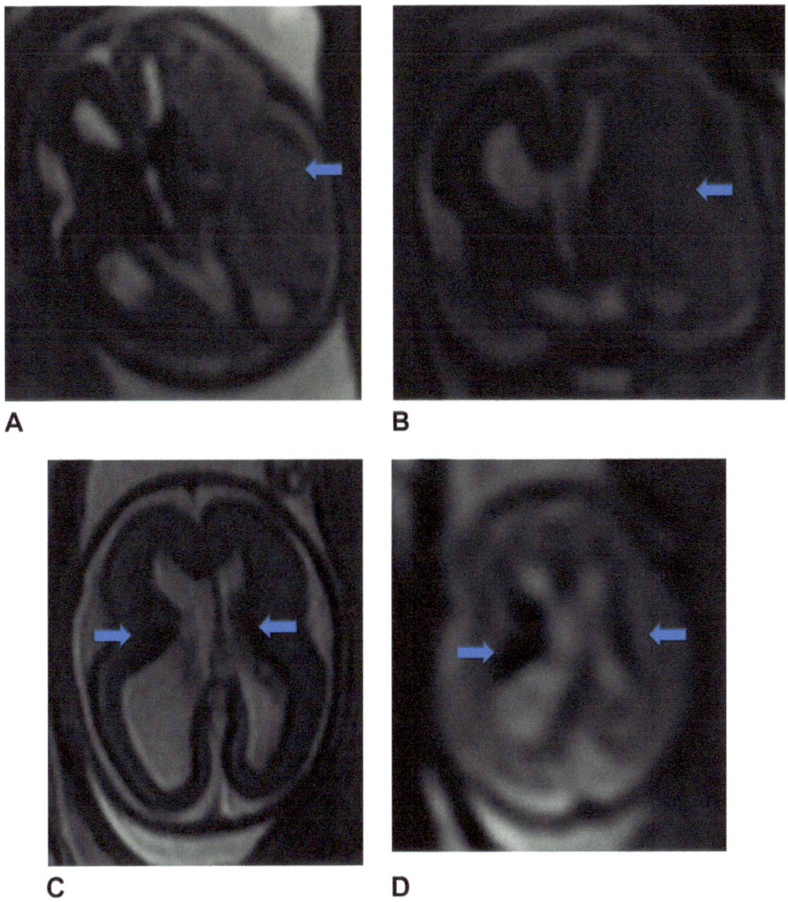

Fig. 15.12. Hemorrhage. Axial (A) and coronal (B) gradient echo, at 21 weeks of gestation, showing hemorrhage of left cerebral hemisphere with mass effect, compression of left lateral ventricle, and mild left to right midline shift. Axial gradient echo (C) and axial diffusion-weighted imaging (D), at 22 weeks of gestation, showing bilateral germinal matrix hemorrhages, right > left, and with enlargement of right lateral ventricle, which may be secondary to entrapment from layering of hemorrhage at level of foramen of Monro.

gene with subsequent microangiopathy (Vermeulen et al., 2011; Risson et al., 2012). Maternal trauma can also produce hemorrhage in the intracranial, subdural, subarachnoid, or epidural spaces (Matsushita et al., 2014; Tattoli et al., 2017; Yamasato et al., 2018). Infratentorial hemorrhages are typically associated with cerebellar injury.

Arterial ischemic strokes are less commonly seen unless there is hemorrhagic conversion in an arterial distribution (Amato et al., 1991). Risk factors include placental pathology, cardiac disease, prothrombotic disorders, genetic causes of arteriopathy such as PHACES syndrome (posterior fossa brain malformations, hemangioma, arterial lesions, cardiac abnormalities, and eye abnormalities), and monochorionic twinning with twin-twin transfusion syndrome (Amlie-Lefond et al., 2008; Das et al., 2008; Fernandez-Mayoralas et al., 2014; Redline and Ravishankar, 2018). Along with focal cerebral infarction, ischemic strokes may cause cortical malformations such as schizencephaly or polymicrogyria.

Outcome of hemorrhagic or ischemic stroke depends on the extent and location of the stroke, and if there are associated findings such as encephalomalacia or hydrocephalus. Seizures are a common presenting sign in neonates with stroke, and there may be subsequent effects on cognition, language, or behavior.

CONCLUSION

Fetal neuroimaging has transformed the field of pediatric neurology, as neurologists are now playing an important role in prognostication, medical care, and decision-making during pregnancy. As the field develops, there will be more comprehensive understanding of prognosis, and neuroimaging techniques will also improve, resulting in even more significant neurologic input into maternal/fetal care.

References

Abdelkader MA, Ramadan W, Gabr AA et al. (2017). Fetal intracranial hemorrhage: sonographic criteria and merits of prenatal diagnosis. J Matern Fetal Neonatal Med 30: 2250–2256.

Adzick NS, Thom EA, Spong CY et al. (2011). A randomized trial of prenatal versus postnatal repair of myelomeningocele. N Engl J Med 364: 993–1004.

Agathokleous M, Chaveeva P, Poon LC et al. (2013). Meta-analysis of second-trimester markers for trisomy 21. Ultrasound Obstet Gynecol 41: 247–261.

Aicardi J (2005). Aicardi syndrome. Brain Dev 27: 164–171.

Aldinger KA, Dempsey JC, Tully HM et al. (2018). Rhombencephalosynapsis: fused cerebellum, confused geneticists. Am J Med Genet C Semin Med Genet 178: 432–439.

Amato M, Huppi P, Herschkowitz N et al. (1991). Prenatal stroke suggested by intrauterine ultrasound and confirmed by magnetic resonance imaging. Neuropediatrics 22: 100–102.

Amir T, Poretti A, Boltshauser E et al. (2016). Differential diagnosis of ventriculomegaly and brainstem kinking on fetal MRI. Brain Dev 38: 103–108.

Amlie-Lefond CM, Basir MA, Franciosi RA (2008). Fatal neonatal stroke from a prenatal cardiac thrombus. Pediatr Neurol 38: 140–142.

Arrigo S, Grech R (2017). Rhomboencephalosynapsis. BMJ Case Rep 2017:bcr-2017-220211.

Avagliano L, Massa V, George TM et al. (2019). Overview on neural tube defects: from development to physical characteristics. Birth Defects Res 111 (19): 1455–1467.

Azab WA, Shohoud SA, Elmansoury TM et al. (2014). Blake's pouch cyst. Surg Neurol Int 5: 112.

Bardon-Cancho EJ, Munoz-Jimenez L, Vazquez-Lopez M et al. (2014). Periventricular nodular heterotopia and dystonia due to an ARFGEF2 mutation. Pediatr Neurol 51: 461–464.

Barkovich AJ, Norman D (1989). Absence of the septum pellucidum: a useful sign in the diagnosis of congenital brain malformations. Am J Roentgenol 152: 353–360.

Barr M, Jr. Hanson JW, Currey K et al. (1983). Holoprosencephaly in infants of diabetic mothers. J Pediatr 102: 565–568.

Bar-Yosef O, Barzilay E, Dorembus S et al. (2017). Neurodevelopmental outcome of isolated ventriculomegaly: a prospective cohort study. Prenat Diagn 37: 764–768.

Baxi L, Warren W, Collins MH et al. (1990). Early detection of caudal regression syndrome with transvaginal scanning. Obstet Gynecol 75: 486–489.

Blondiaux E, Sileo C, Nahama-Allouche C et al. (2013). Periventricular nodular heterotopia on prenatal ultrasound and magnetic resonance imaging. Ultrasound Obstet Gynecol 42: 149–155.

Bosemani T, Orman G, Boltshauser E et al. (2015). Congenital abnormalities of the posterior fossa. Radiographics 35: 200–220.

Braga VL, da Costa MDS, Riera R et al. (2018). Schizencephaly: a review of 734 patients. Pediatr Neurol 87: 23–29.

Broix L, Jagline H, Ivanova E et al. (2016). Mutations in the HECT domain of NEDD4L lead to AKT-mTOR pathway deregulation and cause periventricular nodular heterotopia. Nat Genet 48: 1349–1358.

Bronshtein E, Solt I, Bronshtein M et al. (2017). Detection rate and sonographic signs of trisomy 21 fetuses at 14–17 weeks of gestation. Isr Med Assoc J 19: 8–12.

Bui CJ, Tubbs RS, Shannon CN et al. (2007). Institutional experience with cranial vault encephaloceles. J Neurosurg 107: 22–25.

Calloni SF, Caschera L, Triulzi FM (2019). Disorders of ventral induction/spectrum of holoprosencephaly. Neuroimaging Clin N Am 29: 411–421.

Cameron M, Moran P (2009). Prenatal screening and diagnosis of neural tube defects. Prenat Diagn 29: 402–411.

Cardoza JD, Goldstein RB, Filly RA (1988). Exclusion of fetal ventriculomegaly with a single measurement: the width of the lateral ventricular atrium. Radiology 169: 711–714.

Carta S, Kaelin Agten A, Belcaro C et al. (2018). Outcome of fetuses with prenatal diagnosis of isolated severe bilateral ventriculomegaly: systematic review and meta-analysis. Ultrasound Obstet Gynecol 52: 165–173.

Casanova MF, El-Baz A, Elnakib A et al. (2011). Quantitative analysis of the shape of the corpus callosum in patients with autism and comparison individuals. Autism 15: 223–238.

Chen J, Wang J, Sun H et al. (2019). Fetal cardiac tumor: echocardiography, clinical outcome and genetic analysis in 53 cases. Ultrasound Obstet Gynecol 54: 103–109.

Chow KC, Lee CC, Lin TY et al. (2000). Congenital enterovirus 71 infection: a case study with virology and immunohistochemistry. Clin Infect Dis 31: 509–512.

Coleman BG, Langer JE, Horii SC (2015). The diagnostic features of spina bifida: the role of ultrasound. Fetal Diagn Ther 37: 179–196.

Conti V, Carabalona A, Pallesi-Pocachard E et al. (2013). Periventricular heterotopia in 6q terminal deletion syndrome: role of the C6orf70 gene. Brain 136: 3378–3394.

Copp AJ, Stanier P, Greene ND (2013). Neural tube defects: recent advances, unsolved questions, and controversies. Lancet Neurol 12: 799–810.

Crawford JR, Isaacs H Jr. (2019). Perinatal (fetal and neonatal) choroid plexus tumors: a review. Childs Nerv Syst 35: 937–944.

Croen LA, Shaw GM, Lammer EJ (1996). Holoprosencephaly: epidemiologic and clinical characteristics of a California population. Am J Med Genet 64: 465–472.

Croen LA, Shaw GM, Lammer EJ (2000). Risk factors for cytogenetically normal holoprosencephaly in California: a population-based case–control study. Am J Med Genet 90: 320–325.

Curatolo P, Nabbout R, Lagae L et al. (2018). Management of epilepsy associated with tuberous sclerosis complex: updated clinical recommendations. Eur J Paediatr Neurol 22: 738–748.

Da Silva SL, Jeelani Y, Dang H et al. (2015). Risk factors for hydrocephalus and neurological deficit in children born with an encephalocele. J Neurosurg Pediatr 15: 392–398.

Damaj L, Bruneau B, Ferry M et al. (2010). Pediatric outcome of children with the prenatal diagnosis of isolated septal agenesis. Prenat Diagn 30: 1143–1150.

D'Antonio F, Khalil A, Garel C et al. (2016a). Systematic review and meta-analysis of isolated posterior fossa malformations on prenatal imaging (part 2): neurodevelopmental outcome. Ultrasound Obstet Gynecol 48: 28–37.

D'Antonio F, Khalil A, Garel C et al. (2016b). Systematic review and meta-analysis of isolated posterior fossa malformations on prenatal ultrasound imaging (part 1): nomenclature, diagnostic accuracy and associated anomalies. Ultrasound Obstet Gynecol 47: 690–697.

D'Antonio F, Pagani G, Familiari A et al. (2016c). Outcomes associated with isolated agenesis of the corpus callosum: a meta-analysis. Pediatrics 138: e20160445.

Das S, Ankola P, Chiechi M et al. (2008). Perinatal cerebral arterial infarction associated with a placental chorioangioma. Am J Perinatol 25: 381–383.

De Morsier G (1956). Studies on malformation of cranio-encephalic sutures. III. Agenesis of the septum lucidum with malformation of the optic tract. Schweiz Arch Neurol Psychiatr 77: 267–292.

Deloison B, Sonigo P, Millischer-Bellaiche AE et al. (2018). Prenatally diagnosed periventricular nodular heterotopia: further delineation of the imaging phenotype and outcome. Eur J Med Genet 61: 773–782.

Dennis M, Barnes MA (2010). The cognitive phenotype of spina bifida meningomyelocele. Dev Disabil Res Rev 16: 31–39.

Dewan MC, Wellons JC (2019). Fetal surgery for spina bifida. J Neurosurg Pediatr 24: 105–114.

Dragoumi P, O'Callaghan F, Zafeiriou DI (2018). Diagnosis of tuberous sclerosis complex in the fetus. Eur J Paediatr Neurol 22: 1027–1034.

Dreier JW, Andersen AM, Berg-Beckhoff G (2014). Systematic review and meta-analyses: fever in pregnancy and health impacts in the offspring. Pediatrics 133: e674–e688.

Edison RJ, Muenke M (2005). Gestational exposure to lovastatin followed by cardiac malformation misclassified as holoprosencephaly. N Engl J Med 352: 2759.

Edwards JH (1958). Congenital malformations of the central nervous system in Scotland. Br J Prev Soc Med 12: 115–130.

Edwards TJ, Sherr EH, Barkovich AJ et al. (2014). Clinical, genetic and imaging findings identify new causes for corpus callosum development syndromes. Brain 137: 1579–1613.

English L, Barnes MA, Fletcher JM et al. (2010). Effects of reading goals on reading comprehension, reading rate, and allocation of working memory in children and adolescents with spina bifida meningomyelocele. J Int Neuropsychol Soc 16: 517–525.

Falip C, Blanc N, Maes E et al. (2007). Postnatal clinical and imaging follow-up of infants with prenatal isolated mild ventriculomegaly: a series of 101 cases. Pediatr Radiol 37: 981–989.

Fernandez-Mayoralas DM, Recio-Rodriguez M, Fernandez-Perrone AL et al. (2014). In utero diagnosis of PHACE syndrome by fetal magnetic resonance imaging (MRI). J Child Neurol 29: 118–121.

Fluss J, Blaser S, Chitayat D et al. (2006). Molar tooth sign in fetal brain magnetic resonance imaging leading to the prenatal diagnosis of Joubert syndrome and related disorders. J Child Neurol 21: 320–324.

Fox NS, Monteagudo A, Kuller JA et al. (2018). Mild fetal ventriculomegaly: diagnosis, evaluation, and management. Am J Obstet Gynecol 219: B2–B9.

Franz DN, Krueger DA (2018). mTOR inhibitor therapy as a disease modifying therapy for tuberous sclerosis complex. Am J Med Genet C Semin Med Genet 178: 365–373.

Fry AE, Kerr MP, Gibbon F et al. (2013). Neuropsychiatric disease in patients with periventricular heterotopia. J Neuropsychiatry Clin Neurosci 25: 26–31.

Gaglioti P, Danelon D, Bontempo S et al. (2005). Fetal cerebral ventriculomegaly: outcome in 176 cases. Ultrasound Obstet Gynecol 25: 372–377.

Garel C, Luton D, Oury JF et al. (2003). Ventricular dilatations. Childs Nerv Syst 19: 517–523.

Garg A, Gorla SR, Kardon RE et al. (2019). Rapid involution of large cardiac rhabdomyomas with everolimus therapy. World J Pediatr Congenit Heart Surg 10: 215013 5118822711.

Glass HC, Shaw GM, Ma C et al. (2008). Agenesis of the corpus callosum in California 1983–2003: a population-based study. Am J Med Genet A 146a: 2495–2500.

Goldstein RB, Filly RA (1988). Prenatal diagnosis of anencephaly: spectrum of sonographic appearances and distinction from the amniotic band syndrome. Am J Roentgenol 151: 547–550.

Griffiths PD (2018). Schizencephaly revisited. Neuroradiology 60: 945–960.

Griffiths PD, Reeves MJ, Morris JE et al. (2010). A prospective study of fetuses with isolated ventriculomegaly investigated by antenatal sonography and in utero MR imaging. Am J Neuroradiol 31: 106–111.

Griffiths PD, Brackley K, Bradburn M et al. (2017a). Anatomical subgroup analysis of the MERIDIAN cohort: failed commissuration. Ultrasound Obstet Gynecol 50: 753–760.

Griffiths PD, Brackley K, Bradburn M et al. (2017b). Anatomical subgroup analysis of the MERIDIAN cohort: ventriculomegaly. Ultrasound Obstet Gynecol 50: 736–744.

Guibaud L (2004). Practical approach to prenatal posterior fossa abnormalities using MRI. Pediatr Radiol 34: 700–711.

Guibaud L, des Portes V (2006). Plea for an anatomical approach to abnormalities of the posterior fossa in prenatal diagnosis. Ultrasound Obstet Gynecol 27: 477–481.

Guibaud L, Larroque A, Ville D et al. (2012). Prenatal diagnosis of 'isolated' Dandy-Walker malformation: imaging findings and prenatal counselling. Prenat Diagn 32: 185–193.

Gutierrez-Castillo A, Jimenez-Ruiz A, Chavez-Castillo M et al. (2018). Septo-optic dysplasia plus syndrome. Cureus 10: e3727.

Hinkley LB, Marco EJ, Findlay AM et al. (2012). The role of corpus callosum development in functional connectivity and cognitive processing. PLoS One 7: e39804.

Huang YF, Chen WC, Tseng JJ et al. (2006). Fetal intracranial hemorrhage (fetal stroke): report of four antenatally diagnosed cases and review of the literature. Taiwan J Obstet Gynecol 45: 135–141.

Ivashchuk G, Loukas M, Blount JP et al. (2015). Chiari III malformation: a comprehensive review of this enigmatic anomaly. Childs Nerv Syst 31: 2035–2040.

Izzo G, Conte G, Cesaretti C et al. (2015). Prenatal magnetic resonance imaging of atypical partial rhombencephalosynapsis with involvement of the anterior vermis: two case reports. Neuropediatrics 46: 416–419.

Jamieson DJ, Kourtis AP, Bell M et al. (2006). Lymphocytic choriomeningitis virus: an emerging obstetric pathogen? Am J Obstet Gynecol 194: 1532–1536.

Johnson CY, Rasmussen SA (2010). Non-genetic risk factors for holoprosencephaly. Am J Med Genet C Semin Med Genet 154c: 73–85.

Jouan L, Ouled Amar Bencheikh B, Daoud H et al. (2016). Exome sequencing identifies recessive CDK5RAP2 variants in patients with isolated agenesis of corpus callosum. Eur J Hum Genet 24: 607–610.

Katorza E, Gat I, Duvdevani N et al. (2018). Fetal brain anomalies detection during the first trimester: expanding the scope of antenatal sonography. J Matern Fetal Neonatal Med 31: 506–512.

Katz VL, McCoy MC, Kuller JA et al. (1996). An association between fetal parvovirus B19 infection and fetal anomalies: a report of two cases. Am J Perinatol 13: 43–45.

Kirkham FJ, Zafeiriou D, Howe D et al. (2018). Fetal stroke and cerebrovascular disease: advances in understanding from lenticulostriate and venous imaging, alloimmune thrombocytopaenia and monochorionic twins. Eur J Paediatr Neurol 22: 989–1005.

Klein O, Pierre-Kahn A, Boddaert N et al. (2003). Dandy-Walker malformation: prenatal diagnosis and prognosis. Childs Nerv Syst 19: 484–489.

Kline-Fath BM, Arroyo MS, Calvo-Garcia MA et al. (2018). Prenatal aqueduct stenosis: association with rhombencephalosynapsis and neonatal outcome. Prenat Diagn 38: 1028–1034.

Kumar R, Jain MK, Chhabra DK (2001). Dandy-Walker syndrome: different modalities of treatment and outcome in 42 cases. Childs Nerv Syst 17: 348–352.

Kwiatkowska J, Waldoch A, Meyer-Szary J et al. (2017). Cardiac tumors in children: a 20-year review of clinical presentation, diagnostics and treatment. Adv Clin Exp Med 26: 319–326.

Larbrisseau A, Vanasse M, Brochu P et al. (1984). The Andermann syndrome: agenesis of the corpus callosum associated with mental retardation and progressive sensorimotor neuronopathy. Can J Neurol Sci 11: 257–261.

Laurence KM, Carter CO, David PA (1968). Major central nervous system malformations in South Wales. I. Incidence, local variations and geographical factors. Br J Prev Soc Med 22: 146–160.

Leibovitz Z, Guibaud L, Garel C et al. (2018). The cerebellar "tilted telephone receiver sign" enables prenatal diagnosis of PHACES syndrome. Eur J Paediatr Neurol 22: 900–909.

Lennon CA, Gray DL (1999). Sensitivity and specificity of ultrasound for the detection of neural tube and ventral wall defects in a high-risk population. Obstet Gynecol 94: 562–566.

Leombroni M, Khalil A, Liberati M et al. (2018a). Fetal midline anomalies: diagnosis and counselling Part 1: corpus callosum anomalies. Eur J Paediatr Neurol 22: 951–962.

Leombroni M, Khalil A, Liberati M et al. (2018b). Fetal midline anomalies: diagnosis and counselling part 2: septal anomalies. Eur J Paediatr Neurol 22: 963–971.

Leoncini E, Baranello G, Orioli IM et al. (2008). Frequency of holoprosencephaly in the International Clearinghouse Birth Defects Surveillance Systems: searching for population variations. Birth Defects Res A Clin Mol Teratol 82: 585–591.

Lerman-Sagie T, Prayer D, Stocklein S et al. (2018). Fetal cerebellar disorders. Handb Clin Neurol 155: 3–23.

Leruez-Ville M, Ghout I, Bussieres L et al. (2016). In utero treatment of congenital cytomegalovirus infection with valacyclovir in a multicenter, open-label, phase II study. Am J Obstet Gynecol 215: 462. e461–462.e410.

Levey EB, Stashinko E, Clegg NJ et al. (2010). Management of children with holoprosencephaly. Am J Med Genet C Semin Med Genet 154c: 183–190.

Li Z, Ren A, Zhang L et al. (2006). Extremely high prevalence of neural tube defects in a 4-county area in Shanxi Province, China. Birth Defects Res A Clin Mol Teratol 76: 237–240.

Liu J, Zhang L, Li Z et al. (2016). Prevalence and trend of neural tube defects in five counties in Shanxi province of Northern China, 2000 to 2014. Birth Defects Res A Clin Mol Teratol 106: 267–274.

Liu Z, Han J, Fu F et al. (2017). Outcome of isolated enlarged cisterna magna identified in utero: experience at a single medical center in mainland China. Prenat Diagn 37: 575–582.

Lo BW, Kulkarni AV, Rutka JT et al. (2008). Clinical predictors of developmental outcome in patients with cephaloceles. J Neurosurg Pediatr 2: 254–257.

Mahapatra AK, Suri A (2002). Anterior encephaloceles: a study of 92 cases. Pediatr Neurosurg 36: 113–118.

Malinger G, Lev D, Kidron D et al. (2005). Differential diagnosis in fetuses with absent septum pellucidum. Ultrasound Obstet Gynecol 25: 42–49.

Malinger G, Lev D, Oren M et al. (2012). Non-visualization of the cavum septi pellucidi is not synonymous with agenesis of the corpus callosum. Ultrasound Obstet Gynecol 40: 165–170.

Massoud M, Guibaud L (2018). Prenatal imaging of posterior fossa disorders. a review. Eur J Paediatr Neurol 22: 972–988.

Massoud M, Cagneaux M, Garel C et al. (2014). Prenatal unilateral cerebellar hypoplasia in a series of 26 cases: significance and implications for prenatal diagnosis. Ultrasound Obstet Gynecol 44: 447–454.

Matsunaga E, Shiota K (1977). Holoprosencephaly in human embryos: epidemiologic studies of 150 cases. Teratology 16: 261–272.

Matsushita H, Harada A, Sato T et al. (2014). Fetal intracranial injuries following motor vehicle accidents with airbag deployment. J Obstet Gynaecol Res 40: 599–602.

McKechnie L, Vasudevan C, Levene M (2012). Neonatal outcome of congenital ventriculomegaly. Semin Fetal Neonatal Med 17: 301–307.

McMahon CL, Braddock SR (2001). Septo-optic dysplasia as a manifestation of valproic acid embryopathy. Teratology 64: 83–86.

Melchiorre K, Bhide A, Gika AD et al. (2009). Counseling in isolated mild fetal ventriculomegaly. Ultrasound Obstet Gynecol 34: 212–224.

Merello E, Swanson E, De Marco P et al. (2008). No major role for the EMX2 gene in schizencephaly. Am J Med Genet A 146a: 1142–1150.

Meyer R, Bar-Yosef O, Barzilay E et al. (2018). Neurodevelopmental outcome of fetal isolated ventricular asymmetry without dilation: a cohort study. Ultrasound Obstet Gynecol 52: 467–472.

Miller SP, Shevell MI, Patenaude Y et al. (2000). Septo-optic dysplasia plus: a spectrum of malformations of cortical development. Neurology 54: 1701–1703.

Miller EA, Rasmussen SA, Siega-Riz AM et al. (2010). Risk factors for non-syndromic holoprosencephaly in the National Birth Defects Prevention Study. Am J Med Genet C Semin Med Genet 154c: 62–72.

Mirsky DM, Schwartz ES, Zarnow DM (2015). Diagnostic features of myelomeningocele: the role of ultrafast fetal MRI. Fetal Diagn Ther 37: 219–225.

Mulkey SB, Bulas DI, Vezina G et al. (2019). Sequential neuroimaging of the fetus and newborn with in utero zika virus exposure. JAMA Pediatr 173: 52–59.

Nabavizadeh SA, Zarnow D, Bilaniuk LT et al. (2014). Correlation of prenatal and postnatal MRI findings in schizencephaly. Am J Neuroradiol 35: 1418–1424.

Nagaraj UD, Bierbrauer KS, Stevenson CB et al. (2018). Myelomeningocele versus myelocele on fetal MR images: are there differences in brain findings? Am J Roentgenol 211: 1376–1380.

Naidich TP, Altman NR, Braffman BH et al. (1992). Cephaloceles and related malformations. Am J Neuroradiol 13: 655–690.

Naidich TP, Griffiths PD, Rosenbloom L (2015). Central nervous system injury in utero: selected entities. Pediatr Radiol 45: S454–S462.

Nellist M, Brouwer RW, Kockx CE et al. (2015). Targeted Next Generation Sequencing reveals previously unidentified TSC1 and TSC2 mutations. BMC Med Genet 16: 10.

Ng KH, Ng SM, Parker A (2015). Annual review of children with tuberous sclerosis. Arch Dis Child Educ Pract Ed 100: 114–121.

Nicolaides KH, Campbell S, Gabbe SG et al. (1986). Ultrasound screening for spina bifida: cranial and cerebellar signs. Lancet 2: 72–74.

Nomura ML, Barini R, De Andrade KC et al. (2010). Congenital hydrocephalus: gestational and neonatal outcomes. Arch Gynecol Obstet 282: 607–611.

Norem CT, Schoen EJ, Walton DL et al. (2005). Routine ultrasonography compared with maternal serum alpha-fetoprotein for neural tube defect screening. Obstet Gynecol 106: 747–752.

Oesch G, Maga AM, Friedman SD et al. (2018). Geometric morphometrics reveal altered corpus callosum shape in pyridoxine-dependent epilepsy. Neurology 91: e78–e86.

Olsen CL, Hughes JP, Youngblood LG et al. (1997). Epidemiology of holoprosencephaly and phenotypic characteristics of affected children: New York State, 1984–1989. Am J Med Genet 73: 217–226.

Ornoy A, Reece EA, Pavlinkova G et al. (2015). Effect of maternal diabetes on the embryo, fetus, and children: congenital anomalies, genetic and epigenetic changes and developmental outcomes. Birth Defects Res C Embryo Today 105: 53–72.

Ostrowsky-Coste K, Neal A, Guenot M et al. (2019). Resective surgery in tuberous sclerosis complex, from penfield to 2018: a critical review. Rev Neurol (Paris) 175: 163–182.

Pagani G, Thilaganathan B, Prefumo F (2014). Neurodevelopmental outcome in isolated mild fetal ventriculomegaly: systematic review and meta-analysis. Ultrasound Obstet Gynecol 44: 254–260.

Paladini D, Volpe P (2006). Posterior fossa and vermian morphometry in the characterization of fetal cerebellar abnormalities: a prospective three-dimensional ultrasound study. Ultrasound Obstet Gynecol 27: 482–489.

Paladini D, Malinger G, Pilu G et al. (2017). The MERIDIAN trial: caution is needed. Lancet 389: 2103.

Paladini D, Donarini G, Rossi A (2019). Indications for MRI in fetal isolated mild ventriculomegaly… 'And then, there were none…'. Ultrasound Obstet Gynecol 54 (2): 151–155.

Palmer EE, Mowat D (2014). Agenesis of the corpus callosum: a clinical approach to diagnosis. Am J Med Genet C Semin Med Genet 166c: 184–197.

Park H, Chang CS, Choi SJ et al. (2019). Sirolimus therapy for fetal cardiac rhabdomyoma in a pregnant woman with tuberous sclerosis. Obstet Gynecol Sci 62: 280–284.

Parrini E, Ramazzotti A, Dobyns WB et al. (2006). Periventricular heterotopia: phenotypic heterogeneity and correlation with filamin A mutations. Brain 129: 1892–1906.

Pashaj S, Merz E, Wellek S (2013). Biometry of the fetal corpus callosum by three-dimensional ultrasound. Ultrasound Obstet Gynecol 42: 691–698.

Pasquini L, Masini G, Gaini C et al. (2014). The utility of infection screening in isolated mild ventriculomegaly: an observational retrospective study on 141 fetuses. Prenat Diagn 34: 1295–1300.

Patel DM, Rocque BG, Hopson B et al. (2015). Sleep-disordered breathing in patients with myelomeningocele. J Neurosurg Pediatr 16: 30–35.

Patel C, Abraham S, Ferdman D (2018). Rapid regression of prenatally identified intrapericardial giant rhabdomyomas with sirolimus. CASE (Phila) 2: 258–261.

Paul LK, Brown WS, Adolphs R et al. (2007). Agenesis of the corpus callosum: genetic, developmental and functional aspects of connectivity. Nat Rev Neurosci 8: 287–299.

Pavlicek J, Klaskova E, Kapralova S et al. (2019). Fetal heart rhabdomyomatosis: a single-center experience. J Matern Fetal Neonatal Med 1–7.

Petryk A, Graf D, Marcucio R (2015). Holoprosencephaly: signaling interactions between the brain and the face, the environment and the genes, and the phenotypic variability in animal models and humans. Wiley Interdiscip Rev Dev Biol 4: 17–32.

Pilu G, Reece EA, Goldstein I et al. (1989). Sonographic evaluation of the normal developmental anatomy of the fetal cerebral ventricles: II. The atria. Obstet Gynecol 73: 250–256.

Pilu G, Segata M, Ghi T et al. (2006). Diagnosis of midline anomalies of the fetal brain with the three-dimensional median view. Ultrasound Obstet Gynecol 27: 522–529.

Poretti A, Boltshauser E (2015). Fetal diagnosis of rhombencephalosynapsis. Neuropediatrics 46: 357–358.

Poretti A, Alber FD, Burki S et al. (2009). Cognitive outcome in children with rhombencephalosynapsis. Eur J Paediatr Neurol 13: 28–33.

Poretti A, Boltshauser E, Doherty D (2014). Cerebellar hypoplasia: differential diagnosis and diagnostic approach. Am J Med Genet C Semin Med Genet 166c: 211–226.

Prabowo AS, Anink JJ, Lammens M et al. (2013). Fetal brain lesions in tuberous sclerosis complex: TORC1 activation and inflammation. Brain Pathol 23: 45–59.

Pugash D, Oh T, Godwin K et al. (2011). Sonographic 'molar tooth' sign in the diagnosis of Joubert syndrome. Ultrasound Obstet Gynecol 38: 598–602.

Putoux A, Thomas S, Coene KL et al. (2011). KIF7 mutations cause fetal hydrolethalus and acrocallosal syndromes. Nat Genet 43: 601–606.

Quarello E, Molho M, Garel C et al. (2014). Prenatal abnormal features of the fourth ventricle in Joubert syndrome and related disorders. Ultrasound Obstet Gynecol 43: 227–232.

Raam MS, Solomon BD, Muenke M (2011). Holoprosencephaly: a guide to diagnosis and clinical management. Indian Pediatr 48: 457–466.

Ramaswamy S, Rangasami R, Suresh S et al. (2013). Spontaneous resolution of Blake's pouch cyst. Radiol Case Rep 8: 877.

Randle SC (2017). Tuberous sclerosis complex: a review. Pediatr Ann 46: e166–e171.

Redline RW, Ravishankar S (2018). Fetal vascular malperfusion, an update. APMIS 126: 561–569.

Rethmann C, Scheer I, Meuli M et al. (2017). Evolution of posterior fossa and brain morphology after in utero repair of open neural tube defects assessed by MRI. Eur Radiol 27: 4571–4580.

Rintoul NE, Sutton LN, Hubbard AM et al. (2002). A new look at myelomeningoceles: functional level, vertebral level, shunting, and the implications for fetal intervention. Pediatrics 109: 409–413.

Risson DC, Davies MW, Williams BA (2012). Review of neonatal alloimmune thrombocytopenia. J Paediatr Child Health 48: 816–822.

Rizzo G, Pietrolucci ME, Capponi A et al. (2011). Assessment of corpus callosum biometric measurements at 18 to 32 weeks' gestation by 3-dimensional sonography. J Ultrasound Med 30: 47–53.

Robinson AJ (2014). Inferior vermian hypoplasia—preconception, misconception. Ultrasound Obstet Gynecol 43: 123–136.

Robinson AJ, Ederies MA (2016). Diagnostic imaging of posterior fossa anomalies in the fetus. Semin Fetal Neonatal Med 21: 312–320.

Ross ME, Mason CE, Finnell RH (2017). Genomic approaches to the assessment of human spina bifida risk. Birth Defects Res 109: 120–128.

Rossi AC, Prefumo F (2014). Additional value of fetal magnetic resonance imaging in the prenatal diagnosis of central nervous system anomalies: a systematic review of the literature. Ultrasound Obstet Gynecol 44: 388–393.

Sahni M, Ohri A (2019). Meningomyelocele. In: StatPearls, StatPearls Publishing, Treasure Island, FL.

Sajan SA, Fernandez L, Nieh SE et al. (2013). Both rare and de novo copy number variants are prevalent in agenesis of the corpus callosum but not in cerebellar hypoplasia or polymicrogyria. PLoS Genet 9: e1003823.

Sanapo L, Whitehead MT, Bulas DI et al. (2017). Fetal intracranial hemorrhage: role of fetal MRI. Prenat Diagn 37: 827–836.

Santo S, D'Antonio F, Homfray T et al. (2012). Counseling in fetal medicine: agenesis of the corpus callosum. Ultrasound Obstet Gynecol 40: 513–521.

Santoro M, Coi A, Barisic I et al. (2019). Epidemiology of Dandy-Walker Malformation in Europe: a EUROCAT population-based registry study. Neuroepidemiology 53: 169–179.

Saranac L, Gucev Z (2014). New insights into septo-optic dysplasia. Pril (Makedon Akad Nauk Umet Odd Med Nauki) 35: 123–127.

Seidman DS, Nass D, Mendelson E et al. (1996). Prenatal ultrasonographic diagnosis of fetal hydrocephalus due to infection with parainfluenza virus type 3. Ultrasound Obstet Gynecol 7: 52–54.

Senat MV, Bernard JP, Schwarzler P et al. (1999). Prenatal diagnosis and follow-up of 14 cases of unilateral ventriculomegaly. Ultrasound Obstet Gynecol 14: 327–332.

Sepulveda W, Corral E, Ayala C et al. (2004). Chromosomal abnormalities in fetuses with open neural tube defects: prenatal identification with ultrasound. Ultrasound Obstet Gynecol 23: 352–356.

Shekdar KV, Schwartz ES (2017). Brain tumors in the neonate. Neuroimaging Clin N Am 27: 69–83.

Shellhaas RA, Kenia PV, Hassan F et al. (2018). Sleep-disordered breathing among newborns with myelomeningocele. J Pediatr 194: 244–247.e241.

Shen O, Gelot AB, Moutard ML et al. (2015). Abnormal shape of the cavum septi pellucidi: an indirect sign of partial agenesis of the corpus callosum. Ultrasound Obstet Gynecol 46: 595–599.

Shinar S, Har-Toov J, Lerman-Sagie T et al. (2016). Thick corpus callosum in the second trimester can be transient and is of uncertain significance. Ultrasound Obstet Gynecol 48: 452–457.

Simon EM, Hevner RF, Pinter JD et al. (2002). The middle interhemispheric variant of holoprosencephaly. Am J Neuroradiol 23: 151–156.

Sole G, Coupry I, Rooryck C et al. (2009). Bilateral periventricular nodular heterotopia in France: frequency of mutations in FLNA, phenotypic heterogeneity and spectrum of mutations. J Neurol Neurosurg Psychiatry 80: 1394–1398.

Solomon BD, Rosenbaum KN, Meck JM et al. (2010). Holoprosencephaly due to numeric chromosome abnormalities. Am J Med Genet C Semin Med Genet 154c: 146–148.

Sotiriadis A, Makrydimas G (2012). Neurodevelopment after prenatal diagnosis of isolated agenesis of the corpus callosum: an integrative review. Am J Obstet Gynecol 206: 337. e331–335.

Stoll C, Dott B, Alembik Y et al. (2011). Associated malformations among infants with neural tube defects. Am J Med Genet A 155a: 565–568.

Stothard KJ, Tennant PW, Bell R et al. (2009). Maternal overweight and obesity and the risk of congenital anomalies: a systematic review and meta-analysis. JAMA 301: 636–650.

Stumpel C, Vos YJ (1993). L1 syndrome. GeneReviews((R)). In: MP Adam, HH Ardinger, RA Pagon et al. (Eds.), University of Washington, Seattle, WA.

Sukanya S, Bay BH, Tay SS et al. (2012). Frontiers in research on maternal diabetes-induced neural tube defects: past, present and future. World J Diabetes 3: 196–200.

Summers AD, Reefhuis J, Taliano J et al. (2018). Nongenetic risk factors for holoprosencephaly: an updated review of the epidemiologic literature. Am J Med Genet C Semin Med Genet 178: 151–164.

Szabo N, Gyurgyinka G, Kobor J et al. (2010). Epidemiology and clinical spectrum of schizencephaly in south-eastern Hungary. J Child Neurol 25: 1335–1339.

Takanashi J, Barkovich AJ, Ferriero DM et al. (2003). Widening spectrum of congenital hemiplegia: periventricular venous infarction in term neonates. Neurology 61: 531–533.

Tang PH, Bartha AI, Norton ME et al. (2009). Agenesis of the corpus callosum: an MR imaging analysis of associated abnormalities in the fetus. Am J Neuroradiol 30: 257–263.

Tarui T, Limperopoulos C, Sullivan NR et al. (2014). Long-term developmental outcome of children with a fetal diagnosis of isolated inferior vermian hypoplasia. Arch Dis Child Fetal Neonatal Ed 99: F54–F58.

Tattoli L, Di Vella G, Solarino B (2017). A case of intrauterine lethal fetal injury after attempted suicide of the mother. Forensic Sci Int 280: e1–e5.

Tietjen I, Bodell A, Apse K et al. (2007). Comprehensive EMX2 genotyping of a large schizencephaly case series. Am J Med Genet A 143a: 1313–1316.

Vachon-Marceau C, Guerra V, Jaeggi E et al. (2019). In-utero treatment of large symptomatic rhabdomyoma with sirolimus. Ultrasound Obstet Gynecol 53: 420–421.

van Dijk T, Baas F, Barth PG et al. (2018). What's new in pontocerebellar hypoplasia? An update on genes and subtypes. Orphanet J Rare Dis 13: 92.

Vawter-Lee MM, Wasserman H, Thomas CW et al. (2018). Outcome of isolated absent septum pellucidum diagnosed by fetal magnetic resonance imaging (MRI) scan. J Child Neurol 33: 693–699.

Vermeulen RJ, Peeters-Scholte C, Van Vugt JJ et al. (2011). Fetal origin of brain damage in 2 infants with a COL4A1 mutation: fetal and neonatal MRI. Neuropediatrics 42: 1–3.

Volpe P, Campobasso G, De Robertis V et al. (2009). Disorders of prosencephalic development. Prenat Diagn 29: 340–354.

Volpe P, Contro E, De Musso F et al. (2012). Brainstem-vermis and brainstem-tentorium angles allow accurate categorization of fetal upward rotation of cerebellar vermis. Ultrasound Obstet Gynecol 39: 632–635.

Vos YJ, Hofstra RM (2010). An updated and upgraded L1CAM mutation database. Hum Mutat 31: E1102–E1109.

Weber MA, Sebire NJ (2010). Genetics and developmental pathology of twinning. Semin Fetal Neonatal Med 15: 313–318.

Weller S, Gartner J (2001). Genetic and clinical aspects of X-linked hydrocephalus (L1 disease): Mutations in the L1CAM gene. Hum Mutat 18: 1–12.

Windham GC, Bjerkedal T, Sever LE (1982). The association of twinning and neural tube defects: studies in Los Angeles, California, and Norway. Acta Genet Med Gemellol (Roma) 31: 165–172.

Wortmann SB, Reimer A, Creemers JW et al. (2008). Prenatal diagnosis of cerebral lesions in Tuberous sclerosis complex (TSC). Case report and review of the literature. Eur J Paediatr Neurol 12: 123–126.

Yamasato K, Kurata N, Towner D (2018). Delayed appearance of a traumatic fetal intracranial hemorrhage. Case Rep Obstet Gynecol 2018: 1465034.

Yu T, Meiners LC, Danielsen K et al. (2013). Deregulated FGF and homeotic gene expression underlies cerebellar vermis hypoplasia in CHARGE syndrome. eLife 2: e01305.

Yun-Hai S, Nan B, Ping-Ping G et al. (2015). Is repair of the protruded meninges sufficient for treatment of meningocele? Childs Nerv Syst 31: 2135–2140.

Zamora EA, Ahmad T (2019). Dandy Walker malformation. In: StatPearls, StatPearls Publishing, Treasure Island, FL.

Zou Z, Huang L, Lin S et al. (2018). Prenatal diagnosis of posterior fossa anomalies: additional value of chromosomal microarray analysis in fetuses with cerebellar hypoplasia. Prenat Diagn 38: 91–98.

Handbook of Clinical Neurology, Vol. 171 (3rd series)
Neurology and Pregnancy: Pathophysiology and Patient Care
E.A.P. Steegers, M.J. Cipolla and E.C. Miller, Editors
https://doi.org/10.1016/B978-0-444-64239-4.00016-3

Chapter 16

Perinatal stroke

EMMETT E. WHITAKER[1,2]* AND MARILYN J. CIPOLLA[2,3,4]

[1]*Department of Anesthesiology, University of Vermont Larner College of Medicine, Burlington, VT, United States*

[2]*Department of Neurological Sciences, University of Vermont Larner College of Medicine, Burlington, VT, United States*

[3]*Department of Obstetrics, Gynecology & Reproductive Sciences, University of Vermont Larner College of Medicine, Burlington, VT, United States*

[4]*Department of Pharmacology, University of Vermont Larner College of Medicine, Burlington, VT, United States*

Abstract

Perinatal stroke is a heterogeneous syndrome resulting from brain injury of vascular origin that occurs between 20 weeks of gestation and 28 days of postnatal life. The incidence of perinatal stroke is estimated to be between 1:1600 and 1:3000 live births (approximately 2500 children per year in the United States), though its actual incidence is difficult to estimate because it is likely underdiagnosed. Perinatal arterial ischemic stroke (PAIS) accounts for approximately 70% of cases of perinatal stroke. Cerebral sinovenous thrombosis, while less common, also accounts for a large proportion of the morbidity and mortality seen with perinatal stroke. Hemorrhagic stroke leads to disruption of neurologic function due to intracerebral hemorrhage that is nontraumatic in origin. While most cases of PAIS fall into one of these three categories, other patterns of injury should also be considered perinatal stroke. In some cases, the etiology of PAIS is not known but is idiopathic. This chapter will review the classification, risk factors, pathogenesis, clinical presentation, management, and long-term sequelae of perinatal stroke.

INTRODUCTION

Perinatal stroke is a heterogeneous syndrome resulting from brain injury of vascular origin that occurs between 20 weeks of gestation and 28 days of postnatal life and is not a rare phenomenon. The incidence of perinatal stroke is estimated to be between 1:1600 and 1:3000 live births (approximately 2500 children per year in the United States), though its actual incidence is difficult to estimate because it is likely underdiagnosed (Nelson and Lynch, 2004; Laugesaar et al., 2007; Agrawal et al., 2009). This focal disruption of blood flow is most often due to one of three main causes. Perinatal arterial ischemic stroke (PAIS) accounts for approximately 70% of cases of perinatal stroke (Domi et al., 2008; Govaert et al., 2009a; deVeber et al., 2017). Cerebral sinovenous thrombosis (CSVT), while less common, also accounts for a large proportion of the morbidity and mortality seen with perinatal stroke. Hemorrhagic stroke refers to disruption of neurologic function due to intracerebral hemorrhage that is nontraumatic in origin. While most cases of PAIS fall into one of these three categories, it is important to note that there is ongoing debate as to whether other patterns of injury should also be considered perinatal stroke. In some cases, the etiology of PAIS is not known and it is classified as idiopathic. Classification, risk factors, pathogenesis, clinical presentation, management, and long-term sequelae of perinatal stroke will be reviewed in this chapter. In addition, how perinatal stroke varies from adult stroke and stroke in childhood will be partially reviewed. Last, some information regarding the physiologic maturation of cerebral circulation and cerebral hemodynamics, and how it relates to perinatal stroke, is presented.

*Correspondence to: Emmett E. Whitaker, M.D., Assistant Professor, Departments of Anesthesiology & Neurological Sciences, University of Vermont Larner College of Medicine, 111 Colchester Ave, Burlington, VT 05401, United States. Tel: +1-802-847-2717, Fax: +1-802-847-5324, E-mail: emmett.whitaker@uvmhealth.org

GESTATIONAL CEREBROVASCULAR DEVELOPMENT

In humans, cerebrovascular maturation beings in the fourth week of gestation and continues throughout the first several years of life. Neural tube development, occurring at 3–4 weeks of gestation, is the first in a sequence of events that comprise in utero origination of the cerebral circulation (DeSesso et al., 1999; Rice and Barone Jr., 2000). Thereafter, during weeks 6–8, the aortic arches develop, creating the framework for the basic arterial structure that is present in the human brain. Beyond this point, the cerebral vascular tree continues to arborize, creating the basis of perfusion for the brain (Rhodes and Hyde, 1965; Harnarine-Singh and Hyde, 1970; Otto and Lierse, 1970).

During gestation, the metabolic needs of burgeoning populations of neurons are the primary factors that regulate extension of blood vessels into the brain parenchyma (Degani, 2009).

NEONATAL CEREBROVASCULAR DEVELOPMENT AND AUTOREGULATION OF CEREBRAL BLOOD FLOW

Vascular endothelium plays a critical role in maintenance of homeostasis in the cerebral circulation. Endothelial cells are involved in the modulation of vascular tone, angiogenesis, hemostasis, and function of the blood–brain barrier (BBB). Vascular endothelial cells initiate development of capillaries via release of vascular growth factors (Inagami et al., 1995). However, the developing vascular endothelium can also act in an angiostatic fashion to inhibit angiogenesis. These two functions are tightly regulated to ensure appropriate blood supply to the rapidly developing brain. The vascular endothelium also acts as a "gatekeeper" for trafficking of substances into and/or out of the brain. This immature BBB is significantly more permeable than the adult BBB, and as a result the immature brain is more susceptible to injury and invasion than the adult brain (Buckley, 1986; Stewart and Hayakawa, 1987; Risau and Wolburg, 1990). Further disruption of the BBB by hypoxia/ischemia may further impair the ability of the BBB to protect the brain.

Vascular tone is an important determinant of blood blow to the brain, and close regulation of tone is required for maintenance of adequate perfusion. Vascular endothelial cells release vascoactive substances that act on vascular smooth muscle to affect tone. Nitric oxide (NO) is one of the most important regulators of cerebrovascular tone, acting on nearby smooth muscle cells to cause vasodilation via guanylate cyclase activation (Moncada et al., 1991). Overall, the functional immaturity of vascular endothelium in neonates limits its ability to influence vascular tone when compared with adults.

Vascular smooth muscle cells exist in a number of different phenotypes, resulting in significant variations in physiology among vascular smooth muscle subtypes. These differences allow for differing capacities of these cells to proliferate, secrete, migrate, and contract (Owens et al., 2004; Rensen et al., 2007). A broad variety of factors are responsible for growth and differentiation of smooth muscle, most notably vascular endothelial growth factor (VEGF) (Hoeben et al., 2004), which is a glycoprotein and a growth factor that has potent proangiogenic activity (Ferrara et al., 1991). It also has potent tropic effects on the growth and differentiation of vascular smooth muscle. Hypoxia is an important regulator of VEGF synthesis and release. In the setting of hypoxia, transcriptional activators that affect the expression of the VEGF gene are released (Semenza, 2004); this allows VEGF expression to closely parallel the degree of tissue hypoxia (Germain et al., 2010). Given its ability to promote angiogenesis and response to hypoxia, it is likely that VEGF is involved in the physiologic response to neonatal stroke, potentially via enhancement of angiogenesis and of endothelial proliferation and amelioration of injury (Dzietko et al., 2013).

The reactivity of the vasculature in the immature brain is a continuum that varies based on size of vessels, brain region, age, and smooth muscle cell phenotype. Similar to the adult, this heterogeneity allows the neonatal brain to maintain adequate regional and global perfusion during a period where the milieu in the brain is constantly changing. Fundamentally, for brain tissue to maintain physiologic homeostasis, it must receive blood flow that meets its metabolic requirements. The brain is reliant, more than other type of tissue, on glucose metabolism. As such, flow and metabolism in the brain must be matched not only for oxygen, but also glucose delivery (Kuschinsky, 1991). Regulation of this flow-metabolism coupling is complex, but unlike many other processes in the neonatal brain, this mechanism is fully developed at birth (Pryds et al., 1990; Ramaekers and Casaer, 1990; Walter et al., 2000).

In a disease state such as stroke, hypoxia occurs in the affected tissue. Even short-term hypoxia causes significant vasodilation in the immature brain. Generally speaking, more severe hypoxia produces the most marked responses in areas of brain with high metabolic activity (Lou et al., 1985; Buckley, 1986; Szymonowicz et al., 1990). Vasodilation is especially important because in hypoxic states, increased blood flow can temporarily compensate for hypoxemia (Koehler et al., 1986). Further compensation (as arterial oxygen content continues to fall) includes maximal vasodilation and maximal extraction of oxygen from the blood. However, anaerobic metabolism, tissue acidosis, and, eventually, irreversible tissue damage occur if hypoxia is not corrected (Richardson et al., 1989).

Autoregulation, the capacity of vascular beds to maintain constant blood flow over a range of cerebral perfusion pressures (CPPs), is intact but immature at birth (Strandgaard and Paulson, 1984). In neonates, the lower limit of autoregulation (LLA) is lower than that seen in adults. As the CPP approaches the LLA (30–40 mmHg), vasodilation occurs to maintain adequate blood flow. However, below the LLA the blood flow falls in direct proportion to the CPP (Mann, 1970; Hernandez et al., 1980; Buckley, 1986; Ramaekers et al., 1990). As a baby develops, the baseline blood pressure rises. During this period, the upper limit of autoregulation (ULA) also rises, reaching a value of approximately 80–90 mmHg in normal neonates (Strandgaard and Paulson, 1984). As the CPP approaches the ULA, vasoconstriction occurs to limit blood flow. However, this mechanism is not fully developed in neonates (Walsh and Cole, 2013). CPPs in excess of the ULA can cause vascular insults in neonates, potentially leading to intracerebral hemorrhage (Pasternak and Groothuis, 1985; Ramaekers et al., 1990; Arnold et al., 1991). Taken together, differences in maturation of neonatal cerebrovascular physiology have direct relevance to the pathophysiology, management, and prognosis of perinatal stroke.

CLASSIFICATION OF PERINATAL STROKE

Perinatal stroke can be classified temporally by the time period during which it occurs, based on clinical findings and neuroimaging characteristics. Fetal stroke, as implied by its name, occurs before birth but after 20 weeks of gestation (Nelson and Lynch, 2004; Raju et al., 2007; Govaert et al., 2009a, b). It is often diagnosed postnatally as a chronic clinical state presenting immediately after birth and is confirmed with imaging. Neonatal stroke occurs between birth and 28 days of life and is usually diagnosed by acute, generalized neurologic findings and confirmed with imaging. Presumed perinatal ischemic stroke (PPIS) is considered to have occurred when a focal, chronic neurologic finding presents sometime during the first year of life and is presumed to have occurred in the perinatal period. In the case of PPIS, no acute neurologic syndrome is present (Kirton et al., 2008).

Similar to adult stroke, perinatal stroke can also be classified clinically into three subtypes: arterial ischemic, venous thrombosis, and hemorrhagic. However, there are a number of important differences in perinatal stroke with respect to presentation, diagnosis, and treatment.

Perinatal arterial ischemic stroke (PAIS)

PAIS is the most common subtype of perinatal stroke, comprising approximately 70% of cases (Armstrong-Wells et al., 2009; Govaert et al., 2009a). In PAIS, the area of injury conforms to a specific vascular territory, most frequently the middle cerebral artery (MCA) (Nelson and Lynch, 2004; Govaert et al., 2009a). This is similar to adult stroke, and arterial occlusion leads to hypoperfusion and resultant infarction. It has been theorized that the origin of the left carotid artery from the aorta allows a more direct vascular route to the brain as a corridor for emboli, though the precise reason that the MCA is disproportionately affected remains unknown (de Vries et al., 1997; Govaert et al., 2000).

Stroke occurring in the fetus (before delivery) likely represents an underreported cause of perinatal stroke as it rarely results in noticeable maternal or fetal symptoms. Additionally, the mean age of detection (by routine prenatal ultrasound) is 31 weeks, implying that fetal stroke may go undetected at the initial (screening) ultrasound (Ozduman et al., 2004). The unique properties of the immature cerebral circulation in the fetus make it particularly vulnerable to stroke (Golden, 1985). Fetal strokes are a subset of perinatal stroke and have two distinct features: (1) fetal strokes can be verified radiologically as having occurred before birth and (2) a static neurologic deficit is present during the neonatal period. Although the cause of fetal stroke is unknown, it is thought that thromboses present in the placenta in late pregnancy may embolize into the fetal circulation (Elbers et al., 2011). Even in the case of venous clots, the patent foramen ovale or left-to-right shunts present in congenital heart disease may allow the clot to cross into the arterial circulation, which has a direct path into the cerebral circulation (Pellicer et al., 1992).

Because physical examination of the fetus is not possible, most fetal strokes are detected on routine prenatal ultrasound. Unlike neonatal stroke, a large proportion of fetal stroke is hemorrhagic in nature, with intraventricular hemorrhage present in six of seven cases presented in one series (Ozduman et al., 2004). Fetal ultrasound findings are typically confirmed with magnetic resonance imaging (MRI). Similar to neonatal stroke, many neonates suffering fetal stroke have long-term neurologic sequelae ranging in severity from mild impairment to significant neurodevelopmental delay.

CLINICAL PRESENTATION

Most cases of PAIS occur in neonates born at term, and 60%–70% of cases present during the neonatal period (between birth and 28 days of life) (Lee et al., 2005a; Curry et al., 2007). The remaining 30%–40% present later, likely due to vagaries in presentation that delay diagnosis. Very little is known about stroke that occurs before birth. At the time of presentation, which may occur weeks to months later, it is nearly impossible to distinguish between stroke that occurred in utero and stroke that occurred during the neonatal period. Unlike ischemic stroke in adults, the presentation of PAIS is more subtle and nonspecific (Fujimoto et al., 1992; Estan and Hope, 1997; Sreenan et al., 2000). The most common presenting manifestation of perinatal stroke is new-onset seizures. Seizures are the presenting symptom in 70%–90% of term newborns with PAIS (Lee et al., 2005b; Curry et al., 2007; Kirton et al., 2011). PAIS may also present with more diffuse symptoms, including altered consciousness, hypotonia, apnea, and feeding problems (Chabrier et al., 2010; Darmency-Stamboul et al., 2012). Focal signs, however, are uncommon in PAIS. Typically, seizures associated with PAIS occur within the first 3 days of life, a feature that can distinguish it from seizures associated with generalized hypoxic ischemia (Lee et al., 2005b; Rafay et al., 2009; Chabrier et al., 2010; Harteman et al., 2012). As a result of the nonspecific findings associated with perinatal stroke, PAIS is almost certainly underdiagnosed. The remainder of PAIS cases is diagnosed later in life, usually due to a focal neurologic deficit. Subsequent neuroimaging and incidental discovery of an old cerebral infarct lead to a diagnosis of PPIS (Kirton et al., 2008), which usually presents after 2 months of age, with the median age of presentation being approximately 6 months (Golomb et al., 2001; Lee et al., 2005b). Clinical features of PPIS tend to differ from that of PAIS, with approximately 80% of infants presenting with an early hand preference or unilateral fisting (Golomb et al., 2001; Lee et al., 2005b).

As most PAIS affects the distribution of the MCA, the arm and face are more likely to be affected than the lower body (Bouza et al., 1994). Unilateral lesions may result in a hemiparesis, but this is usually subtle and difficult to detect. It may present as an asymmetry of spontaneous movements, or in the case of bilateral lesions, a mild quadriparesis. Bilateral lesions, depending on location, can also result in abnormal temperature homeostasis and/or altered sleep–wake cycles (van der Aa et al., 2013).

Diagnosis

Initial diagnosis of PAIS depends on a high degree of clinical suspicion so that appropriate evaluation can be carried out to confirm the diagnosis. Once PAIS is suspected, the formal diagnosis is typically made using some sort of imaging modality. With an experienced operator, transcranial ultrasound can be used to detect PAIS, and incidental diagnosis of PAIS with transcranial ultrasound has been described (Cowan et al., 2005). However, the peripheral nature of most arterial infarcts makes most PAIS difficult to detect with this modality. Although both computed tomography (CT) and MRI can be used to diagnose PAIS, MRI is the gold standard and should be performed in all neonates suspected of having PAIS (Lee et al., 2017). MRI lacks the radiation exposure inherent in the use of CT scanning and is capable of detecting small or early infarcts that may not be apparent on CT (Raju et al., 2007; Lee et al., 2017). In the event of an emergency where MRI is not practical or possible, CT can be used to rule out a space-occupying lesion (subdural hematoma that may need surgical treatment), but MRI will ultimately be required to determine the distribution and extent of the lesion.

Diffusion weighted imaging (DWI) MRI is most often used for the diagnosis and evaluation of PAIS as it can identify lesions that are too early to be detected by other MRI modalities (Cowan et al., 1994; Robertson et al., 1999; Krishnamoorthy et al., 2000). DWI is often used in conjunction with MR angiography to identify the specific location of the arterial occlusion and/or arterial narrowing. However, the location of the occlusion cannot always be identified, and as a result it must be inferred based on the location and characteristics of the infarct itself (Govaert et al., 2009a; Ecury-Goossen et al., 2013). Large, wedge-shaped infarcts usually originate from occlusion of large cerebral arteries. In the case of PAIS, the most frequently affected artery is the MCA. Small, subcortical infarcts in the white matter and other deep structures typically arise from blockage of a smaller artery that branches from a major cerebral artery.

The time period over which PAIS evolves in neonates is distinct from that seen in adults. Lack of evidence of PAIS in the first 24 h after infarct does not necessarily indicate that a stroke has not occurred; most strokes are detectable on DWI after 24 h, and the signal begins to dissipate at approximately 5 days poststroke (McKinstry et al., 2002). As such, imaging results must be interpreted with caution in the setting of a strong clinical suspicion of PAIS.

Further evaluation of the neonate with PAIS

Further investigation as to the cause of the stroke is required. Perhaps the most important element of the evaluation is laboratory screening for the presence of thrombophilia, including protein C, protein S, fibrinogen, antithrombin, lipoprotein(a), anticardiolipin antibodies, lupus anticoagulant, partial thromboplastin time, and international normalized ratio (INR) assessment. Thereafter, a follow-up screen that includes activated protein C resistance, factor V Leiden, prothrombin gene mutation, methyltetrahydrofolate reductase (MTHFR), lipoprotein(a), homocysteine, and factors VIII, IX, and X should be performed at age 3–6 months (Nowak-Gottl et al., 1997). Additional laboratory studies include a complete blood count with differential, which may reveal polycythemia. Echocardiography is indicated to confirm or rule out the presence of congenital cardiac lesions, in particular patent foramen ovale or other lesions that allow right-to-left shunting. Finally, electroencephalography (EEG) may be helpful in identifying focal or generalized seizures and may be a useful guide to prognosis in the post-PAIS period (Mercuri et al., 1999).

Etiology and pathogenesis

In many cases, the cause of PAIS is not clear. However, the causes are generally classified into one of three categories: (1) emboli of cardiac or transcardiac origin; (2) disorders of the cerebral arteries (arteriopathy); and (3) thrombosis due to coagulopathy. The most common theory as to the cause of the majority of PAIS with an embolic source is that an embolus (most likely originating from the placenta) enters the fetal or neonatal circulation (peripartum), crosses the patent foramen ovale, and travels to the cerebral circulation leading to infarct.

Arteriopathies are the most common identifiable cause of PAIS in a previously healthy neonate. They have been found to be present in as many as 64% of patients (Danchaivijitr et al., 2006; Fullerton et al., 2007; Zimmer et al., 2007; Amlie-Lefond et al., 2009). Arteriopathy, a term generally used to refer to any disease of the arteries, takes on a more specific meaning in the case of PAIS. Arteriopathies associated with PAIS are usually classified as one of four subtypes that include arterial dissection, focal cerebral arteriopathy-inflammatory type (FCA-i), moyamoya disease, and diffuse multifocal vasculitis (Wintermark et al., 2017). Unlike in adults, arteriopathies in children are nonatherosclerotic and steno-occlusive (Biller, 2009). Arteriopathy in children can be intrinsic or acquired and has been associated with perinatal infection, birth trauma, congenital vascular malformations, and several genetic syndromes (Table 16.1).

FCA-i refers to focal intracranial arterial stenosis that is idiopathic in nature. It accounts for approximately 25% of childhood arteriopathy, making it the most common cause overall (Amlie-Lefond et al., 2009). FCA-i is likely associated with recent infection, although its actual etiology remains unclear (Amlie-Lefond et al., 2009). FCA encompasses two previously named conditions that share significant overlap: transient cerebral arteriopathy (TCA) of childhood and nonprogressive central nervous system vasculitis. TCA is transient in the sense that on follow-up imaging and 6 months poststroke there is no evidence of additional stenoses or progression of the original stenosis. It is postulated that TCA is also precipitated by a viral infection (Sebire et al., 2004). Nonprogressive central nervous system vasculitis it typically seen in children and is not common in neonates (Benseler et al., 2006).

Table 16.1

Genetic syndromes associated with arteriopathy

Neurofibromatosis type 1 (Rea et al., 2009)
Alagille syndrome (Emerick et al., 2005)
Williams syndrome (Wollack et al., 1996)
PHACES syndrome (Heyer et al., 2008)
Trisomy 21 (Benseler et al., 2006)

Postvaricella angiopathy refers to a focal cerebral artery stenosis in the setting of varicella infection in the year preceding the stroke (Sebire et al., 1999, 2004). The pathophysiology of varicella-induced arteriopathy is poorly understood, but it is theorized that invasion of the artery wall by the virus leads to vascular endothelial dysfunction (Linnemann Jr. and Alvira, 1980).

Primary CNS vasculitis is diagnosed when no other condition that could cause inflammation of the blood vessels is present. Vasculitis is termed secondary when it results from another intracranial process, such as malignancy, infection, connective tissue disease, or exposure to certain medications or toxins. Diagnosis of vasculitis can be extremely challenging, especially when primarily small vessels are involved. In these cases, angiography can be negative despite the presence of disease (Benseler et al., 2005). Primary or secondary vasculitis should be considered as part of the differential diagnosis in any case where one or more of the following is present: protracted clinical course, a known risk factor for vasculitis (such as autoimmune disease), diffuse neurologic deficits, or systemic signs such as fever (Benseler et al., 2005).

Primary moyamoya disease, the second most common type of arteriopathy in children, is a progressive occlusive disease of the distal internal carotid arteries or, less frequently, the posterior circulation (Roach et al., 2008). Both types of moyamoya may lead to significant collateralization. Moyamoya disease is idiopathic. If there is a known cause (cranial radiation or genetic syndrome), then it is termed moyamoya syndrome (secondary moyamoya) (Emerick et al., 2005). One unique feature of moyamoya disease is that in many cases surgical revascularization can be performed to bypass the lesion. When surgical revascularization is performed in a timely fashion, subsequent morbidity and mortality can be avoided (Amlie-Lefond et al., 2011).

Cervicocephalic arterial dissection is a type of arteriopathy that usually results from birth trauma. This phenomenon accounts for approximately 7.5%–20% of perinatal strokes (Rafay et al., 2006). Consistent with the mechanism of injury (birth trauma), the anterior circulation is more frequently affected than the posterior circulation (Rafay et al., 2006).

Management and treatment

The primary focus of treatment for PAIS is supportive care, which is directed at treating any underlying conditions and preventing progression of disease or further strokes. As with any critical illness, additional goals include correcting acid–base imbalance and electrolyte abnormalities, ensuring adequate oxygenation and ventilation, treating anemia, and providing antibiotic therapy if bacterial infection is present (Roach et al., 2008). Seizures should be treated with anticonvulsants once they are adequately characterized using EEG. For the most part, once the aforementioned measures are undertaken, therapy consists of close observation and monitoring. Systemic anticoagulation carries a significant risk of hemorrhage, and it remains unclear whether the potential benefit of anticoagulation outweighs the risk. However, anticoagulants may be warranted on a case-by-case basis. The American College of Chest Physicians recommends supportive care unless there is a documented cardioembolic source. In the case of a neonate with a first arterial stroke and a documented thromboembolic source, they recommend aspirin or unfractionated or low molecular weight heparin. This approach is also recommended for neonates who have or are at risk for recurrent stroke (Monagle et al., 2012). Thrombolytic therapy is rarely indicated in neonates, and the American Heart Association specifically recommends against use of thrombolytic agents in neonates (Roach et al., 2008).

Prognosis

A number of sequelae are common after PAIS during the neonatal period. However, because the cause, presentation, and patient characteristics vary so widely among individual cases, the prognosis is also highly variable. Of children who suffer PAIS in the neonatal period, 20%–40% go on to develop normally (deVeber et al., 2000; Sreenan et al., 2000; Nelson and Lynch, 2004; Lee et al., 2005b; Chabrier et al., 2016). In children who do manifest abnormal neurodevelopment, 75% exhibit multiple disabilities. These may include cerebral palsy, cognitive impairment, behavioral abnormalities, epilepsy, and language delay. The relative incidence of these sequelae is summarized in Table 16.2. Recurrence of stroke is considered unlikely (1%–3%) (Kurnik et al., 2003; Fullerton et al., 2007).

Table 16.2

Sequelae after PAIS in the neonate

Disability	Incidence (%)	References
Cerebral palsy/motor disability	20–30	Mercuri et al. (2004), Lehman and Rivkin (2014), and Lee et al. (2005b)
Epilepsy	10–40	Lee et al. (2005b), Golomb et al. (2001), and Billinghurst et al. (2017)
Cognitive impairment	20–25	Bosenbark et al. (2017) and Westmacott et al. (2009)

Cerebral sinovenous thrombosis (CSVT)

CSVT accounts for approximately 10% of cases of pediatric stroke with an approximate incidence of 0.6 per 100,000 (Grunt et al., 2010). Of these cases, 30%–50% are newborns (Mallick et al., 2009). CSVT is associated with thrombosis of at least one major cerebral sinus, leading to venous outflow obstruction, which may or may not result in infarction and/or hemorrhage. The brain's venous system is a complex network of cortical, medullary, and deep veins that all drain into one of several venous sinuses: those that are superficial (sagittal, transverse, and sigmoid) and those that are deep (straight sinus and the vein of Galen) (Cipolla, 2016). Thrombotic occlusion of one of these can impair venous outflow leading to venous congestion, increased central venous pressure, and intracranial hypertension. Intracranial hypertension, in turn, impedes arterial perfusion and can lead to cerebral ischemia.

Risk factors for CSVT are diverse, age-dependent, and similar to those for other types of perinatal stroke. However, the most consistent in newborns appear to be acute systemic illness and/or thrombophilia (Berfelo et al., 2010; Grunt et al., 2010; Jordan et al., 2010; Moharir, 2010; Moharir et al., 2010; Ichord et al., 2015).

Presentation

The presentation and clinical characteristics of CSVT in the neonate are similar to those of PAIS and include seizures and altered mental status (Lehman and Rivkin, 2014). Because of increased intracranial pressure, bulging fontanelles may be present on physical examination. Acute intracranial hypertension is a medical emergency. In the absence of treatment, herniation and death can occur.

Table 16.3

Recommendations for treatment of CSVT in neonates

Source	Recommendations	References
American Heart Association Scientific Statement	• Supportive measures • Consider anticoagulant therapy in cases with prothrombotic disorders, multiple thrombi, or progressive cerebral thrombus	Roach et al. (2008)
American College of Chest Physicians	• Consider anticoagulant therapy for neonates without intracranial hemorrhage for a period of 6 weeks to 3 months • For neonates with intracranial hemorrhage, defer treatment until repeat imaging after 5–7 days shows clot enlargement	Monagle et al. (2012)
British Committee for Standards in Hematology	• Neonates without intracranial hemorrhage should be treated with anticoagulant therapy for o 3 months if provoking illness is reversible o 6 months when no provoking illness can be identified o >6 months for ongoing thrombophilia of persistent venous outflow obstruction • Repeat imaging before stopping anticoagulant therapy • Children with intracranial hemorrhage may be treated after repeat imaging demonstrates enlarging clot	Chalmers et al. (2011)

Diagnosis

Making the diagnosis of CSVT is challenging, because as with PAIS, symptoms are nonspecific and may be confused with characteristics of comorbid conditions. However, timely diagnosis is critically important to guide therapy and optimize outcomes. Acute neurologic change and/or seizures in any high-risk neonate should prompt immediate concern for perinatal stroke, in which case urgent imaging is indicated (Lee et al., 2017). A range of imaging modalities can be used. MRI and magnetic resonance venography should be considered to evaluate clot burden, location, and extent of parenchymal damage. When MRI is impractical or impossible, CT scanning with venography can be used. Although this modality provides a sensitive method of identifying thrombosis, it is considerably less sensitive in identifying and characterizing an infarct (Raju et al., 2007).

Management and treatment

Due to the relatively low incidence and difficulty in studying CSVT, there have been no clinical trials designed to determine the optimal treatment(s) for CSVT in neonates. However, treatment recommendations do exist for adults. Treatment strategies for neonates with CSVT are generally based on adult treatment protocols. Treatment guidelines for CSVT in neonates exist, but they are controversial due to the mixed quality of evidence on which they are based. As a result, supportive therapy is the primary recommendation for most cases, with some exceptions. Table 16.3 summarizes the current recommendations that are available.

Though endovascular (mechanical) intervention has gained attention in the past few years, there are obvious technical challenges/limitations to its use in very small neonates. Nonetheless, case reports of successful use of endovascular therapy have been published, even in very small neonates (Drofa et al., 2016; Sun et al., 2018).

Prognosis

The prognosis for neonates who suffer from CSVT is variable. Most neonates (>90%) survive the acute period of CSVT (deVeber et al., 2001; Jordan et al., 2010). However, a study found that at follow-up (19 months median age), the mortality rate was high (19%) (Berfelo et al., 2010). In survivors of CSVT, the neurologic outcome is highly variable.

Hemorrhagic stroke

Hemorrhagic stroke refers to tissue ischemia and death that result from intracranial hemorrhage. It can be primary (intraparenchymal hemorrhage) or secondary to hemorrhagic conversion of an arterial or venous infarction. In most cases, the etiology of primary hemorrhagic stroke is unknown, but it may be associated with coagulopathy (thrombocytopenia, hemophilia) or vascular abnormality (aneurysm, arteriovenous malformation—AVM).

Clinical presentation and diagnosis

Clinical signs of hemorrhagic stroke are similar to those seen with other types of perinatal stroke, including seizures, hypotonia, lethargy, apnea, poor feeding, or encephalopathy (Armstrong-Wells et al., 2009; Bruno et al., 2014; Cole et al., 2017). Periventricular hemorrhage, a subtype of hemorrhagic stroke, affects primarily premature neonates. It can present insidiously, catastrophically, or have no presenting signs at all. As with other types of perinatal stroke, periventricular hemorrhage is characterized by nonspecific findings, including an altered level of consciousness, hypotonia, decreased spontaneous movements, and subtle changes in eye movements. Because the syndrome can be "silent" in up to 50% of cases, many advocate for daily transcranial ultrasonographic screenings in high-risk neonates.

Etiology and pathogenesis

The cause of hemorrhagic stroke in neonates is poorly understood. In one study of over 323,000 live births, hemorrhagic stroke was identified in only 20. While 75% of cases were idiopathic in nature, 5 of the babies had thrombocytopenia and one was found to have a cavernous malformation (Armstrong-Wells et al., 2009). Another study of 86 infants with hemorrhagic stroke found that primary intraparenchymal hemorrhage accounted for 59% of cases and hemorrhagic conversion accounted for 35%. In most cases of primary intraparenchymal hemorrhage the etiology was never identified (Cole et al., 2017). The cause of intracranial hemorrhage in neonates is frequently elusive, despite extensive evaluation.

In most cases, a definitive cause for hemorrhage leading to stroke in neonates is usually not determined. However, case–control studies have identified several predictors of perinatal intracerebral hemorrhage, including fetal distress, emergency cesarean delivery, prematurity, and postmaturity. Additional risk factors include extracorporeal membrane oxygenation, coarctation of the aorta, venous thrombosis, and vitamin K deficiency (Miller, 2000; Schulte et al., 2014).

AVM is a rare cause of hemorrhagic stroke in the perinatal period. In neonates, AVMs most commonly involve the vein of Galen, but can also be found in the cerebral hemispheres, thalamus and third ventricle, and choroid plexus. Most infants with vein of Galen malformations present with heart failure (De Rosa et al., 2019). The prognosis for neonates with intracerebral hemorrhage from AVMs depends on the location of the lesion, the severity of bleeding, and the degree of mass effect. Definitive therapy for bleeding AVMs is endovascular embolization of the feeding vessel(s), which may or may not be possible in a small neonate.

Management and treatment

Treatment for hemorrhagic stroke is largely supportive, but when possible, underlying causes should be addressed. Guidelines from the American Heart Association/American Stroke Association recommend that thrombocytopenia should be corrected, coagulation factors should be replaced if depleted, and cerebrospinal fluid drainage should be performed if hydrocephalus develops (Ferriero et al., 2019). All neonates should receive vitamin K at birth, as vitamin K deficiency can lead to bleeding. In addition, any bleeding infant with a prolonged prothrombin time or INR should be thought to be deficient in vitamin K until proven otherwise. In cases of severe bleeding, plasma or prothrombin complex may be administered in addition to vitamin K. Although surgical evacuation of the hemorrhage in neonates with elevated intracranial pressure may be performed, it is unclear if this improves outcomes (Ferriero et al., 2019).

Intra/periventricular hemorrhage (seen most frequently in premature neonates) is treated supportively. Goals of management include maintaining cerebral perfusion, preventing further insult, and treating the underlying cause, if any. In most cases, surveillance is performed with daily ultrasonography of the head to assess for evolution of the hemorrhage and ventricular dilation.

Prognosis

Prognosis for neonates who suffer hemorrhagic stroke is highly variable. Mortality rates that range from 4% to 25% have been reported (Jocelyn and Casiro, 1992; Brouwer et al., 2010; Bruno et al., 2014; Cole et al., 2017). Of patients who suffer hemorrhagic stroke during the neonatal period, 45%–75% will have long-term sequelae, which may include cerebral palsy, epilepsy, and developmental delay (Lee et al., 2005b; Brouwer et al., 2010; Bruno et al., 2014). In one study of 86 infants with neonatal hemorrhagic stroke, there were 3 deaths, all of which occurred beyond the neonatal period. Poor neurologic outcome was observed in 42% of the 50 patients that were followed. Impairments included sensorimotor deficits, language delay, and epilepsy. There was no recurrence of hemorrhagic stroke. Interventions beyond the neonatal period included ventriculoperitoneal shunt placement, epilepsy surgery, and AVM embolization (Cole et al., 2017).

RISK FACTORS FOR PERINATAL STROKE

Unlike adult stroke, risk factors for perinatal stroke remain controversial. Although a direct cause–effect relationship has not been established for any risk factor or combination of risk factors, it has been suggested that the presence of more than one risk factor increases the likelihood of perinatal stroke from 1:3000 to 1:200 (Lee et al., 2005a). It is useful to subdivide risk factors for perinatal stroke into maternal/obstetric factors, intrapartum events, neonatal factors, and prothrombotic factors, but it is important to note that most of the risk factors discussed herein were identified based upon case–control studies and that perinatal stroke is likely multifactorial in nature.

Maternal/Obstetric factors

Maternal pathology is associated with an increased risk of perinatal stroke. The risk of perinatal stroke increases as the number of maternal risk factors rises (Lee et al., 2005a). Although the direct correlation between different maternal risk factors and perinatal stroke remains controversial, there are a number of case–control studies that suggest a potential link. Two maternal traits have been found to be independent risk factors for perinatal stroke: a history of maternal infertility and preeclampsia. Table 16.4 summarizes these factors.

Intrapartum events

Events during the course of delivery have been found to be associated with an increased risk of perinatal stroke. In particular, birth asphyxia (OR 44.04) and abnormal fetal heart rate tracing (OR 6.87) have been seen in multiple studies to be associated with an increase in the incidence of PAIS. Table 16.5 summarizes known intrapartum risk factors for perinatal stroke.

Table 16.4

Maternal risk factors for perinatal stroke
Cocaine abuse (Chasnoff et al., 1986; Dominguez et al., 1991; Heier et al., 1991)
Placental abruption (Wu et al., 2005)
Maternal diabetes (Gunther et al., 2000)
Preeclampsia (Lee et al., 2005a; Chabrier et al., 2010; Li et al., 2017)
Maternal smoking (Darmency-Stamboul et al., 2012)
History of infertility (Lee et al., 2005a)
Chorioamnionitis (Lee et al., 2005a)
Twin–twin transfusion (Cheong and Cowan, 2009)
Maternal fever (Harteman et al., 2012; Martinez-Biarge et al., 2016)
Maternal autoimmune or prothrombotic disease (Akanli et al., 1998)
Primiparity (Lee et al., 2005b)
Previous miscarriage (Lee et al., 2005b; Harteman et al., 2012)
Oligohydramnios (Lee et al., 2005b; Darmency-Stamboul et al., 2012)

Neonatal factors

Certain characteristics of the newborn are associated with an increased risk of PAIS. Of particular note are umbilical artery pH at birth <7.0 (OR 8.41), hypoglycemia (OR 6.85), and 5-min Apgar score <7. Known neonatal risk factors are summarized in Table 16.6.

Prothrombotic factors

Prothrombotic states warrant specific mention as a group of risk factors that may predispose a neonate to perinatal stroke. Prothrombotic coagulation factors are present in

Table 16.5

Intrapartum risk factors for perinatal stroke
Birth asphyxia (Wu et al., 2005; Lee et al., 2005b)
Abnormal fetal heart rate tracing (Estan and Hope, 1997; Martinez-Biarge et al., 2016)
Umbilical cord abnormalities (Lee et al., 2005b; Darmency-Stamboul et al., 2012)
Vacuum delivery (Wu et al., 2004)
Prolonged second stage of labor (Lee et al., 2005b)
Prolonged rupture of membranes (Lee et al., 2005a; Martinez-Biarge et al., 2016)
Emergency cesarean section (Wu et al., 2004; Lee et al., 2005a; Benders et al., 2007; Darmency-Stamboul et al., 2012; Harteman et al., 2012; Tuckuviene et al., 2012; Martinez-Biarge et al., 2016)
Meconium-stained amniotic fluid (Estan and Hope, 1997; Lee et al., 2005a; Benders et al., 2007; Darmency-Stamboul et al., 2012; Harteman et al., 2012; Tuckuviene et al., 2012)

Table 16.6

Neonatal risk factors for perinatal stroke
Thrombophilia (Kenet et al., 2010)
Meningitis/sepsis/infection (Gunther et al., 2000; Fitzgerald and Golomb, 2007; Hernandez et al., 2011; Harteman et al., 2012)
5-min Apgar score <7 (Harteman et al., 2012)
Disseminated intravascular coagulation (Gunther et al., 2000)
Congenital heart disease (McQuillen et al., 2007; Chen et al., 2009)
Hypoglycemia (Harteman et al., 2012)
Polycythemia (Gunther et al., 2000)
Hypoxia–ischemia (Gunther et al., 2000; Ecury-Goossen et al., 2013)
Perinatal resuscitation (Kirton et al., 2011)
Male sex (Nowak-Gottl et al., 1997; Fitzgerald and Golomb, 2007; Nwosu et al., 2008; Chabrier et al., 2010; Yang et al., 2010)

more than half of neonates diagnosed with perinatal stroke (Gunther et al., 2000; Cnossen et al., 2009). Elevated lipoprotein(a), MTHFR C677T mutation, factor V Leiden, prothrombin gene mutation, presence of antiphospholipid antibodies, and protein deficiency have all been associated with perinatal stroke (Nowak-Gottl et al., 1997; Gunther et al., 2000). The presence of prothrombotic states likely contributes to development of perinatal stroke and may also influence outcome (Govaert et al., 2009b; Simchen et al., 2009; Kenet et al., 2010). Pregnancy itself is known to be a hypercoagulable state, even in normal parturients. Certain conditions may exacerbate this hypercoagulability, such as preeclampsia (Mello et al., 2005). In addition, inherited or acquired maternal thrombophilic state can be passed from the mother to the neonate via the placenta. As a result, the neonate is also in a prothrombotic state at birth (Lao et al., 1990).

CONCLUSION

Perinatal stroke is not rare. It presents several challenges to the practitioner who cares for neonates, particularly premature neonates, neonates with severe coexisting disease, and neonates born to mothers with disorders of pregnancy. Awareness of the risk factors, presentation, management, and prognosis of this heterogeneous syndrome is critical to providing the best possible outcomes for children affected by perinatal stroke.

References

Agrawal N, Johnston SC, Wu YW et al. (2009). Imaging data reveal a higher pediatric stroke incidence than prior US estimates. Stroke 40: 3415–3421.

Akanli LF, Trasi SS, Thuraisamy K et al. (1998). Neonatal middle cerebral artery infarction: association with elevated maternal anticardiolipin antibodies. Am J Perinatol 15: 399–402.

Amlie-Lefond C, Bernard TJ, Sebire G et al. (2009). Predictors of cerebral arteriopathy in children with arterial ischemic stroke: results of the International Pediatric Stroke Study. Circulation 119: 1417–1423.

Amlie-Lefond C, Zaidat OO, Lew SM (2011). Moyamoya disease in early infancy: case report and literature review. Pediatr Neurol 44: 299–302.

Armstrong-Wells J, Johnston SC, Wu YW et al. (2009). Prevalence and predictors of perinatal hemorrhagic stroke: results from the Kaiser pediatric stroke study. Pediatrics 123: 823–828.

Arnold BW, Martin CG, Alexander BJ et al. (1991). Autoregulation of brain blood flow during hypotension and hypertension in infant lambs. Pediatr Res 29: 110–115.

Benders MJ, Groenendaal F, Uiterwaal CS et al. (2007). Maternal and infant characteristics associated with perinatal arterial stroke in the preterm infant. Stroke 38: 1759–1765.

Benseler SM, deVeber G, Hawkins C et al. (2005). Angiography-negative primary central nervous system vasculitis in children: a newly recognized inflammatory central nervous system disease. Arthritis Rheum 52: 2159–2167.

Benseler SM, Silverman E, Aviv RI et al. (2006). Primary central nervous system vasculitis in children. Arthritis Rheum 54: 1291–1297.

Berfelo FJ, Kersbergen KJ, van Ommen CH et al. (2010). Neonatal cerebral sinovenous thrombosis from symptom to outcome. Stroke 41: 1382–1388.

Biller J (2009). Stroke: improving characterization of childhood cerebral arteriopathies. Nat Rev Cardiol 6: 395–397.

Billinghurst LL, Beslow LA, Abend NS et al. (2017). Incidence and predictors of epilepsy after pediatric arterial ischemic stroke. Neurology 88: 630–637.

Bosenbark DD, Krivitzky L, Ichord R et al. (2017). Clinical predictors of attention and executive functioning outcomes in children after perinatal arterial ischemic stroke. Pediatr Neurol 69: 79–86.

Bouza H, Rutherford M, Acolet D et al. (1994). Evolution of early hemiplegic signs in full-term infants with unilateral brain lesions in the neonatal period: a prospective study. Neuropediatrics 25: 201–207.

Brouwer AJ, Groenendaal F, Koopman C et al. (2010). Intracranial hemorrhage in full-term newborns: a hospital-based cohort study. Neuroradiology 52: 567–576.

Bruno CJ, Beslow LA, Witmer CM et al. (2014). Haemorrhagic stroke in term and late preterm neonates. Arch Dis Child Fetal Neonatal Ed 99: F48–F53.

Buckley NM (1986). Maturation of circulatory system in three mammalian models of human development. Comp Biochem Physiol A Comp Physiol 83: 1–7.

Chabrier S, Saliba E, Sylvie Nguyen The Tich et al. (2010). Obstetrical and neonatal characteristics vary with birthweight in a cohort of 100 term newborns with symptomatic arterial ischemic stroke. Eur J Paediatr Neurol 14: 206–213.

Chabrier S, Peyric E, Drutel L et al. (2016). Multimodal outcome at 7 years of age after neonatal arterial ischemic stroke. J Pediatr 172: 156–161.e153.

Chalmers E, Ganesen V, Liesner R et al. (2011). Guideline on the investigation, management and prevention of venous thrombosis in children. Br J Haematol 154: 196–207.

Chasnoff IJ, Bussey ME, Savich R et al. (1986). Perinatal cerebral infarction and maternal cocaine use. J Pediatr 108: 456–459.

Chen J, Zimmerman RA, Jarvik GP et al. (2009). Perioperative stroke in infants undergoing open heart operations for congenital heart disease. Ann Thorac Surg 88: 823–829.

Cheong JL, Cowan FM (2009). Neonatal arterial ischaemic stroke: obstetric issues. Semin Fetal Neonatal Med 14: 267–271.

Cipolla M (2016). The cerebral circulation, Morgan & Claypool Life Sciences.

Cnossen MH, van Ommen CH, Appel IM (2009). Etiology and treatment of perinatal stroke; a role for prothrombotic coagulation factors? Semin Fetal Neonatal Med 14: 311–317.

Cole L, Dewey D, Letourneau N et al. (2017). Clinical characteristics, risk factors, and outcomes associated with neonatal hemorrhagic stroke: a population-based case-control study. JAMA Pediatr 171: 230–238.

Cowan FM, Pennock JM, Hanrahan JD et al. (1994). Early detection of cerebral infarction and hypoxic ischemic encephalopathy in neonates using diffusion-weighted magnetic resonance imaging. Neuropediatrics 25: 172–175.

Cowan F, Mercuri E, Groenendaal F et al. (2005). Does cranial ultrasound imaging identify arterial cerebral infarction in term neonates? Arch Dis Child Fetal Neonatal Ed 90: F252–F256.

Curry CJ, Bhullar S, Holmes J et al. (2007). Risk factors for perinatal arterial stroke: a study of 60 mother–child pairs. Pediatr Neurol 37: 99–107.

Danchaivijitr N, Cox TC, Saunders DE et al. (2006). Evolution of cerebral arteriopathies in childhood arterial ischemic stroke. Ann Neurol 59: 620–626.

Darmency-Stamboul V, Chantegret C, Ferdynus C et al. (2012). Antenatal factors associated with perinatal arterial ischemic stroke. Stroke 43: 2307–2312.

De Rosa G, De Carolis MP, Tempera A et al. (2019). Outcome of neonates with vein of galen malformation presenting with severe heart failure: a case series. Am J Perinatol 36: 169–175.

de Vries LS, Groenendaal F, Eken P et al. (1997). Infarcts in the vascular distribution of the middle cerebral artery in preterm and fullterm infants. Neuropediatrics 28: 88–96.

Degani S (2009). Evaluation of fetal cerebrovascular circulation and brain development: the role of ultrasound and Doppler. Semin Perinatol 33: 259–269.

DeSesso JM, Scialli AR, Holson JF (1999). Apparent lability of neural tube closure in laboratory animals and humans. Am J Med Genet 87: 143–162.

deVeber GA, MacGregor D, Curtis R et al. (2000). Neurologic outcome in survivors of childhood arterial ischemic stroke and sinovenous thrombosis. J Child Neurol 15: 316–324.

deVeber G, Andrew M, Adams C et al. (2001). Cerebral sinovenous thrombosis in children. N Engl J Med 345: 417–423.

deVeber GA, Kirton A, Booth FA et al. (2017). Epidemiology and outcomes of arterial ischemic stroke in children: the Canadian Pediatric Ischemic Stroke Registry. Pediatr Neurol 69: 58–70.

Domi T, Edgell DS, McCrindle BW et al. (2008). Frequency, predictors, and neurologic outcomes of vaso-occlusive strokes associated with cardiac surgery in children. Pediatrics 122: 1292–1298.

Dominguez R, Aguirre Vila-Coro A, Slopis JM et al. (1991). Brain and ocular abnormalities in infants with in utero exposure to cocaine and other street drugs. Am J Dis Child 145: 688–695.

Drofa A, Kouznetsov E, Tomek S et al. (2016). Successful endovascular management of massive pansinus thrombosis: case report and review of literature. Pediatr Neurosurg 51: 318–324.

Dzietko M, Derugin N, Wendland MF et al. (2013). Delayed VEGF treatment enhances angiogenesis and recovery after neonatal focal rodent stroke. Transl Stroke Res 4: 189–200.

Ecury-Goossen GM, Raets MM, Lequin M et al. (2013). Risk factors, clinical presentation, and neuroimaging findings of neonatal perforator stroke. Stroke 44: 2115–2120.

Elbers J, Viero S, MacGregor D et al. (2011). Placental pathology in neonatal stroke. Pediatrics 127: e722–e729.

Emerick KM, Krantz ID, Kamath BM et al. (2005). Intracranial vascular abnormalities in patients with Alagille syndrome. J Pediatr Gastroenterol Nutr 41: 99–107.

Estan J, Hope P (1997). Unilateral neonatal cerebral infarction in full term infants. Arch Dis Child Fetal Neonatal Ed 76: F88–F93.

Ferrara N, Houck KA, Jakeman LB et al. (1991). The vascular endothelial growth factor family of polypeptides. J Cell Biochem 47: 211–218.

Ferriero DM, Fullerton HJ, Bernard TJ et al. (2019). Management of stroke in neonates and children: a scientific statement from the American Heart Association/American Stroke Association. Stroke 50: e51–e96.

Fitzgerald KC, Golomb MR (2007). Neonatal arterial ischemic stroke and sinovenous thrombosis associated with meningitis. J Child Neurol 22: 818–822.

Fujimoto S, Yokochi K, Togari H et al. (1992). Neonatal cerebral infarction: symptoms, CT findings and prognosis. Brain Dev 14: 48–52.

Fullerton HJ, Wu YW, Sidney S et al. (2007). Risk of recurrent childhood arterial ischemic stroke in a population-based cohort: the importance of cerebrovascular imaging. Pediatrics 119: 495–501.

Germain S, Monnot C, Muller L et al. (2010). Hypoxia-driven angiogenesis: role of tip cells and extracellular matrix scaffolding. Curr Opin Hematol 17: 245–251.

Golden GS (1985). Stroke syndromes in childhood. Neurol Clin 3: 59–75.

Golomb MR, MacGregor DL, Domi T et al. (2001). Presumed pre- or perinatal arterial ischemic stroke: risk factors and outcomes. Ann Neurol 50: 163–168.

Govaert P, Matthys E, Zecic A et al. (2000). Perinatal cortical infarction within middle cerebral artery trunks. Arch Dis Child Fetal Neonatal Ed 82: F59–F63.

Govaert P, Ramenghi L, Taal R et al. (2009a). Diagnosis of perinatal stroke I: definitions, differential diagnosis and registration. Acta Paediatr 98: 1556–1567.

Govaert P, Ramenghi L, Taal R et al. (2009b). Diagnosis of perinatal stroke II: mechanisms and clinical phenotypes. Acta Paediatr 98: 1720–1726.

Grunt S, Wingeier K, Wehrli E et al. (2010). Cerebral sinus venous thrombosis in Swiss children. Dev Med Child Neurol 52: 1145–1150.

Gunther G, Junker R, Strater R et al. (2000). Symptomatic ischemic stroke in full-term neonates: role of acquired and genetic prothrombotic risk factors. Stroke 31: 2437–2441.

Harnarine-Singh D, Hyde JB (1970). Post-natal growth of the arterial net in the human cerebral pia mater. Nature 225: 86–87.

Harteman JC, Groenendaal F, Kwee A et al. (2012). Risk factors for perinatal arterial ischaemic stroke in full-term infants: a case–control study. Arch Dis Child Fetal Neonatal Ed 97: F411–F416.

Heier LA, Carpanzano CR, Mast J et al. (1991). Maternal cocaine abuse: the spectrum of radiologic abnormalities in the neonatal CNS. Am J Neuroradiol 12: 951–956.

Hernandez MJ, Brennan RW, Bowman GS (1980). Autoregulation of cerebral blood flow in the newborn dog. Brain Res 184: 199–202.

Hernandez MI, Sandoval CC, Tapia JL et al. (2011). Stroke patterns in neonatal group B streptococcal meningitis. Pediatr Neurol 44: 282–288.

Heyer GL, Dowling MM, Licht DJ et al. (2008). The cerebral vasculopathy of PHACES syndrome. Stroke 39: 308–316.

Hoeben A, Landuyt B, Highley MS et al. (2004). Vascular endothelial growth factor and angiogenesis. Pharmacol Rev 56: 549–580.

Ichord RN, Benedict SL, Chan AK et al. (2015). Paediatric cerebral sinovenous thrombosis: findings of the International Paediatric Stroke Study. Arch Dis Child 100: 174–179.

Inagami T, Naruse M, Hoover R (1995). Endothelium as an endocrine organ. Annu Rev Physiol 57: 171–189.

Jocelyn LJ, Casiro OG (1992). Neurodevelopmental outcome of term infants with intraventricular hemorrhage. Am J Dis Child 146: 194–197.

Jordan LC, Rafay MF, Smith SE et al. (2010). Antithrombotic treatment in neonatal cerebral sinovenous thrombosis: results of the International Pediatric Stroke Study. J Pediatr 156: 704–710 710.e701–e702.

Kenet G, Lutkhoff LK, Albisetti M et al. (2010). Impact of thrombophilia on risk of arterial ischemic stroke or cerebral sinovenous thrombosis in neonates and children: a systematic review and meta-analysis of observational studies. Circulation 121: 1838–1847.

Kirton A, Deveber G, Pontigon AM et al. (2008). Presumed perinatal ischemic stroke: vascular classification predicts outcomes. Ann Neurol 63: 436–443.

Kirton A, Armstrong-Wells J, Chang T et al. (2011). Symptomatic neonatal arterial ischemic stroke: the International Pediatric Stroke Study. Pediatrics 128: e1402–e1410.

Koehler RC, Traystman RJ, Jones Jr MD (1986). Influence of reduced oxyhemoglobin affinity on cerebrovascular response to hypoxic hypoxia. Am J Physiol 251: H756–H763.

Krishnamoorthy KS, Soman TB, Takeoka M et al. (2000). Diffusion-weighted imaging in neonatal cerebral infarction: clinical utility and follow-up. J Child Neurol 15: 592–602.

Kurnik K, Kosch A, Strater R et al. (2003). Recurrent thromboembolism in infants and children suffering from symptomatic neonatal arterial stroke: a prospective follow-up study. Stroke 34: 2887–2892.

Kuschinsky W (1991). Coupling of function, metabolism, and blood flow in the brain. Neurosurg Rev 14: 163–168.

Lao TT, Yin JA, Yuen PM (1990). Coagulation and anticoagulation systems in newborns—correlation with their mothers at delivery. Lower levels of anticoagulants and fibrinolytic activity in the newborn. Gynecol Obstet Invest 29: 181–184.

Laugesaar R, Kolk A, Tomberg T et al. (2007). Acutely and retrospectively diagnosed perinatal stroke: a population-based study. Stroke 38: 2234–2240.

Lee J, Croen LA, Backstrand KH et al. (2005a). Maternal and infant characteristics associated with perinatal arterial stroke in the infant. JAMA 293: 723–729.

Lee J, Croen LA, Lindan C et al. (2005b). Predictors of outcome in perinatal arterial stroke: a population-based study. Ann Neurol 58: 303–308.

Lee S, Mirsky DM, Beslow LA et al. (2017). Pathways for neuroimaging of neonatal stroke. Pediatr Neurol 69: 37–48.

Lehman LL, Rivkin MJ (2014). Perinatal arterial ischemic stroke: presentation, risk factors, evaluation, and outcome. Pediatr Neurol 51: 760–768.

Li C, Miao JK, Xu Y et al. (2017). Prenatal, perinatal and neonatal risk factors for perinatal arterial ischaemic stroke: a systematic review and meta-analysis. Eur J Neurol 24: 1006–1015.

Linnemann Jr CC, Alvira MM (1980). Pathogenesis of varicella-zoster angiitis in the CNS. Arch Neurol 37: 239–240.

Lou HC, Tweed WA, Davies JM (1985). Preferential blood flow increase to the brain stem in moderate neonatal hypoxia: reversal by naloxone. Eur J Pediatr 144: 225–227.

Mallick AA, Sharples PM, Calvert SE et al. (2009). Cerebral venous sinus thrombosis: a case series including thrombolysis. Arch Dis Child 94: 790–794.

Mann LI (1970). Fetal brain metabolism and function. Clin Obstet Gynecol 13: 638–651.

Martinez-Biarge M, Cheong JL, Diez-Sebastian J et al. (2016). Risk factors for neonatal arterial ischemic stroke: the importance of the intrapartum period. J Pediatr 173: 62–68.e61.

McKinstry RC, Miller JH, Snyder AZ et al. (2002). A prospective, longitudinal diffusion tensor imaging study of brain injury in newborns. Neurology 59: 824–833.

McQuillen PS, Barkovich AJ, Hamrick SE et al. (2007). Temporal and anatomic risk profile of brain injury with neonatal repair of congenital heart defects. Stroke 38: 736–741.

Mello G, Parretti E, Marozio L et al. (2005). Thrombophilia is significantly associated with severe preeclampsia: results of a large-scale, case-controlled study. Hypertension 46: 1270–1274.

Mercuri E, Rutherford M, Cowan F et al. (1999). Early prognostic indicators of outcome in infants with neonatal cerebral infarction: a clinical, electroencephalogram, and magnetic resonance imaging study. Pediatrics 103: 39–46.

Mercuri E, Barnett A, Rutherford M et al. (2004). Neonatal cerebral infarction and neuromotor outcome at school age. Pediatrics 113: 95–100.

Miller V (2000). Neonatal cerebral infarction. Semin Pediatr Neurol 7: 278–288.

Moharir MD (2010). Cerebral sinus venous thrombosis in children: an urgent need for multicentre trials. Dev Med Child Neurol 52: 1080–1081.

Moharir MD, Shroff M, Stephens D et al. (2010). Anticoagulants in pediatric cerebral sinovenous thrombosis: a safety and outcome study. Ann Neurol 67: 590–599.

Monagle P, Chan AKC, Goldenberg NA et al. (2012). Antithrombotic therapy in neonates and children: antithrombotic therapy and prevention of thrombosis, 9th ed: American College of Chest Physicians Evidence-Based Clinical Practice Guidelines. Chest 141: e737S–e801S.

Moncada S, Palmer RM, Higgs EA (1991). Nitric oxide: physiology, pathophysiology, and pharmacology. Pharmacol Rev 43: 109–142.

Nelson KB, Lynch JK (2004). Stroke in newborn infants. Lancet Neurol 3: 150–158.

Nowak-Gottl U, Debus O, Findeisen M et al. (1997). Lipoprotein (a): its role in childhood thromboembolism. Pediatrics 99: E11.

Nwosu ME, Williams LS, Edwards-Brown M et al. (2008). Neonatal sinovenous thrombosis: presentation and association with imaging. Pediatr Neurol 39: 155–161.

Otto KB, Lierse W (1970). The capillaries of various parts of the human brain in the fetal period and during the first years of life. Acta Anat (Basel) 77: 25–36.

Owens GK, Kumar MS, Wamhoff BR (2004). Molecular regulation of vascular smooth muscle cell differentiation in development and disease. Physiol Rev 84: 767–801.

Ozduman K, Pober BR, Barnes P et al. (2004). Fetal stroke. Pediatr Neurol 30: 151–162.

Pasternak JF, Groothuis DR (1985). Autoregulation of cerebral blood flow in the newborn beagle puppy. Biol Neonate 48: 100–109.

Pellicer A, Cabanas F, Garcia-Alix A et al. (1992). Stroke in neonates with cardiac right-to-left shunt. Brain Dev 14: 381–385.

Pryds O, Andersen GE, Friis-Hansen B (1990). Cerebral blood flow reactivity in spontaneously breathing, preterm infants shortly after birth. Acta Paediatr Scand 79: 391–396.

Rafay MF, Armstrong D, Deveber G et al. (2006). Craniocervical arterial dissection in children: clinical and radiographic presentation and outcome. J Child Neurol 21: 8–16.

Rafay MF, Cortez MA, de Veber GA et al. (2009). Predictive value of clinical and EEG features in the diagnosis of stroke and hypoxic ischemic encephalopathy in neonates with seizures. Stroke 40: 2402–2407.

Raju TN, Nelson KB, Ferriero D et al. (2007). Ischemic perinatal stroke: summary of a workshop sponsored by the National Institute of Child Health and Human Development and the National Institute of Neurological Disorders and Stroke. Pediatrics 120: 609–616.

Ramaekers VT, Casaer P (1990). Defective regulation of cerebral oxygen transport after severe birth asphyxia. Dev Med Child Neurol 32: 56–62.

Ramaekers VT, Casaer P, Daniels H et al. (1990). Upper limits of brain blood flow autoregulation in stable infants of various conceptional age. Early Hum Dev 24: 249–258.

Rea D, Brandsema JF, Armstrong D et al. (2009). Cerebral arteriopathy in children with neurofibromatosis type 1. Pediatrics 124: e476–e483.

Rensen SS, Doevendans PA, van Eys GJ (2007). Regulation and characteristics of vascular smooth muscle cell phenotypic diversity. Neth Heart J 15: 100–108.

Rhodes AJ, Hyde JB (1965). Postnatal growth of arterioles in the human cerebral cortex. Growth 29: 173–182.

Rice D, Barone Jr S (2000). Critical periods of vulnerability for the developing nervous system: evidence from humans and animal models. Environ Health Perspect 108: 511–533.

Richardson BS, Rurak D, Patrick JE et al. (1989). Cerebral oxidative metabolism during sustained hypoxaemia in fetal sheep. J Dev Physiol 11: 37–43.

Risau W, Wolburg H (1990). Development of the blood–brain barrier. Trends Neurosci 13: 174–178.

Roach ES, Golomb MR, Adams R et al. (2008). Management of stroke in infants and children: a scientific statement from a Special Writing Group of the American Heart Association Stroke Council and the Council on Cardiovascular Disease in the Young. Stroke 39: 2644–2691.

Robertson RL, Ben-Sira L, Barnes PD et al. (1999). MR line-scan diffusion-weighted imaging of term neonates with perinatal brain ischemia. Am J Neuroradiol 20: 1658–1670.

Schulte R, Jordan LC, Morad A et al. (2014). Rise in late onset vitamin K deficiency bleeding in young infants because of omission or refusal of prophylaxis at birth. Pediatr Neurol 50: 564–568.

Sebire G, Meyer L, Chabrier S (1999). Varicella as a risk factor for cerebral infarction in childhood: a case–control study. Ann Neurol 45: 679–680.

Sebire G, Fullerton H, Riou E et al. (2004). Toward the definition of cerebral arteriopathies of childhood. Curr Opin Pediatr 16: 617–622.

Semenza GL (2004). O_2-regulated gene expression: transcriptional control of cardiorespiratory physiology by HIF-1. J Appl Physiol (1985) 96: 1173–1177. discussion 1170–1172.

Simchen MJ, Goldstein G, Lubetsky A et al. (2009). Factor v Leiden and antiphospholipid antibodies in either mothers or infants increase the risk for perinatal arterial ischemic stroke. Stroke 40: 65–70.

Sreenan C, Bhargava R, Robertson CM (2000). Cerebral infarction in the term newborn: clinical presentation and long-term outcome. J Pediatr 137: 351–355.

Stewart PA, Hayakawa EM (1987). Interendothelial junctional changes underlie the developmental 'tightening' of the blood-brain barrier. Brain Res 429: 271–281.

Strandgaard S, Paulson OB (1984). Cerebral autoregulation. Stroke 15: 413–416.

Sun LR, Pearl M, Bahouth MN et al. (2018). Mechanical thrombectomy in an infant with acute embolic stroke. Pediatr Neurol 82: 53–54.

Szymonowicz W, Walker AM, Yu VY et al. (1990). Regional cerebral blood flow after hemorrhagic hypotension in the preterm, near-term, and newborn lamb. Pediatr Res 28: 361–366.

Tuckuviene R, Christensen AL, Helgested J et al. (2012). Infant, obstetrical and maternal characteristics associated with thromboembolism in infancy: a nationwide population-based case–control study. Arch Dis Child Fetal Neonatal Ed 97: F417–F422.

van der Aa NE, Dudink J, Benders MJ et al. (2013). Neonatal posterior cerebral artery stroke: clinical presentation, MRI findings, and outcome. Dev Med Child Neurol 55: 283–290.

Walsh MP, Cole WC (2013). The role of actin filament dynamics in the myogenic response of cerebral resistance arteries. J Cereb Blood Flow Metab 33: 1–12.

Walter B, Bauer R, Kuhnen G et al. (2000). Coupling of cerebral blood flow and oxygen metabolism in infant pigs during selective brain hypothermia. J Cereb Blood Flow Metab 20: 1215–1224.

Westmacott R, MacGregor D, Askalan R et al. (2009). Late emergence of cognitive deficits after unilateral neonatal stroke. Stroke 40: 2012–2019.

Wintermark M, Hills NK, DeVeber GA et al. (2017). Clinical and imaging characteristics of arteriopathy subtypes in children with arterial ischemic stroke: results of the VIPS study. Am J Neuroradiol 38: 2172–2179.

Wollack JB, Kaifer M, LaMonte MP et al. (1996). Stroke in Williams syndrome. Stroke 27: 143–146.

Wu YW, March WM, Croen LA et al. (2004). Perinatal stroke in children with motor impairment: a population-based study. Pediatrics 114: 612–619.

Wu YW, Lynch JK, Nelson KB (2005). Perinatal arterial stroke: understanding mechanisms and outcomes. Semin Neurol 25: 424–434.

Yang JY, Chan AK, Callen DJ et al. (2010). Neonatal cerebral sinovenous thrombosis: sifting the evidence for a diagnostic plan and treatment strategy. Pediatrics 126: e693–e700.

Zimmer JA, Garg BP, Williams LS et al. (2007). Age-related variation in presenting signs of childhood arterial ischemic stroke. Pediatr Neurol 37: 171–175.

Index

NB: Page numbers in *italics* refer to figures and tables.

B

D

I

O

P

T

X

Z